Methods in Enzymology

Volume 328

APPLICATIONS OF CHIMERIC GENES AND HYBRID PROTEINS

Part C

Protein–Protein Interactions and Genomics

METHODS IN ENZYMOLOGY

Methods in Enzymology

Volume 328

Applications of Chimeric Genes and Hybrid Proteins

Part C
Protein–Protein Interactions and Genomics

EDITED BY

Jeremy Thorner

UNIVERSITY OF CALIFORNIA
BERKELEY, CALIFORNIA

Scott D. Emr

HOWARD HUGHES MEDICAL INSTITUTE
AND SCHOOL OF MEDICINE
UNIVERSITY OF CALIFORNIA, SAN DIEGO
LA JOLLA, CALIFORNIA

John N. Abelson

CALIFORNIA INSTITUTE OF TECHNOLOGY
PASADENA, CALIFORNIA

ACADEMIC PRESS
San Diego London Boston New York Sydney Tokyo Toronto

This book is printed on acid-free paper. 

Academic Press
A Harcourt Science and Technology Company
525 B Street, Suite 1900, San Diego, California 92101-4495, USA

http://www.academicpress.com

Academic Press Limited
32 Jamestown Road, London NW1 7BY, UK

International Standard Book Number: 0-12-182229-X

PRINTED IN THE UNITED STATES OF AMERICA
00 01 02 03 04 05 06 MM 9 8 7 6 5 4 3 2 1

Table of Contents

Section I. Applications of Two-Hybrid Methods and Related Techniques for Analysis of Protein–Protein Interactions

Section II. Other Approaches Using Chimeras for Identification and Analysis of Interactions of Proteins and Nucleic Acids

Section III. Phage Display and Its Applications

Section IV. Construction of Hybrid Molecules by DNA Shuffling and Other Methods

Section V. Applications of Fusions in Functional Genomics

Contributors to Volume 328

Article numbers are in parentheses following the names of contributors.
Affiliations listed are current.

TOM ALBER (17), *Department of Molecular and Cell Biology, University of California, Berkeley, California 94720-3206*

KATJA M. ARNDT (17, 22), *Department of Molecular and Cell Biology, University of California, Berkeley, California 94720-3206*

FRANCES H. ARNOLD (26, 27), *Division of Chemistry and Chemical Engineering, California Institute of Technology, Pasadena, California 91125*

AMI ARONHEIM (4), *Department of Molecular Genetics and Rappaport Family Institute for Research in Medical Sciences, Technion Israel Institute of Technology, Haifa 31096, Israel*

TERRANCE M. ARTHUR (11), *USDA-ARS-RLHUSMARC, Clay Center, Nebraska 68933*

MARK BERGSEID (31), *Structural Genomix, San Diego, California 92121*

BRUCE T. BLAKELY (15), *Department of Molecular Pharmacology, Stanford University School of Medicine, Stanford, California 94305*

ULRICH K. BLASCHKE (29), *Laboratory of Synthetic Protein Chemistry, Rockefeller University, New York, New York 10021*

HELEN M. BLAU (15), *Department of Molecular Pharmacology, Stanford University School of Medicine, Stanford, California 94305*

ERIC T. BODER (25), *Department of Chemical Engineering, University of Pennsylvania, Philadelphia, Pennsylvania 19104*

MICHAEL A. BRASCH (34), *Life Technologies, Inc., Rockville, Maryland 20850*

ROGER BRENT (13), *The Molecular Sciences Institute, Berkeley, California 94704*

RICHARD R. BURGESS (11), *McArdle Laboratory for Cancer Research, University of Wisconsin, Madison, Wisconsin 53706*

GERARD CAGNEY (1), *Banting and Best Department of Medical Research, University of Toronto, Ontario M5G 1L6, Canada*

FRANÇOIS-X. CAMPBELL-VALOIS (14), *Département de Biochimie, Université de Montréal, Montréal, Québec H3C 3J7, Canada*

LEWIS C. CANTLEY (12), *Division of Signal Transduction, Harvard Institutes of Medicine, Department of Medicine, Beth Israel Deaconess Medical Center, and Department of Cell Biology, Harvard Medical School, Boston, Massachusetts 02215*

JANNETTE CAREY (30), *Chemistry Department, Princeton University, Princeton, New Jersey 08544-1009*

CAROL A. CHARLTON (15), *Department of Molecular Pharmacology, Stanford University School of Medicine, Stanford, California 94305*

PAULO S. R. COELHO (33), *Department of Molecular, Cellular, and Developmental Biology, Yale University, New Haven, Connecticut 06520-8103*

BRYAN R. CULLEN (20), *Howard Hughes Medical Institute and Department of Genetics, Duke University Medical Center, Durham, North Carolina 27710*

BRIAN C. CUNNINGHAM (21), *Sunesis Pharmaceuticals, Inc., Redwood City, California 94063*

ANTHONY J. DEMAGGIO (10), *Icos Corporation, Bothell, Washington 98021*

SHELLEY ANN DES ETAGES (33), *Department of Molecular, Cellular, and Developmental Biology, Yale University, New Haven, Connecticut 06520-8103*

STEPHEN J. ELLEDGE (32), *Department of Biochemistry and Molecular Biology, Howard Hughes Medical Institute, Baylor College of Medicine, Houston, Texas 77030*

HIDEKI ENDOH (6), *Laboratory of Molecular Oncology, MGH Cancer Center, Charlestown, Massachusetts 02129*

SARAH J. FASHENA (2), *Division of Basic Science, Fox Chase Cancer Center, Philadelphia, Pennsylvania 19111*

STANLEY FIELDS (1, 19), *Departments of Genetics and Medicine and Howard Hughes Medical Institute, University of Washington, Seattle, Washington 98195-7360*

RUSSELL L. FINLEY, JR. (3) *Center for Molecular Medicine and Genetics and Karmanos Cancer Institute, Wayne State University School of Medicine, Detroit, Michigan 48201*

C. RONALD GEYER (13), *The Molecular Sciences Institute, Berkeley, California 94704*

PHYLLIS GOLDMAN (10), *Icos Corporation, Bothell, Washington 98021*

ERICA A. GOLEMIS (2), *Division of Basic Science, Fox Chase Cancer Center, Philadelphia, Pennsylvania 19111*

RICHARD H. GOODMAN (10), *The Vollum Institute, Oregon Health Sciences University, Portland, Oregon 97201*

ERIC C. GRIFFITH (7), *Center for Cancer Research, Departments of Biology and Chemistry, Massachusetts Institute of Technology, Cambridge, Massachusetts 02139*

JOZEF HANES (24), *Department of Biochemistry, University of Zurich, Zurich CH-8057, Switzerland*

LARS O. HANSSON (23, 28), *Department of Biochemistry, Uppsala University, Uppsala SE-75123, Sweden*

JAMES L. HARTLEY (34), *Life Technologies, Inc., Rockville, Maryland 20850*

JOHN A. HEYMAN (31), *Invitrogen Corporation, Carlsbad, California 92008*

JAMES P. HOEFFLER (31), *Crosswinds Consulting, Anchorage, Alaska 99516*

MERL F. HOEKSTRA (10), *Qbiogene, Inc., Carlsbad, California 92008*

JAMES C. HU (18), *Department of Biochemistry and Biophysics, Texas A&M University, College Station, Texas 77843-2128*

LUTZ JERMUTUS (24), *Department of Biochemistry, University of Zurich, Zurich CH-8057, Switzerland*

SABINE JUNG (22), *Department of Biochemistry, University of Zurich, Zurich CH-8057, Switzerland*

GOUZEL KARIMOVA (5), *Unité de Biochimie Cellulaire, CNRS URA 2185, Institut Pasteur, Paris, Cedex 15, France*

MICHAEL KARIN (4), *Department of Pharmacology, Laboratory of Gene Regulation and Signal Transduction, University of California San Diego, La Jolla, California 92093-0636*

JAREMA P. KOCHAN (9), *Department of Metabolic Diseases, Hoffmann-La Roche, Inc., Nutley, New Jersey 07110*

MIKHAIL G. KOLONIN (3), *Center for Molecular Medicine and Genetics, Wayne State University School of Medicine, Detroit, Michigan 48201*

BRIAN KRAEMER (19), *Department of Biochemistry, University of Wisconsin, Madison, Wisconsin 53706*

CLAUS KREBBER (22), *Maxygen, Inc., Redwood City, California 94063*

ANUJ KUMAR (33), *Department of Molecular, Cellular, and Developmental Biology, Yale University, New Haven, Connecticut 06520-8103*

DANIEL LADANT (5), *Unité de Biochimie Cellulaire, CNRS URA 2185, Institut Pasteur, Paris, Cedex 15, France*

MAMIE Z. LI (32), *Department of Biochemistry and Molecular Biology, Howard Hughes Medical Institute, Baylor College of Medicine, Houston, Texas 77030*

EDWARD J. LICITRA (7), *Center for Cancer Research, Departments of Biology and Chemistry, Massachusetts Institute of Technology, Cambridge, Massachusetts 02139*

DOU LIU (32), *Department of Biochemistry and Molecular Biology, Howard Hughes Medical Institute, Baylor College of Medicine, Houston, Texas 77030*

JUN O. LIU (7), *Center for Cancer Research, Departments of Biology and Chemistry, Massachusetts Institute of Technology, Cambridge, Massachusetts 02139*

QINGHUA LIU (32), *Department of Biochemistry and Molecular Biology, Howard Hughes Medical Institute, Baylor College of Medicine, Houston, Texas 77030*

MONIQUE A. LORSON (34), *Laboratory of Molecular Oncology, MGH Cancer Center, Charlestown, Massachusetts 02129*

HENRY B. LOWMAN (21), *Department of Protein Engineering, Genentech, Inc., South San Francisco, California 94080*

BENGT MANNERVIK (23, 28), *Department of Biochemistry, Uppsala University, Uppsala SE-75123, Sweden*

STEPHEN W. MICHNICK (14), *Département de Biochimie, Université de Montréal, Montréal, Québec H3C 3J7, Canada*

TOM W. MUIR (29), *Laboratory of Synthetic Protein Chemistry, Rockefeller University, New York, New York 10021*

KRISTIAN M. MÜLLER (17), *Department of Molecular and Cell Biology, University of California at Berkeley, Berkeley, California 94720-3206*

MARC S. NASOFF (31), *Genomics Institute of the Novartis Research Foundation, San Diego, California 92121-1125*

MARK A. OSBORNE (9), *Department of Human Genetics, Genome Therapeutics Corporation, Waltham, Massachusetts 02154*

SANG-HYUN PARK (16), *Department of Cellular and Molecular Pharmacology, University of California San Francisco, San Francisco, California 94143-0450*

JOELLE N. PELLETIER (14), *Département de Chimie, Université de Montréal, Montréal, Québec H3C 3J7, Canada*

BRADLEY C. PIETZ (11), *McArdle Laboratory for Cancer Research, University of Wisconsin, Madison, Wisconsin 53706*

ANDREAS PLÜCKTHUN (22, 24), *Department of Biochemistry, University of Zurich, Zurich CH-8057, Switzerland*

RONALD T. RAINES (16), *Departments of Biochemistry and Chemistry, University of Wisconsin, Madison, Wisconsin 53706*

INGRID REMY (14), *Département de Biochimie, Université de Montréal, Montréal, Québec H3C 3J7, Canada*

JENNIFER D. RIEKER (18), *Department of Biology, Massachusetts Institute of Technology, Cambridge, Massachusetts 02139*

G. SHIRLEEN ROEDER (33), *Department of Molecular, Cellular, and Developmental Biology and Department of Genetics, Howard Hughes Medical Institute, Yale University, New Haven, Connecticut 06520-8005*

FABIO M. V. ROSSI (15), *Department of Molecular Pharmacology, Stanford University School of Medicine, Stanford, California 94305*

DHRUBA SENGUPTA (19), *Departments of Genetics and Medical Genetics, University of Washington, Seattle, Washington 98195-7360*

ILYA G. SEREBRIISKII (2), *Division of Basic Science, Fox Chase Cancer Center, Philadelphia, Pennsylvania 19111*

ZHIXIN SHAO (27), *Roche Diagnostics, Werk Penzberg, 82372 Penzberg, Germany*

HSIU-MING SHIH (10), *The Vollum Institute, Oregon Health Sciences University, Portland, Oregon 97201*

SACHDEV S. SIDHU (21), *Department of Protein Engineering, Genentech, Inc., South San Francisco, California 94080*

JONATHAN SILBERSTEIN (29), *Laboratory of Synthetic Protein Chemistry, Rockefeller University, New York, New York 10021*

WILLIAM C. SKARNES (35), *Division of Genetics and Development, Department of Molecular and Cell Biology, University of California at Berkeley, Berkeley, California 94720-3200*

MICHAEL SNYDER (33), *Departments of Molecular, Cellular, and Developmental Biology and Molecular Biophysics and Biochemistry, Yale University, New Haven, Connecticut 06520-8103*

GARY F. TEMPLE (34), *Life Technologies, Inc., Rockville, Maryland 20850*

LISA TRONSTAD (23), *Department of Biochemistry, Uppsala University, Uppsala SE-75123, Sweden*

PETER UETZ (1), *Departments of Genetics and Medicine, University of Washington, Seattle, Washington 98195-7360*

AGNES ULLMANN (5), *Unité de Biochimie Cellulaire, CNRS URA 2185, Institut Pasteur, Paris, Cedex 15, France*

ALEXIS VALLÉE-BÉLISLE (14), *Département de Biochimie, Université de Montréal, Montréal, Québec H3C 3J7, Canada*

SANDER VAN DEN HEUVEL (34), *Laboratory of Molecular Oncology, MGH Cancer Center, Charlestown, Massachusetts 02129*

MARC VIDAL (6, 34), *Dana-Farber Cancer Institute and Department of Genetics, Harvard Medical School, Boston, Massachusetts 02115*

ALEXANDER A. VOLKOV (26, 27), *Genecor International, Palo Alto, California 94304-1013*

CHRISTOPH VOLPERS (9), *Center for Molecular Medicine* (ZMMK), *University of Cologne, Cologne 50931, Germany*

ALBERTHA J. M. WALHOUT (6, 34), *Dana-Farber Cancer Institute and Department of Genetics, Harvard Medical School, Boston, Massachusetts 02115*

JAMES A. WELLS (21), *Sunesis Pharmaceuticals, Inc., Redwood City, California 94063*

MARVIN WICKENS (19), *Department of Biochemistry, University of Wisconsin, Madison, Wisconsin 53706*

MIKAEL WIDERSTEN (23), *Department of Biochemistry, Uppsala University, Uppsala SE-75123, Sweden*

K. DANE WITTRUP (25), *Department of Chemical Engineering and Division of Bioengineering, Massachusetts Institute of Technology, Cambridge, Massachusetts 02139*

MICHAEL B. YAFFE (12), *Division of Signal Transduction, Harvard Institutes of Medicine, Departments of Medicine and Surgery, Beth Israel Deaconess Medical Center, Boston, Massachusetts 02215*

BEILIN ZHANG (19), *Department of Biochemistry, University of Wisconsin, Madison, Wisconsin 53706*

JIE ZHANG (8), *Guilford Pharmaceuticals, Inc., Baltimore, Maryland 21224*

JINHUI ZHONG (3), *Center for Molecular Medicine and Genetics, Wayne State University School of Medicine, Detroit, Michigan 48201*

Preface

The modern biologist takes almost for granted the rich repertoire of tools currently available for manipulating virtually any gene or protein of interest. Paramount among these operations is the construction of fusions. The tactic of generating gene fusions to facilitate analysis of gene expression has its origins in the work of Jacob and Monod more than 35 years ago. The fact that gene fusions can create functional chimeric proteins was demonstrated shortly thereafter. Since that time, the number of tricks for splicing or inserting into a gene product various markers, tags, antigenic epitopes, structural probes, and other elements has increased explosively. Hence, when we undertook assembling a volume on the applications of chimeric genes and hybrid proteins in modern biological research, we considered the job a daunting task.

To assist us with producing a coherent work, we first enlisted the aid of an Advisory Committee, consisting of Joe Falke, Stan Fields, Brian Seed, Tom Silhavy, and Roger Tsien. We benefited enormously from their ideas, suggestions, and breadth of knowledge. We are grateful to them all for their willingness to participate at the planning stage and for contributing excellent and highly pertinent articles.

A large measure of the success of this project is due to the enthusiastic responses we received from nearly all of the prospective authors we approached. Many contributors made additional suggestions, and quite a number contributed more than one article. Hence, it became clear early on that given the huge number of applications of gene fusion and hybrid protein technology—for studies of the regulation of gene expression, for lineage tracing, for protein purification and detection, for analysis of protein localization and dynamic movement, and a plethora of other uses—it would not be possible for us to cover this subject comprehensively in a single volume, but in the resulting three volumes, 326, 327, and 328.

Volume 326 is devoted to methods useful for monitoring gene expression, for facilitating protein purification, and for generating novel antigens and antibodies. Also in this volume is an introductory article describing the genesis of the concept of gene fusions and the early foundations of this whole approach. We would like to express our special appreciation to Jon Beckwith for preparing this historical overview. Jon's description is particularly illuminating because he was among the first to exploit gene and protein fusions. Moreover, over the years, he and his colleagues have continued to develop the methodology that has propelled the use of fusion-based techniques from bacteria to eukaryotic organisms. Volume 327 is focused on procedures for tagging proteins for immunodetection, for using chimeric proteins for cytological purposes, especially the analysis of membrane proteins and intracellular protein trafficking, and for monitoring and manipulating various aspects of cell signaling and cell physiology. Included in this volume is a rather extensive section on the green fluorescent protein (GFP) that deals with applications not covered in Volume 302. Volume 328 describes protocols for using hybrid genes and proteins to identify and analyze protein–protein and protein–nucleic interactions, for mapping molecular recognition domains, for directed molecular evolution, and for functional genomics.

We want to take this opportunity to thank again all the authors who generously contributed and whose conscientious efforts to maintain the high standards of the *Methods in Enzymology* series will make these volumes of practical use to a broad spectrum of investigators for many years to come. We have to admit, however, that, despite our best efforts, we could not include each and every method that involves the use of a gene fusion or a hybrid protein. In part, our task was a bit like trying to bottle smoke because brilliant new methods that exploit the fundamental strategy of using a chimeric gene or protein are being devised and published daily. We hope, however, that we have been able to capture many of the most salient and generally applicable procedures. Nonetheless, we take full responsibility for any oversights or omissions, and apologize to any researcher whose method was overlooked.

Finally, we would especially like to acknowledge the expert assistance of Joyce Kato at Caltech, whose administrative skills were essential in organizing these books.

Jeremy Thorner
Scott D. Emr
John N. Abelson

METHODS IN ENZYMOLOGY

VOLUME I. Preparation and Assay of Enzymes
Edited by SIDNEY P. COLOWICK AND NATHAN O. KAPLAN

VOLUME II. Preparation and Assay of Enzymes
Edited by SIDNEY P. COLOWICK AND NATHAN O. KAPLAN

VOLUME III. Preparation and Assay of Substrates
Edited by SIDNEY P. COLOWICK AND NATHAN O. KAPLAN

VOLUME IV. Special Techniques for the Enzymologist
Edited by SIDNEY P. COLOWICK AND NATHAN O. KAPLAN

VOLUME V. Preparation and Assay of Enzymes
Edited by SIDNEY P. COLOWICK AND NATHAN O. KAPLAN

VOLUME VI. Preparation and Assay of Enzymes (*Continued*)
Preparation and Assay of Substrates
Special Techniques
Edited by SIDNEY P. COLOWICK AND NATHAN O. KAPLAN

VOLUME VII. Cumulative Subject Index
Edited by SIDNEY P. COLOWICK AND NATHAN O. KAPLAN

VOLUME VIII. Complex Carbohydrates
Edited by ELIZABETH F. NEUFELD AND VICTOR GINSBURG

VOLUME IX. Carbohydrate Metabolism
Edited by WILLIS A. WOOD

VOLUME X. Oxidation and Phosphorylation
Edited by RONALD W. ESTABROOK AND MAYNARD E. PULLMAN

VOLUME XI. Enzyme Structure
Edited by C. H. W. HIRS

VOLUME XII. Nucleic Acids (Parts A and B)
Edited by LAWRENCE GROSSMAN AND KIVIE MOLDAVE

VOLUME XIII. Citric Acid Cycle
Edited by J. M. LOWENSTEIN

VOLUME XIV. Lipids
Edited by J. M. LOWENSTEIN

VOLUME XV. Steroids and Terpenoids
Edited by RAYMOND B. CLAYTON

VOLUME XVI. Fast Reactions
Edited by KENNETH KUSTIN

Volume LV. Biomembranes (Part F: Bioenergetics)
Edited by Sidney Fleischer and Lester Packer

Volume LVI. Biomembranes (Part G: Bioenergetics)
Edited by Sidney Fleischer and Lester Packer

Volume LVII. Bioluminescence and Chemiluminescence
Edited by Marlene A. DeLuca

Volume LVIII. Cell Culture
Edited by William B. Jakoby and Ira Pastan

Volume LIX. Nucleic Acids and Protein Synthesis (Part G)
Edited by Kivie Moldave and Lawrence Grossman

Volume LX. Nucleic Acids and Protein Synthesis (Part H)
Edited by Kivie Moldave and Lawrence Grossman

Volume 61. Enzyme Structure (Part H)
Edited by C. H. W. Hirs and Serge N. Timasheff

Volume 62. Vitamins and Coenzymes (Part D)
Edited by Donald B. McCormick and Lemuel D. Wright

Volume 63. Enzyme Kinetics and Mechanism (Part A: Initial Rate and Inhibitor Methods)
Edited by Daniel L. Purich

Volume 64. Enzyme Kinetics and Mechanism (Part B: Isotopic Probes and Complex Enzyme Systems)
Edited by Daniel L. Purich

Volume 65. Nucleic Acids (Part I)
Edited by Lawrence Grossman and Kivie Moldave

Volume 66. Vitamins and Coenzymes (Part E)
Edited by Donald B. McCormick and Lemuel D. Wright

Volume 67. Vitamins and Coenzymes (Part F)
Edited by Donald B. McCormick and Lemuel D. Wright

Volume 68. Recombinant DNA
Edited by Ray Wu

Volume 69. Photosynthesis and Nitrogen Fixation (Part C)
Edited by Anthony San Pietro

Volume 70. Immunochemical Techniques (Part A)
Edited by Helen Van Vunakis and John J. Langone

Volume 71. Lipids (Part C)
Edited by John M. Lowenstein

Volume 72. Lipids (Part D)
Edited by John M. Lowenstein

VOLUME 73. Immunochemical Techniques (Part B)
Edited by JOHN J. LANGONE AND HELEN VAN VUNAKIS

VOLUME 74. Immunochemical Techniques (Part C)
Edited by JOHN J. LANGONE AND HELEN VAN VUNAKIS

VOLUME 75. Cumulative Subject Index Volumes XXXI, XXXII, XXXIV–LX
Edited by EDWARD A. DENNIS AND MARTHA G. DENNIS

VOLUME 76. Hemoglobins
Edited by ERALDO ANTONINI, LUIGI ROSSI-BERNARDI, AND EMILIA CHIANCONE

VOLUME 77. Detoxication and Drug Metabolism
Edited by WILLIAM B. JAKOBY

VOLUME 78. Interferons (Part A)
Edited by SIDNEY PESTKA

VOLUME 79. Interferons (Part B)
Edited by SIDNEY PESTKA

VOLUME 80. Proteolytic Enzymes (Part C)
Edited by LASZLO LORAND

VOLUME 81. Biomembranes (Part H: Visual Pigments and Purple Membranes, I)
Edited by LESTER PACKER

VOLUME 82. Structural and Contractile Proteins (Part A: Extracellular Matrix)
Edited by LEON W. CUNNINGHAM AND DIXIE W. FREDERIKSEN

VOLUME 83. Complex Carbohydrates (Part D)
Edited by VICTOR GINSBURG

VOLUME 84. Immunochemical Techniques (Part D: Selected Immunoassays)
Edited by JOHN J. LANGONE AND HELEN VAN VUNAKIS

VOLUME 85. Structural and Contractile Proteins (Part B: The Contractile Apparatus and the Cytoskeleton)
Edited by DIXIE W. FREDERIKSEN AND LEON W. CUNNINGHAM

VOLUME 86. Prostaglandins and Arachidonate Metabolites
Edited by WILLIAM E. M. LANDS AND WILLIAM L. SMITH

VOLUME 87. Enzyme Kinetics and Mechanism (Part C: Intermediates, Stereochemistry, and Rate Studies)
Edited by DANIEL L. PURICH

VOLUME 88. Biomembranes (Part I: Visual Pigments and Purple Membranes, II)
Edited by LESTER PACKER

VOLUME 89. Carbohydrate Metabolism (Part D)
Edited by WILLIS A. WOOD

VOLUME 90. Carbohydrate Metabolism (Part E)
Edited by WILLIS A. WOOD

VOLUME 227. Metallobiochemistry (Part D: Physical and Spectroscopic Methods for Probing Metal Ion Environments in Metalloproteins)
Edited by JAMES F. RIORDAN AND BERT L. VALLEE

VOLUME 228. Aqueous Two-Phase Systems
Edited by HARRY WALTER AND GÖTE JOHANSSON

VOLUME 229. Cumulative Subject Index Volumes 195–198, 200–227

VOLUME 230. Guide to Techniques in Glycobiology
Edited by WILLIAM J. LENNARZ AND GERALD W. HART

VOLUME 231. Hemoglobins (Part B: Biochemical and Analytical Methods)
Edited by JOHANNES EVERSE, KIM D. VANDEGRIFF, AND ROBERT M. WINSLOW

VOLUME 232. Hemoglobins (Part C: Biophysical Methods)
Edited by JOHANNES EVERSE, KIM D. VANDEGRIFF, AND ROBERT M. WINSLOW

VOLUME 233. Oxygen Radicals in Biological Systems (Part C)
Edited by LESTER PACKER

VOLUME 234. Oxygen Radicals in Biological Systems (Part D)
Edited by LESTER PACKER

VOLUME 235. Bacterial Pathogenesis (Part A: Identification and Regulation of Virulence Factors)
Edited by VIRGINIA L. CLARK AND PATRIK M. BAVOIL

VOLUME 236. Bacterial Pathogenesis (Part B: Integration of Pathogenic Bacteria with Host Cells)
Edited by VIRGINIA L. CLARK AND PATRIK M. BAVOIL

VOLUME 237. Heterotrimeric G Proteins
Edited by RAVI IYENGAR

VOLUME 238. Heterotrimeric G-Protein Effectors
Edited by RAVI IYENGAR

VOLUME 239. Nuclear Magnetic Resonance (Part C)
Edited by THOMAS L. JAMES AND NORMAN J. OPPENHEIMER

VOLUME 240. Numerical Computer Methods (Part B)
Edited by MICHAEL L. JOHNSON AND LUDWIG BRAND

VOLUME 241. Retroviral Proteases
Edited by LAWRENCE C. KUO AND JULES A. SHAFER

VOLUME 242. Neoglycoconjugates (Part A)
Edited by Y. C. LEE AND REIKO T. LEE

VOLUME 243. Inorganic Microbial Sulfur Metabolism
Edited by HARRY D. PECK, JR., AND JEAN LEGALL

VOLUME 244. Proteolytic Enzymes: Serine and Cysteine Peptidases
Edited by ALAN J. BARRETT

VOLUME 336. Microbial Growth in Biofilms (Part A: Developmental and Molecular Biological Aspects) (in preparation)
Edited by RONALD J. DOYLE

VOLUME 337. Microbial Growth in Biofilms (Part B: Special Environments and Physicochemical Aspects) (in preparation)
Edited by RONALD J. DOYLE

Section I

Applications of Two-Hybrid Methods and Related Techniques for Analysis of Protein–Protein Interactions

[1] High-Throughput Screening for Protein–Protein Interactions Using Two-Hybrid Assay

By GERARD CAGNEY, PETER UETZ, and STANLEY FIELDS

Introduction

The accumulation of large amounts of genomic sequence data has prompted studies in protein biology on an unprecedented scale.[1] In 1996, the DNA sequence of the yeast *Saccharomyces cerevisiae* became the first eukaryotic genome to be completed.[2] Although this event was a milestone in the history of genetics, at least 40% of the predicted open reading frames (ORFs) for this relatively small genome have no known function.[3] Determining the functions of uncharacterized ORFs, from yeast as well as from other organisms, is a major challenge, and traditional biochemical and genetic approaches are struggling to keep up with sequence data. One strategy to help determine function is to identify protein–protein interactions using the two-hybrid system, a genetic assay that takes place within living yeast cells.[4] Two proteins, X and Y, expressed as fusions with, respectively, a DNA-binding domain (DBD) and a transcriptional activation domain (AD), will generate a transcriptional signal in the assay if X and Y physically interact in the yeast nucleus (Fig. 1). Large-scale analysis of protein interactions by this assay may help assign functions to uncharacterized proteins by identifying associations with proteins of known function; subsequent genetic or biochemical investigations can confirm hypotheses based on these associations.

We have developed an array format to increase the throughput of proteins that can be screened by the two-hybrid method. By an array, we mean a spatially ordered set of separated elements, in which each element is a unique protein potentially available for interaction. When the array is screened against a test protein, the identities of positives can be rapidly determined by their location in the array. The binding of any protein in the array is independent of the binding of all the others, so that weak signals

[1] S. Fields, *Nature Genet.* **15,** 325 (1997).

[2] A. Goffeau, B. G. Barrell, H. Bussey, R. W. Davis, B. Dujon, H. Feldmann, F. Galibert, J. D. Hoheisel, C. Jacq, M. Johnston, E. J. Louis, H. W. Mewes, Y. Murakami, P. Philippsen, H. Tettelin, and S. G. Oliver, *Science* **274,** 563 (1996).

[3] Yeast Protein Database, http://www.proteome.com/YPDhome.html

[4] S. Fields and O. Song, *Nature* (*London*) **340,** 245 (1989); P. Bartel and S. Fields (eds.), "The Yeast Two-Hybrid System." Oxford University Press, New York, 1997.

0076-6879/00 $30.00

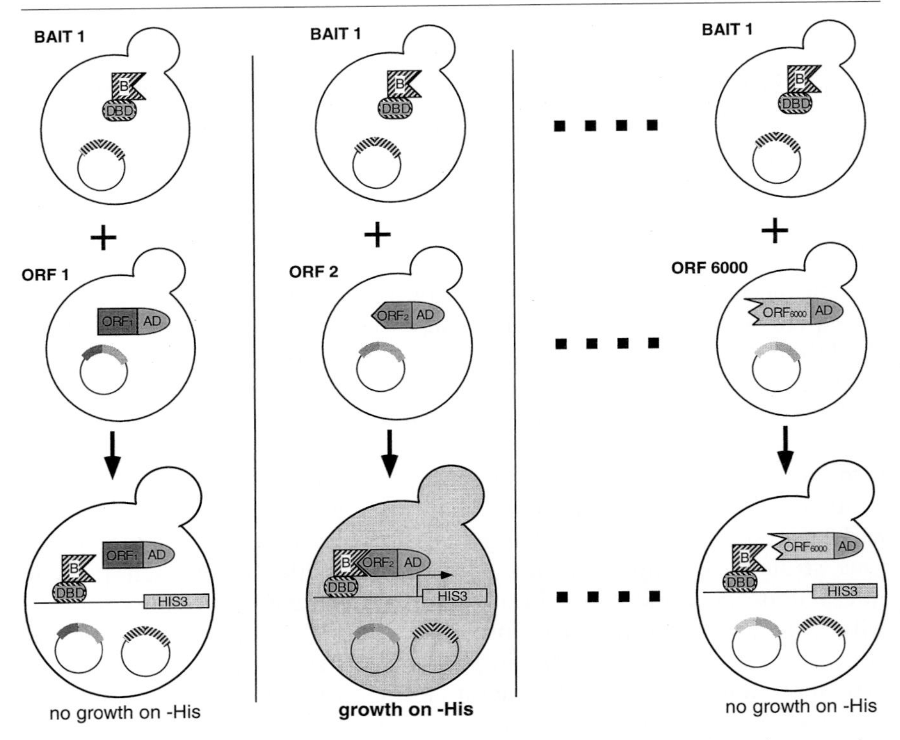

FIG. 1. Schematic of the genome-wide protocol for two-hybrid tests. A strain expressing a Gal4 DNA-binding domain–ORF fusion protein (*top row*) is mated to 6000 different yeast strains of the opposite mating type that express one of the yeast ORFs fused to the Gal4 activation domain (*second row*). If the hybrid proteins interact, a transcriptional activator is reconstituted and a reporter gene (*HIS3*) is expressed. This expression allows the diploid cell to grow on selective media lacking histidine. Each column therefore represents one individual two-hybrid test.

can be readily observed. Moreover, each element serves as a reference for its neighbors, so that a large number of assays can be compared side-by-side. As each potential protein partner exists only once in the array, the problem of multiple positives corresponding to a single interaction does not occur as it does in screens of a random library. In this chapter, we describe methods for constructing and screening a protein array that contains most of the proteins present in *S. cerevisiae*. However, the array may represent any group of proteins: all the expressed proteins present in an organism, a family of related proteins, or even a set of proteins or peptides not found in nature.

Generation of Protein Array Suitable for High-Throughput Screening

The array that we have generated comprises most of the ORFs predicted by the *S. cerevisiae* sequence as a set of approximately 6000 yeast colonies. Each colony expresses a unique ORF fused to the activation domain of the Gal4 transcription factor. In order for large numbers of fusion genes to be constructed, we generated DNAs encoding the proteins of interest with flanking sequences that permit cloning into yeast vectors by homologous recombination.[5,6] These flanking sequences are added by two rounds of polymerase chain reaction (PCR)[7] (Fig. 2). The first-round PCR uses primers specific for each gene. In our case primers were designed to amplify full-length ORF sequences; however, they may be designed to amplify any protein-coding regions, such as individual domains. In addition, both upstream and downstream first-round primers contain ~20-nucleotide tails that are identical to sequences in the two-hybrid vectors pOAD and pOBD2 (Fig. 3). The tail for each of the forward primers is in common, and the tail for each of the back primers is in common. Thus, the second-round PCR step, with a single pair of 70-mer primers terminating in the common sequences, adds an additional 50 nucleotides of homology to the insert DNAs. The recombination event takes place inside the cell after cotransformation of vector and insert DNA by a modification of the lithium acetate–polyethylene glycol technique.[8] An aliquot of each transformant is stored individually and spotted into a position in the array on an agar plate.

Preparation of DNA

First-Round Primers

Forward primer: 5′ AA TTC CAG CTG ACC ACC ATG XXX_{20-30} 3′; where *XXX* represents ORF-specific sequences of 20 to 30 bases and the ATG represents the start codon of the yeast ORF

Reverse primer: 5′ GA TCC CCG GGA ATT GCC ATG *** XXX_{20-30} 3′; where XXX_{20-30} represents 20 to 30 bases of the reverse complement of ORF-specific sequences at the terminus of the reading frame and the triple asterisks (***) represent the reverse complement of one of the three stop codons

[5] T. L. Orr-Weaver, J. W. Szostak, and R. J. Rothstein, *Proc. Natl. Acad. Sci. U.S.A.* **78,** 6354 (1981).

[6] K. R. Oldenburg, K. T. Vo, S. Michaelis, and C. Paddon, *Nucleic Acids Res.* **25,** 451 (1997).

[7] J. R. Hudson, Jr., E. P. Dawson, K. L. Rushing, C. H. Jackson, D. Lockshon, D. Conover, C. Lanciault, J. R. Harris, S. J. Simmons, R. Rothstein, and S. Fields, *Genome Res.* **7,** 1169 (1997).

[8] R. D. Gietz, R. H. Schiestl, A. R. Willems, and R. A. Woods, *Yeast* **11,** 355 (1995).

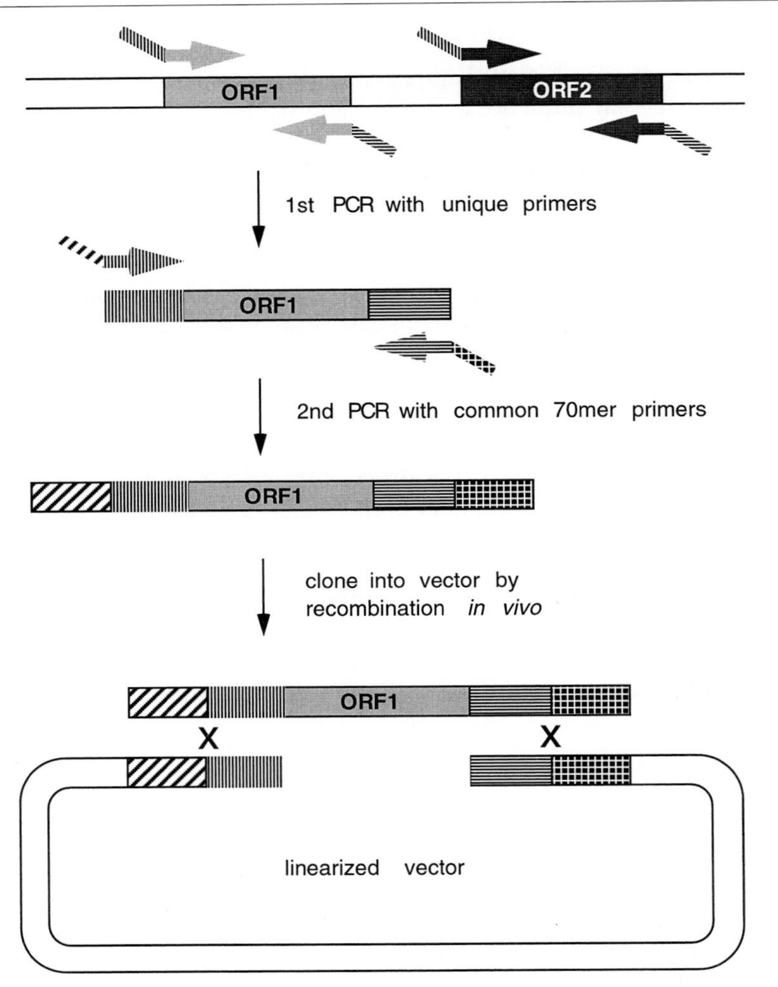

FIG. 2. Construction of a genome-wide protein array for yeast. Yeast ORFs were amplified (first PCR) with specific primers that generated products with common 5′ and common 3′ 20-nucleotide tails. A second PCR generated products with common 5′ and common 3′ ~70-nucleotide tails. The common ~70-nucleotide ends allow cloning into linearized two-hybrid expression vectors by recombination.

Second-Round Primers

Forward primer: 5′ C TAT CTA TTC GAT GAT GAA GAT ACC CCA CCA AAC CCA AAA AAA GAG ATC GAA TTC CAG CTG ACC ACC ATG 3′

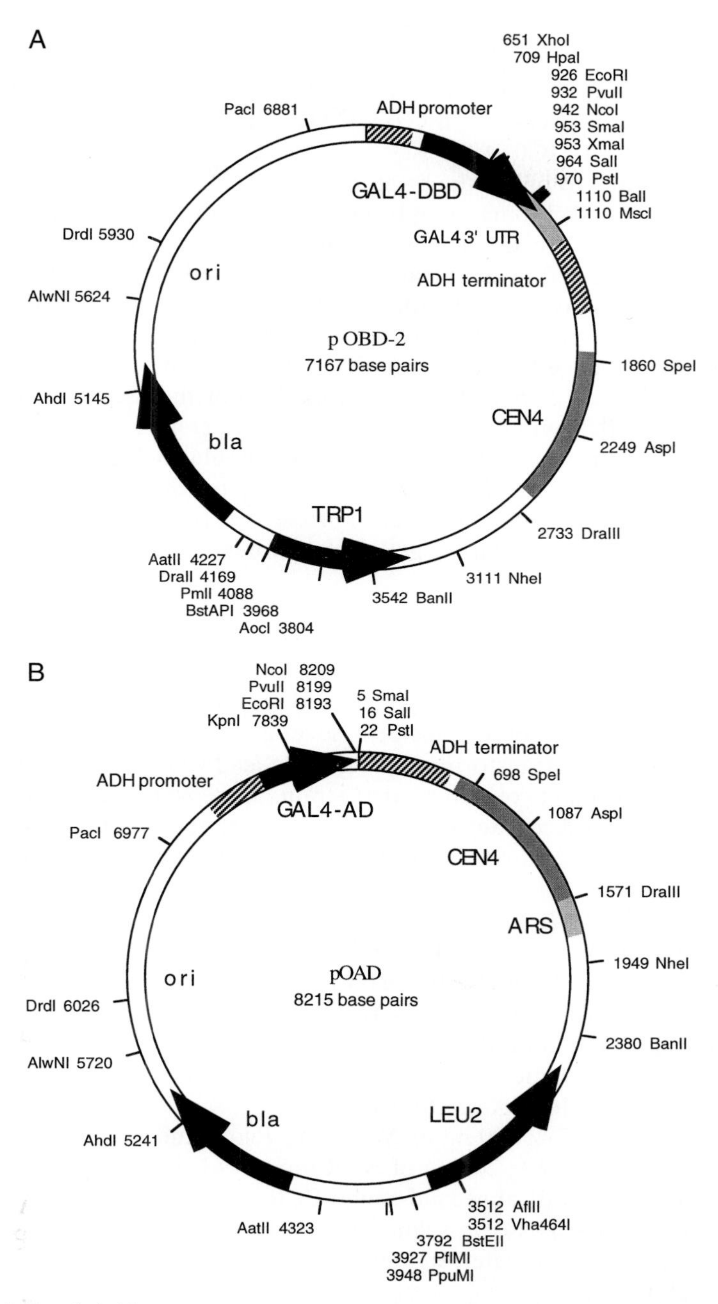

FIG. 3. Two-hybrid vectors used for expression of open reading frames as fusions to the Gal4 DNA-binding domain (A) and the Gal4 activation domain (B). A *CEN4* element is present for stable maintenance of the plasmids. The *TRP1* and *LEU2* genes permit maintenance of the plasmids, and selection for diploids on media lacking leucine and tryptophan.

Reverse primer: 5′ C TTG CGG GGT TTT TCA GTA TCT ACG ATT CAT AGA TCT CTG CAG GTC GAC GGA TCC CCG GGA ATT GCC ATG 3′

It is advisable to purify the primers to the highest level practical [high-performance liquid chromatography (HPLC) or polyacrylamide gel electrophoresis (PAGE)] because deletions or insertions in the primers may result in fusion proteins being out of frame when expressed in the yeast. Standard PCR conditions may be used, although it is convenient to group the PCRs according to the approximate product size, and to vary the extension times accordingly. A suggested set of conditions for the first round is as follows: 30 ng of template, 1.5 m*M* Mg^{2+}, 5 units of *Taq* polymerase (Perkin-Elmer, Norwalk, CT), 0.02 unit of *Pfu* polymerase (Stratagene, La Jolla, CA), and 20 pmol of each primer; the second round should include 5 ng of first-round template, 1.5 m*M* Mg^{2+}, 0.6 unit of *Taq* polymerase, 0.003 unit of *Pfu* polymerase, and 2.5 pmol of each primer.[7] A fraction of each product should be analyzed by agarose gel electrophoresis.

Cloning by Recombination

Cloning by recombination is suitable for the generation of 96 yeast clones in a 96-well microtiter format, and may be scaled up or down as needed. One person may transform several microtiter plates in 1 day. Selection of pOAD transformants requires a *leu2* yeast strain; selection of pOBD2 transformants requires a *trp1* strain. Moreover, at least one of the haploid strains must contain a two-hybrid reporter gene under Gal4 control. We use two strains, PJ69-4A and PJ69-4α, which are identical except for mating type.

PJ69-4A[9]: *MAT***a** *trp1-901 leu2-3,112 ura3-52 his3-200 gal4Δ gal80Δ LYS2::GAL1-HIS3 GAL2-ADE2 met2::GAL7-lacZ*

PJ69-4α[7]: *MATα trp1-901 leu2-3,112 ura3-52 his3-200 gal4Δ gal80Δ LYS2::GAL1-HIS3 GAL2-ADE2 met2::GAL7-lacZ*

The following materials are required:

Yeast synthetic media lacking appropriate nutrients
Host strain: PJ69-4A or PJ69-4α
AD vector DNA: 200 ng of *Nco*I- and *Pvu*II-linearized pOAD
DBD vector DNA: 200 ng of *Nco*I- and *Pvu*I-linearized pOBD2
Insert DNA: 3 μl of second-round PCR products
Carrier DNA: Dissolve salmon sperm DNA (7.75 mg/ml; Sigma, St. Louis, MO) in water and store at −20° after autoclaving (15 min, 121°)
Dimethyl sulfoxide (DMSO)

[9] P. James, J. Halladay, and E. A. Craig, *Genetics* **144,** 1425 (1996)

Lithium acetate, 0.1M

96PEG solution: Add the following reagents and bring the volume to 100 ml with water: 45.6 g of polyethylene glycol (PEG; Sigma), 6.1 ml of 2 M lithium acetate, 1.14 ml of 1 M Tris-HCl (pH 7.5), 232 μl of 0.5 M EDTA

Procedure

1. Inoculate 50 ml of yeast extract–peptone–dextrose (YEPD) medium[10] with 0.3 ml of a host strain in a 250-ml conical flask and grow overnight at 30°.
2. The following morning, harvest the yeast by centrifugation (3500 rpm, 3 min, room temperature on a benchtop centrifuge), pour off the supernatant, and resuspend the yeast in 2 ml of 0.1 M lithium acetate.
3. Boil 0.58 ml of carrier DNA for 5 min, and plunge it into ice water.
4. Add the following reagents in order to a 50-ml tube and immediately shake for 30 sec followed by vortexing for 1 min: 20.73 ml of 96PEG, 0.58 ml of carrier DNA, 200 ng of vector DNA, and 2.62 ml of DMSO.
5. Add the 2 ml of yeast suspension to the mixture and mix well by hand for 1 min.
6. Pour the mixture into a sterile trough and pipette 245 μl into each well of a 96-well microtiter dish.
7. Add 3 μl of insert DNA to each well; seal with secure plastic or aluminium tape and vortex for 4 min.
8. Incubate for 30 min at 42°.
9. Recover the yeast by centrifugation (2000 rpm, 7 min, room temperature in a benchtop centrifuge) and aspirate the supernatant with an eight-channel pipetter.
10. Add 200 μl of water to each well and resuspend the yeast.
11. Plate each of the yeast suspensions onto a 35-mm plate containing selective medium lacking leucine or tryptophan (–Leu or –Trp) and incubate at 30° for 2–3 days.

Nature of Array

Two related issues must be addressed when designing an array suitable for genome-wide screens. First, what constitutes an array element, and second, to what extent can each array element be validated? An array element may constitute a single transformant whose plasmid has been validated by DNA sequencing; alternatively, it may constitute a pool of several transformants that are uncharacterized. We chose to pool two inde-

[10] M. D. Rose, F. Winston, and P. Hieter, "Methods in Yeast Genetics." Cold Spring Harbor Laboratory Press, Cold Spring Harbor, New York, 1990.

pendent colonies per transformation for each array element. The pooling of too many colonies risks the possibility of variants outcompeting the desired one in culture. We determined experimentally that the sum of all errors introduced in the cloning process was approximately 10% (oligonucleotide synthesis, PCR, recombination cloning) and reasoned that the presence of two independent colonies would increase the probability that at least one correctly synthesized fusion protein would be present per array element. In practice, our array probably contains 85–90% of the correct inserts. The most desirable validation is confirmation of protein expression via a Western blot. Failing this, sequencing of the ORF DNA and flanking sequences from a plasmid can indicate that it is likely to express a full-length in-frame fusion protein. Sequencing at least a short region around the recombination sites can indicate that the insert is in the correct reading frame. Neither Western blot nor sequencing reactions are trivial to carry out on large numbers of yeast transformants. A genetic test of correct protein expression (such as encoding a gene downstream of the ORF that must be translated in-frame to allow the transformants to grow) would be helpful but would not detect either substitutions or in-frame deletion or insertion events. An array containing Gal4 AD fusion proteins is preferable to one containing DBD fusions because DBD fusions with the ability to activate the reporter gene independent of a protein interaction result in a high background on the two-hybrid selection plates.

The array of yeast colonies should be of high density, easy to manipulate, and ordered in such a way that the identity of individual colonies may be readily determined. As the "384" format has become an industry standard, we grew the ~6000 yeast colonies as liquid cultures (YEPD or –Leu medium) in sixteen 384-well microtiter plates (Nalge Nunc International; Rochester, NY) before resuspending the pelleted cells in 20% (v/v) glycerol and storing them at −20°. The liquid cultures can be thawed and inoculated onto solid medium with a robotic replicating tool (see the next section). This procedure generates an array of AD transformants that is suitable for probing with DBD fusion proteins, using the two-hybrid assay. The array can be stored on solid medium lacking leucine for up to 4 months at 4°. A working copy of the array grown on complete medium (YEPD) can be generated from the –Leu copy every 4 weeks, for use in the screens.

Screening of Large Protein Array for Protein–Protein Interactions

Interaction Mating Using Robotic Procedures

In order for the ORF array to be screened for protein–protein interactions, it is mated to a strain expressing Gal4 DBD fused to the protein of interest (Fig. 4). Yeast carrying the AD fusions are haploids of mating type **a**, opposite to that of the haploid cells carrying a fusion to the Gal4 DBD (mating type α), such that both fusions can be introduced into the same diploid cell by mating. The use of mating for the screens is more convenient than cotransformation of the two plasmids into the same cell. After 1 to 2 days, the colonies are transferred to medium selective for diploid cells (–Leu –Trp). This step serves to indicate how successful the mating has been. Some DBD–ORF fusions appear to inhibit mating by unknown mechanisms. After an additional 2 days of growth, the diploid cells are transferred to medium that allows growth only of cells expressing a reporter gene under the control of Gal4-binding sites. In strain PJ69-4A and PJ69-4α, the *GAL4* gene is deleted, and the reporter genes *ADE2* and *HIS3* are under the control of two different Gal4-dependent promoters. Growth dependent on *ADE2* expression demands more Gal4 activity than does growth dependent on *HIS3* expression. If the two-hybrid selection is too stringent (adenine selection), many interacting protein combinations will not be detected; thus, our experiments have used only histidine selection. In addition, as strains containing different DBD–ORF fusion proteins vary in basal reporter gene activation, we routinely supplement the –His medium with a competitive inhibitor of His3, 3-aminotriazole (3-AT). Prescreening diploid colonies (colonies resulting from the mating of haploid cells containing the DBD–ORF plasmid with haploid cells containing the parent AD vector) containing each DBD–ORF fusion protein on plates containing different concentrations of 3-AT can establish the level required to eliminate growth due to basal reporter gene expression.

A robotic workstation (Biomek 2000; Beckman Coulter, Fullerton, CA) can be used to carry out all the colony transfer steps of the screen. Thus, in a single screening experiment, approximately 6000 individual two-hybrid assays can be carried out *en masse,* and a single workstation operating for 12 hr a day can process 20 such screens weekly. Throughput may be increased further by using additional workstations, by increasing the density of the array, or by using pooling strategies. Although the upper limit on the density of arrays composed of living yeast colonies is significantly lower than that for oligonucleotide arrays, we have successfully experimented with 768 and 1536 colonies per 86 $\times$ 128 mm plate.

The following procedure describes a genome-wide screen of the yeast ORFs (Fig. 4), using the *HIS3* gene as a reporter for protein–protein interactions. The array is present on 16 OmniTray plates in 384-colony format. The screen can also be carried out manually with a hand-held 384-pin replicating tool such as that manufactured by Nalge Nunc.

The following materials are required:

Single-well OmniTrays (Nalge Nunc)

Yeast synthetic media lacking appropriate nutrients

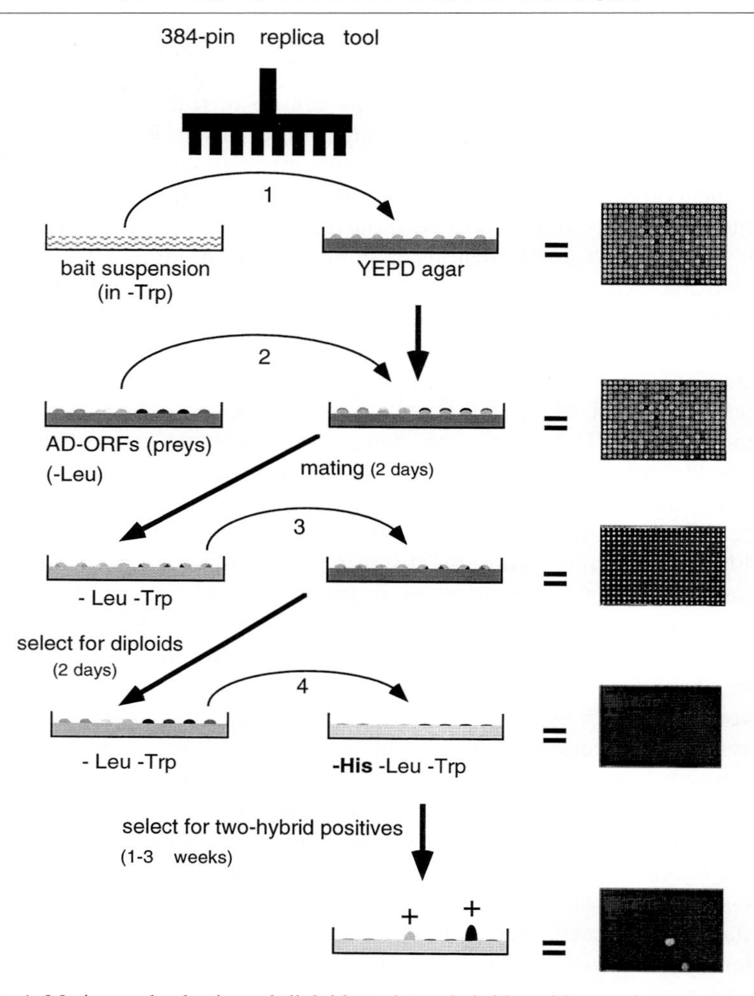

FIG. 4. Mating and selection of diploids and two-hybrid positives, using an automated pinning device. Steps using the replica pinning tool are numbered 1–4. In step 1, haploid yeast expressing the DNA-binding domain–ORF fusion protein are transferred from a liquid culture to solid medium. In step 2, haploid cells from the activation domain array are transferred with the same tool onto the DBD–ORF fusion-expressing cells for mating. After 2 days of growth, the colonies are transferred (step 3) to –Leu –Trp medium to allow growth of diploids. After another 2 days of growth, the diploid cells are transferred in step 4 to two-hybrid selective plates (–Trp –Leu –His), on which visible colonies start to grow after a few days. Weak two-hybrid positives may, however, take up to 3 weeks or more to form colonies.

384-pin replicating tool [Beckman Coulter (robotic) or Nalge Nunc (manual)]

Yeast strains expressing Gal4 AD–ORF fusion proteins arrayed in 384-colony format Yeast strain expressing Gal4 DBD–ORF fusion

Procedure

1. Inoculate 20-ml of complete yeast medium (YEPD) with a 0.3-ml liquid preculture of the strain expressing a Gal4 DBD–ORF fusion into a 250-ml conical flask. The culture is suitable for use for several days if stored at 4°.
2. Grow overnight with shaking at 30°.
3. Pour the DBD–ORF culture into an OmniTray before pinning. Transfer the DBD–ORF culture onto OmniTrays containing solid YEPD medium by using a 384-pin replicating tool (in our hands, the use of a tool for inoculating colonies is superior to spreading of lawns). Allow the yeast to dry onto the plates for 10–20 min.
4. Transfer the yeast AD–ORF colonies from the array directly onto the dry DBD–ORF cultures, using the same tool.
5. Incubate for 1–2 days at 30° to allow mating of the strains.
6. Transfer the colonies to OmniTrays containing medium lacking leucine and tryptophan.
7. Incubate for 2 days at 30° to allow diploid cells to grow.
8. Transfer the colonies to OmniTrays containing selective medium lacking leucine, tryptophan, and histidine, supplemented with an appropriate amount of 3-AT (for nonactivating DBD–ORF strains, we routinely use 3 m*M* 3-AT).
9. Incubate for 1–3 weeks at 30°.
10. Score and document the screen.

Interpreting Results

Colonies on the two-hybrid selective plates that clearly outgrow the background of the majority of colonies are scored as positive. However, some DBD–ORF fusions generate many nonreproducible two-hybrid positives in this procedure. The origin of these false positives is not well understood. We therefore screen each DBD–ORF fusion strain in duplicate, and consider only positives that appear twice. Using this criterion, we screened 192 yeast ORFs against the yeast protein array, and obtained 281 positives from 87 ORFs yielding reproducible results.[11] Many of these interactions

[11] P. Uetz, L. Giot, G. Cagney, T. A. Mansfield, R. S. Judson, V, Narayan, D. Lockshon, M. Srinivasan, P. Pochan, A. Qureshi-Emili, Y. Li, B. Godwin, D. Conover, T. Kalbfleisch, G. Vijayadamodar, M. Yang, M. Johnston, S. Fields, and J. M. Rothberg, *Nature* **403,** 623 (2000). http://depts.washington.edu/sfields/

have been observed previously. In cases where previously observed interactions were not detected, several explanations are possible: the AD–ORF or DBD–ORF fusion proteins may not be expressed, they may localize to an extranuclear location, or full-length ORFs may preclude some interactions. The expression of ORF fragments or domains under ~60 kDa (the upper size limit for passage through the yeast nuclear pore), or the exclusion of hydrophobic membrane domains may overcome these problems to some extent.

The application of genome-wide two-hybrid screening to organisms whose protein complements are considerably larger and more complex than yeast is daunting. However, the increased efficiency of solid-phase oligonucleotide technologies and the reduction in oligonucleotide cost permit the synthesis of the large numbers of primers necessary for the PCR steps. In addition, the increasing use of robotics to handle the array steps makes the protocol feasible for complements of genes numbering more than 10,000.

Acknowledgments

This work was supported by NIH Grants GM54415 and RR11823 and by a grant from the Merck Genome Research Institute. P.U. is supported by a grant from the Deutsche Akademische Austauschdienst. S.F. is an investigator of the Howard Hughes Medical Institute.

[2] LexA-Based Two-Hybrid Systems

By SARAH J. FASHENA, ILYA G. SEREBRIISKII, and ERICA A. GOLEMIS

Introduction

The two-hybrid system, a genetic screening strategy which utilizes yeast, facilitates the rapid identification of novel interacting proteins.[1] The repertoire of two-hybrid components has expanded to include a variety of distinct and generally noninterchangeable sets of reagents (Table I). The availability of different two-hybrid systems provides researchers with options that may serve to optimize an interaction screen for a particular protein. This chapter outlines the LexA-based variants of the basic two-hybrid approach and

[1] C. T. Chien, P. L. Bartel, R. Sternglanz, and S. Fields, *Proc. Natl. Acad. Sci. U.S.A.* **88,** 9578 (1991).

0076-6879/00 $30.00

TABLE I
COMPARISON OF PLASMIDS AND STRAINS FOR USE IN LexA TWO-HYBRID SYSTEMS[a]

Characteristic	I	II
Basic bait plasmids		
Typical representative	pEG202, pMW series	pBTM116
Yeast markers	HIS3	ADE2/TRP1
Bacterial markers	$Ap^R/Cm^R/Km^R$	Ap^R
Expression	Constitutive	Constitutive
Alternative bait plasmids		
NLS enhanced	pJK202	pBTM116-NLS
Inducible expression	pGilda	N/A
COOH-terminal LexA	pNLexA	N/A
Reporters		
1. Auxotrophic	*lexAop*-LEU2 (integrated)	*lexAop*-HIS3 (integrated)
Typical representative	EGY48, EGY191	L40
Sensitivity regulation	Number of ops in series of strains	Addition of 3-AT
Yeast markers	None	LYS2
2. Colorimetric	*lexAop*-lacZ, GFP (episomal)	*lexAop*-LacZ (integrated)
Typical representative	pSH18-34, pJK103	L40
Sensitivity regulation	Number of ops in series of plasmids	N/A
Yeast markers	URA3	URA3
Libraries		
Yeast markers	TRP1	LEU2
Expression	Inducible	Constitutive
Activation domain	B42 (moderate)	VP16 (very strong)
Oligo(dT) primed	Yes	No
Random primed	Yes	Yes
Epitope tagged	Yes	No
In vitro transcription	No	Yes
Commercially available	Yes	No
GAL4 based	No	Yes
Other features		
Compatible with dual bait system	Fully	Partially
Elimination of some false positives by Gal dependence	Yes	No

[a] The yeast host strains used for these systems are generally EGY48 (or related strains EGY191, 195) (*Matα trp1 his3 ura3 6ops-LEU2*) for system I, and L40 [*MAT***a** *his3Δ200 trp1-901 leu2-3,112 ade2 LYS2::(lexAop)$_4$-HIS3 URA3::(lexAop)$_8$-lacZ GAL4,* constructed by R. Sternglanz] for system II. Strains and plasmids are commercially available from a variety of sources (Invitrogen, Clontech, Stratagene, and Display Systems); some reagents can be acquired by request from the Golemis laboratory at the Fox Chase Cancer Center: phone (215) 728-2860, fax (215) 728-3616, e-mail ea_golemis@fccc.edu

compares two systems that are frequently used to screen for novel protein interactions.[2,3]

Briefly, two-hybrid screens require two basic classes of chimeric proteins: (1) a fusion of a particular protein (protein x) to a DNA-binding

[2] A. B. Vojtek, S. M. Hollenberg, and J. A. Cooper, *Cell* **74,** 205 (1993).
[3] J. Gyuris, E. A. Golemis, H. Chertkov, and R. Brent, *Cell* **75,** 791 (1993).

domain (DBD–x) and (2) a fusion of a cDNA library to a transcriptional activation domain (AD–y). The DBD–x fusion protein (or Bait) is transformed into an appropriate strain of *Saccharomyces cerevisiae* containing a dual reporter system. For the screens described here, the DBD used is the bacterial protein LexA. The dual reporters contain *lexA* operator sites (DNA-binding sites) upstream of the *lacZ* gene, which encodes β-galactosidase, and a gene that confers the ability to grow in selective media lacking specific amino acids. The dual reporter system can be introduced into yeast either by stable integration (LexA system II) or by a combination of transformation and stable integration (LexA system I). After transformation of the AD–cDNA library (or Prey) into such yeast, Bait–Prey interactions are revealed by colonies that can grow on selective media and are blue when assayed with the chromogenic substrate 5-bromo-4-chloro-3-indolyl-β-D-galactopyranoside (X-Gal). For the purposes of clarity, we discuss the specifics of LexA systems I and II separately, underscoring the strengths and weaknesses of each approach at the end of the chapter.

LexA System I

In the LexA system I,[3] the selective reporter contains *lexA* operators upstream of *LEU2,* a gene required in the yeast leucine biosynthetic pathway. A number of yeast strains into which the *lexAop-LEU2* reporter gene has been chromosomally integrated are available. The two most commonly used strains are EGY48 and EGY191, which provide differential sensitivity for *trans*-activation.[4] The *lexAop-lacZ* reporter plasmid is introduced into the yeast by cotransformation with the bait construct. The activation domain of AD–y is provided by the "acid-blob" B42.[5] Yeast containing Bait alone cannot transcriptionally activate either of the dual reporters, so they appear white on medium containing X-Gal and fail to grow in the absence of leucine. When such yeast are retransformed with an AD–cDNA library (or Prey), Bait–Prey protein interactions are revealed by colonies that can grow in the absence of leucine and are blue when assayed on X-Gal plates (Fig. 1).

LexA System II

The LexA system II utilizes the *S. cerevisiae* L40 reporter strain, which contains two integrated reporters, the bacterial *lacZ* gene and the yeast *HIS3* gene, which are both under the transcriptional control of *lexA* opera-

[4] J. Estojak, R. Brent, and E. A. Golemis, *Mol. Cell. Biol.* **15,** 5820 (1995).
[5] D. M. Ruden, J. Ma, Y. Li, K. Wood, and M. Ptashne, *Nature* (*London*) **350,** 250 (1991).

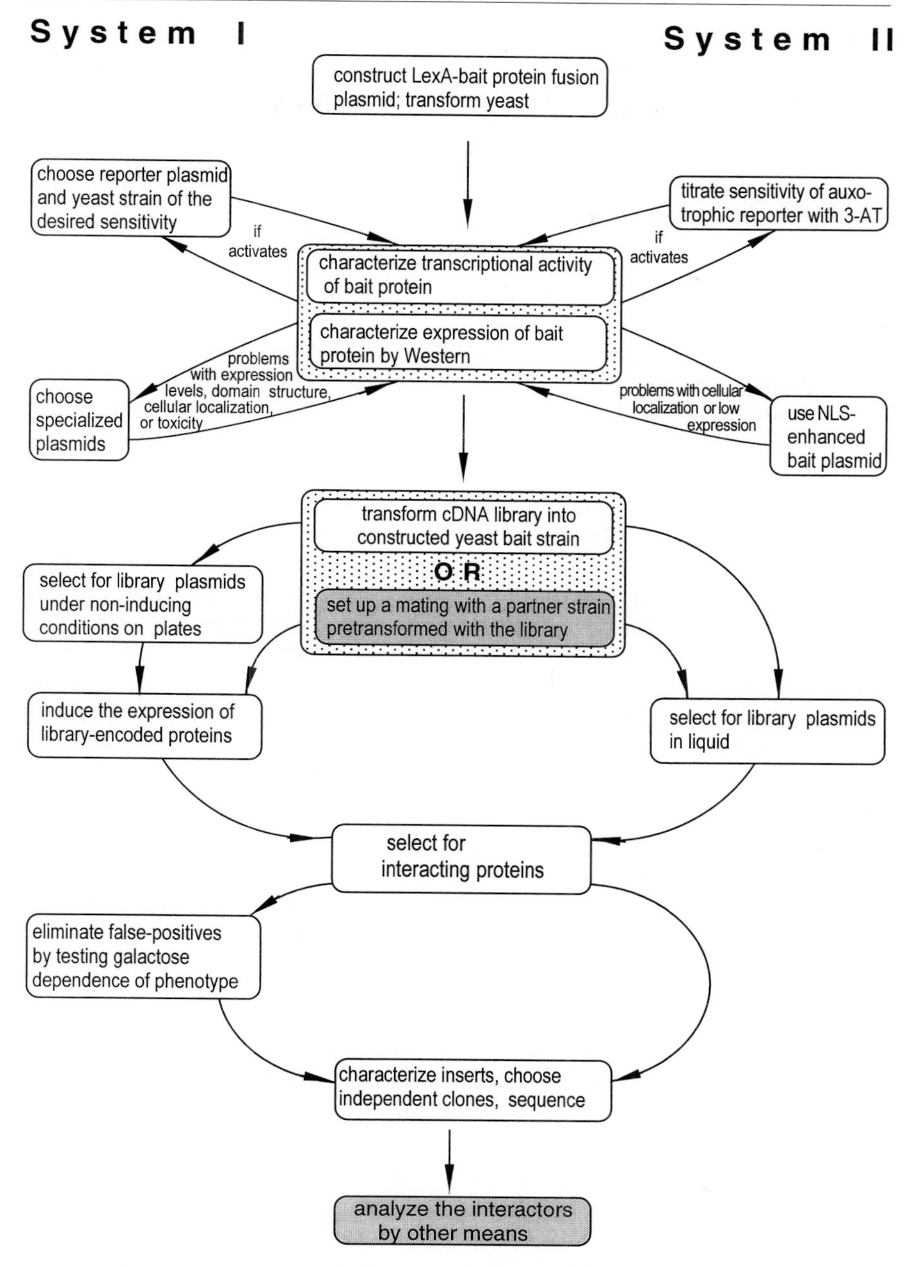

FIG. 1. Flow chart showing progression of Bait development and characterization, and sequence of events for subsequent library screen. See text for details.

tors.[6] The *HIS3* gene is required in the yeast histidine biosynthetic pathway and its expression facilitates colony growth in media lacking histidine. The L40 strain is transformed with both the LexA fusion protein DBD–x and the AD–cDNA library, which encodes fusion proteins containing the herpes simplex virus VP16 activation domain. Bait–Prey interactions are, therefore, revealed by yeast colonies that can grow in the absence of histidine and turn blue on media containing X-Gal (Fig. 1).

Bait Protein

Because the two-hybrid assay requires that the bait protein is transcriptionally silent and can enter the nucleus, its design should take into consideration a number of features. First, any "protein x" sequences that could confer transcriptional activation or membrane attachment should be excised. Second, a decision should be made as to whether the full-length or truncated domains of protein x should be included. If the proposed fusion protein is predicted to be ~60–80 kDa or larger, it may be synthesized at low levels or posttranslationally digested by yeast proteases. Proteins expressed at low levels are frequently inactive in the initial transcriptional activation assays, but may be upregulated during selection and consequently demonstrate a high background for transcriptional activation. Undiagnosed proteolytic processing of the Bait protein, on the other hand, can lead to screens performed with truncated forms of the larger intended Bait. Of note, truncated domains of larger proteins can display high levels of *trans*-activation mediated by Bait alone. Moreover, truncated protein domains can display promiscuity with respect to the spectrum of proteins with which they interact. To evaluate if the Bait suffers from any of the preceding problems, appropriate controls should be performed as outlined and lysates from yeast transformed with the Bait construct should be analyzed by Western transfer and immunoblotting (see below).

LexA System I

The construction of a plasmid that encodes LexA fused to protein x (Bait) is the first step in any interactor hunt. The Bait construct (pBait) is then cotransformed with a *lexAop-lacZ* reporter plasmid into a yeast strain containing a chromosomally integrated *lexAop-LEU2* reporter gene. Control experiments should be performed to demonstrate that the Bait protein is stable, can enter the nucleus and bind *lexA* operator sites, and does

[6] S. M. Hollenberg, R. Sternglanz, P. F. Cheng, and H. Weintraub, *Mol. Cell. Biol.* **15,** 3813 (1995).

not appreciably activate transcription of the *lexA* operator-based reporter genes. Results from these control experiments will dictate if the Bait can be used to screen a library, using the initial test conditions, or if modifications of the Bait and/or different combinations of reporter strains/plasmids must be used.

Methods. The following protocols utilize basic yeast media and transformation procedures, described in greater detail in Refs. 7 and 8. The LexA screen should be performed expeditiously (within 10 days of the Bait/reporter transformation), because fusion protein expression levels decline in yeast stored at 4°. Alternatively, a frozen stock of yeast transformed with Bait and reporter can be prepared and maintained stably at −70°.

1. Using standard subcloning techniques, insert the DNA encoding protein x into the polylinker of pMW103 (Table I[9]) to make an in-frame LexA fusion protein, incorporating a translational stop sequence at the carboxy-terminal end of the Bait sequence.

2. To evaluate Bait-mediated transcriptional activation, transform the *S. cerevisiae lexAop-LEU2* reporter strain EGY48, using the following combinations of LexA fusion and *lexAop-lacZ* plasmids:
   a. pBait + pMW112 (test for background activation)
   b. pSH17-4 + pMW112 (positive control for activation)
   c. pRFHM1 + pMW112 (negative control for activation)

Use 100–500 ng of plasmid DNA of each construct/transformation. See Refs. 10 and 11 for a small-scale transformation protocol. A description of the properties of each of the plasmids is provided in Table I. *Optional:* The Bait can also be evaluated for transcriptional repression activity as described elsewhere.[7,8]

3. Plate each transformation mixture on Glu/CM –Ura, –His dropout plates (i.e., plates containing complete medium with glucose and lacking uracil and histidine). Incubate for 2–3 days at 30° to select for yeast that contain plasmids. Each transformation should yield ~200 colonies.

4. Make a master plate from which specific colonies can be assayed for activation of *lacZ* and *LEU2* reporters. Several independent colonies are assayed for each combination of plasmids because Bait expression levels

[7] E. A. Golemis, I. Serebriiskii, J. Gyuris, and R. Brent, *in* "Current Protocols in Molecular Biology" (F. M. Ausubel, ed.), pp. 20.1.1–20.1.35. John Wiley & Sons, New York, 1997.

[8] E. Golemis and I. Serebriiskii, *in* "Cells: A Laboratory Manual" (D. L. Spector, R. Goldman, and L. Leinward, eds.). Cold Spring Harbor Laboratory Press, Cold Spring Harbor, New York, 1997.

[9] M. A. Watson, R. Buckholz, and M. P. Weiner, *BioTechniques* **21,** 255 (1996).

[10] D. Gietz, A. St. Jean, R. A. Woods, and R. H. Schiestl, *Nucleic Acids Res.* **20,** 1425 (1992).

[11] J. Hill, K. A. Ian, G. Donald, and D. E. Griffiths, *Nucleic Acids Res.* **19,** 5791 (1991).

can vary among colonies, thereby producing an apparent (and misleading) heterogeneity in Bait-mediated transcriptional activation of the two reporters. Alternative procedures for making master plates have been described elsewhere.[7,8]

Use sterilized flat-edged toothpicks to pick six colonies and restreak in a grid on a fresh Glu/CM –Ura –His plate. Incubate at 30° overnight.

5. On the second day, restreak from the master plate to fresh Gal/CM –Ura X-Gal and Gal/CM –Ura –His –Leu plates.

Make a relatively thick streak on X-Gal plates (buffered to pH 7.0 to optimize β-galactosidase activity) because yeast do not grow well at neutral pH; hence it is easier to determine a blue–white phenotype when starting with a large inoculum. Conversely, make a relatively thin streak on –Leu plates to prevent cross-feeding among colonies from occurring.

Use of the two LexA fusions (pSH17-4, pRFHM1) as positive and negative controls allows a rough assessment of the transcriptional activation profile of LexA Bait proteins. pMW103 itself (or related plasmid pEG202) is not a good negative control because the peptide encoded by the uninterrupted polylinker sequence can minimally activate transcription.

6. Incubate the plates at 30° for up to 4 days. Blue colonies should be evaluated repeatedly over the course of 3 days poststreaking, while the LEU2 phenotype should be assayed between days 2 and 3.

Optimally, at 2 days poststreaking to Gal/CM –Ura X-Gal, **b** should be bright blue, **c** should be white, and **a** should be white or very light blue. At 2 days poststreaking to Gal/CM –Ura –His –Leu, **b** growth should equal that of the Glu/CM –Ura –His master plate, while **a** and **c** growth should be negligible. Growth of the **a** transformants indicates that the Bait is transcriptionally active and should be structurally modified (a truncated form of full-length protein x could be fused to LexA). Alternatively, this system provides the option of reducing such background *trans*-activation by using different reporter constructs and/or yeast strains that alter the sensitivity of the transcriptional assays.

LexA System II

The construction of pBait, which encodes LexA fused to protein x, is the first step in any interactor hunt. The pBait construct is then transformed into the *S. cerevisiae* L40 dual reporter strain, which contains integrated *lexAop-lacZ* and *lexAop-HIS3* genes.[6] Control experiments should be performed to demonstrate that the Bait protein is stable, can enter the nucleus and bind *lexA* operator sites, and does not appreciably activate transcription of the *lexA* operator-based reporter genes. Results from these control experiments will dictate if the Bait can be used to screen a library under the initial test conditions or if modifications of the Bait and/or test conditions are required.

Methods. The following protocols utilize basic yeast media and transformation protocols.[7,8]

1. Using standard subcloning techniques, insert the DNA encoding protein x into the polylinker of pBTM116 (Table I[12]) to make an in-frame LexA fusion protein, incorporating a translational stop sequence at the carboxy-terminal end of protein x.
2. To evaluate Bait-mediated transcriptional activation, transform the L40 reporter strain with the following combinations of LexA fusion and VP16 plasmid:
   a. pBait* + pVP16 (empty library plasmid†; test for background activation)
   b. pLexA-Ras + pVP16Raf (positive control for activation)
   c. pLexA-Lamin + pVP16 (negative control for activation)
3. Plate each transformation mix to Glu/CM –Leu –Trp dropout plates and incubate at 30° for 2 to 3 days.
4. To test for Bait-mediated *trans*-activation, pick six colonies and restreak in a grid to Glu/CM –Trp –His –Ura –Leu –Lys and Glu/CM –Leu –Trp plates. Incubate at 30° overnight.

The following results are expected: **c** transformants should grow on Glu/CM –Leu –Trp, but not Glu/CM –Trp –His –Ura –Leu –Lys plates; **b** transformants should grow on both Glu/CM –Leu –Trp and Glu/CM –Trp –His –Ura –Leu –Lys plates; and **a** transformants should grow on Glu/CM –Leu –Trp, but not Glu/CM –Trp –His –Ura –Leu –Lys plates. Growth of the **a** transformants indicates that the Bait is transcriptionally active and should be structurally modified (a truncated form of full-length protein x could be fused to LexA). Alternatively, this system provides the option of reducing such background *trans*-activation by including 3-amino-1,2,4-triazole (3-AT) in the medium. 3-AT is an inhibitor of the dehydratase encoded by the *HIS3* gene.[13] The amount of supplemental 3-AT must be titrated to optimize for inhibition of Bait-mediated *trans*-activation, while

[12] P. Bartel, C. Chien, R. Sternglanz, and S. Fields, *in* "Cellular Interactions in Development: A Practical Approach" (D. A. Hartley, ed.), pp. 153–179. Oxford University Press, Oxford, 1993.

* Expression of LexA alone minimally activates transciption from the dual reporters, therefore pBTM116 is not an appropriate negative control.

† See Refs. 10 and 11 for small-scale transformation protocol.

[13] E. W. Jones and G. R. Fink, *in* "The Molecular Biology of the Yeast *Saccharomyces cerevisiae*" (J. N. Strathern, E. W. Jones, and J. R. Broach, eds.), pp. 181–299. Cold Spring Harbor Press, Cold Spring Harbor, New York, 1982.

minimizing the 3-AT-mediated reduction in plating efficiency (for protocol see Ref. 14).

Detection of Bait Protein Expression

Initial characterization of a Bait protein requires an assessment of Bait expression levels and molecular weight. For this purpose, Western analysis of lysates from yeast transformed with pBait is performed.

1. Inoculate two primary transformants of each pBait assayed and one positive control for protein expression. Sterilely transfer colonies from the master plate into the appropriate selective liquid medium. Grow overnight cultures on a roller drum or other shaker at 30°. The next morning, dilute saturated cultures into fresh tubes containing 3–5 ml of selective medium to a starting density of OD_{600} ~0.15 and grow on a shaker (~4–6 hr).

2. When the culture reaches OD_{600} ~0.45–0.7, centrifuge 1.5 ml of culture in an Eppendorf tube at full speed for 3–5 min. Aspirate the supernatant and resuspend the pellet in 50 μl of 2× Laemmli sample buffer, vortex, and boil the samples for 5 min. Such samples may be analyzed immediately or stably frozen −70°.

Laemmli sample buffer (2×): Stable for ~2 months at room temperature:
2-Mercaptoethanol, 10% (v/v)
Sodium dodecyl sulfate (SDS), 6% (w/v)
Glycerol, 20% (v/v)
Bromphenol blue (0.2 mg/ml)

3. Pulse centrifuge to pellet debris and chill on ice. Load 20–50 μl per well and analyze by SDS–polyacrylamide gel electrophoresis (SDS–PAGE), Western transfer, and immunoblotting according to standard protocols.[15,16] Bait can be visualized with antibody to LexA or protein x.

Troubleshooting and Modifications of Baits

There are three basic problems that can be identified and potentially corrected before screening.

1. *The Bait activates transcription:* If the Bait weakly or moderately activates transcription (LexA system I), the control experiments can be

[14] A. B. Vojtek, J. A. Cooper, and S. M. Hollenberg, *in* "The Yeast Two Hybrid System: A Practical Approach" (P. Bartel and S. Fields, eds.), pp. 29–42. Oxford University Press, New York, 1997.

[15] E. Harlow and D. Lane, "Antibodies: A Laboratory Manual." Cold Spring Harbor Laboratory Press, Cold Spring Harbor, New York, 1988.

[16] J. Sambrook, E. F. Fritsch, and T. Maniatis, "Molecular Cloning: A Laboratory Manual." Cold Spring Harbor Laboratory Press, Cold Spring Harbor, New York, 1989.

repeated using more stringent reporter plasmids and strains (see Table I) to reduce background. Alternatively, an integrating form of the Bait vector could be used (see Table I), which would result in a stable reduction of protein levels. To address high Bait-mediated transcriptional activity in either LexA system I or II, it is also possible to make a series of truncated protein x domains fused to LexA. The truncated fusion proteins will, it is hoped, retain many binding characteristics of the full-length protein, but lose the high transcriptional activity. The truncated fusion protein approach is particularly useful if Bait alone vigorously activates transcription. LexA system II also provides a potentially significant advantage in that inclusion of 3-AT in the media, an inhibitor of the *HIS3*-encoded dehydratase, can reduce Bait-mediated background *trans*-activation to workable levels.[13]

2. *The Bait is proteolytically cleaved or is expressed at low levels:* If Bait is predicted to be ~60–80 kDa or larger and/or is expressed at low levels, a truncated fusion protein approach may be the best strategy (LexA system I or II). Alternatively, the vectors pJK202 (LexA system I) and pBTM116-NLS (LexA system II), which incorporate a nuclear localization sequence into the fusion protein, can be utilized to address low expression levels. The nuclear localization signal should provide a means by which the Bait is concentrated in the nucleus, thereby achieving assayable levels of activity.

3. *Low frequency and/or reduced growth rate of transformants expressing the bait protein:* Either of these results would suggest that the Bait is toxic to the yeast. Because Bait-mediated toxicity can make library screening difficult or impossible, protein x could be recloned into pGilda (credited to D. A. Shaywitz, MIT Center for Cancer Research, Cambridge, MA), a GAL1-inducible LexA fusion protein vector (LexA system I). This should facilitate regulated protein expression that can be optimized for assaying protein interactions while minimizing Bait-related toxicity.

Note: All of the alternative Bait expression vectors (LexA system I) confer ampicillin resistance (Amp^R) to bacteria. Consequently, passage through KC8 bacteria is necessary to facilitate isolation of the library plasmid after a library screen.[7,8]

Selecting an Interactor

LexA System I

In "traditional" usage, LexA system I requires two successive transformations, first with pBait and reporters and second with the cDNA library. Libraries are generally cloned into the pJG4-5 vector, in which the cDNA expression cassette is under the control of the *GAL1* galactose-inducible promoter. A number of libraries compatible with LexA system I are available and protocols detailing their use have been described elsewhere.[7,8]

In brief, yeast are initially plated to select for the library plasmid under noninducing conditions, Glu/CM –Ura –His –Trp. Such transformed yeast are then replated onto Gal/Raff/CM –Ura, –His, –Trp, –Leu dropout plates to induce expression of cDNA encoded proteins and select for Bait–Prey protein interactions. Library plasmids from colonies identified in the second plating are then purified by passage through bacteria and used to transform yeast cells in a final specificity screen.[7,8]

More recently, an alternative means of introducing the library into yeast, via mating, has become popular. This means of performing an interactor hunt is described in the following section on interaction mating.

LexA System II

LexA system II requires cotransformation of pBait and the cDNA library into the L40 strain, which contains the stably integrated dual reporters. Libraries are generally cloned into the pVP16 vector. A number of libraries compatible with LexA system II are available and protocols detailing their use have been described elsewhere.[2,6,14]

In brief, the L40 dual reporter strain is transformed with pBait and plated on Glu/CM –Trp plates. Inoculate transformed yeast into Glu/CM –Trp –Ura medium, expand liquid culture, and prepare competent pBait L40 yeast. Transform a pVP16 library into competent pBait L40 yeast, let the yeast recover in Glu/CM –Trp –Leu –Ura medium (to select for pBait, pVP16, and the HIS3 reporter gene), and plate onto Glu/CM –Trp –Leu –Ura –Lys –His to select for Bait–Prey interactions. His^+ colonies are then assayed on X-Gal plates to provide independent confirmation of protein interaction. Pick and regrow each His^+LacZ^+ colony on Glu/CM –Trp –Leu –Ura –Lys –His plates to isolate single library plasmids encoding interacting proteins.

Mapping Interaction Domains

The two-hybrid system can also be modified to delineate the domains necessary for protein–protein interactions. Such directed analyses have been performed for pairs of proteins known to interact, such as simian virus 40 (SV40) large T antigen–p53[17] and Ras–Raf.[2] Moreover, a directed analysis of novel interacting pairs of proteins identified in either LexA system can be informative for directing future research. Therefore, once an interacting protein(s) has been identified, it is possible to delineate the domains necessary for the protein interaction by making a series of mutants

[17] K. Iwabuchi, B. Li, P. Bartel, and S. Fields, *Oncogene* **8,** 1693 (1993).

in the DNA encoding the Bait protein. The domain structure of the Bait protein, based on sequence alignment or structural data if available, can be used as a guide for designing truncated forms of the Bait. A systematic approach entailing the use of convenient restriction enzyme sites located within the DNA encoding the Bait is also effective. Once a series of pBait truncations has been made, they can be tested for transcriptional activity by cotransformation with the specific interacting protein–AD construct (Prey) and a *lexAop-lacZ* reporter plasmid into an appropriate yeast strain. Analysis of blue–white colonies on X-Gal plates should define the domain(s) of the Bait protein that confer interaction. Point mutations can also be introduced to define specific residues within domains that are essential for binding. A similar approach can be used to delineate the critical domains/residues of the Prey that are required for interaction with the Bait. As detailed above, appropriate controls should always be performed and lysates from yeast transformed with the truncated protein constructs should be analyzed by Western transfer and immunoblotting.

Benefits of LexA Systems I and II

Both LexA systems have been used successfully to isolate proteins that interact with a wide variety of proteins. The decision to use either LexA system I or II should, therefore, be made on the basis of intrinsic properties of the Bait, the availability of compatible libraries derived from tissues/animals of interest, and the level of proficiency of the researcher with yeast techniques. It should be noted that the reagents available for use with either system are not generally compatible with those of the other system, and therefore changing systems is not trivial.

Although more complicated, LexA system I provides the greatest flexibility because of the array of Bait and reporter constructs available. When these constructs are used in the appropriate combinations, it is possible to modulate the readout of the transcriptional assays to suit the Bait, thereby overcoming a host of problems including Bait-mediated *trans*-activation and/or toxicity. For example, Bait-mediated *trans*-activation can be addressed by using different reporter constructs that alter the sensitivity of the transcriptional assays. Moreover, problems due to Bait-mediated toxicity can be sidestepped by using expression constructs in which the Bait is under the control of the inducible *GAL1* promoter. Utilizing a galactose-inducible promoter is also helpful when confronted with false positives resulting from mutations in the yeast strain. This system also provides a means by which Prey-mediated toxicity can be managed because LexA system I-compatible libraries are under the control of the inducible *GAL* promoter. Finally, LexA system I reagents have been adapted for use in dual Bait screens, which facilitate the analysis of more complicated protein interactions.[18] To take full advantage of LexA system I, however, it is necessary to be familiar with the constructs and yeast strains available and fairly proficient in yeast biology.

LexA system II is potentially less labor intensive and should facilitate the rapid identification of interacting proteins. This system provides a potentially significant advantage in that inclusion of 3-AT in the medium can reduce Bait-mediated background *trans*-activation to workable levels. The system is, however, limited in that the sensitivity of the integrated dual reporter system is fixed. Consequently, if the inclusion of 3-AT in the medium does not reduce Bait-mediated *trans*-activation to acceptable levels, a modified Bait must be constructed. Moreover, inducible Bait expression constructs are not available to address Bait-mediated toxicity. Notably, LexA system II does benefit from a larger number of available compatible libraries, including the *GAL4*-based two-hybrid libraries.[1,19]

Acknowledgments

E.A.G. and I.S. are supported in two-hybrid studies by a grant from the Merck Genome Research Institute, and by core funds CA-06927 to the Fox Chase Cancer Center. S.J.F. is supported by NIH T32 CA-09035.

[18] I. Serebriiskii, V. Khazak, and E. A. Golemis, *J. Biol. Chem.* **274,** 17080 (1999).
[19] S. Fields and O. Song, *Nature* (*London*) **340,** 245 (1989).

[3] Interaction Mating Methods in Two-Hybrid Systems

By MIKHAIL G. KOLONIN, JINHUI ZHONG, and RUSSELL L. FINLEY, JR.

Introduction

The yeast two-hybrid system is a powerful assay for protein–protein interactions.[1] As described in chapters 1, 2, 6, and 10 of this volume, several versions of the two-hybrid system have been developed. Most versions have the following features. The two proteins to be tested for interaction are expressed as hybrids in the nucleus of a yeast cell. One of the proteins is fused to the DNA-binding domain (DBD) of a transcription factor and the other is fused to a transcription activation domain (AD). If the two hybrid

[1] S. Fields and O. Song, *Nature* (*London*) **340,** 245 (1989).

0076-6879/00 $30.00

proteins interact, they reconstitute a functional transcription factor that activates one or more reporter genes that contain binding sites for the DBD. This simple assay has been widely used to identify new interacting proteins from libraries, to test interactions between small and large sets of proteins, to map protein networks, and to address the functions of individual proteins and protein interactions.[2–5] Here we describe interaction mating, a two-hybrid variation that can be adapted to most versions of the system and that can simplify and facilitate most two-hybrid experiments.[6,7]

In interaction mating, the AD and DBD fusion proteins begin in two different haploid yeast strains with opposite mating types. To test for interaction, the hybrid proteins are brought together by mating, a process in which two haploid cells fuse to form a single diploid cell. The technique is fairly simple, requiring only that the two haploid strains be mixed together and incubated overnight on rich medium. The diploids that form are then tested for reporter activation as in a conventional two-hybrid experiment. In this chapter we describe methods for interaction mating to facilitate several routine two-hybrid experiments. We begin with a protocol for screening a library for new interacting proteins or peptides. We then present a simple cross-mating assay to test interactions between small sets of proteins. This assay is particularly useful when testing the specificity of proteins isolated in an interactor hunt. We also present methods for identification and characterization of whole networks of proteins by reiterative interactor hunts and by screening arrayed libraries by mating. Finally, we present two mating approaches to study the functions of individual protein interactions. One approach is to isolate interaction mutants and their suppressors, and the other is to isolate peptides that can disrupt specific protein interactions *in vivo*.

The interaction mating methods can be adapted for use with most of the currently popular two-hybrid systems. The key to choosing a combination of strains and plasmids is to ensure that the two strains to be mated are of opposite mating type (*MAT***a** and *MAT*α) and that both have auxotrophies to allow selection for the appropriate plasmids and reporter genes. Here we present detailed procedures for using interaction mating with the version

[2] A. R. Mendelsohn and R. Brent, *Curr. Opin. Biotechnol.* **5,** 482 (1994).

[3] J. B. Allen, M. W. Walberg, M. C. Edwards, and S. J. Elledge, *Trends Biochem.* **20,** 511 (1995).

[4] P. L. Bartel and S. Fields, *in* "The Yeast Two-Hybrid System" (P. L. Bartel and S. Fields, eds.), pp. 3–7. Oxford University Press, Oxford, 1997.

[5] R. Brent and R. L. Finley, Jr., *Annu. Rev. Genet.* **31,** 663 (1997).

[6] R. L. Finley, Jr. and R. Brent, *Proc. Natl. Acad. Sci. U.S.A.* **91,** 12980 (1994).

[7] C. Bendixen, S. Gangloff, and R. Rothstein, *Nucleic Acids Res.* **22,** 1778 (1994).

TABLE I
YEAST STRAINS AND PLASMIDS

Strain	Genotype	Refs.[a]
MATα strains		
EGY48	*MATα his3 trp1 ura3 3LexAop-LEU2::leu2*	1
RFY231	*MATα his3 trp1Δ::hisG ura3 3LexAop-LEU2::leu2*	2
RFY251	*MATα ura3 his3 trp1Δ::hisG lys2Δ201 3LexAop-LEU2::leu2*	Unpublished
*MAT***a** strains		
RFY206	*MAT***a** trp1Δ::*hisG his3Δ200 leu2-3 lys2Δ201 ura3-52*	3
YPH499	*MAT***a** ura3-52 lys2-801 ade2-101 trp1-Δ63 his3-Δ200 leu2-Δ1	4

Plasmids	Marker	Expression cassette[b]	Refs.[a]
Bait plasmids			
pEG202	*HIS3*	*ADH1*p-LexA (DBD)	1
pJK202	*HIS3*	*ADH1*p-LexA (DBD)	5
AD fusion plasmids			
pJG4-5	*TRP1*	*GAL1*p-B42 (AD)	6
pMK2	*URA3*	*ADH1*p-B42 (AD)	Unpublished
Reporter plasmids			
pSH18-34	*URA3*	8LexAop-*lacZ*	5
pCWX24	*LYS2*	8LexAop-*lacZ*	7

[a] Key to references: (1) J. Estojak, R. Brent, and E. A. Golemis, *Mol. Cell. Biol.* **15,** 5820 (1995); (2) M. G. Kolonin and R. L. Finley, Jr., *Proc. Natl. Acad. Sci. U.S.A.* **95,** 14266 (1998); (3) R. L. Finley, Jr. and R. Brent, *Proc. Natl. Acad. Sci. U.S.A.* **91,** 12980 (1994); (4) R. S. Sikorski and P. Hieter, *Genetics* **122,** 19 (1989); (5) E. A. Golemis, *et al., in* "Current Protocols in Molecular Biology" (F. M. Ausubel, *et al.,* eds.), Vol. 20.1. John Wiley & Sons, New York, 1999; (6) J. Gyuris, E. Golemis, H. Chertkov, and R. Brent, *Cell* **75,** 791 (1993); (7) C. W. Xu, A. R. Mendelsohn, and R. Brent, *Proc. Natl. Acad. Sci. U.S.A.* **94,** 12473 (1997).

[b] Expression cassette indicates the promoter (from the *ADH1* or *GAL1* gene, or a minimal promoter with eight LexA-binding sites), followed by the fusion moiety (LexA DBD or B42 AD), or reporter (*lacZ*).

of the two-hybrid system developed by Brent and colleagues.[8,9] In this system the DBD is LexA and the reporters are usually *lacZ* and *LEU2*. The AD is typically a bacterial sequence called B42, and is expressed conditionally from the yeast *GAL1* promoter, which is induced on galactose media and repressed on glucose. The protocols presented here were developed with the yeast strains and plasmids indicated in Table I. Recipes for

[8] J. Gyuris, E. Golemis, H. Chertkov, and R. Brent, *Cell* **75,** 791 (1993).

[9] R. L. Finley, Jr. and R. Brent, *in* "DNA Cloning, Expression Systems: A Practical Approach" (B. D. Hames and D. M. Glover, eds.), pp. 169–203. Oxford University Press, Oxford, 1995.

yeast media can be found elsewhere in this volume[10] and are available at our Web site.[11]

Isolating New Interactors from Libraries

The interaction mating two-hybrid hunt is conducted by mating a haploid strain that expresses the LexA fusion protein, or "bait," with a haploid strain of the opposite mating type that has been pretransformed with the library DNA expressing AD fusions. The resulting diploids are then screened for interactors. This approach can save considerable time and materials when one library is to be screened with two or more bait proteins. Hunts with different baits can be performed by mating each bait-expressing strain with a thawed aliquot of yeast that had been transformed with library DNA in a single large-scale transformation. The interaction mating approach is also useful for bait proteins that interfere with yeast viability because it avoids the difficulty associated with transforming a sick strain expressing such a bait. Finally, because the reporters are less sensitive to transcription activation in diploids than they are in haploids, interaction mating provides a way to reduce the background from baits that activate transcription.

The interaction mating hunt can be divided into four tasks. First, the bait strain is constructed and characterized. The characterization will include a test of how much the bait protein activates the reporters on its own, using a method that mimicks the library screen. The second task is to create the pretransformed library by high-efficiency yeast transformation with the library plasmid DNA. Yeast transformed with the library can be frozen in many aliquots and thawed individually for each interactor hunt. The third task is to mate the bait strain with an aliquot of the pretransformed library strain and allow diploids to form on solid yeast extract–peptone–dextrose (YPD) medium overnight. The resulting diploid yeast are then screened for interactors as in a conventional two-hybrid hunt by testing for galactose-dependent activation of the reporters. The final task is to isolate and characterize the cDNAs from the positives. The initial characterization should include a demonstration that each cDNA encodes a protein that interacts specifically with the original bait. This can be achieved with a cross-mating assay (see Testing Interactions between Small Sets of Proteins by Cross-Mating Assay).

[10] S. J. Fashena, I. G. Serebriiskii, and E. A. Golemis, *Methods Enzymol.* **328,** Chap. 2, 2000 (this volume).

[11] R. L. Finley, Jr., Web site: http://cmmg.biosci.wayne.edu/rfinley/lab.html

Constructing and Characterizing Bait Strain

1. Construct the bait plasmid (pBait) by inserting a cDNA encoding the protein of interest in-frame with LexA into an appropriate bait vector (see Table I), using standard cloning methods or by recombination cloning[12] (e.g., see Constructing New Bait Strains by Recombination Cloning). The bait vectors for the system used here have the *HIS3* marker and 2 μm origin of replication (Table I).

2. Transform RFY206 (or another appropriate strain; see Table I) with pBait and with a *lacZ* reporter plasmid such as pSH18-34. This and most *lacZ* reporter plasmids in this system have the *URA3* marker and 2 μm origin of replication (Table I). This and all yeast transformations can be performed by the lithium acetate method.[13] Select transformants on Glu/CM –Ura, –His medium (glucose/complete medium lacking uracil and histidine).

3. Characterize the bait strain. This may include performing a Western blot to show that full-length stable bait protein is synthesized, and a test for reporter activation by the bait.[9,10,14] The most useful approach to test whether the bait activates the *LEU2* reporter is to mate the bait strain with a control strain containing empty library vector. The resulting diploids can then be plated on medium lacking leucine, which will mimic the library hunt. This test can be done concurrently with the actual hunt as described below.

Preparing Pretransformed Library and Control Strains

1. Perform a large-scale transformation of RFY231 (or another appropriate strain; see Table I) with library plasmid DNA. The library plasmid in this system is derived from pJG4-5, which contains the *TRP1* marker and 2 μm origin of replication, in addition to the *GAL1* promoter driving expression of the AD fusions.[8] Select library transformants on Glu/CM–Trp plates. Collect library transformants by scraping plates, washing yeast, and resuspending in 1 pellet volume of glycerol solution [65% (v/v) glycerol, 25 m*M* Tris (pH 7.5)]. Freeze 0.5-ml aliquots at −70 to −80°.

2. To make a control strain, transform RFY231 with the empty library vector, pJG4-5, and select transformants on Glu/CM –Trp plates. Combine several colonies to inoculate 30 ml of Glu/CM –Trp liquid medium and

[12] H. Ma, S. Kunes, P. J. Schatz, and D. Botstein, *Gene* **58,** 201 (1987).

[13] R. H. Schiestl and R. D. Gietz, *Curr. Genet.* **16,** 339 (1989).

[14] E. A. Golemis, I. Serebriiskii, R. L. Finley, Jr., M. G. Kolonin, J. Gyuris, and R. Brent, *in* "Current Protocols in Molecular Biology" (F. M. Ausubel, *et al.*, eds.), Vol. 20.1. John Wiley & Sons, New York, 1999.

grow at 30° with shaking to OD_{600} ~3. Collect the cells by centrifugation and resuspend in 1 pellet volume of glycerol solution. Freeze in several 0.5-ml aliquots at −70 to −80°.

3. Determine the plating efficiency of the frozen cells by thawing an aliquot of each pretransformed strain from step 1 and step 2 and making 10-fold serial dilutions in sterile water. Plate 100 μl of each of 10^{-5}, 10^{-6}, and 10^{-7} dilutions on Glu/CM –Trp plates and incubate for 2 to 3 days. Count the colonies and determine the number of colony-forming units (CFU) per unit volume of frozen yeast. The plating efficiency for a typical library transformation and for the control strain will be ~1×10^8 CFU/100 μl.

Screening for Interactors by Mating

In this section, the bait strain is mated with a frozen aliquot of the pretransformed library strain. At the same time, the bait strain is mated with a frozen aliquot of the control strain to test for background activation of the *LEU2* reporter by the bait itself.

1. Grow a 30-ml culture of the bait strain in Glu/CM –Ura, –His liquid medium to ~3×10^7 cells/ml (OD_{600} ~1.5). Collect the cells by centrifugation at ~1100*g* for 5 min at room temperature. Resuspend the cell pellet in sterile water to a final volume of 1 ml. This will correspond to about 10^9 CFU/ml.

2. Mix 200 μl of the bait strain with ~1×10^8 CFU (~0.1 to 1 ml) of the pretransformed library strain in a microcentrifuge tube. In a second tube mix 200 μl of the bait strain with 200 μl of the pretransformed control strain. This should approximate a twofold or greater excess of bait strain over pretransformed library strain. Under these conditions, ~10% of the colony-forming units in the pretransformed library strain will mate with the bait strain. This mating efficiency should be considered when calculating how much of the library is being screened.

3. Collect the cells by centrifugation as described in step 1 and resuspend in 200 μl of YPD medium. Plate each suspension on a 100-mm YPD plate. Incubate for 12 to 15 hr at 30° to allow mating.

4. Add ~1 ml of liquid Gal/CM –Ura, –His, –Trp medium to the lawns of mated yeast on each plate and suspend the cells with a sterile applicator stick or glass rod. To induce expression of the library proteins with galactose, dilute each cell slurry into 100 ml of Gal/CM –Ura, –His, –Trp liquid medium in a 500-ml flask and incubate with shaking for 6 hr at 30°.

5. Collect the cells by centrifugation as described in step 1 and wash by resuspending in 30 ml of sterile water and centrifuging again. Resuspend each pellet in 5 ml of sterile water. Measure the OD_{600} and dilute with water to ~1×10^8 cells/ml (OD_{600} ~5.0).

6. For each mating make a series of dilutions from 10^{-1} to 10^{-6} in sterile water. To determine the titer of diploids, plate 100 μl each of the 10^{-4}, 10^{-5}, and 10^{-6} dilutions on 100-mm Gal/Raff/CM –Ura, –His, –Trp plates. To determine the level of activation of the *LEU2* reporter by the bait itself (here called the *trans*-activation potential; see below), plate 100 μl each of the six dilutions on 100-mm Gal/Raff/CM –Ura, –His, –Trp, –Leu plates. To select for interactors from the library mating, plate 100 μl of the 10^{-1} dilution on each of twenty 100-mm Gal/Raff/CM –Ura, –His, –Trp, –Leu plates and 100 μl of the undiluted cells on each of twenty 100-mm Gal/Raff/CM –Ura, –His, –Trp, –Leu plates. Incubate the plates at 30° for 2 to 5 days.

7. Count the colonies. The *trans*-activation potential of the bait can be represented as the number of Leu^+ colonies per colony-forming unit (Leu^+/CFU). This is determined by the ratio of the number of colonies growing on Gal/Raff/CM –Ura, –His, –Trp, –Leu to the number of colonies growing on Gal/Raff/CM –Ura, –His, –Trp medium for a particular dilution of the control mating. A bait with essentially no *trans*-activation potential will produce less than 10^{-6} Leu^+/CFU. To identify all the interactors in the pretransformed library, it may be necessary to pick and characterize all the background Leu^+ colonies produced by the *trans*-activation potential of the bait itself, particularly if the background is high and the frequency of interactors is low. The minimum number of Leu^+ colonies that should be picked is given by

(*trans*-Activation potential, Leu^+/CFU)
× (no. of library transformants screened)

8. Test the Leu^+ clones that grew on the selection plates for galactose-dependent Leu^+ and $LacZ^+$ activity as follows. Streak Leu^+ colonies to 1-cm^2 patches on Glu/CM –Ura, –His, –Trp plates to turn off the *GAL1* promoter. If the Leu^+ colonies were close together it may be necessary to streak purify to single colonies first on Gal/Raff/CM –Ura, –His, –Trp, –Leu plates, and then patch the single colonies onto Glu/CM –Ura, –His, –Trp plates. Include patches of strains that will serve as controls for the interaction phenotype. Grow for 1–2 days at 30°.

9. Transfer the patches to velvet and replicate to five plates in the order shown below. The first four are indicator plates; the last plate serves as a transfer control and a source of yeast for plasmid isolation or polymerase chain reaction (PCR). Incubate for 1 to 5 days at 30° and record the Leu^+ and $LacZ^+$ phenotypes.

1. Glu/CM –Ura, –His, –Trp, X-Gal (5-bromo-4-chloro-3-indolyl-β-D-galactopyranoside)

2. Gal/Raff/CM –Ura, –His, –Trp, X-Gal
3. Glu/CM –Ura, –His, –Trp, –Leu
4. Gal/Raff/CM –Ura, –His, –Trp, –Leu
5. Glu/CM –Ura, –His, –Trp

Alternatively, the LacZ phenotype can be determined by filter lift, X-Gal overlay, or liquid assays.[15–17] Pick for further analysis yeast that require galactose to grow on the –Leu plates. Yeast that grow on the –Leu glucose plates, where the AD fusion is not expressed, are false positives. Because the *LEU2* reporter is more sensitive than the *lacZ* reporter it is possible for a weak interaction to activate only *LEU2* and not *lacZ*.

Characterizing Interactors

To distinguish true positives from false positives, which may arise, for example, from mutations in the reporter genes, it is important to demonstrate that the reporter phenotypes depend on the cDNA. It is also important to show that each cDNA encodes a protein that interacts specifically with the bait. Both of these objectives can be accomplished by isolating the cDNA and reintroducing it into a haploid yeast strain. Interactions can then be assayed by mating the new strain with a number of bait strains. The bait strains should include the original bait strain used for the hunt and strains expressing control baits. The interaction specificity test is performed in the cross-mating assay described in the next section.

If many positives are identified it is useful to restrict the analysis to nonredundant cDNA isolates. The quickest way to identify nonredundant cDNAs is to restriction digest PCR products generated with vector-specific primers[9] (e.g., for pJG4-5 use forward primer FP-J, 5′-GCTGAAATCGAATGGTTTTCATG-3′, and reverse primer RP-J, 5′-GAGTCACTTTAAAATTTGTATACAC-3′). As a template source, fresh yeast cells can be used. For example, cells from the Glu/CM –Ura, –His, –Trp plates in step 9 of the previous section can be added directly to a PCR; the reaction should be preincubated at 95° for 5 min to release DNA. Alternatively, yeast DNA minipreparations can be used as the source of template DNA.[9,14] The amplified products can be digested directly with a restriction enzyme *Alu*I or *Hae*III) and then analyzed by gel electrophoresis. The isolates that produce matching digestion patterns correspond to the same cDNA.

Once the unique cDNA classes have been identified the corresponding

[15] R. W. West, Jr., R. R. Yocum, and M. Ptashne, *Mol. Cell. Biol.* **4,** 2467 (1984).
[16] A. B. Vojtek, S. M. Hollenberg, and J. A. Cooper, *Cell* **74,** 205 (1993).
[17] H. M. Duttweiler, *Trends Genet.* **12,** 340 (1996).

plasmids or cDNAs can be retrieved and introduced into another strain to mate with different bait strains. To isolate the library plasmids, yeast DNA minipreparations can be prepared and used to transform *Escherichia coli,* and then plasmid DNA can be amplified and purified from individual *E. coli* clones. However, there are two potentially faster alternatives to isolating the plasmid from *E. coli.* In one approach the yeast DNA minipreparation is used to transform yeast directly. Transformants are selected on –Trp plates. Colonies that have taken up the *TRP1* plasmid but not the *URA3 lacZ* reporter plasmid or the *HIS3* bait plasmid can be identified by replicating transformants to Glu/CM –His and Glu/CM –Ura plates. Trp^+, Ura^- His^- yeast can then be used in the cross-mating assay described below to test specificity. In the other approach each cDNA is PCR amplified directly from the positive yeast colonies, using FP-J and RP-J, and then subcloned directly into pJG4-5 in a new strain by recombination cloning (Fig. 1). In this procedure RFY231 or a similar strain is cotransformed with the crude PCR product and the library vector (pJG4-5) linearized at the cloning site. The linearized vector will be repaired in the yeast cells by homologous recombination, resulting in insertion of the cDNA into the vector. This recombination cloning approach is discussed in more detail in the context of making new baits (see Constructing New Bait Strains by Recombination Cloning).

Testing Interactions between Small Sets of Proteins by Cross-Mating Assay

The cross-mating assay is a quick method to test interactions between sets of AD- and DBD-fused proteins (Fig. 2). The AD fusions are expressed in individual strains of one mating type, and the DBD fusions are expressed in individual strains of the opposite mating type. Binary interactions between the AD and DBD fusion proteins can be easily sampled by mating the two sets of strains on one plate and replicating to indicator plates (Fig. 2). This approach reduces the number of yeast transformations needed to test interaction between two sets of proteins, because otherwise a separate transformation would be necessary for each binary interaction to be tested. Cross-mating is particularly useful for testing the specificity of new interactors isolated in an interactor hunt. In this case the AD fusion strains express new cDNAs, and the DBD fusion strains include the original bait strain used for the screen, plus a number of strains expressing unrelated baits.

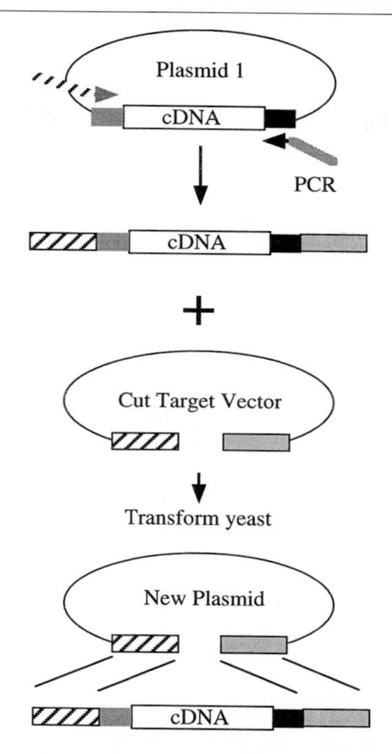

FIG. 1. Recombination subcloning in yeast to make new plasmids. A DNA fragment from one vector is PCR amplified with primers that each include about 60 bases of sequence from the region flanking the cloning site of a target vector. Yeast are then cotransformed with the PCR product and with the target vector, which has been linearized with a restriction enzyme that cuts within the cloning site. The target vector will be repaired in the yeast cell by homologous recombination, resulting in a new plasmid containing the DNA fragment in the target vector [H. Ma, S. Kunes, P. J. Schatz, and D. Botstein, *Gene* **58,** 201 (1987)]. When the starting vector (Plasmid 1) is the same vector as the target vector, the PCR primers should correspond to sequences 60 bp upstream and downstream of the cloning site. For example, when subcloning cDNAs from pJG4-5 to pJ-G4-5 in a new strain, use primers FP-J and RP-J, as described in text.

Procedure

1. Streak parallel lines (at least 2 mm wide) of AD fusion strains (e.g., RFY231 transformed with various pJG4-5 derivatives) onto a Glu/CM –Trp plate. Incubate at 30° for 1–3 days.

2. Streak parallel lines (at least 2 mm wide) of bait strains (e.g., RFY206 transformed with pSH18-34 and various bait plasmids) to a Glu/CM –Ura, –His plate. Incubate at 30° for 1–3 days.

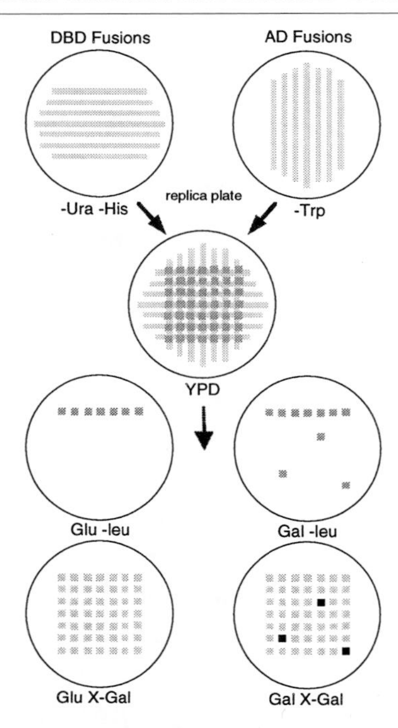

FIG. 2. Interaction mating assay for protein interactions. Several strains of one mating type expressing different DBD fusions are streaked in parallel lines on one plate. Strains of the opposite mating type expressing AD fusions are streaked on another plate. The two strain types are crossed onto the same replica velvet and lifted with a single YPD plate. After growth overnight the YPD plate is replicated to indicator plates. All the indicator plates are –Ura –His –Trp to select for diploid growth. Two of the indicator plates (–Leu) test for expression of the *LEU2* reporter and two test for *lacZ* expression (X-Gal). In this version of the two-hybrid system [J. Gyuris, E. Golemis, H. Chertkov, and R. Brent, *Cell* **75,** 791 (1993)], the AD fusions are expressed only in galactose medium. This allows detection of false positives in which reporter activation is independent of the AD fusion, as in the bait strain on the top row. [Adapted from R. L. Finley, Jr. and R. Brent, *Proc. Natl. Acad. Sci. U.S.A.* **91,** 12980 (1994).]

3. Press the bait strains and AD fusion strains to the same replica velvet so that the lines of bait and AD yeast strains intersect. Replicate the impression onto a YPD plate and incubate at 30° overnight.

4. Test the reporter phenotypes by replica plating from the YPD plate to indicator plates as described in step 9 of Screening for Interactors by Mating (above). The phenotypes are interpreted as before: galactose-dependent growth on –Leu plates (Leu^+) and blue color on X-Gal plates ($LacZ^+$) of the diploids that grow at the intersections of the two mated strains indicate an interaction.

Mapping Networks of Interacting Proteins

Many important regulatory pathways consist of networks of interacting proteins. Protein interaction data derived from two-hybrid experiments can suggest the functions for individual genes and assist in assembling proteins into regulatory pathways. Here we present two approaches for elaborating protein networks using the yeast two-hybrid system and interaction mating. The first approach involves sequential (reiterative) library screening in which newly isolated cDNAs are used as baits for subsequent library screens. By streamlining the interactor hunt protocol and introducing a rapid way to make new bait strains, this approach is a time- and cost-effective way to elaborate large protein networks. The second approach involves systematic screening of arrayed yeast expressing DBD or AD fusion proteins.

Elaborating Protein Networks by Iterative Interactor Hunts

Starting with one or more proteins, a protein interaction network can be generated by exhaustive and sequential screening of AD fusion libraries (see, e.g., Ref. 18). New AD fusion proteins isolated in each hunt are converted to baits for use in subsequent hunts. The strategy is outlined in Fig. 3. A potential rate-limiting step in conducting iterative interactor hunts is the step of subcloning newly isolated cDNAs into the bait vector followed by transformation and characterization of the bait strain. This process can be expedited dramatically by using recombination cloning (Fig. 1; see protocol below). In this approach the new bait plasmid is constructed by recombination in yeast, so that the new bait strain is created in the same step. Each new bait strain can be easily tested by mating it with a test strain expressing an AD fusion that should interact with the new bait. For most steps in an iterative hunt, the test AD fusion strain will be available from the previous screen (see Fig. 3). This test will confirm that the new bait strains contain cDNAs in the correct reading frame and that the expressed fusion protein can enter the nucleus and bind the reporters. A reiterative series of interactor hunts might proceed as follows.

1. Construct a strain expressing the initial protein of interest as a bait as described above. Construct a strain expressing an AD fusion version of the protein of interest by inserting the cDNA into a library vector (e.g., pJG4-5), and transforming RFY231.
2. Conduct an interactor hunt by mating the bait strain with an aliquot of frozen library as described above in the interactor hunt protocol. Identify

[18] M. Fromont-Racine, J. C. Rain, and P. Legrain, *Nature Genet.* **16,** 277 (1997).

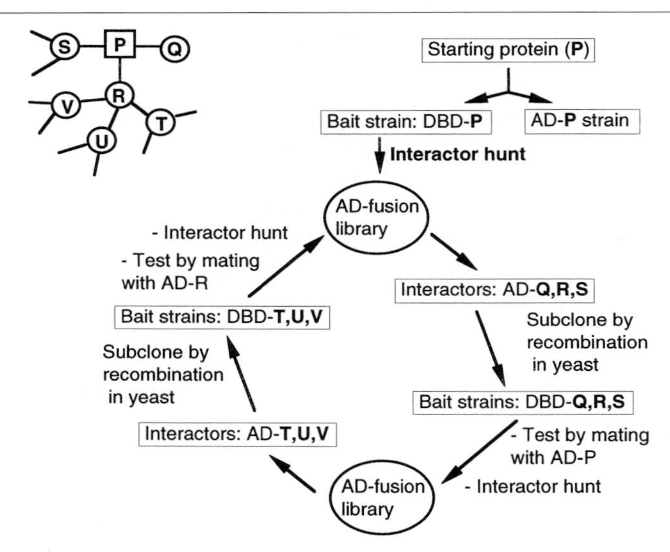

FIG. 3. Iterative interactor hunts. New clones isolated in interactor hunts are converted to baits for subsequent hunts. The fastest approach is to construct new bait strains by recombination cloning from the AD fusion library vector to the bait vector (see Fig. 2 and text). New bait strains can then be quickly tested by mating with a strain expressing an AD-fused interactor, for example, from the previous hunt. See text for more details.

specific interactors as described above. If many hunts are planned, the fastest approach to isolating the positive cDNAs should be used (e.g., PCR amplification of cDNAs from yeast, as described above).

3. Make new bait strains from the positive interactors by recombination cloning (see protocol below). Test the new bait strains for interaction with the AD fusion to the original protein of interest (from step 1 above) in a cross-mating assay.

4. Use the new bait strains to conduct interactor hunts by mating. Again, make new bait strains from the interactors. This time, and for all subsequent hunts, test the new bait strains by mating with the AD fusion isolated in the previous hunt.

Constructing New Bait Strains by Recombination Cloning

In this protocol, the cDNAs encoding new interactors are PCR amplified directly from the library clone (pJG4-5 derivative), using primers that include sequences from the bait vector (Fig. 1). The new bait strain is then made by cotransforming yeast with the PCR product and the bait vector linearized at the cloning site. The yeast cells will repair the double-strand break in the bait vector by homologous recombination, using the PCR product as a template. The result is that the cDNA is inserted into the bait vector in-frame with the DBD.

Procedure

1. Amplify the cDNA of an interactor clone by PCR, using the following primers: If making a pJK202 bait vector (Table I), use forward primer FP-N (5-GAC TGG CTG GAA TTG GCC CCC AAG AAA AAG AGA AAG GTG CCA GAT TAT GCC TCT CCC G 3′); if making a pEG202 bait vector (Table I), use forward primer FP-E (5′ GGG CTG GCG GTT GGG GTT ATT CGC AAC GGC GAC TGG CTG GTG CCA GAT TAT GCC TCT CCC G 3′). The underlined portion of each is from pJG4-5; the remainder is homologous to the appropriate bait vector. The same reverse primer can be used for both vectors (RP-J; see above). The bait vectors share nearly the same terminator region as pJG4-5 so that the total length of homology with pJG4-5 at the 3′ end of the PCR product is 80 bp. Use a high-fidelity polymerase such as Vent (New England BioLabs, Beverly, MA) or *Pfu* (Stratagene; La Jolla, CA). For template DNA use either whole yeast cells containing the pJG4-5 clone (add a 5-min 95° incubation step in the beginning of the PCR program), or yeast DNA minipreparations.
2. Digest the bait vector with *Eco*RI and *Xho*I, and purify over a Centricon-100 concentrator (Amicon, Beverly, MA). Adjust the concentration to 50 ng/μl.
3. Make the bait strain by transforming RFY206/pSH18-34 with 200 ng of linearized vector and 5–10 μl (10–200 ng) of unpurified PCR. Perform a control transformation with no PCR product. The presence of the PCR product should increase the transformation efficiency by greater than five-fold. More than 90% of the transformants will contain inserts.
4. Test several independent transformants by mating with a strain (e.g., RFY231) expressing an AD fusion to a known interactor by performing a cross-mating assay as described above.

Screening Two-Hybrid Arrays

Interaction data generated by testing interactions between sets of proteins encoded by cloned cDNAs are often easier to interpret than results from screening libraries. This is in part because a library screen includes a selection for false positives. These arise, for example, from yeast mutations or library proteins that increase the reporter readout.[19] Library screens are

[19] E. Golemis, Web site: http://www.fccc.edu/research/labs/golemis/InteractionTrapInWork.html

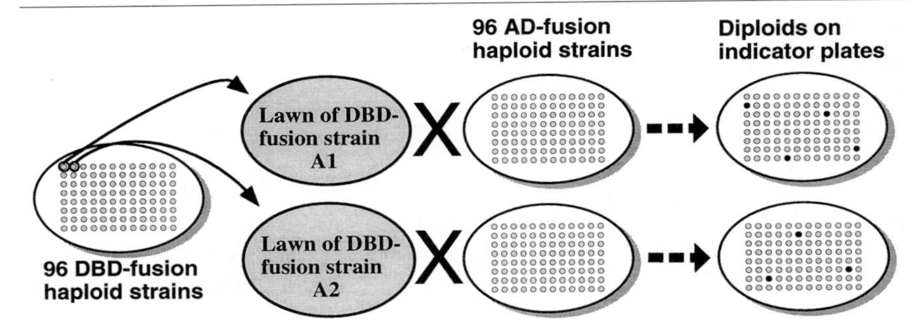

FIG. 4. Interaction mating with arrayed libraries. Arrayed strains expressing DBD fusions and strains expressing AD fusions are collected and tested for interaction by mating. In the approach shown, individual DBD fusion strains are first spread on lawns and then mated with plates containing 96 arrayed AD fusion strains. As in the cross-mating assay (Fig. 1), the two strains are mated by mixing them on a YPD plate, growing them overnight, and then replicating them onto indicator plates, in this case X-Gal plates.

also subject to false negatives, for example, due to the absence or low frequency of a particular cDNA in a library, or due to loss of library clones encoding proteins that reduce yeast viability. These problems are avoided when two strains, each expressing a single defined hybrid protein, are mated and the reporter readout is assayed in the diploids. There is no selection for false positives in such a binary assay, and clones encoding mildly toxic proteins can be maintained in haploids under repressing conditions. Finally, while transcription activators make poor baits in library screens, it is possible to directly test them for interactions, either by expressing them as AD fusions, or by looking for increases in reporter activation over the background generated by the bait itself.

The quickest way to test for interactions between small sets (tens) of proteins is the cross-mating assay presented above.[6] Interactions between large sets of proteins can be assayed by mating arrays of DBD fusion strains and AD fusion strains[5,20] (Fig. 4). Strains expressing individual hybrid proteins are collected and arrayed in the standard 48-well or 96-well format. Individual proteins are then tested against each array by mating. For example, a lawn of yeast expressing a DBD version of the protein can be mated with the array of strains expressing AD fusions. Large arrays can be systematically mated with each other as shown in Fig. 4. For very large arrays, a pooling scheme may be useful for mating (J. Zhong and R. L. Finley, Jr., unpublished, 1999).

[20] S. B. Hua, L. Ying, M. Qiu, E. Chan, H. Zhou, and L. Zhu, *Gene* **215,** 143 (1998).

Prospects for Genome-Wide Protein Interaction Maps

The numerous individual successes of two-hybrid experiments, combined with the potential for scale-up, have opened the prospect of generating genome-wide interaction maps by two-hybrid screening. The first attempt at generating a genome-wide interaction map was the work of Bartel and colleagues to map the interactions between the 55 proteins encoded by the bacteriophage T7 genome.[21] They used a combination of approaches, including mating individual and small pools of bait strains with AD fusion libraries, and numerous pairwise matings of individual DBD and AD fusion strains. The interaction map that they derived from their exhaustive two-hybrid analysis included a wealth of new functional information and insights about T7 biology. For example, they discovered a number of unsuspected interactions, and they confirmed many interactions that were previously suspected on genetic grounds. While the T7 genome is relatively small as genomes go, their success in mapping its protein interactions was nevertheless a landmark accomplishment because it demonstrated that large-scale interaction maps generated by two-hybrid technology can have immense utility as we try to decipher the functions of many genes and whole genomes.

The approaches used for larger genomes may differ depending on the size of the genome and whether cDNAs are available. For example, random mating of strains expressing DBD and AD fusion libraries may be useful for mapping the interactions encoded by small genomes, although the false-positive and false-negative frequency may interfere with analysis. Alternatively, arrayed libraries could readily be made from the relatively small microbial genomes.[22] Arrayed libraries are also likely to be used in any approach that maps interactions for a larger complex genome. Arrayed libraries may be generated from collections of expressed sequence tags (ESTs), for example, by recombination cloning into the DBD and AD fusion plasmids.[20] For genomes not represented by significant EST collections, arrayed yeast two-hybrid libraries can be generated by constructing normalized libraries.

Exploring Functions of Individual Protein Interactions

As new protein–protein interactions are identified, two-hybrid technology can be used to begin to study the functions of specific interactions. Interaction domains and contacts between proteins can be mapped by

[21] P. L. Bartel, J. A. Roecklein, D. SenGupta, and S. Fields, *Nature Genet.* **12,** 72 (1996).

[22] J. R. Hudson, Jr., E. P. Dawson, K. L. Rushing, C. H. Jackson, D. Lockshon, D. Conover, C. Lanciault, J. R. Harris, S. J. Simmons, R. Rothstein, and S. Fields, *Genome Res.* **7,** 1169 (1997).

screening libraries of mutations of one or the other protein.[23] Mutant versions of a protein that fail to interact with a potential partner protein could be used to explore the function of that interaction *in vivo.* For example, in some organisms the function of specific interactions can be addressed by expressing interaction mutant alleles of either partner protein in a null mutant background. Furthermore, starting with a noninteracting mutant version of one protein it is a relatively simple task to isolate a potential suppressor mutation in one of its partner proteins that can restore the interaction, by screening a library of mutant partner proteins. The ability of a suppressor to restore wild-type activity to an interaction mutant when both are expressed *in vivo* can provide strong evidence of the function of the specific interaction. This approach is akin to the use of classic suppressor genetics to suggest functional interactions, but has the advantage of using designed and characterized mutant alleles so that the nature of the mutation is known in advance. Another approach to testing the functions of specific interactions *in vivo* is to use reagents that disrupt them. The yeast two-hybrid system is a useful assay for isolating and characterizing such reagents.

Isolating Interaction Mutants and Suppressors

Interaction mutants of a protein can be isolated by creating a library strain expressing its mutant versions as AD fusions and screening the library for noninteractors with a partner protein as a bait. A number of efficient methods for making a mutant library have been described.[24,25] For example, one convenient approach is to perform low-fidelity mutagenic PCR[25] and subclone the PCR products into the AD fusion vector. By using PCR primers with sequences corresponding to the AD fusion vector (e.g., primers FP-J and RP-J; see above), the mutagenized PCR products can be readily inserted into the vector by recombination cloning (Fig. 1). This approach simultaneously creates the library strain that can be collected and frozen in aliquots to be screened with any number of interacting bait proteins. For many studies, the goal will be to identify single amino acid substitutions that abolish interaction, suggesting residues and domains important for interaction. However, random mutagenesis will lead to many noninteracting clones that result from nonsense mutations encoding truncated AD fusions. Such truncated proteins could be identified by performing Western blots on yeast clones containing noninteractors. Alternatively, stable full-length noninteractors can be isolated directly by including an easily detectable

[23] R. Jiang and M. Carlson, *Genes Dev.* **10,** 3105 (1996).
[24] A. Greener, M. Callahan, and B. Jerpseth, *Methods Mol. Biol.* **57,** 375 (1996).
[25] R. C. Cadwell and G. F. Joyce, *PCR Methods Appl.* **2,** 28 (1992).

carboxy-terminal moiety, such as green fluorescent protein (GFP) or β-galactosidase, on the AD fusion.[26] In this case, missense mutants can be identified by screening the mutant library for noninteractors that express GFP or *lacZ*. Another alternative is to use a two-hybrid system with two different reporter systems responsive to different DBD fusions.[27,28] This would allow isolation of mutant versions of an AD fusion protein that selectively abolish interactions with one DBD fusion but not the other.

Once a mutant library has been made it can be screened for noninteractors by interaction mating with a bait strain. The first part of this screen can be conducted as in an interactor hunt. First, an aliquot of the mutant library is thawed and mixed with fresh bait strain. The mixture is then plated on a YPD plate. To screen the diploids for noninteractors, the mated yeast could be plated onto diploid selection plates, and the colonies could then be replicated to the reporter indicator plates. The number of colonies to screen depends on the size of the target protein and the efficiency of the mutagenesis, but generally will be fewer than 10,000 to cover all possible mutations. An alternative to screening through this many colonies is to use a toxic reporter gene. For example, expression of the yeast *URA3* gene is toxic to yeast grown on medium containing 5-fluoroorotic acid (5-FOA). Strains containing a *URA3* reporter have been described for systems using the LexA or Gal4 DBD.[27,29]

Breaking Interactions with Proteins and Peptides

Another way to determine the function of a specific protein interaction is to disrupt it *in vivo* using a *trans*-acting reagent. The yeast two-hybrid system provides an assay to identify such reagents. For example, random peptide libraries can be screened to isolate peptide aptamers, which are peptides that bind tightly and specifically to a given protein.[30,31] Some aptamers will interact with surfaces of their target proteins that normally make contacts with other proteins, and thus will disrupt the corresponding interactions. Like dominant-negative mutant proteins,[32] peptides that disrupt specific protein interactions may be useful for genetic analysis when

[26] H. M. Shih, P. S. Goldman, A. J. DeMaggio, S. M. Hollenberg, R. H. Goodman, and M. F. Hoekstra, *Proc. Natl. Acad. Sci. U.S.A.* **93,** 13896 (1996).

[27] C. W. Xu, A. R. Mendelsohn, and R. Brent, *Proc. Natl. Acad. Sci. U.S.A.* **94,** 12473 (1997).

[28] I. Serebriiskii, V. Khazak, and E. A. Golemis, *J. Biol. Chem.* **274,** 17080 (1999).

[29] M. Vidal, R. K. Brachmann, A. Fattaey, E. Harlow, and J. D. Boeke, *Proc. Natl. Acad. Sci. U.S.A.* **93,** 10315 (1996).

[30] M. Yang, Z. Wu, and S. Fields, *Nucleic Acids Res.* **23,** 1152 (1995).

[31] P. Colas, B. Cohen, T. Jessen, I. Grishina, J. McCoy, and R. Brent, *Nature* (*London*) **380,** 548 (1996).

[32] I. Herskowitz, *Nature* (*London*) **329,** 219 (1987).

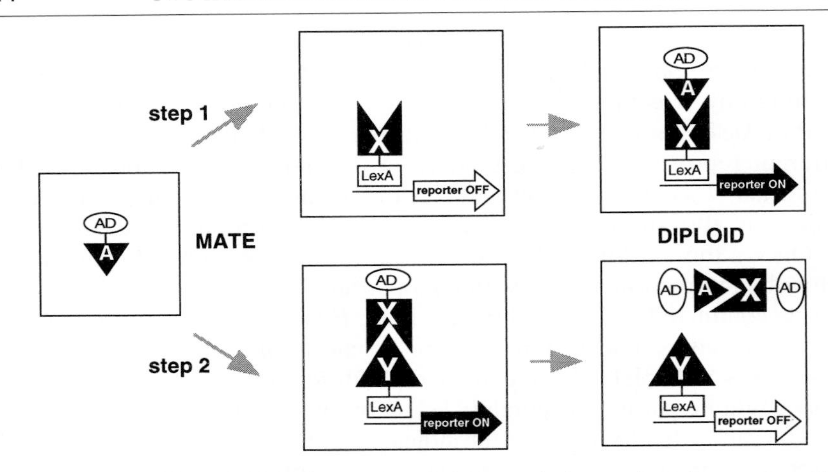

FIG. 5. Identifying peptides that disrupt specific protein interactions. In step 1 peptide aptamers that bind to protein X are isolated in an interactor hunt. In step 2 a strain expressing X as an AD fusion, and a LexA-fused interacting protein, Y, is mated with a strain expressing the peptide aptamer. This can be done by the cross-mating assay (Fig. 2). If the aptamer disrupts the X–Y interaction the reporter expression will be decreased or turned off. Note that the peptide aptamer is expressed as an AD fusion in both steps so that subcloning is not necessary. Note that "A" could be a peptide aptamer or any protein that interacts with X.

loss-of-function mutations in a gene cannot be obtained. For example, peptide aptamers can be expressed in a model organism in predictable spatial and temporal patterns, or injected into specific cells or embryos, to target specific proteins[33,34] (M. G. Kolonin and R. L. Finley, Jr., unpublished, 1999). Although here we focus on isolation of inhibitory peptide aptamers, the same approach could be taken to identify cellular proteins or their derivatives that disrupt specific protein contacts.

Peptide aptamers that disrupt specific protein interactions could be isolated from combinatorial peptide libraries by screening for loss of interaction using a counterselectable reporter such as *URA3*.[27,29] In this case it is important to include controls to ensure that the loss of reporter expression is due to disruption of the specific protein interaction. An alternative approach is to isolate the disruptive peptide aptamers in two steps (Fig. 5). In the first step, a combinatorial library is screened for peptide aptamers that strongly and specifically bind one of the interacting proteins. In the

[33] B. A. Cohen, P. Colas, and R. Brent, *Proc. Natl. Acad. Sci. U.S.A.* **95,** 14272 (1998).
[34] M. G. Kolonin and R. L. Finley, Jr., *Proc. Natl. Acad. Sci. U.S.A.* **95,** 14266 (1998).

second step, individual peptide aptamers are assayed for their ability to disrupt a two-hybrid interaction. This second step is done by mating one strain that expresses both the AD and DBD fusions with strains expressing potentially disruptive peptide aptamers. Disruption of the interaction between the AD- and DBD-fused proteins will result in the loss of growth or color on the indicator plates. An advantage to this two-step approach is that each peptide aptamer can be simultaneously tested for the ability to disrupt interactions between the target protein and any number of its partners.

Isolating Disruptive Peptide Aptamers. The following protocol outlines a two-step approach to isolating peptide aptamers (A) that bind specifically to a bait protein, X, and that disrupt interactions between X and another protein, Y (Fig. 5).

1. Screen a combinatorial peptide library for peptide aptamers that bind protein X bait, using the mating protocol described above for screening cDNA libraries. Random peptide libraries expressed from a *TRP1*-marked plasmid similar to pJG4-5 have been described elsewhere.[31]
2. Transform yeast strain RFY251 (Table I) with library plasmids expressing peptides. Simultaneously, transform RFY251 with pJG4-5 to construct a control strain. These RFY251 transformants can be used to test the specificity of the peptides (step 3 below) and to perform the disrupter assay (steps 4–7 below).
3. Test the specificity of peptides for bait X in the corss-mating assay (Fig. 2), using the RFY251/peptide transformants. Use the original bait strain and an appropriate set of nonspecific bait strains, which should possibly include baits closely related to protein X. Peptide aptamers are those peptides that interact specifically with protein X.
4. Make a plasmid for constitutive expression of the AD–X fusion. To do this, subclone a protein X cDNA in-frame with the AD moiety in plasmid pMK2. pMK2 (M. G. Kolonin and R. L. Finley, Jr., unpublished, 1999) is a plasmid that carries the *URA3* selectable marker and expresses AD fusion proteins from the yeast *ADH1* promoter.
5. Make a plasmid for constitutive expression of the DBD-Y fusion (pBait-Y). To do this, subclone a protein Y cDNA in-frame with LexA into a pBait vector (Table I).
6. Cotransform RFY206/pCWX24, the host strain that carries the *lacZ* reporter on a *LYS2*-marked plasmid (Table I), with pMK2-X and the pBait-Y. It is also useful to make a control strain expressing an interacting pair of proteins unrelated to X and Y. To create another control strain, cotransform RFY206/pCWX24 with pMK2 (no cDNA insert) and the pBait-Y. Select transformants on Glu/CM –Ura, –His, –Lys medium (~3 days).

7. Mate the strains created in step 2 above with the strains created in step 6 above by cross-mating as described previously (Fig. 2). In this case the RFY206 derivatives should be streaked on Glu/CM –Ura, –His, –Lys, and all the indicator plates must lack lysine as well.

Interpreting Results. On the Gal/Raff plates (aptamers expressed) compare the X–Y interaction in the presence of different peptide-expressing plasmids and pJG4-5. A reduction in reporter expression indicates that a corresponding aptamer disrupts the interaction. To distinguish peptides that disrupt the X–Y interaction from peptides that may be mildly toxic to yeast it is useful to test for disruption of an unrelated pair of interacting proteins. True disrupters will specifically decrease reporter expression in the strain with X and Y. If the reporters are active in the control strain (RFY206/pCWX24/pMK2/pBait-Y) it indicates that the Y bait activates transcription. In such cases disrupters may sometimes still be identified if the X–Y interaction activates the reporters above the *trans*-activation background. Some peptides that bind protein X without disrupting the X–Y contact might actually enhance the X–Y interaction phenotype because they bring in an additional AD.

Concluding Remarks

The two-hybrid system has evolved from an elegant assay for protein interactions to a robust technology for genetic analysis and functional genomics. Introducing the hybrid proteins to one another by interaction mating facilitates most two-hybrid experiments. Mating methods may also become useful for other yeast hybrid systems including those that involve bridging RNAs or small organic compounds,[35,36] as well as systems that operate in the cytoplasm.[37]

Acknowledgments

We thank Eugenius S. B. C. Ang, Jr. and Kory Levine for critical comments on this manuscript, and Jennifer Fonfara and other members of the Finley laboratory for help in developing and testing these procedures. This work was supported by grants to R.L.F. from the National Institutes of Health Center for Human Genome Research, and from the Merck Genome Research Institute.

[35] D. J. SenGupta, B. Zhang, B. Kraemer, P. Pochart, S. Fields, and M. Wickens, *Proc. Natl. Acad. Sci. U.S.A.* **93,** 8496 (1996).

[36] G. R. Crabtree and S. L. Schreiber, *Trends Biochem. Sci.* **21,** 418 (1996).

[37] N. Johnsson and A. Varshavsky, *Proc. Natl. Acad. Sci. U.S.A.* **91,** 10340 (1994).

[4] Analysis and Identification of Protein–Protein Interactions Using Protein Recruitment Systems

By AMI ARONHEIM and MICHAEL KARIN

Introduction

The immense output of various genome sequencing projects has been providing the scientific community with numerous proteins with no assigned function. It is well accepted that there is no single protein that functions in isolation within the cell, and protein–protein interaction serves as a general means to provide functional specificity and control the activity of enzymes and signaling molecules. To achieve a better functional understanding of known and novel gene products, much current research activity in molecular cell biology is focused on identification of interacting proteins and characterization of these interactions. To date, multiple methods exist to identify and characterize protein–protein interactions; however, few of them employ a genetic selection system *in vivo.* One that does is the yeast two-hybrid system.[1] The two-hybrid system is as an excellent research tool, and is commonly used to identify and characterize novel and known interaction partners for proteins of interest.[2–8] Although powerful, the two-hybrid system, which is based on a transcriptional readout, exhibits several limitations and inherent problems. The two-hybrid system cannot be used with transcription activators or proteins with transcription repression activity. In addition, multiple proteins, although not considered transcription factors, exhibit intrinsic transcription activity and, therefore, cannot be used in this assay. Moreover, the fact that protein interaction underlying this method occurs in the yeast nucleus may result in problems of toxicity due to nuclear expression of DNA-binding proteins and cell cycle regulators. If kept out of the nucleus such proteins should not exert a toxic effect. Alternatively, proteins of nonnuclear origin may be inappropriately folded or modified while expressed in the nucleus. Finally, the two-hybrid system,

[1] S. Fields and O. K. Song, *Nature* (*London*) **340,** 245 (1989).
[2] J. B. Allen, M. W. Walberg, M. C. Edwards, and S. J. Elledge, *Trends Biochem. Sci.* **20,** 511 (1995).
[3] J. Boeke and R. K. Brachmann, *Curr. Biol.* **8,** 561 (1997).
[4] R. K. Brachmann and J. D. Boeke, *Curr. Opin. Biotechnol.* **8,** 561 (1997).
[5] C. Evangelista, D. Lockshon, and S. Fields, *Trends Cell Biol.* **6,** 196 (1996).
[6] R. M. Fredrickson, *Curr. Opin. Biotechnol.* **9,** 90 (1998).
[7] K. Hopkin, *J. NIH Res.* **8,** 27 (1996).
[8] K. H. Young, *Biol. Reprod.* **58,** 302 (1998).

0076-6879/00 $30.00

used for almost a decade with different baits, has generated data regarding the repetitive isolation of prey proteins as a result of a library screening approach. While numerous proteins may pass bait-specificity tests, eventually they turn out to be "false positives" after all, resulting in wasted effort and confusion.

To overcome some of the problems and limitations mentioned above, we have developed a cytoplasmic protein recruitment system designated the Sos recruitment system (SRS).[9] This and the related Ras recruitment system (RRS)[10] take advantage of a general phenomenon in signal transduction, that generation of local high concentrations of a signaling intermediate can result in a dramatic increase in signaling activity. This phenomenon was demonstrated for several signaling molecules, including son of sevenless (hSos),[11,12] Ras GTPase,[13] Raf kinase,[14,15] phosphatidylinositol 3-kinase,[16] GTPase-activating protein,[17] and more. In many cases, the localization effect initially identified in mammalian cells functions just as well in a lower eukaryote, the yeast *Saccharomyces cerevisiae.*[9,11] This allowed the use of a corresponding yeast mutant strain whose deficiency can be complemented by the activated mammalian protein to develop efficient screening assays based on recruitment of an active reactant via protein–protein interaction to the plasma membrane and its resultant activation, collectively termed "protein recruitment systems."

A case in point is the recruitment of the Ras guanyl nucleotide exchange factor hSos to the plasma membrane. In mammalian cells, membrane localization of hSos is achieved after the interaction of an activated tyrosine kinase receptor (a membrane protein) and the adaptor protein Grb2.[18] Constitutive localization of hSos through covalent attachment of either farnesyl or myristic acid moieties results in activation of the Ras signaling

[9] A. Aronheim, E. Zandi, H. Hennemann, S. Elledge, and M. Karin, *Mol. Cell. Biol.* **17,** 3094 (1997).

[10] Y. C. Broder, S. Katz, and A. Aronheim, *Curr. Biol.* **8,** 1121 (1998).

[11] A. Aronheim, D. Engelberg, N. Li, N. Al-Alawi, J. Schlessinger, and M. Karin, *Cell* **78,** 949 (1994).

[12] L. A. Quilliam, S. Y. Huff, K. M. Rabun, W. Wei, W. Park, D. Broek, and C. J. Der, *Proc. Natl. Acad. Sci. U.S.A.* **91,** 8512 (1994).

[13] J. F. Hancock, A. I. Magee, J. Childs, and C. J. Marshall, *Cell* **57,** 1167 (1989).

[14] S. J. Leevers, H. F. Paterson, and C. J. Marshall, *Nature* (*London*) **369,** 411 (1994).

[15] D. Stokoe, S. G. Macdonald, K. Cadwallader, M. Symons, and J. F. Hancock, *Science* **264,** 1463 (1994).

[16] A. Klippel, C. Reinhard, M. Kavanaugh, G. Apell, M. A. Escobedo, and L. T. Williams, *Mol. Cell. Biol.* **16,** 4117 (1996).

[17] D. C. S. Huang, C. J. Marshall, and J. F. Hancock, *Mol. Cell. Biol.* **13,** 2420 (1993).

[18] L. Buday and J. Downward, *Cell* **73,** 611 (1993).

pathway.[11,12,19,20] In yeast, the Ras signaling pathway is required for cell proliferation. Fortunately, a temperature-sensitive yeast strain that is mutated in its Ras guanyl nucleotide exchange factor, Cdc25, was identified.[21] Using this strain we have demonstrated that expression of membrane-bound hSos can activate the yeast Ras pathway and confer cell growth at the restrictive temperature.[11] Recruitment of hSos to the plasma membrane can also be achieved via protein–protein interaction sufficient for Ras activation and complementation of Cdc25 temperature-sensitive mutation.[9] This system, designated "SRS," was shown to be suitable for identification and isolation of known and novel protein interactions.[9,22] The SRS approach has been improved by replacing the hSos effector molecule with Ras, designated "RRS,"[10,23] thus overcoming several technical problems that arose while using the SRS. Both systems employ the yeast cytoplasm as a milieu for protein–protein interaction and therefore overcome some of the limitations and problems of the two-hybrid assay. Therefore, SRS and RRS serve as attractive alternatives for the isolation and characterization of protein–protein interaction *in vivo*. The use of either system for the analysis of a bait of interest is similar, and is described below.

Description of hSos Recruitment System/Ras Recruitment System Selection Approaches

In general, the cDNA encoding a protein of interest for which protein partners are to be isolated is fused in frame with the effector molecule (the "bait"). The effector molecule could be either Ras or hSos. At the restrictive temperature, 36°, in the *cdc25-2* yeast strain, the yeast Ras is found in the GDP-bound inactive form, resulting in growth arrest (Fig. 1A). On the other hand, a protein partner for the bait or a cDNA library is fused in frame with a membrane localization signal (the "prey"). These signals could be either a myristoylation or farnesylation sequence to be fused at the amino or carboxy terminal relative to the prey cDNA, respectively. cDNAs encoding proteins different from Ras or hSos are not expected to confer cell growth of the *cdc25-2* yeast strain at the restrictive temperature (Fig. 1B). When protein–protein interaction occurs between an hSos bait and

[19] L. A. Quilliam, R. Khosravi, S. Y. Huff, and C. J. Der, *BioEssays* **17,** 395 (1995).

[20] L. J. Holsinger, D. M. Spencer, D. J. Austin, S. L. Schreiber, and G. R. Crabtree, *Proc. Natl. Acad. Sci. U.S.A.* **92,** 9810 (1995).

[21] A. Petitjean, F. Higler, and K. Tatchell, *Genetics* **124,** 797 (1990).

[22] X. Yu, L. C. Wu, A. M. Bowcock, A. Aronheim, and R. Baer, *J. Biol. Chem.* **273,** 25388 (1998).

[23] A. Aronheim, Y. C. Broder, A. Cohen, A. Fritsch, B. Belisle, and A. Abo, *Curr. Biol.* **8,** 1125 (1998).

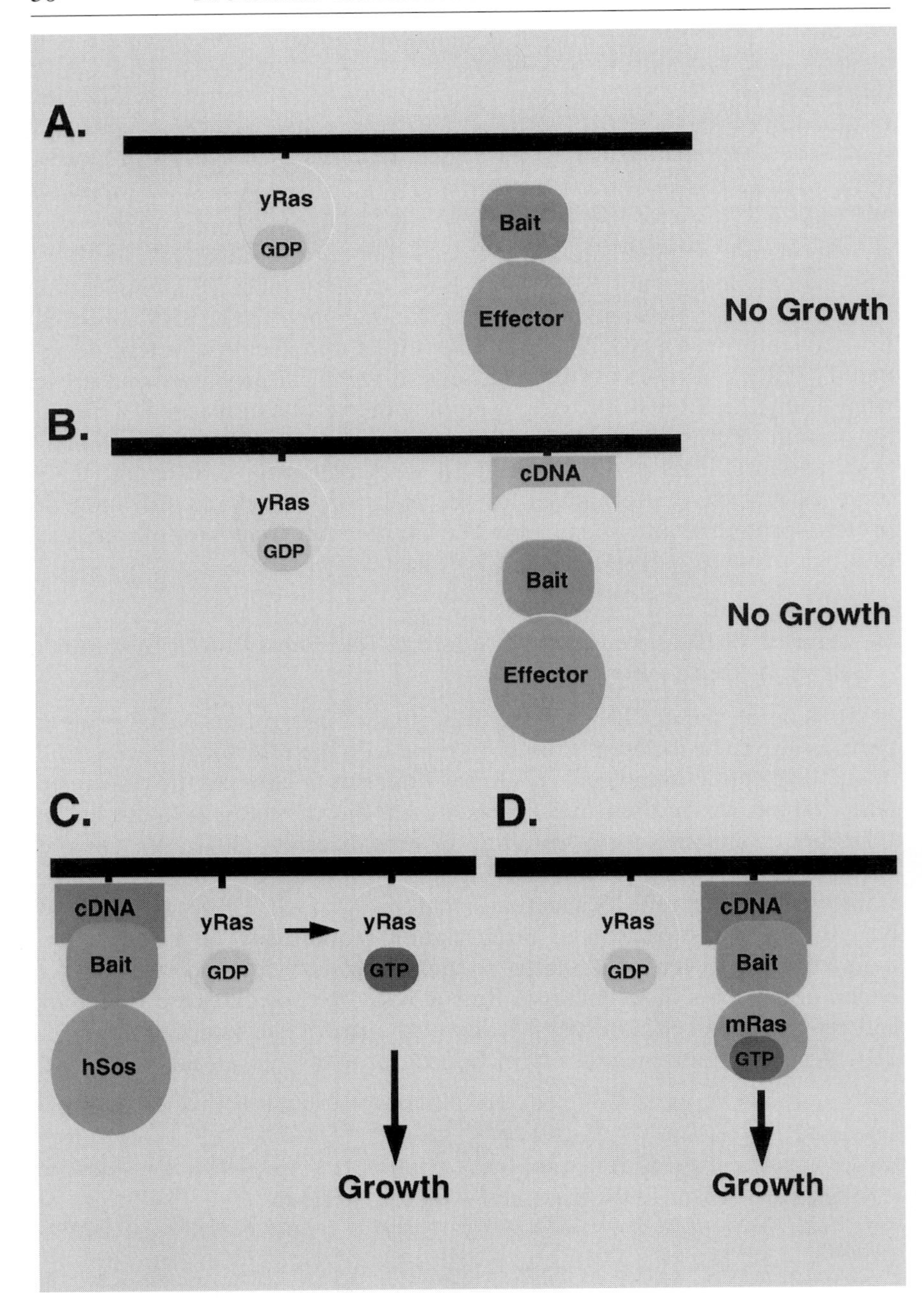
A.
yRas
GDP
Bait
Effector
No Growth
B.
cDNA
yRas
GDP
Bait
Effector
No Growth
C.
cDNA
Bait
hSos
yRas
GDP
yRas
GTP
Growth
D.
yRas
GDP
cDNA
Bait
mRas
GTP
Growth

its corresponding prey it results in exchange of GDP-bound yeast Ras to GTP-bound yeast Ras, which allows growth at the restrictive temperature (Fig. 1C). As for a Ras bait, its localization to the plasma membrane via protein–protein interaction by the prey protein is sufficient to allow cell growth at the restrictive temperature, rendering the yeast Ras in the GDP-bound inactive form at all stages of the assay.

Designing Bait for hSos Recruitment System

The bait is a hybrid protein constructed by fusing the cDNA for the protein of interest or a protein fragment to the 5′ hSos cDNA fragment that encodes the catalytic domain of hSos without its C-terminal regulatory domain. This truncation mutant exhibits increased exchange activity relative to full-length hSos.[11] The bait-5′ hSos hybrid expression cassette is placed under the control of the alcohol dehydrogenase (ADH) promoter in the pADNS expression vector,[24] which includes the *LEU2* gene as a yeast nutrient selection marker. The bait can be fused in frame to either the 5′ (N) or 3′ (C) termini of 5′ hSos. While testing known protein interactions, we observed no difference in the effectiveness of either configuration. However, more successful screens have been conducted so far with baits fused to the 3′ (C) terminus of 5′ hSos. Nevertheless, it is advised to test novel baits fused to 5′ hSos in both configurations. In the event that a positive control for the newly designed bait is available, it is highly recommended that the corresponding prey be designed and its interaction with the bait be tested. The prey cDNA is placed under the control of the *GAL1* promoter, using a pYes2 (Invitrogen, San Diego, CA)-based expression

[24] J. Colicelli, C. Birchmeier, T. Michaeli, K. O'Neill, M. Riggs, and M. Wigler, *Proc. Natl. Acad. Sci. U.S.A.* **86,** 3599 (1989).

FIG. 1. Schematic diagram describing the SRS and RRS. (A) The protein of interest is fused in frame with an effector molecule that can be either Ras or hSos (the "bait"). In both cases the bait fusion is found in the cytoplasm and when expressed in temperature-sensitive yeast cells cdc25-2 is unable to confer cell growth at the restrictive temperature. (B) The cDNA encoding a protein partner for the bait or a cDNA library is fused to membrane localization signals, such as a myristoylation sequence. Unless the cDNA encodes either Ras or hSos its expression is not expected to result in growth of the *cdc25-2* cells at the restrictive temperature. (C) The localization of the hSos fusion bait to the plasma membrane via protein–protein interaction results in exchange of GDP-bound yeast Ras for GTP-bound yeast Ras and thereby activation of the Ras viability pathway at the restrictive temperature conferring efficient cell growth. (D) The localization of the Ras fusion bait via protein–protein interaction directly activates the Ras viability pathway of the *cdc25-2* yeast strain and thereby allows efficient cell growth at the restrictive temperature.

vector, which provides the *URA* gene as a yeast selection marker. Alternatively, if a positive control for an interaction with the bait does not exist, it is possible to test the bait with a plasmid encoding a Ras protein (the Rit homolog).[25] Rit was found to cause efficient growth of *cdc25-2* transformants only when coexpressed with 5′ hSos independent of the bait fused.[26] This provides a test for the catalytic activity of hSos as part of the hybrid bait protein.

Designing Bait for Ras Recruitment System

The bait is designed as a hybrid protein between the protein (or protein fragment) of interest and the mammalian Ras protein lacking its farnesylation (CAAX) box. In principle, both wild-type Ras and activated Ras can be used (A. Aronheim, unpublished, 1998) for testing known protein–protein interactions. However, for identification of protein–protein interactions following a library screen, it is necessary to use the activated form of Ras(L61), because it is insensitive to the mammalian GTPase-activating protein that is used in the screen.[26] The bait can be fused to either the N or C terminus of Ras, exactly as described for the SRS bait. Expression in yeast is controlled as described above.

Designing Prey for hSos Recruitment System/Ras Recruitment System

The prey cDNA is placed under control of the *GAL1* promoter, using a pYes2 (Invitrogen)-based expression vector, which provides the *URA* gene as a yeast selection marker. The cDNA can be placed amino terminal to a Ras farnesylation sequence or, alternatively, carboxy terminal to a myristoylation sequence. The fusion of a cDNA library is designed in the same vector by using the v-Src myristoylation sequence, which provides and ensures a common translation initiation sequence and membrane modification for all cDNAs fused independent of the existence of a stop codon within the cDNAs. The *GAL1* promoter provides the ability to induce/repress the expression level of the resulting prey protein hybrid, depending on the sugar-containing medium used. The ability to control the expression levels of the prey provides an additional level of selection during a library screening approach for cDNAs that encode proteins that contribute directly to phenotypic growth (see below).

Testing Bait

Although the expression from the *GAL1* promoter is highly induced when cells are grown on galactose-containing medium, some interactions

[25] C. Lee, N. G. Della, C. E. Chew, and D. J. Zack, *Neuroscience* **16,** 6784 (1996).
[26] A. Aronheim, *Nucleic Acids Res.* **25,** 3373 (1997).

are readily detectable when cells are grown on glucose, for example, c-Jun and c-Fos heterodimerization, using the SRS.[27] The ability of a protein pair to confer growth at the restrictive temperature on galactose or glucose will depend on the strength of interaction between the bait and prey, their level of expression, and the ability of the membrane localization signal to actually place the prey protein at the inner leaflet of the plasma membrane. In contrast, Jun–Fos heterodimerization is galactose dependent when c-Jun is fused to Ras. Assuming that the affinity of interaction is not affected by either hSos or Ras, these results indicate that the RRS requires more of the protein prey to be made as compared with the SRS, suggesting that it is a more efficient approach for isolation of strong binding interactions, while the SRS is a more suitable approach for isolation of weaker interactions.

The fusion of a bait sometime results in self-activation of the SRS, for example, the regulatory domain of Pak65[28] fused to hSos, 5′ hSos–PakR. Therefore, it is essential to test each bait for its inability to allow growth of the *cdc25-2* strain at the restrictive temperature. Using an identical Pak65 fragment fused to Ras that lacks its farnesylation signal (CAAX box), self-activation was eliminated and further characterization of the PakR bait was made possible.[10,23]

In addition, because nutritional supplements from different suppliers may contain various contaminants that might affect the activity of the *GAL1* promoter, it is important to test a known galactose-dependent interaction that cannot be detected on glucose, on the plates intended for use in the screen. The growth of *cdc25-2* cells that coexpress Rit and 5′ hSos–bait is commonly galactose dependent and therefore can serve as a general positive control for both the functionality of the hSos part of the bait fusion protein and the quality of the culture media. No such positive control for Ras–baits exists.

Materials and Methods

Culture Media

YNB galactose medium (500 ml)

Yeast nitrogen base without amino acids (Difco, Detroit, MI)	0.85 g
Ammonium sulfate (Fisher, Pittsburgh, PA)	2.5 g
Galactose (Sigma, St. Louis, MO)	15 g

[27] A. Aronhein, submitted (2000).

[28] E. Manser, T. Leung, H. Salihuddin, Z. S. Zhao, and L. Lim, *Nature* (*London*) **367,** 40 (1994).

D-Raffinose (Sigma)	10 g
Glycerol (Fisher)	10 g
Bacto-agar (Difco)	>20 g
Distilled water	to 500 ml

It is important to use a high concentration of agar for efficient replica plating.

YNB glucose medium (500 ml)

Yeast nitrogen base without amino acids	0.85 g
Ammonium sulfate	2.5 g
Glucose (Sigma)	10 g
Bacto-agar	>20 g
Distilled water	to 500 ml

The following amino acids and bases should be added to a final concentration of 50 mg/liter, excluding those amino acids that serve for selection of the transfected plasmids: leucine, uracil, tryptophan, methionine, lysine, adenine, and histidine. All amino acids and bases are obtained from Sigma.

YPD nonselection medium (500 ml)

Yeast extract (Difco)	1% (w/v)
Bacto-peptone (Difco)	2% (w/v)
Glucose	2% (w/v)
Bacto-agar	>20 g
Distilled water	to 500 ml

Transfection of Yeast Cells

1. A single *cdc25-2* yeast colony is grown in 200 ml of YDP medium overnight at 24° to 2–10 × 10^6 cells/ml.
2. Cells are pelleted for 5 min at 2500 rpm and resuspended in 20 ml of LISORB [100 m*M* lithium acetate, 1 *M* sorbitol in TE (10 m*M* Tris-HCl, pH 8.0; 1 m*M* EDTA)]. After two washes with LISORB, cells are resuspended with LISORB to 2–5 × 10^8 cells/ml and rotated for 30 min at 24°.
3. For each transfection 10 μl of preboiled sheared salmon sperm DNA (20 mg/ml) and 2–3 μg of plasmid DNA are added.
4. DNA is mixed by vortexing and 200 μl of prewashed cells is added to each transfection tube. The cell–DNA mixture is mixed briefly and 1.2 ml of LIPEG [40% (w/v) polyethylene glycol (PEG) 3350 in 100 m*M* lithium acetate in TE] is added followed by rigorous mixing. The mixture is incubated for 30 min at room temperature with constant rotation. To increase transfection efficiency, 100 μl of dimethyl sulfoxide (DMSO) is added followed by a 10-min heat shock at 42°.

5. The transfection mixture is pelleted for 1 min and the supernatant is discarded.
6. The remaining PEG is completely removed, using a pipettor with a 200 μl tip. The pellet is recovered in 150 μl of 1 *M* sorbitol and the cells are plated on the appropriate medium followed by incubation at 24° in a humidified incubator for 4 days. Individual colonies are plated on glucose medium containing the appropriate amino acids and bases and incubated for an additional 2 days at 24°, followed by replica plating to both galactose- and glucose-containing plates incubated at 36°.
7. The following two controls should be included in every transfection.
   a. Plating ~10^6 cells (100 μl of culture ready for transfection) into a YPD plate incubated directly at 36°; this control tests the original *cdc25-2* culture and provides an estimate of the rate of reversion of *cdc25-2* mutation. Typically 10–20 colonies per 10^6 cells will grow at 36°.
   b. A control transfection tube should include both pYes2 and pADNS expression plasmids. After transfection the control transformants are plated on a YNB glucose (–Leu –Ura) plate and incubated directly at 36°. This control gives an estimate regarding the *cdc25-2* reversion rate and possible contamination of the culture during the transfection procedure. No colony is expected to grow on this plate.

hSos Recruitment System/Ras Recruitment System Library Screening Protocol

Library screening using the SRS/RRS approaches requires specific cDNA libraries (see flow chart, Fig. 2). The cDNAs are routinely inserted fused to a DNA fragment encoding the Src myristoylation signal (M) through the *Eco*RI–*Xho*I restriction sites in the pYes2-(URA)-derived expression vector. To reduce the isolation of mammalian Ras cDNAs as false positives, a plasmid encoding mGAP (GTPase-activating protein) is coexpressed with the bait and the cDNA library expression plasmids.[26] The mGAP protein, which inhibits Ras activity, is expressed under the control of the *GAL1* promoter, using the pYes2 (TRP)-based expression vector. Efficient elimination of mammalian Ras false positives requires expression of mGAP from a multicopy expression plasmid, because a single-copy plasmid encoding mGAP can eliminate only part of the mRas false positives when expressed in *cdc25-2* cells, due to low expression levels (A. Aronheim, unpublished results, 1997).

To obtain high transformation efficiency, the bait and mGAP expression plasmids are first introduced into the *cdc25-2* strain. Transformants are

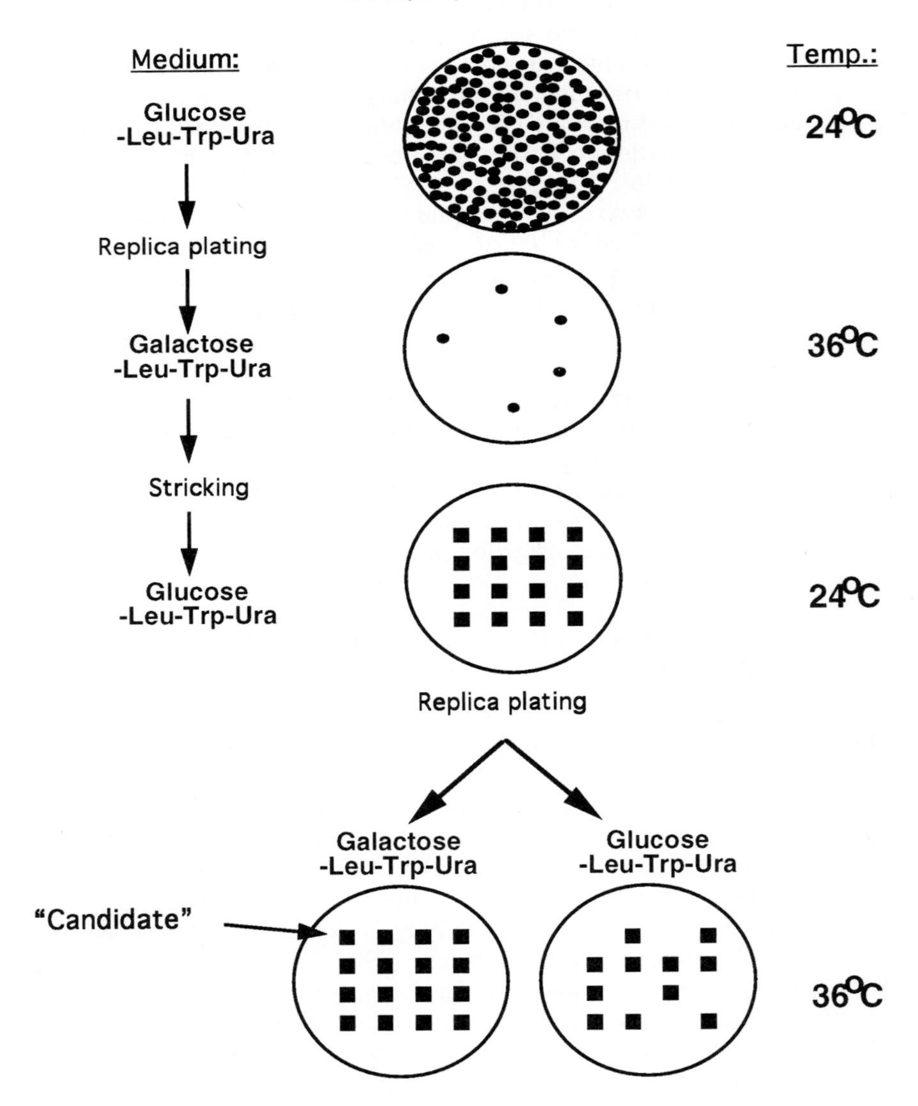
Transfected plasmids:
1. ADNS(LEU)-Ras(61)ΔF-Bait/ 5'Sos-Bait
2. Yes(TRP)-mGAP
3. Yes(URA)-M-cDNA
Medium:
Temp.:
Glucose -Leu-Trp-Ura
24°C
Replica plating
Galactose -Leu-Trp-Ura
36°C
Stricking
Glucose -Leu-Trp-Ura
24°C
Replica plating
Galactose -Leu-Trp-Ura
Glucose -Leu-Trp-Ura
"Candidate"
36°C

isolated and used to inoculate a 3-ml liquid culture for overnight growth at 24°. The culture is subsequently transferred to 200 ml of liquid culture for additional overnight growth at 24°. The culture is than pelleted and transferred to 200 ml of YPD medium for a recovery period of 3–5 hr. Subsequently the cells are used to transform 20 tubes with 3 μg of library plasmids resulting in 5000–10,000 transformants on each 10-cm plate. After 4–5 days at 24°, plates are used for replica plating onto galactose (–Leu –Ura –Trp) plates incubated for 3–4 days at 36°. Colonies that grow are selected and placed on a glucose plate containing the appropriate amino acids and bases, using a grid, and are incubated at 24° for an additional 2 days. These clones are tested for their ability to grow at 36° depending on the presence of galactose in the medium. Those clones that show preferential growth when grown on galactose medium compared with the growth obtained on YPD glucose medium are considered candidates. To test the specificity of the library plasmids, plasmid DNA is extracted from candidate clones[29] and is used to cotransform *cdc25-2* cells with either the specific bait or a nonrelevant bait. Candidate clones that exhibit bait-specific growth are further analyzed.

Perspective

SRS and RRS are efficient methods for the identification and characterization of protein–protein interactions. The obvious difference between the two systems is the effector size. This makes the RRS, using Ras as the effector protein (20 kDa), more convenient in all DNA manipulations and more attractive for bait fusions either a few amino acids long or full-length proteins as well. On the other hand, the relatively large size of hSos (150 kDa) might provide a preferential cytoplasmic localization of a bait fusion

[29] K. Robzyk and Y. Kassir, *Nucleic Acids Res.* **20,** 3790 (1992).

FIG. 2. Flow chart describing the screening of cDNA libraries by the SRS or RRS. Cdc25-2 yeast cells transformed with bait and GAP expression plasmids are transfected with cDNA library expression plasmids. Cells are plated on glucose minimal medium lacking tryptophan, uracyl, and leucine, resulting in 5000–10,000 transformants per 10-cm plate incubated at 24°. Transformants are allowed to grow for 4–5 days at 24° before replica plating onto galactose-containing plates that are incubated at 36° for 3–4 days. Clones that exhibit efficient growth at 36° are isolated, placed on a glucose-containing plate, and grown for an additional 2 days at 24°. Subsequently, the isolated clones are tested for their galactose-dependent growth at 36°. Clones exhibiting efficient growth on galactose-containing medium but slower growth on glucose-containing medium are considered promising candidates and are further analyzed.

that contains a strong nuclear localization signal. Thus far, protein–protein interactions that could be observed using the SRS were also detectable using the RRS. Nevertheless, several differences between the two systems were observed that should be taken into consideration when deciding on whether to use the SRS or RRS. First, when a specific bait exhibited efficient growth at the restrictive temperature, even in the absence of a specific prey, using the SRS, this background growth was eliminated by using the RRS. Second, protein–protein interactions that could be detected on both glucose and galactose plates when using the SRS were detected only on galactose plates when using the RRS. Because the expression levels provided by the *GAL1* promoter on glucose plates are not sufficient to activate the RRS, indicate that the RRS system would be preferable for detecting interaction between protein pairs with stronger binding affinities as compared to the SRS. Therefore, when strong interactions are expected it is better to use the RRS approach, because only clones exhibiting efficient growth on galactose but not on glucose will be further analyzed. Third, the RRS approach is advantagous over the SRS approach in preventing the isolation of cDNAs encoding the Ras protein during a cDNA library screening effort. However, neither system completely prevents the isolation of cDNAs encoding hSos homologs. A yeast strain that would eliminate the isolation of Ras guanyl nucleotide exchange factors would greatly improve the reliability of both systems. So far, the ability of *cdc25-2* cells to grow at the restrictive temperature serves as the only readout parameter for a functional protein–protein interaction. Because this parameter is mainly qualitative and not quantitative, the addition of Ras-responsive reporter genes would greatly improve the usefulness of both systems. The RRS system has been shown to function in mammalian host cells as well.[30] This enables a novel interaction to be tested directly, initially identified in yeast, *in vivo* in tissue culture cells, and provides a useful quantitative readout to compare the association with different protein pairs. In addition, it provides the possibility of using a reverse hybrid approach directly in target cells to screen for drugs that inhibit specific protein–protein interaction.

In summary, both the SRS and RRS approaches represent efficient means for the identification and characterization of protein–protein interactions and thus serve as attractive alternatives to the two-hybrid system, by overcoming some of the problems and limitations associated with this classic assay, and also provide novel features.

[30] M. Maroun and A. Aronheim, *Nucleic Acids Res.* **27,** e4 (1999).

Acknowledgments

A.A. thanks Y. C. Broder and S. Katz for fruitful discussions. Research was supported by the Israel Science Foundation, the Israel Academy of Sciences, the Humanities-Charles H. Revson Foundation, the Israel Cancer Research Foundation (ICRF), and the National Institutes of Health (NIH). A.A. is a recipient of an academic lectureship from Samuel and Miriam Wein. M.K. is the Frank and Else Schilling-American Cancer Society Research Professor.

[5] A Bacterial Two-Hybrid System that Exploits a cAMP Signaling Cascade in *Escherichia coli*

By GOUZEL KARIMOVA, AGNES ULLMANN, and DANIEL LADANT

Introduction

Most biological processes involve specific protein–protein interactions. The yeast two-hybrid system[1] represents a powerful *in vivo* approach to analyze interactions between macromolecules and screen for polypeptides that bind to a given "bait" protein. Bacterial equivalents to the yeast two-hybrid system have not been developed until recently. In this chapter we describe a novel bacterial two-hybrid system that allows easy *in vivo* screening and selection of functional interactions between two proteins.[2] This system, because of its sensitivity and simplicity, could have broad application in the studies of structure–function relationships in biological macromolecules, in the functional analysis of genomes, as well as in high-throughput screening of interacting ligands or new therapeutic agents.

Principle of the Bacterial Two-Hybrid System Based on a cAMP Signaling Cascade

The two-hybrid system that we have set up is based on the reconstitution of an artificial cAMP signal transduction pathway in an *Escherichia coli* adenylate cyclase-deficient strain (*cya*). It takes advantage of the modular structure of the catalytic domain of *Bordetella pertussis* adenylate cyclase, which consists of two complementary fragments.[3,4] When they are expressed

[1] S. Fields and O. Song, *Nature (London)* **340,** 245 (1989).

[2] G. Karimova, J. Pidoux, A. Ullmann, and D. Ladant, *Proc. Natl. Acad. Sci. U.S.A.* **95,** 5752 (1998).

[3] D. Ladant, *J. Biol. Chem.* **363,** 2612 (1988).

[4] D. Ladant, S. Michelson, R. Sarfati, A. M. Gilles, R. Predeleanu, and O. Barzu, *J. Biol. Chem.* **264,** 4015 (1989).

0076-6879/00 $30.00

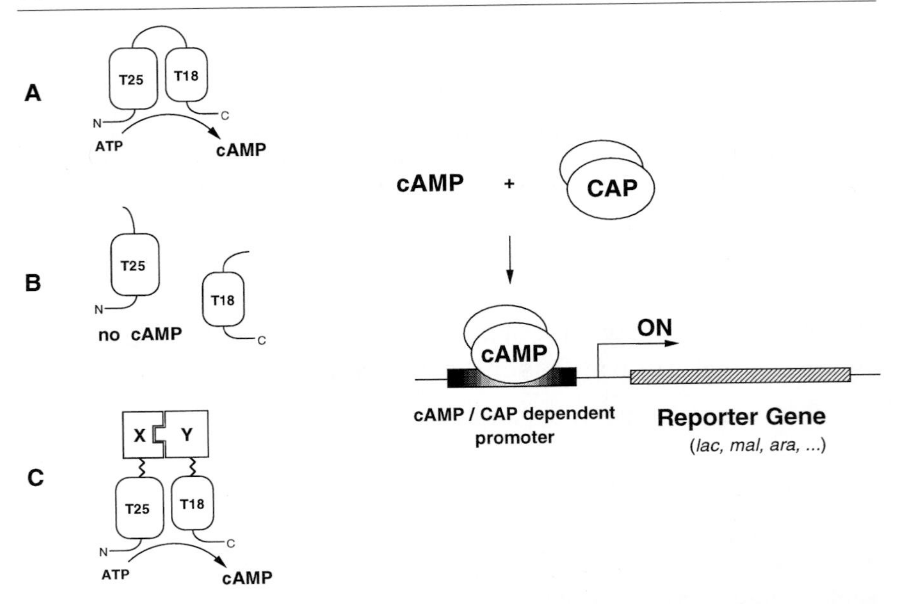

FIG. 1. Principle of the bacterial two-hybrid system. T25 and T18 (boxed), the two fragments of the catalytic domain of *B. pertussis* adenylate cyclase; X and Y, the interacting polypeptides fused to T25 and T18; CAP, the catabolite activator protein.

separately in *E. coli,* they cannot be converted to an active enzyme, unless interacting polypeptides are genetically fused to these fragments (Fig. 1). In the bacterial two-hybrid system, interaction between the two chimeric proteins results in functional complementation between the two adenylate cyclase fragments and restoration of enzymatic activity. The resulting cAMP synthesis triggers the expression of several *E. coli* resident genes, thus giving rise to a selectable phenotype.

cAMP Signaling in Escherichia coli

In *E. coli,* cAMP is a key signaling molecule that binds to the pleiotropic transcriptional activator, CAP (catabolite activator protein). The cAMP–CAP complex controls the expression of a large number of genes, including genes involved in the catabolism of carbohydrates such as lactose or maltose.[5,6] Hence, *E. coli* strains deficient in their endogenous adenylate cyclase

[5] A. Ullmann and A. Danchin, *in* "Advances in Cyclic Nucleotide Research" (P. Greengard and G. A. Robinson, eds.), p. 1. Raven Press, New York, 1983.

[6] M. H. J. Saier, T. M. Ramseier, and J. Reizer, *in* "*Escherichia coli* and *Salmonella:* Cellular and Molecular Biology" (F. C. Neidhart, ed.), p. 1325. ASM Press, Washington, D.C., 1996.

(*cya*) are unable to ferment lactose or maltose, in contrast to *cya*$^+$ bacteria. The *cya* and *cya*$^+$ cells can be discriminated easily, either on indicator media or on selective media.

Modular Structure of the Bordetella pertussis Adenylate Cyclase Catalytic Domain

Bordetella pertussis, the causative agent of whooping cough, synthesizes a calmodulin-dependent adenylate cyclase toxin encoded by the *cyaA* gene.[7,8] Its catalytic domain, located within the first 400 amino acids of the 1706-residue protein,[9] exhibits a high catalytic activity (k_{cat} = 2000 sec^{-1}) in the presence of calmodulin (CaM), and a residual but detectable activity (k_{cat} = 2 sec^{-1}) in the absence of this activator.[3,10] When the catalytic domain is expressed in *E. coli cya,* which does not produce calmodulin, its residual (i.e., CaM-independent) activity is sufficient to render Cya$^+$ an adenylate cyclase-deficient strain.[11]

The catalytic domain has a modular structure: it consists of two complementary fragments, T25 and T18, originally identified by limited proteolyis studies.[3,4] These two fragments, when expressed in *E. coli cya* as separated entities, are unable to recognize each other and cannot reconstitute a functional enzyme. However, when the T25 and T18 fragments are fused to peptides or proteins that can interact, heterodimerization of these chimeric polypeptides results in a functional complementation and, therefore, in cAMP synthesis.[2] Bacteria that express chimeric proteins that are able to heterodimerize will exhibit a Cya$^+$ phenotype that can be easily detected either on indicator media or on selective media.

Materials

Bacterial Strain

Adenylate cyclase-deficient (*cya*) *E. coli* strains harboring point mutations, deletions, or insertions within the *cya* gene (adenylate cyclase struc-

[7] M. Mock and A. Ullmann, *Trends Microbiol.* **1,** 187 (1993).
[8] P. Guermonprez, C. Fayolle, C. Leclerc, G. Karimova, A. Ullmann, and D. Ladant, *Methods Enzymol.* **326,** Chap. 32 (2000).
[9] P. Glaser, D. Ladant, O. Sezer, F. Pichot, A. Ullmann, and A. Danchin, *Mol. Microbiol.* **2,** 19 (1988).
[10] J. Wolff, G. H. Cook, A. R. Goldhammer, and S. A. Berkowitz, *Proc. Natl. Acad. Sci. U.S.A.* **77,** 3841 (1980).
[11] D. Ladant, P. Glaser, and A. Ullmann, *J. Biol. Chem.* **267,** 2244 (1992).

tural gene) have been described.[5,6] In most of our studies we have used the DHP1 strain, which is a spontaneous *cya* derivative of DH1 [F^- *glnV44(AS) recA1 endA1 gyrA96* (*Nal*^{r}) *thil hsdR17 spoT1 rfbD1*], isolated by a modified phosphomycin selection procedure.[2] Functional complementation between hybrid proteins was found to be more efficient in DHP1 than in other *cya* strains tested, possibly because of the higher stability of chimeric proteins. Transformation of DHP1 bacteria is performed by standard techniques ($CaCl_2$ or electroporation) as described in Sambrook *et al.*[12]

Plasmids

Genetic screening of protein interaction requires coexpression of the two hybrid proteins within the same recipient *cya* bacteria. For this purpose two compatible vectors were constructed[2] (Fig. 2).

1. pT25 encodes the T25 fragment of *B. pertussis* adenylate cyclase, corresponding to the first 224 amino acids of CyaA.[4,13] This vector is a derivative of low copy-number plasmid pACYC184,[14] and carries the chloramphenicol acetyltransferase gene that confers resistance to chloramphenicol. A multicloning sequence was inserted at the 3′ end of the gene encoding T25 to facilitate construction of fusions in frame at the C-terminal end of the T25 polypeptide.
2. pT18 encodes the T18 fragment (amino acid 225 to 399 of CyaA)[4,13] and is a derivative of pBluescript II KS (Stratagene, La Jolla, CA), which expresses the β-lactamase selectable marker (ampicillin resistance). Its ColE1 origin of replication is compatible with that of pT25, carrying the p15A replicon. Hence, both plasmids, pT18 and pT25, can be maintained in the same cell. The T18 coding region is fused in frame to the multicloning sequence of pBluescript II KS. In both vectors the expression of fusion proteins is under the control of the *lac* promoter.

Media

Analysis of complementation can be scored either on indicator plates [i.e., LB–X-Gal, MacConkey, or EMB (eosin–methylene blue) medium supplemented with maltose or lactose] or on selective media (minimal medium supplemented with lactose or maltose as unique carbon sources).

[12] J. Sambrook, E. F. Fritsch, and T. Maniatis, "Molecular Cloning: A Laboratory Manual." Cold Spring Harbor Laboratory Press, Cold Spring Harbor, New York, 1989.

[13] H. Munier, A. M. Gilles, P. Glaser, E. Krin, A. Danchin, R. Sarfati, and O. Barzu, *Eur. J. Biochem.* **196,** 469 (1991).

[14] J. H. Miller, "A Short Course in Bacterial Genetics." Cold Spring Harbor Laboratory Press, Cold Spring Harbor, New York, 1992.

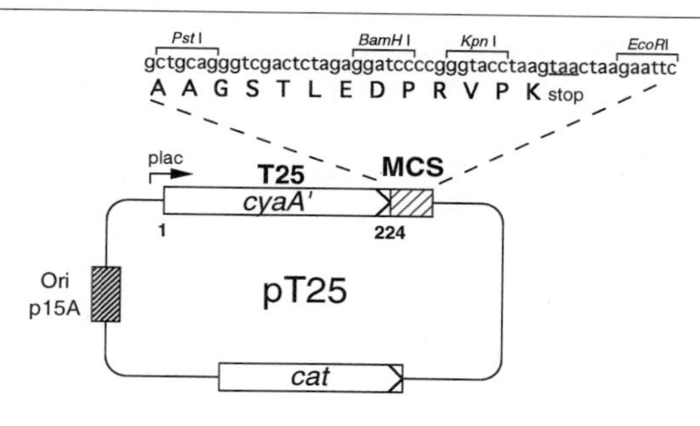

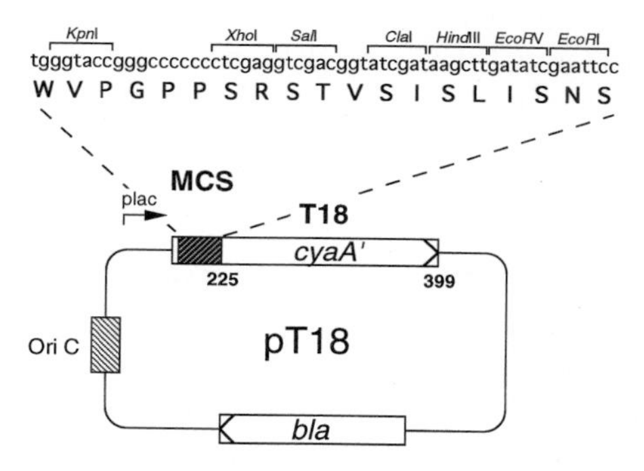

FIG. 2. Cloning vectors used for chimeric protein expression. Schematic representation of plasmids pT25 and pT18. Rectangles represent the open reading frames of T25 and T18 fragments, and of the chloramphenicol acetyltransferase (*cat*) and β-lactamase (*bla*) genes. The multicloning sequences (MCSs), which allow insertion of foreign genes, are displayed with the corresponding open reading frames.

LB–X-Gal Medium. The β-galactosidase chromogenic substrate 5-bromo-4-chloro-3-indolyl-β-D-galactopyranoside (X-Gal; final concentration, 40 μg/ml) and antibiotics (ampicillin and chloramphenicol at 100 and 30 μg/ml, respectively) are added to the autoclaved rich Luria–Bertani (LB) medium just before pouring plates.[14] Isopropyl-β-D-thiogalactopyranoside (IPTG) (final concentration, 0.5 mM) can be added to increase β-galactosidase expression.

MacConkey Medium. MacConkey base medium can be purchased from Difco Laboratories (Detroit, MI). Maltose [1% (w/v) final concentration; stock solution of 20% (w/v) in water, sterilized by filtration] and antibiotics (ampicillin and chloramphenicol at 100 and 30 μg/ml, respectively) are added to the autoclaved MacConkey medium just before pouring the plates.[14]

EMB Medium. The EMB base contains 10 g of Bacto-tryptone, 10 g of yeast extract, 5 g of NaCl, 2 g of KH_2PO_4, and 15 g of Difco agar in 1 liter of distilled water and adjusted to pH 7.4. After autoclaving, 10 ml each of sterile solutions of 4.5% (w/v) eosin yellow and 0.7% (w/v) methylene blue are added. The solutions containing the dyes should be autoclaved separately. Maltose and antibiotics are added as described above, just before pouring the plates.

Standard Synthetic Medium M63 Supplemented with 0.4% (w/v) Lactose. A stock 5× concentrated M63 medium[14] is prepared by adding 10 g of $(NH_4)_2SO_4$, 68 g of KH_2PO_4, and 2.5 mg of $FeSO_4 \cdot 7H_2O$ to 1 liter of water (adjust to pH 7.0 with KOH). Two hundred milliliters of sterile 5× M63 medium containing 1% (w/v) lactose, vitamin B_1 (1 μg/ml), and antibiotics (chloramphenicol at 15 μg/ml, ampicillin at 50 μg/ml), are added to autoclaved agar (15 g in 800 ml of water) before pouring the plates.

cAMP Assays

Commercial kits for cAMP determination [enzyme-linked immunosorbent assay (ELISA) or radioimmunoassays] can be purchased from various companies. In our laboratory we use a homemade ELISA.[2] Briefly, a cAMP–biotinylated bovine serum albumin (BSA) conjugate, prepared by coupling 2′-*O*-monosuccinyladenosine 3′,5′-monophosphate (Sigma, St. Louis, MO) to biotinylated BSA (Sigma) in the presence of N-Ethoxycarbonyl-2-ethoxy-1,2-dihydroquinoline (EEDQ) according to Joseph and Guesdon,[15] is coated on ELISA plates (50 μl/well) and incubated overnight at 4°. Nonspecific protein-binding sites are saturated with 50 m*M* HEPES (pH 7.5), 150 m*M* NaCl, and 0.1% (v/v) Tween 20 (HBST buffer) containing BSA at 20 mg/ml (250 μl/well) for 1 hr at room temperature. After washing with HBST, 100 μl of boiled (then cooled) bacterial culture, or standard dilutions (ranging from 3 μ*M* to 3 n*M*) of cAMP in LB medium, are added to the wells, followed by 50 μl of a rabbit anti-cAMP antiserum diluted 5000-fold in HBST buffer containing BSA at 10 mg/ml. After overnight incubation at 4°, the plates are washed extensively with HBST, and then goat anti-rabbit IgG coupled to alkaline phosphatase (AP) is added (100 μl/well) and incubated for 1 hr at 30°. After washing with HBST, the AP activity is revealed by 5′-*p*-nitrophenyl phosphate (100 μl/well of AP buffer, 100 m*M* NaCl, 5 m*M* $MgCl_2$, 50 m*M* Tris-HCl, pH 9.5, containing 10 mg (2 tablets) of Sigma 104 phosphatase substrate per 11 ml). Cyclic AMP concentrations in bacterial extracts are determined from the standard curve established with known concentrations of cAMP diluted in LB medium.

[15] E. Joseph and J. L. Guesdon, *Anal. Biochem.* **119,** 335 (1982).

β-Galactosidase Assay

β-Galactosidase assays are carried out on exponentially growing cells (overnight cultures can also be used) according to Pardee *et al.*[16] Bacteria are permeabilized by adding 1 drop of toluene and 1 drop of a 0.1% (w/v) sodium dodecyl sulfate (SDS) solution per 2–3 ml of cell suspension. The tubes are vortexed for 10 sec and placed in a shaker at 37° for 30 min, lightly plugged with cotton (to allow the toluene to evaporate). Aliquots of the permeabilized cells (0.1 to 1 ml) are added to the assay medium (PM2 buffer, which contains 70 m*M* $Na_2HPO_4 \cdot 12H_2O$, 30 m*M* $NaHPO_4 \cdot H_2O$, 1 m*M* $MgSO_4$, and 0.2 m*M* $MnSO_4$, pH 7.0, supplemented just before use with 100 m*M* 2-mercaptoethanol) to a final volume of 2 ml. The tubes are equilibrated at 28° and the reaction is started by adding 0.5 ml per tube of the substrate O-nitrophenyl-β-D-galactoside (ONPG) (4 mg/ml in PM2 buffer without 2-mercaptoethanol). After sufficient yellow color has developed, the reaction is stopped by adding 1 ml of a 1 *M* Na_2CO_3 solution. The optical density at 420 nm is then recorded for each tube. The reaction is linear up to an absorbance of 1.6. The reading at 420 nm is a combination of absorbance by the *o*-nitrophenol and light scattering of the cell debris. This latter can be neglected if small volumes of cell suspensions are used and the reading at 420 nm is above 0.3. At lower absorbance the light scattering can be corrected for by obtaining the absorbance at 600 nm from the same reaction mixture and using a correction factor: $OD_{420} - 1.5 \times OD_{600}$, which then compensates for light scattering. Enzymatic activities are calculated by using a molar absorption coefficient of 5 for *o*-nitrophenol at pH 11. One unit of β-galactosidase activity corresponds to 1 nmol of ONPG hydrolyzed per minute at 28°.[16] Results are generally given as units per milligram (dry weight) of bacteria. One milligram (dry weight) of bacteria corresponds to about 2×10^9 bacteria per milliliter, determined from the reading at 600 nm. Other methods are described in Miller[14] and Sambrook *et al.*[12]

Analysis of Protein–Protein Interactions with the Bacterial Two-Hybrid System: General Procedure

The general procedure to analyze *in vivo* interactions between two proteins (e.g., X and Y) is outlined in Fig. 3. In a first step, the genes encoding the two proteins of interest are cloned into pT25 and pT18 vectors, in frame with the adenylate cyclase fragment open reading frames. In a second step, the resulting plasmids, which express the hybrid proteins (T25-X and Y-T18), are cotransformed in competent DHP1 cells. The cotransfor-

[16] A. B. Pardee, F. Jacob, and J. Monod, *J. Mol. Biol.* **1,** 165 (1959).

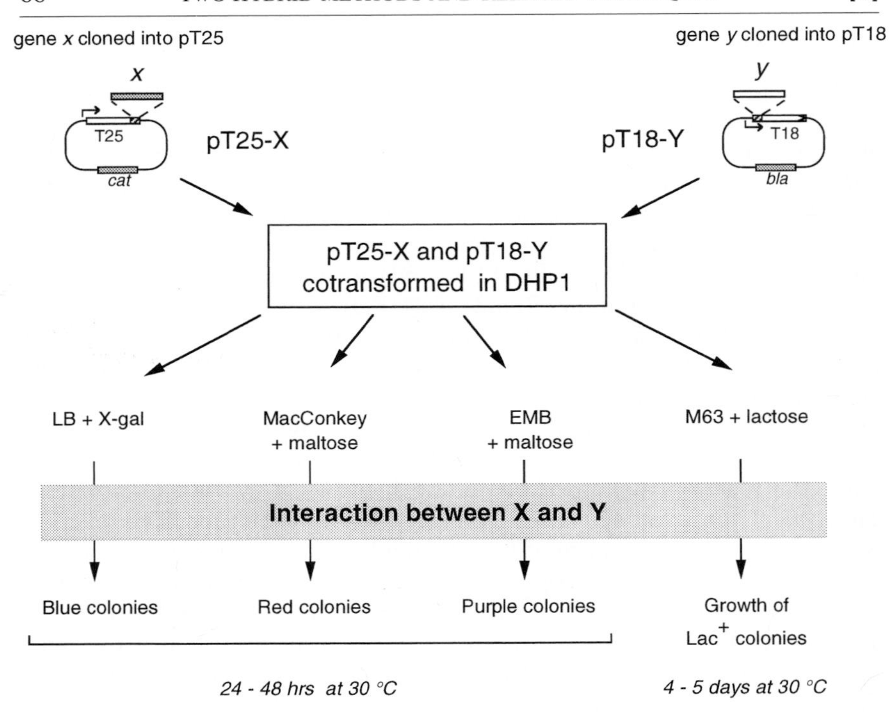

FIG. 3. Analysis of protein–protein interactions using the bacterial two-hybrid system. Cloning strategy and screening/selection procedures are depicted. For detailed explanations, see text.

mants are plated either on indicator plates or on selective media, depending on the kind of experiments performed. Interaction between the two hybrid proteins results in functional complementation between the T25 and T18 fragments, leading to cAMP synthesis, which then confers a Cya^+ phenotype to the recipient *cya* bacteria.

Screening Procedure

To determine whether two particular proteins (X and Y) interact, a screening procedure on indicator plates is particularly appropriate. On these rich media, the cells are growing faster and results can be obtained after 1–2 days of incubation at 30°, but only a small number of cotransformants (usually fewer than 500 per plate) can be plated. Two different types of indicator plates can be used.

LB–X-Gal Medium. In *E. coli* the expression of the *lacZ* gene encoding β-galactosidase is positively controlled by cAMP–CAP. Therefore, bacteria expressing interacting hybrid proteins will form blue colonies on rich Luria–Bertani (LB) medium[14] in the presence of the chromogenic substrate X-Gal. Cells expressing noninteracting proteins should remain colorless; however, they will frequently form pale blue colonies because DHP1 cells express a basal level of β-galactosidase activity even in the absence of cAMP. This sometimes renders the discrimination between complementing and noncomplementing clones difficult.

MacConkey and EMB Media. MacConkey and EMB media are used as fermentation indicators that make use of the pH differences between colonies that metabolize specific carbohydrates and colonies that do not. Adenylate cyclase-deficient bacteria are unable to ferment lactose or maltose[5,6]: they form white–pink colonies on MacConkey indicator medium containing lactose or maltose, while Cya^+ bacteria form red colonies on the same media (fermentation of the added sugar results in the acidification of the medium, which is revealed by a color change of the dye).[14] On EMB medium the ability to ferment the added carbohydrate yields dark purple colonies, which sometimes have a green sheen. Failure to metabolize the added carbohydrate results in white–light pink colonies. As noted above, because of the background expression of β-galactosidase in the absence of cAMP, noncomplementing *cya* bacteria will weakly ferment lactose and will display an intermediate Lac^+ phenotype. It is therefore highly preferable to use maltose as the added sugar. Because the expression of the maltose regulon is more stringently dependent on cAMP, no fermentation of maltose occurs in bacteria that do not synthesize cAMP, that is, bacteria that express noninteracting chimeric proteins. Hence, a clear, unambiguous difference between complementing and noncomplementing bacteria will be seen on MacConkey or EMB–maltose.

Selection Procedure

To analyze a large number of colonies, as in a library screening, it will be advantageous to plate the transformants on a synthetic medium. As Cya^+ cells are Lac^+/Mal^+, they are able to grow on a minimal medium supplemented with lactose/maltose as unique carbon sources.[14] A powerful selection procedure to isolate bacteria that express interacting hybrid proteins is to use a standard synthetic medium, M63,[14] supplemented with 0.4% (w/v) lactose. Up to 10^6 cells can be plated on a single dish on this selective minimal medium containing lactose. Growth of Lac^+ colonies will be detected after 4–5 days of incubation at 30°. The X-Gal substrate can also be added to the medium to facilitate the early visualization of Lac^+ complementing colonies. The growing Lac^+ colonies are, in fact, of two different types: truly positive cells expressing interacting hybrid proteins and false-positive cells having acquired a Lac^+ phenotype due to endogenous mutations (see Discussion). However, for unknown reasons, most of these false-positive clones (which appear at a frequency of about 10^{-6}) exhibit a mucoid phenotype and, therefore, they can be easily differentiated from truly positive clones.

Quantification of the functional complementation between chimeric proteins can be obtained by measuring cAMP levels and β-galactosidase activities in liquid cultures.

It is worth noting that, in most cases, the functional complementation was found to be more efficient at 30° than at 37°, although the precise reason for this is still unclear.

Applications to Model Proteins

We have shown that this bacterial two-hybrid system is able to reveal interactions between small peptides (GCN4 leucine zipper), bacterial proteins (tyrosyl-tRNA synthetase, tyrosine–tRNA ligase), and eukaryotic proteins (yeast Prp11–Prp21 complex).

The 35-amino acid leucine zipper motif of GCN4, a yeast transcriptional activator,[17] dimerizes with high affinity. Its DNA sequence was amplified by polymerase chain reaction (PCR) and subcloned into pT25 and pT18[2] to yield plasmids pT25-zip and pT18-zip. The tyrosyl-tRNA synthetase from *Bacillus stearothermophilus* (TyrRS) is a dimer with a monomer–dimer equilibrium constant of about 10 n*M*.[18] A DNA fragment encoding the N-terminal part of the synthetase, which forms a dimer in solution (residues 1 to 302), was subcloned into pT25 and into pT18 to construct plasmids pT25-Tyr and pT18-Tyr, respectively. Similarily, the interacting yeast splicing factors Prp11 and Prp21, previously characterized with the yeast two-hybrid assay,[19] were genetically fused to T25 and T18, respectively (encoded by plasmids pT25-Prp11 and pT18-Prp21). Table I indicates the results of different types of DHP1 cotransformations with the above-described plasmids. These experiments demonstrate that the functional complementation between the T25 and T18 fusion proteins is dictated by the specificity of recognition of the polypeptides fused to the two adenylate cyclase fragments.

We have extended this analysis and studied the interactions between various subdomains of the TyrRS protein.[19a] In addition, we checked that

[17] E. K. O'Shea, R. Rutkowski, and P. S. Kim, *Science* **243,** 538 (1989).

[18] W. H. Ward, D. H. Jones, and A. R. Fersht, *Biochemistry* **26,** 4131 (1987).

[19] P. Legrain and C. Chapon, *Science* **262,** 108 (1993).

[19a] G. Karimova, A. Ullmann, and D. Ladant, submitted (2000).

TABLE I
ANALYSIS OF COMPLEMENTATION IN DHP1 STRAIN

Plasmids	Phenotype on MacConkey–maltose	cAMP[a] (pmol/mg dry weight)	β-Galactosidase activity[a] (units/mg dry weight)
None	White	<10	179
pT25 + pT18	White/72 hr	<10	130
pT25 + pT18-zip	White/72 hr	<10	183
pT25-zip + pT18	White/72 hr	<10	178
pT25-zip + pT18-zip	Red/30 hr	1100	4750
pT25-Tyr + pT18-Tyr	Red/40 hr	580	2800
pT25-Tyr + pT18	White/96 hr	<10	193
pT25 + pT18-Tyr	White/96 hr	<10	183
pT25-Tyr + pT18-zip	White/96 hr	<10	134
pT25-zip + pT18-Tyr	White/96 hr	<10	126
pT25-prp11 + pT18-prp21	Red/40 hr	65	850

[a] Bacteria were grown in LB at 30° in the presence of 0.5 m*M* IPTG plus appropriate antibiotics. The results represent the average values obtained for at least five independent cultures, which differed by less than 10%.

a point mutation, previously shown[18] to abolish dimerization of TyrRS (Phe 164 → Glu), also abolishes functional complementation between the chimeric T25–Tyr and T18–Tyr proteins. We have also used this bacterial two-hybrid system to analyze the dimerization capabilities of a DNA-binding protein, BvgA, which is the main transcriptional activator of *B. pertussis* virulence-associated genes[20] and belongs to the large family of bacterial two-component signal transduction regulators. Our results showed that BvgA has the capacity to dimerize (our unpublished results, 1999). Experiments are in progress to delineate the amino acids that are directly involved in the dimerization of BvgA.

In addition, we performed a "model" library screen on minimal medium containing lactose, to show that this system can be used to identify a few interacting proteins among an excess of noninteracting proteins.[2] DHP1 bacteria expressing interacting proteins (T25-zip and T18-zip) were mixed with a 10^5 excess of DHP1 bacteria expressing noninteracting proteins (T25 and T18) and about 10^7 cells were plated on minimal medium supplemented with lactose plus antibiotics. After 4–5 days at 30°, about 100 Lac^+ colonies appeared. Plasmid DNA analysis indicated that 18 of 20 of these colonies tested harbored pT25-zip and pT18-zip. The other two false-positive colonies (i.e., expressing noninteracting T25 and T18 fragments) appeared to

[20] V. Scarlato, B. Arico, M. Domenighini, and R. Rappuoli, *BioEssays* **15,** 99 (1993).

represent spontaneous revertants of DHP1 to a Lac$^+$ phenotype (see Discussion).

Discussion: Advantages and Drawbacks

Although this bacterial two-hybrid system has been developed only recently, it is expected to be useful in studying *in vivo* protein–protein interactions and in identifying potential partners of a given polypeptide. Below, we examine briefly some advantages of this technique as well as its potential drawbacks.

Advantages

1. The bacterial two-hybrid system relies on a signaling cascade that utilizes the diffusable regulatory molecule cAMP. As a consequence, the physical association of the two interacting chimeric proteins can be spatially separated from the transcription activation readout. Hence, it is possible to analyze protein–protein interactions that occur either in the cytosol, at the inner membrane level, or on the DNA. This system should offer the possibility of studying bacterial proteins in their native context and eventually to analyze simultaneously both the functional activity and the association state of given proteins (provided that the hybrid proteins retain their function).

2. As this genetic test is carried out in *E. coli,* the screening and the characterization of the interacting proteins should be facilitated. First, as *E. coli* is the basic "tool" of molecular biologists, this technique can be exploited immediately by most researchers. Second, the high efficiency of transformation that can be achieved in *E. coli* will allow screening of libraries of high complexity. Third, the same plasmid constructs used in library screening to identify a putative binding partner to a given "bait" can be employed to express the chimeric proteins in order to characterize their interaction by *in vitro* binding assays.

3. This two-hybrid system offers the possibility not only to select colonies that express interacting proteins (positive selection) but also to select colonies that express noninteracting proteins among a background of cells expressing interacting hybrids (negative selection). To select for bacteria that no longer synthesize cAMP, two properties of the *cya* cells can be used: (1) resistance to phosphomycin and (2) resistance to bacteriophage lambda (λ) infection[5,6] (the bacterial receptor for phage λ, LamB, is positively regulated by cAMP and therefore *cya* strains express a low level of LamB). Such a negative selection procedure will be of great interest to finely characterize interacting proteins: it will allow a search for point

mutations that abolish the interaction under study. Compensatory mutations that restore the interaction could then be selected for by using the positive selection procedure. The negative selection procedure should also prove to be useful in high-throughput screening for molecules that could disrupt a given interaction between two proteins of interest.

Pitfalls and Potential Limitations

False-negative results that correspond to hybrid proteins that interact but do not yield functional complementation. As complementation requires the spatial proximity between the T25 and T18 fragments, it is likely that in a number of cases, steric constraints imposed by the three-dimensional structure of the hybrid protein complex will affect the adenylate cyclase enzymatic activity of the reconstituted complex. This could lead to a lack of functional complementation between two chimeric proteins despite their physical interaction. In other words, the absence of phenotypic complementation between two hybrid proteins will not necessarily mean that they do not associate. In fact, this rule applies similarly to all two-hybrid methodologies including the yeast system, although this latter is probably much more tolerant to large hybrid proteins. Future work with this bacterial system will give more insight into its flexibility and tolerance. Yet, it is noteworthy that many proteins are made of separate modules with autonomous interacting abilities. In many cases functional domains that are able to specifically interact with each other are in the range of 100–200 amino acid residues. It is likely that most of the interactions mediated by domains of this size can be detected with this bacterial two-hybrid technique.

False-positive results that correspond to clones that exhibit a Cya$^+$ phenotype (*i.e., Lac$^+$ or Mal$^+$*) *although they harbor noninteracting hybrid proteins.* False-positive colonies in this two-hybrid technique can arise from two different origins.

1. The recipient cells acquire a Lac$^+$ or Mal$^+$ phenotype independently of the expressed hybrid proteins. This can be due to (a) the spontaneous reversion of the Cya$^-$ to Cya$^+$ phenotype that occurs in DHP1, at a frequency of 10^{-6}. This problem can be eliminated by using *E. coli* strains deleted of the endogenous adenylate cyclase gene; or (b) spontaneous mutations leading to a cAMP-independent Lac$^+$ or Mal$^+$ phenotype as a result of mutations of the CAP gene toward a cAMP-independent phenotype (the so-called CRP* mutants occur at frequencies of about $10^{-8}/10^{-9}$) or mutations in the *lac* promoter (which occur at frequencies of about $10^{-7}/10^{-8}$). The probability of obtaining cAMP-independent Mal$^+$ cells should be about 10^{-21}.

In all cases, these false-positive colonies can be eliminated on transformation of the plasmids into new competent *cya* cells.

2. The cotransformed plasmids confer a Cya^{+} phenotype although they encode noninteracting hybrid proteins. This can result from (a) the insertion in one of the cotransformed plasmids (e.g., during a library screen) of a gene encoding a fragment of an adenylate cyclase that is catalytically active in *E. coli* or (b) the insertion in one of the cotransformed plasmids of a gene encoding a protein that is able to bind directly to the T25 and/or T18 fragments and restore their adenylate cyclase activity. This could be the case for the eukaryotic calmodulin or possibly for other calmodulin-like proteins.

In both cases, these false-positive plasmids might be easily detected on cotransformation with the complementary plasmid expressing only the adenylate cyclase fragment (T25 or T18, not fused to the bait protein).

Conclusions and Future Developments

We are currently working to improve the basic version of this bacterial two-hybrid system in order to render it more sensitive, more efficient, and more versatile. Some of the possibilities that we are currently examining are listed below.

Strains

To facilitate library screening, we are currently constructing an *E. coli cya* strain that harbors an antibiotic resistance gene under the control of a promoter dependent on cAMP–CAP. Hence, colonies that express interacting hybrids could be selected much more easily (and rapidly) on LB medium supplemented with the appropriate antibiotic. Also, we are designing an *E. coli cya* strain that harbors a reporter gene, driven by cAMP–CAP, that encodes a toxic product that will be helpful in searching for chemical compounds or mutations that abolish a given interaction between the two studied proteins.

Vectors

We have constructed new plasmids that permit fusions at the C terminus of T18. We have also constructed a plasmid that expresses a T25 fragment with a hexahistidine (His_6) tag at its N terminus; this construct will facilitate the *in vitro* characterization of the interactions between the two hybrid proteins. Hence, during the course of a library screen, it would be possible to check directly if positive clones encode truly interacting polypeptides (the His_6-tagged T25 hybrid protein will be specifically retained on a nickel column; coretention of the T18 chimeric proteins will demonstrate undoubtedly a direct interaction between the hybrid molecules). Alternatively, different immunological tags could also be introduced in both fragments and the physical interactions could be established by a coimmunoprecipitation approach.

New Design for Increased Sensitivity

The present system makes use of only the basal enzymatic activity of *B. pertussis* adenylate cyclase, i.e., in the absence of calmodulin. This system could be rendered exquisitely sensitive by using the full catalytic potency of *B. pertussis* adenylate cyclase, in the presence of its natural activator calmodulin. In this case, it is anticipated that reconstitution of only a few hybrid molecules per cell will be sufficient to elicit a detectable signal (compared to 1000–5000 in the present design).

Concluding Remarks

This bacterial two-hybrid technique should find wide applications for the analysis of *in vivo* interactions, in functional genomics, and in drug discovery.[21] Also, this technique could be used to design selection schemes in which the fitness of the bacterial population would be determined by the affinity and/or specificity of interactions between bybrid proteins.

Acknowledgments

Financial support for the authors' laboratory came from the Institut Pasteur, the Centre National de la Recherche Scientifique (URA 1129), the Ministry of National Education, Research, and Technology, and a collaborative Research and Development Contract with Hybrigenics SA (Paris).

[21] S. Fields, *Nature Genet.* **15,** 325 (1997).

[6] A Green Fluorescent Protein-Based Reverse Two-Hybrid System: Application to the Characterization of Large Numbers of Potential Protein-Protein Interactions

By HIDEKI ENDOH, ALBERTHA J. M. WALHOUT, and MARC VIDAL

Introduction

Complete genome sequences reveal high numbers of predicted open reading frames (ORFs) for which no functional information is yet available. Consequently, high-throughput standardized assays are being developed to functionally annotate the corresponding predicted proteins. One of the challenges of this emerging field of "functional genomics" is to convert these annotations into useful biological hypotheses. Considering the amount of information, it is important to design efficient methods to test these hypotheses in the relevant biological systems. For example, protein interaction mapping projects result in the identification of large numbers of potential interactions and it has been argued that interaction-defective alleles can be used to test the validity of these interactions. Here, we describe the protocols of a novel system that can be used for the convenient identification of interaction-defective alleles. The system is based on a modified version of the reverse two-hybrid system and is compatible with high-throughput settings.

"Postfunctional Genomics" Era

The information generated by complete genome sequences is expected to revolutionize the way biological questions can be formulated. Biological questions can now be approached on a genome-wide scale rather than gene by gene or protein by protein. The emerging field of "functional genomics" characterizes large numbers of ORFs predicted from complete genome sequences by the use of standardized functional assays that can be applied in high-throughput settings. Examples of such projects include gene knockouts,[1] microarray or chip analyses,[2] protein–protein interaction maps,[3–6]

[1] E. A. Winzeler, D. D. Shoemaker, A. Astromoff, H. Liang, K. Anderson, B. Andre, R. Bangham, R. Benito, J. D. Boeke, H. Bussey, A. M. Chu, C. Connelly, K. Davis, F. Dietrich, S. Whelen Dow, M. El Bakkoury, F. Foury, S. H. Friend, E. Gentalen, G. Giaever, J. H. Hegemann, T. Jones, M. Laub, H. Liao, N. Liebundguth, D. J. Lockhart, A. Lucau-Danila, M. Lussier, N. M'Rabet, P. Menard, M. Mittmann, C. Pai, C. Rebischung, J. L. Revuelta,

METHODS IN ENZYMOLOGY, VOL. 328
0076-6879/00 $30.00

and projects under development such as protein localization maps, spatial and temporal expression patterns, and protein structure analyses (Fig. 1A).

Novel strategies need to be developed downstream of these standardized functional assays to validate the large amount of information generated. Thus, bioinformatic tools are being made increasingly available to organize the data (Fig. 1A). For example, clustering programs have been used to examine the outcome of microarray analysis.[7] In addition, the data are increasingly released on the internet rather than in the pages of a scientific journal. This allows Boolean searches to be performed and the information to be retrieved conveniently. Finally, the information produced by functional genomics projects will have to be integrated to formulate the most appropriate hypotheses.[8] Thus, hyperlinks are being created between different databases to generate a backbone of functional relationships between otherwise unannotated predicted ORFs.[6]

One of the challenges in what could be referred to as the "postfunctional genomics" era will be to generate large numbers of biological reagents to test the hypotheses generated by the combination of standardized functional assays and bioinformatics (Fig. 1A).

Interaction-Defective Alleles

Because many proteins require physical interactions with other proteins to function appropriately, a number of laboratories have initiated protein interaction mapping projects as a way to annotate predicted ORFs. The yeast two-hybrid system[9] is most commonly used as a standardized functional assay for protein interaction mapping because it has been demon-

L. Riles, C. J. Roberts, P. Ross-MacDonald, B. Scherens, M. Snyder, S. Sookhai-Mahadeo, R. K. Storms, S. Véronneau, M. Voet, G. Volckaert, T. R. Ward, R. Wysocki, G. S. Yen, K. Yu, K. Zimmermann, P. Philippsen, M. Johnston, and R. W. Davis, *Science* **285,** 901 (1999).

[2] M. Schena, D. Shalon, R. W. Davis, and P. O. Brown, *Science* **270,** 467 (1995).

[3] P. L. Bartel, J. A. Roecklein, D. SenGupta, and S. Fields, *Nature Genet.* **12,** 72 (1996).

[4] M. Fromont-Racine, J. C. Rain, and P. Legrain, *Nature Genet.* **16,** 277 (1997).

[5] A. Flores, J.-F. Briand, O. Gadal, J.-C. Andrau, L. Rubbi, V. Van Mullem, C. Boschiero, M. Goussot, C. Marck, C. Carles, P. Thuriaux, A. Sentenac, and M. Werner, *Proc. Natl. Acad. U.S.A.* **96,** 7815 (1999).

[6] A. J. M. Walhout, R. Sordella, X. Lu, J. L. Hartley, G. F. Temple, M. A. Brasch, N. Thierry-Mieg, and M. Vidal, *Science* **287,** 116 (2000).

[7] M. B. Eisen, P. T. Spellman, P. O. Brown, and D. Botstein, *Proc. Natl. Acad. Sci. U.S.A.* **95,** 14863 (1998).

[8] A. J. M. Walhout, H. Endoh, N. Thierry-Mieg, W. Wong, and M. Vidal, *Am. J. Hum. Genet.* **63,** 955 (1998).

[9] S. Fields and O. Song, *Nature (London)* **340,** 245 (1989).

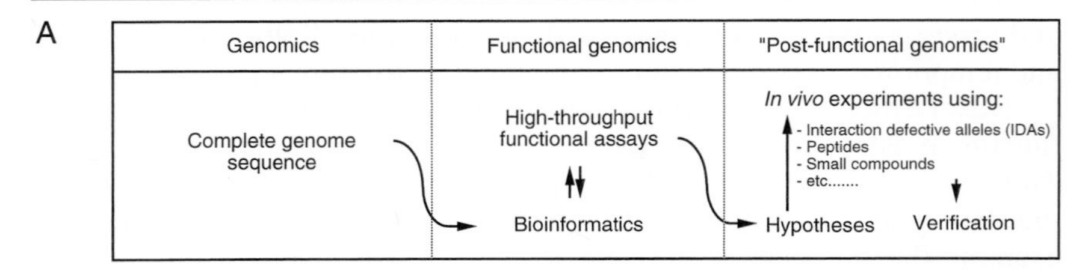

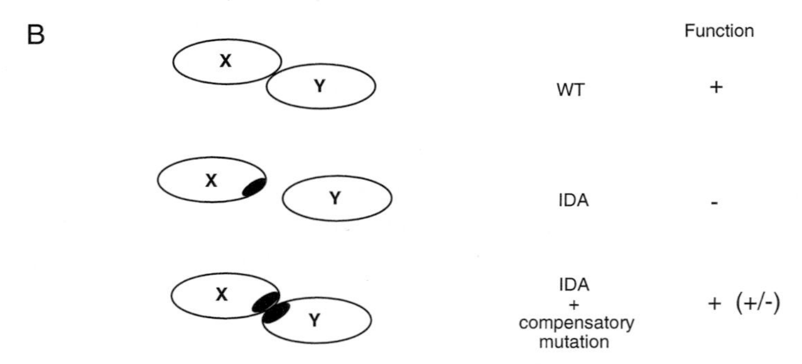

FIG. 1. The "postfunctional genomics" era. (A) The functional genomics field translates genomic information into functional annotations by using a combination of standardized functional assays and bioinformatic tools. It is expected that in the near future large numbers of hypotheses will derive from this work. Thus in a "postfunctional genomics" era, it will be important to design efficient ways to generate reagents such as interaction-defective alleles and others to validate these hypotheses. (B) Interaction-defective alleles (IDAs) of interactions that are relevant for a particular function should alter that function in either genetic assays *in vivo* or in biochemical assays *in vitro*. The use of compensatory mutations in the potential partner should then restore that function.

strated to be useful for the identification of protein–protein interactions and is amenable to automation (see, e.g., Ref. 10). As a result, an increasing number of potential protein–protein interactions [referred to here as interaction sequence tags (ISTs)[6]] is already available in several databases. Thus it is important to design strategies to efficiently generate reagents to test the validity of ISTs back *in vivo* (Fig. 1A).

Several strategies currently used to study protein–protein interactions *in vivo* are amenable to high-throughput settings. For example, reagents that dissociate or prevent interactions can be generated and tested in the relevant biological system. In this strategy, it may be expected that if an IST is critical for a hypothetical function, the dissociation of the corresponding

[10] M. Vidal and P. Legrain, *Nucleic Acids Res.* **27,** 919 (1999).

interaction would impair this function *in vivo*. Protein–protein interactions can be dissociated by *trans*-acting molecules such as peptides[11] and small compounds[12] or by *cis*-acting mutations in one partner that prevent the interaction with the other partner (Fig. 1B).[13] Such alleles, referred to here as interaction-defective alleles (IDAs), can be compared with their wild-type counterparts for their ability to function in genetic or biochemical assays (Fig. 1B).[10] For example, IDAs can be tested *in vivo* for the complementation of a loss-of-function phenotype in model organisms such as yeast, worms, and fruit flies. Alternatively, the corresponding proteins can be expressed and purified and subsequently tested in biochemical assays.[14] In both the genetic and the biochemical approach, IDAs can be tested along with compensatory alleles of the potential partner that restore the interaction (Fig. 1B). This control experiment can be used to demonstrate that the ability to interact correlates with the ability to function properly. For example, one report described the use of IDAs and corresponding compensatory mutations to demonstrate that the interaction between LIN-7 and LET-23 is essential for proper LET-23 receptor localization.[15]

Reverse Two-Hybrid System and C-Terminal Tags

Because the three-dimensional structure of the interacting domains of most ISTs is unknown, it is often not possible to design IDAs rationally. Thus, genetic selections are needed to identify IDAs from large libraries of randomly generated alleles. The reverse two-hybrid system provides such a genetic selection *against* protein–protein interactions.[16] In this system, two proteins X and Y are expressed in yeast cells as fusions to a DNA-binding (DB) domain and a transcriptional activation (AD) domain, respectively (DB–X and AD–Y). The interaction between X and Y (DB–X/AD–Y) results in activation of the expression of a toxic gene such as *URA3*. IDAs can be genetically selected because they reduce the expression of *URA3* and thus provide a growth advantage on media containing 5-fluoroorotic acid (5-FOA).[16]

A convenient way to generate a library of alleles suitable for the reverse two-hybrid system is by polymerase chain reaction (PCR) amplification of

[11] M. Vidal, R. Brachmann, A. Fattaey, E. Harlow, and J. D. Boeke, *Proc. Natl. Acad. Sci. U.S.A.* **93,** 10315 (1996).

[12] M. Vidal and H. Endoh, *Trends Biotechnol.* **17,** 374 (1999).

[13] M. Vidal, P. Braun, E. Chen, J. D. Boeke, and E. Harlow, *Proc. Natl. Acad. Sci. U.S.A.* **93,** 10321 (1996).

[14] T. Yasugi, M. Vidal, H. Sakai, P. M. Howley, and D. Benson, *J. Virol.* **71,** 5942 (1997).

[15] J. S. Simske, S. M. Kaech, S. A. Harp, and S. K. Kim, *Cell* **85,** 195 (1996).

[16] M. Vidal, *in* "The Yeast Two-Hybrid System" (P. Bartels and S. Fields, eds.), pp. 109–147. Oxford University Press, New York, 1997.

the X- or Y-encoding sequence under conditions that favor misincorporations. This is followed by gap repair transformation of the products into recipient yeast cells expressing an AD–Y or DB–X fusion protein, respectively.[16] Transformants are subsequently replica plated onto plates containing 5-FOA. 5-FOA-resistant colonies (5-FOA^R) are expected to express X or Y alleles that have lost the ability to interact with the partner protein of interest. Despite the apparent convenience of such a genetic selection, the reverse two-hybrid system is complicated by a high rate of nonsense or frameshift mutations that lead to truncated proteins. Such mutants are not informative because they are often unstable and are likely to be deficient in the interaction with multiple partner proteins (Fig. 2A). Therefore, one way to avoid the recovery of truncated proteins is to select for alleles of X that affect a particular interaction, e.g., with Y, but not others (Z, W, etc.).[17] This strategy is complicated by the fact that different combinations of interactors are needed in each case and is thus not amenable to high-throughput settings. An alternative approach consists in expressing the protein to be mutagenized, DB–X or AD–Y, in fusion with a scorable C-terminal tag. In this configuration, missense mutations allow expression of the C-terminal tag whereas truncated alleles do not (Fig. 2A). This provides an opportunity to distinguish between the two classes of alleles. Such a strategy should be amenable to high-throughput settings because it is based on a standardized approach, i.e., the same C-terminal tag can be used for large numbers of ISTs to be tested.

Ideally, C-terminal tags should not interfere with protein–protein interactions and their scoring should be convenient, automatable, and nondestructive so that positive colonies can be retrieved and analyzed. For example, the use of antibody-based epitope tags is limited by the fact that current Western-blotting methods are both time-consuming and destructive for the cells. C-terminal tags based on fusions to enzymes such as β-galactosidase (β-Gal) are easily testable but usually require killing of the yeast cells. Here, we describe fusions to the green fluorescent protein (GFP) and its derivatives. Detection of GFP is automatable and can be performed with living cells.

Green Fluorescent Protein-Based Reverse Two-Hybrid System

For most DB–X/AD–Y ISTs identified, it should be possible to recover interaction–defective alleles of X (X*), or Y (Y*), that cannot interact with Y or X, respectively. We designed a GFP-based reverse two-hybrid system that allows the use of DB–X–GFP (Fig. 2B) or AD–Y–GFP[18] fusion

[17] C. Inouye, N. Dhillon, T. Durfee, P. C. Zambryski, and J. Thorner, *Genetics* **147,** 479 (1997).

[18] H. Endoh, Y. Jacob, A. J. M. Walhout, and M. Vidal, submitted (2000).

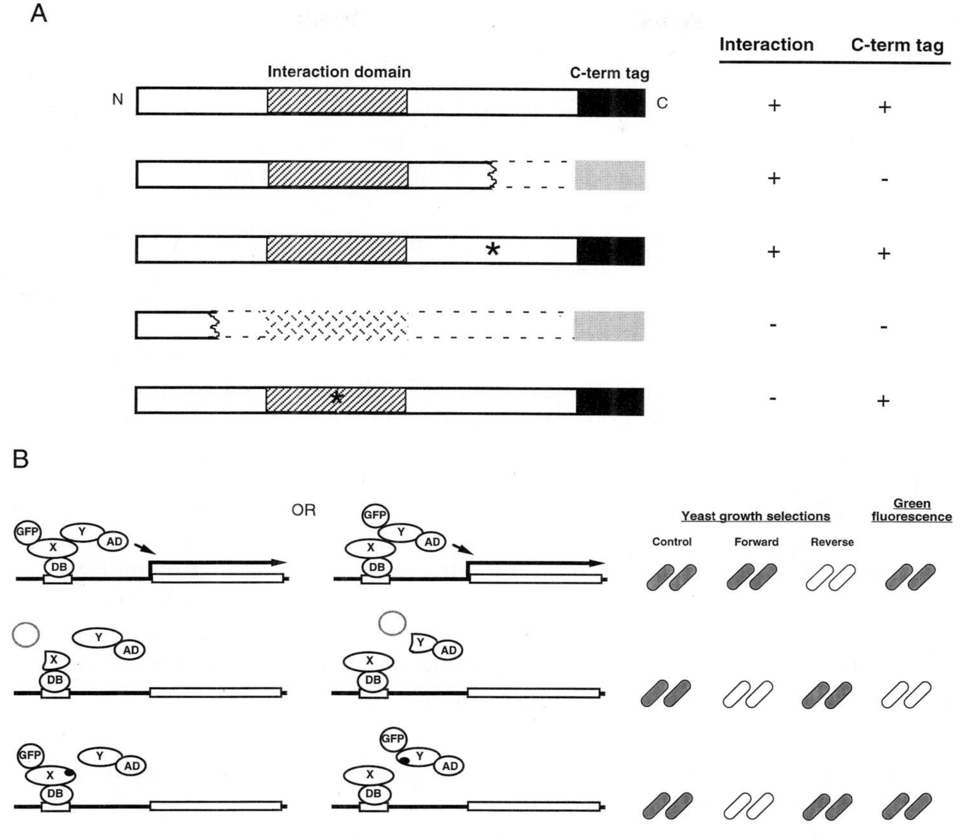

FIG. 2. C-terminal tags and reverse two-hybrid system. (A) C-terminal tags. By combining the conventional reverse two-hybrid system and the use of C-terminal tags (C-term tags), it is possible to select for IDAs corresponding to single amino acid changes. In contrast to alleles leading to truncations, single amino acid substitutions are more likely to become useful tools for biological experiments. (B) GFP-based reverse two-hybrid system. Derivatives of the green fluorescent protein (GFP) can be fused to the C-terminal end of DB–X (*left*) or AD–Y (*right*) fusion proteins. The expected readouts shown on the right should allow the distinction between subtle amino acid changes and truncations.

proteins, together with AD–Y or DB–X, respectively. To test the feasibility of the GFP-based reverse two-hybrid system and the range of its applicability, we asked two basic questions for each configuration, DB–X–GFP (Fig. 2B) and AD–Y–GFP.[18] First, is there any effect of expressing DB–X or AD–Y in fusion with GFP on their ability to interact with Y or X? Conversely, is the GFP activity altered by a fusion to DB–X or AD–Y? Among more than 20 protein–protein interactions tested so far, only 1, DB–pRB/

AD–E2F1,[18] is incompatible with the GFP-based reverse two-hybrid system, suggesting that the system can be used for the majority of potential interactions (19/20 = 95%).

In summary, the GFP-based reverse two-hybrid system is designed as follows. Multicopy plasmids (2 μm) are used because they confer relatively high expression levels of the protein to be characterized. This allows convenient detection of the GFP activity of 5-FOA^R yeast colonies (GFP$^+$). For the visualization of the yeast colonies expressing GFP, a fluorescence dissecting microscope is used. The yeast strains used, MaV103 and MaV203,[16] contain three different reporter genes: *GAL1*::*lacZ* for quantitation of the interaction readout on 5-bromo-4-chloro-3-indolyl-β-D-galactopyranoside (X-Gal) medium, *GAL1*::*HIS3* for forward two-hybrid selections on media containing 3-aminotriazole (3-AT), and *SPAL10*::*URA3* for reverse two-hybrid selections on plates containing 5-FOA.[16] We demonstrated that 5-FOA^R colonies can be selected on 5-FOA plates for both DB–X–GFP/AD–Y or DB–X/AD–Y–GFP interactions. As exemplified by Figs. 3A and 4 (see color inserts), it is possible to manually pick colonies that correspond to the expected readouts for missense interaction-defective alleles: 5-FOA^R and fluorescent GFP$^+$. It should be noted that neither X-Gal nor 3-AT plates (Fig. 4, see color insert) could be used to pick these alleles from a large background of undesired colonies. We envision that an automated colony-picker device could be designed to recover fluorescent GFP$^+$ from nonfluorescent 5-FOA^R colonies.

Nature of Interaction-Defective Alleles Obtained from Green Fluorescent Protein-Based Reverse Selections

The nature of the IDAs that can be recovered from GFP-based reverse two-hybrid selections has been investigated in detail, using the interaction between FKBP12 and the cytoplasmic domain of the type I transforming growth factor β (TGF-β) receptor (RIC).[18] The three-dimensional structure of this interaction has been resolved.[19] The majority of FKBP12 IDAs selected using the GFP-based reverse two-hybrid system contained substitutions in amino acids that are clustered within the FKBP12 interaction domain.[18] This suggests that the GFP-based reverse two-hybrid system can specifically target amino acids involved in a protein–protein interaction rather than generally affecting folding and/or stability.

This notion was further demonstrated in the context of the protein interaction mapping project for the *Caenorhabditis elegans* proteome initi-

[19] M. Huse, Y. G. Chen, J. Massague, and J. Kuriyan, *Cell* **96,** 425 (1999).

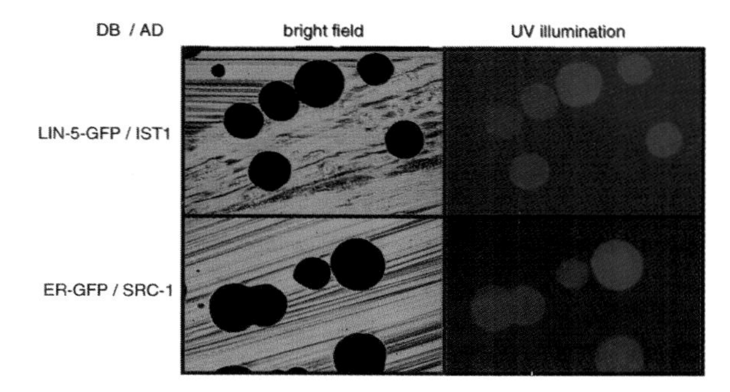

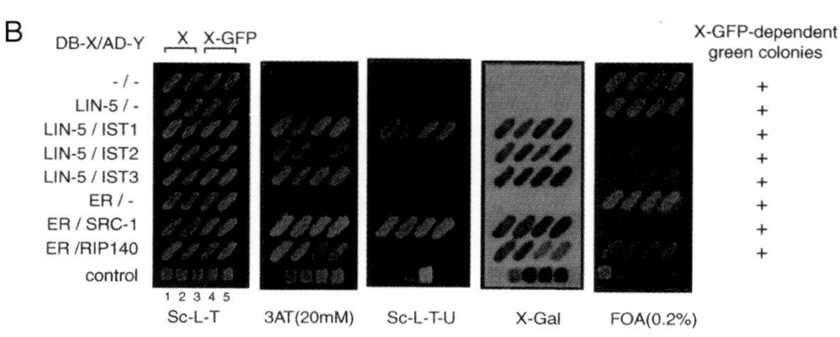

FIG. 3. The GFP-reverse two-hybrid system. (A) Fluorescence read-out of DB–GFP and DB–X–GFP fusions. The left side shows light microscopy 5-FOA^R colonies while the right side shows the same colonies under UV illumination. (B) Two-hybrid read-outs of DB–X/AD–Y and DB–X–GFP/AD–Y interactions. Leu$^+$ Trp$^+$ transformants expressing the indicated fusion proteins were patched on Sc–L–T (*left*) and then replica-plated on the plates photographed here. For each panel, the two right patches express GFP while the left patches do not. The five controls at the bottom were described elsewhere[16] and provide an indication of the quality of the plates and incubation conditions. The plates contained 10^7 *M* of 17 β-estradiol to facilitate binding between DB-ER (estrogen receptor) and AD–SRC-1 or AD–RIP140.

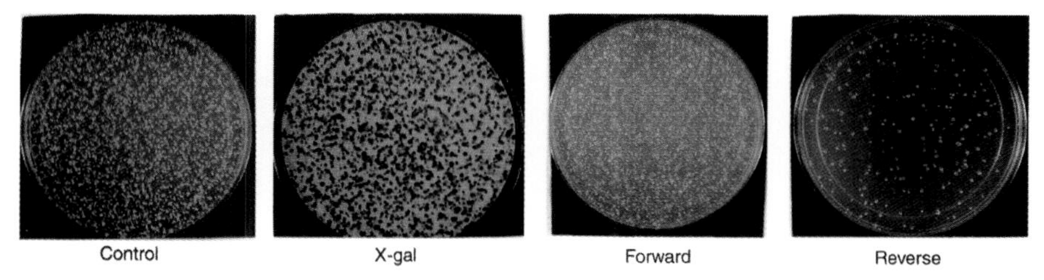

FIG. 4. GFP-reverse versus forward two-hybrid selections. This figure compares the colony read-outs of the forward two-hybrid system and the GFP-reverse two-hybrid selections. The "X-Gal" plate detects the expression of *GAL1::lacZ*, the "Forward" plate (3-AT 10 m*M*) provides a growth selection for cells that express *GAL1::HIS3*, and the "Reverse" plate (5-FOA 0.2%) provides a growth selection for cells that fail to express *SPAL10::URA3*. Note that it is not convenient to detect white and nongrowing colonies on the "X-Gal" and "Forward" plate, respectively. In contrast, colonies growing on the "Reverse" plate are readily detectable.

ated in our laboratory.[8] Using large-scale two-hybrid analysis, we have identified ~150 potential interactions mediated by proteins involved in vulval development[6] and in other cell cycle-related functions.[19a] So far, a number of conclusions can be drawn from the analysis of half a dozen *C. elegans* ISTs (Table I). First, a substantial number of 5-FOA^R GFP$^+$ alleles recovered after the initial selection/screening procedure did indeed retest when reintroduced in fresh recipient cells (see Protocols). Second, the IDAs usually contained single amino acid substitutions. Third, the number of independent alleles recovered suggested that the mutagenesis protocol did not introduce any major bias for nucleotide substitutions. Finally, the *lin-53* IDAs exemplify the specificity of the interaction mediated by different alleles selected against different interactors. This suggests that these alleles could be used to dissect the different functions of LIN-53.

Protocols

Green Fluorescent Protein-Based Reverse Plasmids

Plasmids used to express DB–X–GFP and AD–Y–GFP fusion proteins were designed as originally described.[18] Multicopy plasmids used in the conventional two-hybrid system were adapted to the GFP-based reverse two-hybrid system by cloning a sequence encoding one of the GFP derivatives into their multiple cloning site (MCS).

As described below, the gap repair method[20] is used to introduce PCR-generated libraries into these plasmids. This method can be readily automated. However, one of the disadvantages is that a relatively high frequency of transformants obtained is derived from the recircularization of the plasmid. In the context of the reverse two-hybrid system, these molecules give rise to a 5-FOA^R phenotype and thus complicate the recovery of genuine IDAs. To provide a way to eliminate these false positives, the GFP-encoding sequence was cloned in a different frame than that of DB or AD. Thus, on recircularization of the plasmid, GFP is not expressed and the corresponding transformants can be eliminated (5-FOA^R GFP$^-$).

The AD–GFP plasmid (pACT2-viGFP) was designed as follows. A sequence encoding the GFP mutant derivative Phe64Leu, Ser65Thr, Val68Ile ("viGFP"), which gives rise to stronger fluorescence than wild-type GFP, was inserted into the MCS of the two-hybrid vector pACT2 (Clontech, Palo Alto, CA). The coding frame of GFP was designed to correspond to the +1 frame of AD.[18]

[19a] M. Lorson, A. J. M. Walhout, M. Vidal, and S. van den Heuvel, in preparation (2000).
[20] T. L. Orr-Weaver, J. W. Szostack, and R. J. Rothstein, *Methods Enzymol.* **101,** 228 (1983).

TABLE I
RELATIVELY SPECIFIC INTERACTION-DEFECTIVE ALLELES READILY FOUND USING GFP-BASED REVERSE TWO-HYBRID SYSTEM[a]

DB–X	AD–Y	Number of colonies (without/with PCR products)	5-FOA^R/ green colonies	Number of retested alleles	Number of alleles with single amino acid substitution in target (underlined X or Y)	Number of independent alleles	LIN-53 IDA	IST: LIN-37	IST: MO3C11.4	IST: T27C4.4	IST: C47E8.5
TGF-β RIC	FKBP12	420/4700	2060/15	9	9	9					
IST1	LIN-5	1300/4500	2100/13	1	1	1					
IST2	LIN-5	624/6720	3840/7	0	0	0					
IST3	LIN-5	480/6800	2640/26	2	2	2					
LIN-53	LIN-37	300/3300	320/12	6	6	3					
LIN-53	LIN-37	180/1600	340/26	3	3	2	IDA1	−	−	−	−
							IDA2	−	+	−	+
LIN-53	MO3C11.4	200/2400	560/35	2	2	2	IDA3	+	−	−	−
							IDA4	+	−	−	+
LIN-53	T27C4.4	150/1440	264/1	1	1	1	IDA5	+	+	−	−
							IDA6	+	+	+	−
LIN-53	C47E8.5	180/2200	600/32	5	5	2	IDA7	+	+	+	−

[a] For each IST characterized so far (DB–X, AD–Y), the ratio of colony numbers obtained from transformation of linearized vectors without and with PCR product, the number of 5-FOA^R and 5-FOA^R GFP$^+$ colonies, the number of genuine interaction-defective alleles obtained after retesting in fresh recipient cells, the number of single amino acid substitutions, and the number of independent alleles obtained are shown. For the LIN-53, ISTs, the specificity of interaction is compared between different IDAs.

The DB–GFP plasmid (pAS2-EGFP) was designed as follows. A sequence encoding another GFP derivative (EGFP) was subcloned from the plasmid pEGFP (Clontech) into the *Bgl*II–*Not*I sites of the two-hybrid vector pPC97,[16] such that the coding frame of GFP corresponds to the +1 frame of DB. The resulting plasmid was used as template DNA in a PCR with DB and Term primers[16] (see below) and the amplified fragment was introduced into the two-hybrid vector pAS2-1 (Clontech) by gap repair.

Primers for Yeast Gap Repair

Yeast transformation by gap repair is a convenient method for the introduction of large numbers of PCR-generated alleles in frame with the GFP-encoding sequence. The fragment to be cloned requires a minimum of 40 nucleotides identical to the extremities of the linearized vector. Thus we designed primers for the amplification of DNA fragments to be mutagenized with 40-nucleotide tags that are identical to either Gal4pDB, Gal4pAD, viGFP, or EGFP-encoding sequences. The primer sequences are as follows:

Forward primers used to generate DB–X fusions:

5′-AGAGAGTAGTAACAAAGGTCAAAGACAGTTGACTG-TATCGNNNNNNNNN-3′

The underlined sequence corresponds to Gal4pDB and the Ns correspond to 20–25 nucleotides (nt) of X N terminus-encoding sequence [melting temperature (T_m), ~60°]

Forward primers used to generate AD–Y fusions:

5′-ATTCGATGATGAAGATACCCCACCAAACCCAAAAA-AAGAGNNNNNNNNN-3′

The underlined sequence corresponds to Gal4pAD and the Ns correspond to 20–25 nt of Y N terminus-encoding sequence (T_m ~60°)

Reverse primers to generate X–EGFP fusions:

5′-GCACCACCCCGGTGAACAGCTCCTCGCCCTTGCTCAC-CATNNNNNNNNN-3′

The underlined sequence corresponds to EGFP and the Ns correspond to 20–25 nt of X C terminus-encoding sequence (T_m ~60 ± 4°). The stop codon is omitted from the sequence

Reverse primers to generate Y–viGFP fusions:

5′-TTGGGACAACTCCAGTGAAAAGTTCTTCTCCTTTACC-CATNNNNNNNNN-3′

The underlined sequence corresponds to viGFP and the Ns correspond to 20–25 nt of Y C terminus-encoding sequence (T_m ~60 ± 4°). The stop codon is omitted from the sequence

These primers are specific for each X or Y sequence to be mutagenized. In the context of the "postfunctional genomics" era, we expect to apply the GFP-based reverse two-hybrid selections to large numbers of ORFs. It would be advantageous to use a single pair of primers rather than large numbers of specific pairs for such large-scale projects. Our laboratory is constructing a near complete array of *C. elegans* ORFs in which each ORF is flanked by sequences referred to as B1 and B2.[21] Thus we designed universal primers that can be used in PCR with these constructs as template DNA.

Forward primer used to generate DB–B1–X fusions:

5′-AGAGAGTAGTAACAAAGGTCAAAGACAGTTGACTG-TATCG*ACAAGTTTGTACAAAAAAGCAGGCT*-3′

The underlined sequence corresponds to Gal4pDB and the italicized sequence corresponds to B1. Note that the B1 and X sequences are in frame in the template DNA

Forward primer used to generate AD–B1–Y fusions:

5′-ATTCGATGATGAAGATACCCCACCAAACCCAAAAA-AAGAG*ACAAGTTTGTACAAAAAAGCAGGCT*-3′

The underlined sequence corresponds to Gal4pAD and the italicized sequence corresponds to B1. Note that the B1 and Y sequences are in frame in the template DNA

Reverse primer used to generate X–B2–EGFP fusions:

5′-GCACCACCCCGGTGAACAGCTCCTCGCCCTTGCTCAC-CAT*CCACTTTGTACAAGAAAGCTGGG*-3′

The underlined sequence corresponds to EGFP and the italicized sequence corresponds to B2. Note that the B2 and X sequences are in frame in the template DNA

Reverse primer used to generate Y–B2–viGFP fusions:

5′-TTGGGACAACTCCAGTGAAAAGTTCTTCTCCTTTACC-CAT*CCACTTTGTACAAGAAAGCTGGG*-3′

[21] A. J. M. Walhout, G. F. Temple, M. A. Brasch, J. L. Hartley, M. A. Lorson, S. van den Heuvel, and M. Vidal, *Methods Enzymol.* 328, Chap. 35, 2000 (this volume).

The underlined sequence corresponds to viGFP and the italicized sequence corresponds to B2. Note that the B2 and Y sequences are in frame in the template DNA

Polymerase Chain Reaction Conditions

Previously, we have described various parameters that must be considered for mutagenic PCRs for the generation of libraries of mutant alleles.[16] Briefly, concentrations of manganese and individual nucleotides should be titrated to obtain a mutation rate of about 10^{-3}. The optimal condition will depend on the length and sequence of the fragment to be mutagenized. For the GFP-based reverse two-hybrid system, we typically use the following protocols.

1. For sequences to be mutagenized that are equal to or shorter than 1 kb:
   a. For each PCR prepare a 100-μl mixture containing 100 ng of DNA template, 0.3 μM primers, 50 μM each dNTP, 50 mM KCl, 10 mM Tris-HCl (pH 9.0), 0.1% (v/v) Triton X-100, 1.5 mM $MgCl_2$, bovine serum albumin (BSA, 1 mg/ml), and 2.5 units of *Taq* DNA polymerase.
   b. First perform a set of 10 elongation cycles: step 1, 94° for 1 min; step 2, 45° for 1 min; step 3, 72° for 2 min.
   c. Add 100 μM $MnCl_2$ to the reaction.
   d. Perform another set of 30 amplification cycles as follows: step 1, 94° for 1 min; step 2, 45° for 1 min; step 3, 72° for 2 min.
2. Because the misincorporation rate of *Taq* polymerase is sufficient to generate approximately one mutation in sequences longer than 1 kb, conventional PCR conditions should be used to prevent the occurrence of multiple mutations per molecule:
   a. For each PCR prepare a 100-μl mixture containing 100 ng of DNA template, 0.3 μM primers, 50 μM each dNTP, 50 mM KCl, 10 mM Tris-HCl (pH 8.3), 0.1% (v/v) Triton X-100, 2.5 mM $MgCl_2$, BSA (1 mg/ml), and 2.5 units of *Taq* DNA polymerase.
   b. Perform a set of 35 elongation cycles: step 1, 94° for 1 min; step 2, 45° for 1 min; step 3, 72° for 2 min.

Yeast Transformations and gap Repair Cloning

The different steps of the protocol for yeast gap repair transformation are as follows[16]: preparation of recipient cells that contain a plasmid expressing the unmutagenized partner, preparation of linearized vector DNA, mutagenic PCRs (see above), yeast transformation, and selection of transformants.

1. The pAS2-EGFP plasmid harbors the *TRP1* selectable marker. Thus, the recipient cells of mutagenized DB–X–GFP PCR products should contain unmutagenized AD–Y plasmids that harbor *LEU2*.[16] For example, pGAD10, pGAD424, or pACT2 (Clontech) can be used for this purpose.

2. Linearization of the pAS2-EGFP plasmid can be performed with *Sal*I because the corresponding site is unique on the plasmid and located between the coding region of Gal4pDB and EGFP.

3. The pACT2-viGFP plasmid harbors the *LEU2* selectable marker. Thus, the recipient cells of mutagenized AD–Y–GFP PCR products should contain unmutagenized DB–X plasmids that harbor *TRP1*. For example, pGBT9 (Clontech) or pMAB20 (a centromeric DB vector based on pPC97) can be used for this purpose.

4. Linearization of the pACT2-viGFP plasmid can be performed with *Bam*HI because the corresponding site is unique on the plasmid and located between the coding region of Gal4pAD and viGFP.

5. Approximately 100 ng of linearized pAS2-EGFP or pACT2-viGFP plasmid should be mixed with 100 ng of PCR product and transformed into MaV103 cells carrying the relevant AD–Y or DB–X plasmid, respectively.

6. Yeast transformation protocols that are optimal for this yeast strain have been described.[16] Using this lithium acetate-based protocol, we usually obtain 2–5 $\times$ 10^3 transformants.

7. Controls for gap repair should include transformation with a reference circular plasmid, no DNA, and circularized vector alone. The number of transformants using linearized vector together with the PCR fragments is usually 3–10 times higher compared with linearized vector only.

8. After transformation, the cells are plated on a synthetic complete (Sc) plate lacking leucine and tryptophan (Sc –L –T). Transformants are incubated at 30° for 2 days. The Sc –L –T plates are then replica plated onto Sc –L –T plus 5-FOA. As previously reported the optimal concentration of 5-FOA should be determined.[16] Most experiments can be performed at 5-FOA concentrations between 0.15 and 0.2% (w/v). The 5-FOA plates are then immediately replica cleaned[16] and incubated for 24 hr at 30°. 5-FOA^R yeast colonies can then be scored for GFP activity by fluorescence microscopy (Fig. 3A, see color insert).

Fluorescence Microscopy

Light and fluorescence microscopy can be used to detect 5-FOA^R colonies directly on plates containing 2% (w/v) agar. By switching back and forth from light to fluorescence, GFP$^+$ colonies can conventiently be discriminated from GFP$^-$ colonies. After detection, GFP$^+$ colonies should be picked with sterile toothpicks. It is conceivable that automated devices could be designed for large-scale detection and retrieval of 5-FOA^R GFP$^+$ colonies. We currently use a Leica MZ8 microscope (Kramer Scientific, Burlington, MA). We usually detect a few 5-FOA^R GFP$^+$ per hundred 5-FOA^R colonies.

Recovery of Alleles and Verification

Subsequent to the selection and screening of 5-FOA^R GFP$^+$ colonies, it is crucial to reintroduce the DNA fragments encoding potential IDAs into fresh recipient cells (see above) as a verification step. Genuine interaction-defective alleles should reproducibly confer a 5-FOA^R GFP$^+$ phenotype. The following primers can be used to PCR amplify the sequences directly from yeast colonies.

Forward primers:
- DB primer: 5′-GGCTTCAGTGGAGACTGATATGCCTC-3′
- AD primer: 5′-CGCGTTTGGAATCACTACAGGG-3′

Reverse primers:
- Term primer: 5′-GGAGACTTGACCAAACCTCTGGCG-3′
- Reverse viGFP: 5′-GAATTGGGACAACTCCAGTG-3′
- Reverse eGFP: 5′-TGGGCACCACCCCGGTGAAC-3′

The conditions for yeast colony PCRs are identical to what is described above for longer DNA fragments, with the exception that the template DNA is prepared as follows. A volume of approximately 3 μl of yeast cells is resuspended in 10 μl of 0.02 *N* NaOH and incubated for 30 sec in a microwave at maximal power. The transformation conditions of the resulting PCR products are identical to what is described above.

Downstream Steps

The experimental steps that should be performed downstream of the reverse two-hybrid selections have already been described. These include sequencing of the alleles and Western-blotting analysis of the fusion proteins from yeast extracts.[16] Other steps, such as cloning of the alleles into appropriate expression vectors, and biological experiments that can be performed with interaction-defective alleles are described elsewhere.[18]

Conclusions

We have shown that the addition of derivatives of GFP in fusion with DB–X or AD–Y fusion proteins does not significantly modify the reverse two-hybrid readout. Thus the GFP-based reverse two-hybrid system described here should be versatile and applicable to many protein–protein interactions. Using the GFP-based reverse system, we have identified IDAs for eight potential protein–protein interactions.

Our preliminary experiments suggest that most IDAs selected contain single amino acid changes resulting from missense mutations. For FKBP12, it has been shown that the altered amino acids are embedded in the interacting domain.[18] From the other interactions characterized here, it is becoming increasingly clear that most IDAs affect specific interactions while leaving other interactions intact. This strongly suggests that those alleles will become valuable tools to functionally dissect proteins mediating multiple contacts, such as those involved in cross-talk between different pathways of regulatory networks.

Finally, it is important to reiterate that the GFP-based reverse two-hybrid system can be automated. The mutagenic PCRs can be performed on 96-well plates, using automated pipetting devices. The picking of 5-FOA^R GFP$^+$ colonies should be easily automatable using colony-picker devices that can detect fluorescence. Subsequent yeast colony PCRs, retransformation, and sequencing steps can also be performed on 96-well plates. We are currently working on automatable methods to clone interaction-defective alleles into various expression vectors. Altogether the methods described here suggest that in the near future, large numbers of ISTs derived from protein–protein interaction mapping projects could be characterized using IDAs.

Acknowledgments

We thank M. Ewen, Y. Jacob, J. Lamb, S. van den Heuvel, and T. Wang for reagents and/or unpublished information; Yamanouchi Pharmaceutical Co., Ltd., for support; and S. van den Heuvel and A. Hart for generous help with the microscopy. This work is supported by grants from the National Human Genome Research Institute (1 RO1 HG01715-01), the National Cancer Institute (1 R21 CA81658A01), and the Merck Genome Research Institute awarded to M.V.

[7] Yeast Three-Hybrid System for Detecting Ligand–Receptor Interactions

By ERIC C. GRIFFITH, EDWARD J. LICITRA, and JUN O. LIU

Interactions between small ligands and proteins are involved in many fundamental biological processes and form the molecular basis for pharmacological intervention in human disease. Small molecule ligands often manifest their effects by noncovalent interactions with high-affinity binding proteins within a specific cell type, blocking the functions of genes at the protein level.[1] As such, small ligands have also served as valuable probes for the elucidation of regulatory mechanisms of cellular processes ranging from transcription to signal transduction.[2–5] The majority of drugs used in the clinic today are small molecule agonists or antagonists of protein targets. Therefore, the identification of specific small ligands for various enzymes and receptors is a prerequisite for drug discovery. Advances in both chemistry and biology have led to an explosive growth in the number of small ligands and known genes. The use of combinatorial chemistry is generating millions of different chemical structures within a relatively short time.[6] In addition, the rapid progress of the human and related genome projects will eventually provide the cDNA sequences of all genes for various organisms. These developments heighten the need to develop simple and rapid methods for detection of ligand–receptor interactions.

The classic approach for detecting ligand–protein receptor interactions in the context of receptor discovery relies primarily on *in vitro* biochemical methods, including photo-cross-linking, radiolabeled ligand binding, and affinity chromatography.[7] These methods are often laborious and time-consuming, requiring purification of sufficient quantities of target receptors from complex mixtures of cellular proteins, peptide sequencing, and subsequent cloning of the cDNAs encoding the receptors (Fig. 1). Similarly,

[1] J. G. Hardman, L. E. Limbird, P. B. Molinoff, R. W. Ruddon, and A. G. Gilman, "The Pharmacological Basis of Therapeutics." McGraw-Hill, New York, 1996.

[2] K. R. Yamamoto, *Annu. Rev. Genet.* **19,** 209 (1985).

[3] R. M. Evans, *Science* **240,** 889 (1988).

[4] S. L. Schreiber, *Science* **251,** 283 (1991).

[5] R. D. Klausner, J. G. Donaldson, and J. Lippincott-Schwartz, *J. Cell Biol.* **116,** 1071 (1992).

[6] M. A. Gallop, R. W. Barrett, W. J. Dower, S. P. A. Fodor, and E. M. Gordon, *J. Med. Chem.* **37,** 1233 (1994).

[7] W. B. Jakoby and M. Wilcheck, *Methods Enzymol.* **46,** 1 (1974).

METHODS IN ENZYMOLOGY, VOL. 328
0076-6879/00 $30.00

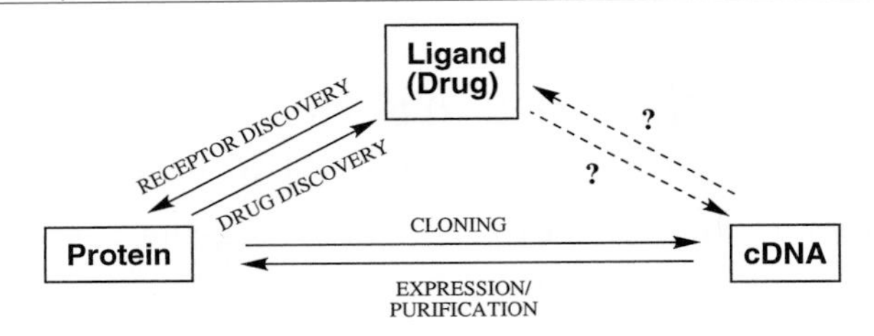

FIG. 1. Classic biochemical approaches for studying small ligand–protein interactions.

screening assays for drugs using known protein targets are dependent on the expression and purification of sufficient quantities of recombinant target proteins and establishment of appropriate assays with suitable readout. Not all the lead compounds identified in such *in vitro* screening assays are useful due to lack of either cell permeability or stability inside mammalian cells. To date, no general genetic method has been available for the study of small molecule–protein interactions *in vivo*. In an attempt to speed up the receptor discovery process, we developed the yeast three-hybrid system that now provides a general and rapid method for detecting ligand–protein interactions in yeast cells.[8]

The yeast three-hybrid system is an extension of the two-hybrid system.[9,10] In addition to the two hybrid fusion proteins needed in the two-hybrid system, the three-hybrid system requires a third hybrid ligand, which acts as a chemical inducer of dimerization.[11] The heterodimeric small molecule ligand (or "bait"), the third hybrid molecule, acts to bring together a DNA-binding domain fused to the receptor for one ligand with an activation domain fused to the receptor for the second ligand, reconstituting a functional transcriptional activator in yeast cells (Fig. 2). One-half of the hybrid ligand consists of dexamethasone, a high-affinity ligand for the glucocorticoid receptor. Binding of dexamethasone to the glucocorticoid receptor fused to a DNA-binding domain (or "hook") serves to anchor the ligand to the promoter of a reporter gene. Interaction of the tethered orphan ligand with a protein receptor fused to an activation domain (or "fish") initiates transcription of the reporter gene whose activation is coupled to growth or color selection similar to that used for the traditional two-hybrid system. The plasmid encoding the fish protein can then be retrieved and

[8] E. J. Licitra and J. O. Liu, *Proc. Natl. Acad. Sci. U.S.A.* **93,** 12817 (1996).

[9] S. Fields and O.-K. Song, *Nature* (*London*) **340,** 245 (1989).

[10] C. -T. Chien, P. L. Bartel, R. Sternglanz, and S. Fields, *Proc. Natl. Acad. Sci. U.S.A.* **88,** 9578 (1991).

[11] D. M. Spencer, T. J. Wandless, S. L. Schreiber, and G. R. Crabtree, *Science* **262,** 1019 (1993).

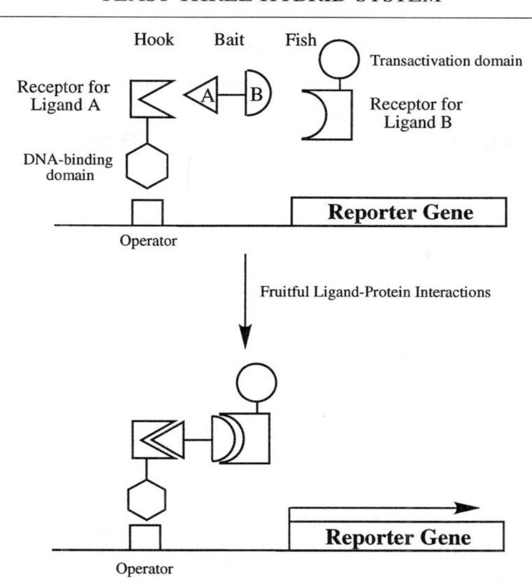

FIG. 2. Components of the yeast three-hybrid system. The triangle and the semicircle represent ligands A and B, respectively. [Taken from E. J. Licitra and J. O. Liu, *Proc. Natl. Acad. Sci. U.S.A.* **93,** 12817 (1996), with permission. Copyright (1996) National Academy of Sciences, U.S.A.]

sequenced, enabling the rapid detection of protein receptors for small molecules without recourse to traditional protein biochemistry.

The yeast three-hybrid system consists of two pairs of ligand–receptor interactions. As such, it can be conceptually broken into two parts: a conserved ligand–receptor pair common to all three-hybrid screens and a second ligand–receptor pair that represents the interaction of interest. The first conserved pair represents a known and well-established high-affinity interaction. For the second ligand–receptor pair, one component, either the ligand or the receptor, can be unknown. Application of the three-hybrid system to screen a library of either small molecules or receptors would allow for the identification of the missing component. Thus, the yeast three-hybrid system can be used to identify new ligands that bind to a known receptor or to identify new receptors for an "orphan" ligand, the latter of which is of more general interest to those who are trying to identify targets for natural products or synthetic ligands.

This chapter is intended to provide a practical guide to the yeast three-hybrid system, using as an example the interaction between the immunosuppressive drug FK506 and its binding protein FKBP12. Given the limitations of space, methods conserved between the yeast two-hybrid and the three-hybrid systems are either covered briefly or omitted completely where possible, with the assumption that the reader has had prior experience with the yeast two-hybrid method. Several excellent review articles are available on the yeast two-hybrid system[12] (and see Gerard *et al.*,[12a] Geyer and Brent,[12b] and Liu *et al.*[12c] in this volume.

Key Reagents

Many of the components required to perform a three-hybrid screen are widely available because of the widespread use of the two-hybrid system. Traditional yeast strains and cDNA libraries used for the identification of protein–protein interactions are all compatible with this new system. However, some additional reagents are required.

Vectors

A number of different DNA-binding domain and transcriptional activation domain vectors have been used for the two-hybrid system. In principle, any of these systems could be adapted for use in a three-hybrid screen; however, we confine this discussion to vectors developed by R. Brent (Molecular Sciences Institute, Berkeley, CA) based on fusion to the DNA-binding protein LexA.[13] Fish cDNA libraries constructed in the pJG4-5 vector as fusions to the bacterial protein B42 activation domain are completely adaptable for use in this system. The pEG202 vector, however, has been modified to express the glucocorticoid receptor–LexA fusion protein.[8] Several variants of the hook vector have been generated with the hormone-binding domain of the rat glucocorticoid receptor fused to LexA (Fig. 3). In addition, we have introduced two point mutations in an attempt to increase efficiency. The C656G mutation has been shown to increase the affinity of the hormone-binding domain for dexamethasone.[14] A second point mutation, F620S, is also known to increase the availability of ectopically expressed glucocorticoid receptor in yeast for interaction with the dexamethasone ligand.[15] Testing of the various hook constructs in yeast,

[12] P. L. Bartel and S. Fields, "The Yeast Two-Hybrid System." Oxford University Press, New York, 1997.

[12a] G. Cagney, P. Vetz, and S. Fields, *Methods Enzymol.* **328,** Chap. 1, 2000 (this volume).

[12b] C. R. Geyer and R. Brent, *Methods Enzymol.* **328,** Chap. 13, 2000 (this volume).

[12c] Q. Liu, M. Z. Li, D. Liu, and S. J. Elledge *Methods Enzymol.* **328,** Chap. 32, 2000 (this volume).

[13] J. Gyuris, E. A. Golemis, H. Chertkov, and R. Brent, *Cell* **75,** 791 (1993).

[14] P. K. Chakraborti, M. J. Garabedian, K. R. Yamamoto, and S. S. J. Simons, *J. Biol. Chem.* **266,** 22075 (1991).

[15] M. J. Garabedian and K. R. Yamamoto, *Mol. Biol. Cell* **3,** 1245 (1992).

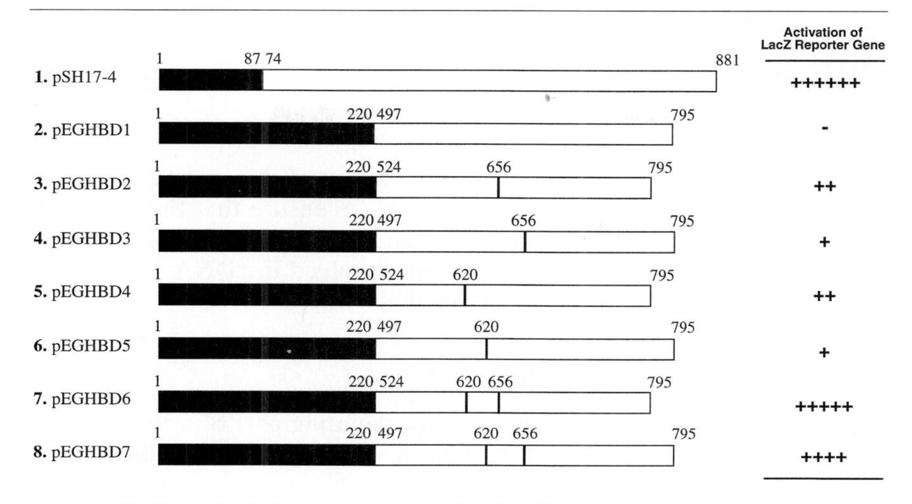

FIG. 3. Different hook fusion constructs and their ability to activate the *lacZ* reporter gene on reconstitution of the yeast three-hybrid assay, using dexamethasone–FK506 as the bait. The control fusion protein (pSH17-4) between the LexA DNA-binding domain (filled) and the Gal4 transactivation domain (shaded) is indicated along with variants of hook fusion proteins between the LexA DNA-binding domain and the hormone-binding domain of the rat glucocorticoid receptor (empty). The first and the last amino acid of the LexA DNA-binding protein, the Gal4 transactivation domain, and the hormone-binding domain of the rat glococorticoid receptor as well as the relative positions of the two mutations in the hormone-binding domain of rat glucocorticoid receptor are indicated above each construct. [Taken in part from E. J. Licitra and J. O. Liu, *Proc. Natl. Acad. Sci. U.S.A.* **93,** 12817 (1996), with permission. Copyright (1996) National Academy of Sciences, U.S.A.]

using a dexamethasone–FK506 bait and a pJG4-5-FKBP12 fish construct, indicated that hook constructs possessing both point mutations give much higher sensitivity for detection of ligand–receptor interactions (Fig. 3). Thus, the pEGHBD6 and pEGHBD7 vectors currently represent the best hook constructs for use in three-hybrid screens. The pEGHBD6 insert has also been cloned into the pAS2 vector based on the *GAL4* DNA-binding domain and also shows good sensitivity (E. C. Griffith and J. O. Liu, unpublished observations, 1999).

Strains

The strain should contain nonreverting mutations in genes present as selectable markers on the hook and fish vectors. In addition, the strain must contain a reporter gene under the control of an upstream activating sequence corresponding to the DNA-binding domain. Several strains exist that can be used to conduct three-hybrid screens, including EGY48 (*ura3 trpl his3 lexA* operator-*LEU2*), which is compatible with the pEG202 and pJG4-5 vectors.

cDNA Libraries

A number of cDNA libraries from a variety of tissues and organisms exist and are commercially available. Check to ensure that the selectable markers are compatible with the strain and vectors. It may be necessary to construct a library; however, protocols involved in cDNA library construction have been previously described elsewhere.[16]

Preparation of a Suitable Hybrid Bait

For many biologists perhaps the most daunting part of a three-hybrid screen will be preparation of a suitable small molecule hybrid ligand bait. Although the procedures involved are relatively simple, it requires a basic setup for organic synthesis and access to modern analytical instrumentation such as a nuclear magnetic resonance (NMR) and mass spectrophotometer to characterize the synthetic intermediates. For those who do not have a synthesis laboratory, it is best carried out with a group engaged in organic synthesis.

For a ligand to be usable in the three-hybrid system, it must possess a suitable functional group that may be modified without significant loss in activity. This determination requires sufficient structure–activity data for the ligand. The hybrid ligand should be prepared by linking two ligands via their "viable" or tolerable positions. For the current version of the three-hybrid system, we have employed as the conserved ligand–receptor pair dexamethasone and the glucocorticoid receptor. This ligand–receptor pair was chosen because of the large amount of structure–activity data available on the high-affinity interaction of dexamethasone with the glucocorticoid receptor. In fact, dexamethasone has been modified into an affinity reagent by synthesizing a tether to biotin, which retains high affinity for the receptor.[17] The glucocorticoid receptor is a "natural" fusion protein between a DNA-binding domain and a ligand-binding domain, making it possible to fuse the ligand-binding domain to the LexA DNA-binding domain without losing ability to bind ligands such as dexamethasone. Moreover, dexamethasone is relatively stable and hydrophobic, making it an attractive choice as the conserved portion of the hybrid ligand.

[16] L. Zhu, D. Gunn and S. Kuchibhatla, *in* "The Yeast Two-Hybrid System" (P. L. Bartel and S. Fields, eds.), pp. 73–96. Oxford University Press, New York, 1997.

[17] B. Manz, A. Heubner, I. Kohler, H. J. Grill, and K. Pollow, *Eur. J. Biochem.* **131,** 333 (1983).

For conjugation with a second ligand, dexamethasone can be converted into either an amine or activated carboxylic acid for reaction with either nucleophilic or electrophilic groups on the ligand. The chemistry will vary for individual ligands, but the preparation of activated dexamethasone will be the same. Here, we include synthetic procedures used to prepare the dexamethasone–FK506 hybrid ligand that is used for experiments described in this chapter.

Preparation of Dexamethasone–FK506 Hybrid Ligand

The dexamethasone–FK506 heterodimer was synthesized in five chemical steps from dexamethasone and FK506 (Fig. 4, structures **1–5**). Dexamethasone is first converted into an activated ester that can be coupled to a second ligand containing a primary amino group. Another conjugate between dexamethasone and the natural product pateamine A has also been made by following a similar procedure.[18]

Reagents and General Synthetic Manipulations

Dexamethasone is purchased from Research Biochemicals (Seattle, WA). FK506 is a gift from Fujisawa Pharmaceuticals (Deerfield, IL). All other reagents and solvents are commercial grade and purchased from Aldrich (Milwaukee, WI), EM Science (Cherry Hill, NJ), or Mallinckrodt Chemical (St. Louis, MO) and are used without further purification except for tetrahydrofuran and dichloromethane, which are distilled under nitrogen or argon from sodium benzophenone ketyl and calcium hydride, respectively.

Analytical thin-layer chromatography (TLC) is performed on EM Science precoated glass-backed silica gel 60 F-254 0.25-mm plates with staining, using an ethanolic solution of 3% (v/v) *p*-anisaldehyde containing 0.5% (v/v) concentrated sulfuric acid. Preparative chromatography is performed on Analtech (Newark, DE) precoated glass-backed silica gel GF-254 1.0-mm plates. Flash chromatography is performed with 230–400 mesh Merck (Rahway, NJ) silica gel.

Procedures

Dexamethasone 17β-*N*-hydroxysuccinimidylcarboxy ester (*N*-hydroxysuccinimidyl-9-fluoro-16α-methyl-11β,17-dihydroxy-3-oxo-1,4-androstadiene-17β-carboxylic acid) (**1**) and 24,32-bis(*tert*-butyldimethylsilyloxy)–

[18] D. Romo, R. M. Rzasa, K. Park, H. A. Shea, J. M. Langenhan, L. Sun, A. Akhiezer, and J. O. Liu, *J. Am. Chem. Soc.* **120,** 12237 (1998).

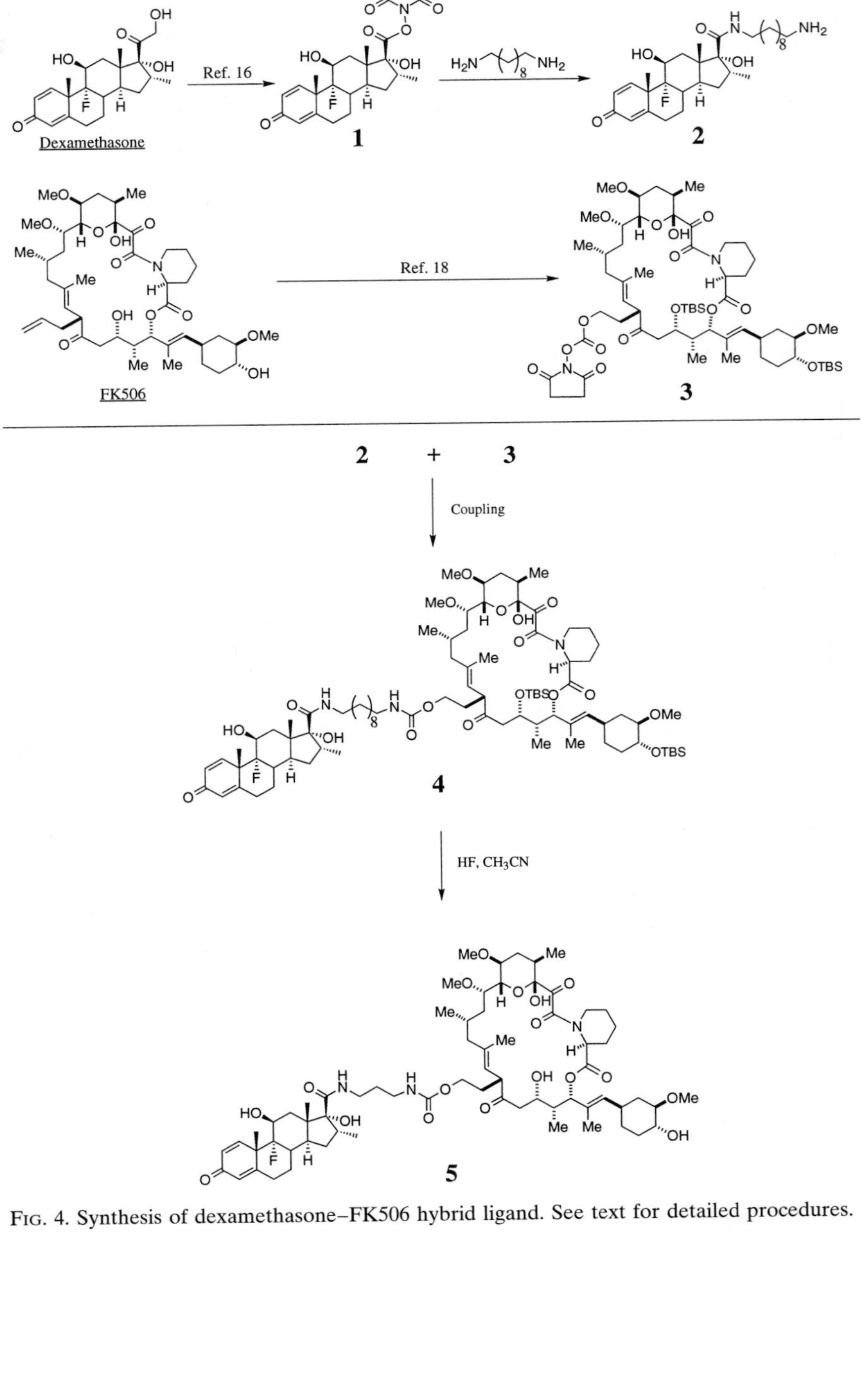

FIG. 4. Synthesis of dexamethasone–FK506 hybrid ligand. See text for detailed procedures.

FK506 *N*-hydroxysuccinimidyl mixed carbonate (**3**) are synthesized according to procedures described previously.[8,19] The procedures for synthesizing the intermediates **2**, **4**, and the hybrid ligand **5** follow.

Intermediate **2**—dexamethasone 17β-*N*-(10-amino)decylcarboxamide [*N*-(10-amino)decyl-9-fluoro-16α-methyl-11β,17-dihydroxy-3-oxo-1,4-androstadiene-17β-carboxamide]. Dexamethasone 17β-*N*-hydroxysuccinimidylcarboxy ester (0.090 g, 0.190 mmol) is dissolved in tetrahydrofuran (3 ml) and to this solution is added 1,10-diaminodecane (0.97 g, 0.562 mmol, 2.95 equivalents). The reaction is allowed to stir at room temperature for 12 hr before concentration *in vacuo.* The product is purified by preparative thin-layer chromatography [SiO_2, methanol–chloroform (10%, v/v), triethylamine (1%, v/v)] to afford 0.090 g of a white solid (89%).

Intermediate **4**—dexamethasone 17β-*N*-(10-amino)decylcarboxamide-24,32-bis(*tert*-butyldimethylsilyloxy)–FK506. 24,32-Bis(*tert*-butyldimethylsilyloxy)–FK506 *N*-hydroxysuccinimidyl mixed carbonate (0.008 g, 0.0067 mmol) and dexamethasone 17β-*N*-(10-amino)decylcarboxamide (0.020 g, 0.037 mmol, 5.5 equivalents) are placed in a 5-ml round-bottom flask and dissolved in anhydrous tetrahydrofuran (1 ml) and excess triethylamine. The reaction is allowed to proceed at room temperature for 3 hr. At this time the reaction is quenched with saturated sodium bicarbonate, and the aqueous phase is extracted with ethyl acetate. The organic layer is dried by filtration through $MgSO_4$ and concentrated *in vacuo.* The product is carried on to the next step without further purification.

Hybrid ligand **5**—dexamethasone 17β-*N*-(10-amino)decylcarboxamide–FK506. Dexamethasone 17β-*N*-(10-amino)decylcarboxamide-24,32-bis(*tert*-butyldimethylsilyloxy)–FK506 (approximately 0.010 g, 0.0062 mmol) is dissolved in acetonitrile (1 ml) in a 15-ml polypropylene tube. To this solution is added 49% (v/v) hydrofluoric acid (100 μl) and the reaction is allowed to proceed at room temperature for 24 hr. The reaction is quenched with saturated sodium bicarbonate, and the aqueous phase is extracted with ethyl acetate. The organic phase is dried by filtration through $MgSO_4$ and concentrated *in vacuo.* The product is purified by preparative thin-layer chromatography [SiO_2, methanol–dichloromethane (5%, v/v)] to afford 0.0037 g of product (40% for two steps). ^{1}H NMR (500 MHz, $CDCl_3$) δ 7.19 (d, 1H, J = 9.8 Hz), 6.56 (t, 1H), 6.33 (d, 1H, J = 9.8 Hz), 6.11 (s, 1H), 5.38 and 5.17 (rotamers, br s, 1H), 5.24 (1 rotamer, br d, J = 8.6 Hz), 5.03 (m, 1H), 4.42 (br d, 1H, J = 11 Hz), 4.38 (br d, 1H, J = 5.4 Hz), 3.73 (d, 1H, J = 7.7 Hz), 3.61 (d, 1H, J = 8.1 Hz), 3.45, 3.43, 3.40, 3.39, 3.31, 3.30 (rotamers of 3 methoxyls s, 9H) 2.84 and 2.68 (rotamers, d,

[19] M. N. Pruschy, D. M. Spencer, T. M. Kapoor, H. Miyaki, J. Crabtree, and S. L. Schreiber, *Chem. Biol.* **1,** 163 (1994).

1H, $J = 14.2$ Hz). PDMS (m/z) calculated for $C_{75}H_{116}N_3O_{18}F$: 1365.8. Found: 1365.9 $(M + H)^+$, 1389.9 $(M + Na)^+$.

After the last chromatographic step, dexamethasone–FK506 is dried under vacuum overnight. The hybrid ligand can be stored at −20° for more than 1 year without significant decomposition as judged by both TLC and the activity of the dimer in three-hybrid assays.

Preparation of Synthetic Complete Dropout Plates Containing Dexamethasone–FK506 Dimer

Dropout plates containing dexamethasone–FK506 are prepared slightly differently from a typical dropout plate. It has been found that dexamethasone–FK506 is not particularly stable at pH values below 5. The hybrid ligand is, however, quite stable at a pH close to or slightly above neutrality. Therefore, 70 m*M* phosphate buffer is included to adjust the medium to pH 6.5. To avoid potential decomposition, exposure of dexamethasone–FK506 to temperatures higher than 65° is also avoided.

Reagents

Synthetic complete dropout plate and medium (500 ml)

Synthetic minimal media (2×)	250 ml
Agar, 4% (w/v)	250 ml
Glucose or galactose, 40% (w/v)	25 ml
Amino acid stock solution (20×) (see below)	25 ml
Adenine, 0.2% (w/v)	5.0 ml
Uridine, 0.2% (w/v)	5.0 ml
Leucine, 1% (w/v)	4.5 ml
Lysine, 1% (w/v)	1.5 ml
Arginine, 1% (w/v)	1.0 ml
Histidine, 1% (w/v)	1.0 ml
Tryptophan, 1% (w/v)	1.0 ml

Amino acid stock solution (20×)

Double-distilled H_2O	700 ml
Methionine	0.28 g
Tyrosine	0.42 g
Isoleucine	0.42 g
Phenylalanine	0.70 g
Glutamic acid	1.40 g
Aspartic acid	1.40 g
Valine	2.10 g
Threonine	2.80 g
Serine	5.60 g

5-Bromo-4-chloro-3-indolyl-β-D-galactopyranoside (X-Gal): 20-mg/ml solution in dimethylformamide (DMF): Use at a 1:250 dilution
Potassium phosphate buffer (10×): 0.7 *M*, pH 6.5
Dexamethasone–FK506, 10 m*M* in dimethyl sulfoxide (DMSO) or ethanol: Store at −20°

Procedures

For 20 plates:

1. Melt 4% (w/v) agar in a microwave. Combine 250 ml of 4% (w/v) agar and 250 ml of 2× S solution and mix by swirling. Allow to cool to about 65° in a water bath.
2. Add glucose or galactose and appropriate amino acid stock solutions as well as 5 ml of 10× potassium phosphate buffer.
3. Add 50 μl of 10 m*M* dexamethasone–FK506 in ethanol. Mix quickly.
4. Pour 25 ml of the mixed agarose solution per plate. Allow the agarose to solidify and dry for 1 day.
5. The plates can be used immediately or stored at 4° for at least 1 week.

Checking Feasibility of a Hybrid Bait

Once a hybrid bait molecule has been synthesized, it is important to verify that it is suitable for a fruitful three-hybrid screen. First, it should be assayed to determine if the molecule retains biological activity in the original assay; even if previous structure–activity data suggest that the hybrid ligand should retain activity, it is important to verify that conjugation with dexamethasone does not abrogate activity. A relatively small decrease in potency should be tolerable, but decreases in potency that are likely to increase the dissociation constant for the interaction above 5–10 n*M* are likely to be undetectable with the current system.

Second, we have found that hybrid ligands must be tested to ensure their stability in yeast media. For example, we have found that dexamethasone–FK506 (Dex–FK506) is relatively unstable in normal yeast media. Much better results are obtained, however, if the pH of yeast plates is first adjusted with phosphate buffer (pH 6.5). Different hybrid ligands will necessarily possess different sensitivities to acidic and basic conditions. Therefore, it is important to check the hybrid bait stability in liquid yeast media, following the decomposition of the compound by high-performance liquid chromatography (HPLC) or some other method.

Finally, the new hybrid ligand should be tested for yeast permeability. The precise rules for permeability of small molecules in yeast are still unclear. Smaller and more hydrophobic compounds penetrate easier, but even some small hydrophobic compounds are impermeable. Molecules permeable to mammalian cells are also sometimes unable to penetrate yeast, presumably because of the presence of the yeast cell wall or the presence of different drug transporters. We are still working on methods to increase the permeability of yeast, using yeast strains deficient in certain drug transporters to further increase the generality of the system. Permeability of the hybrid ligand can be assayed on the basis of the ability to compete a standard three-hybrid interaction, using, for example, Dex–FK506 and FKBP12 fused to the transcriptional activation domain, as described below. The presence of excess Dex–ligand in the yeast cell will compete with Dex–FK506 for binding to the glucocorticoid receptor, blocking reporter gene activation. This competition test will ensure that the bait is capable of penetrating the yeast cell. However, it is necessary to be cautious in the interpretation of this result; because competition is through the dexamethasone portion of the molecule, it will still be unclear if the orphan ligand part of the molecule has been in some way metabolized and inactivated once taken up by the yeast.

Media and Reagents

Glu/SC –His, –Ura, –Trp liquid media: Glucose-containing dropout media (minus histidine, uracil, and tryptophan)
Gal/SC –His, –Ura, –Trp, –Leu liquid media
Dexamethasone–FK506 (10 m*M* in DMSO or ethanol)
Dex–ligand dissolved in DMSO or ethanol
EGY48 yeast strain transformed with pEGHBD7, pSH18-34, and pJG-FKBP12

Procedure for Dex–FK506 Competition

1. Grow the EGY48 yeast strain transformed with the appropriate vectors in Glu/SC –His, –Ura, –Trp liquid medium at 30°.
2. Inoculate overnight cultures of Gal/SC –His, –Ura, –Trp, –Leu with the yeast and add 1 μM Dex–FK506 dissolved in ethanol [final ethanol concentration, 0.1% (v/v)].
3. Add increasing concentrations of Dex–ligand stock solution to the cultures, from 1 to 200 μM final concentration [final ethanol or DMSO concentration, <0.5% (v/v)]. Make sure to do a positive control with no Dex–ligand. Also determine if Dex–ligand alone affects yeast growth. Incubate at 30° overnight.
4. Measure Dex–FK506-dependent yeast growth by absorbance at 600 nm.

Library Transformation and Selection

Once a feasible bait ligand has been constructed, it can now be used for screening of an activation domain-fused cDNA library. The EGY48 strain transformed with pEGHBD7 and the pSH18-34 reporter plasmid is grown and transformed with the fish library according to standard two-hybrid methods, using a high-efficiency lithium acetate protocol.[20] The yeast are then plated on Gal/SC –His, –Leu, –Ura, –Trp plus Dex–ligand (1–2 μM final concentration) and incubated at 30°. Colonies that grow within the first 10–14 days are restreaked and checked for specificity. Note that unlike in a two-hybrid screen, the fastest growing colonies are likely to represent false positives, such as direct protein interactors with the glucocorticoid receptor.

Eliminating False Positives

Currently the three-hybrid system suffers from the problem of numerous false positives. Such clones could arise for multiple reasons, including the presence of glucocorticoid receptor-interacting proteins in the fish cDNA library. To identify relevant interactors, several specificity tests must be performed. Colonies should be tested for bait-dependent growth by plating in the absence of hybrid ligand. This step usually eliminates the vast majority of false positives and should be performed prior to any other tests. In addition, a relevant interaction should exhibit galactose dependence because the fish library expression is under the control of the inducible *GAL1* promoter. Finally, an exquisite test for bona fide orphan ligand receptors is competition of the interaction with an excess of free ligand. Addition of 10- to 50-fold excess free ligand should block the fish–bait interaction because of competition with the hybrid ligand for fish binding. Alternatively, reporter gene activation can be assayed by a β-galactosidase assay using standard protocols instead of yeast growth on auxotrophic media. The combination of these three specificity tests has been sufficient to eliminate false positives in our experience.

Media and Reagents

Gal/SC –His, –Ura, –Trp, –Leu + Dex–ligand plates
Gal/SC –His, –Ura, –Trp, –Leu plates
Glu/SC –His, –Ura, –Trp, –Leu + Dex–ligand plates
Gal/SC –His, –Ura, –Trp, –Leu + Dex–ligand + ligand plates

[20] D. Gietz, A. St. Jean, R. A. Woods, and R. H. Schiestl, *Nucleic Acids Res.* **20,** 1425 (1992).

Procedure for Specificity Test

1. Streak colonies on Gal/SC –His, –Ura, –Trp, –Leu + Dex–ligand plates and grow at 30°.
2. Restreak or replica plate colonies onto Gal/SC –His, –Ura, –Trp, –Leu plates and incubate at 30° for 2–3 days.
3. Clones that fail to grow on Gal/SC –His, –Ura, –Trp, –Leu plates are streaked to Glu/SC –His, –Ura, –Trp, –Leu + Dex–ligand plates and incubated at 30° for 2–3 days.
4. Galactose-dependent clones are streaked to Gal/SC –His, –Ura, –Trp, –Leu + Dex–ligand plates and incubated at 30° for 2–3 days. Colonies that fail to grow in the presence of excess free ligand are taken for retrieval of the fish plasmid.

Recovery and Characterization of Three-Hybrid Clones

Once specific interactors have been identified, the library plasmid can be recovered through bacterial transformation of DNA isolated from the yeast clones as in a yeast two-hybrid screen. Such procedures have been covered previously.[21] Because false positives could occur even if the clones pass all of the above-described specificity tests, it is important to have an independent test to confirm the interaction. Therefore, an independent biochemical assay for the ligand–protein interaction should be used. Examples include the use of a ligand–biotin conjugate that can be used to determine interaction with recombinant bait protein[22] or the other, more traditional biochemical approaches.

Other Three-Hybrid Applications

Of course, the three-hybrid system is not limited to the identification of novel receptors for orphan ligands. The system is also suitable for use to define the receptor domains/residues required for interaction. Similarly, the use of several related hybrid ligands allows the examination of the structural components of the ligand required for interaction with the receptor. Similar to the reverse two-hybrid system, the three-hybrid system could be used to identify small molecules or peptides capable of blocking a ligand–protein interaction.[23] In addition, with the advent of combinatorial

[21] E. A. Golemis and R. Brent, *in* "The Yeast Two-Hybrid System" (P. L. Bartel and S. Fields, eds.), pp. 43–72. Oxford University Press, New York, 1997.

[22] E. C. Griffith, Z. Su, B. E. Turk, S. Chen, Y. H. Chang, Z. Wu, K. Biemann, and J. O. Liu, *Chem. Biol.* **4,** 461 (1997).

[23] M. Vidal, P. Braun, E. Chen, J. D. Boeke, and E. Harlow, *Proc. Natl. Acad. Sci. U.S.A.* **93,** 10321 (1996).

chemistry, the system could also be used to screen for novel small molecule ligands for a given protein. Screening could be used to identify lead compounds, using a library of ligands tethered to dexamethasone, or used to optimize a lead compound by competition experiments against a lower potency lead compound. This approach has advantages over traditional screening methods because of its versatility and generality. Several different protein targets can be rapidly screened with the same library without requiring time-consuming protein biochemistry.

As it stands, the yeast three-hybrid system requires further improvement before it can be applied to ligands with a K_d value higher than 5 nM. This limitation is understandable because the three-hybrid system depends on two ligand–receptor interactions as opposed to the single protein–protein interaction in the two-hybrid system. We are in the process of trying to improve the sensitivity of the system so that ligands with lower affinity for their receptors than the FK506–FKBP12 interaction can be detected.

[8] Use of a Yeast Three-Hybrid System to Clone Bridging Proteins

By JIE ZHANG

Introduction

The original two-hybrid system was designed to test protein interactions in pairs.[1] Many protein complexes, however, involve multiple components. The conventional two-hybrid system is limited in efficiency to analyze complex protein assemblies. Expanding the two-hybrid system to a multiple-hybrid system would facilitate dissecting the protein linkage network. For example, in a ternary protein complex, there are cases that protein X does not directly contact protein Y, but rather requires a third protein Z to mediate the interactions (Fig. 1A). In another situation, a protein Y may bind only to the composite site created as a result of association between proteins X and Z (Fig. 1B). Both cases demand introducing a "bridging" protein on a third shuttle plasmid in order to form a ternary complex in yeast. The three-hybrid system described here permits expressing a third component in the traditional two-hybrid system. It can be used to analyze ternary protein complexes and to isolate genes encoding components of the complex. In addition, the shuttle plasmid also allows testing of proteins

[1] S. Fields and O.-K. Son, *Nature* (*London*) **340,** 245 (1989).

0076-6879/00 $30.00

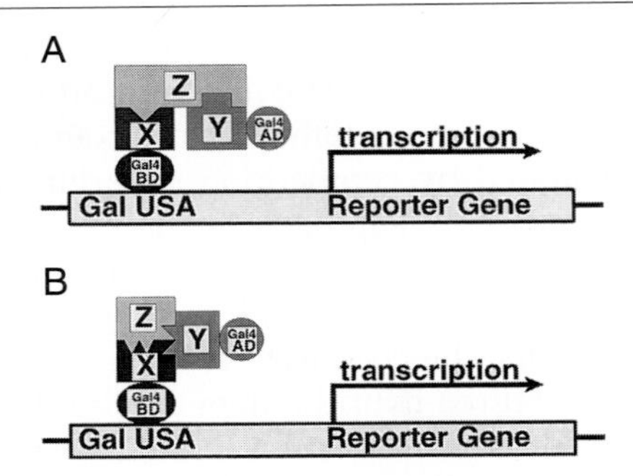

FIG. 1. Uses of the yeast three-hybrid system. The three-hybrid system may be useful for detecting ternary complex formation or for cloning the missing third component required for complex scenarios such as in (A), where interaction of proteins X and Y must be mediated by Z; or in (B), where Y binds only to a composite contour created by the combination of X and Z. Z can be a protein/peptide or an oligonucleotide. (Reproduced with permission from Ref. 4.)

assembled on RNA. A peptide library can be constructed in the shuttle plasmid by using degenerative oligonucleotides. Such a library, when used in conjunction with the reverse two-hybrid system,[2,3] makes it possible to screen for peptides that dissociate protein interactions.

In most cases, before applying the three-hybrid system, there should be circumstantial evidence that more than two components are involved in complex formation. Typical indications of multiple unit interactions are listed in Table I. The expanded hybrid system is useful either to test if a known protein is a component of a ternary protein complex, or to screen a library for cloning the component. At least two of the components in the complex should have been cloned. The two components should have been shown not to interact directly in the two-hybrid system. They form the basis for further analysis in the three-hybrid system.

Here the EGFR/Grb2/Sos complex is used as a model system to demonstrate the principles of the three-hybrid system. It has been established that the dimerization of the EGFR (epidermal growth factor receptor) activates its tyrosine kinase in the cytoplasmic domain. As a result, autophosphorylation of tyrosines in the EGFR creates high-affinity sites for binding coherent SH2 domains. Grb2, an adaptor protein, contains a central SH2 domain flanked by two SH3 domains. The SH2 domain of Grb2 binds to auto-

[2] M. Vidal, R. K. Brachmann, A. Fattaey, E. Halow, and J. D. Boeke, *Proc. Natl. Acad. Sci. U.S.A.* **93,** 10315 (1996).

[3] H. M. Shih, P. S. Goldman, A. J. DeMaggio, S. M. Hollenberg, R. H. Goodman, and M. F. Hoekstra, *Proc. Natl. Acad. Sci. U.S.A.* **93,** 13896 (1996).

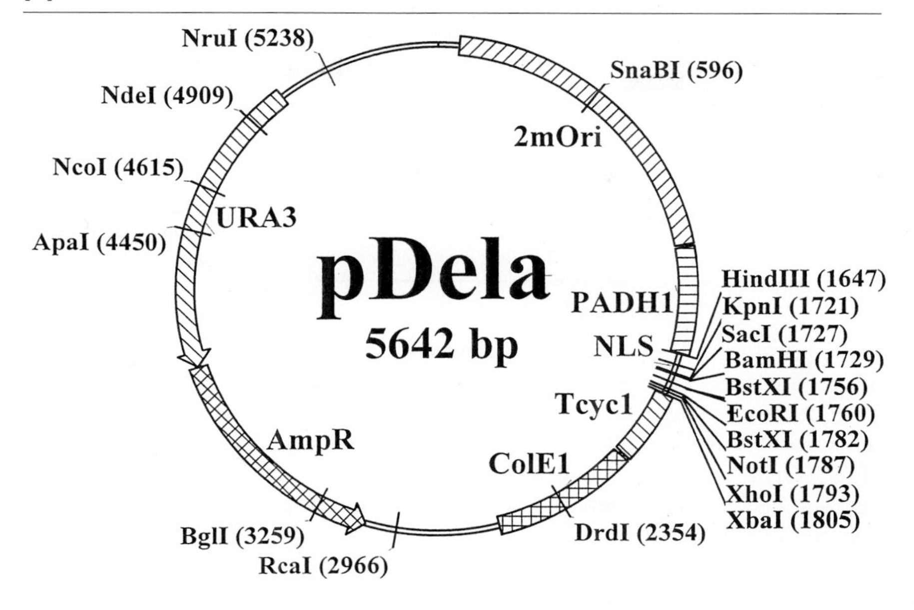


```
A AGC TTT GCA AAG ATG GAT AAA GCG GAA TTA ATT CCC GAG CCT
Hind III          Met Asp Lys Ala Glu Leu Ile Pro Glu Pro

  CCA AAA AAG AAG AGA AAG GTC GAA TTG GGT ACC GAG CTC GGA
  Pro Lys Lys Lys Arg Lys Val             Kpn I   Sac I   Bam

  TCC ACT AGT AAC GGC CGC CAG TGT GCT GGA ATT CTG CAG ATA
  HI   Spe I                  Bst XI       EcoRI

  TCC ATC ACA CTG GCG GCC GCT CGA GCA TGC ATC TAG A
       Bst XI         Not I   Xho I           Xba I
```

FIG. 2. Map of pDela. The shuttle vector plasmid contains an ampicillin resistance gene and a ColE1 replication sequence for ampicillin resistance selection and propagation in *E. coli,* a 2 μ Ori replication sequence, and a $Ura3^+$ gene for propagation and selection in a Ura^- yeast strain. There are multiple cloning sites (MCSs) for inserting proteins or protein domains of interest to be expressed in frame as a fusion with the SV40 T antigen nuclear localization sequence (NLS) at the amino terminus. The DNA sequence in the MSC region is shown with the NLS and cloning sites listed. The expression of the hybrid protein is under the control of the alcohol dehydrogenase promoter (*PADH1*) with cyc transcription termination sequence (Tcyc1).

TABLE I
EVIDENCE THAT IMPLIES POSSIBLE MULTIPLE PROTEIN INTERACTIONS

Experiment	Result
Immunoprecipitation	Multiple protein bands on SDS gel
Protein purification	Copurification of protein subunits
Enzymatic assays	Addition of fractions/factors for maximal activity
Signal transduction	Missing links between an upstream and downstream event
Genetic analysis	Multiple genes compensate/affect one phenotype

phosphorylated EGFR. The SH3 domains of Grb2 bind to the proline-rich motifs such as those in the carboxyl-terminal region of Sos, a guanine-nucleotide exchange factor for Ras proteins. The interacting domains of the EGFR, Grb2, and Sos have been subcloned to create pGBT9-EGFR, pDela-Grb2, and pGAD424-Sos plasmids, respectively. These plasmids can be used as positive control for the experiments using the three-hybrid system described below.

Only the salient features of the three-hybrid system are recapitulated here. The rest of the standard procedures for the yeast two-hybrid system can be found in the related chapters of this series.

Plasmids and Yeast Strains

Yeast strain BY3161 (*MAT***a** *leu2-3 trp1-901 his3-200 ura3-52 ade2-101 gal4-542 gal80-538* GAL1-lacZ GAL1-His3) is from J. Boeke (Johns Hopkins University School of Medicine, Baltimore, MD). pGBT9, pGAD424, pTD1, and pVA3 are from Clontech (Palo Alto, CA). pDela (Fig. 2) is constructed by ligating a 1.7-kb *Msp*A1I–*Kpn*I fragment of pGAD424 (nucleotides 5420 to 479) and a 3.9-kb *Hpa*I–*Kpn*I fragment of pYes2 (nucleotides 2284 to 354 (pYes2 is from InVitrogen, San Diego, CA) in an orientation such that the two *Kpn*I complementary ends join together, and the blunt ends of *Msp*A1I and *Hpa*I join together. Construction of positive control plasmids pGBT9-EGFR, pDela-Grb2, and pGAD424-Sos has been reported.[4]

Testing a Known Protein to Mediate Hybrid Interactions

The following procedure is for testing whether a cloned protein Z can bridge interactions between cloned proteins X and Y, which do not associate with each other directly.

[4] J. Zhang and S. Lautar, *Anal. Biochem.* **242,** 68 (1996).

Constructing Three Fusion Genes

Standard molecular cloning techniques should be followed to insert cDNA encoding appropriate proteins or protein domains X, Y, or Z into pGBT9, pGAD424, or other compatible plasmids, and pDela, respectively, to create pGBT9-X, pGAD424-Y, and pDela-Z. To express the fusion proteins, X and Y should be in frame with Gal4-BD and Gal4-AD. For the third protein Z, if only a partial sequence is to be expressed, its cDNA should be cloned into the multiple cloning sites in frame with the nuclear localization signal (NLS) in pDela. If a complete nuclear protein Z is to be expressed, it can be subcloned either into the *Hin*dIII site 5′-upstream of the NLS or into the multiple cloning site (MCS) in frame.

Triple Transformation

One-step transformation can be carried out to introduce all three plasmids, pGBT9-X/pGAD424-Y/pDela-Z, into the BY3161 yeast strain. Separate transformation with pGBT9/pGAD424-Y/pDela-Z, pGBT9-X/pGAD424/pDela-Z, and pGBT9-X/pGAD424-Y/pDela could serve later as a control to eliminate false interaction in the β-galactosidase assay. Transformation efficiency could be improved if sequential transformations are performed to introduce plasmids one by one. Triple transformants should be selected on yeast media without tryptophan, leucine, and uracil. Typically, triple transformants would emerge in 1 week while single transformants would appear in 2–3 days.

Assaying the Transformants for β-Galactosidase Activity

A positive signal for pGBT9-X/pGAD424-Y/pDela-Z but negative for pGBT9/pGAD424-Y/pDela-Z, pGBT9-X/pGAD424/pDela-Z, and pGBT9-X/pGAD424-Y/pDela would be a strong indication of a Z protein-dependent X and Y interaction. In separate *in vitro* binding experiments, the X/Y/Z complex should be confirmed with the purified or recombinant components. Together, the *in vivo* and *in vitro* binding data would establish the formation of an X/Y/Z complex. A typical 5-bromo-4-chloro-3-indolyl-β-D-galactopyranoside (X-Gal) filter assay result for the EGFR/Grb2/Sos interaction is shown in Fig. 3. A liquid β-galactosidase assay can also be used to quantify the relative strength of complex formation.

Constructing cDNA Library

For cloning a component of a "bridging" protein Z, a cDNA library needs to be constructed from the most abundant source of mRNA encoding

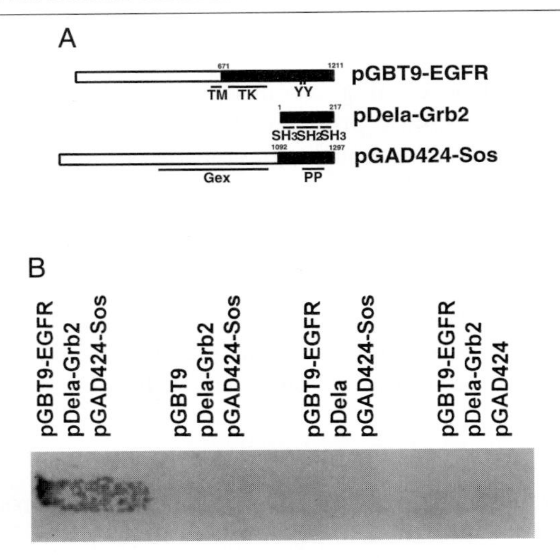

FIG. 3. Detecting the formation of a ternary complex of EGFR, Grb2, and Sos in the three-hybrid system. (A) Construction of three-hybrid expressing plasmids. DNA fragments encoding part of or a whole protein (filled bars) were obtained to make pGBT9-EGFR for expressing GAL4-BD and the cytoplasmic domain of the EGFR fusion protein, pDela-Grb2 for expressing the SV40 T-antigen nuclear localization domain and Grb2 fusion protein, and pGAD424-Sos for expressing GAL4-AD and the carboxyl-terminal domain of Sos2 fusion protein. TM, Transmembrane domain; TK, tyrosine kinase domain; Gex, guanine-nucleotide exchange catalytic domain; PP, proline-rich domain. (B) Expression of β-galactosidase as a result of interactions among the EGFR, Grb2, and Sos. Yeast transformed with different combinations of plasmids as indicated were grown on SD plates without tryptophan, uracil, and leucine, and their β-galactosidase activities were determined by filter X-Gal assay. (Reproduced with permission from Ref. 4.)

the putative Z protein. First, cDNA is made from the mRNA of the tissues or cells. Appropriate linker-adaptors compatible with the MCS in pDela are then ligated to the cDNA. Sizing selection for the long fragments of the cDNA is recommended. In-frame fusion between the NLS and the cDNA in pDela may not be necessary, because association of Z protein with either BD–X or AD–Y can usually take place in the cytoplasm first, and then the NLS in BD or AD will direct translocation of the complex to the nucleus. Therefore, it is desirable to select long cDNAs to increase the complexity of the library. This is different from an AD–cDNA fusion library, in which sizing fractionation for full-length cDNA could be counterproductive because potential stop codons in the 5′-untranslated sequence would abort the translation of AD fusion protein.

Used as an example here is a pDela-cDNA library synthesized from adult mouse brain poly(A)-selected mRNA using random and oligo(dT) primers. *Bst*XI linkers are added to the cDNA. The cDNA is sized and fragments longer than 500 bp are inserted into two *Bst*XI sites in pDela. The primary library consists of 6×10^5 independent colonies. More than 90% of the plasmids contain inserts with an average size of 2 kb.

Sequential Library Screening

For library screening, 500 μg of the plasmid library DNA is used to transform BY3161 yeast harboring pGBT9-X and pGAD424-Y by the polyethylene glycol (PEG)–lithium method. Yeast transformants are selected on SD plates with 50 m*M* 3-amino-1,2,4-triazole (AT), and without histidine, tryptophan, leucine, and uracil at 30° for 2 weeks. His^+ transformants are tested for β-galactosidase activity by the filter X-Gal assay. X-Gal-positive colonies are restreaked to obtain single colonies for further characterization. Restreaking is usually performed on –His, +AT, –Trp, –Leu, –Ura plates in order to maintain histidine selection pressure on weak or transient ternary complex formation.

Identifying Positive Clones

1. Prepare plasmid DNA from positive yeast colony and use the DNA to transform *Escherichia coli.*

2. Pick a dozen ampicillin-resistant *E. coli* transformants and use the polymerase chain reaction (PCR) to identify pDela-clone-containing transformants. A pair of primers, 5Dela3007 and 3Dela3551, can be used to specifically amplify a 545-bp fragment from pDela-clones. 5Dela3007 has the sequence 5′-CCTTCATCTCTTCCACCCACACACC-3′, and

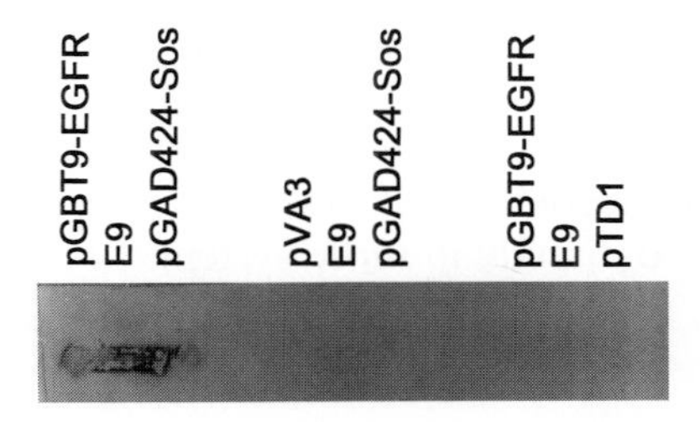

FIG. 4. Isolating the E9 clone that mediates the EGFR and Sos interaction. The yeast BY3161 strain containing pGBT9-EGFR and pGAD424-Sos was used to screen a mouse brain cDNA library in pDela. The pDela plasmid retrieved from the positive clone E9 was used to retransform yeast BY3161 together with pGBT9-EGFR, pGAD424-Sos, or their cognate control plasmids pVA3 and pTD1. Shown here are X-Gal filter assay results of the triple transformants. (Reproduced with permission from Ref. 4.)

3Dela3551 has the sequence 5′-CATCCTAGTCCTGTTGCTGCCAAGC-3′. Alternatively, primers bracketing the MSC in pDela (Fig. 2) can be used to amplify the insert by PCR, which can eliminate redundant clones.

3. Make a large preparation of pDela-clone plasmid from the PCR-positive colony.

Confirming Positive Clones

1. Transform BY3161 with pGBT9-X/pGAD424-Y/pDela-clone, and as two negative controls transform with pGBT9/pGAD424-Y/pDela-clone and pGBT9-X/pGAD424/pDela-clone.
2. Perform the β-galactosidase assay on triple transformants.
3. Sequence the pDela-clone. Express and purify recombinant cloned protein and test its *in vitro* interaction with X and Y.

As an example, a clone named E9 was isolated when the pDela-mouse brain cDNA library was screened with BY3161 containing pGBT9-EGFR and pGAD424-Sos to search for protein links in the EGFR/Sos signal transduction pathway (Fig. 4). The DNA sequence revealed that the 1.3-kb insert contained the entire coding sequence for mouse Grb2, as well as a 279-bp noncoding sequence at the 5′ end and a 346-bp noncoding sequence at the 3′ end.[4]

Summary

The three-hybrid system described here should facilitate screening cDNA libraries to clone "bridging" proteins in ternary complex. The pDela constructs were initially designed to be compatible with the Gal4-based two-hybrid system, such as pGAD424/pGBT9. The Ura selection offered by pDela makes it easy to introduce the "bridging" protein into the yeast containing the other two hybrids, the BD fusion protein and the AD fusion protein. Because many of the widely used two-hybrid systems employ Trp^-, Leu^-, His^-, or ziocin selection in the shuttle vectors, the Ura^+-based pDela constructs can be used directly in those systems.

Acknowledgment

I thank Academic Press for permission to reproduce Figs. 1, 3, and 4 and adapt experimental procedures for this chapter, and Eaton Publishing for permission to reproduce Fig. 2 and Table I.

[9] The Yeast Tribrid System: cDNA Expression Cloning of Protein Interactions Dependent on Posttranslational Modifications

By JAREMA P. KOCHAN, CHRISTOPH VOLPERS, and MARK A. OSBORNE

Introduction

In the study of cellular signaling, modifications of proteins by various enzymes such as protein tyrosine kinases, serine/threonine kinases, glycosyltransferases, and acylating enzymes often play critical roles. These modifications permit additional protein–protein interactions and often initiate a series of biochemical cascades that follow extracellular and physiological stimuli. One of the best-studied posttranslational modifications is the phosphorylation of tyrosine residues within the intracellular domains of cell surface receptors and other signaling proteins following activation by extracellular ligand binding.[1,2] The elucidation of protein–protein interaction networks via the yeast two-hybrid system has demonstrated the power and adaptability of the method. As experience with the yeast two-hybrid system has grown, so have the needs to introduce various improvements and modifications to the system originally described by Fields and Song in 1989.[3] In the case of mammalian proteins that participate in intracellular signaling, phosphotyrosine is a common modification that has consequences for protein–protein interactions. However, because there is minimal endogenous protein tyrosine phosphorylation in *Saccharomyces cerevisiae,* the traditional yeast two-hybrid approach would not be expected to identify protein–protein interactions that depend on phosphotyrosine. Because the two-hybrid system utilizes *S. cerevisiae* as its host, only those processes that occur within the yeast cytosol or nucleus will be able to modify the two fusion partners. In addition, these enzymes must be able to specifically modify the proteins of interest. Because of enzyme specificity, even modifications that occur endogenously in yeast may not be able to modify exogenous proteins in the two-hybrid arrangement. Therefore the ability to specifically modify sites on proteins provides a tremendous advantage.

We have introduced a third component into the yeast two-hybrid system, a protein tyrosine kinase, which *trans*-phosphorylates protein substrates in

[1] T. Hunter, *Cell* **80,** 225 (1995).
[2] C.-H. Heldin, *Cell* **80,** 213 (1995).
[3] S. Fields and O.-K. Song, *Nature* (*London*) **340,** 245 (1989).

0076-6879/00 $30.00

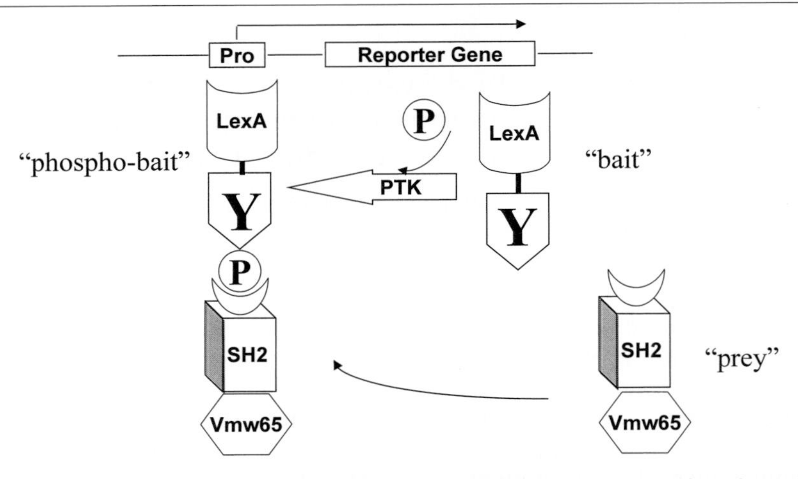

FIG. 1. Phosphorylation of bait protein on tyrosine residue permits detection of protein–protein interaction via two-hybrid analysis. The LexA fusion ("bait") protein is phosphorylated by the protein tyrosine kinase (PTK), yielding the "phospho-bait." This is then able to bind to the SH2 domain fused to the Vmw65 transcription activation domain, resulting in the reconstitution of the active transcription factor complex and transcription of the reporter gene.

the yeast cell (see Fig. 1). The series of vectors that has been developed permits the investigation of phosphotyrosine-dependent signal transduction pathways, which are often critical in higher eukaryotic cell function. The design of the vectors will also accommodate virtually any other enzyme that is involved in posttranslational modification that is necessary to facilitate protein interactions, or other proteins that form multisubunit complexes. Here, we describe the three-component approach to investigate tyrosine phosphorylation (Fig. 1). This modification is only an initial example, as its design is amenable to incorporate many posttranslational modifications or allosteric regulations that are necessary to facilitate protein interactions. The vectors described for the yeast tribrid system are available to interested researchers.

We describe here the steps in our method that differ from those typically used in two-hybrid screening as noted in other chapters in this volume and elsewhere.[4]

[4] P. L. Bartel and S. Fields, "The Yeast Two Hybrid System." Oxford University Press, New York, 1997.

1. **Yeast Selectable Marker**
2. **Yeast promoter**
3. **Gene for DNA binding domain (LexA)**
4. **Sites for cloning gene of interest**
5. **Transcriptional Termination signal**

LexA fusion("bait"); p4402

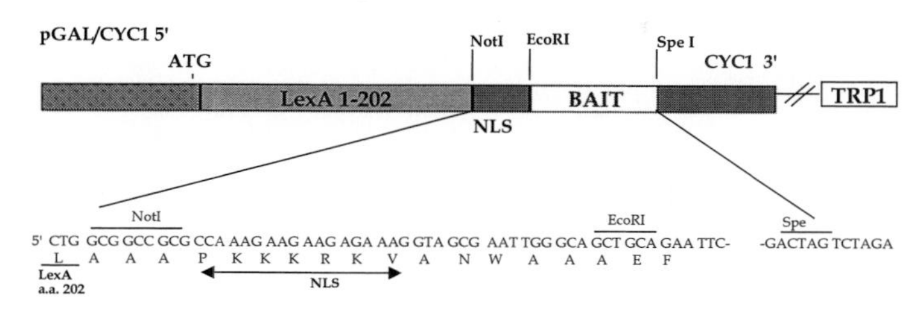

FIG. 2. p4402 DNA-binding domain fusion plasmid vector. Construction of an in-frame fusion requires the use of the *Eco*RI and *Spe*I sites. The resulting protein fusion contains the complete LexA protein sequence followed by the SV40 T nuclear localization sequence (NLS) and the gene product of interest.

Vectors and Reagents

Vectors

The tribrid system uses three different plasmid vectors, which direct the expression of the bait and prey fusion proteins and a tyrosine kinase (see Figs. 2–4). The DNA-binding domain fusion vector (Fig. 2) contains the following key features: the DNA-binding domain of the LexA bacterial protein (amino acids 1–202), the *TRP1* selectable marker, the strong inducible *pGAL10/CYC1* hybrid promoter,[5,6] the 3′ transcriptional termination signal of *CYC1,*[6,7] and a multiple cloning site. The DNA activation domain fusion vector (Fig. 3) contains the following key features: the transcriptional activation domain from the Vmw65 protein of herpes simplex virus type 1 (amino acids 410–490[8]; similar to the VP16 protein of HSV-2 protein), the *URA3* selectable marker, the strong inducible *pGAL10/CYC1* hybrid promoter, the 3′ transcriptional termination signal of *CYC1,* and a multiple cloning site. The tyrosine kinase vector (also known as the posttranslational modification vector) (Fig. 4) contains the following key features: a suitable

[5] M. Johnston, *Microbiol. Rev.* **51,** 458 (1987).
[6] S. Dalton, and R. Treisman, *Cell* **68,** 597 (1992).
[7] P. Russo and F. Sherman, *Proc. Natl. Acad. Sci. U.S.A.* **86,** 8348 (1989).
[8] M. A. Dalrymple, D. J. Mcgeoch, A. J. Davison, and C. M. Preston, *Nucleic Acids Res.* **13,** 7865 (1985).

1. **Yeast Selectable Marker**
2. **Yeast Promoter**
3. **Gene for Trans-activation domain (TAD), VP16(Vmw65) aa810-890**
4. **Sites for cloning in library inserts or specific cDNA**
5. **Transcriptional Termination signal**

Vmw65-cDNA fusion ("prey"); p4064

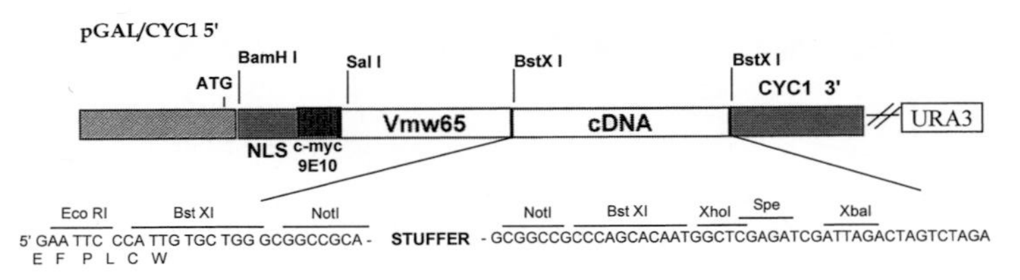

FIG. 3. p4064 transcription activation domain fusion plasmid vector. Construction of a cDNA library in this plasmid is facilitated by the noncomplementary *Bst*XI restriction sites. The ~1-kb stuffer fragment is derived from the mouse p53 gene. The protein product will include a nuclear localization sequence (NLS), the 9E10 c-Myc epitope tag, and the protein of interest. Stop codons in all three reading frames are present after the second *Bst*XI site.

tyrosine kinase or tyrosine kinase domain such as Lck or Lyn, the *LEU2* selectable marker, the strong inducible *pGAL10/CYC1* hybrid promoter, the 3′ transcriptional termination signal of *CYC1*, and a multiple cloning site. The details of the construction of all plasmids have been published

1. **Yeast Selectable Marker**
2. **Yeast Promoter**
3. **Tyrosine kinase gene (Lck 1-509)**
4. **Transcriptional Termination signal**

Tyrosine kinase Kinase; Lck p4140

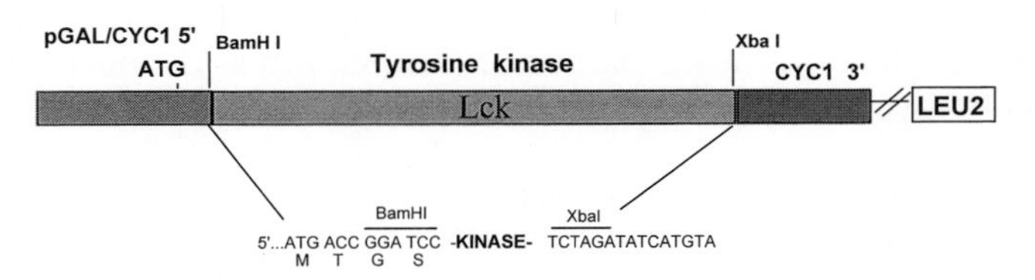

FIG. 4. p4140 posttranslational modification vector. Contains the same promoter and terminator as the two other plasmids but no epitope tag or nuclear localization sequence.

previously.[6,9] Both the LexA fusion vector and the Vmw65 fusion vector contain the nuclear localization signal of simian virus 40 (SV40) T antigen,[10] and the Vmw65 fusion vector also contains the 9E10 c-Myc epitope for fusion protein detection. The tyrosine kinase plasmid does not have a nuclear localization sequence.

There are a number of significant advantages to using the tribrid system over those available elsewhere. These include the inducible promoters, which make it possible to conditionally express all three of the proteins, thereby eliminating any growth disadvantages that might be conferred on the host cell by overexpression of certain proteins when using constitutive promoters. The Vmw65 fusion protein vector, in which cDNA libraries are generated, is amenable to the use of *Bst*XI linkers, which greatly reduce the background of non-insert-containing library members.[11] The plasmids are all CEN based, so copy number fluctuations within the host cell are minimized.

Microbial Strains Used

The *S. cerevisiae* reporter strain used in our studies is S-260 (*MATα ura3*::*ColE1* operator (×6)-*lacZ leu2–3,112 trp1-1 ade2-1 can1-100 ho*). For mating assays, W303-1a (*MAT***a** *leu2-3,112 trp1-1 his3-11,15 ura3-52 can1-100 ho*) is used. Rescue of library plasmids from S-260 is accomplished by electroporation into *Escherichia coli* strain KC8 (*pyrF*::Tn*5 hsdR leuB600 trpC9830 LacΔ74 strA galK hisB436*). The *E. coli* strain DH10B [F⁻ *mcrA* Δ(*mrr-hsdRMS-mcrBC*) *ϕ80dlacZΔM15 ΔlacX74 deoR recA1 endA1 araD139* Δ(*ara, leu*)*7697 galU galK rpsL nupG*] is used for propagation of the cDNA library. For a high-yield plasmid DNA preparation, *E. coli* strain XL-2 (*recA1 endA1 gyrA96 thi-1 hsdR17 supE44 relA1 lac*[F′ *pro AB lacI*^{q} *ZΔM15* Tn*10* (Tet^r) Amy Cam^r]) is used.

Polymerase Chain Reaction and Sequencing Primers

The following primers are suitable for sequencing and PCR of DNAs subcloned into the three plasmids:

LexA (anneals ~75 bp 5′ to the polylinker):

5′-TCG TTG ACC TTC GTC AGC AGA GCT TCA-3′

Vmw65 (anneals ~50 bp 5′ to the polylinker):

5′-TCG AGT TTG AGC AGA TGT TTA CCG ATG-3′

[9] M. A. Osborne, S. Dalton, and J. P. Kochan, *Bio/Technology* **13,** 1474 (1995).
[10] D. Kalderon, B. L. Roberts, W. D. Richardson, and A. E. Smith, *Cell* **39,** 499 (1984).
[11] A. Aruffo and B. Seed, *Proc. Natl. Acad. Sci. U.S.A.* **84,** 8573 (1987).

pGAL10/CYC1 5′ (for kinase inserts; anneals 30 bp 5′ to *Bam*HI site):
5′-TTA CTA TAC TTC TAT AGA CAC GCA-3′

CYC1 3′ UTR (will sequence from the 3′ end of the polylinker of all three plasmids; anneals 25 bp 3′ to the *Xba*I site):
5′-GAG GGC GTG AAT GTA AGC GTG AC-3′

Reagents Purchased from Commercial Sources

Reinforced nitrocellulose must be used for the β-galactosidase filter assays, because unsupported nitrocellulose shatters when immersed in liquid nitrogen. Schleicher & Schuell (Keene, NH; grade BA-S), Micron Separations (Westboro, MA; NitroPure), and Sartorius (Göttingen, Germany; reinforced cellulose nitrate) supported nitrocellulose products have been used successfully. Large Bio-assay dishes (245 × 245 mm) used for cDNA library construction and primary screening are manufactured by Nunc (Roskilde, Denmark). Glass beads (425–600 μm, acid washed) for yeast DNA isolation are purchased from Sigma (St. Louis, MO). Anti-phosphotyrosine antibodies are purchased from Upstate Biotechnology (Lake Placid, NY). Anti-FLAG antibodies are purchased from Eastman Kodak (Rochester, NY).

Media Recipes

YPD (rich medium) contains 10 g of yeast extract, 20 g of Bacto-peptone (Difco, Detroit, MI), 20 g of glucose, and 30 mg of adenine sulfate per liter. For plates, add 15 g of agar. Plates for KC8 uracil selection are prepared by mixing 6 g of Na_2HPO_4, 3 g of KH_2PO_4, 0.5 g of NaCl, 1 g of NH_4Cl, 2 g of Casamino acids, and 15 g of agar in 1 liter of water. Autoclave for 20 min, and add 1 ml of 1 *M* $MgSO_4$, 10 ml of 20% (w/v) glucose, 0.1 ml of 1 *M* $CaCl_2$, and 5 ml of tryptophan 4 mg/ml (all filter sterilized). Ampicillin is added to 100 μg/ml. Yeast selective plates are prepared by standard methods.[11a] with the addition of 0.2 ml of 10 *N* NaOH per liter to raise the pH to ~6, which greatly assists with the hardening of agar.

Solutions

Solutions for Yeast Transformation. Sterilize the following by filtration.

Lithium acetate buffer: 1× Tris–EDTA (TE), lithium acetate made fresh: TE (10×): 0.1 *M* Tris-HCl, 0.01 *M* EDTA, pH 7.5. Lithium acetate (10×): 1 *M* lithium acetate, pH 7.5

[11a] Guthrie and Fink (1991).

Polyethylene glycol (PEG) solution: 40% (w/v) PEG, 1× TE, and 1× lithium acetate made fresh from 50% (w/v) PEG 4000, 10× TE, and 10× lithium acetate

TPBS: 150 m*M* NaCl, 16 m*M* NaH_2PO_4, 4 m*M* Na_2HPO_4, 0.1% (v/v) Tween 20, pH 7.3

Enzymatic lysis solution: 1.2 *M* sorbitol, 20 m*M* HEPES (pH 7.4) containing 0.25 mg of yeast lytic enzyme (ICN, Costa Mesa, CA; also known as Zymolyase) and 15 μl of glusulase (New England Nuclear, Boston, MA) per milliliter

Lysis buffer for plasmid preparation from yeast cells: 2% (v/v) Triton X-100, 1% (w/v) sodiumdodecyl sulfate (SDS), 100 m*M* NaCl, 10 m*M* Tris-HCl (pH 8.0), 1 m*M* EDTA

X-Gal solution: 5-Bromo-4-chloro-3-indolyl-β-D-galactoside, 20 mg/ml in dimethyl formamide (store in glass or polypropylene tube at −20°)

Z buffer: mix 16 g of Na_2HPO_4, 5.5 g of NaH_2PO_4, 0.75 g of KCl, 0.25 g of $MgSO_4$, and 2.7 ml of 2-mercaptoethanol in 1 liter of water, adjust to pH 7.0 (store at 4°). Dilute X-Gal 1:20

Lysis buffer for COS cells: 10 m*M* 3-[(3-cholamidopropyl)-dimethyl-ammonio]-1-propanesulfonate (CHAPS) in phosphate-buffered saline (PBS) plus protease inhibitors [aprotinin (30 μg/ml), phenylmethylsulfonyl fluoride (PMSF, 200 μg/ml), leupeptin (10 μg/ml), pepstatin (10 μg/ml)] and phosphatase inhibitors (1 m*M* sodium orthovanadate, 50 μ*M* pervanadate, 1 m*M* NaF)

Methods

Bait Suitability for Tribrid Screening

The bait cDNA is subcloned into the LexA fusion vector (p4402) to make an in-frame fusion protein. The sequence of the fusion insert should be confirmed, particularly if it was generated by polymerase chain reaction (PCR), in order to ensure that the two reading frames mesh properly and that PCR-induced sequence errors have not been generated.

Next, as in the traditional two-hybrid method, the bait must be tested to ensure that the LexA–bait fusion protein does not confer the ability to activate transcription of the reporter gene in the absence of an interactor. If the interactor screen that is being considered makes use of the *trans*-phosphorylation by a protein tyrosine kinase (PTK), it is critical to ensure that the self-activation test is performed in the presence of the PTK. We have found some examples of LexA fusion proteins that do not activate transcription unless they are coexpressed with a PTK. Thus, it is important to verify that the LexA fusion protein does not activate transcription on its own both in the presence and absence of kinase. If it does not activate transcription, the expression of the LexA fusion protein and the PTK must be confirmed. Both steps are accomplished in the following experiments.

Method 1: Testing the Bait for Use in the Tribrid System

1. Transform the yeast S-260 with the LexA fusion protein expression vector along with the plasmid directing the expression of the PTK. Select colonies on SC Trp^- Leu^- glucose plates.

2. Pick a colony and inoculate 5 ml of Trp^- Leu^- glucose liquid medium and grow overnight at 30° with vigorous shaking (300 rpm).

3. Overlay the transformation plate with a reinforced nitrocellulose filter. Allow the filter to become completely wetted from beneath.

4. Remove the filter and place, colony side up, on a Trp^- Leu^- galactose plate. Incubate the plate at 30° overnight (18 hr).

5. The next morning, do a β-galactosidase filter assay on the nitrocellulose filter. If the colonies do not turn blue within ~30 min, proceed to step 6.

6. Measure the OD_{600} of the overnight culture started in step 2. Centrifuge the remaining cells for 5 min at 3000*g* at room temperature.

7. Wash the cell pellet in 5 ml of H_2O, and recentrifuge. Resuspend the pellet in 1 ml of Trp^- Leu^- liquid medium lacking carbon source.

8. Prepare two tubes for each transformant to be analyzed with 4.5 ml of Trp^- Leu^- liquid medium in each. To one tube, add 0.5 ml of 20% (w/v) glucose and to the other, add 0.5 ml of 20% (w/v) galactose.

9. Add a portion of the washed cell culture to each of the tubes containing 5 ml of medium to obtain a final OD_{600} of 0.2. Shake the cultures at 30° for 4–6 hr.

10. Centrifuge the cell cultures for 5 min at 3000*g* at room temperature to pellet the cells. Wash the pellet in 1 ml of sterile H_2O and transfer the culture to a microcentrifuge tube. Spin again (15 sec, maximum speed) to pellet the cells and remove the supernatant.

11. Add 100–300 μl of 1× Laemmli sample buffer to each pellet, resuspend by pipetting up and down, and boil for 5 min.

12. Centrifuge the lysates for 2–5 minutes at maximum speed at room temperature to precipitate unlysed cells and debris. Load the lysates onto SDS–polyacrylamide gels and analyze the products for LexA fusion protein and tyrosine kinase expression by Western blotting.

Considerations for Method 1. It is necessary to wash the yeast cells in step 7 to remove all the residual glucose, because glucose will interfere with induction of the *pGAL* promoter by galactose through the catabolite repression pathway.[5] Although other protocols for yeast cell lysis may call for the use of glass beads to increase the number of lysed cells, we have found that this dramatically increases the proteolysis observed as well. Controls to include in the expression test include using a transformant containing LexA protein alone (relative mobility about 25,000 Da) to verify the reduced mobility of the LexA fusion protein (when using anti-LexA antibodies). Running lysates from both the glucose- and galactose-grown cultures will provide a relative level of protein expression induction.

A second immunoblot with anti-phosphotyrosine antibodies will confirm the expression and activity of the tyrosine kinase, as most PTKs will phosphorylate endogenous yeast proteins. We have observed that different PTKs give different patterns of tyrosine phosphorylation of yeast proteins. For example, the Src family PTKs Lck and Lyn phosphorylate a large number of endogenous yeast proteins.[9] Jak3, in contrast, gives rise to only a few phosphotyrosine-containing bands (our unpublished results, 1999). Others have observed similar results with c-Src and v-Src.[12,13] Although v-*src* expression is lethal in *S. cerevisiae,*[14] we have not observed any growth impairment due to expression of any PTKs expressed in yeast. The control extract of uninduced cells grown in parallel (glucose as a carbon source) eliminates confusion that may be due to yeast proteins that immunoreact with the anti-LexA and/or anti-phosphotyrosine antibodies.

Tyrosine Phosphorylation of LexA Fusion Proteins. If the LexA fusion protein is expressed (and does not activate transcription on its own) and endogenous yeast proteins are phosphorylated by the PTK, it is also important to verify that the PTK is indeed phosphorylating the LexA fusion protein. Immunoprecipitation with anti-LexA antibodies followed by immunoblotting with anti-phosphotyrosine antibodies will show that the LexA fusion protein is tyrosine phosphorylated. Control strains grown with a LexA nonfusion protein and without the PTK plasmid should also be used for completeness. Alternatively, phosphorylation of the "prey" may be important for the interaction. If this is the case, skip to method 3.

Method 2: Phosphorylation of LexA Fusion Proteins

1. Inoculate 5 ml of Trp^- Leu^- glucose liquid medium with a yeast colony bearing the appropriate LexA fusion plasmid and kinase plasmid. Grow overnight at 30° with vigorous shaking.

[12] S. M. Murphy, M. Bergman, and D. O. Morgan, *Mol. Cell. Biol.* **13,** 5290 (1993).

[13] M. Okada, B. W. Howell, M. A. Broome, and J. A. Cooper, *J. Biol. Chem.* **268,** 18070 (1993).

[14] F. Boschelli, S. M. Uptain, and J. J. Lightbody, *J. Cell Sci.* **105,** 519 (1993).

2. The next morning, inoculate 10 ml of Trp^- Leu^- galactose liquid medium with an appropriate volume of H_2O-washed yeast cells (see method 1, steps 6–10) to give an OD_{600} of 0.2. Grow at 30° for 4–6 hr.

3. Centrifuge the cells to pellet (3000*g*, 5 min, room temperature), wash with sterile H_2O, respin, wash with 1 ml of sterile H_2O, and spin down in preweighed sterile microcentrifuge tubes.

4. Remove the supernatant and weigh the tubes, recording the mass of the wet cell pellet.

5. Resuspend the pellet in 2 ml of 0.1 *M* Tris-HCl (pH 9.4)–10 m*M* dithiothreitol (DTT) per gram wet weight. Incubate at 30° for 15 min.

6. Centrifuge for 2 min at 6000 rpm at room temperature and resuspend the pellet in 5 ml (per gram wet weight) of 1.2 *M* sorbitol–20 m*M* HEPES (pH 7.4) containing 0.25 mg of yeast lytic enzyme (ICN; also known as Zymolyase) and 15 μl of glusulase (NEN) per milliliter. Incubate at 30° with occasional rocking for 15–45 min.

7. Check for spheroplasting by microscopic inspection—mix 10 μl of cell suspension with 10 μl of H_2O and look to see if there are any cells remaining. Membrane ghosts should be plentiful and unlysed cells should be few. If not, continue incubating for up to 1 hr.

8. Spin down the spheroplasts for 5 min at 6000 rpm in a microcentrifuge.

9. Resuspend the pellet in PBS containing 1% (v/v) Triton X-100 (protease inhibitors may help as well as 1 m*M* $NaVO_4$, a phosphatase inhibitor), and pipette up and down to fully resuspend the spheroplasts. Incubate on ice for 5 min. The suspension should be clear, not cloudy like a suspension of cells. Check for lysis by microscopic examination if desired.

10. Perform an immunoprecipitation, using standard methods. Analyze immunoprecipitates by SDS–PAGE followed by immunoblotting.

Considerations for Method 2. We have investigated the tyrosine phosphorylation of LexA fusion proteins by immunoprecipitating with anti-LexA antibodies and then immunoblotting with anti-phosphotyrosine antibodies.[9] It should also be possible to use bait-specific antibodies instead of anti-LexA antibodies, or to reverse the order of the reagents—immunoprecipitation with anti-phosphotyrosine first, followed by anti-bait antibodies. We have found that preparing spheroplasts, although time consuming, is worth the effort because proteolysis is reduced relative to glass bead lysis. It is important to include several controls—glucose induction to control for endogenous yeast proteins immunoreacting with the antibody used, a no-kinase control (transforming the LexA fusion plasmid along with the kinase vector, pRS415, allows growth in Trp^- Leu^- medium), and a control using the LexA fusion vector without an insert.

LexA has no tyrosine residues, so it should not be modified by a tyrosine kinase.

The tyrosine phosphorylation of proteins is often dependent on the specific PTK used. It is best to test several different PTKs to determine which is best for phosphorylating a specific bait. Once the phosphorylation of the LexA–bait fusion protein has been demonstrated, the interactor screen can be performed as described below.

cDNA Library Transformation

The construction of cDNA libraries in the Vmw65 fusion vector has been described.[15] Even though there are several libraries constructed in the Vmw65 fusion protein vector [HeLa,[6] rat mast cells,[9] and human peripheral blood lymphocytes and hypothalamus (C. Volpers, M. A. Osborne, and J. P. Kochan, unpublished, 1999)], it is important to select a library from which it is reasonable to expect to identify interactors. One way to examine the suitability of a specific library is to determine whether the bait cDNA is present by Northern blot or reverse transcriptase (RT)-PCR analysis. If the bait is expressed in the cells from which the library is derived, it is not unreasonable to assume that its binding partners will also be expressed. In addition, it may also be best to use a PTK that is known to phosphorylate the bait of interest, if this information is available.

A pilot transformation of the cDNA library into S-260 carrying the LexA fusion and kinase plasmids will give an estimate of how many colonies to expect per microgram of library DNA. This number can then be used to scale up the transformation for the screen. Generally, we plate about 20,000 colonies per large plate and deal with about 20 plates at one time. If more than 400,000 colonies are to be screened, transformations performed on successive days may be less problematic.

Method 3: Large-Scale Library Transformation

1. Inoculate a culture of 25 ml of Trp^- Leu^- glucose liquid medium with a transformant containing the LexA–bait fusion and the tyrosine kinase plasmid (e.g., a colony from the plate used in method 1, step 1). Grow the culture at 30° with shaking overnight.

2. Dilute the overnight culture 1 : 10 in a final volume of 200 ml of Trp^- Leu^- glucose liquid medium and grow overnight at 30° with shaking.

3. The next morning, count the cells in the overnight culture from step 2 and inoculate three separate 2-liter flasks containing 300 ml of Trp^- Leu^-

[15] M. Osborne, M. Lubinus, and J. P. Kochan, The tribrid system—detection of protein–protein interactions dependent on posttranslational modifications. *In* "The Yeast Two Hybrid System," pp. 233–258. Oxford University Press, New York, 1997.

glucose liquid with the overnight culture to a final cell density of 2×10^6 cells/ml. Incubate at 30° for 5 hr with shaking.

4. Centrifuge the cells and process for lithium acetate transformation as usual. Transform with an appropriate amount of cDNA library to generate 20,000 colonies per plate × 15 plates = 300,000 colony-forming units.

5. After heat shock, plate the cells on 245 × 245 mm (Nunc) Trp^- Leu^- Ura^- glucose plates. In addition, plate a small (0.0002%) amount of the transformation mixture on small (10-cm) Trp^- Leu^- Ura^- glucose plates, in order to estimate the total number of colonies screened. Incubate the plates at 30° for 24–36 hr.

6. When colonies are clearly visible but not touching, overlay with reinforced nitrocellulose (22 × 22 cm) sheets. Allow to completely wet from beneath, then transfer (cell side up) to Trp^- Leu^- Ura^- galactose plates for 18 hr at 30°.

7. Perform the β-galactosidase filter assay.

8. With a sterile toothpick, pick blue colonies as they appear onto a small Trp^- Leu^- Ura^- glucose plate, and incubate this plate for 2–3 days at 30°.

Consideration for Method 3. The nitrocellulose should not be autoclaved before using it to overlay the transformation plates. Autoclaving causes wrinkles, which interfere with the ability of the nitrocellulose to lie flat on the agar surface, and causes smearing of the colonies. It is also helpful to ensure that the Trp^- Leu^- Ura^- glucose plates are not wet when plating the transformation mixture. A volume of cell suspension of 3–4 ml per plate has worked well in our hands. The transformation mixture must be spread carefully to ensure even distribution of colonies, as a highly dense area of cells will give rise to small colonies that may be difficult to detect when blue. The liquid nitrogen permeabilization step is best accomplished in a large, glass baking dish, followed by allowing the filters to equilibrate to room temperature while sitting on a piece of dry blotting paper [such as Whatman (Clifton, NJ) 3MM]. This allows the colonies, which tend to "sweat" as they warm up, to dry a bit before becoming wet with Z buffer in the assay. This step minimizes the possibility of colonies smearing and running into each other during the β-galactosidase color development step. Pieces of blotting paper are then cut to fit inside the lid of the inverted Nunc bioassay dishes. Add ~25 ml of Z buffer to saturate. The Whatman 3MM paper must be perfectly flat in the lid, to ensure even contact with the nitrocellulose filter. Once the filters are placed on the Whatman paper saturated with Z buffer, additional Z buffer may be needed in order to maintain an appropriate level of moisture. Too much buffer results in runny colonies; too little results in poor color development. The filters should be covered with the bottom part of the plates for two reasons: (1) so that the filters do not dry out, and (2) so that other laboratory members are not offended by the pungent aroma of 2-mercaptoethanol in the Z buffer.

Considerations for Library Transformations

The transformation mixture may be plated in several different ways. One is to plate it directly on Trp^- Leu^- Ura^- glucose plates, allow the transformants to grow for 24–36 hr, and then transfer them to reinforced nitrocellulose filters, which are then placed, colony side up, on Trp^- Ura^- Leu^- galactose plates for 18 hr. The filter is then processed for a β-galactosidase assay as usual. Blue colonies may be scraped directly from the nitrocellulose filter, or, alternatively, identified and picked from the glucose "master" plate. Picking from the master plate can be tricky, as the colonies on the glucose plate are often small and easily smeared by the process of lifting to nitrocellulose and the positives identified on the filter after assay may be difficult to match against the master glucose plate later. The ease of the former method (scraping directly from the nitrocellulose) is tempered by the fact that the cells that constitute the colony being picked from the nitrocellulose have been grown on galactose. Thus, if one of the plasmids encodes a toxic protein, this may make the rescue of living cells from a colony exposed to galactose difficult. We have not observed this phenomenon, but it bears keeping in mind. Picking from the master plate as well as from the nitrocellulose filter is probably the safest alternative. Another method to consider is to plate the transformation mixture directly onto nylon membranes (such as Hybond; Amersham, Arlington Heights, IL) placed on the Trp^- Ura^- Leu^- glucose plates. The colonies will then grow directly on the filters, which are then transferred to Trp^- Ura^- Leu^- galactose plates after 24–36 hr. The β-galactosidase assay is then performed. Although this method involves the fewest manipulations, we find that the overall transformation efficiency for plating directly on nylon filters is reduced by fivefold relative to plating onto agar plates directly.

Recovery of Blue Colonies

There are several technical considerations to consider when performing large-scale screenings of yeast colonies. It is important to keep the filters moist with Z buffer during the time it takes for blue colonies to appear (typically 15 to 90 min from the start of the assay), and to ensure that the filter paper placed under the nitrocellulose is flat so that the nitrocellulose can make contact with it uniformly. If bubbles form between the nitrocellulose filter and the filter paper, the X-Gal solution will not reach the colonies, and positives may not be identified in this region. Blue colonies are then picked by either scraping the filter (or the corresponding master plate) and patching them onto a Trp^- Ura^- Leu^- glucose plate (as some cells that have been dipped in liquid nitrogen do live). Typically, the patches will show growth in 2–3 days at 30°.

Toward the end of the assay, it becomes increasingly difficult to discern "true" positives from the emerging background of blue colonies. The intensity of the blue background is dependent on the particular bait; some seem to display higher incidences of background than others. It is often helpful to note the time elapsed since the start of the assay when identifying blue colonies, so that they can be ranked according to intensity during subsequent analysis. We have not found any true positives that take longer than 90 min to turn blue during an initial screen.

Purification of Positive Colonies

Because it is not generally feasible to scrape the blue colony from the filter or master plate without touching neighboring colonies, it is important to obtain a pure population before proceeding with analysis. Each patch contained on the plate generated in step 8 of method 3 is therefore a mixed population of both the desired positive cells and contaminating noninteractors. The following protocol directs the isolation of pure positives.

Method 4: Colony Purification

1. After the patches of cells from step 8 of method 3 have grown, scrape the cells with a toothpick into 1 ml of Trp^- Ura^- Leu^- medium lacking carbon source. If there are a number of single, isolated colonies in the patch, scrape them all together.
2. Dilute 100 μl of the cell suspension in 900 μl of H_2O and determine the OD_{600}. Assume that 1 $OD_{600} = 1.5 \times 10^6$ colony-forming units/ml, and calculate the volume needed to deliver 200 cells.
3. Dilute the cell suspension to 200 colony-forming units/100 μl, and plate onto a 10-cm Trp^- Ura^- Leu^- glucose plate. Incubate at 30° for 1–2 days.
4. Overlay each plate with a nitrocellulose filter, and transfer to a Trp^- Ura^- Leu^- galactose plate for 18 hr. Perform a β-galactosidase assay.
5. Identify a well-isolated blue colony on each filter, and pick the corresponding colony from the glucose plate. Patch the colony onto a fresh Trp^- Ura^- Leu^- glucose plate. Incubate at 30° for 1–2 days.
6. Scrape a portion of the patch into 1 ml of 50% (v/v) glycerol and freeze at −70° as a backup.

Considerations for Method 4. The factor required to convert OD_{600} to colony-forming units must be determined empirically, and may vary depending on the bait or other considerations. Once pure populations of each positive have been obtained, they can be used to explore two important questions: (1) Does the Vmw65 fusion protein ("prey") interact specifically with the LexA fusion protein used? (2) Is the interaction dependent on protein tyrosine kinase activity?

Determining Specificity and Tyrosine Kinase Dependence

Once a pure population of yeast colonies bearing interacting cDNAs exists, many researchers may be tempted to obtain sequence information as quickly as possible. Although this may be the most immediately satisfying approach, it is not always the most time- and cost-effective, but will depend on the resources that are available in different laboratories. The rescue of cDNA library plasmids from yeast via *E. coli* is time consuming, and, because some of the plasmids may prove uninteresting to the investigator (e.g., false positives and nonspecific interactors), it may be more efficient to eliminate these clones quickly. Alternatively, high-quality PCR products can be obtained directly from the yeast colonies. This method eliminates the *E. coli* rescue step.

The two questions posed at the end of the previous section require two different transformants. The first question (whether the interaction is specific) can be addressed by curing the host strain of the LexA fusion plasmid (p4402; *TRP1* marker) and then mating the resulting strain (still carrying the Vmw65–cDNA fusion and tyrosine kinase plamids) to W303-1a (*MAT***a**) transformed with a number of different LexA fusion protein transformants.[16] The resulting diploids can then be tested individually by nitrocellulose filter β-galactosidase assay following galactose induction.

To identify whether the interactions observed require the tyrosine kinase, the host strain should be screened for transformants that have lost the tyrosine kinase plasmid (*LEU2* marker) and the resulting cells will be Trp^+ Ura^+ Leu^-. These double transformants can then be screened by filter assay to determine which, if any, transformants retain β-galactosidase activity. Those that do will not require the kinase for interaction.

The two different transformant profiles required for these two determinations can be identified in the same experiment. To do this, triple-transformant yeast (which are blue in the β-galactosidase assay) are grown under nonselective conditions (in rich media such as YPD) for several generations to allow for the loss of plasmids, which will occur randomly because they

[16] C. Bai and S. J. Elledge, Searching for interacting proteins with the two-hybrid system. *In* "The Yeast Two Hybrid System," pp. 11–28. Oxford University Press, New York, 1997.

are no longer being subjected to nutritional selection (see Fig. 3). In our experience, following the protocol provided allows for identification of all classes of transformants.

Method 5: Plasmid Segregation

1. For each positive interactor, inoculate a 2-ml culture in YPD medium. Shake at 30° for 24–48 hr.
2. Dilute the culture 1:300 and plate 75 μl on a fresh YPD agar plate. Incubate at 30° overnight.
3. The next day, replica plate each YPD plate to two plates: one Trp⁻ Ura⁻ glucose (plate 1) and one Ura⁻ Leu⁻ glucose (plate 2). Incubate at 30° overnight.
4. Compare the two replicas. Identify two colonies on each plate that do not grow on the other replica. Patch these colonies to a fresh agar plate (with the appropriate supplemental amino acids) for further analysis.

Considerations for Method 5. Comparison of the two replica plates will reveal the two types of colonies sought. If a colony grows on plate 1 (LexA⁺ Vmw65–cDNA⁺) but does not grow on plate 2 (Vmw65–cDNA⁺ kinase⁺), then it must have lost the kinase plasmid, because it will not grow on the plate lacking leucine (plate 2). This colony may be picked for analysis of kinase dependence, and should be patched or streaked onto a Trp⁻ Ura⁻ glucose plate. It is easiest and most convincing to patch several colonies onto one plate along with positive and negative controls and then assay for β-galactosidase activity. Trp⁺ Ura⁺ Leu⁻ colonies that remain blue after β-galactosidase assay do not require the tyrosine kinase for an interaction. This also permits evaluation of the relative intensities of the interactors with each other. If a colony grows on plate 2 but not on plate 1, then it has lost the LexA fusion plasmid, because it will not grow on a plate lacking tryptophan. Patch this colony onto a Ura⁻ Leu⁻ glucose plate for use in mating assays to dermine the specificity of its interaction.

Mating Assay to Test for Specificity

We find it most convenient to maintain glycerol stocks of several LexA fusion protein bait plasmid transformants in the strain W303–1a. These can then be individually mated to the Ura⁻ Leu⁻ interactors to test for specificity. Mating is accomplished by mixing the LexA fusion bait in strain W303–1a with the library/kinase transformant (Ura⁺ Leu⁺ Trp⁻) in S-260 on a YPD plate. After incubating for several hours to overnight at 30°, replica plate to a Trp⁻ Leu⁻ Ura⁻ glucose plate and incubate at 30° overnight. Only the diploids will grow. This plate is then overlaid with nitrocellulose and the filter induced on a Trp^- Leu^- Ura^- galactose plate overnight, followed by β-galactosidase filter assay. If blue colonies are observed only for the bait of interest and the library/kinase diploid but not with the extraneous baits, it is now appropriate to rescue the library plasmid in *E. coli* for sequencing and further manipulation.

At the end of the preceding analysis, it will be known how specific the interaction is for a particular LexA fusion, in addition to whether the tyrosine kinase is required for the interaction to be detected. The Vmw65–cDNA fusion plasmid can now be rescued from these cells into *E. coli* by standard methods and analyzed by restriction endonucleases, DNA sequencing, and retransformation back into S-260 to verify the interaction. The latter step, retransforming the rescued plasmid back into S-260 with the appropriate bait and kinase to verify its interaction, is particularly important, as we have observed blue colonies whose β-galactosidase activity is not linked to the plasmid (and is therefore an artifact). β-Galactosidase units may also be measured to quantitate the interaction. We commonly observe 1000 units of activity for Src homology 2 (SH2)–phosphotyrosine interactions with a background of less than 10 units.[9]

Conclusion

In this chapter we have provided a detailed overview of the methods and techniques that are utilized in the yeast tribrid system for the identification and characterization of protein interactors that require *trans*-phosphorylation on tyrosine. Various peculiarities and benefits have been described to help the reader perform the experiments. There are a variety of modifications that can be adapted to the tribrid system that we have described.[15] Moreover the tribrid system can be utilized with just about any posttranslational modification or allosteric interaction available. There are probably other modifications that can be developed to the tribrid system, and we encourage innovation in the pursuit of discovery.

[10] The Yeast Split-Hybrid System

By ANTHONY J. DEMAGGIO, PHYLLIS GOLDMAN, HSIU-MING SHIH, RICHARD H. GOODMAN, and MERL F. HOEKSTRA

Introduction

Considering the large number of intricate and highly regulated processes occurring in the cell, it is not surprising that the multiprotein networks mediating these processes are themselves quite complex. Characterization of the different networks of protein–protein interactions has contributed to virtually all aspects of cellular biology including signal transduction, transcription, translation, protein processing, protein translocation, differentiation, apoptosis, and the cell cycle. As research in these fields continues to advance and as genome and mass spectrometry sequencing efforts identify novel uncharacterized proteins, the study of protein–protein interactions will remain paramount to understanding cellular events.

Many techniques, including immunoprecipitation, cross-linking, and fluorescence energy transfer, are useful in characterizing protein–protein interactions. In 1989, Fields and Song[1] introduced the yeast two-hybrid assay, which allows investigators to study known protein–protein interactions and identify novel protein-binding partners. This yeast-based genetic assay capitalizes on the fact that many eukaryotic transcription factors consist of two separable functional domains, a DNA-binding domain (DBD) that localizes the transcription factor to a site-specific DNA sequence and a transcriptional activation domain (TAD) that interacts with the basal transcriptional machinery. In the yeast two-hybrid assay (Fig. 1), one gene expressing a protein of interest is fused to a DBD such as GAL4 or LexA, while a second gene (or a cDNA library) expressing another protein (or a library of proteins) is fused to a TAD such as GAL4 or VP16. Interaction of the two proteins reconstitutes a functional transcription factor that can activate the expression of reporter genes under the control of GAL4 -or LexA-binding sites. The reporter genes generally encode enzymes required for yeast growth (*HIS3, LEU2,* or *URA3*), enzymes that can be detected colorimetrically (*lacZ*), or those detected visually or luminometrically (green fluorescent protein, GFP). Identifying protein–protein interactions by growth selection, color, or light is fast, simple, inexpensive, and is likely why the yeast two-hybrid system has become a method of choice for studying protein–protein interactions.

[1] S. Fields and O. Song, *Nature* (*London*) **340,** 245 (1989).

METHODS IN ENZYMOLOGY, VOL. 328
0076-6879/00 $30.00

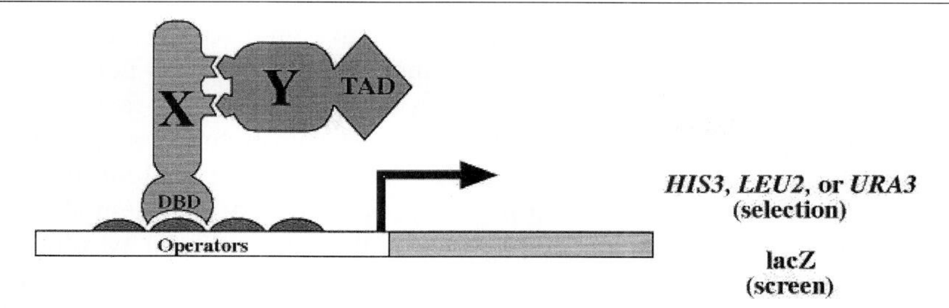

FIG. 1. The yeast two-hybrid system. Interaction of protein X fused to a DBD with protein Y fused to a TAD activates expression of nutritional (*HIS3, LEU2, URA3*) and/or colorimetric (*lacZ*) reporter genes.

Since its inception, many modifications have been made to the yeast two-hybrid system. False-positive clones identified in library screens have been decreased by incorporating a two-reporter gene system,[2] and methods to screen libraries of proteins fused to the DBD have been developed.[3] Two-hybrid technology has also become widely used in bacterial and mammalian cells.

Many creative variations of the basic two-hybrid theme have been devised to study interaction scenarios. For instance, in one-hybrid systems, proteins that bind to specific DNA sequences can be identified by screening cDNA–TAD fusion libraries against a reporter gene driven by that particular DNA sequence.[4,5] One three-hybrid assay screens for small ligands that interact with known receptors or for receptors that interact with known ligands,[6] while another three-hybrid system detects RNA–protein interactions.[7] The yeast two-hybrid system can also be used to identify peptide aptamers that bind to a given protein.[8,9] Large combinatorial peptide libraries can be screened rapidly in an *in vivo* setting, proving advantages over other *in vitro* peptide-screening methods. Identified peptides may provide insight into protein-binding domains and, importantly, may also be useful as inhibitors of protein–protein interactions. While most of the current hybrid technology concentrates on characterizing protein–protein interac-

[2] P. Bartel, C.-T. Chien, R. Sternglanz, and S. Fields, *Biotechniques* **14,** 920 (1993).

[3] W. Du, M. Vidal, J. E. Xie, and N. Dyson, *Genes Dev.* **10,** 1206 (1996).

[4] J. J. Li and I. Herskowitz, *Science* **262,** 1870 (1993).

[5] M. M. Wang and R. R. Reed, *Nature* (*London*) **364,** 121 (1993).

[6] E. J. Licitra and J. O. Liu, *Proc. Natl. Acad. Sci. U.S.A.* **93,** 12817 (1996).

[7] D. J. SenGupta, B. Zhang, B. Kraemer, P. Pochart, S. Fields, and M. Wickens, *Proc. Natl. Acad. Sci. U.S.A.* **93,** 8496 (1996).

[8] M. Yang, Z. Wu, and S. Fields, *Nucleic Acids, Res.* **23,** 1152 (1995).

[9] P. Colas, B. Cohen, T. Jessen, I. Grishina, J. McCoy, and R. Brent, *Nature* (*London*) **380,** 548 (1996).

tions, the ability to study the disruption of protein–protein interactions should provide additional valuable information about protein networks.

Two technologies take a direct approach in identifying factors that disrupt protein–protein interactions. The yeast split-hybrid[10] and reverse two-hybrid systems[11–14] both convert the disruption of a protein–protein interaction into a positive selection. The yeast split-hybrid system (Fig. 2) employs many of the same components of the conventional two-hybrid system; however, it is a binary system that also incorporates the *Escherichia coli tet* repressor (TetR) operator system. In this system, LexA operator sites are located in a promoter that drives expression of the TetR gene. The interaction of one protein fused to the LexA DBD with a second protein fused to the VP16 TAD activates the expression of TetR, which subsequently binds to TetR-binding sites (*tet* operators) located upstream of the nutritional reporter gene *HIS3*. Binding of TetR to the *tet* operators represses the expression of the *HIS3* gene, preventing yeast growth in the absence of histidine. If the protein–protein interaction is disrupted, no TetR protein is made, the *HIS3* gene is transcribed, histidine is produced, and yeast grow in the absence of histidine.

General Procedure

Yeast Strains

Yeast strains are grown in yeast extract–peptone–dextrose (YEPD) or selective minimal medium under standard conditions.[15,16] Selective minimal medium is dropout or omission medium in which all amino acids and nucleosides are added except for those required to select for specific auxotrophies/prototrophies. YEPD and synthetic medium are purchased from Bio 101 (Carlsbad, CA). Yeast strains are derived from AMR69 and AMR70.[17] Two strains are commonly used: YI584 and YI671. The genotype

[10] H. M. Shih, P. S. Goldman, A. J. DeMaggio, S. M. Hollenberg, R. H. Goodman, and M. F. Hoekstra, *Proc. Natl. Acad. Sci. U.S.A.* **93,** 13896 (1996).

[11] M. Vidal, R. K. Brachmann, A. Fattaey, E. Harlow, and J. D. Boeke, *Proc. Natl. Acad. Sci. U.S.A.* **93,** 10315 (1996).

[12] M. Vidal, P. Braun, E. Chen, J. D. Boeke, and E. Harlow, *Proc. Natl. Acad. Sci. U.S.A.* **93,** 10321 (1996).

[13] J. Huang and S. L. Schreiber, *Proc. Natl. Acad. Sci. U.S.A.* **94,** 13396 (1997).

[14] A. Borchardt, S. D. Liberles, S. R. Biggar, G. R. Crabtree, and S. L. Schreiber, *Chem. Biol.* **4,** 961 (1997).

[15] F. Sherman, *Methods Enzymol.* **194,** 3 (1991).

[16] R. S. Sikorski and J. D. Boeke, *Methods Enzymol.* **194,** 302 (1991).

[17] S. M. Hollenberg, R. Sternglanz, P. F. Cheng, and H. Weintraub, *Mol. Cell. Biol.* **15,** 3813 (1995).

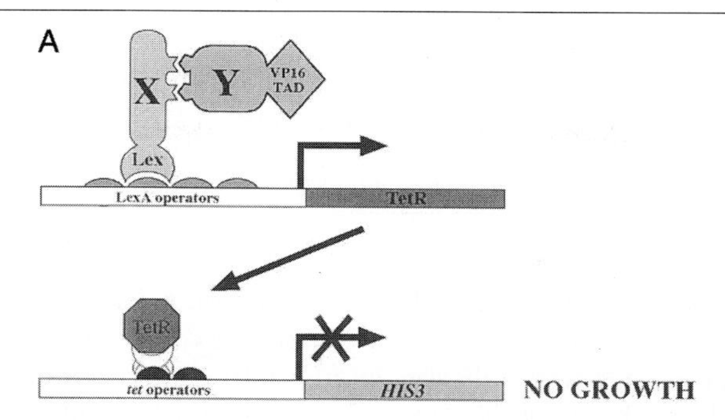

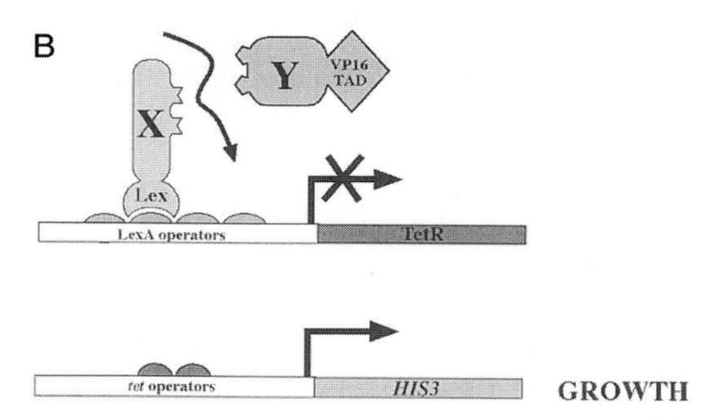

FIG. 2. The yeast split-hybrid system. (A) Interaction of protein X fused to LexA with protein Y fused to the VP16 TAD activates the expression of TetR. By binding to *tet* operators located in the promoter of the *HIS3* gene, TetR prevents the expression of the *HIS3* gene product, blocking the growth of yeast in media lacking histidine. (B) Disruption of the LexA–X and VP16–Y interaction prevents TetR expression, allowing *HIS3* expression and growth of yeast in media lacking histidine.

of YI584 is *MAT***a***/MATα, his3Δ200/his3Δ200 trp1-901/trp1-901 leu2-3, 112/leu2-3, 112 ade2/ade2 URA3*::(LexA operator)8-*Tet*R *LYS2*::(Tet operator)2::*HIS3*. YI671 is a *MATα* haploid derivative of YI154 that contains an *ADH1* construct driving expression of TetR. YI671 picked up a cryptic methionine mutation during construction and it is recommended that synthetic complete-based dropout medium be used rather than "add back" synthetic medium, which contains only the specific nutritional requirements for the strains.

Plasmids

The split-hybrid system is designed for use with *lexA* DNA-binding domain constructs. Various plasmids, such as pBTM116,[1] are compatible with the split-hybrid system. Plasmid pVP16-lacZ is used for mutagenesis experiments. Both VP16 and *GAL4* activation domains are amenable to use in the split-hybrid system. Plasmids and yeast strains are available from Bio 101.

General Methods

Yeast strains are transformed[10] and plated on selective medium plates. After 3 days of growth at 30°, transformants are diluted in 5 ml of selective medium, vortexed, and mildly sonicated for 10 sec to disrupt clumps. Cells are counted in a hemocytometer and seeded at 1000 cells/ml in selective medium. Tetracycline (Tc), 3-aminotriazole (3-AT), and histidine are supplemented as appropriate. Samples are incubated with shaking for 2 days at 30° and are quantitated by measuring the optical density at 600 nm (OD_{600}). The β-galactosidase (β-Gal) liquid assays are performed as described.[18]

Use of the System

The split-hybrid system was developed with several possible uses in mind: mutagenesis screening, screening for factors that abrogate protein associations, and identifying low molecular weight compounds that modulate specific protein interactions. To date the greatest utility of the system has been for mutagenesis studies[10,19] and for this reason we describe a general mutagenesis screen as an example of how to use the split-hybrid system.

Step 1: Plasmid Construction. The gene of interest should be cloned into a *lexA* fusion protein vector. If a two-hybrid screen was performed with the gene of interest, this may have already been performed. If a new construct is being prepared for split-hybrid use, we recommend pVP16-LacZ as the backbone for target gene constructs as the *lacZ* gene provides a screen for truncation or stop codon mutations.

Step 2: Identifying Optimal Growth Conditions for Yeast Transformants. YI584 and YI671 should both be transformed with target and bait plasmids. YI671 has a strong promoter driving TetR and is used for greater growth differences with weak interacting proteins. We recommend sequential transformations: i.e., introduce target plasmid and bait plasmids in independent experiments. This allows controlled isolation of strains so that all plasmid permutations can be obtained with minimal manipulations. Controls are

[18] H. J. Himmelfarb, J. Pearlberg, D. H. Last, and M. Ptashne, *Cell* **63,** 1299 (1990).
[19] J. D. Crispino, M. B. Lodish, J. P. MacKay, and S. H. Orkin, *Mol. Cell* **3,** 219 (1999).

essential for the split-hybrid characterization and should include transformations with empty vectors. Table I lists the desired plasmid combinations.

Transformants must be tested for the possibility of intrinsic *trans*-activation by the LexA fusion protein. Transformants listed in Table I should be tested on SD –His –Leu –Lys –Trp –Ura diagnostic medium. In the optimum situation, pLexA-YFG/T + pVP16-LacZ should show wild-type growth, whereas pLexA-YFG/T + pVP16-LacZ-YFG/B should not grow. Liquid cultures are recommended and wild-type growth is measured by using complete medium containing all nutritional requirements. If growth is compromised for pLexA-YFG/T + pVP16-LacZ, this indicates that pLexA-YFG/T can affect the *tet* gene and inhibit growth without disruption of pVP16-LacZ-YFG/B. To dampen this effect, YI584 should be compared with YI671 strains or system modulation with Tc and 3-AT is required. If pLexA-YFG/T + pVP16-LacZ-YFG/B shows growth in diagnostic medium, background must be suppressed with 3-AT. For optimum use of the split hybrid, growth of this pair must be kept low.

Step 3: System Modulation (Optional). If system modulation is required, transformants (Table I) are picked, suspended in SD –His –Leu –Lys–Trp –Ura medium, and diluted to 1000 cells/ml, and 100 μl is seeded into 96-well microtiter plates. Increasing amounts of tetracycline (0–25 μg/ml) and 3-AT (0–100 m*M*) are added (Fig. 3) and cultures with added histidine without drug are included. Cells are grown with gentle rocking and are examined at 1 and 2 days for growth. It is recommended that Ura^+ and Ura^- media be tested here, as differences are observed for certain interactions.

Step 4: Screening. After mutagenesis, the pLexA-YFG/T-containing strain is transformed with a mutagenized pool of pVP16-LacZ-YFG/B. Transformants are selected on SD –His –Leu –Lys –Trp –Ura medium

TABLE I

SPLIT-HYBRID PLASMID COMBINATIONS[a]

Transformant	Transformation medium
pLexA + pVP16-LacZ	SD –Leu –Lys –Trp –Ura
pLexA-YFG/T + pVP16-LacZ	SD –Leu –Lys –Trp –Ura
pLexA + pVP16-LacZ-YFG/B	SD –Leu –Lys –Trp –Ura
pLexA-YFG/T + pVP16-LacZ-YFG/B	SD –Leu –Lys –Trp –Ura

[a] YFG/T and YFG/B refer to your favorite gene target and bait, respectively. The medium described here is recommended for identifying transformants because the LexA operator is integrated at *URA3* and the *tet* operator is integrated at *LYS2*. Medium lacking uracil and lysine result in minimal background growth. SD –Leu –Trp can be employed but background can be higher.

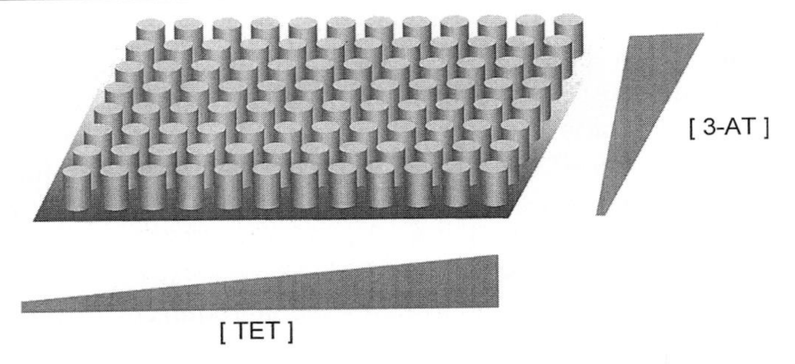

FIG. 3. Optimization grid. Yeast cells are plated into 96-well microtiter plates at a low cell concentration. Tetracycline and 3-AT are added at increasing concentrations on the X and Y grids and optimum growth is monitored visually, microscopically with a hemocytometer, or spectrophotometrically.

containing the appropriate concentration of tetracycline and 3-AT and grown for up to 5 days for colonies to form. Resulting colonies are screened for β-Gal activity and $LacZ^-$ colonies are excluded from further analysis. Plasmids are rescued into *Escherichia coli* from positive colonies and are retransformed into both a traditional two-hybrid system (e.g., using strain L40) and the split-hybrid strain. The nonmutagenized parental plasmids should grow in the two-hybrid system but not in the split-hybrid system, whereas the mutagenized plasmid should not grow in the traditional two-hybrid but grow in the split-hybrid system.

Discussion

The TetR protein and *HIS3* components provide two important points for regulation of the split-hybrid system. First, binding of the TetR protein to *tet* operators can be blocked with tetracycline (Tc). The ability to modulate TetR activity with Tc is useful when a LexA fusion protein is able to activate the expression of TetR (and thus inhibit yeast growth) in the absence of the VP16 fusion protein. In these cases, the LexA fusion protein is probably interacting directly with the yeast basal transcriptional machinery. However, it is unlikely that the basal *trans*-activating capabilities of these LexA fusion proteins are stronger than VP16. Therefore, the LexA fusion protein alone will produce lower amounts of TetR protein than is produced by the actual interaction of the LexA fusion protein with the VP16 fusion protein. Tc concentration can be titrated to suppress the activity of the TetR produced by the LexA fusion protein alone without suppressing the activity of the TetR produced by the actual LexA and VP16 interacting fusion proteins.

A second regulatory mechanism utilizes the drug 3-aminotriazole, an inhibitor of the *Saccharomyces cerevisiae* histidine pathway. When the interaction of two proteins is relatively weak, detection can be difficult. Because of the extreme sensitivity of yeast nutritional reporter genes (particularly *HIS3*), weak protein interactions often can be detected in the conventional yeast two-hybrid assay but not in other binding assays. In the split-hybrid system, weak interacting proteins may not produce sufficient TetR to fully shut down *HIS3* gene expression, resulting in background growth. In these situations, 3-AT can be used to supplement the TetR effects to shut off histidine production completely. This dual modulation of the system by Tc and 3-AT greatly expands the number of protein interactions that can be studied in the split-hybrid system.

There are several different classes of factors that might disrupt protein–protein interactions. Mutations that disrupt the binding of one protein with its interacting protein can be screened efficiently in the split-hybrid system. CREB was randomly mutagenized to identify residues involved in the binding to its coactivator Creb-binding protein (CBP).[10] The majority of mutations identified in this screen are located in the CREB Ser-133 PKA phosphorylation motif, which had previously been demonstrated to be necessary for CBP binding.[20–22] The other mutations identified are located at hydrophobic residues 137, 138, and 141. Subsequent nuclear magnetic resonance (NMR) studies have identified these hydrophobic CREB residues as points of contact with CBP.[22] These results highlight the extraordinary precision of using the split-hybrid system for mutagenesis screening.

The reverse two-hybrid system has also been used successfully for mutagenesis screens. Mutations in E2F1 were identified that disrupted its interaction with its dimerization partner DP1.[12] In addition, the reverse two-hybrid assay was used to identify dominant negative mutations of p53[23] whereas the split-hybrid system was used to identify altered specificity mutant in the GAT-1 : FOG complex.[19]

The split-hybrid and reverse two-hybrid systems can also be used to screen cDNA libraries for disrupters of protein–protein interactions. These screens may identify enzymes such as kinases, phosphatases, or proteases that disrupt a protein–protein interaction by covalently modifying one

[20] J. C. Chrivia, R. P. S. Kwok, N. Lamb, M. Hagiwara, M. R. Montminy, and R. H. Goodman, *Nature (London)* **365,** 223 (1993).

[21] R. P. S. Kwok, J. R. Lundblad, J. C. Chrivia, J. P. Richards, H. P. Bñchinger, R. G. Brennan, S. G. E. Roberts, M. R. Green, and R. H. Goodman, *Nature* (*London*) **370,** 223 (1994).

[22] I. Radhakrishnan, G. C. Pçrez-Alvardo, D. Parker., H. J. Dyson, M. R. Montminy, and P. E. Wright, *Cell* **91,** 741 (1997).

[23] R. K. Brachmann, M. Vidal, and J. D. Boeke, *Proc. Natl. Acad. Sci. U.S.A.* **93,** 4091 (1996).

of the interacting proteins. Proteins that act as competitive inhibitors or inhibitors that require the presence of both interacting proteins may also be identified. The utility of using the reverse two-hybrid system to screen for protein inhibitors of protein–protein interactions was demonstrated by the ability of E1A to inhibit the interaction of pRB and p107 with E2F.[11] Large combinatorial peptide libraries can also be screened in the split-hybrid or reverse two-hybrid systems to identify peptides capable of disrupting a protein–protein interaction. Because the peptide inhibitors are screened to block a specific protein–protein interaction, this method has significant advantages over using the two-hybrid system for identifying peptide inhibitors.

Another potentially important use of the split-hybrid and reverse two-hybrid systems is as a screen for small molecules that disrupt clinically important protein–protein interactions. Because the disruption event occurs *in vivo,* small molecules that are permeable and stable are preferentially selected. Furthermore, because the readout of the disruption is yeast growth, toxic small molecules are screened against. The ability to "weed out" impermeable, unstable, and toxic small molecules in the primary screen reduces the time consumed performing secondary screens in living cells that eliminate undesirable small molecule candidates.

Tetracycline can be used in split-hybrid small molecule screens as a positive control. For instance, predetermined concentrations of Tc can establish maximal yeast growth levels. In such a screen, as a small molecule inhibits an interaction between the LexA and VP16 fusion proteins, the level of TetR expression will be decreased, allowing an increase in yeast growth. If the screen is performed in liquid culture, the rate and extent of this yeast growth can be measured spectrophotometrically and compared with yeast growth in the presence of Tc. Small molecules that may nonspecifically promote yeast growth (Tc or histidine analogs; LexA or VP16 blockers) can be readily identified by rescreening in split-hybrid yeast expressing an unrelated protein interaction.

As an example, the reverse two-hybrid assay was modified to accommodate screens for small molecules that disrupt protein–protein interactions.[14] In this system, the expression of two interacting proteins is controlled with an inducible promoter (*GAL1*), ensuring that the small molecules are present in the yeast before the interacting proteins activate the expression of the toxic reporter gene. Inducible expression of the interacting proteins is not necessary in the split-hybrid system because TetR produced prior to the presence of drug can be suppressed with Tc (our unpublished observations, 1999).

Over time, the split-hybrid and reverse two-hybrid systems will likely undergo improvements and modifications in much the same way as the original yeast two-hybrid system. Logical extensions include one-hybrid systems that screen for factors that disrupt protein–DNA binding and adapting the systems to mammalian cells. These screens should provide valuable improvements to the understanding of cellular protein networks.

Section II

Other Approaches Using Chimeras for Identification and Analysis of Interactions of Proteins and Nucleic Acids

[11] Mapping Protein–Protein Interaction Domains Using Ordered Fragment Ladder Far-Western Analysis of Hexahistidine-Tagged Fusion Proteins

By RICHARD R. BURGESS, TERRANCE M. ARTHUR, and BRADLEY C. PIETZ

Transferring materials out of sodium dodecyl sulfate (SDS) gels (blotting) onto a nitrocellulose membrane has become a widely used technique. It not only takes advantage of the high resolving power of polyacrylamide gel electrophoresis but also allows ready access to the blotted target material by a variety of interaction probes. This approach started with the transfer of DNA[1] (termed a Southern blot after its inventor, Ed Southern), then the transfer of RNA (termed a Northern blot), and finally the transfer of protein[2] (termed a Western blot). It is now generally called a Western blot if the blotted protein is probed or detected with an antibody, and a south-Western blot if it is probed with a labeled DNA. More recently a technique called a far-Western blot[3] (or sometimes a west-Western) has gained increasing use. In a far-Western blot, instead of probing with an antibody, the probing is performed with another protein, taking advantage of specific protein–protein interactions. This approach requires that at least some region (the interaction domain) of a fraction of the blotted target protein be able to refold on the membrane and form a three-dimensional structure containing the interaction site. This approach is particularly useful in determining which subunit of a multisubunit complex is involved in an interaction with the probe protein.

In this chapter we describe our methods for determining interaction domains along a polypeptide chain by "ordered fragment ladder" far-Western analysis. Whereas far-Western analysis allows one to determine "What?" interacts with the probe, this method allows one to determine "Where?" This method evolved from our earlier work, in which a similar method was used to map monoclonal antibody epitopes by ordered fragment ladder Western analysis.[4] More recently we have published our first results using the ordered fragment ladder far-Western approach described

[1] E. M. Southern, *J. Mol. Biol.* **98,** 503 (1975).
[2] H. Towbin, T. Staehelin, and J. Gordon, *Proc. Natl. Acad. Sci. U.S.A.* **76,** 4350 (1979).
[3] P. M. Lieberman and A. J. Berk, *Genes Dev.* **5,** 2441 (1991).
[4] L. Rao, D. P. Jones, L. H. Nguyen, S. A. McMahan, and R. R. Burgess, *Anal. Biochem.* **241,** 173 (1996).

METHODS IN ENZYMOLOGY, VOL. 328
0076-6879/00 $30.00

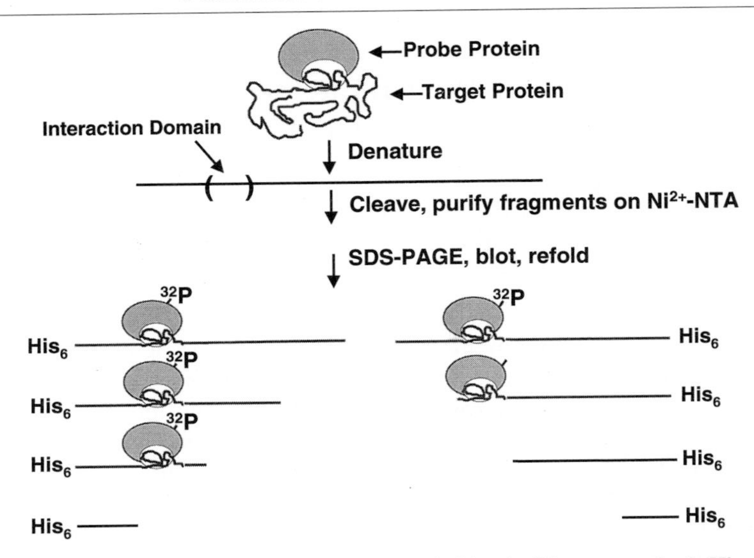

FIG. 1. A schematic of the ordered fragment ladder far-Western method. His_6-tagged target protein is cleaved, and the fragments are purified on an Ni^{2+}-NTA column, fractionated by SDS–PAGE, and electroblotted onto nitrocellulose. The denaturant is washed away from the blotted protein fragments, and the interaction domains on fragments are allowed to refold. These interaction domains can be identified by probing with a radioactively labeled protein. The interaction domain is mapped by identifying fragments that have part of their interaction domain missing and can no longer bind the probe. [Reproduced with permission from R. R. Burgess, T. M. Arthur, and P. C. Pietz, "Cold Spring Harbor Symposium," Vol. 63, p. 277. Cold Spring Harbor Laboratory Press, Cold Spring Harbor, New York, 1998.]

here.[5,6] This approach takes advantage of the ease of constructing hybrid proteins containing hexahistidine tags (His_6 tags) at either terminus and the unusual property of His_6-tagged proteins to bind to Ni-chelate columns, even in the presence of denaturants. To our knowledge this is the first description of a systematic approach to mapping interaction domains along a polypeptide chain by far-Western analysis. This method is potentially applicable to any protein–protein interaction study. A schematic that illustrates the principle of this method is shown in Fig. 1.

Overall Strategy

Ordered fragment ladder far-Western analysis is a general method for mapping an interaction domain on one protein that is necessary for binding

[5] T. M. Arthur and R. R. Burgess, *J. Biol. Chem.* **273,** 31381 (1998).

[6] R. R. Burgess, T. M. Arthur, and B. C. Pietz, *in* "Cold Spring Harbor Symposium," Vol. 63, p. 277. Cold Spring Harbor Press, Cold Spring Harbor, New York, 1998.

another protein. It involves several steps that are listed below. The italicized steps are described in detail in the following material.

A. Cloning and purification of a protein of interest with a His_6 tag fused to either the N terminus or C terminus
B. *Chemical or enzymatic partial cleavage* of the His_6-tagged protein to create a series of fragments
C. *Purification of His_6-tagged fragments on a Ni-chelate affinity column* under denaturing conditions to obtain a set of fragments all containing the His_6-tagged end
D. Fractionating these fragments on the basis of size by SDS–polyacrylamide gel electrophoresis (SDS–PAGE) to form an "ordered fragment ladder"
E. *Transferring the protein fragments out of the gel* and onto a nitrocellulose membrane and allowing these protein fragments to refold on the membrane
F. *Preparing a ^{32}P-labeled protein probe* by labeling a heart muscle protein kinase (HMK) recognition site-tagged protein with [γ-^{32}P]ATP and heart muscle protein kinase
G. *Probing the membrane with the labeled probe,* washing, and detection
H. Validating and characterizing the far-Western complex

A. Cloning and Purification of Protein with a Hexahistidine Tag

Standard cloning methods are used to place the gene of interest into an overproducing vector that puts a His_6 tag on either the N or C terminus of the protein. A variety of such vectors is available. We typically use pET vectors,[7] a T7 polymerase-based expression system, available from Novagen (Madison, WI). The appropriate overexpression strain is induced with isopropyl-β-D-thiogalactopyranoside (IPTG), the cells are lysed, and inclusion bodies are prepared as described.[5] The resuspended inclusion bodies are aliquoted into 1-mg portions and frozen at $-70°$ until use. It is not necessary that the target protein be refolded into its native conformation, because the next step is chemical or enzymatic cleavage, often under denaturing conditions.

B. Chemical and Enzymatic Cleavage of Proteins

The basic approach is to predict the chemical and enzymatic cleavage sites of an overproduced, His_6-tagged target protein and to generate partial

[7] F. W. Studier, A. H. Rosenberg, J. J. Dunn, and J. W. Dubendorff, *Methods Enzymol.* **185,** 60 (1990).

digests with chemical reagents or proteases. The amino acid sequence of a target protein can be entered into a computer program that predicts protein cleavage sites [e.g., there are such programs as part of the MacVector (Oxford Molecular, London, UK) or DNASTAR (Madison, WI) packages]. On the basis of the predicted pattern of cleavage, select one or more cleavage protocols. An example of the predicted cleavage sites for *Escherichia coli* RNA polymerase β' subunit for two chemical cleavage agents is shown in Fig. 2A.

Materials

Buffer B
- Tris-HCl (pH 7.9), 20 m*M*
- NaCl, 500 m*M*
- Imidazole (Fisher, Pittsburgh, PA), 5 m*M*
- Tween 20, 0.1% (v/v)
- Glycerol, 10% (v/v)

Urea buffer
- Buffer B + 8 *M* urea

Hydroxylamine buffer; For 15 ml, combine the following:

2-(*N*-cyclohexylamino)ethanesulfonic acid (CHES) buffer (pH 9.5)	(1 *M* final)
Hydroxylamine hydrochloride	(4 *M* final)
Adjust to pH 9.5 with 10 *M* NaOH	

Hydroxylamine hydrochloride (Aldrich, Milwaukee, WI)
Iodosobenzoic acid (IBA; Sigma, St. Louis, MO)
2-Nitro-5-thiocyanobenzoic acid (NTCB; Sigma)
Dithiothreitol (DTT)
Thermolysin from *Bacillus thermoproteolyticus* (Boehringer Mannheim, Indianapolis, IN)
Trypsin, tolylsulfonyl phenylalanyl chloromethyl ketone (TPCK) treated (Worthington Biochemicals, Freehold, NJ)

Cleavage Protocols

The following protocols for cleavage are routinely used in our laboratory. The conditions given work for most of the proteins we have studied and can be varied, by varying the time of cleavage or the dose of cleavage reagent, to accommodate proteins that are particularly easy or difficult to cleave. We aim for cleavage conditions generating as even as possible a distribution of fragments. This often is a reaction that leaves 10–30% of the polypeptide uncleaved.

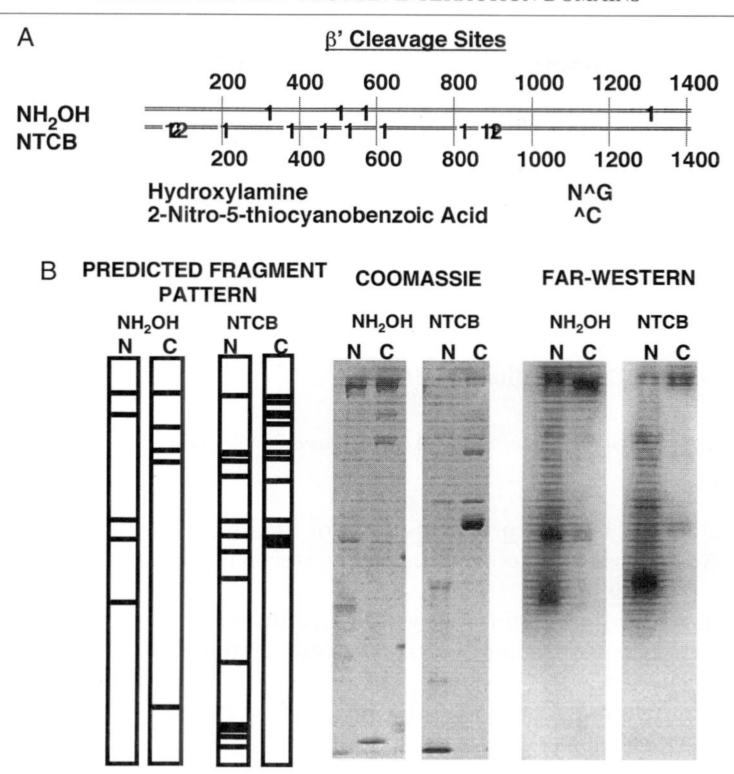

FIG. 2. Example of an order fragment ladder far-Western analysis. (A) A schematic indicating the positions of the chemical cleavage sites on the *E. coli* RNA polymerase β' subunit for chemical cleavage agents hydroxylamine (NH_2OH) and 2-nitro-5-thiocyanobenzoic acid (NTCB). These cleavage sites were predicted from the amino acid sequence using the MacVector program (Oxford Molecular Group). The numbers refer to the amino acid positions from the N terminus at the left. A "1" indicates the position of a cleavage site, whereas a "2" indicates two sites close together. (B) The ordered fragment ladder of N- and C-terminal His_6-tagged β' subunit cleaved with the two chemicals in (A). On the left is shown a schematic of the expected bands on an SDS gel. In the middle is the actual Coomassie-stained gel, and on the right is an identical gel blotted onto nitrocellulose and probed with ^{32}P-labeled σ^{70}. It can be seen from the NH_2OH cleavage fragment results that most of the N-terminal His_6-tagged fragments bind the probe, while only the full-length C-terminal His_6-tagged protein binds the probe. These results allow us to infer that the interaction domain is within the region from amino acid 1 to 309 of β'.[5]

Iodosobenzoic Acid Cleavage after Tryptophan

See Ref. 8.

1. Dissolve 1 mg of protein in 200 μl of 8 *M* guanidine hydrochloride.
2. Add 800 μl of 100% acetic acid, 3 μl of *p*-cresol, and 2 mg of IBA.
3. Incubate at room temperature for 20 hr.
4. Dry in a SpeedVac (Savant, Holbrook, NY) (about 1 hr).
5. Resuspend in 1 ml of urea buffer and adjust to pH 8.0 with 10 *M* NaOH (50–100 μl).
6. Load on an Ni column.

2-Nitro-5-thiocyanobenzoic Acid Cleavage before Cysteine

See Ref. 9.

1. Dissolve 1 mg of protein in 1 ml of urea buffer without glycerol.
2. Add a fivefold molar excess (to cysteine in protein) of DTT made fresh (1 *M* stock).
3. Incubate for 15 min at 37° to reduce disulfides.
4. Add a fivefold molar excess (over total cysteine) of NTCB and adjust to pH 9.5 with NaOH.
5. Incubate at room temperature for 2–6 hr for partial cleavage or for 24–30 hr for total cleavage.
6. Dilute 1 : 10 in urea buffer and load on an Ni column.

Hydroxylamine Cleavage between Asparagine and Glycine

See Ref. 10.

1. Dissolve 1 mg of protein in 1 ml of urea buffer.
2. Incubate for 15 min at 37°.
3. To 500 μl of urea-solubilized protein, add 500 μl of hydroxylamine buffer and incubate for 2 hr at 42°.
4. Add 7 μl of 2-mercaptoethanol (to 0.1 *M*), mix, and incubate for 15 min at 37°.
5. Dilute 1 : 10 in urea buffer and load on an Ni column.

[8] A. Fontana, D. Dalzoppo, C. Grandi, and M. Zambonin, *Methods Enzymol.* **91,** 311 (1983).

[9] G. R. Jacobsen, M. H. Schaffer, G. R. Stark, and T. C. Vanaman, *J. Biol. Chem.* **248,** 6583 (1973).

[10] P. Bornstein and G. Bolian, *Methods Enzymol.* **47,** 132 (1970).

Thermolysin Cleavage before Hydrophobic Amino Acids

See Ref. 4.

1. Resuspend 1 mg of inclusion body protein in 100 μl of urea buffer.
2. Incubate for 15 min at 37°.
3. Add thermolysin at protein:protease ratios of 4000:1, 8000:1, and 16,000:1 (w/w).
4. Digest for 30 min at room temperature.
5. Load on an Ni column.

Trypsin Cleavage after Arginine and Lysine

See Ref. 4.

1. Resuspend 1 mg of inclusion body protein in 1 ml of urea buffer.
2. Incubate for 15 min at 37°.
3. Dilute to 4 *M* urea by adding an equal volume of buffer B.
4. Add trypsin at protein:protease ratios of 4000:1, 8000:1, and 16,000:1 (w/w).
5. Digest for 30 min at room temperature.
6. Load on an Ni column.

Comments

1. Chemical cleavage ladders are useful in determining the precise size of fragments because it is known exactly where the cleavage occurs. This is particularly important because many proteins migrate abnormally on SDS–polyacrylamide gel electrophoresis.[4,11] The use of standard molecular weight markers would lead to serious errors in fragment size determination.

2. Because most chemical cleavage reagents produce only a few cuts per polypeptide chain, the ordered fragment ladder generated by chemical cleavage has only a few "rungs" on the ladder. Therefore, partial cleavage with one or more proteases can be used to create a ladder with more rungs that is capable of higher resolution mapping of interaction domains. Light, moderate, and heavy cleavage reactions can be performed with a given protease, and the resulting cleavage reactions mixed together before or after purification on the Ni-chelate column. This helps to produce a ladder containing similar amounts of each fragment size.

3. Sometimes it is difficult to generate a good ladder by partial cleavage methods, either because of a scarcity or uneven distribution of chemical cleavage sites or because the protein is relatively resistant to proteolysis.

[11] M. S. Strickland, N. E. Thompson, and R. R. Burgess, *Biochemistry* **27,** 5755 (1988).

In these cases it is also possible to produce ladders by cloning individual truncated fragments. This has proved necessary for our mapping of sites on *E. coli* RNA polymerase σ^{70}.[4,6]

C. Nickel-Chelate Column Purification of Hexahistidine Tagged Fragments

Materials

Ni^{2+}-NTA agarose (Qiagen, Chatsworth, CA)
Buffer B and urea buffer (see above)

Procedure

1. Load Ni^{2+}-NTA resin slurry into a Bio-Rad (Hercules, CA) mini-column to generate a 300-μl column bed.
2. Wash with 5 column volumes of Milli-Q (Millipore, Bedford, MA) water. All column operations are carried out at room temperature.
3. Wash with 5 column volumes of urea buffer.
4. Load the cleavage reaction (see above) and let it drain to the top of the resin.
5. Wash with 10 column volumes of urea buffer to remove non-His_6-tagged fragments.
6. Wash with 10 column volumes of buffer B to remove urea.
7. Elute with 500 μl of buffer B with 200 m*M* imidazole.
8. Check extent of cleavage by SDS–PAGE.
9. Store fragments as 50-μl aliquots frozen at $-20°$.

Comments

1. A key feature of Ni-chelate column purification is the ability of His_6-tagged proteins or fragments to bind to a Ni^{2+}-NTA column, even in the presence of 8 *M* urea or 6 *M* guanidine hydrochloride. Washing with a solution containing a denaturant prevents interactions between hydrophobic protein fragments and ensures that only His_6-tagged fragments are purified.
2. Once a set of ordered fragments is produced, they may be stored at -20 or $-70°$ for more than 1 year, until they are needed to map the binding of a monoclonal antibody or interacting protein.

D. Gel Electrophoresis

Use standard SDS–polyacrylamide gel electrophoresis procedures. We often use colored molecular weight markers [such as the Novex (San Diego,

CA) MultiMark multicolor standards] to aid in determining whether the transfer to nitrocellulose is efficient and to aid in cutting the nitrocellulose filter if probing with several different radioactive probes or antibodies. We often use prepoured 8–16% gradient polyacrylamide Tris–glycine gels (Novex) so that both large polypeptides, like the *E. coli* RNA polymerase β' subunit and its much smaller partial proteolysis fragments, can be visualized on the same gel.

E. Transfer of Protein Fragments from a Sodium Dodecyl Sulfate Gel to a Nitrocellulose Membrane

The proteins or peptides separated by SDS gel electrophoresis are electrophoretically transferred to nitrocellulose membrane as described below prior to either Western analysis or far-Western analysis.

Materials

Nitrocellulose membrane: Schleicher & Schuell (Keene, NH) Protran, 0.05 μm

Whatman paper: Whatman (Clifton, NJ) grade 3MM chromatography paper (Fisher)

Tris–Glycine (4×), pH 8.5: To make 1 liter, combine the following:

Glycine	57.6 g
Tris base	12.0 g

Towbin buffer (TB): To make 2 liters, combine the following:

Methanol	400 ml [20% (v/v) final]
Tris–glycine (4×)	500 ml (1× final)
SDS (10%, w/v)	10 ml [0.05% (w/v)]

TBST: To make 1 liter, combine the following:

Tris-HCl (pH 7.9), 1 *M*	10 ml (10 m*M* final)
NaCl, 4 *M*	37.5 ml (150 m*M* final)
Tween 20	1 ml [0.1% (v/v) final]

Blotto: Carnation nonfat dry milk, 2% (w/v) in TBST

Procedure

1. Cut one piece of nitrocellulose and two pieces of Whatman paper slightly larger than the gel.
2. Prewet one sponge and one piece of Whatman paper with TB.
3. Place the Whatman paper on top of the sponge, and then place the gel on top of the paper.
4. Wet the nitrocellulose and place it on the gel (avoid bubbles between gel and nitrocellulose) followed by wet Whatman paper and two sponges.

5. Place the resulting sandwich in the cage and put it into the transfer box with the nitrocellulose membrane toward the positive terminal (follow the "run it toward the red" rule).

6. Fill the transfer box with TB and transfer for 3 hr at a constant current of 200 mA (about 60 V).

7. Remove the nitrocellulose, place it protein side up in a petri dish, add 10–25 ml of Blotto to cover the blot, and block the membrane for 1–2 hr at room temperature with shaking or overnight at 4°.

8. For Western analysis: Wash once with TBST for about 30 sec at room temperature; incubate for 1 hr at room temperature in 10 ml of Blotto with a 1 : 1000 dilution of primary antibody; wash three times with 10 ml of TBST for 5 min each; incubate for 1 hr at room temperature in 10 ml of Blotto with a 1 : 1000 dilution of secondary antibody conjugated with horseradish peroxidase (HP) or alkaline phosphatase (AP); wash three times with 10 ml of TBST for 5 min each; and develop with an appropriate colorimetric or chemiluminescent detection reagent.

9. For far-Western analysis: proceed as described in Section G below.

F. Using Protein Kinase A to Label Proteins with ^{32}P for Far-Western Probing

We have found it convenient and satisfactory to use ^{32}P-labeled protein probes. If the five-amino acid recognition site for the catalytic subunit of the cAMP-dependent protein kinase A from heart muscle (RRASV) is attached to the terminus of a cloned protein, the protein can readily be labeled by reaction with [γ-^{32}P]ATP and protein kinase A.[12–14] Previously we constructed a cloning vector, based on the pET vector pET-28b(+) (Novagen), which contains the HMK recognition site and results in an N-terminal addition to a cloned protein of 25 amino acids.[15] The vector is available from Novagen as pET-33b(+). While pET-33b(+) has proved useful, we have constructed several additional vectors for the purpose of producing HMK site-tagged protein probes. These constructs contain either an *Nde*I or *Nco*I cloning site that allows the N-terminal methionine of the probe protein to be fused to a 13-amino acid N-terminal HMK-His_6 tag and provides the choice of either a Kan^R or an Amp^R antibiotic resistance marker. The relevant information about these vectors is summarized in the tabulation.

[12] B.-L. Li, J. A. Langer, B. Schwartz, and S. Pestka, *Proc. Natl. Acad. Sci. U.S.A.* **86,** 558 (1989).

[13] S. Pestka, L. Lin, W. Wu, and L. Izotova, *Protein Express. Purif.* **17,** 203 (1999).

[14] M. A. Blanar and W. J. Rutter, *Science* **256,** 1014 (1992).

[15] M. deArruda and R. R. Burgess, *inNovations* **4a,** 7 (1996). [Novagen Newsletter]

Vector name	Derived from:	N-Terminal tag	Cloning site	Resistance marker
pAP1	pET-28b	MA**RRASV**HHHHHH	*Nde*I	Kan^R
pAP2	pET-21a	MA**RRASV**HHHHHH	*Nde*I	Amp^R
pAP3	pET-32b	M**RRASV**HHHHHHA	*Nco*I	Amp^R

Preparation of HMK Recognition Site-Tagged Probe Protein

A BL21(DE3) *E. coli* strain,[7] containing the probe protein cloned into a suitable vector such as one of those described above, is cultured, induced, and the inclusion bodies purified as described.[5] The washed inclusion body may be solubilized with guanidine hydrochloride or the detergent sodium *N*-lauroylsarcosine (Sarkosyl) and refolded as described elsewhere.[16,17] Often the washed inclusion bodies are solubilized with 8 *M* urea and purified either before or after refolding by affinity chromatography on an Ni-chelate column.[6]

Materials

Protein kinase A (PKA): Catalytic subunit of bovine heart muscle protein kinase A (PKace kit; Novagen)

PKA buffer (4×; Novagen): 200 m*M* Tris-HCl (pH 8.0), 1.5 *M* NaCl, 200 m*M* $MgCl_2$, 100 μM ATP

[γ-^{32}P]ATP (NEN/Du Pont, Boston, MA): 6000 Ci/mmol, 5 mCi/33 μl; store at $-20°$

Labeling buffer (1× LB): 25% (v/v) glycerol, 40 m*M* Tris-HCl (pH 7.4), 100 m*M* NaCl, 12 m*M* $MgCl_2$, 0.1 m*M* DTT (added fresh)

Bio-Rad spin column (Bio-Spin P6): Just before use, vortex the column to resuspend the resin, remove the column bottom, and allow the column to drain. Add 1 ml of 1× labeling buffer and allow it to flow through by gravity. Spin in a Beckman (Fullerton, CA) TJ-6 centrifuge (TH-4 swinging bucket rotor) in a 50-ml conical plastic tube at 1000*g* for 2 min at room temperature and discard the flowthrough

Procedure

1. Add 5 μl of 10× PKA buffer to a 1.5-ml microcentrifuge tube.
2. Add 20–40 μg (about 500 pmol) of protein to be labeled [often stored

[16] R. R. Burgess, *Methods Enzymol.* **273,** 145 (1996).

[17] D. Marshak, J. Kadonaga, R. Burgess, M. Knuth, S.-H. Lin, and W. Brennan, "Strategies for Protein Purification and Characterization: A Laboratory Manual," pp. 205–274. Cold Spring Harbor Laboratory Press, Cold Spring Harbor, New York, 1996.

in 50% (v/v) glycerol] and bring the total volume to 43 μl with Milli-Q water. The final glycerol concentration should be 20–25%.

3. Add 5 μl of PKA (20-U/μl stock) and 2 μl of [γ-^{32}P]ATP (300 μCi = 6.6 $\times$ 10^8 dpm), mix, and incubate for 60 min at room temperature.

4. Add 50 μl of 1$\times$ LB to the reaction, add the resulting 100 μl of diluted reaction to a washed spin column, and spin for 4 min at 1000*g* at 20°.

5. Collect the flowthrough in a microcentrifuge tube.

6. Store the labeled probe frozen at −20° until use. It can be stored for up to 30 days.

Comments

1. Approximately 30–50% of the label is incorporated into protein.
2. After centrifugation in the spin column, about 90% of the label is in protein.
3. A typical labeling of HMK–σ^{70} at 35 μg in a 50-μl reaction gives about 1–4 $\times$ 10^6 cpm/μg.
4. The preceding protocol yields about 100 μl of material, suitable for probing 10–20 far-Western blots. The procedure can easily be scaled down twofold if desired.
5. It is also possible to label with [γ-^{33}P]ATP.[15] While this gives a lower specific activity, and thus a lower detection sensitivity, it does result in a labeled probe that has a longer half-life and that gives sharper bands on imaging.

G. Probing Far-Western Blots with ^{32}P-Labeled Protein

Materials

^{32}P-Labeled protein probe: Make according to the preceding protocol
Probe buffer (ProB): Store at 4° for at most 1 or 2 days

Stock	Final concentration	100 ml	500 ml
HEPES (pH 7.2), 1 *M*	20 m*M*	2 ml	10 ml
KCl, 2.5 *M*	200 m*M*	8 ml	40 ml
$MgCl_2 \cdot 6H_2O$, 1 *M*	2 m*M*	0.2 ml	1.0 ml
$ZnCl_2$ 0.25 *M*	0.1 m*M*	40 μl	0.2 ml
DTT, 1 *M*	1 m*M*	0.1 ml	0.5 ml
Tween 20, 100%	0.5% (v/v)	0.5 ml	2.5 ml
Nonfat dried milk	1% (w/v)	1 g	5 g
Glycerol, 50% (v/v)	10% (v/v)	20	100
Milli-Q Water		to 100 ml	to 500 ml

Procedure

1. Transfer proteins or fragments from the gel or spot proteins onto nitrocellulose membrane.
2. Block the membrane for 2 hr in ProB at room temperature with shaking (or overnight at 4°).
3. Add 5–10 μl of labeled probe (^{32}P-labeled protein) solution to 15 ml of ProB and incubate for 2 hr with the membrane at room temperature with shaking.
4. Wash the membrane three times for 3 min each with 10 ml of ProB.
5. Air dry the membrane (about 15 min), wrap in Saran wrap, and expose it to film or a PhosphorImager screen (Molecular Dynamics, Sunnyvale, CA).

Comments

1. This method is more powerful if ordered fragment ladders of both N- and C-terminally His$_6$-tagged versions of the target protein can be generated. In that way the interaction domain from both directions can be mapped. Figure 2B shows a predicted fragment pattern and a Coomassie blue-stained SDS gel for the ordered fragment ladders that result from cleaving both N-terminal (N) and C-terminal (C) His$_6$-tagged *E. coli* RNA polymerase β' subunit with hydroxylamine (NH_2OH) or with NTCB. The right-hand part of Fig. 2B shows the results of a far-Western analysis of an identical gel, probed with ^{32}P-labeled σ^{70}.

2. Ordered fragment ladder far-Western analysis requires that at least a fraction of the molecules in a blotted fragment band be able to refold at least that part of the polypeptide (the interaction domain) needed to create the three-dimensional interaction surface or interaction site. A number of reports have reported increased refolding and thus sensitivity in far-Western analysis when the blotted target protein is subjected to denaturation followed by renaturation prior to probing.[3,18] We have not yet found conditions under which we obtain significant improvement over merely blocking the membrane with probe buffer for at least 2 hr as described above. Presumably the SDS transferred with the blotted protein is removed (by becoming bound to the large excess of casein in the probe buffer), allowing the target protein to refold, at least partially, on the membrane.

3. Nitrocellulose pore sizes larger than 0.05 μm can be used; however, we find that 0.05-μm pores result in better retention of small protein fragments.

[18] C. R. Vinson, K. L. LaMarco, P. F. Johnson, W. H. Landschulz, and S. L. McKnight, *Genes Dev.* **2,** 801 (1988).

4. We have consistently observed that several different labeled probes bind nonspecifically to colored molecular weight markers, most likely an interaction between the His_6 tag or the HMK tag and the dyes that are attached to the markers. This, however, provides a useful set of labeled markers to help orient the resulting data.

H. Validating and Characterizing the Far-Western Complex

If a positive signal is detected in a far-Western analysis, it may still be necessary to provide additional evidence that the binding observed is due to a specific, relevant interaction and not merely to a nonspecific ionic or hydrophobic interaction. We have used several means to validate our results and to give us confidence that we are studying meaningful interactions. The major method has been the coimmobilization assay,[5,6] in which the putative interaction domain is cloned with a His_6 tag on one terminus and bound to an Ni-chelate column. If the probe (lacking a His_6 tag) binds to the immobilized target domain and elutes from the column when the target is eluted with imidazole, then it can be inferred that the two proteins interact. We have confirmed many far-Western results by coimmobilization assays. We have noticed that we can observe smaller target domains using the coimmobilization assay.[5,6] We attribute this effect to the fact that the refolded target domain is attached to the Ni-chelate column through interaction of its terminal His_6 tag and so can display the smallest functional interaction domain. In contrast, the blotted fragment must be attached by at least one or more contacts between the protein and the membrane. It requires that extra amino acids be present to allow binding without interfering with refolding of the minimal interaction domain on the membrane.

Another line of evidence to support the validity of the results of an ordered fragment ladder far-Western analysis is site-directed mutagenesis of the interaction domain on the target protein. We confirmed that the amino acid 260–309 region of the *E. coli* RNA polymerase β' subunit, the σ^{70} interaction domain identified by far-Western analysis, is involved in σ^{70} binding *in vivo*. We made mutations in this region that no longer show interaction in the far-Western analysis and that also fail to bind σ^{70}, while retaining the ability to assemble β' into core polymerase.[18a]

Of course, significant nonspecific interaction can be ruled out by demonstrating that the labeled probe will not give a signal above background when used to probe a blot of a number of proteins such as bovine serum albumin (BSA) or major bacterial proteins in a bacterial extract.

Finally, the nature of the probe–target complex can be partially charac-

[18a] T. Arthur, L. Anthony, and R. R. Burgess, *J. Biol. Chem.* **275,** 23113 (2000).

terized in an ordered fragment ladder far-Western analysis. We do this by probing as described in Section G and then washing the membrane with probe buffer for varying lengths of time and measuring the amount of labeled probe remaining bound to the membrane. In this way, the approximate half-life of the complex on the membrane can be determined. For example, the complexes between σ^{70} and fragments of β' dissociate with a half-life of about 2.5 hr. Similarly, the salt used during such a wash can be varied to determine the effect of salt on the rate of dissociation (T. Arthur and B. Pietz, unpublished results, 1999).

Conclusions

Requirements

To carry out the ordered fragment ladder far-Western analysis described in this chapter, it is necessary to have a clone that allows overexpression of a His_6-tagged target protein, the target gene sequence, and a suitable, labeled probe. The first two conditions are commonly satisfied due to the popularity of using His_6-tagged proteins and the rapid increase in genome sequence information available. The cloning of an HMK recognition site-tagged protein probe is now routine and rapid, employing the vectors described above in Section F.

Advantages of the Method

A positive result is useful provided it is shown to be specific. This is a rapid means of detecting binding and locating the region containing an interaction domain. It can focus more tedious mapping approaches, such as cloning individual truncated fragments or making multiple mutations, to a relatively small segment of the target polypeptide. The use of the ^{32}P-labeled protein probe allows relatively weak interactions to be detected. The probe can easily be labeled to more than 10^6 cpm/μg and the final wash after incubation of the blot with the probe and before exposure to PhosphorImaging can take as little as 5–10 min. In contrast, the detection of bound probe by immunological methods requires incubation with primary and secondary antibodies that can take several hours or more. The extended incubation and wash time can allow the probe to dissociate from the target.

Limitations of the Method

This method does not map the interaction site, but rather the whole region (the interaction domain) needed to form the contact surface or interaction site. A negative result is not very useful. An important interaction might not be detected if the interaction domain is difficult to refold or binds the probe protein too weakly. To detect binding, it is necessary to be within the "window of the assay," i.e., the half-life must be longer than the time of the wash. Convincing results are not obtained if the binding is weak, on the order of the nonspecific binding to random proteins or background. This method is ineffective if the interaction domain involves regions of two different-sized polypeptides and might not work if it involves two distant regions of the same polypeptide. It also would work poorly if there were a strong membrane-binding site in the middle of the interaction domain that prevented refolding on the membrane.

Applications and Potential Applications

1. *Epitope mapping of monoclonal antibodies:* This is an excellent method for moderate resolution mapping of epitopes. A suitable ordered fragment ladder is blotted to a membrane as described above and then probed with the monoclonal antibody (MAb) to be mapped (Refs. 4 and 19; T. Arthur and B. Pietz, unpublished results, 1999).

2. *Protein–protein interaction domain mapping* (described in this chapter): We have successfully used this method to map the binding site for *E. coli* RNA polymerase σ^{70} subunit to the amino acid 260–309 region of core polymerase subunit β'.[5] We have also used the β'_{1-309} fragment as a probe and mapped its binding to an interaction domain in region 2 of σ^{70}.[6,20] Preliminary positive results have been obtained in mapping the interaction domain for several other σ factors on the N-terminal region of β' (B. Pietz and T. Arthur, unpublished, 1999) and for GreB on the C terminus of β' (T. Arthur, unpublished, 1999). Far-Western analysis using labeled probes has detected interactions of human transcription factors TFIIB and RAP30 with several subunits of calf thymus and yeast RNA polymerase II (N. Thompson, unpublished results, 1999) and of several yeast RNA polymerase II subunits with other RNA polymerase II subunits (M. de Arruda and V. Svetlov, unpublished results, 1999).

3. *DNA- or RNA-binding site mapping:* While we have not yet done this, it seems likely that the probe used could be a labeled DNA or RNA molecule, instead of a protein. This would allow mapping of a specific nucleic acid-binding site along a polypeptide. The necessary precautions would have to be taken to determine that the binding is specific and not just nonspecific ionic interaction.

[19] L. Ellgaard, T. Holtet, S. Moestrup, M. Etzerodt, and H. Thogersen, *J. Immunol. Methods* **180,** 53 (1995).

[20] B. Pietz, in preparation (2000).

4. *Mapping sites of modification:* A slight variation of this approach should allow mapping sites of radioactive modification, such as phosphorylation or the labeled tag from a tag-transfer cleavable cross-linker,[21] along a polypeptide. The procedure would start with a His_6-tagged protein, allow the modification to occur, chemically or enzymatically cleave the target protein, isolate the His_6-tagged fragments on an Ni-chelate column, fractionate by SDS gel electrophoresis, transfer to a membrane, and expose to film or a PhosphorImager screen to determine at which point, moving down the ordered fragment ladder, the label is no longer detected.

Acknowledgments

We thank Monika deArruda, Nancy Thompson, Dale Jones, Lin Rao, Scott McMahan, and Anne Dunn for contributing to the early development of the present methods. Aspects of this work were refined during the Cold Spring Harbor Course on Protein Purification and Characterization. This work was supported by NIH Grants CA07175, CA23076, and GM28575, and by NIGMS predoctoral biotechnology training grant fellowships to T.A. and B.P.

[21] Y. Chen, Y. W. Ebright, and R. H. Ebright, *Science* **265,** 90 (1994).

[12] Mapping Specificity Determinants for Protein–Protein Association Using Protein Fusions and Random Peptide Libraries

By Michael B. Yaffe and Lewis C. Cantley

Introduction

Many signaling proteins and small modular domains such as SH2, SH3, 14-3-3, WW, PTB, and PDZ domains interact with protein ligands by recognizing short linear stretches of amino acids. Determining the precise amino acid sequence that constitutes the binding motif for these modular signaling elements can be a technical challenge, and a variety of techniques are available for this purpose including phage display,[1] one-peptide/one-pin type techniques[2,3] and alanine-scanning mutagenesis of the target proteins.

[1] G. P. Smith, *Science* **228,** 1315 (1985).

[2] H. M. Geysen, R. H. Meloen, and S. J. Barteling, *Proc. Natl. Acad. Sci. U.S.A.* **81,** 3998 (1984).

[3] S. P. Fodor, J. L. Read, M. C. Pirrung, L. Stryer, A. T. Lu, and D. Solas, *Science* **251,** 767 (1991).

0076-6879/00 $30.00

Oriented peptide library screening is a technique that differs from those described above primarily by the rapidity with which sequence motifs can be determined for a protein or domain of interest.[4–9] Once an immobilized form of the protein or domain is available, usually as a glutathione *S*-transferase (GST) or maltose-binding protein (MBP) fusion, soluble peptide library screening can be performed in an afternoon, and a motif determined within 1–2 days. An additional advantage of the soluble peptide library technique is that the results provide a quantitative measure of the binding affinity for any amino acid within the deduced binding motif. This information can then be used to (1) identify the binding site within a known protein target for the protein or domain that was screened, (2) search a protein or nucleic acid database to identify unrecognized proteins or open reading frames (ORFs) that are likely to interact with the domain of interest, (3) design optimal peptide ligands for biophysical studies and structural analysis of protein : peptide complexes using multidimensional nuclear magnetic resonance (NMR) or X-ray crystallography, (4) construct peptidomimetic inhibitors of the protein or domain for general use in cell biological studies and as lead compounds for drug design, and (5) rationalize the effects of known or engineered mutations on interactions between the domain and target proteins. This latter application may be particularly relevant to understanding the genetic basis for certain human diseases.

Principles

The theoretical basis for oriented peptide library screening is depicted in Fig. 1. To apply the technique, it is helpful to know the identity of a single amino acid required for high-affinity binding to the domain of interest, although this can ultimately be deduced by testing libraries oriented

[4] S. Zhou, B. Margolis, M. Chaudhuri, S. E. Shoelson, and L. C. Cantley, *J. Biol. Chem.* **270,** 14863 (1995).

[5] M. B. Yaffe, M. Schutkowski, M. Shen, X. Z. Zhou, P. T. Stukenberg, J. U. Rahfeld, J. Xu, J. Kuang, M. W. Kirschner, G. Fischer, L. C. Cantley, and K. P. Lu, *Science* **278,** 1957 (1997).

[6] M. B. Yaffe, K. Rittinger, S. Volinia, P. R. Caron, A. Aitken, H. Leffers, S. J. Gamblin, S. J. Smerdon, and L. C. Cantley, *Cell* **91,** 961 (1997).

[7] Z. Songyang, A. S. Fanning, C. Fu, J. Xu, S. M. Marfatia, A. H. Chishti, A. Crompton, A. C. Chan, J. M. Anderson, and L. C. Cantley, *Science* **275,** 73 (1997).

[8] Z. Songyang, S. E. Shoelson, J. McGlade, P. Olivier, T. Pawson, X. R. Bustelo, M. Barbacid, H. Sabe, H. Hanafusa, T. Yi, R. Ren, D. Baltimore, S. Ratnofsky, R. A. Feldman, and L. C. Cantley, *Mol. Cell. Biol.* **14,** 2777 (1994).

[9] Z. Songyang, S. E. Shoelson, M. Chaudhuri, G. Gish, T. Pawson, W. G. Haser, F. King, T. Roberts, S. Ratnofsky, R. J. Lechleider, B. G. Neel, R. B. Birge, J. E. Fajardo, M. M. Chou, H. Hidesaburo, B. Schaffhausen, and L. C. Cantley, *Cell* **72,** 767 (1993).

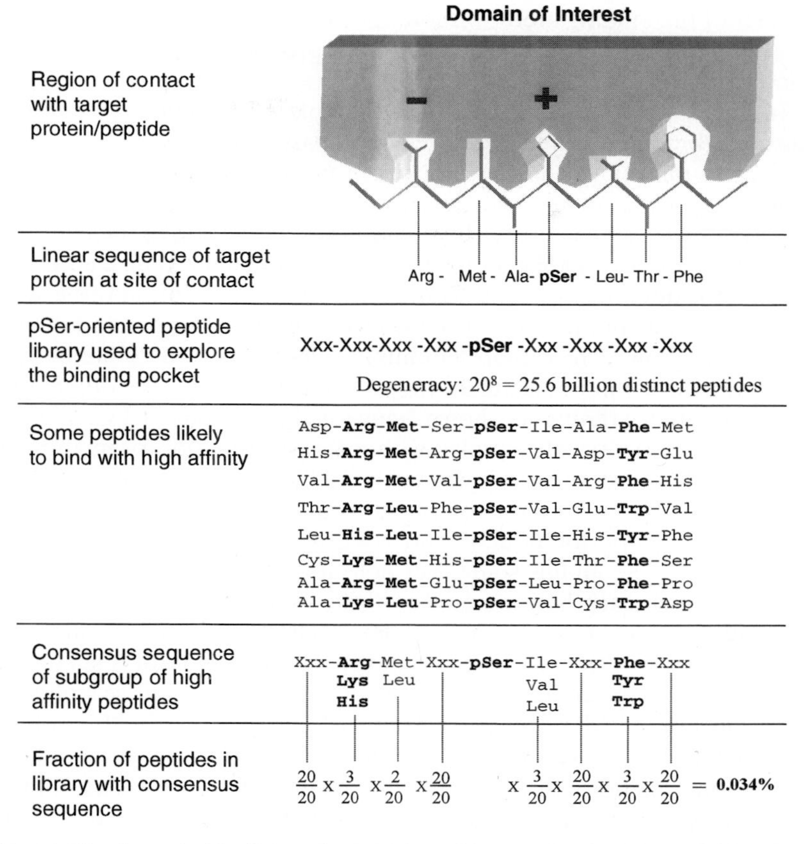

FIG. 1. The theoretical basis for selection of peptides contained within a soluble oriented peptide library by a binding protein or modular signaling domain. Note that the orienting phosphoserine residue in this example maintains all of the bound peptides in register for evaluation by successive rounds of N-terminal sequencing. In this example 1 mg of peptide library (~1 μmol of total peptide) would contain about 340 pmol of peptides with high affinity for the domain of interest.

with individual amino acids. This information might be obtained by site-directed mutagenesis experiments, genetic screening studies, or sequence comparisons between known ligands prior to peptide library screening. In some cases, such as 14-3-3 proteins and SH2 domains, it is sufficient to know only in a general sense that a single amino acid is critical for the interaction, particularly if that amino acid binds to the domain only after a posttranslational modification such as phosphorylation. In such cases, phosphotyrosine or phosphoserine libraries can be used to derive the remainder of the binding motif, without the necessity of knowing beforehand the sequence of even a single bona fide protein target of the domain of interest.

In Fig. 1 we show a hypothetical example of a domain composed of individual binding pockets that recognize basic, linear and branched aliphatic, phosphoacidic, and aromatic contributions on the ligand. This information, of course, is not known beforehand, but emerges only after the analysis; however, it is useful for illustrating the general principle of the technique. Consider a collection of peptides in which the single amino acid essential for interaction with the domain is always present, in this example a phosphoserine, but positions both C terminal and N terminal to the fixed residue are allowed to contain equimolar amounts of all 20 amino acids. A subset of these peptides might be expected to bind to the domain with reasonably high affinity, as shown beneath the domain in Fig. 1. In this example, the fraction of peptides within the starting library mixture that interact with the domain of interest would be ~0.034% of all possible peptides present. Observed experimental values typically fall within 0.02–0.4%, depending on the domain and orienting residue(s) selected.

Each peptide contains some, but not all, of the residues preferred for interaction with each binding pocket of the domain. The entire ensemble of peptides, however, is sufficient to define all of the optimal amino acids in each position within the binding motif. The inclusion of a single fixed residue that is sufficient to dominate the interaction, in this case a phosphoserine, ensures that all peptides bind in register with respect to this orienting amino acid. Thus, by sequentially degrading the entire collection of bound peptides one position at a time, starting from the N terminus, using standard Edman degradation sequencing, it is possible to determine the relative amounts of each individual amino acid within each position in the bound peptides (Fig 2). By normalizing the amount of each amino acid at a particular sequencing cycle in the bound material to that present in the initial peptide mixture, preference values for each amino acid within each position of a binding motif are obtained.

Experimental Procedures

Materials

Chemicals

- Standard reagents for peptide synthesis and protein sequencing.
- Glutathione beads (Sigma, St. Louis, MO) or amylose beads (New England BioLabs, Beverly, MA)
- Acetic acid, 30% (v/v)

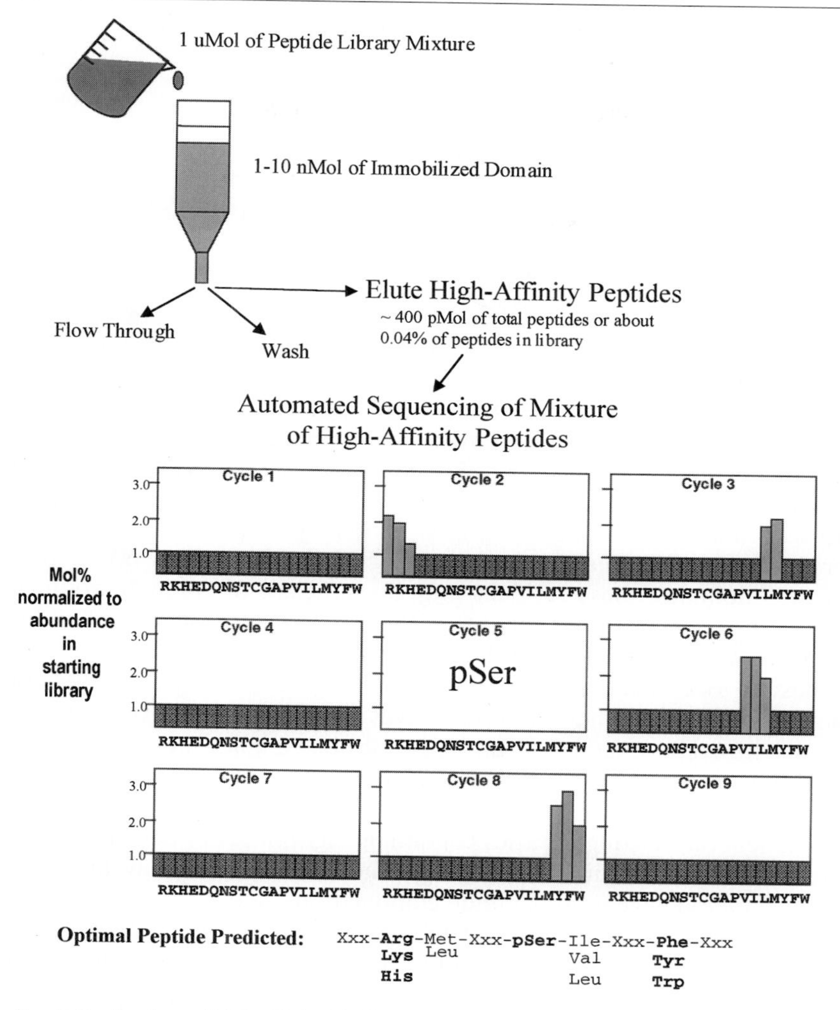

FIG. 2. Purification of high affinity binding peptides using immobilized proteins or modular signaling domains, and determination of the binding motif by successive cycles of Edman sequencing. The example shown is the same as that in Fig. 1.

Phosphate-buffered saline (150 m*M* NaCl, 3 m*M* KCl, 10 m*M* Na_2HPO_4, 2 m*M* KH_2PO_4) with and without 0.5% (v/v) nonidet P-40 (NP-40)

Equipment

Peptide synthesizer (431A; Applied Biosystems, Foster City, CA)

and protein sequencer (Procise, single or multicartridge; Applied Biosystems)
Polypropylene columns or syringes with caps
Glass wool
SpeedVac (Savant, Holbrook, NY)
Microcentrifuge
Microcentrifuge tubes containing nylon filters (Costar Spin-X filters)

Design and Synthesis of Degenerate Peptide Libraries

The selection of one or more residues around which to fix and orient a degenerate peptide library is the most critical decision in the experimental design, and ultimately will determine the success or failure of the technique. Libraries oriented on single residues often suffice when that residue is both critical for the interaction, and makes a dominant contribution to binding through strong electrostatic or ionic contributions, such as phosphotyrosine residues in the case of peptide binding to SH2 domains. In other cases, where unusual amino acids show strong selection by virtue of conformational preferences, such as proline in the case of peptide binding to WW domains or SH3 domains, these residues may be sufficient to uniquely orient the library. For domains where neither of these criteria is satisfied, i.e., for those domains whose binding results from several distributed weak interactions, it is best to lock in several fixed amino acids spread among the degenerate positions to ensure that all peptides will bind in register. For example, a domain thought to bind to an amphipathic helix might contain an orienting Leu, Ile, or Val residue placed every three or four amino acids within the sequence.[10]

In general, starting libraries containing eight or fewer degenerate positions work best for initial screening. These libraries already have more than 25 billion degenerate peptide combinations, and the fixed amino acid residue can be positioned within the center of the degenerate stretch, or along either end. Incorporating more than 10 degenerate positions in the starting library will increase the initial complexity of the peptide mixture; however, it can simultaneously decrease the relative fraction of peptides within the starting mixture that bind with high affinity, yielding inadequate amounts for sequencing. Longer stretches of degenerate positions are possible, however, in secondary libraries (see below). Libraries are routinely constructed beginning with the sequence Met-Ala, and terminated after the last degenerate position with the sequence Ala-Lys-Lys-Lys. A typical library used to deduce interactions with novel SH2 domains, for example, contained the sequence Met-Ala-Xxx-Xxx-Xxx-Xxx-pTyr-Xxx-Xxx-Xxx-

[10] D. Parker, U. S. Jhala, I. Radhakrishnan, M. B. Yaffe, C. Reyes, A. I. Shulman, L. C. Cantley, P. E. Wright, and M. Montminy, *Mol. Cell* **2,** 353 (1998).

Xxx-Ala-Lys-Lys-Lys,[10a] where Xxx indicates all amino acids except Cys. The Met-Ala sequence at the N terminus of the peptide libraries allows verification that peptides from this mixture are being sequenced, and provides a quantitative estimate of peptide binding. The N- and C-terminal Ala residues bracketing the degenerate positions allow estimation of how much peptide has been lost during sequencing (i.e., repetitive yield) and give an indication of cleavage lag and carryover between sequencing cycles. The polylysine tail ensures solubility of the libraries and aids in retention of the peptides on the Biobrene (Applied Biosystems)-coated filters during the sequencing steps.

Once an initial binding motif has been determined, it is often helpful to construct secondary libraries to further refine and expand individual amino acid preferences. For example, an initial screening using a phosphoserine (pSer)-based library might reveal a strong preference for Pro in the pSer + 1 position. A library in which both the pSer and Pro were fixed could then be constructed. Because the general affinity for any random peptide within this pSer-Pro library is higher than that for a peptide from a library oriented on pSer alone, weaker interactions from residues bracketing the pSer-Pro core can now be used to provide the selectivity for higher affinity peptides. In the primary screening performed with the pSer-only library, information about these interactions was lost, because the binding was largely dominated by selection for Pro in the pSer + 1 position.

Solid-phase synthesis of the degenerate peptide libraries is performed with a Rink amide resin matrix (American Peptide, Sunnyvale, CA) and an automated peptide synthesizer (Applied Biosystems 431A) according to standard benzotriazole-1-yl-oxy-tris(dimethylamino)-phosphoniumhexafluorophosphate/1-hydroxybenzotriazole (BOP/HOBt) coupling protocols with commercially available 9-fluorenylmethoxycarbonyl (Fmoc)-protected amino acids. Side-chain protected forms of phosphotyrosine, phosphothreonine, and phosphoserine are now available, as are a variety of interesting and unusual nonnatural amino acids that can be used for orienting the library. When nonnatural amino acids are placed in a fixed position within a library, we extend the coupling time to ensure good incorporation. At the degenerate positions, equal amounts of Fmoc-blocked amino acids (except Cys) are weighed out and loaded into the synthesizer cartridges so that the mixture is in fourfold molar excess to the coupling resin. The ratio of Fmoc-blocked amino acids needs to be adjusted slightly on different synthesizers in order to obtain a roughly even distribution of degenerate amino acids. Once the synthesis is complete, the degenerate peptide libraries are deprotected and cleaved from the resin for 3 hr at room temperature,

[10a] F. Poy, M. B. Yaffe, J. Sayos, K. Saxena, M. Morra, J. Sumegi, L. C. Cantley, C. Terhorst, and M. J. Eck, *Mol. Cell* **4,** 555 (1999).

using 4 ml of a cleavage reagent prepared by dissolving 0.75 g of phenol in 10 ml of trifluoroacetic acid, and adding 0.5 ml of H_2O, 0.5 ml of thioanisole, and 0.25 ml of ethane dithiol. The crude library mixture is precipitated by slow addition into 50 ml of diethyl ether followed by cooling to −20° for 1 hr. The precipitate is filtered with a fretted funnel, washed six times with chilled diethyl ether, dissolved in H_2O, and lyophilized. The final peptide libraries are stored as dry powders at −20°. Working stocks of the libraries are prepared at 10–30 mg/ml in phosphate-buffered saline (PBS), adjusted by addition of dry HEPES or Tris to pH ~7.0, and a small amount of the library sequenced to ensure that all amino acids are present at similar amounts (within a factor of 3) at each of the degenerate positions.

Fusion Protein Domain Construction, Preparation of Immobilized Protein Beads, and Library Screening

Fusion proteins with the domain of interest placed C terminal to a gluthathione *S*-transferase or maltose-binding protein tag for immobilization are generated by standard recombinant DNA procedures, using pGex (Pharmacia, Piscataway, NJ) or pMal (New England BioLabs) expression vectors. As a general starting point for expression, the constructs are introduced into DH5α or BL21(DE3) strains of *Escherichia coli,* and 500- to 2000-ml cultures are induced by the addition of isopropylthiogalactoside (IPTG) to 0.4 m*M* when the OD_{600} has reached 0.6–0.8. Induction times of 3–5 hr at 37° suffice for most proteins. Sodium dodecyl sulfate–polyacrylamide gel electrophoresis (SDS–PAGE) analysis of total bacterial extracts from 1-ml aliquots before and after induction is used as a guide to adjust the induction times, temperatures, and bacterial strains to optimize protein production. After induction, the bacteria are pelleted and stored frozen at −20°.

It is essential that the immobilized protein be present at high concentrations on the bead matrix in order for peptide library screening to be successful, because the major determinants of specificity between high- and low-affinity peptides is the off rate.[6] Consequently, the higher the density of immobilized proteins, the better one can discriminate between weak- and tight-binding peptides during the wash steps. We routinely aim for immobilizing 1 mg of a GST fusion protein on 100 μl of glutathione beads, or 1 mg of MBP fusion protein on 300 μl of amylose beads. The bacterial pellet is resuspended in 7.5 ml of a 50 m*M* Tris (pH 7.5), 150 m*M* NaCl, 1 m*M* EDTA, 5 m*M* dithiothreitol (DTT) solution supplemented to 4 μg/ml with leupeptin and aminoethylbenzene sulfonic acid (AEBSF) along with other protease inhibitors as necessary. Forty microliters of a 50-mg/ml solution of lysozyme is added, and the suspension is allowed to incubate for 20 min on ice. The cells are disrupted either by repeated rounds of sonication on ice, or by the addition of 10 mg of deoxycholic acid (sodium salt) and incubation at room temperature for 15 min. The lysate is then supplemented to 10 m*M* $MgCl_2$, 200 units of DNase I is added, and the sample is incubated for an additional 10 min at room temperature. The suspension is clarified by centrifugation at 12,000*g* for 15 min at 4°, and the supernatant is incubated with 100 μl of GSH beads or 300 μl of amylose beads at 4° in 15- or 50-ml disposable centrifuge tubes (Corning, Corning, NY) with continual mixing. The binding of MBP fusion proteins to amylose beads is rapid, requiring only a 30-min incubation. In contrast, maximal binding of GST fusion proteins to GSH beads requires 60–180 min (Fig. 3).

After binding, the beads are collected by centrifugation at ~300*g* for 1 min at 4°, and washed three times with 10–25 ml of PBS + 0.5% (v/v) NP-40 and then twice with PBS in the absence of detergent. After the final wash, the beads are resuspended in a minimal volume of PBS and loaded into small disposable minicolumns (Bio-Rad, Hercules, CA) or 1-cm^3 TB syringes plugged with glass wool. The beads are washed 3 times with 1 ml of PBS and allowed to drain by gravity. Using long gel loading tips and a pipettor, 0.5–1 mg of the peptide library solution is applied and allowed to enter the bead bed, and the column is then capped and left at room temperature for 10 min. The beads are then rapidly washed twice with 1 ml of ice-cold PBS + 0.5% (v/v) NP-40 followed by two washes with 1 ml of ice-cold PBS. It is important that the washes be done quickly, generally

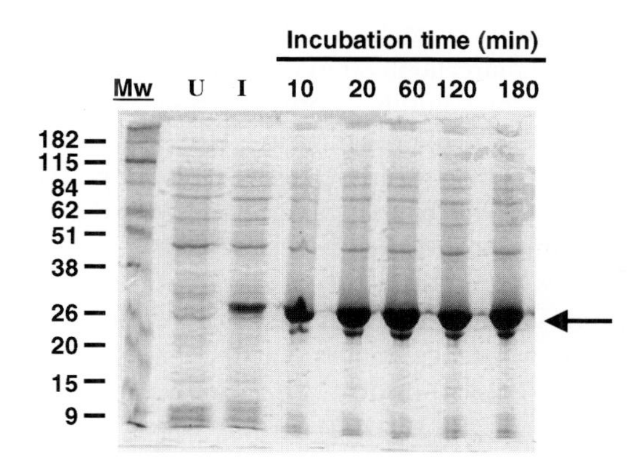

FIG. 3. Time dependence of binding of GST fusion proteins to glutathione beads. Lanes U and I show crude bacterial lysates before and after IPTG induction, respectively. The protein, in this case a 30-kDa WW domain–GST fusion, is evident at 30 kDa in the induced sample. Maximal binding to GSH beads requires 60–180 min of incubation. The amount shown in each lane represents approximately 1/50 of the total amount immobilized on the 100 μl of beads used for library screening.

within 60–120 sec total elapsed time, in order to retain enough peptide for sequencing, particularly during the primary screening as the binding constants for first-round peptides are likely to be relatively large. To facilitate rapid washing we use an air-filled 60-cm^3 syringe whose tip has been fashioned to make a gas-tight seal with the top of the minicolumn, using the air cushion to drive the wash solution through the beads. Once the final wash has been completed, the column is capped and the beads are resuspended in 200–600 μl of the eluent for 10 min at room temperature. The eluent is collected and evaporated to dryness on a SpeedVac apparatus (Savant). We routinely perform a control experiment simultaneously by performing library screening on beads containing immobilized GST alone. The entire procedure, from lysis of the bacterial pellet until sample dry-down on the SpeedVac, can be accomplished within a single afternoon.

The choice of elution agent depends on the domain to be screened. For SH2 domains, specific elution can be obtained with a 20 m*M* sodium phenylphosphate solution adjusted to pH 7.2 to specifically compete with phosphotyrosine binding.[8] In these cases, after evaporation in the SpeedVac, the recovered peptides can be resuspended in 80 μl of H_2O, and 20–40 μl used directly for amino acid sequencing.

No alternative specific eluent has proved suitable for most other domains, including the inability of phosphoserine to compete effectively for binding to domains that recognize phosphoserine-based peptides. Therefore, a general strategy is to elute both the peptides and the fusion protein from the beads, using 30% (v/v) acetic acid in water. This treatment results in irreversible denaturation of the fusion protein. Resuspension of the SpeedVac-evaporated material in 80 μl of water for 5 min at room temperature with gentle vortexing solubilizes only the bound peptides. After water reconstitution, the tubes are microcentrifuged for 2 min and the supernatant carefully removed from the insoluble fusion protein pellet. If small flecks of the fusion protein prove troublesome to remove, the supernatants can be recentrifuged in microcentrifuge tubes fitted with 0.22-μm nylon filters (Costar Spin-X tubes) for 1 min to collect a fusion protein-free filtrate. Control experiments have verified that this filter treatment results in negligible peptide loss with no change in amino acid composition before and after filtration. Twenty- to 40-μl aliquots are routinely used for sequencing.

Peptide Sequencing and Data Analysis

The recovered peptides are analyzed by standard automated amino acid sequencing. It is also important to sequence the initial peptide library mixture. Samples can be sequenced by any commercial sequencing facility, using any sequencing protocol sufficient to detect individual amino acids at the 5- to 10-pmol range. In our own laboratory, we spot the peptides on Biobrene-coated glass-fiber filter disks and load the samples into an Applied Biosystems Procise sequencer. A pulsed-liquid solvent delivery cycle is used with standard times for phenylisothiocyanate (PITC) coupling, extraction, cleavage, and transfer along with a 9-min phenylthiohydantoin (PTH) conversion at 64°. The PTH-derivatized amino acids are then detected by high-performance liquid chromatography (HPLC) with a gradient optimized for separation between amino acids, particularly between Phe, Ile, Lys, and Leu, which elute late. For best accuracy, amino acid standards should be prepared weekly, and the calibration run preceding each sequencing reaction performed with amounts of each PTH amino acid reasonably close to what is expected in the actual peptide library sequencing. The yield of amino acids per cycle in the actual experiment should typically exceed that in the GST control by at least a factor of 4.

To determine the optimal peptide motif selected by the protein domain that was screened, the abundance of each amino acid at a given cycle in the sequence of the bound peptide mixture is divided by the abundance of the same amino acid in the same cycle of the starting mixture. This corrects for variations in the abundance of particular amino acids at particular residues in the starting library as well as variations in yield of amino acids during sequencing. This can easily be performed in spreadsheet fashion by normalizing each amino acid in a sequencing cycle to its mole percentage and dividing by the corresponding mole percentage of that amino acid in the initial library. To scale the relative preferences among all amino acids present in the degenerate library, these raw preference values are then summed and normalized to the total number of amino acids in the degenerate position (typically 19, as Cys is omitted from the libraries). Graphic plots of normalized preference values versus amino acid for each sequencing cycle (Fig. 2) are useful for revealing the optimal peptide motif. If the domain has no specificity at a particular residue position within the motif, then the mole percentage of all amino acids in this position will be the same as that present in the initial mixture, and the preference values for all amino acids will be ~1. In identifying amino acids that are selected at particular positions within a motif, pay special attention to those amino acids that change from cycle to cycle, rather than those that remain persistently slightly elevated or depressed in relative abundance. This latter effect usually results from small systematic errors in sequencing for a particular residue between the sample and the starting library mixture.

On occasion a domain may give a weak signal on an initial screen that can sometimes be deconvoluted by either subtracting the background binding as estimated from the GST control sample screened simultaneously with the protein of interest, or by normalizing the mole percentages of amino acids in each cycle for the sample to those in the GST control. Because these motifs are more likely to contain errors, they should be independently confirmed by secondary library screening in which additional residues deduced from the primary screened are placed in fixed positions.

Using the Optimal Domain Motif to Elucidate Protein Interactions and Cell Signaling Pathways

Once a motif has been determined for the domain or protein of interest, it can be used to identify the site of interaction on proteins that are known to bind to the target protein, or to identify new potential interacting proteins via database searching. A variety of software available over the internet allows rapid searching for the motif in either protein or translated nucleic acid databases, including such programs as Ross Overbeek's PatScan (www.unix.mcs.anl.gov/compbio/PatScan/HTML/patscan.html), FASTA[11] (www2.ebi.ac.uk/fasta3), and BLAST[12] (www.ncbi.nlm.nih.gov/BLAST). It is also possible to perform a restricted search for the motif in a subset of proteins having homology to a known target (or to any other protein) using Pattern Hit Initiated Blast (*PHI-BLAST*). In database searching it is important to allow for partial matches to the motif in scoring query sequences to avoid missing interacting proteins, because many proteins that bind will match at many but not all residue positions within the motif. A program that will perform a quantitative ranking of likely targets within the SWISS-PROT database using user-entered peptide library data is currently under development.

Comments

The peptide library screening approach offers a rapid quantitative method for evaluating the contribution of each amino acid to a binding motif recognized by a protein or modular domain. The final motif derived from the screening can be extremely helpful in identifying proteins likely to interact with the target protein that was screened, selecting optimal peptides for binding and inhibition studies, and designing reagents for structural studies of the domain of interest. We routinely synthesize several of the optimal peptides predicted by oriented peptide library screening to independently measure their binding constants by surface plasmon resonance or fluorescence polarization anisotropy, and believe that all investigators using this technique should do likewise. A variety of variations on the

[11] W. R. Pearson and D. J. Lipman, *Proc. Natl. Acad. Sci. U.S.A.* **85,** 2444 (1988).
[12] S. F. Altschul, W. Gish, W. Miller, E. W. Myers, and D. J. Lipman, *J. Mol. Biol.* **215,** 403 (1990).

technique can be devised, including the incorporation of nonnatural amino acids and fractionated washes of increasing stringency to separate bound peptides into various classes on the basis of their binding affinities.

Troubleshooting

Failure of the Protein to Bind to Peptides in the Library

Most commonly, failure of the protein to bind to peptides in the library is the result of failure to select the correct orienting residue(s) in designing the peptide library. For the technique to work, the protein or domain being screened must recognize and bind a short linear stretch of amino acids. Proteins that bind to a motif that is distributed throughout the sequence of a target protein, or that bind to extended sequence regions greater than 10–15 amino acids long, are unlikely to give good results by this technique. Proteins that recognize motifs within a particular three-dimensional fold, such as an α helix, can give reasonable results with this technique, but often require the inclusion of fixed amino acids within the library to favor these peptide conformations.[10]

If peptide library screening fails to give a motif, examine the insoluble protein pellet obtained after acetic acid elution and water reconstitution by resuspending the pellet in Laemmli sample buffer followed by SDS–PAGE analysis. This ensures that the target protein was present in sufficient amounts on the beads during screening. The addition of certain cofactors such as metal ions to stabilize the target protein and/or the removal of certain ions or salts that may compete with peptide binding (i.e., high concentrations of phosphate) can occasionally result in peptide library binding that was previously undetectable.

Persistent High Abundance of One Amino Acid during Sequencing, Particularly Glycine

Persistent high abundance of one amino acid during sequencing is occasionally observed even in well-prepared samples. All buffers should be examined for the presence of the contaminant. It is important that all buffers used in the wash steps after peptide binding lack primary amine groups (i.e., exclude Tris, etc.), which is a frequent cause of this problem. Increasing the filter wash step during peptide sequencing prior to the initial PITC coupling can prove helpful. Irregardless of the source, the contaminating amino acid can often be neglected in the analysis, providing useful data from the experiment albeit lacking information about the omitted amino acid. On occasion, impurity peaks eluting in the serine or threonine position arise in late sequencing cycles.

Significant Amounts of Peptide Binding but No Clearly Discernible Motif

The absence of a clearly discernible motif in the context of significant peptide binding usually indicates that the fixed residue within the library is sufficient to confer binding but not strong enough to orient the bound peptides. Consequently, peptides bind but are out of register, obscuring the presentation of the motif when analyzed by N-terminal sequencing. Locking in additional fixed residues within the library to maintain the alignment between different peptides during binding should remedy this problem.

Tailing Effects for Strongly Selected Residues

When a residue is strongly selected in a particular sequencing cycle, it is common for that residue to also show some selectivity in the following cycle. This "tailing" effect results from a combination of carryover of PTH-derivatized material from the previous cycle and cleavage lag of the peptides on the filter during Edman degradation. Consequently, although the selectivity value is still greater than 1 in the cycle following one showing strong selection, it is significantly decreased, and can therefore be neglected in motif analysis. Some domains, however, truly select for the same residue(s) in multiple consecutive cycles, for example, selecting a motif such as pYLLL. This is evident as a persistently high selectivity value that is stable between cycles, and also shows significant elevation when compared with cycles in which it was not selected. Recognizing these types of tandem selectivities can be difficult, and verification of selection should be confirmed by examining the binding constants for individual peptides.

Sequencing Glutathione S-Transferase

On occasion a clear signal emerges from the library screen with a single dominant residue in each sequencing position. A particularly unfortunate result is the "motif" MSPILGYWK. . . since this is the N-terminal sequence of GST! The absence of Ala in the second cycle is an immediate tip-off that this problem has occurred. GST contamination results from inadvertent solubilization of the fusion protein along with the bound peptides following the acetic acid elution and water reconstitution step. Samples with GST contamination can occasionally be rescued by spinning them through microcentrifuge tubes fitted with nylon filters. This problem can usually be avoided by gentle handling of the acetic acid precipitate along with strong centrifugation following resuspension of the peptides in water.

Acknowledgment

We thank German Gaston Leparc for help with figure construction.

[13] Selection of Genetic Agents from Random Peptide Aptamer Expression Libraries

By C. RONALD GEYER and ROGER BRENT

Peptide Aptamers as Genetic Agents

Most cellular processes are regulated by networks of interactions among proteins. Classic "forward" transmission genetic approaches have succeeded in identifying proteins involved in such networks and provide information on the relationships between them. Approaches such as (1) isolation of mutants and mapping the genes responsible for mutant phenotypes, (2) epistasis analysis to determine the order in which gene products might act on each other, (3) dependency analysis to determine the relationship between gene products with respect to the completion of a cellular process,[1] and (4) allele-specific suppression analysis to give clues as to the possible physical interaction between two proteins[2] provide detailed information about genes and their positions and interactions within networks. Although these approaches are often highly informative, they are often difficult to perform, especially in diploid organisms. Their full application is thus generally limited to a few tractable organisms such as phage, bacteria, yeast, worms, and flies.[3]

The ability to manipulate individual genes by recombinant DNA approaches allowed the development of "reverse" genetic techniques, in which investigators mutate the function of individual genes and monitor the resulting phenotypes. Reverse genetic approaches that mutate or knock out the function of genes at the DNA level suffer from the following limitations: (1) if the gene under study is essential, then mutations in it will yield nonviable organisms, but often no other information; (2) gene knockouts provide information about the phenotypic consequences of abolishing all the protein interactions in which the knockout gene product participates; (3) gene knockouts typically provide information related only to those phenotypes caused by the initial function of the gene in the development of the organism; and (4) in diploid organisms, as for forward genetic analysis, the observation of recessive phenotypes requires the generation of organisms homozygous for particular mutations, typically in the second generation.

[1] L. M. Hereford and L. H. Hartwell, *J. Mol. Biol.* **84,** 445 (1974).
[2] P. E. Hartman and J. R. Roth, *Adv. Genet.* **17,** 1 (1973).
[3] J. H. Nadeau and P. J. Dunn, *Curr. Opin. Genet. Dev.* **8,** 311 (1998).

0076-6879/00 $30.00

More recently, dominant reverse approaches have been devised that affect gene products rather than genes. These dominant reverse approaches rely on generating "mutagenic" agents that inactivate gene products in *trans* without affecting their coding DNA. Examples of dominant genetic agents include small molecule inhibitors,[4] dominant negative proteins,[5] injection of antibodies,[6] antisense RNA,[7] ribozymes,[8] and nucleic acid aptamers.[9] These methods are particularly useful for studying gene function in diploids; however, they too have various weaknesses. Dominant negative proteins and small molecule inhibitors may not exist for all gene products, thus restricting their universality. Antibodies, antisense RNA, ribozymes, and nucleic acid aptamers can in principle be generated to inactivate almost any gene product. Antibodies, however, are not cell permeable and their injection into cells is time consuming and not practicable in all organisms. RNA-based agents are generally not stable in intracellular environments. Although RNA stability can be partly overcome by using RNAs with different chemistry, it remains difficult to predict the sites on the target RNA that are exposed for antisense inhibition. Also, the stability of the protein product of the antisense and ribozyme target gene affects the timing (perdurance) and extent (penetrance) of the mutant phenotype. Moreover, like DNA mutagenesis, mutagenesis at the RNA level results in phenotypes that are caused by the abolishment of all protein interactions in which the target gene product is involved.

A new class of dominant agents has been developed to facilitate the analysis of processes in diploid and genetically intractable organisms. We termed these molecules "peptide aptamers," because of their similarity to nucleic acid aptamers.[10] We define peptide aptamers as antibody-like recognition agents that consist of conformationally constrained peptides displayed on the surface of scaffold proteins, and distinguish such molecules from peptides of variable sequence displayed that are not constrained at both ends by the scaffold. Peptide aptamers are designed to inhibit cellular processes by interacting with proteins and disrupting their biological functions. Combinatorial libraries of peptide aptamers in principle contain aptamers that bind almost any protein target. Peptide aptamers specific for numerous proteins have been isolated by using the yeast two-hybrid system.

[4] T. J. Mitchison, *Chem. Biol.* **1,** 3 (1994).
[5] I. Herskowitz, *Nature* (*London*) **329,** 219 (1987).
[6] G. J. Gorbsky, R. H. Chen, and A. W. Murray, *J. Cell Biol.* **141,** 1193 (1998).
[7] A. D. Branch, *Trends Biochem. Sci.* **23,** 45 (1998).
[8] B. Bramlage, E. Luzi, and F. Eckstein, *Trends Biotechnol.* **16,** 434 (1998).
[9] M. Thomas, S. Chedin, C. Carles, M. Riva, M. Famulaok, and A. Sentenac, *J. Biol. Chem.* **272,** 27980 (1997).
[10] A. E. Ellington and J. Szostak, *Nature* (*London*) **346,** 818 (1990).

Using this strategy, we and others have isolated peptide aptamers against Cdk2,[11] Ras,[12] human immunodeficiency virus type 1 (HIV-1) Rev,[13] E2F[14] as well as linear peptides against Rb,[15] and mutagenized PDZ domains against C-terminal peptides.[16] Peptide aptamers recognize their protein targets with K_d values and half-inhibitory concentrations between 10^{-6} and 5×10^{-11} *M*. The fact that peptide aptamers are typically selected *in vivo* by two-hybrid methods increases the likelihood that the molecules will function *in vivo*. In support of this notion, peptide aptamers function as reverse dominant agents in both mammalian cells[17,18] and *Drosophila melanogaster*.[19]

We and others have also used peptide aptamers as dominant agents for the forward "genetic" analysis of cellular processes.[20–22] In this approach, combinatorial libraries of peptide aptamers are used as random "mutagens" that interfere with cellular processes. To make this work, the investigator (1) introduces combinatorial expression libraries of peptide aptamers into cells, (2) selects cells that carry peptide aptamers that cause a desired phenotype, and (3) identifies the targets of the peptide aptamers. Target identification is performed by screening the aptamers for interactions against panels of proteins or products of cDNA libraries, using the two-hybrid system. This combination of aptamer-mediated "mutagenesis" with the two-hybrid system to identify aptamer targets provides a general method for characterizing processes in a variety of hitherto genetically intractable organisms.

Here we describe how to obtain and characterize peptide aptamers for both reverse and forward analysis of cellular processes. We describe methods to isolate aptamers against specific proteins with the yeast two-hybrid system as well as methods to identify the targets of aptamers isolated from

[11] P. Colas, B. Cohen, T. Jessen, I. Grishina, J. McCoy, and R. Brent, *Nature (London)* **380,** 548 (1996).

[12] C. W. Xu, A. Mendelsohn, and R. Brent, *Proc. Natl. Acad. Sci. U.S.A.* **94,** 12473 (1997).

[13] B. Cohen, Ph.D. thesis. Harvard University, Boston, Massachusetts, 1998.

[14] E. Fabbrizio, L. Le Cam, J. Polanowski, R. Brent, and C. Sardet, *Oncogene* **18,** 4357 (1999).

[15] M. Yang, Z. Wu, and S. Fields, *Nucleic Acids Res.* **23,** 1152 (1995).

[16] S. Schneider, M. Buchert, O. Georgiev, B. Catimel, M. Halford, S. A. Stacker, T. Baechi, K. Moelling, and C. M. Hovens, *Nature Biotechnol.* **17,** 170 (1999).

[17] B. A. Cohen, P. Colas, and R. Brent, *Proc. Natl. Acad. Sci. U.S.A.* **95,** 14272 (1998).

[18] Xu *et al.,* submitted (2000).

[19] M. G. Kolonin and R. L. Finley, Jr., *Proc. Natl. Acad. Sci. U.S.A.* **95,** 14266 (1998).

[20] G. Caponigro, K. R. Abedi, A. P. Hurlburt, A. Maxfield, W. Judd, and A. Kamb, *Proc. Natl. Acad. Sci. U.S.A.* **95,** 7508 (1998).

[21] C. R. Geyer, A. Colman-Lerner, and R. Brent, *Proc. Natl. Acad. Sci. U.S.A.* **96,** 8567 (1999).

[22] T. C. Norman, D. L. Smith, P. K. Sorger, B. L. Drees, S. M. O'Rourke, T. R. Hughes, C. J. Roberts, S. H. Friend, S. Fields, and A. W. Murray, *Science* **285,** 591 (1999).

forward selections. We also illustrate how to use the two-hybrid system to define the specificity of aptamers, to isolate mutant aptamers with enhanced affinity for their targets, and to obtain aptamers specific for allelic variants of proteins. We provide protocols that detail the necessary reagents and techniques to perform the above-described work. While a basic knowledge of yeast and molecular biology techniques is required, descriptions of reagents and techniques not described in this review can be obtained from basic molecular biology manuals.[23,24]

Design of Intracellular Peptide Aptamers

In Vitro Selection Methods

A number of *in vitro* selection methods have been developed to express randomly encoded peptides (and peptide aptamers) from combinatorial libraries in order to isolate peptides that bind specific targets. These methods range from displaying peptides on the surface of phage,[25] yeast,[26] DNA-binding proteins,[27] polysomes,[28] and covalently attached RNA[29] to displaying bona fide peptide aptamers on the main flagellar protein of *Escherichia coli.*[30] These methods allow isolation of peptides that bind extracellular proteins and proteins immobilized on solid supports. The main advantage of using *in vitro* selection methods to isolate such peptides is the size of the random libraries that can be generated. Currently, libraries containing 10^{10} members are readily obtainable.[31]

The tactics for displaying peptides on the surface of phages, bacteria, or yeast, or in association with the coding RNA, however, all suffer from the following limitations. First, with the exception of the flagellar display of peptide aptamers, the displayed peptides are typically unconstrained

[23] F. M. Ausubel, R. Brent, R. E. Kingston, D. D. Moore, J. G. Seidman, and K. Struhl (eds.), "Current Protocols in Molecular Biology." Greene & Wiley-Interscience, New York, 1992.

[24] J. Sambrook, E. F. Fritsch, and T. Maniatis (eds.), "Molecular Cloning: A Laboratory Manual." Cold Spring Harbor Laboratory Press, Cold Spring Harbor, New York, 1989.

[25] G. P. Smith and V. A. Petrenko, *Chem. Rev.* **97,** 391 (1997).

[26] E. T. Boder and K. D. Wittrup, *Nature Biotechnol.* **15,** 553 (1997).

[27] G. M. Gates, W. P. C. Stemmer, R. Kaptein, and P. J. Schatz, *J. Mol. Biol.* **255,** 373 (1996).

[28] L. C. Mattheakis, R. R. Bhatt, and W. J. Dower, *Proc. Natl. Acad. Sci. U.S.A.* **91,** 9022 (1994).

[29] R. Roberts and J. W. Szostak, *Proc. Natl. Acad. Sci. U.S.A.* (1997).

[30] Z. Lu, K. S. Murray, V. Van Cleave, E. R. LaVallie, M. L. Stahl, and J. M. McCoy, *BioTechnology* **13,** 366 (1994).

[31] J. L. Harrison, S. C. Williams, G. Winter, and A. Nissim, *Methods Enzymol* **267,** 83 (1996).

and of low affinity. Second, none of these selection methods guarantee that the peptide or peptide aptamer will be functional in the cellular environment. For example, peptides may be toxic, they may not bind their target under cellular conditions, and the target may not have the proper conformation or display the same surfaces available for binding. Finally, most *in vitro* selections take place in oxidizing environments, in which disulfide bonds, which are sometimes used to constrain peptides *in vitro,* can form. The cytoplasm and the nucleus are reducing environments, in which disulfide bonds do not typically form.

In Vivo Selection Methods

Peptide aptamers can be directly selected to interact with proteins within cells by the two-hybrid system[11,32] or by genetically screening for aptamers that cause particular phenotypes.[20–22] Because these selections take place inside the cell, the aptamers are by definition active and nontoxic. However, when the selection is performed in yeast or in mammalian cells, the number of aptamer-expressing cells that can be easily surveyed ($\sim 10^6$–10^7) is typically smaller than the number of peptides that can be surveyed by *in vitro* methods. Libraries of this lower complexity still possess significant biochemical diversity[33] and have been successfully used in both phage display[34,35] and *in vivo* selections.[11,15]

Properties of Intracellular Aptamers

The selection of peptide aptamers within cells restricts the types of scaffold proteins and variable region constraints that can be used. Constrained peptide libraries generally produce peptide aptamers with higher binding affinities and greater specificity relative to unconstrained peptide libraries.[11,36] For *in vitro* selections, the variable region peptide is often constrained either by cyclization through the formation of disulfide bonds[37] or by its immobilization on a protein scaffold.[38] For selections in the reducing environment of a cell, peptides must either be unconstrained or constrained by means other than disulfide bonds. At least in *E. coli,* unconstrained peptides tend to be unstable. For example, only 5% of the members of a random unconstrained peptide library are able to direct the synthesis

[32] Deleted in proof.

[33] M. E. DeGraaf, R. M. Miceli, J. E. Mott, and H. D. Fischer, *Gene* **128,** 13 (1993).

[34] M. A. McLafferty, K. A. Kent, R. C. Ladner, and W. Markland, *Gene* **128,** 29 (1993).

[35] G. P. Smith, D. A. Schultz, and J. E. Ladbury, *Gene* **128,** 37 (1993).

[36] R. Ladner, *Trends Biotechnol.* **13,** 426 (1995).

[37] E. Koivunen, B. Wang, and E. Rouslahti, *BioTechnology* **13,** 265 (1995).

[38] N. Per-Ake and M. Uhlen, *Curr. Biol.* **7,** 463 (1997).

TABLE I
SCAFFOLDS USED FOR DISPLAYING RANDOM PEPTIDES

Scaffold	Target	K_d	IC_{50} (nM)	Ref.
Thioredoxin	Cdk2	30–120 nM	5	11
	Ras	—	100–500	12
	E2F	—	0.05	14
PDZ domain	Peptide	200 nM	—	16
Linear peptide	Rb	10–75 μM	—	15

of observable protein in *E. coli*.[39] Moreover, higher proteolytic stability is expected for peptides that are displayed from a folded structure relative to a linear or unordered peptide. For these reasons, we prefer to use constrained peptide libraries for intracellular selections.

Other desirable properties for aptamer scaffolds include the following: (1) The scaffold should possess good solubility properties, in order to help prevent aggregation of aptamers that contain hydrophobic variable regions; (2) the scaffold should be small, stable, easily purified, and expressed at high levels without toxicity; and (3) the scaffold should be tolerant of modifications such as localization sequences, epitope tags, purification tags, and other protein moieties that allow the aptamer valency to be increased [e.g., aptamers can be dimerized by fusing them to glutathione *S*-transferase (GST)].[18] In the future, we imagine that individual selections will benefit from the use of promiscuous protein scaffolds that can provide additional functional groups for enhanced binding interactions, and through the development of scaffolds carefully designed to display variable regions of different lengths, with different distances between the constrained ends of the variable region, and with different distances between the variable region and the body of the scaffold protein.

Scaffolds Used for Intracellular Aptamers

To date, only a few different scaffolds have been used to display constrained or unconstrained variable regions inside cells. In two-hybrid selections, *E. coli* thioredoxin[11] and PDZ domains[16] have been used as scaffolds, and the Gal4 activation domain has been used to display linear peptides.[15] A comparison of the binding constants shows that the aptamers, with their constrained variable regions, bind their targets between 100- and 10,000-fold better than linear peptides (see Table I). For direct phenotypic selec-

[39] A. R. Davidson and R. T. Sauer, *Proc. Natl. Acad. Sci. U.S.A.* **91,** 2146 (1994).
[40] Deleted in proof.

tions, thioredoxin,[21] green fluorescent protein (GFP),[20] and staphylococcal nuclease[22] have been used as aptamer scaffolds. Because the dissociation constants for these aptamers were not measured, it is difficult to compare them. However, the fact that the selected aptamers function in two-hybrid experiments suggests that binding constants for all these aptamers are likely to be tighter than the 10^{-6} *M* threshold for these reporters.

Properties of Thioredoxin Aptamer Proteins

In this review we focus on the properties and uses of *E. coli* thioredoxin (TrxA) as a scaffold protein for constrained variable regions. TrxA was originally used as a scaffold by J. McCoy and co-workers to display peptides on the surface of *E. coli* as a fusion to flagellin.[30] Using this approach, TrxA aptamers were obtained against a number of monoclonal antibodies[41] and against protein phosphatase 1.[42]

TrxA possesses many desirable properties that make it an effective intracellular scaffold protein. TrxA is a small (12-kDa) stable cytoplasmic protein that can be expressed at high levels in cells without toxicity.[43] Structural studies have shown that the TrxA active site consists of a short disulfide-constrained peptide loop (–CGPC–).[44] This loop is tolerant to insertion of peptides[43] and thus provides a site to introduce peptides of random sequence. TrxA enhances the solubility of proteins fused to it,[43] suggesting that it will enhance the solubility of peptides that would normally aggregate. TrxA can interact with many different protein substrates, so long as those contain disulfide bonds.[45] The ability of TrxA to interact with many different proteins suggests that it may bear functional groups that make some contacts with the proteins recognized by different aptamer variable regions.

Method 1: Construction of a Thioredoxin Peptide Aptamer Library

Reagents

Vectors

pJM-1: Peptide aptamer vector for use with the interaction trap
pJM-2: Peptide aptamer vector for use in yeast genetic selections

[41] Lu *et al.,* 1995.
[42] S. Zhao and E. Y. Lee, *J. Biol. Chem.* **7,** 28368 (1997).
[43] E. R. LaVallie, E. A. Diblasio, S. Kovacic, K. L. Grant, P. F. Schendel, and J. M. McCoy, *BioTechnology* **11,** 187 (1993).
[44] S. K. Katti, D. M. LeMaster, and H. Eklund, *J. Mol. Biol.* **212,** 167 (1990).
[45] B. Wetterauer, M. Veron, M. Miginiac-Maslow, P. Decottignies, and J. P. Jacquot, *Eur. J. Biochem.* **209,** 643 (1992).

pJM-3: Peptide aptamer vector containing a nuclear localization signal for use in yeast genetic selections

See Fig. 1 for a detailed description of the above-described vectors.

Construction of a Random Thioredoxin Peptide Aptamer Library. Thioredoxin peptide aptamer libraries are constructed by inserting an oligonucleotide encoding a random 20-amino acid peptide into an *Rsr*II restriction site that occurs naturally in the DNA encoding the TrxA active site loop. This generates a solvent-exposed random 20-amino acid peptide constrained by the active site of TrxA.

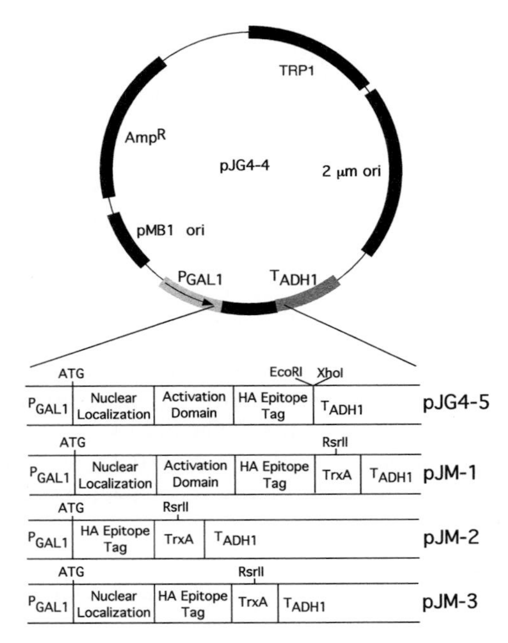

FIG. 1. Expression vectors used for interaction trap and forward genetic selection of peptide aptamers. The yeast–*E. coli* shuttle vectors are derived from pJG4-4.[47] They contain an *E. coli* origin of replication (pMB1 ori), an ampicillin resistance gene (Amp^R), a *TRP1* marker gene, a yeast 2 μm origin of replication, and one of the following expression cassettes: pJG4-5, yeast *GAL1* promoter (P_{GAL1}), SV40 nuclear localization signal, B42 activation domain and hemagglutinin epitope tag, *Eco*RI and *Xho*I cloning sites, and yeast *ADH1* transcription terminator (T_{ADH1})[47]; pJM1, yeast *GAL1* promoter (P_{GAL1}), SV40 nuclear localization signal, B42 activation domain and hemagglutinin epitope tag, TrxA, and yeast *ADH1* transcription terminator (T_{ADH1})[11]; pJM2, yeast *GAL1* promoter (P_{GAL1}) and hemagglutinin epitope tag, TrxA, and yeast *ADH1* transcription terminator (T_{ADH1}) (J. McCoy, personal communication, 1999; and Ref. 21); pJM3, yeast *GAL1* promoter (P_{GAL1}), SV40 nuclear localization signal and hemagglutinin epitope tag, TrxA, and yeast *ADH1* transcription terminator (T_{ADH1}) (J. McCoy, personal communication, 1999; and Ref. 21).

PROTOCOL 1-1: CONSTRUCTION OF RANDOM PEPTIDE APTAMER LIBRARY

1. Choose one of the three TrxA peptide expression vectors (pJM-1, pJM-2, or pJM-3) shown in Fig. 1. pJM-1 is designed for use in the yeast interaction trap two-hybrid system.[11] pJM-2 and pJM-3 are, respectively, designed for genetic selections of nuclear-localized and nonlocalized peptide aptamers in yeast.[21]

2. Digest the appropriate pJM vector (Fig. 1) with *Rsr*II and dephosphorylate with calf intestine phosphatase.

3. Prepare the following oligonucleotide on an automated DNA synthesizer: 5′-GACTGACTGGTCCG $(NNG/C)_{20}$GGTCCTCAGTCAGTCAG, where N represents G, C, A, or T.

4. Add 200 μg (a 10-fold excess) of the primer 5′-CTGACTGACTGAGGACC to 100 μg of the above-described random oligonucleotide to give a total volume of 890 μl in 1× Klenow polymerase buffer. Anneal the primer by heating the sample to 94° for 5 min and slowly cool to room temperature.

5. Add 90 μl of a 5 m*M* mixture of the four dNTPs and 20 μl of Klenow polymerase (5 U/μl) and incubate the reaction for 3 hr at 37°.

6. Phenol–chloroform extract the mixture. Follow this with a chloroform extraction.

7. Ethanol precipitate and wash the pellet with 80% (v/v) ethanol. Dissolve the pellet in 0.9 ml of H_2O.

8. Add 0.1 ml of *Ava*II buffer and digest the DNA with 1000 units of *Ava*II for 4 hr. *Optional:* Digest the DNA from step 7 with 1000 units of *Bfa*I overnight. *Bfa*I cleaves DNA containing CTAG and eliminates TAG stop codons following C.

9. Phenol–chloroform extract the mixture, followed by a chloroform extraction.

10. Ethanol precipitate the restriction digest and dissolve the DNA in 10 m*M* Tris (pH 8) and a 1/5 volume of nondenaturing loading buffer.

11. Purify the DNA on a 10% nondenaturing polyacrylamide gel. Expose the DNA by UV shadowing. Cut out the DNA band and elute in 10 m*M* Tris, pH 8, overnight. Ethanol precipitate the supernatant and dissolve in 10 m*M* Tris, pH 8.

12. Ligate 8 μg of the random sequence DNA insert into 12 μg of the *Rsr*II-cut dephosphorylated pJM vector from step 1 with 80,000 units [New England BioLab (Beverly, MA) units] of ligase in a 1-ml reaction.

13. Purify the DNA with a QIAquick gel extraction kit from Qiagen (Chatsworth, CA) and elute with 30 μl of ultrapure H_2O.

14. Add 30 μl of DNA to 350 μl of competent *E. coli* (e.g., MC1061; Bio-Rad, Hercules, CA) in 0.2-cm gap electroporation cuvettes. Electroporate,

using the following settings: 2.5 kV, 200 Ω, 25 μF. Recover the cells in 25 ml of SOC for 1.5 hr.

15. Take a small sample to determine the transformation efficiency and transfer the cells to 1 liter of LB medium with ampicillin (50 μg/ml). Incubate overnight at 37°.

16. Purify the plasmid DNA with a plasmid megakit from Qiagen, or with successive ethidium bromide–CsCl gradients.

Reverse Genetics with Peptide Aptamers

Selection of Aptamers to Specific Proteins

Two-hybrid systems have been used to detect protein–protein interactions[46,47] and are effective for screening proteins for interactions with cDNA and genomic libraries.[48,49] The two-hybrid system has been used to screen libraries of peptide aptamers[11] and linear peptides[15] for those that interact with given baits. A number of working two-hybrid systems have been developed,[46,47,50–52] all of which have the following features: (1) a target protein of interest fused to a DNA-binding domain referred to as the "bait," (2) a library of proteins or peptide aptamers fused to a transcription activation domain, sometimes referred to as the "prey," and (3) one or more reporter genes that detect interactions between the bait and prey proteins.

In this review we describe the use of the interaction trap two-hybrid system[47] to isolate peptide aptamers (illustrated in Fig. 2). In the interaction trap, the protein of interest is fused to the LexA DNA-binding domain (bait). The bait protein is expressed constitutively from the yeast *ADH1* promoter. The bait protein binds to the LexA operators upstream of the two reporter genes, an integrated LexA operator–*LEU2* gene and a LexA operater–*lacZ* gene, but does not activate transcription of the reporters. The peptide aptamer library is fused to the amino-terminal moiety consisting of three domains: nuclear localization signal, transcriptional activation domain, and an epitope tag. The peptide aptamer library is expressed from the *GAL1* promoter, allowing high levels of expression in the presence of

[46] C. T. Chien, P. L. Bartel, R. Sternglanz, and S. Fields, *Proc. Natl. Acad. Sci. U.S.A.* **88,** 9578 (1991).

[47] J. Gyuris, E. Golemis, H. Chertkov, and R. Brent, *Cell* **75,** 791 (1993).

[48] C. Bai and S. J. M. Elledge, *Methods Enzymol.* **273,** 331 (1996).

[49] Finley and Brent, 1996.

[50] T. Durfee, K. Becherer, P. L. Chen, S. H. Yeh, Y. Yang, A. E. Kilburn, W. H. Lee, and S. J. Elledge, *Genes Dev.* **7,** 555 (1993).

[51] A. B. Vojtek, S. M. Hollenberg, and J. A. Cooper, *Cell* **74,** 205 (1993).

[52] S. Dalton and R. Treisman, *Cell* **68,** 59 (1992).

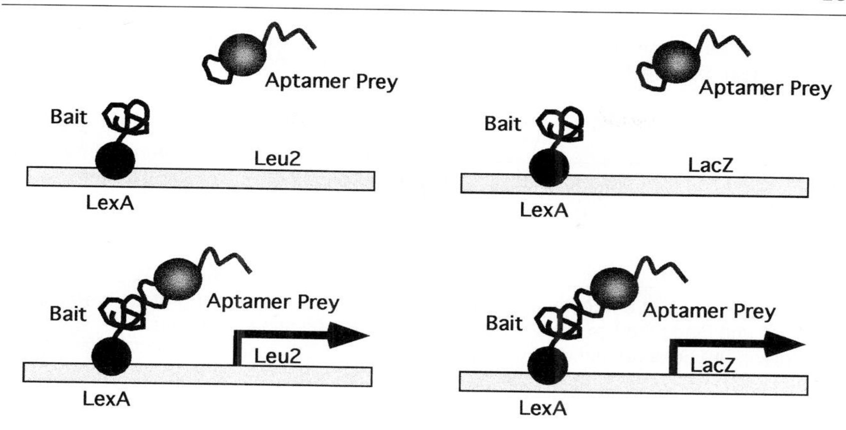

FIG. 2. Isolation of peptide aptamers with the interaction trap. The yeast strain (EGY48) used in the interaction trap contains an integrated *LexAop–LEU2* reporter (required for growth on Leu^- medium) and a plasmid (pSH18-34) containing a *LexAop–lacZ* reporter (required for blue color on X-Gal medium). The bait proteins are constitutively expressed from a plasmid (pEG202) as fusions to LexA (represented by black circles). The library of random peptide aptamer preys is expressed as fusions to an activation domain from a plasmid (pJM-1) containing a galactose-inducible promoter (P_{GAL1}). In the presence of glucose the peptide aptamer preys are not expressed and no interactions occur. In the presence of galactose the peptide aptamer preys are expressed. Interactions between the bait protein and peptide aptamer prey activate the *LexAop–LEU2* and *LexAop–lacZ* reporters, allowing the yeast to grow on Leu^- medium and turn blue on X-Gal.

galactose and repression in the presence of glucose. Interaction between the bait protein and a peptide aptamer prey is detected by specific activation of two reporter genes in the presence of galactose but not in the presence of glucose. A general flow chart for performing the interaction trap is outlined in Fig. 3.

Peptide aptamers have been isolated against Cdk2,[11] Ras,[12] E2F,[14] and HIV-1 Rev[13] with the interaction trap. These aptamers bind their targets with K_d values and half-inhibitory concentrations ranging from 10^{-8} to 5×10^{-11} *M*. In general, approximately 1 of every 10^5 aptamers interacts with the target protein.[11,14,18] For the most part, the variable regions of peptide aptamers isolated against target proteins with the interaction trap (Cdk2, Ras, and HIV-1 Rev) show no resemblance to any known proteins. However, in one case, the variable region of an anti-E2F aptamer resembled DP, a natural protein that interacts with E2F. The anti-E2F aptamer contains a –WIGL– motif, which is present in the DP family of proteins and is responsible for the heterodimerization of E2F–DP.[14]

[53] Deleted in proof.

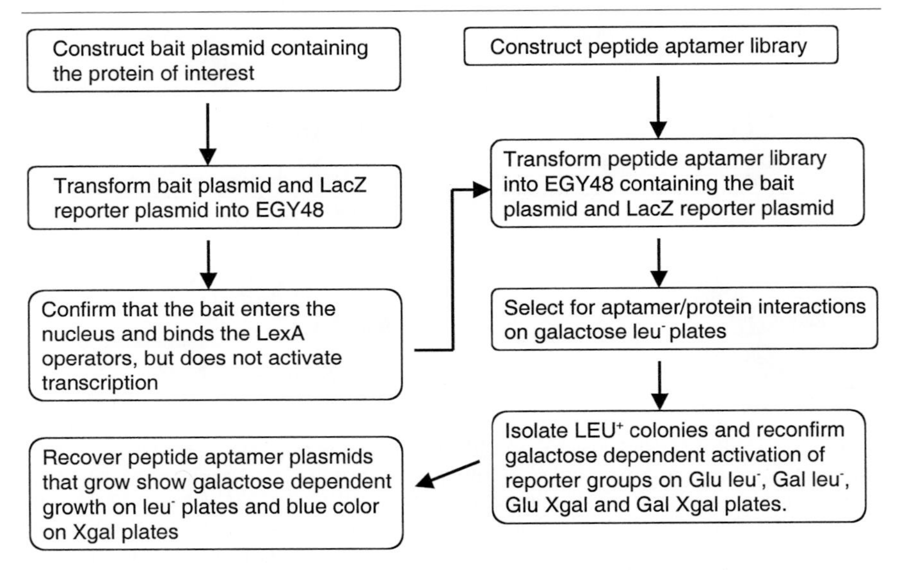

FIG. 3. Flow chart for isolating peptide aptamers with the interaction trap.

The majority of peptide aptamers isolated with the interaction trap function effectively *in vivo*. The anti-Cdk2 aptamer inhibits the progression of human cells through the G_1 phase of the cell cycle.[54] The anti-Cdk2 aptamer also functions in *Drosophila melanogaster* by inhibiting the cell cycle during organogenesis, resulting in eye defects.[19] Anti-Ras aptamers function effectively in mammalian cells by blocking EGG-stimulated Raf kinase activation.[18] The anti-E2F aptamers function in mammalian cells as growth inhibitors by impeding the entry of cells into S phase.[14] In summary, the interaction trap provides an effective method for obtaining high-affinity peptide aptamers that bind specific protein targets. These peptide aptamers function effectively under a variety of *in vivo* conditions.

Method 2: Isolation of Peptide Aptamers with the Interaction Trap

Reagents

Strain
- EGY48: *Matα his3 trp1 ura3-52 leu2::LexA6op-LEU2*

Vectors
- pEG202: LexA fusion plasmid for the construction of bait proteins (pBait) (see Fig. 4 for details)

[54] B. A. Cohen, P. Colas, and R. Brent, *Proc. Natl. Acad. Sci. U.S.A.* **95,** 14272 (1998).

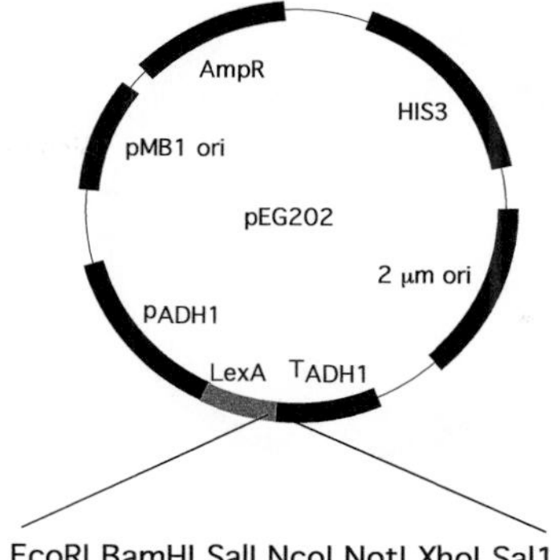

FIG. 4. Bait protein expression plasmid. pEG202[47] is a yeast shuttle vector consisting of an *E. coli* origin of replication (pMB1 ori), an ampicillin resistance gene (Amp^R), a *HIS3* marker gene, a yeast 2 μm origin of replication, and an expression cassette that contains a yeast *ADH1* promoter (P_{ADH1}), DNA encoding amino acids 1–202 of the bacterial repressor protein LexA, a polylinker for creating in-frame fusions to LexA, and a yeast *ADH1* transcription terminator (T_{ADH1}).

pJM-1 peptide aptamer library: Peptide aptamer prey library for use in the interaction trap (see Fig. 1 for details)
pSH18-34: *lacZ* reporter plasmid (see Fig. 5[54a–54d] for details)
pJK101: Repression assay plasmid (see Fig. 5 for details)

Media

YPD plates and liquid media

1. Combine the following:

Yeast extract	10 g
Peptone	20 g
Agar (only for plates)	20 g
H_2O	950 ml

2. Autoclave and add 50 ml of 40% (w/v) glucose

Dropout plates and liquid media

1. Combine the following:

[54a] R. R. Yocum, S. Hanley, R. J. West, and M. Ptashne, *Mol. Cell. Biol.* **4,** 1985 (1984).
[54b] J. Kamens, P. Richardson, G. Mosialos, R. Brent, and T. Gilmore, *Mol. Cell. Biol.* **10,** 2840 (1990).
[54c] R. W. West, Jr., R. R. Yocum, and M. Ptashne, *Mol. Cell. Biol.* **4,** 2467 (1984).
[54d] J. Kamens and R. Brent, *N. Biol.* **3,** 1005 (1991).

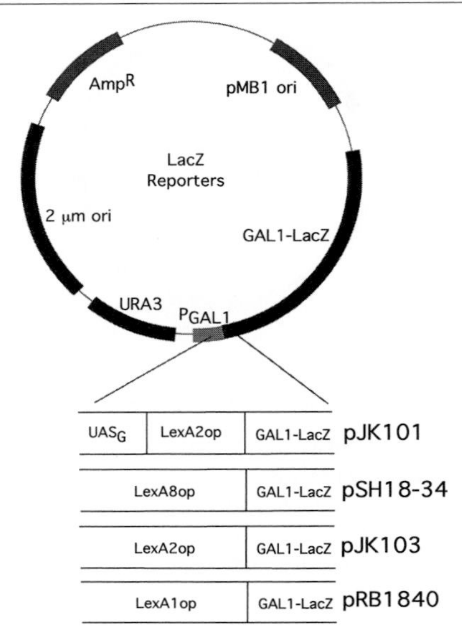

FIG. 5. *lacZ* reporter plasmids. The *lacZ* reporters are derived from a plasmid containing an *E. coli* origin of replication (pMB1 ori), an ampicillin resistance gene (Amp^R), *URA3* marker gene, yeast 2 μm origin of replication, and a *GAL1* promoter fused to a *lacZ* reporter.[54a] pJK101 is used for the repression assay and contains two LexA operators between the UAS_G and the *GAL1* TATA.[54b] *lacZ* reporter plasmids are derived from pLR1Δ1, in which the UAS_G is deleted.[54c] pSH18-34, pJK103,[54d] and pRB1840[64] contain binding sites for eight, two, and one LexA dimers, respectively.

Yeast nitrogen base without amino acids	6.7 g
Dropout mixture minus appropriate amino acids (see below)	2 g
Agar (plates only)	20 g
H_2O	950 ml (glucose plates or media) *or* 925 ml (galactose/raffinose plates or media)

2. Autoclave and add either 50 ml of 40% (w/v) glucose or 50 ml of 40% (w/v) galactose and 25 ml of 40% (w/v) raffinose

Dropout mixture: Combine the required nutrients listed in the tabulation to make the appropriate selection plates. Grind the components of the mixture into a fine powder, using a mortar and pestle

Nutrient	Amount (g)	Concentration in medium (μg/ml)
Adenine	2.5	40
L-Arginine	1.2	20
L-Aspartic acid	6.0	100
L-Glutamic acid	6.0	100
L-Histidine	1.2	20
L-Isoleucine	1.8	30
L-Leucine	3.6	60
L-Lysine	1.8	30
L-Methionine	1.2	20
L-Phenylalanine	3.0	50
L-Serine	22.5	375
L-Threonine	12.0	200
L-Tryptophan	2.4	40
L-Tyrosine	12.0	30
L-Valine	9.0	150
Uracil	1.2	20

X-Gal plates

1. Combine the following:

Yeast nitrogen base without amino acids	6.7 g
Dropout mixture	1.5 g
Agar	20 g
H_2O	850 ml (glucose plates or media) *or* 825 ml (galactose/raffinose plates or media)

2. Autoclave and add either 50 ml of 40% (w/v) glucose or 50 ml of 40% (w/v) galactose and 25 ml of 40% (w/v) raffinose.
3. Cool to 55°.
4. Add 100 ml of 10× BU salts (see below).
5. Add 4 ml of 20 mg/ml of 5-bromo-4-chloro-3-indolyl-β-D-galactopyranoside (X-Gal) dissolved in dimethylformamide.

BU salts (10×):

1. Combine the following:

$Na_2PO_4 \cdot 7H_2O$	70 g
NaH_2PO_4	30 g

2. Adjust to pH 7 and autoclave.

High-Efficiency Lithium Acetate Yeast Transformation. (After Geitz and Schiestl.[55]) Transformations with lithium salts typically result in approximately 10^5 transformants per microgram of plasmid DNA. The protocol described below may need to be optimized for individual yeast strains. Factors such as the final concentration of the cells prior to transformation and the heat shock time should be optimized. In general, the highest transformation efficiencies per microgram of plasmid are obtained when 1 μg of plasmid is used per 50-μl aliquot of competent yeast cells.

PROTOCOL 2-1: YEAST TRANSFORMATION

1. Inoculate a 10-ml culture with the yeast to be transformed in appropriate media and incubate overnight at 30° with shaking.
2. Determine the OD_{600} and dilute colonies in 50 ml of appropriate medium to a concentration of 5×10^6 cells/ml of culture.
3. Continue to grow the cells at 30° with shaking until an OD_{600} of 0.6–0.8 is reached. This will provide enough cells for ten 50-μl transformations.
4. Centrifuge the yeast cells at 4000 rpm for 5 min at room temperature.
5. Remove the supernatant and resuspend the yeast pellet in 25 ml of H_2O. Centrifuge again as in step 4.
6. Remove the supernatant and resuspend the yeast in 1 ml of 100 m*M* lithium acetate. Centrifuge at 13,000 rpm for 15 sec.
7. Remove the supernatant and resuspend the yeast pellet in 100 m*M* lithium acetate to give a final volume of ~500 μl.
8. Split the samples in 50-μl aliquots. Centrifuge at 13,000 rpm for 15 sec.
9. Remove the supernatant and add the following ingredients in the order listed to one yeast aliquot from step 8.

Polyethylene glycol (50%, w/v)	240 μl
Lithium acetate (1 *M*)	36 μl
Single-stranded DNA (ssDNA, 2 mg/ml)	25 μl
Plasmid DNA (1.0 μg)	50 μl

Note: Single-stranded carrier DNA (ssDNA) must be boiled for 5 min and cooled rapidly on ice prior to use.

10. Vortex the mixture vigorously until the pellet is completely resuspended.
11. Incubate the sample at 30° for 30 min.
12. Heat shock the sample at 42° for 20 min.
13. Centrifuge the sample at 8000 rpm for 15 sec and remove the supernatant.

[55] R. D. Gietz and R. H. Schiestl, *Mol. Cell. Biol.* **2,** 255 (1995).

14. Resuspend the pellet in 500 μl of H_2O and plate 200 μl onto a 10-cm-diameter plate containing the appropriate dropout medium.

Construction and Characterization of Bait Plasmid (pBait). (After Ausubel *et al.*[23]) Insert DNA encoding protein that the peptide aptamers are to target into the bait plasmid pEG202 (Fig. 4). Using standard subcloning techniques, insert the coding region for the protein of interest into the polylinker of pEG202 in frame with LexA to create pBait—plasmid that directs the synthesis of a chimeric protein LexA fusion protein that contains the bait moiety. Introduce pBait with pSH18-34 and pEG202 with pSH18-34 (control) into the yeast selection strain EGY48, using the lithium acetate method (protocol 2-1). Select transformants on Glu His^- Ura^- plates. Test the baits for their ability to self-activate the *LEU2* and *lacZ* reporters.

PROTOCOL 2-2: TESTING BAITS FOR SELF-ACTIVATION OF REPORTERS

1. Take four colonies from both pBait with pSH18-34 and pEG202 with pSH18-34 transformations and inoculate 5 ml of Glu Ura^- His^- cultures. Grow the cultures overnight at 30° with shaking. Dilute cultures to an OD_{600} of 0.2 and continue to grow the cultures at 30° with shaking until they are in midlog phase (OD_{600} ~0.6).
2. Make 100- and 1000-fold dilutions of the culture in H_2O.
3. Spot 10 μl of the stock culture and each dilution onto the following plates and incubate at 30°: Gal/Raf Ura^- His^-, Gal/Raf Ura^- His^- Leu^-, and Gal/Raf Ura^- His^- X-Gal.
4. Check the growth of the yeast after 3 days. The yeast should all grow on Gal/Raf Ura^- His^- plates, indicating that the baits are not toxic to the yeast. Yeast should not grow on Gal/Raf Ura^- His^- Leu^- plates or have blue color on Gal/Raf Ura^- His^- X-Gal plates, indicating that the baits do not activate the *LEU2* or *lacZ* reporters.

After confirming that the bait(s) do not activate the reporter genes, it is useful to determine whether the baits enter the nucleus and bind the LexA operators. This is accomplished with a repression assay that tests whether the baits can repress transcription of a *lacZ* reporter on plasmid pJK101 (Fig. 5), which contains LexA operators between the TATA and the *GAL1* upstream activating sequence. *lacZ* expression is induced in the presence of galactose. Baits that enter the nucleus and do not activate transcription bind to the *lexA* operators and repress the transcription of *lacZ*.

PROTOCOL 2-3: REPRESSION ASSAY

1. Transform pBait with pJK101 and pEG202 with pJK101 into EGY48 as described in protocol 2-1. Select transformants on Glu Ura^- His^- plates.

2. Take four transformants from each transformation and streak them onto Glu Ura$^-$ His$^-$ X-Gal and Gal/Raf Ura$^-$ His$^-$ X-Gal plates. Incubate the plates at 30°.

3. Monitor the plates for 3 days. Yeast with pEG202 and pJK101 should turn blue on Gal/Raf Ura$^-$ His$^-$ X-Gal plates after 1 day and should be light blue on Glu Ura$^-$ His$^-$ X-Gal after 2–3 days. Yeast expressing baits that enter the nucleus should turn blue more slowly than the controls.

4. Baits that do not cause a conspicuous diminution in blue color on X-Gal plates relative to the controls should be measured by liquid β-galactosidase assays (see protocol 2-4).

5. If no differences are detected by either plate or liquid β-galactosidase assay, then the baits may not be expressed correctly, or they may be unable to enter the nucleus and bind the operator. Expression of the full-length protein can be verified by standard Western blot methods.[23] If the protein is expressed, a nuclear localization signal may need to be added to the bait to aid its entry into the nucleus (Breitwieser and Ephrussi, unpublished, 1999).

PROTOCOL 2-4: β-GALACTOSIDASE ASSAY. (After Stern *et al.*[56] and Miller.[57])

1. Isolate three individual colonies from each sample and grow cells to log phase in liquid culture (OD_{600} ~0.6).

2. Centrifuge 1 ml of culture (in triplicate) at 13,000 rpm for 5 min at room temperature.

3. Remove the supernatant and resuspend pellet in 1 ml of Z buffer* without 2-mercaptoethanol.

4. Centrifuge again as in step 2 and resuspend the pellet in 150 μl of Z buffer with 2-mercaptoethanol (27 μl/10 ml), 50 μl of chloroform, and 20 μl of 0.1% (w/v) sodium dodecyl sulfate (SDS) in the order listed.

5. Vortex the mixture vigorously for 15 sec.

6. Start the reaction by adding 700 μl of *o*-nitrophenyl-β-galactopyranoside (ONPG) prewarmed to 30° (1 mg/ml in Z buffer plus 2-mercaptoethanol).

7. Incubate the reaction at 30° until the reaction turns a medium yellow color (20 min to 3 hr).

8. Quench the reaction with 0.5 ml of 1 *M* $NaCO_3$.

[56] M. Stern, R. Jensen, and I. Herskowitz, *J. Mol. Biol.* **178,** 853 (1994).

[57] J. H. Miller, *in* "Experiments in Molecular Genetics," p. 352. Cold Spring Harbor Laboratory Press, Cold Spring Harbor, New York, 1972.

* Z buffer (1 liter): $Na_2HPO_4 \cdot 7H_2O$, 16.1 g (8.5 g anhydrous); $NaH_2PO_4 \cdot H_2O$, 5.5 g; KCl, 0.75 g; $MgSO_4 \cdot 7H_2O$, 0.246 g.

9. Centrifuge the reaction mixture at 13,000 rpm for 10 min at room temperature.

10. Measure the OD_{420} of the supernatant and calculate the β-galacosidase activity in Miller units.

$$\text{Miller unit} = (A_{420})\ (1000)/(A_{600})\ (\text{time})\ (\text{volume})$$

where time is reaction time (minutes) and volume is reaction volume (milliliters).

Selection of Aptamers that Bind the Bait Protein. The pJM-1-derived peptide aptamer library expresses proteins that contain a nuclear localization signal and a transcription activation domain. This library is used in the interaction trap to select aptamers that bind a specific protein target. The library is introduced into a selection strain that contains the bait plasmid and LexA fusion-responsive *LEU2* and *lacZ* reporters and transformants are selected on dropout plates. Transformants are collected by scraping the plates, and stored as frozen stocks. In this procedure, each yeast cell containing an aptamer is allowed to grow as a colony and the colonies are pooled when they are approximately the same size. Selection of transformants on plates prior to selection for expression of the reporter genes results in a more uniform representation of yeast that contain unique peptide aptamers. Aliquots of the frozen stock are then plated onto galactose Leu^- plates. Galactose induces the expression of the peptide aptamer preys. The absence of leucine selects for peptide aptamer preys that interact with baits and activate the *LEU2* reporter. Colonies that grow on the galactose Leu^- plates are subsequently tested for galactose-dependent growth and blue color on Leu^- and X-Gal plates, respectively.

Protocol 2-5: Selection of Peptide Aptamers

1. Introduce 50–100 μg of pJM-1 aptamer library into EGY48 containing pBait and pSH18-34 to give between 10^6 and 10^7 transformants (protocol 2-1). Plate the transformation on Glu Ura^- His^- Trp^- plates and incubate at 30° for 2–3 days (colonies should be ~1 mm in diameter).

2. Scrape the yeast cells from the plate by adding 1–2 ml of H_2O to a 24 × 24 cm plate, collect the yeast with a glass spreader, and transfer them to a 50-ml Falcon tube. Pool the transformants from all plates and centrifuge at 3500 rpm for 4 min at room temperature. Remove the supernatant and resuspend the pellet in 25 ml of H_2O. Centrifuge again and wash the pellet once more in H_2O.

3. Resuspend the pellet in H_2O. Add an equal volume of 65% (v/v) glycerol, 0.1 *M* $MgSO_4$, 25 m*M* Tris-HCl, pH 7.4. Divide into 1-ml aliquots and freeze at −70°.

4. Determine the plating efficiency of the frozen aliquot. Inoculate 10 library equivalents in 1 ml of Gal/Raf Ura^- His^- Trp^- liquid medium. Incubate at 30° for 4 hr with shaking.

5. Pellet the yeast cells by centrifuging at 3500 rpm for 4 min at room temperature. Remove the supernatant and resuspend the pellet in H_2O.

6. Spread yeast onto Gal/Raf Ura^- His^- Trp^- Leu^- plates at a density of 10^6 yeast per 10-cm-diameter plate and incubate at 30°. Colonies should begin to appear between 2 and 5 days.

7. Streak colonies onto a Glu Ura^- His^- Trp^- master plate and incubate at 30° for 1 day.

8. Replica plate the master plate on the following four plates: Glu Ura^- His^- Trp^- X-Gal, Gal/Raf Ura^- His^- Trp^- X-Gal, Glu Ura^- His^- Trp^- Leu^-, and Gal/Raf Ura^- His^- Trp^- Leu^- and incubate at 30°. Monitor the plates for blue color on X-Gal plates and growth on Leu^- plates.

9. Select colonies that show galactose-dependent growth and blue color on Leu^- and X-Gal plates, respectively, for further characterization.

Recover peptide aptamer plasmids from the two-hybrid selection strain by a yeast minipreparation (protocol 2-6). Use the plasmid DNA from the yeast minipreparation to transform *E. coli* so that the aptamer plasmids can be separated from the pBait and pSH18-34. Plasmids isolated with this yeast minipreparation can also be used as templates for sequencing.

PROTOCOL 2-6: YEAST MINIPREPARATION. (After the "smash and grab" procedure of Hoffman and Winston.[58])

1. Inoculate 2 ml of Glu Trp^- medium with a single colony from the two-hybrid selection and grow overnight at 30° with shaking.

2. Centrifuge 1.5 ml of the saturated overnight culture at 13,000 rpm for 15 sec. Discard the supernatant.

3. Resuspend the yeast pellet in 200 μl of breaking buffer [2% (v/v) Triton X-100, 1% (v/v) SDS, 100 m*M* NaCl, 10 m*M* Tris-HCl (pH 8), 1 m*M* EDTA]. Add ~0.3 g of glass beads and 200 μl of phenol–chloroform–isoamyl alcohol (25:24:1, v/v/v). Vortex the mixture vigorously for 2 min.

4. Centrifuge at 13,000 rpm for 5 min at room temperature. Remove 50 μl of the aqueous layer. The aqueous layer can be stored at −20° or 1–5 μl can be used directly for transforming *E. coli.*

5. *Escherichia coli* transformants that contain the peptide aptamer plasmid can be identified by performing the polymerase chain reaction (PCR) directly from the colonies, using primers that flank the thioredoxin insert. Perform a PCR for 20 cycles. Colonies that contain the aptamer will give

[58] C. S. Hoffman and F. Winston, *Gene* **57,** 267 (1987).

a bright band on an ethidium bromide–agarose gel. Colonies that do not contain the thioredoxin plasmid will not give a band. Colonies that contain a library plasmid that does not contain a variable region insert will give a band that runs 60 bp slower than the thioredoxin aptamer.

Determining Peptide Aptamer Specificity

The specificity of peptide aptamers isolated with the interaction trap can be evaluated by reintroducing the peptide aptamers into yeast (EGY48) that contain the original target protein bait as well as related proteins. The procedure for determining the specificity of the aptamers is identical to that used in the peptide aptamer library screen (protocol 2-5) except that single peptide aptamers, rather than libraries of aptamers, are screened for interactions with chosen baits. Variations of this two-hybrid method are also used to identify regions or specific amino acids of the target protein that are involved in the aptamer–protein interaction (see below).

Determining Peptide Aptamer Specificity Using Interaction Mating

An alternative method for determining peptide aptamer specificity against a large panel of proteins is to use interaction mating.[59] Haploid yeast exist in one of two mating types (**a** or α) and opposite mating types mate to form diploids (reviewed in Herskowitz[60]). Haploid yeast that contain the bait or prey in strains of the opposite mating types can be mated to generate diploids that carry both the bait and prey. Use of interaction mating to test specificity is outlined in Fig. 6.

For example, Colas *et al.*[11] used interaction mating to study the specificity of anti-Cdk2 peptide aptamers (Fig. 7). All the aptamers isolated against Cdk2 interact with Cdk2 and not with unrelated proteins such as Max and Rb. Some of the aptamers isolated against Cdk2 interacted with related cyclin-dependent kinases, whereas other aptamers were specific for Cdk2. The results suggest that aptamers are capable of recognizing different epitopes on Cdk2, some of which are conserved among different subsets of the cyclin-dependent kinases.

Interaction mating assays can also determine the amino acids and/or regions of the target protein that are required for the interaction. For example, Cohen *et al.*[54] showed that one anti-Cdk2 aptamer is a competitive inhibitor of the Cdk2–cyclin E phosphorylation of histone H1. They used interaction mating against a panel of mutant Cdk2 proteins to show that

[59] R. L. Finley, Jr., and R. Brent, *Proc. Natl. Acad. Sci. U.S.A.* **91,** 12980 (1994).
[60] I. Herskowitz, *Microbiol. Rev.* **52,** 536 (1988).

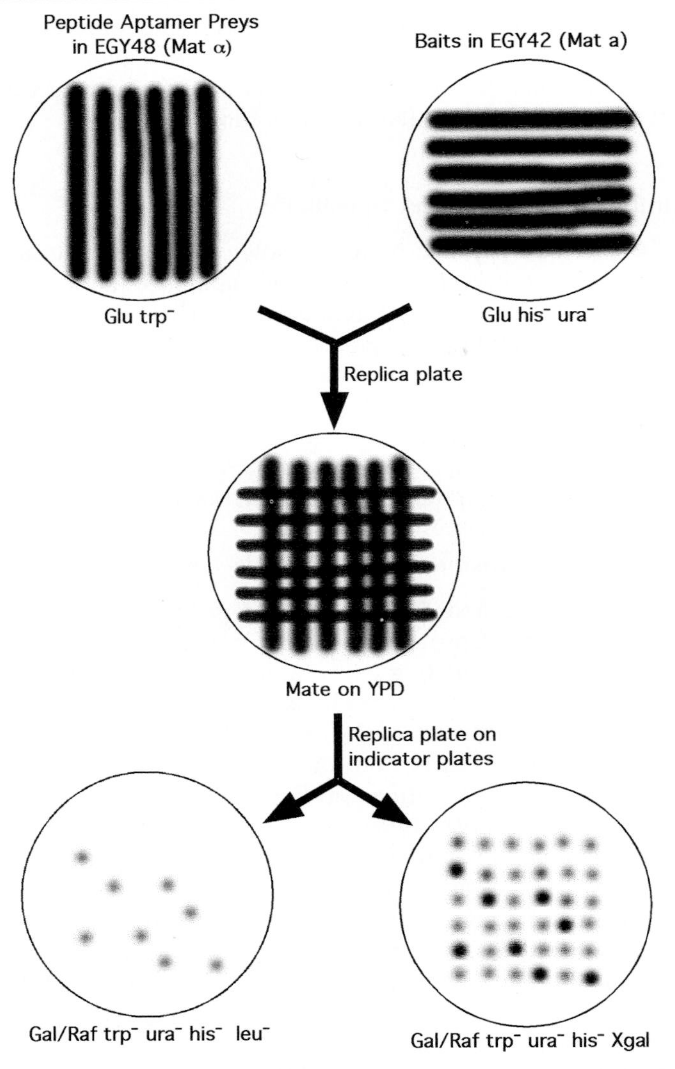

FIG. 6. Interaction mating assay.[59] Yeast strain EGY48 (*Mat*α) containing the individual peptide aptamer preys is streaked horizontally on Glu Trp$^-$ plates. Yeast strain EGY42 (*Mat***a**) containing individual baits and a *lacZ* reporter is streaked vertically on Glu His$^-$ Ura$^-$ plates. The two strains are mated by replica plating on YPD plates. The strains mate at the intersections, forming **a**/α diploids that contain the peptide aptamer prey, bait, and the *lacZ* reporter. The YPD plate is then replica plated onto the mating interaction scoring plates: Glu Ura$^-$ His$^-$ Trp$^-$ Leu$^-$, Gal/Raf Ura$^-$ His$^-$ Trp$^-$ Leu$^-$, Glu Ura$^-$ His$^-$ Trp$^-$ X-Gal, Gal/Raf Ura$^-$ His$^-$ Trp$^-$ X-Gal. Peptide aptamer preys and baits that interact grow on Gal/Raf Leu$^-$ plates and turn blue on Gal/Raf X-Gal plates.

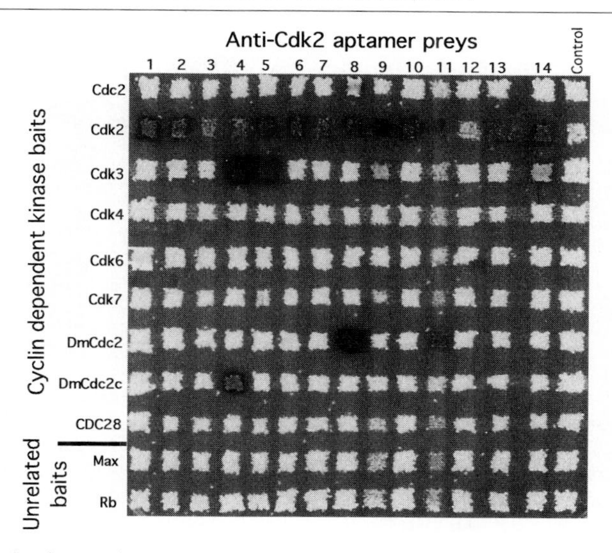

FIG. 7. Mating interaction specificity test (from Ref. 11). Specificity of anti-Cdk2 peptide aptamers against a panel of cyclin-dependent kinases and two unrelated proteins (Max and Rb). The specificity is assessed on galactose X-Gal plates. Interactions between the peptide aptamer preys and the baits result in the activation of the *lacZ* reporter, giving blue color on galactose X-Gal plates.

residues in the active site of Cdk2 are required for the aptamer–Cdk2 interaction (Fig. 8). Specifically, two point mutations within the active site (Cdk2-145: D145N and Cdk2-30: V30A) and two point mutations outside the active site (Cdk2-38: D38A, E40A and Cdk2-150: R150A, A151F, F152A) diminished the interaction between the peptide aptamer and Cdk2. Interestingly, these results showed that the Cdk2 residues recognized by this anti-Cdk2 aptamer differed from those recognized by other Cdk2-binding proteins tested (Cdi1, Cks1, p27, and p21). For example, the Cdk2-30 V30A mutation affects the binding of both the aptamer and p21 but not p27. This aptamer also inhibits the ability of Cdk2 to phosphorylate one substrate but not another, suggesting that it can block specific protein–protein interactions while leaving other interactions uninterrupted. Similarly, this strategy can also be expanded to determine the domains or regions of protein targets where the aptamers interact.[21]

In summary, use of interaction mating together with appropriately chosen panels of wild-type and mutant proteins can give insight into the details of aptamer binding. The assay is able to discriminate between aptamers that recognize different but related proteins, as well as variants of individual proteins. This two-step approach to isolating peptide aptamers with the

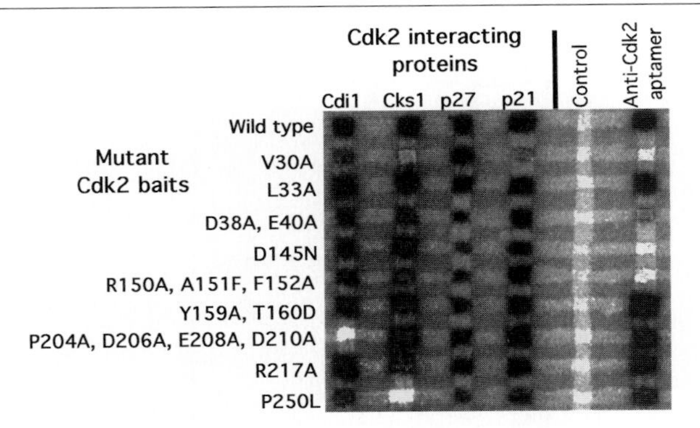

FIG. 8. Interaction of anti-Cdk2 aptamers with mutant alleles of Cdk2 (see Ref. 54). Mating interaction assay involves Cdk2-interacting protein preys (Cdi1, Cks1, p27, p21, and anti-Cdk2 aptamer) and mutant allele baits of Cdk2. The mutant alleles are listed in the columns. Interactions are scored on a galactose X-Gal plate. Interactions between the Cdk2-interacting protein preys and the mutant allele baits result in the activation of the *lacZ* reporter, giving blue color on galactose X-Gal plates.

interaction trap and their subsequent characterization by the mating assay provides a simple and effective method for obtaining aptamers with different specificities toward the same target.

Method 3: Defining Recognition Specificity with Interaction Mating

Reagents

Strains
- EGY48: *Mat*α *his3 trp1 ura3-52 leu2::LexA6op-LEU2*
- EGY42: *Mat***a** *his3 trp1 ura3 leu2*

Vectors
- pEG202: LexA fusion plasmid for the construction of bait proteins (pBait) (see Fig. 4 for details)
- pBait: pEG202 containing bait protein of interest inserted into the polylinker in frame with LexA
- pJG4-5: Nuclear localization, activation domain, and hemagglutinin (HA) epitope tag fusion for the construction of prey proteins (see Fig. 1 for details)
- Peptide aptamer prey: pJM-1 containing peptide aptamer (see Fig. 1 for details)
- pSH18-34: *lacZ* reporter plasmid (see Fig. 5 for details)

Media: See method 2 for recipes

Mating Interaction Assay. Here, the peptide aptamer prey is introduced into yeast of the α-mating type and the bait protein is transformed into yeast of the **a**-mating type. The two yeast strains are mated by replica plating streaks of yeast of opposite mating types on top of each other (see Fig. 6). The resulting diploid yeast contain both the bait and peptide aptamer prey. Interaction between the bait and prey is detected by the activation of *LexAop–LEU2* and *LexAop–lacZ* reporters. This assay allows peptide aptamers to be simultaneously screened for interactions against panels of bait proteins.

Protocol 3-1: Interaction Mating Assay

1. Introduce the peptide aptamer prey plasmid into EGY48 (*Matα*) (protocol 2-1) and select for transformants on Glu Trp$^-$ plates. At the same time, perform an identical transformation with pJG4-5, a control plasmid with no peptide aptamer.

2. Introduce the bait plasmid (pBait) that expresses the protein target, together with pSH18-34, a *lacZ* reporter, into the yeast strain EGY42 (Mat**a**) (protocol 2-1). Select transformants on Glu His$^-$ Ura$^-$ plates. At the same time, perform an identical transformation with pEG202, a control plasmid with no bait protein with pSH18-34.

3. Streak individual peptide aptamer strains and pJG4-5 control strain (from step 1) in parallel lines on Glu Trp$^-$ plates. Streak individual bait strains and control strain (from step 2) in parallel lines on Glu His$^-$ Ura$^-$ plates. Incubate the plates at 30° overnight or until the streaks show heavy growth.

4. Replica plate the bait and prey stains on the same replica velvet by pressing the replica block onto the bait plate, rotating the block by 90°, and pressing it again onto the prey plate. Transfer the imprint on the replica velvet to a YPD plate and incubate overnight at 30°. Mating occurs at patches where the two strains intersect.

5. Replica plate the YPD plate on a replica velvet. Use this velvet to transfer the strains to the following selection indicator plates: Glu Ura$^-$ His$^-$ Trp$^-$ Leu$^-$, Gal/Raf Ura$^-$ His$^-$ Trp$^-$ Leu$^-$, Glu Ura$^-$ His$^-$ Trp$^-$ X-Gal, and Gal/Raf Ura$^-$ His$^-$ Trp$^-$ X-Gal. Incubate the selection plates at 30° and examine for results daily. Diploid colonies should grow on the X-Gal plates. Interaction between the bait and prey will result in growth on the Leu$^-$ plates and blue color on the X-Gal plates.

One-Step Selection for Peptide Aptamers Specific for a Desired Protein

A "two-bait" version of the interaction trap has been developed that enables a one-step selection of peptide aptamers that bind one protein but

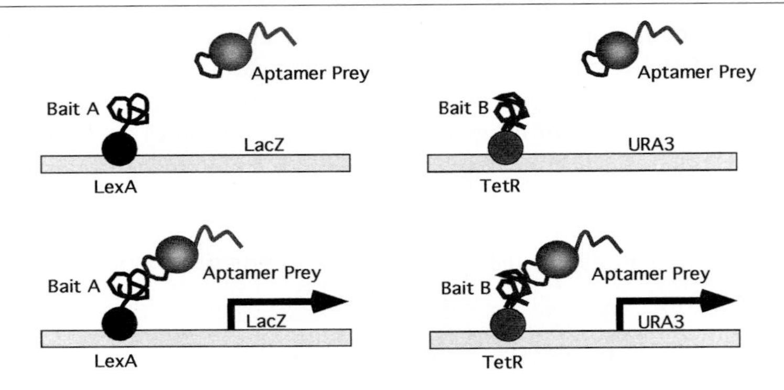

LexA-Protein A	TetR-Protein B	Output
On	Off	Blue, ura⁻
Off	On	White, URA⁺
On	On	Blue, URA⁺
Off	Off	White, ura⁻

FIG. 9. Two-bait interaction trap. The yeast strain (pWX200) contains two reporters, an integrated *Tetop–URA3* reporter and a *LexAop–lacZ* reporter on a plasmid (pCWX24). The bait proteins (LexA and TetR fusions) are constitutively expressed from plasmids (pEG202 or pCWX200). The library of random peptide aptamer preys is expressed as fusions to an activation domain from a plasmid (pJM-1) containing a galactose-inducible promoter (P_{GAL1}). In the presence of glucose the peptide aptamer preys are not expressed and no interactions occur. In the presence of galactose the peptide aptamer preys are expressed. Interactions between the bait proteins and peptide aptamer prey activate the *TetRop–URA3* and *LexAop–lacZ* reporters. The specificity of the peptide aptamer preys toward the two baits is observed by the differential activation of the reporters.

not another, even if the second protein is a close variant.[12] The two-bait interaction trap is outlined in Fig. 9 and uses two DNA-binding domains, Tet repressor (TetR) and LexA, to localize protein targets of interest upstream of two reporter genes. The Tet repressor binds to integrated Tet operator upstream of a *URA3* reporter. The LexA DNA-binding domain binds to LexA operator upstream of the *lacZ* gene on a 2 μm plasmid. Peptide aptamers that interact with either of the proteins are detected by differential activation of the reporters (Fig. 9).

The two-bait interaction trap has been used successfully to isolate allele-specific Ras aptamers.[12,18] Ras proteins cycle between two states, GTP bound (active) or GDP bound (inactive). This variable conformation allows Ras to function as a molecular switch in signal transduction pathways.[61]

[61] M. S. Boguski and F. McCormick, *Nature* (*London*) **366,** 643 (1993).

Allelic variations of Ras exist that are either locked in the GTP-bound (RasV12) or GDP-bound (RasA15) state. Using the two-bait interaction trap, two classes of peptide aptamers were isolated that displayed different interactions with the allelic variants of Ras. The first class of aptamers interacted with RasV12 but not RasA15, and the second class interacted with both RasV12 and RasA15. In addition, some of the aptamers that interacted with RasV12 (GTP bound) preferentially bound this allelic variant compared with wild-type GTP-bound Ras (Ras^+). These results demonstrate that peptide aptamers are capable of discriminating between epitopes on GTP-bound active Ras (RasV12), GDP-bound inactive Ras (RasA15), and allelic variants of the GTP-bound active Ras (RasV12 and Ras^+).

The two-bait interaction trap provides a rapid and efficient method for obtaining aptamers that interact differently with two related proteins. This strategy should be especially useful for isolating peptide aptamers that are specific for polymorphic variants of proteins.

Method 4: One-Step Determination of Peptide Aptamers Specific for One of Two Chosen Proteins

Reagents

Strain
- CWXY2: *Matα ura3 his3 trp1 leu2 lys2::Tetop-URA3*

Vectors
- pEG202: LexA fusion plasmid for the construction of bait proteins (pBait) (see Fig. 4 for details)
- pCWX200: TetR fusion plasmid for the construction of bait proteins (pBait). The plasmid is similar to pEG202 except for a *LEU2* marker and a TetR operator
- pJM-1 peptide aptamer library: Peptide aptamer prey library for use in the interaction trap (see Fig. 1 for details)
- pCWX24: *lacZ* reporter plasmid similar to pSH18-34, except that it contains a *LYS2* marker

Media: See method 2 for recipes

Selection of Aptamers with the Two-Bait Interaction Trap. The two-bait interaction trap can be used to select peptide aptamers specific for one of two chosen proteins. In this system, the target protein to which the peptide aptamer preys are selected to bind are constructed as TetR fusions (TetR bait). A second target protein that the peptide aptamers can be screened to bind or not to bind is constructed as LexA fusion (LexA bait). In one use of this method, peptide aptamers are first selected for interactions with the TetR bait. The TetR bait binds to the Tet operator upstream of the *URA3* reporter. Peptide aptamer preys that interact with the bait are selected on the basis of their ability to grow on Ura^- media. Second, the URA^+ aptamers are screened for interactions with the LexA bait. The LexA bait binds to the LexA operator, which is upstream of the *lacZ* reporter. Aptamers that interact with the LexA bait are blue on X-Gal plates. Aptamers that do not interact with the LexA bait are white.

Protocol 4-1: Two-Bait Interaction Trap

1. Create the desired target bait protein to isolate aptamers against in pCWX200. Create related target protein or allelic variant in pEG202. Confirm that the bait proteins (1) do not activate transcription and (2) are properly expressed, and (3) that the LexA fusions enter the nucleus and bind operator as described previously in protocols 2-2 and 2-3.
2. Transform the bait proteins (pCWX200 and pEG202 fusions) with pCWX24 (*lacZ* reporter) into CWXY2 as described in protocol 2-1. Plate transformants on Glu Leu^- His^- Lys^- dropout plates.
3. Transform 50–100 μg of pJM-1 library into CWXY2 with baits and pCWX24 to give between 10^6 and 10^7 transformants (protocol 2-1). Plate transformation on Glu Leu^- His^- Lys^- Trp^- plates and incubate at 30° for 2–3 days (colonies should be ~1 mm in diameter).
4. Scrape the yeast cells with a glass spreader from the plate by adding 1–2 ml of H_2O to a 24 × 24 cm plate and collect the yeast in a 50-ml Falcon tube. Pool the transformants from all plates and centrifuge at 3500 rpm for 4 min at room temperature. Remove the supernatant and resuspend the pellet in 25 ml of H_2O. Centrifuge again and wash the pellet once more in H_2O.
5. Resuspend the pellet in H_2O and add an equal volume of 65% (v/v) glycerol, 0.1 *M* $MgSO_4$, 25 m*M* Tris-HCl, pH 7.4. Divide into 1-ml aliquots and freeze at −70°.
6. Determine the plating efficiency of the frozen aliquot. Inoculate 10 library equivalents in 1 ml of Gal/Raf Leu^- His^- Lys^- Trp^- liquid medium. Incubate at 30° for 4 hr with shaking.
7. Pellet the yeast cells by centrifuging at 3500 rpm for 4 min at room temperature. Remove the supernatant and resuspend the pellet in H_2O.
8. Spread yeast onto Gal/Raf Leu^- His^- Lys^- Trp^- Ura^- plates at a density of 10^6 yeast per 10-cm-diameter plate and incubate at 30°. Colonies should begin to appear between 2 and 5 days.
9. Streak colonies onto Glu Leu^- His^- Lys^- Trp^- master plates and incubate at 30° for 1 day.
10. Replica plate the master plate on the following four plates: Glu Leu^- His^- Lys^- Trp^- X-Gal, Gal/Raf Leu^- His^- Lys^- Trp^- X-Gal, Glu Leu^- His^- Lys^- Trp^- Ura^-, and Gal/Raf Ura^- Leu^- His^- Lys^- Trp^- and

incubate at 30°. Monitor the plates for blue color on X-Gal plates and growth on Ura^- plates.

11. Select colonies that are URA^+ and $LacZ^+$ or URA^+ and $LacZ^-$ on galactose for further characterization.

12. Isolate and sequence selected aptamer plasmids as described in protocol 2-6.

Affinity Maturation of Peptide Aptamers

Aptamer selections do not take advantage of the entire space of possible variable regions. There are 20^{20} (10^{27}) possible 20-mer variable regions. Libraries of aptamers contain from 10^9 to 10^{10} members, of which $\sim 10^7$ peptides can be looked at in any given selection. In addition, depending on the construction and the length of the random peptide insert, many of the aptamers will contain stop codons within the random region. Together, these limitations reduce the sequence space that can be sampled by any one interaction trap selection. In fact, for the 20-mer variable regions described above only $\sim 9 \times 10^{-6}\%$ of the sequence space is represented. As a result, peptide aptamers isolated with the interaction trap may not represent the highest affinity aptamers from all the possible sequences in a random 20-amino acid thioredoxin library.

However, in strict analogy to the recognition molecules of the immune systems, mutation of the aptamer variable regions, followed by selection for higher affinity aptamers, allows isolation of tighter binding aptamers[13,62] (see Fig. 10). Mutations can be introduced into individual aptamer sequences by random PCR mutagenesis or by synthesizing degenerate oligonucleotides with varying degrees of randomness. The stringency of selection in the interaction trap is manipulated by the choice of reporter genes that have LexA operators in different numbers, affinity, and distance upstream of the reporter genes. The *LexA–lacZ* (pSH18-34) and *LexA–LEU2* (in EGY48) reporters used in the "standard" interaction trap detect dissociation constants as weak as $<1\ \mu M$.[63] A series of *lacZ* reporter genes[64] that contain fewer and lower affinity LexA operators (Fig. 5) is used to allow selection of peptide aptamers that show increased affinity for their targets. Reporters containing one and two operators allow the detection of interactions from 20 n*M* to $<1\ \mu M$.[63] Mutagenesis of DNA encoding the aptamer variable region, followed by selection of mutant aptamers in a strain with a low-sensitivity reporter, allows selection of peptide aptamers with higher affinity.

[62] Colas *et al.*, submitted (2000).

[63] J. Estojak, R. Brent and E. A. Golemis, *Mol. Cell. Biol.* **15**, 5820 (1995).

[64] R. Brent, and M. Ptashne, *Cell* **43**, 729 (1985).

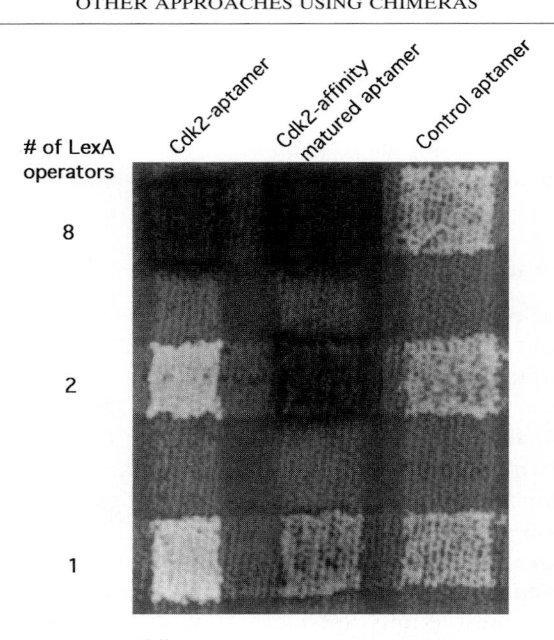

FIG. 10. Affinity maturation.[13,62] Interaction mating assays comparing the affinities of Cdk2 bait for the original Cdk2 aptamer prey and the affinity-matured Cdk2 aptamer prey. Yeast strains (EGY48 *Matα*) containing either the original Cdk2 aptamer, affinity-matured Cdk2 aptamer, or a control aptamer were mated to yeast strains (EGY42 *Mat***a**) containing a Cdk2 bait with a *lacZ* reporter carrying either eight (pSH18-34), two (pJK103), or one (pRB1840) LexA operator. Interactions between the Cdk2 aptamer preys and the Cdk2 bait result in the activation of the *lacZ* reporter. The strength of the interaction is related to the intensity of the blue color on galactose X-Gal plates.

For example, the affinity of an anti-Cdk2 aptamer has been enhanced by using PCR to randomly mutate DNA encoding the variable region, followed by selecting those that bound Cdk2 in a bait strain with a more stringent reporter.[13,62] In this work, a library of 15,000 mutant aptamers was screened for interactions with Cdk2, using a strain that contained a single LexA operator upstream of a *lacZ* reporter gene. A peptide aptamer was recovered with two amino acid changes that activated this *lacZ* reporter. Evanescent wave measurements revealed that the K_d of the aptamer was reduced 20-fold from 10.5 μM to 5 nM. It should be possible to further increase the binding constants of peptide aptamers by using more stringent *lacZ* reporters. Moreover, it should be possible to isolate aptamers of higher affinity by screening larger numbers of mutagenized aptamers, by using more stringent *LEU2* reporters, rather than *lacZ* reporters, as the primary screen, or by using *URA3* reporters in the presence of increasing concentrations of the *URA3* inhibitor 6-azauracil. Finally, it is possible to decrease bait and prey concentrations by changing the copy number of the plasmid encoding them [e.g., by changing the origin of replication from a 2 μm to a CEN autonomous replication sequence (ARS)]. We imagine that the use of such reduced copy bait and prey plasmids may facilitate the isolation of increased affinity aptamers.

Method 5: Affinity Maturation

Reagents

Strain
- EGY48: *Matα his3 trp1 ura3-52 leu2::LexA6op-LEU2*

Vectors
- pBait: pEG202 (Fig. 4) containing the bait protein of interest inserted into the polylinker in frame with LexA
- pJG4-5: Nuclear localization, activation domain, and HA epitope tag fusion for the construction of prey proteins (see Fig. 1 for details)
- Peptide aptamer prey: pJM-1 containing peptide aptamer (see Fig. 1 for details)
- pRB1840: *lacZ* reporter plasmid containing one LexA operator (see Fig. 5 for details)

Media: See method 2 for recipes

Mutagenic Polymerase Chain Reaction. Amplify by PCR the variable region of the aptamer to be mutated in the presence of Mn^{2+}, as described by Cadwell and Joyce.[65]

PROTOCOL 5-1: MUTAGENIC POLYMERASE CHAIN REACTION

1. Add 1 μl of *Taq* polymerase, 12 μl of H_2O, 1 μl of template DNA, and 10 μl of Mg/Mn mixture (45 m*M* $MgCl_2$ and 5 m*M* $MnCl_2$) to 76 μl of PCR premixture.

PCR premixture

Taq polymerase buffer, 10×	500 μl
$MgCl_2$, 1 *M*	5 μl
dATP, 100 m*M*	10 μl
dGTP, 100 m*M*	10 μl
dCTP, 100 m*M*	50 μl
dTTP, 100 m*M*	50 μl
5′ Primer, 20 μ*M*	125 μl
3′ Primer, 20 μ*M*	125 μl
H_2O	2.9 ml

[65] R. C. Cadwell and G. F. Joyce, *PCR Methods Appl.* **2,** 28 (1992).

2. PCR amplify the reaction for 10 cycles. After one cycle remove 13 μl and add it to a solution containing 10 μl of Mg/Mn mixture, 1 μl of *Taq* polymerase, and 76 μl of PCR premixture. Perform the following PCR cycle: 95° for 30 sec, 55° for 1 min, and 72° for 1 min; repeat four times.

Selection of Higher Affinity Aptamers. The PCR-mutagenized variable region is used to construct a small library of mutagenized aptamers. Higher affinity aptamers are identified from this pool with selection strains that contain a low-sensitivity *lacZ* reporter (pRB1840).

Protocol 5-2: Selection of Mutagenized Aptamers

1. Introduce the mutated variable region from protocol 5-1 into *Rsr*II-cut pJM-1, using standard subcloning techniques.
2. Transform the mutagenized peptide aptamer library into EGY48 (protocol 2-1) that contains the pBait of interest and pRB1840. Select transformants on Glu His^- Ura^- Trp^- plates.
3. Replica plate the transformants onto Gal/Raf His^- Ura^- Trp^- X-Gal and Glu His^- Ura^- Trp^- X-Gal plates.
4. Incubate at 30° until galactose-dependent blue colonies are visible.
5. Rescue the plasmids from the galactose-dependent blue colonies (protocol 2-6) and reintroduce the plasmids into EGY48 containing the pBait of interest and pRB1840 to reconfirm the interaction phenotype.
6. Rescue the plasmid from the galactose-dependent blue colonies and sequence their variable regions.

Forward Genetics with Peptide Aptamers

Classic forward genetics involves isolating organisms displaying altered phenotypes and identifying the mutations responsible for the phenotypic changes. Forward genetic analysis is extremely powerful for characterizing cellular processes; however, it is limited by the ability to detect mutations causing the phenotypic change. Mutations causing phenotypic changes are readily identified in haploid organisms. In contrast, only dominant mutations can easily be detected in diploid organisms. Although in some diploid organisms recessive mutations are discernible, their detection requires the generation of organisms homozygous for the recessive mutation. Because creation of these homozyotes requires two generations of breeding, use of homozygous recessive mutations is generally confined to organisms with well-developed genetics.

To alleviate the problem of analyzing recessive mutations in diploid organisms, dominant agents have been developed that mutate gene function without altering the genetic material. Combinatorial peptide aptamer libraries can function as dominant "mutagenic" that randomly inactivate gene function. Forward analysis of cellular processes with peptide aptamers involves expressing combinatorial peptide libraries in cells and screening or selection for aptamer-induced variations of cellular processes. The specific proteins and key points of disruption in the cellular process are subsequently identified by two-hybrid experiments. The advantage of using peptide aptamers for analyzing cellular processes is that they execute their "mutagenic" function at the protein level, eliminating problems associated with analyzing recessive mutations in diploid organisms. This strategy allows traits in diploid organisms and multicopy gene phenotypes to be examined.

For example, we and others have used aptamers in the forward analysis of a complex phenotype, the yeast pheromone response pathway, to identify genes and protein interactions involved in causing the phenotype.[20–22] In the presence of mating pheromone, yeast arrest at the G_1 phase in the cell cycle.[5] Peptide aptamers that inhibited the pheromone response pathway were selected on the basis of their ability to grow in the presence of mating pheromone. The protein targets of the selected peptide aptamers were identified with yeast two-hybrid systems.

Mating interaction assays successfully identified peptide aptamer targets from panels of proteins known to be involved in the pheromone response pathway[20,21] as well as from a larger panel containing proteins from the entire yeast genome.[22] We have also identified peptide aptamer targets by using them as baits in two-hybrid experiments to find interacting proteins in a yeast partial genome interaction library.[21] Interestingly, this interactor hunt identified *CBK1,* a gene not previously known to be involved in the pheromone response pathway.

This result demonstrated the benefits of combining library screening with mating interaction assay panels. However, due in part to the poor quality of the partial yeast genomic library, the library screens were not as effective as the interaction mating panels for identifying the aptamer targets. Many of the peptide aptamer targets that were identified with the mating interaction assay were not observed in the interaction trap hunt using the yeast genomic library.[21] These results illustrate the utility of identifying targets by mating with ordered yeast arrays rather than by library screens. Mating interactions against ordered arrays of proteins have the advantage of (1) presenting the potential interactors as a fully normalized library, (2) easily detecting interactions that result in activation above the basal activation of the bait alone, and (3) detecting interactions that are independent of differences in the plating efficiency due to differences in reporter activation.[63] However, until complete panels of proteins are available for a variety of organisms, mating interactions with limited panels of known proteins will still need to be supplemented with two-hybrid genomic and cDNA library selections to identify peptide aptamer targets.

Method 6: Forward Analysis of Cellular Process in Yeast with Peptide Aptamer Mutagens

Reagents

Vectors

- pJM-2: Peptide aptamer vector for use in yeast genetic selections (see Fig. 1 for details)
- pJM-3: Peptide aptamer vector containing a nuclear localization signal for use in yeast genetic selections (see Fig. 1 for details)
- pEG202: LexA fusion plasmid for the construction of bait proteins (pBait) (see Fig. 4 for details)
- pSH18-34: *lacZ* reporter plasmid (see Fig. 5 for details)
- pJG4-5: Nuclear localization, activation domain, and HA epitope tag fusion for the construction of prey proteins (see Fig. 1 for details)

Media: See method 2 for recipes

Yeast Genetic Selection of Peptide Aptamers. The design of yeast genetic selection strategies is beyond the scope of this review. A typical genetic selection of peptide aptamers involves transforming a peptide aptamer expression library consisting of at least 10^6–10^7 members into the selection strain. On the basis of results obtained with the yeast pheromone response pathway, approximately 1 of every 10^5–10^6 TrxA peptide aptamers with 20-mer variable regions is able to inhibit this process.[21] Therefore, a collection of yeast containing approximately 10^6–10^7 unique 20-mer aptamers should possess enough aptamer diversity to inhibit at least many cellular processes.

In this procedure, the peptide aptamer transformation is plated under nonselective conditions to allow uniform representation of colonies containing individual peptide aptamers. The transformants are then scraped from the plates and pooled as a frozen stock. An aliquot of the stock containing 10 equivalents of the library is subjected to the selection conditions. Plasmids encoding the peptide aptamer from the selected colonies are isolated and retransformed into the selection strain to reconfirm the aptamer phenotype. In the procedure given below, the peptide aptamers are conditionally expressed by the yeast *GAL1* promoter. The inducible promoter allows the confirmation of aptamer-dependent phenotypes by comparing the phenotypes of the yeast with the aptamer in the presence and absence of the inducer.

PROTOCOL 6-1: YEAST GENETIC SELECTION OF PEPTIDE APTAMERS WITH A GALACTOSE-INDUCIBLE PROMOTER

1. Transform selection strain with 50–100 μg of pJM-2 or pJM-3 peptide aptamer library to give 10^6–10^7 transformants (protocol 2-1). Plate transformation on Glu Trp$^-$ plates and incubate at 30° until colonies form.

2. Scrape the yeast cells with a glass spreader from the plate by adding 1–2 ml of H_2O to a 24 × 24 cm plate and collect the yeast in a 50-ml Falcon tube. Pool the transformants from all plates and centrifuge at 3500 rpm for 4 min at room temperature. Remove the supernatant and resuspend the pellet in 25 ml of H_2O. Centrifuge again and wash the pellet once more in H_2O.

3. Resuspend the pellet in H_2O and add an equal volume of 65% (v/v) glycerol, 0.1 *M* $MgSO_4$, 25 m*M* Tris-HCl, pH 7.4. Divide into 1-ml aliquots and freeze at −70°.

4. Determine the plating efficiency of the frozen aliquot. Inoculate 10 library equivalents in 1 ml of Gal/Raf Ura$^-$ His$^-$ Trp$^-$ liquid medium. Incubate at 30° for 4 hr with shaking.

5. Pellet the yeast cells by centrifuging at 3500 rpm for 4 min at room temperature. Remove the supernatant and resuspend the pellet in H_2O.

6. Plate yeast onto Gal/Raf Trp$^-$ selection plates and incubate under selection conditions.

7. Streak positive colonies onto Glu Trp$^-$ plates.

8. Replica plate the master plate onto Glu Trp$^-$ and Gal/Raf Trp$^-$ selections plates and incubate under selection conditions.

9. Isolate aptamer-encoding plasmids from colonies showing the galactose-dependent phenotype (protocol 2-6).

10. Retransform the peptide aptamers into the selection strain and reconfirm the galactose-dependent phenotype.

11. Isolate the galactose-dependent aptamer encoding plasmids from step 10 (protocol 2-6) for further characterization and sequencing.

Identification of Peptide Aptamer Targets. The targets of the peptide aptamers obtained from a genetic selection can be identified by interaction mating experiments[51] or by interaction trap hunts against genomic or cDNA libraries.[47] The peptide aptamer baits are constructed by transferring the aptamer encoding DNA from the expression vector used in the genetic selection to the bait vector, pEG202 (Fig. 4). The aptamer baits are then screened for interactions against panels of known prey proteins or genomic or cDNA libraries of prey proteins. Panels of prey proteins are constructed by inserting the coding region of a desired protein into pJG4-5 (Fig. 1).

The construction of genomic and cDNA libraries is beyond the scope of this review; however, details of their construction can be found elsewhere.[47,66,67]

Identification of peptide aptamer targets by the mating interaction assay is performed as described in method 3, with the following alterations. To identify aptamer targets, aptamers are expressed as baits rather than preys as in method 3, because the same aptamers can also be screened for interactions against genomic and cDNA libraries. Genomic and cDNA libraries are constructed as preys because they contain many sequences capable of activating transcription in the bait configuration. The peptide aptamer targets are identified by transforming the aptamer baits and pSH18-34 (*lacZ* reporter; Fig. 5) into EGY48 (*Mat*α) and the library of preys into EGY42 (*Mat***a**). The strains are mated and interactions between baits and preys are scored on interaction detection plates.

Identification of the peptide aptamer targets by interaction trap library hunts is performed as described in method 2. However, instead of screening a library of peptide aptamers preys for interactions against a specific protein bait, a library of protein preys is screened for interactions against a specific peptide aptamer bait. The peptide aptamer targets are identified in the interaction hunt by transforming the peptide aptamer baits and pSH18-34 into EGY48. A genomic or cDNA library is then introduced into this strain. Cells that contain a prey protein that interacts with the aptamer bait are detected by the activation of their *LexAop–LEU2* reporters, allowing them to form colonies on Leu$^-$ medium. Aptamer–target protein interactions that activate the *LexAop–LEU2* reporter are subsequently tested for their ability to activate a *LexAop–lacZ* reporter.

Once putative peptide aptamer targets are identified with the two-hybrid system, their significance can be further corroborated by a variety of approaches. Immunoprecipitation can be used to ensure that the aptamer–target complex forms under *in vivo* conditions. Epistasis analysis of the peptide aptamer can be used to confirm that it functions in the same "neighborhood" of a known pathway as the protein target. The peptide aptamer target can be deleted or overexpressed and the resulting phenotypic consequences compared with those caused by the peptide aptamer. Finally, we have extended the interaction trap to allow identification of specific protein interactions that the aptamer disrupts.[21]

Concluding Statements

In this review, we have described properties of peptide aptamers and methods to isolate them. We have also provided examples describing the

[66] U. Gubler and B. J. Hoffmann, *Gene* **16,** 263 (1983).
[67] W. D. Huse and C. Hansen, *Stratagene Strategies* **1,** 1 (1988).

use of this new class of dominant "genetic" agent for the analysis of cellular processes. Although peptide aptamers are still in their early stages of development, the pilot studies presented make evident a variety of applications for peptide aptamers in the analysis of cellular processes. Peptide aptamers are particularly well suited for this task, as they can (1) bind a variety of protein targets, (2) bind protein targets with high affinities, (3) specifically recognize their protein targets, and (4) function effectively within cells. Peptide aptamers can inhibit protein function by a variety of mechanisms. Peptide aptamers can disrupt protein interactions within cells[14,18,54] and in two-hybrid assays.[21] In addition, peptide aptamers can competitively compete with substrates for the active site of enzymes.[54] Alternatively, peptide aptamers can be used to modify their protein targets. Peptide aptamers with nuclear localization signals can mislocalize their protein targets to the nucleus.[62] Peptide aptamers have also been used to direct a ubiquitin ligase to a specific protein target.[62]

Peptide aptamers targeted to specific proteins can be used in the reverse analysis of cellular processes. By disrupting the function of their protein targets, peptide aptamers can be used to analyze the phenotypic consequences of individual genes. The high specificity of peptide aptamers enables them to inhibit specific protein–protein interactions while leaving other interactions with the target protein intact.[21] Unlike gene knockout phenotypes, peptide aptamers should enable the analysis of phenotypes caused by the disruption of individual protein interactions in genetic networks. The ability of peptide aptamers to recognize allelic variants of proteins will aid in the functional characterization of polymorphic protein variants now being rapidly identified by genome sequencing projects.[68] In addition, peptide aptamer expression levels and timing can be controlled with inducible promoters. This fact will allow the penetrance and timing of the aptamer phenotype to be controlled.

Random peptide aptamer expression libraries can be used in the forward analysis of cellular processes. Peptide aptamer "mutagens" can be identified that disrupt a cellular process. These aptamers can be subsequently used to identify proteins and protein interactions required for the process. In addition to identifying protein targets, the location of the target protein in a cell can be determined. Peptide aptamer libraries with different localization signals can be used in genetic selections. Further, a comparison of the protein targets obtained from each selection would permit researchers to deconvolute the locations of individual proteins in a genetic pathway.

In higher eukaryotes, various processes such as senescence and metastasis are not well understood. The success of dominant peptide aptamers in

[68] M. D. Adams *et al., Nature* (*London*) **377**(Suppl.), 3 (1995).

the reverse analysis of cellular processes in mammalian cell culture and in *Drosophila* indicates that peptide aptamer expression libraries will be useful for the forward analysis of cellular process in higher eukaryotes. Entire libraries of peptide aptamers have been utilized to "mutagenize" mammalian cells, to try to identify proteins that function in particular processes (A. Colman-Lerner, personal communication, 1999). Such forward "genetic" analysis with random peptide aptamer expression libraries should expand our ability to analyze processes in genetically intractable organisms.

Acknowledgments

We thank Alejandro Colman-Lerner, Andrew Mendelsohn, and Barak Cohen for helpful discussions, and Barak Cohen and Pierre Colas for permission to cite unpublished data. C.R.G. was supported by a postdoctoral fellowship from the National Research Council of Canada. Work on peptide aptamers is supported by a grant to R.B. from the NIH National Institute of General Medical Sciences, while work on interaction mating and the two-bait system was supported by the NIH National Human Genome Research Institute.

[14] Detection of Protein–Protein Interactions by Protein Fragment Complementation Strategies

By Stephen W. Michnick, Ingrid Remy, François-X. Campbell-Valois, Alexis Vallée-Bélisle, and Joelle N. Pelletier

Introduction

Much of modern biological research is concerned with identifying proteins involved in cellular processes, determining their functions and how, when, and where they interact with other proteins involved in specific biochemical pathways. Advances in genome projects have led to rapid progress in the identification of novel genes. In applications to biochemical research, there is now the pressing need to determine the functions of novel gene products. It is in addressing questions of function that genomics-based research becomes bogged down and there is now the need for advances in the development of simple and automatable functional assays. Identifying proteins by functional cloning of novel genes remains a significant experimental challenge. Many ingenious strategies have been devised to simultaneously screen cDNA libraries in the context of assays that allow both

0076-6879/00 $30.00

selection of clones and validation of their biological relevance.[1–4] However, in the absence of an obvious functional assay that can be combined with cDNA library screening, researchers have turned to strategies that use as readout some general functional properties of proteins. A first step in defining the function of a novel gene is to determine its interactions with other gene products in an appropriate context; that is, because proteins make specific interactions with other proteins as part of functional assemblies, an appropriate way to examine the function of the product of a novel gene is to determine its physical relationships with the products of other genes. This is the basis, in part, of the highly successful yeast two-hybrid system.[5,6] In its most commonly used form, interactions detected using the yeast two-hybrid strategy occur in the yeast nucleus. Detection is based on transcriptional activation resulting from these specific interactions and requires cell-specific components in order for interactions to give rise to a detectable signal. The success of this strategy in identifying biologically significant protein–protein interactions has been well documented, whether between two specific partners or between a "bait" and "prey" library.[7,8]

In addition to detecting protein–protein interactions as an expression cloning strategy, there is intense interest in exploring the physical basis of the molecular complementarity of proteins. Phage display and related approaches have been devised to screen linear or structurally constrained peptide libraries to identify characteristic sequence and structural motifs for classes of protein–protein interactions.[9] Here, a library of proteins or peptides is expressed as a fusion to a phage coat protein, such that it is solvent-exposed and can be "panned" *ex vivo* for binding against an immobilized target of interest that may be a protein.[10] Phage display has the advantage of allowing for screening of large libraries (up to 10^{12}) due to the success of infection compared with transformation. Further, the ease of *ex vivo* panning means that full library representation can be achieved. Nonetheless, protein–protein interactions detected by phage display may not be representative of those occurring *in vivo,* as the selection conditions

[1] A. Aruffo and B. Seed, *Proc. Natl. Acad. Sci. U.S.A.* **84,** 8573 (1987).
[2] A. D. D'Andrea, H. F. Lodish, and G. G. Wong, *Cell* **57,** 277 (1989).
[3] H. Y. Lin, X. F. Wang, E. Ng-Eaton, R. A. Weinberg, and H. F. Lodish, *Cell* **68,** 775 (1992).
[4] D. Sako, X. J. Chang, K. M. Barone, G. Vachino, H. M. White, G. Shaw, G. M. Veldman, K. M. Bean, T. J. Ahern, and B. Furie, *Cell* **75,** 1179 (1993).
[5] C. Evangelista, D. Lockshon, and S. Fields, *Trends Cell Biol.* **6,** 196 (1996).
[6] S. Fields and O. Song, *Nature* (*London*) **340,** 245 (1989).
[7] B. L. Drees, *Curr. Opin. Chem. Biol.* **3,** 64 (1999).
[8] M. Vidal and P. Legrain, *Nucleic Acids Res.* **27,** 919 (1999).
[9] G. P. Smith, *Science* **228,** 1315 (1985).
[10] D. J. Rodi and L. Makowski, *Curr. Opin. Biotechnol.* **10,** 87 (1999).

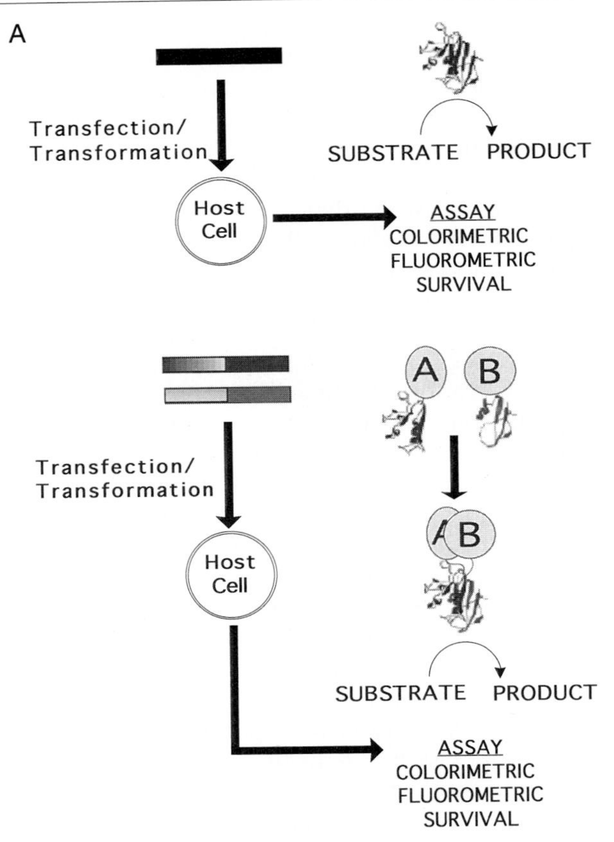

FIG. 1. (A) An enzyme can be transformed/transfected into a host cell and its activity detected by an *in vivo* assay. Oligomerization domains A and B are fused to N- and C-terminal fragments of the gene for that enzyme. Cotransformation/transfection of oligomerization domain–fragment fusions results in reconstitution of enzyme activity by oligomerization domain-assisted reassembly of the enzyme. Reassembly of enzyme will not occur unless oligomerization domains interact. (B) Schematic kinetic model of induced fragment complementation in PCA. K_f and K_{unf} are the folding and unfolding rate constants of the two fragments. It is important to make a clear distinction between the bimolecular rate constant of folding (K_f) and the observed folding rate of the fragments, as the latter is concentration-dependent while the first constant is intrinsic to the nature of the two fragments.

differ from the natural interaction environment. This results in the occurrence of nonspecific binding to the panning supports or to other proteins.

Our laboratory has developed a general strategy for detecting protein–protein interactions in intact cells based on protein fragment complementation assays (PCAs). The gene coding for an enzyme is rationally dissected

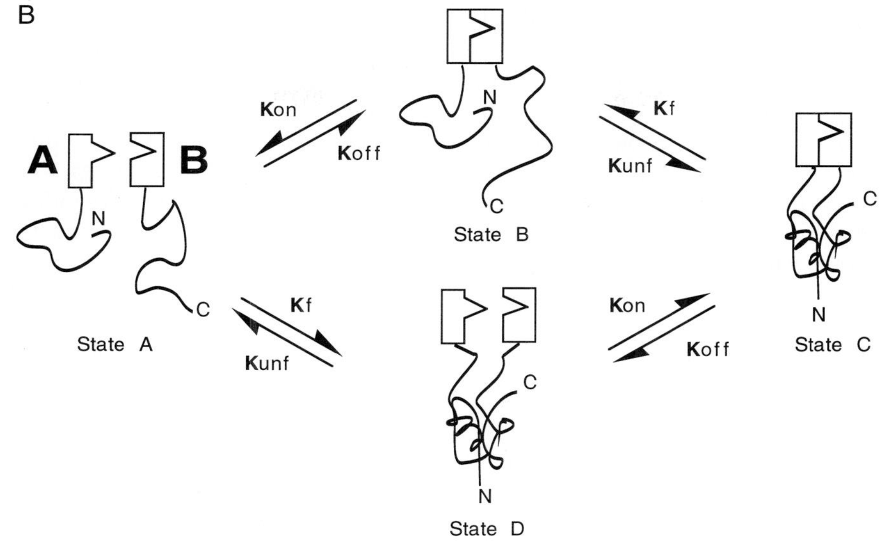

FIG. 1. (*continued*)

into two pieces. Fusion proteins are constructed with two proteins that are thought to bind to each other, fused to either of the two probe fragments. Folding of the probe protein from its fragments is initiated by the binding of the test proteins to each other, and is detected as reconstitution of enzyme activity (Fig. 1A). Our goal in developing the PCA strategy was to devise a way to combine the elegance of protein interaction detection afforded by the yeast two-hybrid approach, with true, context-dependent functional studies as in specific expression cloning strategies. In a general way, such a strategy would be capable of (1) detection of binary protein–protein or higher order interactions *in vivo,* (2) detection of these interactions in appropriate contexts, such as within as specific organism, cell type, cellular compartment, or organelle, (3) distinguishing induced versus constitutive protein–protein interactions (such as by a cell growth or inhibitory factor), (4) distinguishing specific versus nonspecific protein–protein interactions, (5) detection of the kinetics and equilibrium behavior of protein assembly, and (6) being useful for screening cDNA or defined oligonucleotide libraries. In this chapter we discuss how all of these criteria can be met by the PCA strategy.

Here we present the basic concept of PCAs and how they are designed, with particular attention to the first system developed, based on murine dihydrofolate reductase (mDHFR).[11–15] We then present several applications of the assay, including a simple, large-scale library-versus-library screening strategy in *Escherichia coli*.[16,17] We then discuss the implementation of mammalian assays, including applications to the quantitative detection of induced protein interactions and allosteric transitions, in intact cells.[14,15] Finally, we demonstrate the generality of the PCA strategy with examples of assays we have designed on the basis of other enzymes including glycinamide ribonucleotide transformylase, aminoglycoside kinase, and hygromycin B kinase.

Basic Concepts

Polypeptides have evolved to encode all the chemical information necessary to spontaneously fold into a stable, unique three-dimensional structure.[18–20] It logically follows that the folding reaction can be driven by the interaction of two peptides that together contain the entire sequence, and in the correct order of a single peptide that will fold. This was demonstrated in the classic experiments of Richards[21] and Taniuchi and Anfinsen.[22] In practice this does not work easily; the major driving force for protein folding is the hydrophobic effect, but so also is nonspecific aggregation. However, if soluble oligomerization domains are added to the fragments that, by interacting, increase the effective concentration of the fragments, correct folding could be favored over any other nonproductive process.[11,13,23] If the protein that folds from its constitutive fragments is an enzyme, whose activity could be detected *in vivo,* then the reconstitution of its activity can be used as a measure of interaction of the oligomerization domains. Further,

[11] J. N. Pelletier and S. W. Michnick, *Protein Eng.* **10,** 89 (1997).
[12] J. N. Pelletier, I. Remy, and S. W. Michnick., *J. Biomol. Tech. Acc. No. S0012* (1998).
[13] J. N. Pelletier, F.-X. Campbell-Valois, and S. W. Michnick, *Proc. Natl. Acad. Sci. U.S.A.* **95,** 12141 (1998).
[14] I. Remy I., A. Wilson, and S. W. Michnick, *Science* **283,** 990 (1999).
[15] I. Remy and S. W. Michnick, *Proc. Natl. Acad. Sci. U.S.A.* **96,** 5394 (1999).
[16] K. M. Arndt, J. N. Pelletier, K. M. Müller, T. Alber, S. W. Michnick, and A. Plückthun, *J. Mol. Biol.* **295,** 627 (2000).
[17] J. N. Pelletier, K. M. Arndt, P. A. Plückthun, and S. W. Michnick, *Nature Biotechnol.* **17,** 683 (1999).
[18] C. B. Anfinsen, E. Haber, M. Sela, and F. H. White, Jr., *Proc. Natl. Acad. Sci. U.S.A.* **47,** 1309 (1961).
[19] C. B. Anfinsen, *Science* **181,** 223 (1973).
[20] B. Gutte and R. B. Merrifield, *J. Biol. Chem.* **246,** 1922 (1971).
[21] F. M. Richards, *Proc. Natl. Acad. Sci. U.S.A.* **44,** 162 (1958).
[22] H. Taniuchi and C. B. Anfinsen, *J. Biol. Chem.* **216,** 2291 (1971).
[23] N. Johnsson and A. Varshavsky, *Proc. Natl. Acad. Sci. U.S.A.* **91,** 10340 (1994).

this binary, all or none folding event provides for a specific measure of protein interactions dependent not on mere proximity, but on the absolute requirement that the peptides must be oriented precisely in space to allow for folding of the enzyme from the polypeptide chain. As discussed below, we have selected to dissect proteins into fragments that are subdomains of the complete protein but should not be capable of spontaneously folding along with their complementary fragments into functional and complete proteins. If this were true, the PCA strategy would not work, as illustrated in Fig. 1B. Further, to assure that spontaneous fragment complementation does not occur, a set of controls must always be performed with any pair of interacting proteins. These controls should include (1) *noninteracting proteins*—a PCA response should not be observed if a protein that is not known to interact with either of two interacting proteins being tested is used as a PCA partner, nor should overexpression of this protein alone compete for the known interaction; and (2) *partner protein interface mutations*—point or deletion mutations of a partner that are known to disrupt an interaction should also prevent a PCA response. These are the most crucial controls to demonstrate that an observed PCA response is due to specific association of two proteins. Some additional controls could include (1) *competition*—a PCA response should be diminished by simultaneous overexpression of one or the other interacting partners alone; (2) *expression levels*—expression of each of the domain–fragment fusions should be equal and titratable; and (3) *fragment swapping*—an observed interaction between oligomerization domains should occur regardless of whether partner proteins are attached to one or the other PCA fragments.

Selection of Enzymes for a Protein Fragment Complementation Assay

In designing PCAs, we have sought to identify enzymes for which the following is true: (1) The enzyme is relatively small and monomeric. This assures that detected protein–protein interactions are binary, not of higher order; (2) comprehensive literature of structural and functional information is available for the enzyme; (3) simple assays are available both *in vivo* and *in vitro* for the enzyme; and (4) overexpression of the enzyme in eukaryotic and prokaryotic cells has been demonstrated. Murine DHFR, the example we discuss in detail here, meets all these criteria. Prokaryotic and eukaryotic DHFR is central to cellular one-carbon metabolism and is absolutely required for cell survival in both prokaryotes and eukaryotes. Specifically, it catalyzes the reduction of dihydrofolate to tetrahydrofolate for use in transfer of one-carbon units required for biosynthesis of serine, methionine, pantothenate (in prokaryotes), purines, and thymidylate.[24] The DHFRs are small (17 to 21 kDa), monomeric proteins. The crystal structures of DHFR from various bacterial and eukaryotic sources are known and substrate-binding sites and active site residues have been determined,[25–28] allowing for rational design of protein fragments. The folding, catalysis, and kinetics of a number of DHFRs have been studied extensively.[29–32] The enzyme activity can be monitored *in vitro* by a simple spectrophotometric assay[33] or *in vivo* by cell survival in cells grown in the absence of DHFR end products. DHFR is specifically inhibited by the antifolate drug trimethoprim. As mammalian DHFR has a 12,000-fold lower affinity for trimethoprim than does bacterial DHFR,[34] growth of bacteria expressing mDHFR in the presence of trimethoprim levels lethal to bacterial is an efficient means of selecting for reassembly of mDHFR fragments into active enzyme. mDHFR expression in cells can also be monitored by binding of fluorescent high-affinity substrate analogs for DHFR.[35,36] Finally, mDHFR is used routinely to demonstrate heterologous expression of protein in transformed prokaryotic or transfected eukaryotic cells. In both cases expression has been demonstrated to occur at high levels.[37–41]

Protein Fragment Complementation Assay Design Considerations

The design of a PCA starts with the known three-dimensional structure of the enzyme to be used. This is crucial to design, as structural and func-

[24] J. R. Bertino and B. L. Hillcoat, *Adv. Enzyme Regul.* **6,** 335 (1968).
[25] K. W. Volz, D. A. Matthews, R. A. Alden, S. T. Freer, C. Hansch, B. T. Kaufman, and J. Kraut, *J. Biol. Chem.* **257,** 2528 (1982).
[26] C. Bystroff and J. Kraut, *Biochemistry* **30,** 2227 (1991).
[27] D. J. Filman, J. T. Bolin, D. A. Matthews, and J. Kraut, *J. Biol. Chem.* **257,** 13663 (1982).
[28] C. Oefner, A. D'Arcy, and F. K. Winkler, *Eur. J. Biochem.* **174,** 377 (1988).
[29] B. E. Jones and C. R. Matthews, *Protein Sci.* **4,** 167 (1995).
[30] P. A. Jennings, B. E. Finn, B. E. Jones, and C. R. Matthews, *Biochemistry* **32,** 3783 (1993).
[31] J. Thillet, J. A. Adams, and S. J. Benkovic, *Biochemistry* **29,** 5195 (1990).
[32] J. Andrews, C. A. Fierke, B. Birdsall, G. Ostler, J. Feeney, G. C. Roberts, and S. J. Benkovic, *Biochemistry* **28,** 5743 (1989).
[33] B. L. Hillcoat, P. F. Nixon, and R. L. Blakley, *Anal. Biochem.* **21,** 178 (1967).
[34] J. R. Appleman, N. Prendergast, T. J. Delcamp, J. H. Freisheim, and R. L. Blakley, *J. Biol. Chem.* **263,** 10304 (1988).
[35] G. B. Henderson, A. Russell, and J. M. Whiteley, *Arch. Biochem. Biophys.* **202,** 29 (1980).
[36] R. J. Kaufman, J. R. Bertino, and R. T. Schimke, *J. Biol. Chem.* **253,** 5852 (1978).
[37] R. J. Kaufman, *Proc. Natl. Acad. Sci. U.S.A.* **82,** 689 (1985).
[38] H. Hao, M. G. Tyshenko, and V. K. Walker, *J. Biol. Chem.* **269,** 15179 (1994).
[39] P. Loetscher, G. Pratt, and M. Rechsteiner, *J. Biol. Chem.* **266,** 11213 (1991).
[40] R. J. Kaufman, M. V. Davies, V. K. Pathak, and J. W. Hershey, *Mol. Cell. Biol.* **9,** 946 (1989).
[41] T. Grange, F. Kunst, J. Thillet, B. Ribadeau-Dumas, S. Mousseron, A. Hung, J. Jami, and R. Pictet, *Nucleic Acids Res.* **12,** 3585 (1984).

tional constraints must be met in deciding the optimal site at which to cleave the polypeptide. These constraints include selecting sites that are outside of catalytic or cofactor-binding sites, for obvious reasons, and preferably in disordered regions as opposed to within regions of periodic secondary structure.[42] A second set of constraints concerns how best to fuse the fragments to oligomerization domains, specifically what choices of amino or carboxy termini to use. This depends on the conception of how a protein folds from fragments. If the native structure of the enzyme used for the PCA is taken as the starting point, it would seem that the best strategy would be to attach the oligomerization domains, via a flexible peptide linker, in such a way that in the final, folded state, the N and C termini are correctly juxtaposed (Fig. 1B). This is certainly true, as we discuss below in the context of detecting allosteric transitions, but it is not necessarily the best configuration. It could also be imagined that instead of starting from the native structure, it should instead be assumed that the unfolded polypeptide is the starting point. In this case the best configuration of oligomerization domains and fragments would be that in which association of the oligomerization domains brings the peptides into an arrangement in which the peptide is contiguous with the intact, unfolded polypeptide. We do not know that either arrangement is better than the other. For the DHFR assay, we have not observed any difference in how the assays perform with the alternative configuration, whereas for the assays based on aminoglycoside phosphotransferase and hygromycin B phosphotransferase described below, only the second, alternative configuration works.

Design of the Murine Dihydrofolate Reductase Protein Fragment Complementation Assay

mDHFR shares high sequence identity with the human DHFR (hDHFR) sequence (91% identity) and is highly homologous to the *E. coli* enzyme (29% identity, 68% homology), and these sequences share visually superimposable tertiary structure.[25] Comparison of the crystal structures of mDHFR and hDHFR suggests that their active sites are identical.[28,43] DHFR has been described as being formed of three structural fragments forming two domains,[26,44] the adenine-binding domain (residues 47 to 105; fragment [2]) and a discontinuous domain [residues 1 to 46 (fragment [1])

[42] H. Taniuchi, G. R. Parr, and M. A. Juillerat, *Methods Enzymol.* **131,** 185 (1986).

[43] D. K. Stammers, J. N. Champness, C. R. Beddell, J. G. Dann, E. Eliopoulos, A. J. Geddes, D. Ogg, and A. C. North, *FEBS Lett.* **218,** 178 (1987).

[44] C. V. Gegg, K. E. Bowers, and C. R. Matthews, *in* "Techniques in Protein Chemistry" (D. R. Marshak, ed.), p. 439. Academic Press, New York, 1996.

and 106 to 186 (fragment [3]); numbering according to the murine sequence]. The folate-binding pocket and the NADPH-binding groove are formed mainly by residues belonging to fragments [1] and [2]. Fragment [3] has few crystal contacts with the substrates and is not directly implicated in catalysis.

Residues 101 to 108 of hDHFR, at the junction between fragment [2] and fragment [3], form a disordered loop between α-helix E and β-strand E; the loop lies on the same face of the protein as both termini. Cleavage of mDHFR at this loop and fusion of the native termini has produced a circularly permuted protein with physical and kinetic properties similar to the native enzyme.[45] We chose to cleave mDHFR between fragments [1, 2] and [3], just after residue 105, so as to cause minimal disruption of the active site and NADPH cofactor-binding sites. The native N terminus of mDHFR and the novel N terminus created by cleavage occur on the same surface of the enzyme,[28,43] allowing for ease of N-terminal covalent attachment of each fragment to associating proteins. Cleavage of the *E. coli* DHFR into the fragments corresponding to the MDHFR fragments [1,2] and [3] (EcDHFR 1-86 and 87-159) has shown that neither fragment folds independently *in vitro*.[46]

Application of the Dihydrofolate Reductase Protein Fragment Complementation Assay to *in Vivo* Screening of Designed Libraries of Interacting Proteins

We have demonstrated the application of the mDHFR PCA to a simple strategy for large-scale library-versus-library screening of heterodimeric coiled-coil forming sequences in *E. coli.*[16,17] This strategy could have broad applications in protein design, in exploring the details of interactions between proteins or protein subdomains, and in a simple cDNA library screening strategy.

As with any screening technique, it is necessary to introduce stringency of some sort, to select for maximal affinity and/or specificity. We have devised three selection strategies for the DHFR PCA in *E. coli,* each type having different levels of stringency. In the lowest stringency case, we screen two expressed libraries against each other, beginning with a single-step selection, thereby identifying all interacting partners. In the second, we increase the stringency of the selection, by using a mutant DHFR fragment (Ile114Ala) that prevents the stable reassembly of DHFR from the frag-

[45] A. Buchwalder, H. Szadkowski, and K. Kirschner, *Biochemistry* **31,** 1621 (1992).
[46] C. V. Gegg, K. E. Bowers, and C. R. Matthews, *Protein Sci.* **6,** 1885 (1997).

ments and should thus require more efficiently interacting partners to produce enough reconstituted enzyme for propagation. Finally, we introduce competitive metabolic selection, in which clones obtained in the second type of screening are pooled and passaged through several rounds of competition selection in order to enrich for optimally interacting partners.

By simultaneously screening two libraries against each other, we have demonstrated the advantages of screening a large, combinatorial sequence space to identify stably heterodimerizing pairs. We partially sampled a sequence space of 1.72×10^{10} combinations to select novel leucine zipper pairs with characteristics consistent with stable and specific heterodimerization.

Varying the Stringency of Selection

Use of the Ile-114 Mutants. The interface between the two fragments is composed in part of a hydrophobic β sheet formed mostly by fragment [3] but that includes a β strand from fragment [1, 2]. We designed the mutations Ile114Val, Ile114Ala, and Ile114Gly,[13] which sequentially reduce the volume of a side chain that is directly at the fragment interface but distant from the active site, according to the structure of the highly conserved avian enzyme.[25] We hypothesized that these mutations should decrease the association of the fragments by destabilizing the hydrophobic contacts at the interface, with an increasingly severe effect (Val > Ala > Gly). Indeed, the mutations impair growth of bacteria expressing reconstituted mDHFR in the predicted fashion. In conjunction with the homodimerizing GCN4 coiled coil, there was no significant difference in growth rate with the valine mutant, while the rate was 2.5 times slower with the alanine mutant (as quantitated in liquid culture). Again, no growth was observed with the glycine mutant.[13] Thus, the severity of the effect correlates with the difference in side-chain volume.

We hypothesized that, in order to restore the bacterial growth rate to maximal levels when using a mutant fragment, it would be necessary to provide partner proteins that interact more stably. Thus the increased interaction between the partner proteins should improve the reconstitution of the destabilized mDHFR. We tested this hypothesis in the selection of coiled-coil partner proteins from semirandomized libraries. We immediately observed a 50-fold increase in selection stringency when using the Ilel14Ala mutant. The features in the resulting coiled coils were consistent with selection for more stable pairs with a higher requirement for parallel, in-register heterodimerization than in selection using the wild-type fragments.[16,17] Thus, the wild-type and Ilel14 mutant resulted in differing selection stringencies, where destabilization of the mDHFR fragment interface was compensated by more stably and specifically interacting partners. The Ile114Val mutant could also be applied to selection strategies, and should yield a stringency intermediate between the wild-type and the Ile114Ala mutant. In principle, it should also be possible to apply the Ile114Gly mutant to obtain yet higher selection stringency. However, as we have not yet observed enzymatic activity when using this mutant with a variety of partner proteins, it is possible that the interface is too perturbed in this mutant to allow fragment reconstitution.

Application of Metabolic Competition. The identification of the best-performing enzymes from a population, by functional selection, is a straightforward process. We have used this approach in the selection of the best-performing reconstituted mDHFRs. As a result of the absolute requirement of cells for the product of DHFR turnover, a limiting level of DHFR activity determines the growth rate. In a mixed clonal population in which the mDHFR in each clone is reconstituted by a different interacting partner pair, the most efficiently interacting partners should allow for higher DHFR turnover, leading to a higher growth rate for this clone. Therefore, liquid culture can be enriched for the fastest growing bacteria, which express the most efficiently interacting pairs (Fig. 2). We have demonstrated the direct link between the rate of clonal enrichment, and stability and specificity of partner association, by analyzing the results of such a metabolic "competition selection" of the coiled-coil partner proteins from semirandomized libraries. We observe rapid convergence toward predominant populations of clones. Figure 3 shows the results of such a metabolic screening, in which the starting library-versus-library screening would represent 1.7×10^{10} possible interactions. In practice we could screen 2.0×10^{6} of these and at the final passage (P12) there emerged a predominant clone representing 82% of the coiled-coil pairs sequenced. The features in the resulting coiled coils were consistent with selection for more stable pairs than before competition, with a large increase in hetero- versus homospecificity.[16,17]

Mammalian Dihydrofolate Reductase Protein Fragment Complementation Assay

As stated in the introduction, our goal in developing the PCA strategy is to combine screening with quantitative analysis of induced or constitutive protein–protein interactions (true expression cloning). To this end, we have demonstrated how the DHFR PCA can be used, simultaneously, in a strategy for dominant clonal selection and as a universal *in vivo* fluorescence

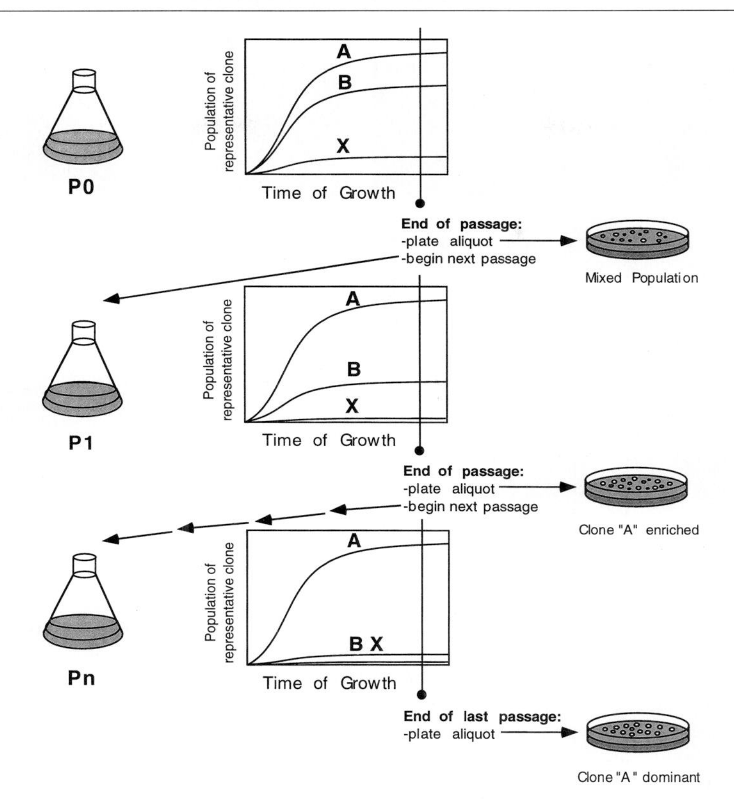

FIG. 2. Schematic representation of metabolic selection strategy: a hypothetical competition between three pairs of interacting proteins labeled A, B, and X. At the first passage (P0) the clones are grown in minimal M9 medium and a portion of the mixed population is plated on agar and minimal medium for clonal analysis. A second portion of the mixed population is used to inoculate the second passage (P1) in liquid medium. This procedure is repeated (P*n*) until homogeneous colonies are obtained. When it is necessary to control precisely the starting number of cells in a competition, the number of viable cells in the starter cultures is quantitated as follows. The appropriate clones are propagated in liquid medium under selective conditions and dilute aliquots are frozen at $-80°$ with 15% (v/v) glycerol. One aliquot for each clone is thawed and plated under selective conditions, and the colonies counted after 45 hr. The volume of cells to use for P0 is then calculated, such that each clone should be overrepresented by a factor of at least 20.

assay for quantitative pharmacological analysis of protein and protein–small molecule binding in mammalian cells.[14,15]

Reconstitution of DHFR activity can be monitored *in vivo* by cell survival in DHFR-negative cells (e.g., CHO-DUKX-B11) grown in the absence of nucleotides. Alternatively, recessive selection can be achieved

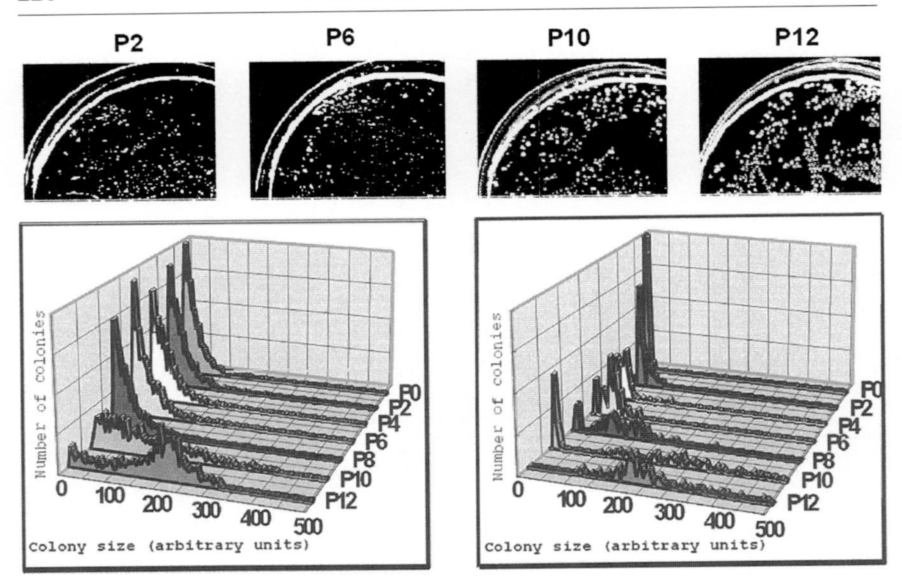

FIG. 3. Competition single-step or library-versus-library selection for optimally interacting leucine zipper-forming peptides. Approximately 1.42×10^4 clones resulting from a single-step or library-versus-library selection of one library fused to the Ile114Ala mutant of fragment [3] (F[3]) were pooled (P0) and competition selection was undertaken as described in Fig. 2. At each passage, some cells were plated and colony sizes were digitally quantitated by the particle analysis facility of the computer program NIH Image, version 1.59. Presented here is quantitation of the colony sizes for a library-versus-library (*left*) or a library-versus-target selection (*right*) at different passages.

by using DHFR fragments containing one or more of several mutations that render the refolded DHFR incapable of binding the antifolate drug methotrexate and growing cells in the absence of nucleotides with selection for methotrexate resistance. The principle of the survival DHFR PCA is that cells simultaneously expressing complementary fragments of DHFR fused to interacting proteins or peptides will survive in media depleted of nucleotides (Fig. 4A). The second approach is a fluorescence assay based on the detection of fluorescein-conjugated methotrexate (fMTX) binding to reconstituted DHFR (Fig. 4B). The basis of this assay is that complementary fragments of DHFR, when expressed and reassembled in cells, will bind with high affinity (K_d of 540 pM) to fMTX in a 1 : 1 complex. fMTX is retained in cells by this complex, while the unbound fMTX is actively and rapidly transported out of the cells.[36,47] In addition, binding of fMTX to

[47] D. I. Israel and R. J. Kaufman, *Proc. Natl. Acad. Sci. U.S.A.* **90,** 4290 (1993).

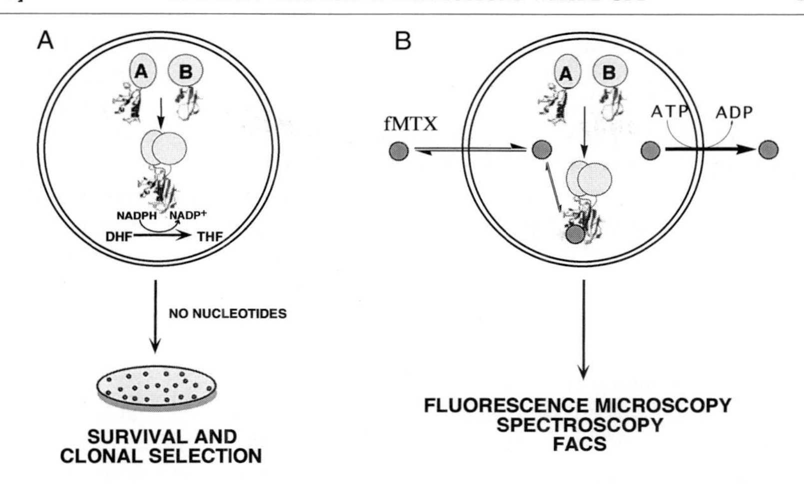

FIG. 4. Schematic representation of the strategy used to study protein–protein interactions in mammalian cells with the DHFR PCA. (A) Interacting proteins A and B are fused to one of two complementary fragments of murine DHFR (F[1,2] or F[3]) to generate A–F[1, 2] and B–F[3]. Association of A and B drives the reconstitution of DHFR (F[1,2] + F[3]), allowing DHFR-negative cells expressing these constructs to grow in medium lacking nucleotides. (B) The fluorescence assay is based on high-affinity binding of the specific DHFR inhibitor fluorescein–methotrexate (fMTX) to reconstituted DHFR. fMTX passively crosses the cell membrane and binds to reconstituted DHFR (F[1, 2] + F[3]) and is thus retained in the cell. Unbound fMTX is rapidly released from the cells by active transport. Bound and retained fMTX can then be detected by fluorescence microscopy, FACS, or fluorescence spectroscopy.

DHFR results in a 4.5-fold increase in quantum yield. Bound fMTX, and by inference reconstituted DHFR, can then be monitored by fluorescence microscopy, fluorescence-activated cell sorting (FACS), or spectroscopy.[14,15] It is important to note that, although fMTX binds to DHFR with high affinity, it does not induce DHFR folding from the fragments in the PCA. This is because the folding of DHFR from its fragments is obligatory; if folding is not induced by binding of the oligomerization domains, no binding sites for fMTX are created. Therefore the number of complexes observed, as measured by the number of fMTX molecules retained in the cell, is a direct measure of the equilibrium number of complexes of oligomerization domain complexes formed, independent of binding of fMTX.[14,15,36]

We chose as our first test system for the mammalian DHFR PCA the pharmacologically well-characterized rapamycin-induced association of FK506-binding protein (FKBP) to its target, the FKBP–rapamycin-binding

(FRB) domain of the FKBP12-rapamycin-associating protein (FRAP).[48] The DHFR-negative CHO-DUKX-B11 cells were stably cotransfected with FRB and FKBP fused to one of the two DHFR complementary fragments (FRB–F[1, 2] and FKBP–F[3]). Cotransfectants were selected for survival in nucleotide-free medium (selection for DHFR activity) and in the presence of rapamycin. Only cells grown in the presence of rapamycin undergo normal cell division and colony formation. We generally observe colony formation after 3 to 5 days. Rapamycin is a known cell cycle inhibitor and so these results would appear to be paradoxical; however, they are not. The concentration range over which survival was induced is 100 times less than the typical median effective concentration (EC_{50}) necessary to arrest cell division (Fig. 5A).[49] Survival is dependent only on the number of molecules of DHFR reassembled, and we have determined that this number is approximately 25 molecules of DHFR per cell.[15] Further, the efficacy of selection was determined by mixing, at ratios ranging from 1 : 100 to 1 : 1,000,000, stably cotransfected with untransfected cells grown in selective medium. After 5 days of incubation, resistant clones could be detected at all dilutions up to 1 in 10^6, although it is likely that further dilutions could be made, certainly to one or two orders of magnitude greater dilution, while still being practical to detect rare interactions in a library.

Formation of the FKBP–rapamycin–FRB complex was also detected in stably and transiently transfected cells with the fluorescence assay described above, based on stoichiometric binding of fluorescein-conjugated methotrexate to reconstituted DHFR *in vivo.* Fluorescence microscopy of unfixed cotransfected cells that had been incubated with fMTX showed high levels of fluorescence when cells were treated with rapamycin at saturating concentrations.[15] The fluorescence response of cell populations was quantified by FACS. The rapamycin-induced formation of FKBP/FRB was monitored by the shift in mean cell population fluorescence compared with noninduced cells (Fig. 5B). Quantitative rapamycin dose dependence of this complex was demonstrated to be consistent with the known pharmacological response. Cell fluorescence versus rapamycin concentration demonstrated single site saturable binding with a calculated K_d of 6 nM (Fig. 5C), compared with a value of 3 nM determined *in vitro.*[50] FK506, a competitive inhibitor of rapamycin for binding to FKBP, diminished rapamycin-induced fluorescence with a K_i of 53 nM, also comparable to the K_i of FK506 to compete for this interaction *in vitro* (Fig. 5D).[50]

[48] J. Chen, X. F. Zheng, E. J. Brown, and S. L. Schreiber, *Proc. Natl. Acad. Sci. U.S.A.* **92,** 4947 (1995).

[49] J. Chung, C. J. Kuo, G. R. Crabtree, and J. Blenis, *Cell* **69,** 1227 (1992).

[50] P. Chen, U. Schulze-Gahmen, E. A. Stura, J. Inglese, D. L. Johnson, A. Marolewski, S. J. Benkovic, and I. A. Wilson, *J. Mol. Biol.* **227,** 283 (1992).

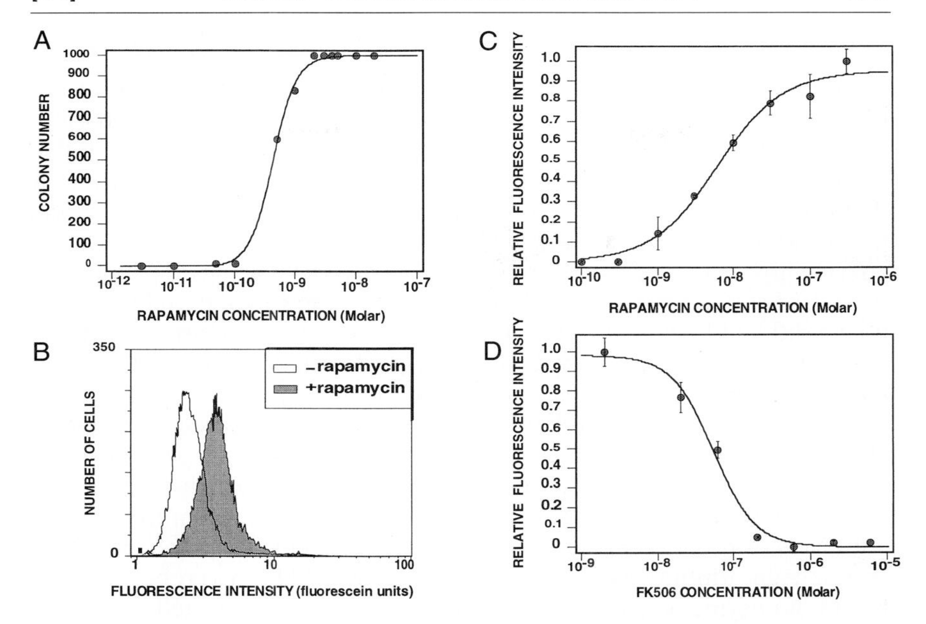

FIG. 5. (A) Survival selection curve showing that colony formation is dependent on the concentration of rapamycin. Stably transfected CHO DUKX-B11 cells expressing FRB–F[1, 2] and FKBP–F[3] were split in different concentrations of rapamycin from 0 to 20 n*M* in selective medium without nucleotides. The number of colonies was established after 4 days of incubation in selective medium. (B) Detection of the formation of the FKBP–rapamycin–FRB complex by fluorescence analysis. CHO-DUKX-B11 cells stably expressing the fusions FRB–F[1, 2] and FKBP–F[3] were incubated with fMTX at a final concentration of 10 μM, with or without addition of 20 n*M* rapamycin, for 22 hr at 37°. Induced formation of FKBP–rapamycin–FRB complex was monitored by fluorescence flow cytometry. Gray histogram corresponds to cells expressing FRB–F[1, 2] and FKBP–F[3] and treated overnight with 20 n*M* rapamycin. White histogram corresponds to untreated cells. (C) Dose–response curve for rapamycin was based on flow cytometric analysis of CHO-DUKX-B11 cells expressing the same fusions. Mean fluorescence intensities were determined for three independent samples at each rapamycin concentration (between 0.1 and 300 n*M*). (D) Competition curve with the inhibitor FK506, an analog of rapamycin. Mean fluorescence intensities were determined for three independent samples at each inhibitor concentration (between 0 and 6 μM, corresponding to a rapamycin: FK506 ratio of 1 : 0 to 1 : 300) (closed circles). The concentration of rapamycin was kept constant at 20 n*M*.

We also show that this strategy can be applied to study membrane protein receptors, demonstrating dose-dependent activation of the erythropoietin receptor by ligands.[15] We have also applied the DHFR PCA strategy to study induced conformational change in a constitutive erythropoietin receptor dimer.[14]

Protein Fragment Complementation Assay Based on Other Enzymes

The design and implementation of PCAs follow from the DHFR example in a logical way. We have developed several other PCAs based on other enzymes. These different systems illustrate the generality of the design principles set out at the beginning of this chapter as well as presenting assays that would themselves have broad applications to studies of protein–protein interactions. We have developed a survival PCA for *E. coli* based on the enzyme glycinamide ribonucleotide transformylase. The advantage of this assay is that cell lines missing this gene have been isolated and therefore no conditional knockout of the intrinsic gene need be performed with the risk of resistance to drugs evolving. We also describe two assays for dominant selection of mammalian cells based on the enzymes: aminoglycoside phosphotransferase (Neo) and hygromycin B phosphotransferase (HygroB). These assays will be of universal utility for library screening because they can be used in virtually any mammalian cell line.

Another reason for describing these assays is that they represent different levels of protein engineering complexity and illustrate that regardless of how counterintuitive to expectation the design of a PCA may be, reconstitution of enzyme activity from fragments can be demonstrated.

Glycinamide Ribonucleotide Transformylase

The glycinamide ribonucleotide transformylase (GAR Tfase) is the third enzyme in the *de novo* purine biosynthesis pathway for *E. coli* and eukaryotes.[51]

The *E. coli* GAR Tfase (EC 2.1.2.2, phosphoribosylglycinamide formyltransferase) is a monomeric enzyme of 23 kDa encoded by the *purN* gene. It has been the subject of numerous crystallographic studies[50,52,53] because of its homology with the C-terminal region of a trifunctional enzyme that contains the GAR Tfase activity in eukaryotic organisms.[54–56] Inhibitors

[51] S. J. Benkovic, *Annu. Rev. Biochem.* **49,** 227 (1980).

[52] R. J. Almassy, C. A. Janson, C. C. Kan, and Z. Hostomska, *Proc. Natl. Acad. Sci. U.S.A.* **89,** 6114 (1992).

[53] C. Klein, P. Chen, J. H. Arevalo, E. A. Stura, A. Marolewski, M. S. Warren, S. J. Benkovic, and I. A. Wilson, *J. Mol. Biol.* **249,** 153 (1995).

[54] C. C. Kan, M. R. Gehring, B. R. Nodes, C. A. Janson, R. J. Almassy, and Z. Hostomska, *J. Protein Chem.* **11,** 467 (1992).

[55] S. C. Daubner, J. L. Schrimsher, F. J. Schendel, M. Young, S. Henikoff, D. Patterson, J. Stubbe, and S. J. Benkovic, *Biochemistry* **24,** 7059 (1985).

[56] C. A. Caperelli, P. A. Benkovic, G. Chettur, and S. J. Benkovic, *J. Biol. Chem.* **255,** 1885 (1980).

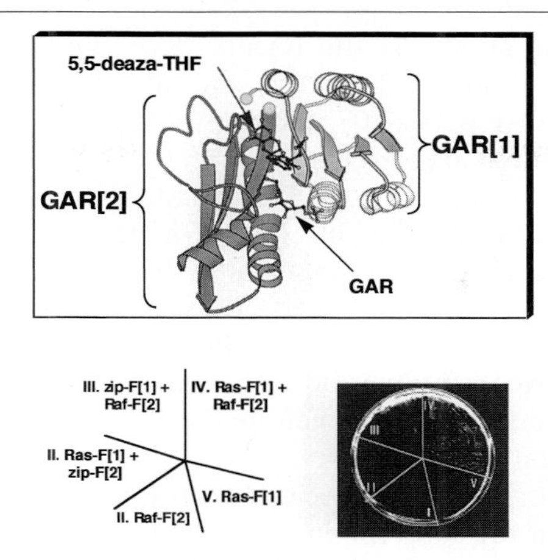

FIG. 6. *Top:* Structure of *E. coli* GAR transformylase with subdomain dissection used in this study (see text). *Bottom:* Survival selection of MW12 (*purN*-negative, *purT*-negative) *E. coli* expressing Ras–GAR[1] and Raf–GAR[2] fusions. Colonies were obtained after 30 hr of growth at 30°. Indicated controls include the individual fusions alone or fusions cotransformed with GCN4 leucine zipper-forming sequences fused to complementary GAR[1] or GAR[2] fragments.

of enzyme activity had been shown to be potential chemotherapeutic agents against certain cancers.[57–60] The enzyme is formed of two distinct domains formed sequentially according to the primary sequence. The amino-terminal domain binds GAR (residues 1–101) and is constituted of a four-stranded parallel β sheet interspersed with four α helices (Fig. 6, top). The carboxy-terminal domain (residues 102–212) binds 10-methylene-tetrahydrofolate (10-CHO-H_4F) and initially continues the parallel β sheet and has two antiparallel stands. The mixed β sheet is covered on one side by a long α helix situated at the interface between the two domains. The two domains are linked by a small loop facing outward from the protein core. The active site is situated in a long crevice approximately 20 Å in diameter and 572 $Å^2$ at the interface between the two domains. The free amino group of GAR is brought into the vicinity of the formyl group of 10-CHO-H_4F to

[57] E. A. Stura, D. L. Johnson, J. Inglese, J. M. Smith, S. J. Benkovic, and I. A. Wilson, *J. Biol. Chem.* **264,** 9703 (1989).

[58] L. L. Habeck, T. A. Leitner, K. A. Sackelford, L. S. Gossett, R. M. Schultz, S. L. Andis C. Shih, G. B. Grindey, and L. G. Mendelsohn, *Cancer Res.* **54,** 1021 (1994).

[59] R. J. Mullin, B. R. Keith, E. C. Bigham, D. S. Duch, R. Ferone, L. S. Heath, S. Singer, K. A. Waters, and H. R. Wilson, *Biochem. Pharmacol.* **43,** 1627 (1992).

[60] G. P. Beardsley, B. A. Moroson, E. C. Taylor, and R. G. Moran, *J. Biol. Chem.* **264,** 328 (1989).

permit nucleophilic attack by the former.[50,52,53] The key residues in the active sites as demonstrated by active site labeling, crystallography, and mutagenesis are Asn-106, His-108, and Asp-144; these are conserved among the prokaryote and eukaryote enzymes.[50,52,53,57,61–63]

The modular structure of GAR Tfase made it a promising prospect for designing a PCA. Further, the conditions for performing a survival assay and the proper *E. coli* auxotrophic strain [MW 12, with PurN and PurT (a second GAR Tfase activity) knocked out] were available.[64] Although the two clear-cut domains made it an attractive target, the fact that the active site is situated between the two domains and that the fragments must be brought together perfectly to reconstitute the correct geometry of the active site presented a challenge.

Based on the structure, we decided to fragment the protein between the two domains in the loop linking α helix 4 and β sheet 5. Specifically, we generated fragment 1 (GAR[1]) by polymerase chain reaction (PCR) from residues 1 to 102, and fragment 2 (GAR[2]) by PCR from residues 103 to 212. The amino terminal of GAR[1] was fused to GCN4 zipper in a pQE 32-derived vector and the amino terminal of GAR[2] was fused to GCN4 zipper in a pQE 32-derived vector. In both cases the two parts of the fusion were separated by a soluble linker of 12 amino acids. Benkovic and co-workers have generated two incremental and complementary truncation libraries of fragments of GAR Tfase (*purN*) by using a strategy in which 5′ or 3′ *purN* gene truncations of different lengths are generated by treatment of double-stranded coding sequence with 3′- or 5′-specific exonucleases.[65] They were able to show fragment complementation by cotransforming an auxotrophic strain (*purN*$^-$ and *purT*$^-$) of *E. coli* with members of each library and observing cell survival. These results illustrate that correct folding and enzymatic activity of the full-length GAR transformylase is possible from some fragments under these conditions. The complementing fragments obtained through this strategy were shown to be clustered mainly in two regions (amino acids 60–70 and 111–114), which do not correspond to the cut site we use in the PCA (between amino acids 102 and 103). Another difference between our fragments and theirs is that,

[61] J. Inglese, J. M. Smith, and S. J. Benkovic, *Biochemistry* **29,** 6678 (1990).

[62] J. Inglese, D. L. Johnson, A. Shiau, J. M. Smith, and S. J. Benkovic, *Biochemistry* **29,** 1436 (1990).

[63] M. S. Warren and S. J. Benkovic, *Protein Eng.* **10,** 63 (1997).

[64] A. E. Marolewski, K. M. Mattia, M. S. Warren, and S. J. Benkovic, *Biochemistry* **36,** 6709 (1997).

[65] M. Ostermeier, A. E. Nixon, J. H. Shim, and S. J. Benkovic, *Proc. Natl. Acad. Sci. U.S.A.* **96,** 3562 (1999).

in most cases, the complementary fragments they observed had significant overlap in amino acid sequence. These results are consistent with numerous cases documenting spontaneously complementing fragments of peptides/proteins.[22,66,67] One could say that our "design" strategy is exactly the opposite of what most researchers interested in protein dissection would seek, in that our goal is to choose fragments incapable of refolding alone!

We have demonstrated the validity of our approach for the assisted complementation of GAR Tfase fragments with Ras and the Ras-binding domain (RBD) of Raf as a model, because a crystallographic structure of the complex was known and we were interested later to use this system as a starting point in binding studies. The RBD of Raf interacts with the GTP-loaded form of Ras, which is likely to be dominant in *E. coli* because of the absence of the proteins regulating the low basal GTPase activity of Ras. Thus, Ras was fused to the GAR[1] fragment and Raf to the GAR[2] fragment and the plasmids expressing the fusion were contransformed. Colonies were obtained after 30 hr of growth at 30° (Fig. 6, bottom). The specificity of the detection of the interaction was demonstrated by the abolition of colony-forming capacity when a point mutation in the RBD of Raf that was previously demonstrated to abolish the interaction with Ras, Arg89Leu,[68] was introduced in the fusion.

Aminoglycoside and Hygromycin B Phosphotransferases

Aminoglycoside phosphotransferase [APH(3′)-IIIa] and hygromycin B phosphotransferase [APH(4)-Ia] genes, corresponding to neomycin and hygromycin B resistance genes, respectively, are commonly used as dominant selectable markers in the selection of stable mammalian cell lines. Neomycin, kanamycin, and similar compounds such as G418 inhibit protein synthesis in prokaryotes and eukaryotes. G418, for example, is an aminoglycoside antibiotic that blocks protein synthesis in eukaryotic cells by interfering with the function of 80S ribosomes.[69,70] The bacterial aminoglycoside phosphotransferases can inactivate these antibiotics by phosphorylation. Hygromycin B, another aminoglycoside antibiotic, has been reported to

[66] A. G. Ladurner, L. S. Itzhaki, G. de Prat Gay, and A. R. Fersht, *J. Mol. Biol.* **273,** 317 (1997).
[67] L. Stewart, G. C. Ireton, and J. J. Champoux, *J. Mol. Biol.* **269,** 355 (1997).
[68] J. R. Fabian, A. B. Vojtek, J. A. Cooper, and D. K. Morrison, *Proc. Natl. Acad. Sci. U.S.A.* **91,** 5982 (1994).
[69] S. Bar-Nun, Y. Shneyour, and J. S. Beckmann, *Biochim. Biophys. Acta* **741,** 123 (1983).
[70] D. C. Eustice and J. M. Wilhelm, *Antimicrob. Agents Chemother.* **26,** 53 (1984).

interfere with translocation[71,72] and to cause mistranslation.[73] The hygromycin B phosphotransferase gene also confers resistance to hygromycin B by encoding a kinase that inactivates the antibiotic through phosphorylation.

The design of PCAs based on APH(3′)-IIa and APH(4)-Ia has been aided by the determination of the structure of the aminoglycoside phosphotransferase subform APH(3′)-IIIa and the realization of its homology to eukaryotic protein kinases.[74] Like the eukaryotic kinases, the general structure of APH(3′)-IIIa consists of two domains: an N-terminal ATP-binding domain and a C-terminal catalytic domain. Further, the primary sequence identity among the aminoglycoside kinases is high (20–40%) and thus it is relatively easy to predict where to dissect one subform versus another based on the structure of APH(3′)-IIIa.[75] We chose to dissect the two enzymes between the ATP-binding and catalytic domains, corresponding to Gly-99 for APH(3′)-IIa and Glu-108 for APH(4)-Ia (Fig. 7, top).

We have examined the two PCAs (called here the Neo and Hygro PCAs) with our test systems, the rapamycin-induced interaction of the FKBP–rapamycin-binding domain of FRAP (FRB) with FKBP and the constitutively interacting GCN4 leucine zipper-forming sequences. In all cases, interacting proteins are separated from the PCA fragments by a 15-amino acid flexible linker consisting of the sequence (Gly-Gly-Gly-Gly-Ser)$_3$ in the final configurations Neo[1]–A, B–Neo[2], Hygro[1]–A, and B–Hygro[2], where A and B correspond to the interacting proteins.

The procedures for dominant selection are not different from how it is normally done. Our procedure in CHO cells is to use lipofectamine to transfect cells in medium enriched with fetal bovine serum and, 48 hr after transfection, to split the cells at approximately 5×10^4 in six-well plates in medium containing G418 (400 μg/ml) or hygromycin B (250 μg/ml) for the neomycin and hygromycin PCAs, respectively. For FRB/FKBP transfection, rapamycin is added to the cells at a final concentration of 10 n*M*. The appearance of distinct colonies occurs after 10 to 14 days (Fig. 7, bottom): colony formation is not observed in the absence of rapamycin. For the constitutively interacting GCN4 leucine zipper-forming sequences, colonies are observed only for clones that simultaneously expressed both

[71] M. J. Cabanas, D. Vazquez, and J. Modolell, *Eur. J. Biochem.* **87,** 21 (1978).

[72] A. Gonzalez, A. Jimenez, D. Vazquez, J. E. Davies, and D. Schindler, *Biochim. Biophys. Acta* **521,** 459 (1978).

[73] A. Singh, D. Ursic, and J. Davies, *Nature* (*London*) **277,** 146 (1979).

[74] W. C. Hon, G. A. McKay, P. R. Thompson, R. M. Sweet, D. S. Yang, G. D. Wright, and A. M. Berghuis, *Cell* **89,** 887 (1997).

[75] K. J. Shaw, P. N. Rather, R. S. Hare, and G. H. Miller, *Microbiol Rev.* **57,** 138 (1993).

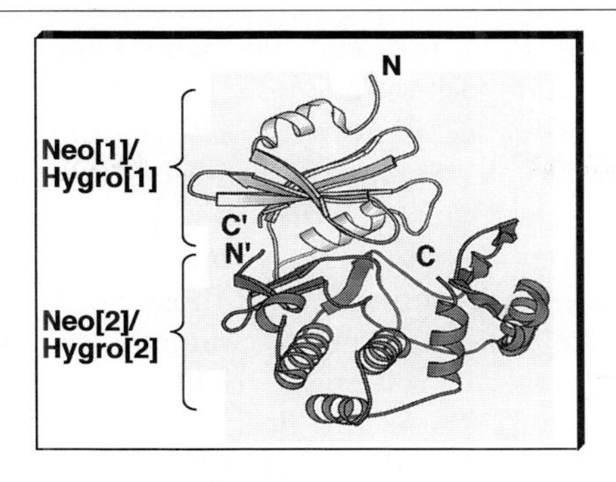

CHO DUKX-B11 Selection in Hygromycin B

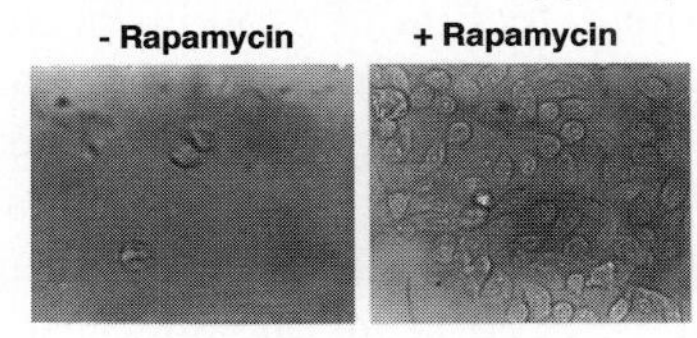

FIG. 7. *Top:* Structure of aminoglycoside phosphotransferase APH(3′)-IIIa with subdomain dissection used in this study (see text). *Bottom:* Survival selection of CHO cells expressing Hygro[1]–FKBP and FRB–Hygro[2]. CHO cells were stably transfected with both fusions and selected on medium containing hygromycin B (250 μg/ml) and in the presence of 10 nM rapamycin to induce the association of FKBP with FRB (reconstitution of hygromycin B phosphotransferase activity). Photographs of cells were taken after 14 days of incubation in the selective medium in the absence (*left*) or in the presence (*right*) of rapamycin.

interacting peptides fused to one or the other complementary DHFR fragments (data not shown).

Conclusions

At the outset of this chapter, we proposed a wish list of features that would make an experimental strategy of general utility for studying protein–protein interactions. The varied applications of the PCA strategy demonstrate that PCA can meet these criteria. Specifically, the features of the PCA strategy that make it uniquely suited to simple *in vivo* studies of protein interactions include the following: (1) It is "complete"; no other cellular activity is necessary and as a result a PCA can be done in any prokaryotic or eukaryotic cell type, or the PCA can be directed to a specific cellular compartment organelle or membrane surface with the inclusion of appropriate signal sequences; (2) the portability of PCAs also means that induced versus constitutive protein–protein interactions can be distinguished by doing the PCA in a cell type where specific protein–protein interactions are thought to be induced by, for example, a growth factor-mediated signal transduction pathway; (3) PCA is not a single assay but a series of assays. The PCA strategy therefore has the added flexibility that an assay can be chosen because it works in a specific cell type appropriate for studying interactions of some class of proteins; (4) PCAs are inexpensive, requiring no specialized reagents beyond those necessary for a particular assay and off-the-shelf materials and technology; (5) PCAs can be automated and high-throughput screening could be done with little human intervention; (6) because PCAs are designed at the level of the atomic structure of the enzymes used, there is additional flexibility in determining the sensitivity and stringencies of the assays; finally, (7) PCAs can be based on enzymes for which the detection of protein–protein interactions can be determined differently. The simplest and most general approaches discussed here were based on dominant selection, in which the reconstituted enzyme complements some missing metabolic enzyme in cells grown under selective pressure. For the purposes of designed or cDNA library screening, we would argue that this strategy will represent the simplest and least technology-intensive approach to detecting interactions between "bait" and "prey" pairs. In practice, the survival DHFR assay is sensitive, as only approximately 25 molecules of DHFR per cell is necessary to support cell proliferation.[15] However, it could prove more rapid to screen on the basis of a fluorescence signal such as the variant of the DHFR PCA described above combined with fluorescence-activated cell sorting. However, enzymes can also be chosen that produce a fluorescent or colored product. The amplification afforded by enzymatic generation of a fluorescent product could be as sensitive as the DHFR survival PCA. We are presently testing several PCAs based on enzymes that produce fluorescent products.

Acknowledgments

We thank Myriam Salem and Mireille Nohra for their preliminary work on development of the GAR Tfase PCA; we also thank Katja Arndt, Christian Müller, and Andreas Plückthun for discussion and fruitful collaboration. The authors gratefully acknowledge Monique Davies for the gift of pMT3 vector and the CHO DUKX-B11 cell line, Joe Heitman for scTOR2, Emil Pai for p21*ras*, and David Waugh for *raf.* Support has been provided by the Burroughs-Wellcome Fund (BWF) and the Medical Research Council of Canada (MOP-42477). S. W. Michnick is a BWF New Investigator Awardee and holds an MRC (Canada) Scientist award. I. Remy and F-X. Campbell-Valois are recipients of predoctoral awards of excellence from le Fonds pour la Formation de Chercheurs et l'Aide à la Recherche (FCAR). J. N. Pelletier was a recipient of a Fellowship from les Fonds de la Recherche en Santé du Québec (FRSQ), and currently holds a Fellowship from le Conseil de Recherche en Sciences Naturelles et en Génie du Canada (CRSNG).

[15] Monitoring Protein–Protein Interactions in Live Mammalian Cells by β-Galactosidase Complementation

By FABIO M. V. ROSSI, BRUCE T. BLAKELY, CAROL A. CHARLTON, and HELEN M. BLAU

Introduction

Conditional protein–protein interactions are critical to a vast number of intracellular regulatory mechanisms. A case in point is the activation and autophosphorylation of growth factor receptors, which is often determined by the ligand-dependent induction of receptor dimerization.[1] Docking sites on the activated receptor become accessible to downstream components of the relevant signaling pathways, leading to further protein–protein interactions.[2] Although efficient genetic methods to identify interacting partners exist and have been successfully applied by a number of laboratories, a technique that allows the monitoring of interactions in real time in the cellular compartment in which they normally take place would extend the types of interactions that could be studied. The β-galactosidase-based intracistronic complementation methodology described here is the first technology that fulfills these requirements and can be applied to live mammalian cells.

The novel β-galactosidase (β-Gal) complementation assay for monitoring protein–protein interactions in intact mammalian cells has many advantages (Fig. 1). Unlike the yeast two-hybrid assay,[3] this assay is independent of transcription. Instead, protein interactions are monitored directly *in situ*. Because it is based on the detection of an enzymatic activity, it provides amplification of the signal. Intracistronic β-Gal complementation, first described in prokaryotes by Jacob and Monod more than 30 years ago,[4,5] is a phenomenon whereby two mutants of the bacterial enzyme β-Gal that have inactivating deletions in different critical domains are capable of recreating an active enzyme by sharing their intact domains.[6] It has long been known in *Escherichia coli* that specific mutants can complement each

[1] A. Ullrich and J. Schlessinger, *Cell* **61,** 203 (1990).
[2] T. Pawson and J. D. Scott, *Science* **278,** 2075 (1997).
[3] C. Bai and S. J. Elledge, *Methods Enzymol.* **273,** 331 (1996).
[4] A. Ullmann, D. Perrin, F. Jacob, and J. Monod, *J. Mol. Biol.* **12,** 918 (1965).
[5] A. Ullmann, F. Jacob, and J. Monod, *J. Mol. Biol.* **24,** 339 (1967).
[6] I. Zabin, *Mol. Cell. Biochem.* **49,** 87 (1982).

0076-6879/00 $30.00

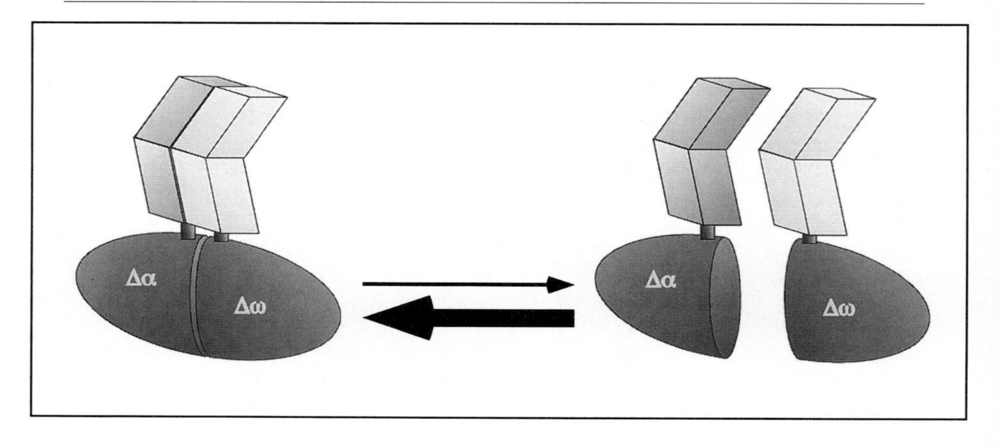

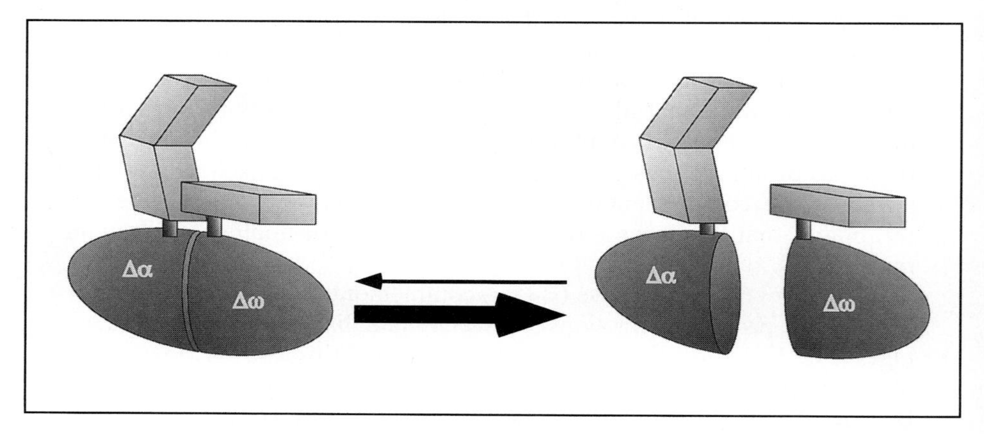

FIG. 1. Schematic of experimental hypothesis. (A) When the $\Delta\alpha$ and $\Delta\omega$ β-Gal mutants are fused to proteins that heterodimerize, their association in the active enzyme conformation is favored. (B) When the two proteins fused to $\Delta\alpha$ and $\Delta\omega$ cannot interact to form a complex, the formation of active β-Gal is not favored.

other more or less efficiently, depending on the nature of the mutations.[7] Our results provide the first demonstration that the same holds true in mammalian cells.[8] Moreover, by using inefficiently complementing mutants to recreate an active β-Gal enzyme, our method allows the interaction of linked protein sequences to be assessed. Thus the β-Gal activity monitors,

[7] M. Villarejo, P. J. Zamenhof, and I. Zabin, *J. Biol. Chem.* **247,** 2212 (1972).
[8] W. A. Mohler and H. M. Blau, *Proc. Natl. Acad. Sci. U.S.A.* **93,** 12423 (1996).

or serves as a readout, for the physical interaction between the non-β-Gal protein components.[9]

There are several significant advantages of the β-Gal complementation method.[10] First, quantitation is possible, as the enzyme that provides the readout of the assay is covalently linked to the interacting proteins. Thus, the number of complemented β-Gal molecules present in the cell correlates with the number of complexes formed by the test proteins, providing a means of quantifying the interaction. Second, the signal is generated directly in the subcellular compartment where the interaction takes place.[9] Finally, the sensitivity of the assay is such that interactions of trace amounts of proteins can be assayed, avoiding perturbations of cellular physiology due to overexpression.[10a] Using this system, we have been able to monitor ligand-induced interactions between two cytoplasmic proteins (FKBP12 and the FRB domain of FRAP),[9] between a cytoplasmic protein and a transmembrane receptor (FADD and FAS), and the formation of homodimers (EGF receptor[10a]) and higher order complexes (FAS) composed of transmembrane proteins.

Development of Mammalian β-Galactosidase Complementation: Historical Perspective

The derivation of a mammalian β-Gal complementation system arose from our interest in muscle cell differentiation. To better understand signaling from cell surface molecules, substrates, or growth factors that have been implicated in the fusion of myoblasts to form multinucleate syncytial myotubes, we sought a rapid, sensitive, highly reproducible biochemical assay to supplant the traditional microscopic scoring of the percentage of nuclei contained inside and outside syncytia. The classic bacterial genetic phenomenon of intracistronic complementation of the *lacZ* gene proved ideal for this purpose.[4] In *E. coli,* deletions of either the N or C terminus of β-Gal produce enzyme that is inactive, but can be complemented by coexpression of an inactive mutant defective in the other terminus. Although in *E. coli,* restoration of enzyme activity results from formation of a heterooctamer of complementing mutant peptides rather than the tetramer that normally constitutes active β-Gal, in mammalian cells the precise

[9] F. Rossi, C. A. Charlton, and H. M. Blau, *Proc. Natl. Acad. Sci. U.S.A.* **94,** 8405 (1997).
[10] F. M. V. Rossi, B. T. Blakely, and H. M. Blau, *Trends Cell Biol.* **10,** 119 (2000).
[10a] B. T. Blakely *et al., Nature Biotech.* **18,** 218 (2000).

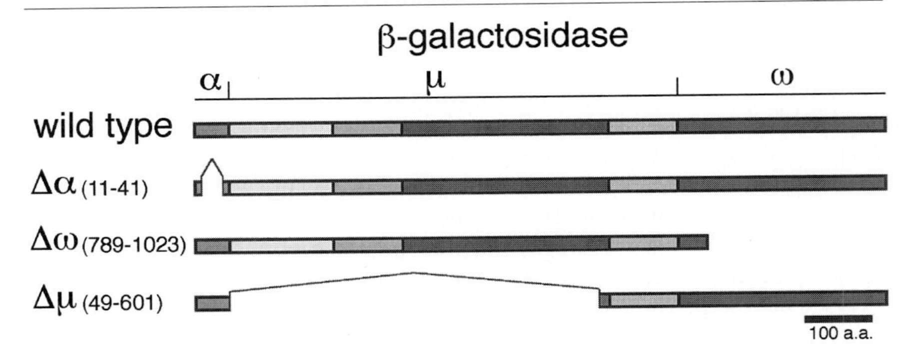

FIG. 3. Schematic representation of the β-Gal mutants. The strongly complementing mutants $\Delta\alpha$ and $\Delta\mu$ are used to monitor cell fusion.[8] In contrast, the weakly complementing $\Delta\alpha$ and $\Delta\omega$ mutants are also used to monitor protein–protein interactions.[9] [Reproduced from W. A. Mohler and H. M. Blau, *Proc. Natl. Acad. Sci. U.S.A.* **93,** 12423 (1996).]

number of monomers in an active complex remains to be determined.[11–13] On the basis of the published bacterial genetic map, biochemical data, 3D structure (Fig. 2; see color insert), and the existence of convenient restriction sites, we constructed vectors expressing more than 20 deleted versions of *E. coli* β-Gal, including $\Delta\alpha$ (N-terminal deletion), $\Delta\omega$ (C-terminal deletion), and $\Delta\mu$ (a large central deletion; Fig. 3).[6–8,11,14] When NIH 3T3 cells were sequentially infected with pairs of potentially complementing constructs and assayed histochemically for β-Gal activity, certain mutant pairs complemented more efficiently than others. The most strongly complementing pair, $\Delta\alpha$ and $\Delta\mu$, was used to develop a fusion assay in which populations of myoblasts independently infected with virus expressing either β-Gal mutant $\Delta\alpha$ or $\Delta\mu$ were mixed in equal proportions and plated in coculture.[8] After different periods of time in differentiation medium [Dulbecco's modified Eagle's medium (DMEM) plus 5% (v/v) equine serum, changed daily], the cells were assayed for β-Gal activity using a sensitive chemiluminescent substrate (see Chemiluminescence Assay, below). The kinetics of myoblast fusion as assayed by β-Gal activity correlated extremely well with the classically used, labor-intensive microscopic analysis of cell fusion.[15]

The β-Gal complementation assay has since been used to investigate

[11] A. Ullmann, F. Jacob, and J. Monod, *J. Mol. Biol.* **32,** 1 (1968).

[12] K. E. Langley and I. Zabin, *Biochemistry* **15,** 4866 (1976).

[13] F. Celada and I. Zabin, *Biochemistry* **18,** 404 (1979).

[14] R. H. Jacobson, X. J. Zhang, R. F. DuBose, and B. W. Matthews, *Nature* (*London*) **369,** 761 (1994).

[15] C. A. Charlton, W. A. Mohler, G. L. Radice, R. O. Hynes, and H. M. Blau, *J. Cell Biol.* **138,** 331 (1997).

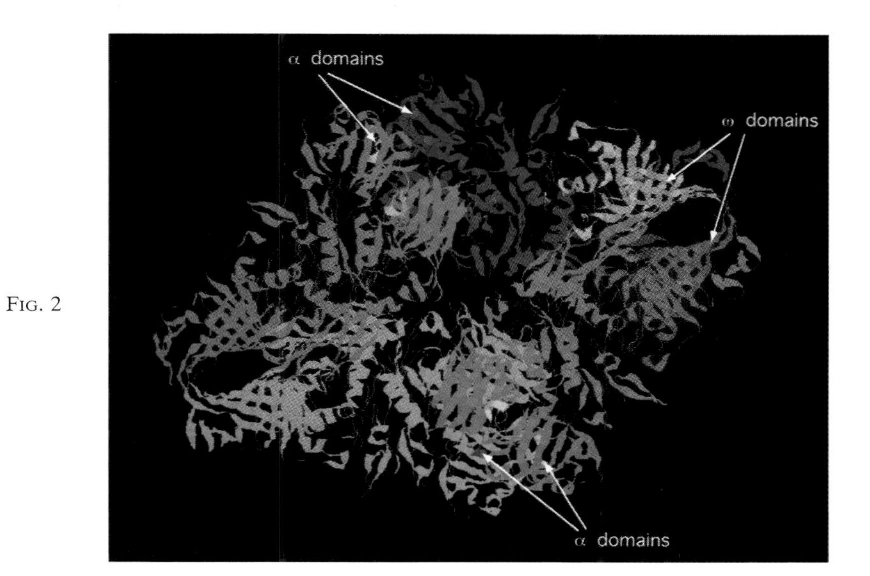

FIG. 2

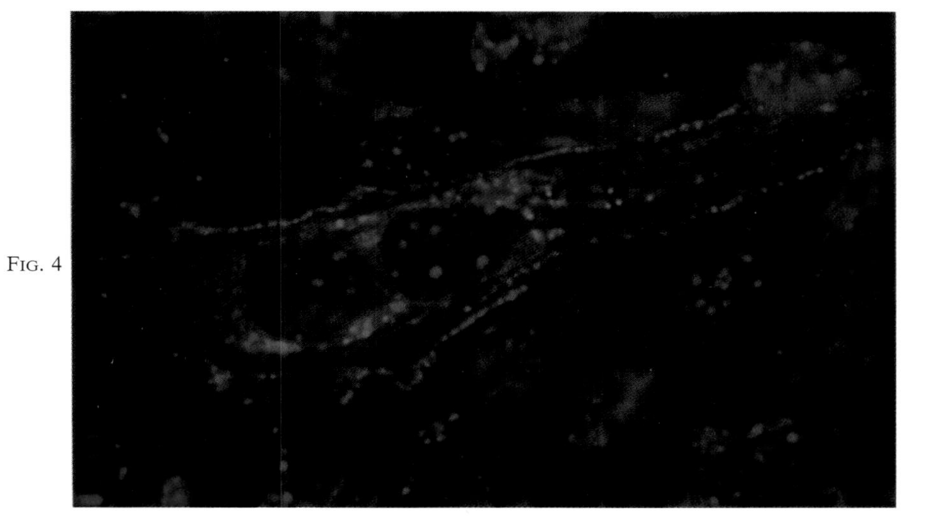

FIG. 4

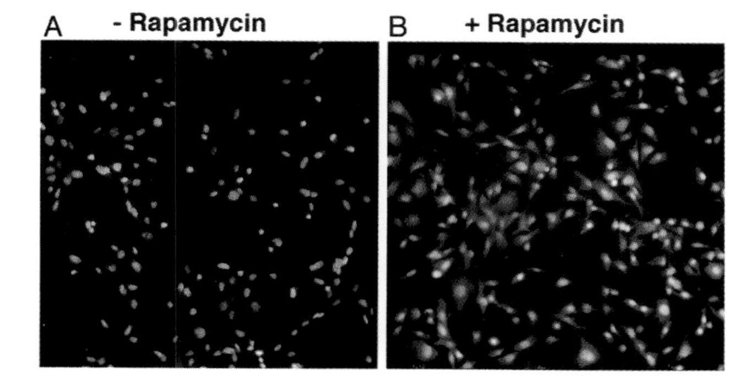

FIG. 7

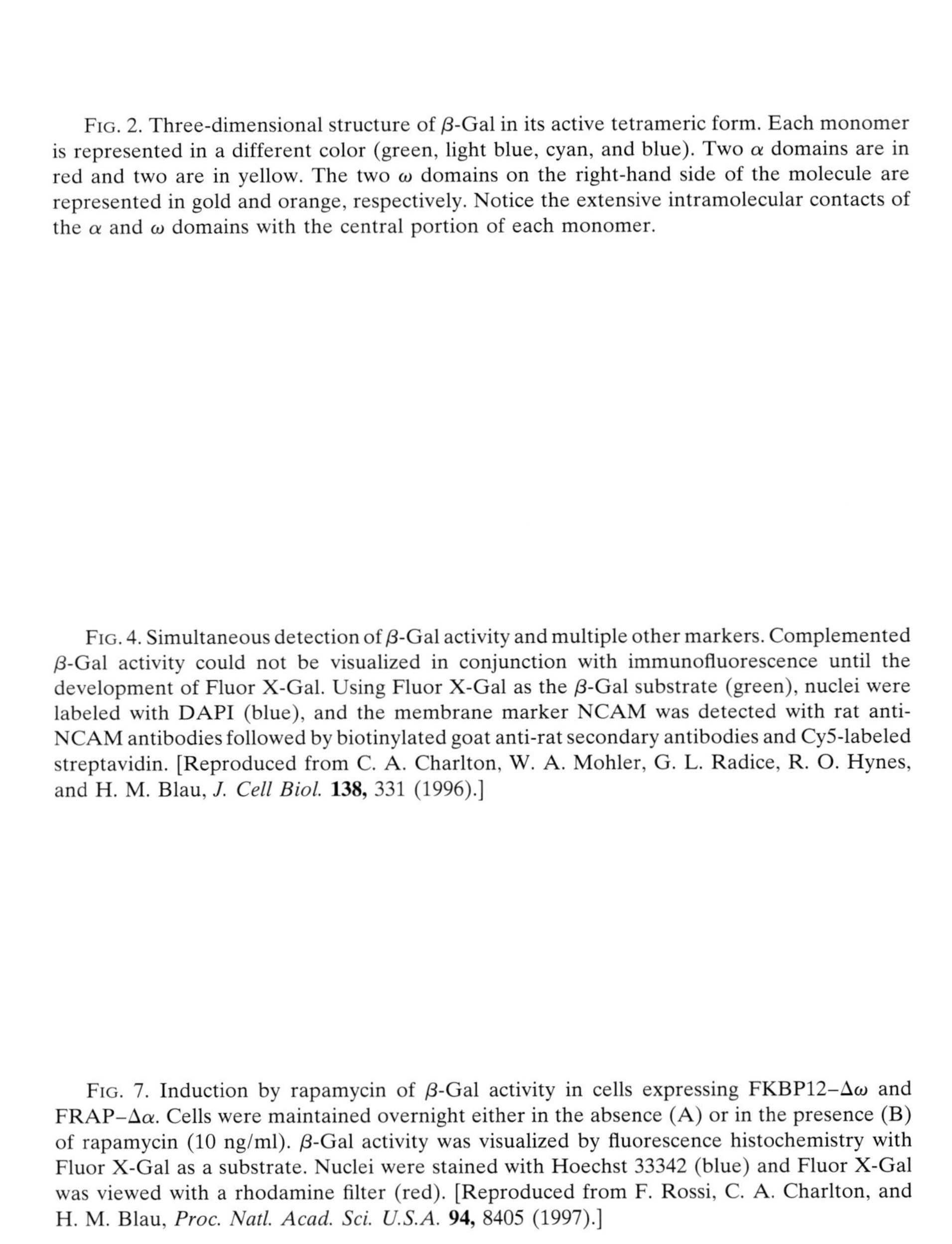

FIG. 2. Three-dimensional structure of β-Gal in its active tetrameric form. Each monomer is represented in a different color (green, light blue, cyan, and blue). Two α domains are in red and two are in yellow. The two ω domains on the right-hand side of the molecule are represented in gold and orange, respectively. Notice the extensive intramolecular contacts of the α and ω domains with the central portion of each monomer.

FIG. 4. Simultaneous detection of β-Gal activity and multiple other markers. Complemented β-Gal activity could not be visualized in conjunction with immunofluorescence until the development of Fluor X-Gal. Using Fluor X-Gal as the β-Gal substrate (green), nuclei were labeled with DAPI (blue), and the membrane marker NCAM was detected with rat anti-NCAM antibodies followed by biotinylated goat anti-rat secondary antibodies and Cy5-labeled streptavidin. [Reproduced from C. A. Charlton, W. A. Mohler, G. L. Radice, R. O. Hynes, and H. M. Blau, *J. Cell Biol.* **138,** 331 (1996).]

FIG. 7. Induction by rapamycin of β-Gal activity in cells expressing FKBP12–$\Delta\omega$ and FRAP–$\Delta\alpha$. Cells were maintained overnight either in the absence (A) or in the presence (B) of rapamycin (10 ng/ml). β-Gal activity was visualized by fluorescence histochemistry with Fluor X-Gal as a substrate. Nuclei were stained with Hoechst 33342 (blue) and Fluor X-Gal was viewed with a rhodamine filter (red). [Reproduced from F. Rossi, C. A. Charlton, and H. M. Blau, *Proc. Natl. Acad. Sci. U.S.A.* **94,** 8405 (1997).]

the effect on fusion of various factors and cell adhesion molecules purported to function in myogenesis. The molecules tested to date include α_4-integrin, N-cadherin, and neural cell adhesion molecule (NCAM), all of which had been implicated in myogenic fusion in previous studies.[16–18] The fusion potential of myoblasts that were rendered homozygous null for these adhesion molecules was compared with wild-type cells using β-Gal complementation. Lack of each of the three cell surface adhesion molecules did not have an effect on myoblast fusion in this rigorous assay, suggesting that if they are involved at all, they are not essential.[15,19] This assay should prove generally useful for monitoring fusion in other cell types such as osteoclasts, trophoblasts, and mitotically inactive syncytia found in tumors.

To examine β-Gal activity simultaneously with expression of different cell surface molecules at the single-cell level, we developed a new histochemical assay for β-Gal (see protocol, below). This was necessary because a drawback of the 5-bromo-4-chloro-3-indolyl-β-D-galactopyranoside (X-Gal) substrate typically used to assay β-Gal activity is that it quenches fluorescence of fluorophores. On the other hand, fluorogenic substrates for β-Gal that are used for live cell sorting [e.g., fluorescein di-β-D-galactopyranoside (FdG)] diffuse readily from the site of enzyme activity.[20] We found that the combination of an azo dye, Fast Red Violet LB, with either X-Gal or with 5-bromo-6-chloro-3-indolyl-β-D-galactopyranoside (5-6 X-Gal) constitutes a substrate (Fluor X-Gal) that is a more sensitive measure of β-Gal activity than the X-Gal substrate alone.[8] The Fluor X-Gal product can be visualized simultaneously with other fluorescent markers (Fig. 4; see color insert), thus allowing for study of the spatial and temporal expression of different molecules in relationship to fusion.[15] This sensitive histochemical assay may also be applicable to monitoring the expression and localization of interacting proteins fused to complementing β-Gal mutants.

β-Galactosidase Complementation Analysis of Protein Interactions

We hypothesized that the less efficient pairs of complementing β-Gal mutants could be used to monitor physical interactions between proteins. This would require that the mutants not drive the reaction. To test this

[16] K. A. Knudsen, S. A. McElwee, and L. Myers, *Dev. Biol.* **138,** 159 (1990).

[17] K. A. Knudsen, L. Myers, and S. A. McElwee, *Exp. Cell Res.* **188,** 175 (1990).

[18] G. D. Rosen, J. R. Sanes, R. LaChance, J. M. Cunningham, J. Roman, and D. C. Dean, *Cell* **69,** 1107 (1992).

[19] J. T. Yang, T. A. Rando, W. A. Mohler, H. Rayburn, H. M. Blau, and R. O. Hynes, *J. Cell Biol.* **135,** 829 (1996).

[20] G. P. Nolan, S. Fiering, J. F. Nicolas, and L. A. Herzenberg, *Proc. Natl. Acad. Sci. U.S.A.* **85,** 2603 (1988).

hypothesis, we took advantage of the well-characterized interaction between the cyclophilin FKBP12 and the FKBP12–rapamycin-binding domain (FRB domain) of the FKBP12–rapamycin-associated protein (FRAP).[21,22] The interaction between FKBP12 and FRAP is particularly well suited for testing a novel methodology, as it takes place only in the presence of rapamycin, a small cell-permeable drug commonly used as an immunosuppressant. The rapamycin molecule is trapped between the two polypeptides within the FKBP12–rapamycin–FRAP complex and supplies most of the contact interface by binding tightly to both partners.[23] Consequently, FKBP12 and FRAP have no tendency to interact in the absence of rapamycin. Indeed, several laboratories that used this inducible interaction to conditionally target a transcriptional activation domain to a promoter detected no transcription above background in the absence of rapamycin.[24–27] We generated two chimeric proteins in which FKBP12 was fused to $\Delta\omega$ and the FRB domain of FRAP was fused to $\Delta\alpha$ (FKBP12–$\Delta\omega$ and FRAP–$\Delta\alpha$, respectively; Fig. 5). The chimeric cDNAs were inserted into retroviral vectors that also contained genes encoding the selectable markers neomycin or hygromycin resistance downstream of an internal ribosome entry sequence (IRES). After infection of C2F3 cells with retrovirus and selection for resistance to neomycin and hygromycin, the activity of β-Gal was assayed (Fig. 6) in the presence or in the absence of rapamycin (10 ng/ml; higher concentrations of the inducer were toxic to the cells). The cells responded to rapamycin with a detectable increase in β-Gal activity that was first observed after more than 30 min of treatment and reached 30-fold above the background activity assayed in untreated cells after 5 hr of treatment (Fig. 7; see color insert). In addition, the cells responded to concentrations of rapamycin between 0.5 and 10 ng/ml with a linear increase in β-Gal activity, providing evidence that this method allows quantitation of the interaction (Fig. 8). Thus, the response to rapamycin was both time

[21] E. J. Brown, M. W. Albers, T. B. Shin, K. Ichikawa, C. T. Keith, W. S. Lane, and S. L. Schreiber, *Nature (London)* **369,** 756 (1994).

[22] J. Chen, X. F. Zheng, E. J. Brown, and S. L. Schreiber, *Proc. Natl. Acad. Sci. U.S.A.* **92,** 4947 (1995).

[23] J. Choi, J. Chen, S. L. Schreiber, and J. Clardy, *Science* **273,** 239 (1996).

[24] P. J. Belshaw, S. N. Ho, G. R. Crabtree, and S. L. Schreiber, *Proc. Natl. Acad. Sci. U.S.A.* **93,** 4604 (1996).

[25] V. M. Rivera, T. Clackson, S. Natesan, R. Pollock, J. F. Amara, T. Keenan, S. R. Magari, T. Phillips, N. L. Courage, F. Cerasoli, Jr., D. A. Holt, and M. Gilman, *Nature Med.* **2,** 1028 (1996).

[26] S. R. Magari, V. M. Rivera, J. D. Iuliucci, M. Gilman, and F. Cerasoli, Jr., *J. Clin. Invest.* **100,** 2865 (1997).

[27] S. N. Ho, S. R. Biggar, D. M. Spencer, S. L. Screiber, and G. R. Crabtree, *Nature (London)* **382,** 822 (1996).

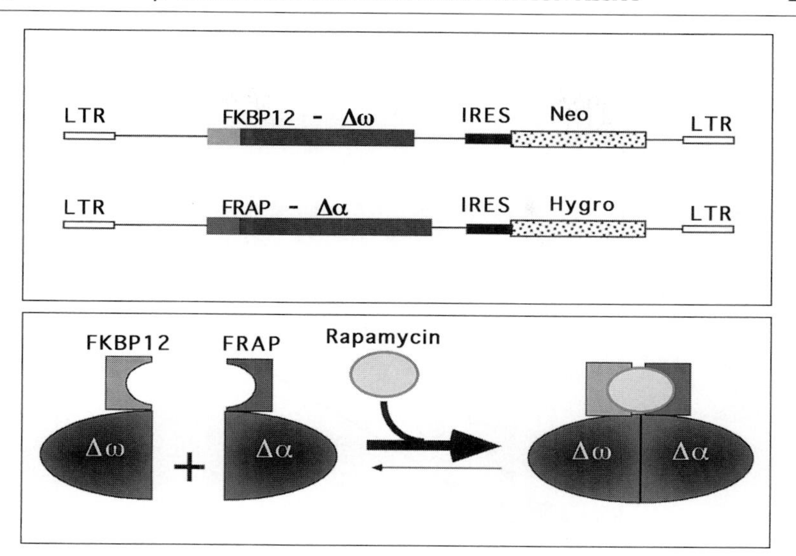

FIG. 5. Use of β-Gal complementation to monitor FKBP12 and FRAP interactions. *Top:* Schematic representation of the FKBP12–Δω–Neo and FRAP–Δα–Hygro retroviral constructs. IRES, Internal ribosome entry sequence; LTR, long terminal repeat. *Bottom:* Rapamycin induces the association of FKBP12 with the FRB domain of FRAP, favoring the reconstitution of functional β-Gal.

and dose dependent.[9] The response was only slightly reduced by treatment of the cells with cytostatic drugs such as mitomycin C. The magnitude of the reduction in β-Gal activity was in good agreement with the expected reduction due to the smaller number of signal-producing cells at given points in time, suggesting that growth-arrested cell populations could be used to facilitate the interpretation of experiments that test protein interactions that affect cell division (Fig. 7).

Application to Membrane Proteins

A major advantage of the β-Gal complementation system for analyzing protein interactions is its ability to detect interactions in the cellular compartment in which they normally occur, in either individual live cells or in mass cultures. The β-Gal complementation assay can be used to detect interactions between two membrane proteins, or between a membrane protein and a cytoplasmic protein. For example, if chimeric proteins consisting of the epidermal growth factor (EGF) receptor linked to two complementing β-Gal proteins are expressed in a single cell, addition of ligand

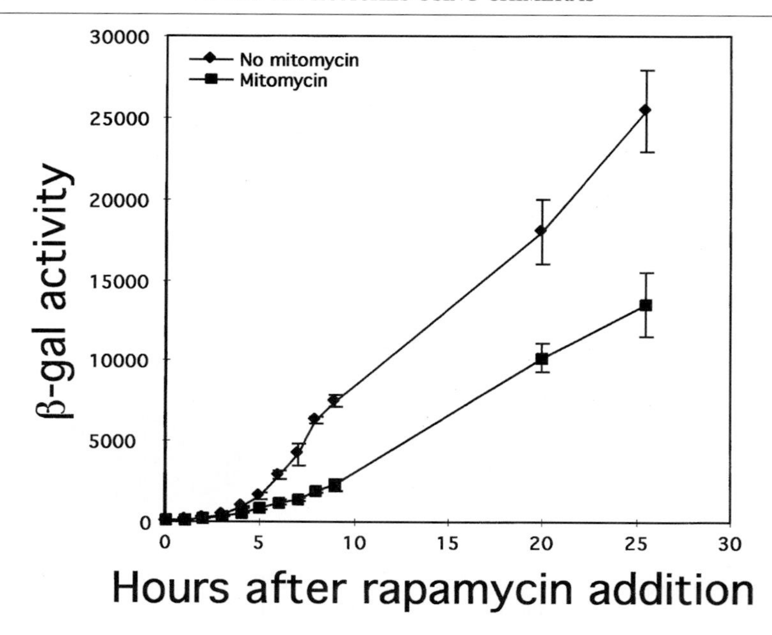

FIG. 6. Kinetics of induction of β-Gal activity on treatment with rapamycin in the presence or absence of the cytostatic drug mitomycin C. C2C12 cells expressing both FKBP12–Δω and FRAP–Δα were either treated with 2 μg of mitomycin overnight or left untreated. Both types of culture were then treated with rapamycin (10 ng/ml). Cells were lysed at different time intervals and the β-Gal activity in the lysates was quantified by chemiluminescence. β-Gal activity is expressed as luminescence counts per second. Each point represents the average of six replicate samples and error bars express standard deviations of the mean.

(EGF or other EGF receptor ligands) results in an increase in β-Gal activity within minutes. β-Gal activity declines after removal of the ligand, and can be induced on addition of fresh ligand. Moreover, an increase in β-Gal activity is prevented by the addition of molecules that inhibit ligand binding.[10a]

This ligand-dependent and ligand-specific β-Gal activity presents several advantages over other methods currently in use or in development: First, it is fast; second, it is direct; third, it occurs *in situ*. Traditionally, dimerization or activation of membrane receptors has been measured by assaying for some downstream event, such as phosphorylation of the receptor or a receptor substrate (in the case of receptor kinases). Such gel electrophoresis-based assays are laborious and time-consuming, and because they permit the analysis of only a few samples at a time, are not adaptable to high-throughput screening technology. Other traditional methods for detecting membrane protein interactions include immunopre-

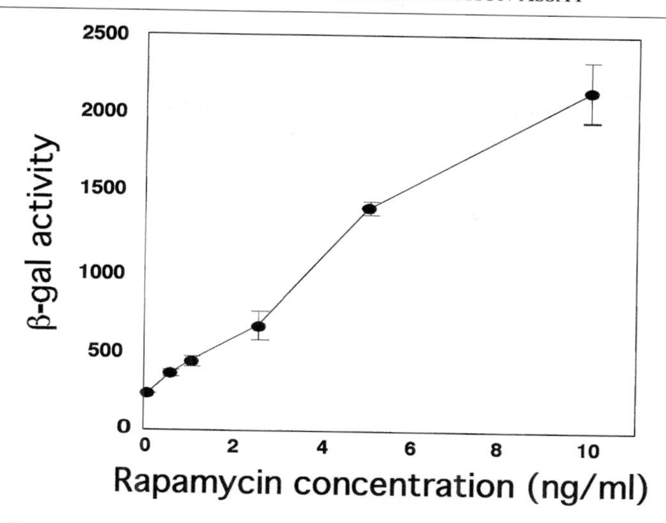

FIG. 8. Dose response to rapamycin of C2C12 cells expressing both FKBP12–$\Delta\omega$ and FRAP–$\Delta\alpha$. Cells were treated for 5 hr with the indicated concentrations of rapamycin, lysed, and β-Gal activity in the lysates was quantified by chemiluminescence. β-Gal activity is expressed as luminescence counts per second. [Reproduced from F. Rossi, C. A. Charlton, and H. M. Blau, *Proc. Natl. Acad. Sci. U.S.A.* **94,** 8405 (1997).]

cipitation, chemical cross-linking followed by immunoprecipitation, or density gradient centrifugation. These methods suffer the same limitations as other gel electrophoresis-based assays, and also detect only a subset of interactions, as some interactions do not survive the disruption of the cell.

Other methods such as fluorescence resonance energy transfer (FRET) assays have been developed to study protein interactions in live mammalian cells.[28] FRET-based assays exist in two forms. In the context of membrane receptors, fluorescent ligands can be added to the cell. This method has been used to analyze ligand binding and receptor dimerization in cells via microscopy.[29] Alternatively, chimeric interacting proteins can be expressed that have a fluorescent molecule, such as variants of green fluorescent protein (GFP), linked to the protein of interest. In this case, FRET analysis can be used to measure the interaction of the fluorescently labeled pro-

[28] S. R. Adams, A. T. Harootunian, Y. J. Buechler, S. S. Taylor, and R. Y. Tsien, *Nature (London)* **349,** 694 (1991).

[29] T. W. Gadella, Jr. and T. M. Jovin, *J. Cell Biol.* **129,** 1543 (1995).

teins.[30,31] FRET analysis is limited by spatial constraints and by the sensitivity of detection because there is no signal amplification.

More recently, protein interactions have been monitored with chimeric proteins that complement and reconstitute a functional protein. One such method is the complementation of β-Gal protein as described here. Another is complementation of dihydrofolate reductase (DHFR). Complementation, which restores the ability of DHFR to bind a fluorescent inhibitor, methotrexate, is measured after a prolonged (22-hr) incubation with the inhibitor.[32–34] As with GFP tagging, a disadvantage of DHFR complementation is that it lacks signal amplification, as the assay does not measure enzymatic activity. Whereas one active β-Gal complex can cleave many fluorogenic molecules, each DHFR-bound methotrexate molecule or GFP-tagged protein constitutes a single fluorescent molecule. Thus, expression levels of the protein must exceed a threshold level before an interaction event can be detected. This can be a problem for some proteins, such as the EGF receptor (EGFR), as high levels of receptor expression are known to lead to ligand-independent dimerization.[29] This problem is overcome with the β-Gal system described here, as proteins in low abundance can be detected, allowing ligand-dependent dimerization of the EGF receptor to be readily monitored.[10a] Furthermore, β-Gal substrates can be introduced into live cells within minutes and results obtained rapidly.[20] The 1-day period required for the DHFR assay suggests that transient interactions may be missed. In addition, it remains unclear whether the DHFR technology can be used in quantitative assays in wild-type cells with endogenous, active DHFR.

Of particular importance was the finding that the kinetics of β-Gal activity in response to an inducer varied with the different protein pairs. When the dimerization of chimeric EGFR–β-Gal constructs was induced with EGF, an increase in β-Gal activity was detected with less than 1 min of treatment with EGF, and enzyme activity continued to increase for another 1 to 2 hr.[10a] In contrast, in the case of the rapamycin-dependent FKBP12–FRAP interaction, β-Gal activity was not detectable until after more than 30 min of treatment with the inducer, rapamycin, and continued to increase over the subsequent 12 hr.[9] These kinetics of association fit well

[30] B. A. Pollok and R. Heim, *Trends Cell Biol.* **9,** 57 (1999).

[31] A. Miyawaki, J. Llopis, R. Heim, J. M. McCaffery, J. A. Adams, M. Ikura, and R. Y. Tsien, *Nature* (*London*) **388,** 882 (1997).

[32] J. N. Pelletier, K. M. Arndt, A. Pluckthun, and S. W. Michnick, *Nature Biotechnol.* **17,** 683 (1999).

[33] I. Remy, I. A. Wilson, and S. W. Michnick, *Science* **283,** 990 (1999).

[34] J. N. Pelletier, F. X. Campbell-Valois, and S. W. Michnick, *Proc. Natl. Acad. Sci. U.S.A.* **95,** 12141 (1998).

with those previously reported for rapamycin-binding proteins.[24,27] These findings indicate that the time course of β-Gal activity reflects the characteristics of the non-β-Gal protein interactions being monitored rather than the kinetics of the complementation reaction. The reconstitution of β-Gal activity by intracistronic complementation is thought to require partial refolding of the mutants and subsequent assembly of the tetrameric active enzyme.[6,11,12] In theory, a delay might be expected between the time of formation of the complex containing the test proteins and the development of a detectable β-Gal signal. However, our experiments suggest that such a lag is on the order of minutes and should therefore be negligible for most applications.

Potential Applications and Future Directions

Because the β-Gal complementation system is a rapid, enzymatic assay, it is amenable to high-throughput screening technologies, such as simple multiwell microplate assays that yield results within 1 hr (see Chemiluminescence Assay, below). The most straightforward and immediate application of this method is likely to be the screening of combinatorial chemical libraries in order to identify compounds that can either block or induce a specific interaction. This method is particularly well suited to the monitoring of interactions between membrane proteins such as receptors, or between receptors and cytoplasmic components of a given downstream signaling pathway. For example, cells expressing EGF receptor chimeras can be plated in multiwell plates and used to screen a library of chemical compounds or proteins for molecules that act as receptor agonists, or antagonists. This could be of particular interest for a molecule such as the EGF/erbB receptor family member erbB2, which has been implicated in a number of cancers, particularly breast cancer.[35] Overexpression of the erbB2 receptor leads to constitutive dimerization and activation of the receptor.[36] Molecules that block the dimerization and thus activation of erbB2 may be useful in anticancer therapy. Furthermore, known ligands of EGF receptor family members can easily be tested on cells expressing different combinations of erbB receptors, which could lead to a better understanding of the homo- and heterodimerization events that occur in this receptor family and underlie normal or neoplastic cell growth.[37–39]

[35] E. Tzahar and Y. Yarden, *Biochim. Biophys. Acta* **1377,** M25 (1998).
[36] R. Worthylake, L. K. Opresko, and H. S. Wiley, *J. Biol. Chem.* **274,** 8865 (1999).
[37] R. R. Beerli and N. E. Hynes, *J. Biol. Chem.* **271,** 6071 (1996).
[38] H. S. Earp, T. L. Dawson, X. Li, and H. Yu, *Breast Cancer Res. Treat.* **35,** 115 (1995).
[39] D. Graus-Porta, R. R. Beerli, J. M. Daly, and N. E. Hynes, *EMBO J.* **16,** 1647 (1997).

We have found that the β-Gal complementation methodology requires only a low level of expression of the chimeric proteins to generate a detectable signal on interaction. Indeed, the lower the expression, the lower the frequency of nonspecific ligand-independent interactions.[10a] Thus it is possible that in most cases, the levels of expression that can be detected will be physiological and comparable to the expression levels of the endogenous proteins. These characteristics of β-Gal complementation may make it applicable to detailed studies of protein interactions involved in signaling pathways. Furthermore, β-Gal complementation may allow direct monitoring of protein interactions in transgenic animals. Specific chimeric proteins could be constructed and tested *in vitro* for their ability to generate β-Gal activity on interaction. The best chimeras could then be recreated *in vivo* by "knock in" of $\Delta\alpha$ and $\Delta\omega$ in frame with the endogenous genes. Animals carrying both transgenes should develop β-Gal activity only in those cells in which the endogenous proteins are expressed and actually interact. Histochemical detection of β-Gal in these animals models should allow the mapping of specific protein interactions both during development and in disease states. Such studies would allow an advance from gene expression to gene function *in vivo.*

Mammalian Two-Hybrid System

Another exciting potential application of β-Gal complementation stems from the possibility of screening a library of cDNAs fused to β-Gal mutants in order to identify genes encoding novel protein interaction partners in mammalian cells. Although the yeast two-hybrid system and its modifications have enabled researchers to screen cDNA libraries for gene products interacting with a known protein, leading to a great increase in the number of known interaction pairs, the β-Gal complementation system may have distinct advantages. In the simplest and most common embodiment of the yeast two-hybrid system, the interaction between a DNA-bound protein (bait) and a polypeptide derived from a cDNA expression library targets a potent transcriptional activation domain to a synthetic promoter and leads to transcription of a detectable marker gene. This interaction-dependent activation of transcription has the advantage of greatly amplifying the signal. Indeed, the yeast two-hybrid system can easily detect low-affinity interactions.[3] However, it also presents disadvantages: for transcription to be activated, the interaction must take place in the nucleus; interactions that take place only in other cellular compartments, such as membranes, will not be detected by this method. Furthermore, attempts to adapt this system for use in mammalian cells have met with limited success.

The adaptation of β-Gal complementation to two-hybrid screens offers several advantages over the yeast system. In theory, a chimeric library obtained by fusing random cDNAs to one of the β-Gal mutants could be constructed. This library could then be screened for gene products that interact with a given "bait" protein fused to the complementary β-Gal mutant. Such a screen would be performed in cultured mammalian cells, allowing the detection of interactions that need to be facilitated by an endogenous protein, or that take place only in a specific cellular compartment. The screen would be facilitated by the availability of flow cytometry-based techniques that allow the isolation of β-Gal-positive cells without compromising their viability[20] (Fig. 9). Furthermore, because the readout of the assay would not be dependent on the nuclear localization of the interacting protein partners, this method could be used to isolate partners of membrane-linked proteins. Such a mammalian two-hybrid screen should lead to the identification of signal transduction components in particular cell contexts (e.g., normal vs neoplastic) and, ultimately, new drug targets.

Practical Issues in Using β-Galactosidase Complementation to Monitor Protein–Protein Interactions

To monitor the interaction between two proteins, each of the proteins is fused to one of a pair of β-Gal mutants (Fig. 1), and the two fusion proteins are expressed at low levels in mammalian cells. The induced interaction between the two test proteins will then facilitate the complementation reaction and result in an increase in β-Gal activity. For this type of application, we have developed a specific pair of β-Gal deletion mutants ($\Delta\alpha$ and $\Delta\omega$) that display a low level of spontaneous complementation

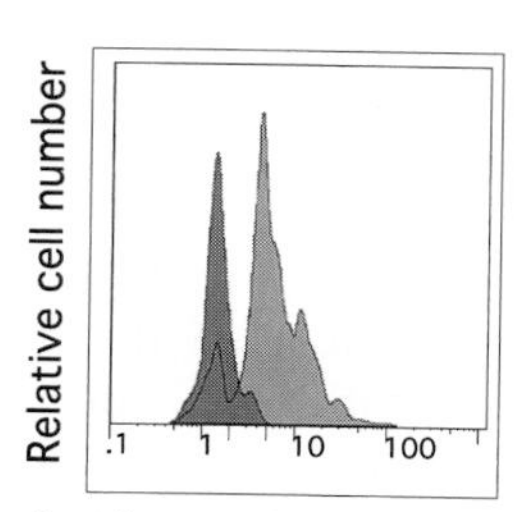

FIG. 9. Fluorescence-activated cell sorter (FACS) analysis of rapamycin-induced β-Gal complementation. The induced and uninduced populations yield essentially nonoverlapping peaks. The vertical axis represents relative cell number, and the horizontal axis depicts intensity of fluorescence on a logarithmic scale. The peaks on the left represent the untreated samples and those on the right represent samples treated with rapamycin (10 ng/ml).

when coexpressed in mammalian cells.[8] Nevertheless, to avoid background β-Gal activity that could mask the interaction-dependent complementation, it is critical that the expression of the chimeric proteins be kept relatively low.[9] This should be an advantage in most cases: Low-level expression is less likely to perturb the balance of proteins in the cell and lead to nonphysiological results than the high levels of expression required by other systems that do not provide any signal amplification. To control the expression levels of the chimeric proteins, we favor the use of IRES-containing retroviral vectors expressing the chimeric protein and a selectable marker from the same bicistronic messenger RNA.[40] Unlike the use of plasmid transfection, which usually results in multiple copies of the vector in the genome of the selected stable clones, infection with retroviral vectors permits the introduction of a single copy of the expression cassette in the target cells, thereby reducing the chances of overexpression. Furthermore, in our experience, bicistronic mRNAs tend to yield lower protein expression levels compared with shorter mRNAs containing a single coding sequence.

A variety of assays are available to detect β-Gal activity, reflecting its extensive use as a marker over three decades. We have successfully detected "complemented" β-Gal in mammalian cells by histochemistry with both the traditional chromogenic substrate X-Gal and a novel, more sensitive fluorogenic substrate (see Histochemical Assays, below). These detection methods, however, are not quantitative and require fixation of the cells. Sensitive quantitative detection of complemented β-Gal can be achieved by assaying cell lysates with commercially available chemiluminescence substrates [e.g., the Gal-Screen kit from Tropix (Bedford, MA); see Chemiluminescence Assays, below]. Chemiluminescence assays in multiwell microplates permit a rapid analysis of the effect of a variety of pharmacological and biological agents on the protein interactions of interest under a range of conditions. Quantitative analysis is also possible in live cells by using the fluorogenic substrate fluorescein di-β-D-galactopyranoside (FdG) followed by detection by flow cytometry (see Flow Cytometry Assay, below, and Fig. 9).[20] FdG can be introduced into live cells by hypotonic loading and the fluorescent cleavage product of the reaction catalyzed by β-Gal (free fluorescein) will remain within the cell as it is unable to cross the plasma membrane. The fluorescence-activated cell sorting (FACS) assay has the further advantage of allowing recovery of live cells in which a given interaction has taken place.

[40] R. A. Morgan, L. Couture, O. Elroy-Stein, J. Ragheb, B. Moss, and W. F. Anderson, *Nucleic Acids Res.* **20,** 1293 (1992).

Molecular Basis of Interaction-Induced β-Galactosidase Complementation

How does forcing the interaction between inefficiently complementing β-Gal mutants increase their ability to recreate an active enzyme? For a hypothesis to be generated, it is necessary first to consider what limits the ability of the $\Delta\alpha$ and $\Delta\omega$ mutants to complement each other spontaneously. For the complementation reaction to take place, one of the two mutants must be able to provide in *trans* the critical domain that the other mutant lacks. For example, in α-complementation in *E. coli,* a peptide containing the α portion of β-Gal "donates" this domain to a β-Gal mutant that harbors a deletion in the α region, a process that requires the establishment of noncovalent interactions between the two proteins.[11,12] Consequently, a critical requirement for the re-creation of enzymatic activity is that the relevant domain is available to be "donated." An analysis of the crystal structure of wild-type β-Gal clearly indicates that both the α domain and the ω domain that are absent from the $\Delta\alpha$ and $\Delta\omega$ mutants are involved in extensive intramolecular interactions with the central portion of the molecule[14] (μ domain; Fig. 2). As both $\Delta\alpha$ and $\Delta\omega$ contain a complete μ domain, it is likely that in these mutants the intact ω and α domains (either of which could in theory be "donated" to the complementary mutant) are sequestered in such intramolecular interactions. Thus, they would not be readily available to establish the intermolecular interactions with the μ domain of the complementary mutant required to reconstitute a functional monomer. Support for this hypothesis comes from the finding that in inefficient α donors, in contrast to efficient α donors, the α domain is not accessible to proteolytic enzymes, suggesting that it is probably masked by intramolecular interactions.[7,13] When the two β-Gal mutants are forced to come in close contact by fusion to interacting proteins their effective local concentration increases greatly, the intermolecular interaction is more likely to occur, and consequently the rate of complementation is increased, generating a detectable signal. Future analysis of crystal structures of the complementing mutants will certainly be instructive.

Troubleshooting

In theory, the use of the β-Gal complementation method for detecting protein–protein interactions may be subject to steric constraint. On interaction of the test proteins, $\Delta\alpha$ and $\Delta\omega$ may need to associate with a certain optimal orientation in order to generate detectable enzymatic activity. This is likely to depend on the orientation of the test proteins within the complex they form. For example, two chimeric proteins fused to the same end of the β-Gal mutants may not lead to efficient complementation if they interact in a head-to-tail fashion, as the β-Gal mutants may project out of the complex in opposite directions and thus never contact each other. The same could happen if the test proteins are too bulky. It is unlikely, however, that these theoretical limitations will preclude the analysis of specific interactions. Given two known polypeptides, it should be possible to generate a range of chimeric proteins *in vitro* and select those that display the best characteristics for later use in, for example, high-throughput screens or transgenic animals. This limitation, however, may preclude the detection of a subset of cDNAs in a library screen, because the cDNA–β-Gal fusion points are necessarily generated at random and cannot be predicted or individually optimized. Studies to optimize interactions and overcome these potential problems are underway, for example, the inclusion of linkers of different lengths between the test protein and β-Gal.

Another limitation affecting this method stems from the fact that available substrates easily localize β-Gal activity to individual cells within a population, but are not efficient in localizing the enzyme at the subcellular level. Whereas nuclear β-Gal can be distinguished from cytoplasmic β-Gal, a more refined localization of the enzyme is prevented by the propensity of the existing substrates to diffuse. Such a problem underscores the need to develop less diffusible substrates, and work is underway to address this need.

Concluding Remarks

Clearly, the application of intracistronic β-Gal complementation to monitoring protein–protein interactions in live cells has tremendous potential, but is still in its infancy. Improvements will certainly be forthcoming as well as increased insights into the underlying molecular mechanisms. Nevertheless, the β-Gal system already allows an investigation of fundamental questions regarding protein interactions involved in cell signaling such as membrane receptor dimerization. High-throughput screens for antagonists and agonists of specific protein interactions can be envisioned. In addition, the development of a "mammalian two-hybrid" system may soon become a reality.

Methods

Cloning, Retrovirus Production, and Infection

To produce the infectious viral particles, the Phoenix-E ecotrophic packaging cell line (a kind gift of G. Nolan, Stanford University) is transiently transfected with the plasmids containing the appropriate proviral genome.

Cells are plated in 60-mm plates the day before transfection at 3×10^6 cells/plate, and transfected with Fugene 6 (Roche, Indianapolis, IN) according to the manufacturer instructions. With this cell line, we obtain optimal transfection with 6 μl of Fugene and 2 μg of plasmid DNA per plate. Although other methods, including the traditional calcium phosphate transfection protocols, can sometimes result in transfection of a higher percentage of the cells, we consistently obtain higher viral titers with Fugene 6, possibly because of the absence of toxic side effects that may affect viral production. Tissue culture supernatant containing the infectious particles is harvested between 36 and 72 hr after transfection and used to infect C2F3 myoblasts plated in 60-mm plates. In some cases, depending on the efficiency of transfection and virus production, we dilute the supernatants 1:10 in fresh medium to limit the number of vector copies introduced in each target cell. Infections are carried out either sequentially or simultaneously by replacing the C2F3 medium with viral supernatant filtered through a 0.45-μm pore size filter. Polybrene is added at a final concentration of 8 μg/ml and the plates containing the target cells are centrifuged in a tabletop Beckman (Fullerton, CA) GS-6 centrifuge equipped with microplate platforms at 2500 rpm for 30 min. The centrifugation step leads to an increase in infection efficiency, but it can be omitted as high multiplicity of infection is not desirable for most applications of this method. Selection with hygromycin and/or neomycin (0.8 mg/ml of each drug) is started 24 hr after infection.

Flow Cytometry Assay

To date, a selection strategy that imparts a growth advantage specifically to cells that express β-Gal is not available. However, β-Gal can be detected in live cells with fluorescent substrates, and fluorescent cells can be efficiently separated from mixed populations rapidly and efficiently by flow cytometry without affecting their viability. This assay is relatively simple and can be performed in less than half an hour when the number of samples to be analyzed is limited. Furthermore, it is quantitative and we use it routinely to monitor the kinetics of a given interaction in response to specific signals. The traditional substrate used for this assay, fluorescein di-β-D-galactopyranoside (FdG), is not cell permeable but it can be introduced into the cells by hypotonic shock.[20] Cleavage by β-Gal results in the production of free fluorescein, which is also unable to cross the plasma membrane and is trapped inside the β-Gal-positive cells. The cells to be analyzed are trypsinized, resuspended in phosphate-buffered saline (PBS) containing 5% (v/v) fetal bovine serum (PBS–FBS), and pelleted in 5-ml polystyrene round-bottom tubes (Falcon 2058; Becton Dickinson Labware, Lincoln

Park, NJ). The cells are then resuspended in 100 μl of PBS–FBS and an equal volume of doubly distilled water containing the substrate at a concentration of 1 m*M* is added. After 3 min at room temperature, the hypotonic conditions are quenched by adding 10 volumes of ice-cold PBS–FBS containing propidium iodine (PI, 1 μg/ml). PI is a red fluorescent compound that is actively excluded from living cells but accumulates in dead cells, allowing their exclusion from the analysis or the sorting. After quenching, the cells are pelleted again, resuspended in approximately 200 μl of ice-cold PBS–FBS, and analyzed on a Becton Dickinson (San Jose, CA) FACScan or sorted on a Becton Dickinson FACStar flow cytometer. In most clonal populations, the cells respond to signals inducing the test interaction with an homogenous increase in β-Gal activity, and the mean fluorescence in each sample can be used as a reliable measure of the interaction. In polyclonal populations, however, we have often noticed that a subset of the cells does not respond to the inducing signal, and the mean fluorescence is less useful. However, the main advantage of the FACS-based assay is that it provides a simple means of selecting for cells in which a given interaction takes place, opening the path to screening cDNA libraries for novel interaction partners of a given "bait" protein in mammalian cells. This rapid method of selection is in our opinion superior to growth-based selection protocols, as many interactions are likely to be transient under physiological conditions and could possibly affect the growth characteristics of the cells.

Histochemical Assays

X-Gal Histochemistry. Chromogenic detection of β-Gal activity can be performed on cells cultured in plastic tissue culture dishes or on glass coverslips. Cells are fixed for 4 min in cold (4°) 4% (w/v) paraformaldehyde in PBS and rinsed for 5 min, two times. A stock solution of X-Gal [Sigma (St. Louis, MO); 40 mg/ml in dimethylformamide, stored at −20°, protected from light] is diluted to a final concentration 1 mg/ml in 5 m*M* $K_3Fe(CN)_6$, 5 m*M* $K_4Fe(CN)_6$, 2 m*M* $MgCl_2$ in PBS, applied to cells, incubated at 37° overnight (shorter times are sufficient for high levels of β-Gal activity), and examined microscopically for blue cells.

Fluor X-Gal Histochemistry. Muscle cells to be labeled for fluorescent histochemical detection of β-Gal activity are cultured on sterile collagen-coated glass coverslips (Becton Dickinson), fixed in 4% (w/v) paraformaldehyde in PBS, and rinsed twice with PBS (in which they can be stored at 4° until staining is performed). If cells are to be labeled with antibody as well as Fluor X-Gal, cells are blocked for 30 min in PBS + 10% (v/v) equine serum, incubated for 2 hr in primary antibody, rinsed four times (10 min each) in blocking buffer, incubated for 1 hr in biotinylated secondary antibody, washed again four times in blocking buffer, incubated for 1 hr in Cy5-labeled streptavidin (diluted 1:100; Amersham, Arlington Heights, IL), and then washed two times in blocking buffer and two times in PBS. All immunolabeling steps are performed at 4°. β-Gal substrate is prepared by diluting a stock solution of Fast Red Violet LB (Sigma; stock is 50 mg/ml in dimethylformamide, stored at −20°, not totally dissolved at this concentration) to a final concentration of 100 μg/ml and a stock solution of 5-6 X-Gal [5-bromo-6-chloro-3-indolyl-β-D-galactopyranoside Fluka (Ronkonkoma, NY); stock at 50 mg/ml in dimethylformamide, stored at −20°, will change from pale blue to yellow after exposure to light, but this does not appear to affect activity] mixed to a final concentration of up to 25 μg/ml in PBS (decrease the concentration if β-Gal activity is strong). A 0.45-μm pore size syringe filter is used to remove any precipitate. The mixture of Fast Red Violet LB and 5-6 X-Gal is added to fixed cells and incubated 60 to 90 min at 37°, and then rinsed in 25 ml of PBS for 30 min at room temperature. Nuclei may be stained by diluting 4′,6-diamidino-2-phenylindole dihydrochloride hydrate (DAPI; Sigma) in PBS to a final concentration of 100 ng/ml, incubating for 10 min at room temperature, and rinsing twice for 5 min. Coverslips are mounted in PBS (no glycerol-based antifade solution) and sealed with nail polish. Fluor X-Gal staining can be viewed with either fluorescein isothiocyanate (FITC) or rhodamine (TRITC) filter sets of an epifluorescence microscope (signal-to-noise peak is 560 nm for weak signals; the rhodamine filter set has a 560-nm bandpass emission filter). The FITC channel gives a better signal-to-background ratio for weak signals, but strong signals appear to be quenched. Therefore, Fluor X-Gal stain is best viewed with TRITC filters (Figs. 4 and 6).

Chemiluminescence Assay

Quantitative assays for β-Gal activity are carried out in 96-well plates with a commercial chemiluminescent assay system with prolonged (glow) emission. Although the FACS-based assay is important for characterization at the single-cell level of live cells expressing β-Gal chimeric proteins, it is difficult to prepare more than about 150 samples at a time for FACS analysis. The chemiluminescent assay described here permits the analysis of thousands of samples in one experiment. Numerical results can be obtained 1 hr after the assay is begun.

Subconfluent tissue culture cells are trypsinized, counted in a cell counter (Z1; Coulter, Hialeah, FL), and replated in white 96-well plates at 10,000 cells per well (for C2F3 mouse myoblasts) in a volume of 100 μl the day before the assay. Plates are maintained in an appropriate medium and incubator conditions [DMEM with 20% (v/v) FBS at 37° in 10% CO_2 for C2F3 cells] up to the addition of the chemiluminescent reagent. We have tested white 96-well plates with white opaque bottoms and clear bottoms (which permit viewing cells on a microscope). Opaque plates yield higher absolute luminescence values; however, the signal-to-background ratio in the two types of plates are similar. In clear bottom plates, luminescence from a strongly luminescent sample can be measured in an adjoining empty or weakly luminescent well. Such cross-talk is prevented in the fully opaque plates.

Cells are treated with agents that stimulate protein interactions of chimeric β-Gal proteins as desired. For example, cells are treated for minutes or hours with a growth factor that induces dimerization of a chimeric growth factor receptor–β-Gal protein. Normally, addition of agents is accomplished by replacing the medium in the well with fresh medium containing the agent, maintaining a volume of 100 μl per well. The medium in untreated wells is also replaced at the same time, as background luminescence (luminescence measured in wells containing only medium and the chemiluminescence reagents) increases the longer the medium (DMEM in our case) is left in the well. Wells are grouped into triplicate or quadruplicate samples for treatment. If wells are to be treated for different times, the start of treatment must vary so that all treatments in a given plate end at the same time, or the different time points must be on separate plates. In the latter case, controls must be present on each plate to control for any variations in the chemiluminescent assay.

At the end of the treatment time, the Tropix Gal-Screen chemiluminescent reagent is added to the plates. Tropix Gal-Screen (GSY200 or GSY1000) consists of two components. The Gal-Screen substrate (1,2-dioxetane compound) is diluted 1:25 immediately before use into Gal-Screen buffer B equilibrated to room temperature. Gal-Screen (100 μl) is added to each well without removing the medium. The plate is then incubated at 26 to 28° for 45 min to 1 hr, and is then read on a plate reader, measuring each well for 1 sec. All readings, especially if multiple plates are used, are made in the period of time in which the chemiluminescence has plateaued, which is typically in the 45-min to 2-hr range.

The following plate readers offer comparable sensitivity: Tropix TR717, EG&G Wallac Berthold (Bad Wildbad, Germany), LB96V, Wallac/EG&G MicroBeta Plus (current version is MicroBeta Trilux). All these machines have optional injectors. On the MicroBeta, addition of an injector to the machine reduces the sensitivity of the instrument. On the TR717 and LB96V, the presence of an injector does not affect the sensitivity of the instrument. The MicroBeta has the advantage of being able to count multiple plates unattended. However, it takes 5 to 10 min to count one plate whereas the TR717 and LB96V, which hold only one plate at a time, require about 2 min to count a plate. The data are then analyzed in a Microsoft Excel spreadsheet. Replicate samples are averaged together, and the luminescence value of wells containing only medium, or containing cells lacking the β-Gal constructs, is subtracted from the measured values.

[16] Green Fluorescent Protein Chimeras to Probe Protein–Protein Interactions

By SANG-HYUN PARK and RONALD T. RAINES

Introduction

Green fluorescent protein (GFP) from the jellyfish *Aequorea victoria* is autofluorescent. The fluorophore of GFP forms on translation of the protein, without the action of any additional factors. GFP can be expressed in a wide variety of nonnative cell types. To date, GFP has been used largely *in vivo* as a marker for gene expression and as a fusion tag to monitor protein localization in living cells.[1–6]

GFP has exceptional physical and chemical properties besides spontaneous fluorescence. These properties include high thermal stability and resistance to detergents, organic solvents, and proteases. These properties endow GFP with enormous potential for biotechnical applications.[7–9] Since the cDNA of GFP was cloned,[10] a variety of GFP variants have been generated that broaden the spectrum of its application.[11–16] Among those

[1] M. Chalfie, Y. Tu, G. Euskirchen, W. W. Ward, and D. C. Prasher, *Science* **263,** 802 (1994).
[2] S. Inouye and F. I. Tsuji, *FEBS Lett.* **341,** 277 (1994).
[3] P. Ren, C.-S. Lim, R. Johnsen, P. S. Albert, D. Pilgrim, and D. L. Riddle, *Science* **274,** 1389 (1996).
[4] M. K. Topham, M. Bunting, G. A. Zimmerman, T. M. McIntyre, P. J. Blackshear, and S. M. Prescott, *Nature* (*London*) **394,** 697 (1998).
[5] M. Maletic-Savatic, R. Malinow, and K. Svoboda, *Science* **283,** 1923 (1999).
[6] F. Perez, G. S. Diamantopoulos, R. Stalder, and T. E. Kreis, *Cell* **96,** 517 (1999).
[7] S. H. Bokman and W. W. Ward, *Biochem. Biophys. Res. Commun.* **101,** 1372 (1981).
[8] W. W. Ward, *in* "Bioluminescence and Chemiluminescence" (M. DeLuca and W. McElroy, eds.), p. 235. Academic Press, New York, 1981.
[9] W. W. Ward and S. H. Bokman, *Biochemistry* **21,** 4535 (1982).
[10] D. C. Prasher, V. K. Eckenrode, W. W. Ward, F. G. Prendergast, and M. J. Cormier, *Gene* **111,** 229 (1992).
[11] A. B. Cubitt, R. Heim, S. R. Adams, A. E. Boyd, L. A. Gross, and R. Y. Tsien, *Trends Biochem. Sci.* **20,** 448 (1995).

0076-6879/00 $30.00

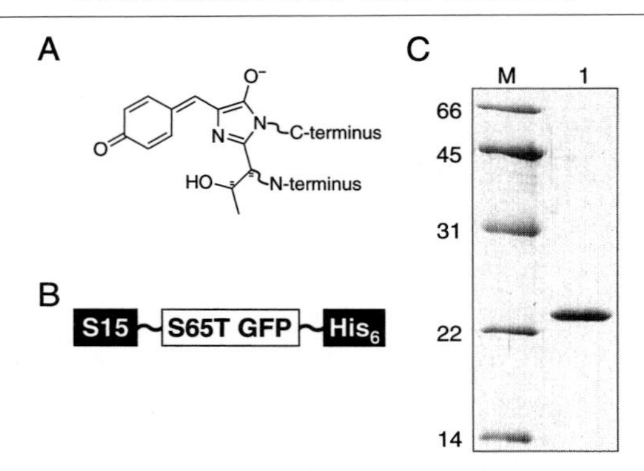

FIG. 1. Green fluorescent protein chimera. (A) Chemical structure of the fluorophore in the core of the S65T variant of green fluorescent protein.[17] This fluorophore forms spontaneously from residues Thr-65, Gly-66, and Tyr-67. The quinone methide resonance form shown here is likely to be responsible for the characteristic fluorescence of S65T GFP. (B) Structure of S15–GFP(S65T)–His_6, in which S15 is fused to the N terminus and six histidine residues are fused to the C terminus of S65T GFP. Two amino acid residues (Pro-Asp and Leu-Glu, respectively) link the tags to S65T GFP. (C) SDS–PAGE analysis of purified GFP chimera. Lane M, Molecular mass markers (in kDa); lane 1, S15–GFP(S65T)–His_6.

variants, S65T GFP (Fig. 1A) is unique in having increased fluorescence intensity, faster fluorophore formation, and altered excitation and emission spectra than that of the wild-type protein.[14,17] The wavelengths of the excitation and emission maxima of S65T GFP (490 and 510 nm, respectively) resemble closely those of fluorescein. The fluorescein-like spectral characteristics of S65T GFP enable its use with instrumentation that has been designed specifically for use with fluorescein.

Here, we describe the use of S65T GFP to probe protein–protein interactions *in vitro*.[18] This method requires fusing GFP to one of the target proteins to create a GFP chimera. The interaction of this fusion protein with another protein can be analyzed by two distinct methods.

[12] S. Delagrave, R. E. Hawtin, C. M. Silva, M. M. Yang, and D. C. Youvan, *Bio/Technology* **13,** 151 (1995).

[13] T. Ehrig, D. J. O'Kane, and F. G. Prendergast, *FEBS Lett.* **367,** 163 (1995).

[14] R. Heim, A. B. Cubitt, and R. Y. Tsien, *Nature* (*London*) **373,** 663 (1995).

[15] A. Crameri, E. A. Whitehorn, E. Tate, and W. P. C. Stemmer, *Nature Biotechnol.* **14,** 315 (1996).

[16] W. W. Ward, "Green Fluorescent Protein: Properties, Applications and Protocols." John Wiley & Sons, New York, 1997.

[17] M. Ormö, A. B. Cubitt, K. Kallio, L. A. Gross, R. Y. Tsien, and S. J. Remington, *Science* **237,** 1392 (1996).

[18] S.-H. Park and R. T. Raines, *Protein Sci.* **6,** 2344 (1997).

The first method is a fluorescence gel retardation assay. The gel retardation assay has been used widely to study protein–DNA interactions.[19] This assay is based on the electrophoretic mobility of a protein–DNA complex being less than that of either molecule alone. In a fluorescence gel retardation assay, electrophoretic mobility is detected by the fluorescent properties of S65T GFP. The fluorescence gel retardation assay is a rapid method to demonstrate the existence of a protein–protein interaction and to estimate the equilibrium dissociation constant (K_d) of the resulting complex.

The second method is a fluorescence polarization assay. The fluorescence polarization assay is an accurate method to evaluate the K_d in a specified homogeneous solution. Fluorescence polarization assays usually rely on fluorescein as an exogenous fluorophore. S65T GFP can likewise serve in this role. Further, the fluorescence polarization assay can be adapted for the high-throughput screening of protein or peptide libraries.

Production, Purification, and Detection of an S65T Green Fluorescent Protein Chimera

To demonstrate the potential of S65T GFP chimeras in exploring protien–protein interactions, we use as a model system the well-characterized interaction of the S-peptide and S-protein fragments of bovine pancreatic ribonuclease A (RNase A; EC 3.1.27.5).[20] Subtilisin treatment of RNase A yields two tightly associated polypeptide chains: S-peptide (residues 1–20) and S-protein (residues 21–124).[21] Residues 1–15 of S-peptide (S15) are necessary and sufficient to form a complex with S-protein.[22] A GFP chimera has also been used to explore the interaction of the BALF-1 protein and Bax/Bak, which are Bcl-2 homologs from Epstein–Barr virus.[23]

Protocol

S15–GFP(S65T)–His_6 (Fig. 1B) is produced by standard recombinant DNA techniques.[18] A cDNA fragment encoding S65T GFP is amplified by the polymerase chain reaction (PCR) and inserted into the *Bgl*II/*Xho*I sites of pET-29b (Novagen, Madison, WI), which encodes S15 and hexahistidine (His_6). The resulting expression vector is used to transform *Escherichia coli* strain BL21(DE3). Cells are grown at 37° in 0.5 liter of Luria–Bertani

[19] J. Carey, *Methods Enzymol.* **208,** 103 (1991).
[20] R. T. Raines, *Chem. Rev.* **98,** 1045 (1998).
[21] F. M. Richards, *C.R. Trav. Lab. Carlsberg Ser. Chim.* **29,** 322 (1955).
[22] J. T. Potts, D. M. Young, and C. B. Anfinsen, *J. Biol. Chem.* **238,** 2593 (1963).
[23] W. L. Marshall, C. Yim, E. Gustafson, T. Graf, D. R. Sage, K. Hanify, L. Williams, J. Fingeroth, and R. W. Finberg, *J. Virol.* **73,** 5181 (1999).

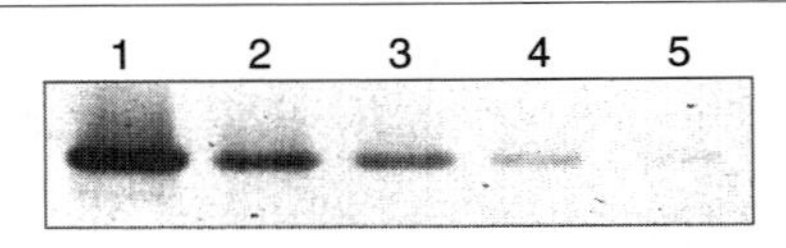

FIG. 2. Fluorimager analysis of purified green fluorescent protein chimera. Lanes 1–5, 10, 3.0, 1.0, 0.3, and 0.1 ng of S15–GFP(S65T)–His_6, respectively.

(LB) medium until the absorbance at 600 nm is 0.5 OD. Isopropylthiogalactoside (IPTG) is then added to a final concentration of 0.5 m*M*, and the cells are grown at 30° for an additional 4 hr. The culture is harvested and resuspended in 25 ml of 50 m*M* HEPES buffer, pH 7.9, containing NaCl (0.3 *M*), dithiothreitol (DTT, 0.5 m*M*), and phenylmethylsulfonyl fluoride (PMSF, 0.2 m*M*), and the cells are lysed by using a French pressure cell. The lysed cells are subjected to centrifugation at 18,000*g*. The supernatant is collected and loaded onto a Ni^{2+}-NTA agarose column (Qiagen, Chatsworth, CA). The column is washed with 50 m*M* HEPES buffer, pH 7.9, containing imidazole (8 m*M*), NaCl (0.3 *M*), and PMSF (0.5 m*M*). S15–GFP(S65T)–His_6 is eluted in the same buffer containing imidazole (0.10 *M*). The green fractions are pooled and purified further by FPLC (fast protein liquid chromatography) on a Superdex 75 gel-filtration column (Pharmacia, Piscataway, NJ) with elution by 50 m*M* HEPES buffer, pH 7.9. Purified S15–GFP(S65T)–His_6 migrates as a single species during sodium dodecyl sulfate–polyacrylamide gel electrophoresis (SDS–PAGE) (Fig. 1C).

S15–GFP(S65T)–His_6 remains fluorescent after electrophoresis in a native polyacrylamide gel (Fig. 2). Further, the altered excitation and emission spectra of S65T GFP are well suited for detection by a fluorimager. The sensitivity of S65T GFP detection in a native polyacrylamide gel is ≥0.1 ng, which is comparable to that of an immunoblot using an anti-GFP antibody.[24]

The regioisomer of S15–GFP(S65T)–His_6, His_6–GFP(S65T)–S15, can also be produced and analyzed by similar protocols.[18] Surprisingly, His_6–GFP(S65T)–S15 migrates as two distinct species during SDS–PAGE, native PAGE, and zymogram electrophoresis.[25,26] Apparently, two isoforms of His_6–GFP(S65T)–S15 exist that migrate differently during electrophoresis, even in the presence of SDS. We therefore recommend constructing GFP chimeras in which the target protein is fused to the N terminus of GFP, rather than to the C terminus.

[24] S. Colby, K. Dohner, D. Gunn, and K. Mayo, *CLONTECHniques* **10,** 16 (1995).
[25] J.-S. Kim and R. T. Raines, *Protein Sci.* **2,** 348 (1993).
[26] J.-S. Kim and R. T. Raines, *Anal. Biochem.* **219,** 165 (1994).

Fluorescence Gel Retardation Assay

Gel mobility retardation is a useful tool for both qualitative and quantitative analyses of protein–nucleic acid interactions.[19] The fluorescence gel retardation assay applies gel retardation of fluorescent species to the study of a protein–protein interaction. In this assay, free and bound S65T GFP chimeras are resolved and visualized in a native polyacrylamide gel.

Example

The fluorescence gel retardation assay is used to quantify the interaction between S-protein and S15–GFP(S65T)–His_6. A fixed quantity of S15–GFP(S65T)–His_6 is incubated with a varying quantity of S-protein prior to electrophoresis in a native polyacrylamide gel. After electrophoresis, the gel is scanned with a fluorimager and the fluorescence intensities of bound and free S15–GFP(S65T)–His_6 are quantified (Fig. 3A). From the relative fluorescence intensities of the bound and free S15–GFP(S65T)–His_6, the binding ratio (R = fluorescence intensity of bound S15–GFP(S65T)–His_6/total fluorescence intensity) at each concentration is obtained. The value

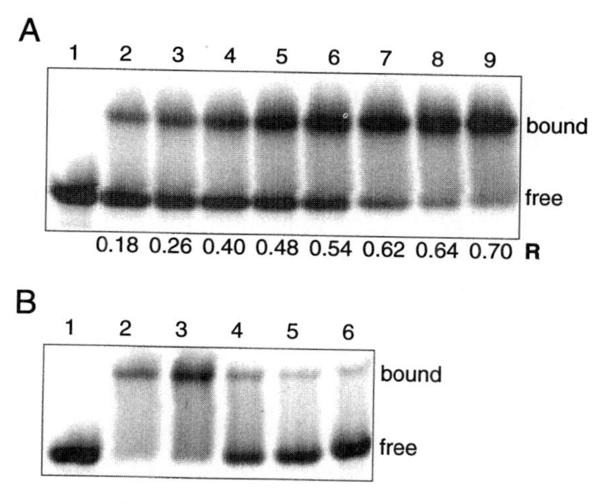

FIG. 3. Gel retardation assay of a protein–protein interaction. (A) Gel retardation assay of the interaction of S15–GFP(S65T)–His_6 with varying amounts of S-Protein. Lanes 1–9, 1 μM S15–GFP(S65T)–His_6 and 0, 0.2, 0.3, 0.4, 0.5, 0.6, 0.7, 0.8, and 0.9 μM S-protein, respectively. The relative mobilities of free and bound S15–GFP(S65T)–His_6 are 0.72 and 0.47, respectively. The value of R is obtained for each lane, and values of K_d are calculated by using Eq. (1) with the average being $K_d = (6 \pm 3) \times 10^{-8}\ M$. (C) Gel retardation assay demonstrating that S-peptide competes with S15–GFP(S65T)–His_6 for interaction with S-protein. Lane 1, 1 μM S15–GFP(S65T)–His_6 and no S-protein or S-peptide. Lanes 2–6, 1 μM S15–GFP(S65T)–His_6, 1 μM S-protein, and 0, 0.3, 1.0, 3.0, and 10 μM S-peptide, respectively.

of K_d for the complex formed in the presence of different S-protein concentrations is calculated from the values of R and the total concentrations of S-protein and S15–GFP(S65T)–His_6 [see Eq. (1)]. The average (±SE) value of K_d is $(6 \pm 3) \times 10^{-8}$ M.

A competition assay is used to probe the specificity of the interaction between S15–GFP(S65T)–His_6 and S-protein. S15–GFP(S65T)–His_6 and S-protein are incubated to allow for complex formation. Varying amounts of S-peptide are added and the resulting mixture is incubated further, and then subjected to native gel electrophoresis. As shown in Fig. 3B, the addition of S-peptide converts bound S15–GFP(S65T)–His_6 to the free state. Thus, S15–GFP(S65T)–His_6 and S-peptide bind to the same region of S-protein.

Protocol

Purified fusion proteins are quantified by using the extinction coefficient (ε = 39.2 mM^{-1} cm^{-1} at 490 nm[14]) of S65T GFP. S-protein (Sigma, St. Louis, MO) is quantified by using its extinction coefficient (ε = 9.56 mM^{-1} cm^{-1} at 280 nm[27]). To begin the gel retardation assay, purified S15–GFP(S65T)–His_6 (1.0 μM) is incubated at 20° with varying amounts of S-protein in 10 μl of 10 mM Tris-HCl buffer, pH 7.5, containing glycerol (5%, v/v). After 20 min, the mixtures are loaded onto a native polyacrylamide (6%, w/v) gel,[28] and the loaded gel is subjected to electrophoresis at 4° at 10 V/cm. For inhibition assays, S-peptide is added and the mixtures are incubated at 20° for another 20 min before electrophoresis. Immediately after electrophoresis, the gel can be scanned with a Fluorimager SI System (Molecular Dynamics, Sunnyvale, CA) at 700 V using a built-in filter set (490 nm for excitation and ≥515 nm for emission).

Calculation of Equilibrium Dissociation Constants

The fluorescence intensities of bound and free S15–GFP(S65T)–His_6 can be quantified by using the program ImageQuaNT 4.1 (Molecular Dynamics). The value of R (the fluorescence intensity of bound S15–GFP(S65T)–His_6/total fluorescence intensity) for each gel lane is determined from the fluorescence intensities, and values of K_d are calculated by using Eq. (1):

$$K_d = \frac{1-R}{R}([\text{S} - \text{protein}]_{\text{total}} - R[\text{S15–GFP(S65T)–His}_6]_{\text{total}}) \qquad (1)$$

[27] P. R. Connelly, R. Varadarajan, J. M. Sturtevant, and F. M. Richards, *Biochemistry* **29,** 6108 (1990).

[28] U. K. Laemmli, *Nature* (*London*) **227,** 680 (1970).

Fluorescence Polarization Assay

The fluorescence gel retardation assay described above is a convenient method by which to visualize a protein–protein interaction as well as to estimate the value of K_d for the resulting complex. Still, gel retardation assays have an intrinsic limitation in evaluating equilibrium dissociation constants. In a gel retardation assay, it is assumed that a receptor–ligand interaction remains at equilibrium during sample loading and electrophoresis. Yet, as samples are loaded and migrate through a gel, complex dissociation is unavoidable and results in an underestimation of the value of K_d. Moreover, if the conditions (e.g., pH, or salt type or concentration) encountered during electrophoresis differ from those in the incubation, then the measured value of K_d may not correspond to the true value.

In a fluorescence polarization assay, the formation of a complex is deduced from an increase in fluorescence polarization, and the equilibrium dissociation constant is determined in a homogeneous aqueous environment. Most applications of fluorescence polarization assay have used fluorescein as a fluorophore.[29–32] Like a free fluorescein-labeled ligand, a free GFP chimera is likely to rotate more rapidly and therefore to have a lower rotational correlation time than does a bound chimera. An increase in rotational correlation time on binding results in an increase in fluorescence polarization, which can be used to assess complex formation.[33] In contrast to a gel retardation assay, the fluorescence polarization assay is performed in a homogeneous solution, in which the conditions can be dictated precisely.

Example

Fluorescence polarization is used to determine the effect of salt concentration on the formation of a complex between S-protein and S15–GFP(S65T)–His_6. The value of K_d increases by fourfold when NaCl is added to a final concentration of 0.10 *M* (Fig. 4). A similar salt dependence for the dissociation of RNase S had been observed previously.[34] The added salt is likely to disturb the water molecules hydrating the hydrophobic patch in the complex between S-peptide and S-protein, resulting in a decrease in the binding affinity.[35] Finally, the value of $K_d = 4.2 \times 10^{-8}$ *M* observed in 20 m*M* Tris-HCl buffer, pH 8.0, containing NaCl (0.10 *M*) is similar (i.e.,

[29] V. LeTilly and C. A. Royer, *Biochemistry* **32,** 7753 (1993).
[30] T. Heyduk, Y. Ma, H. Tang, and R. H. Ebright, *Methods Enzymol.* **274,** 492 (1996).
[31] G. Malpeli, C. Folli, and R. Berni, *Biochim, Biophys. Acta* **1294,** 48 (1996).
[32] B. M. Fisher, J.-H. Ha, and R. T. Raines, *Biochemistry* **37,** 12121 (1998).
[33] D. M. Jameson and W. H. Sawyer, *Methods Enzymol.* **246,** 283 (1995).
[34] A. A. Schreier and R. L. Baldwin, *Biochemistry* **16,** 4203 (1977).
[35] R. L. Baldwin, *Biophys. J.* **71,** 2056 (1996).

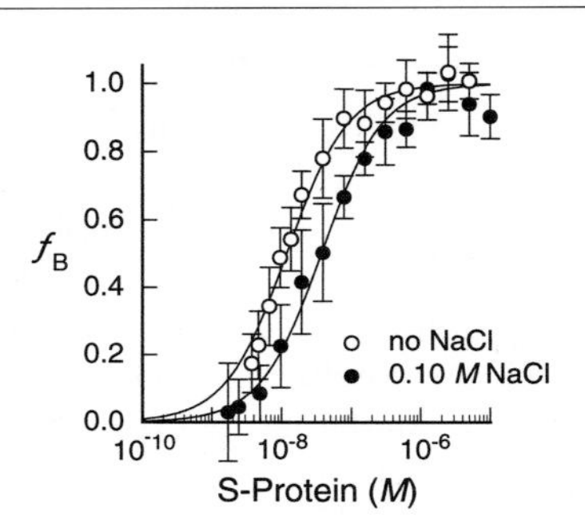

FIG. 4. Fluorescence polarization assay of a protein–protein interaction. S15-GFP(S65T)-His_6 with S-protein. S-protein is added to 20 m*M* Tris-HCl buffer, pH 8.0, in a volume of 1.0 ml. Each data point is an average of five to seven measurements. Curves are obtained by fitting the data to Eq. (3). The values of K_d in the presence of 0 and 0.10 *M* NaCl are 1.1×10^{-8} and 4.2×10^{-8} *M*, respectively.

threefold lower) to that obtained by titration calorimetry in 50 m*M* sodium acetate buffer, pH 6.0, containing NaCl (0.10 m*M*).[27]

Protocol

Fluorescence polarization (P) is defined as

$$P = \frac{I_{\|} - I_{\perp}}{I_{\|} + I_{\perp}} \tag{2}$$

where $I_{\|}$ is the intensity of the emission light parallel to the excitation light plane and $I_{\perp}$ is the intensity of the emission light perpendicular to the excitation light plane. P, being a ratio of light intensities, is a dimensionless number and has a maximum value of 0.5. Fluorescence polarization can be measured with a Beacon fluorescence polarization system (Pan Vera, Madison, WI).[36] Purified S15–GFP(S65T)–His_6 (0.50 n*M*) is incubated at 20° (±2°) with various concentrations of S-protein (20 μM–1.0 n*M*) in 1.0 ml of 20 m*M* Tris-HCl buffer, pH 8.0, containing NaCl (0 or 0.10 *M*). Five to seven polarization measurements are made at each S-protein concentration.

Calculation of Equilibrium Dissociation Constants

Values of K_d can be determined by using the program DeltaGraph 4.0 (DeltaPoint; Monterey, CA) to fit the data to Eq. (3):

[36] P. K. Wittmayer and R. T. Raines, *Biochemistry* **35,** 1076 (1996).

$$P = \frac{\Delta P F}{K_d + F} + P_{min} \tag{3}$$

In Eq. (3), P is the measured polarization, ΔP (= $P_{max} - P_{min}$) is the total change in polarization, and F is the concentration of free S-protein. The fraction of bound S-protein (f_B) is obtained by using Eq. (4):

$$f_B = \frac{P - P_{min}}{\Delta P} = \frac{F}{K_d + F} \tag{4}$$

The binding isotherms (Fig. 4) are obtained by plotting f_B versus F.

Prospectus

Methods to reveal and characterize the noncovalent interaction of one molecule with another are necessary to understand and control such interactions.[37–39] We describe two new methods for probing protein–protein interactions. The first method is a fluorescence gel retardation assay in which one protein is fused to GFP. The GFP fusion protein is incubated with the other protein, and the mixture is separated by native PAGE. The interaction between the two proteins is evident by a decrease in the mobility of the fluorescent fusion protein that results from complex formation.

The fluorescence gel retardation assay is a fast and convenient way to demonstrate interactions between two proteins, and in addition allows for an estimation of the value of K_d for the resulting complex. Conventional methods to demonstrate an interaction between two proteins (e.g., protein A and protein B) are more laborious or less informative (or both).[40–45] In a typical method, protein A is fused to an affinity tag [such as glutathione *S*-transferase (GST)], which is then used to immobilize protein A on a resin. Protein B is applied to the resin to allow for complex formation. The complex is eluted and detected by an immunoblot using an antibody to protein B. In contrast, the fluorescence gel retardation assay requires simply

[37] A. D. Attie and R. T. Raines, *J. Chem. Educ.* **72,** 119 (1995).

[38] D. J. Winzor and W. H. Sawyer, "Quantitative Characterization of Ligand Binding." Wiley-Liss, New York, 1995.

[39] I. M. Klotz, "Ligand–Receptor Energetics: A Guide for the Perplexed." John Wiley & Sons, New York, 1997.

[40] D. W. Carr and J. D. Scott, *Trends Biochem. Sci.* **17,** 246 (1992).

[41] T. Lu, M. Van Dyke, and M. Sawadogo, *Anal. Biochem.* **213,** 318 (1993).

[42] P. Rajagopal, E. B. Waygood, J. Reizer, M. H. Saier, Jr., and R. E. Klevit, *Protein Sci.* **6,** 2624 (1997).

[43] A. Cooper, *Methods Mol. Biol.* **88,** 11 (1998).

[44] I. Kameshita, A. Ishida, and H. Fujisawa, *Anal. Biochem.* **262,** 90 (1998).

[45] J. H. Lakey and E. M. Raggett, *Curr. Opin. Struct. Biol.* **8,** 119 (1998).

mixing a protein A–GFP chimera with protein B, separating the mixture by native PAGE, and scanning the gel with a fluorimager. The interaction between protein A and protein B is apparent from the shift of the protein A–GFP band that results from complex formation. The sensitivity of S65T GFP detection (≥0.1 ng; Fig. 2) approaches that of an immunoblot using an anti-GFP antibody.[24] Brighter GFP variants have become available. These variants show similar excitation and emission spectra but two- to eightfold stronger fluorescence intensity than the S65T variant.[46–48] Using these brighter variants in the gel retardation assay will improve further the sensitivity of the fluorescence gel retardation assay.

The second new method for probing protein–protein interactions, a fluorescence polarization assay, provides a more accurate assessment of the value of K_d. Most applications of fluorescence polarization have focused on analyzing protein–DNA interactions, with fluorescein (linked to DNA) serving as the fluorophore. Here, a GFP fusion protein is titrated with another protein, and the equilibrium dissociation constant is obtained from the increase in fluorescence polarization that accompanies binding. The interaction between the two proteins is detected in a homogeneous solution rather than a gel matrix. The fluorescence polarization assay thereby allows for the determination of accurate values of K_d in a wide range of solution conditions. GFP is particularly well suited to this application because its fluorophore (Fig. 1A) is held rigidly within the protein, as revealed by the three-dimensional structures of wild-type GFP and the S65T variant.[17,49] Such a rigid fluorophore minimizes local rotational motion, thereby ensuring that changes in polarization report on changes to the *global* rotational motion of GFP, as effected by a protein–protein interaction. Finally, it is worth noting that this assay is amenable to the high-throughput screening of protein or peptide libraries for effective ligands.[50]

Another advantage of both of these new methods is the ease with which a protein can be fused to GFP, using recombinant DNA techniques, and the high integrity of the resulting chimera. Traditionally, fluorophores have been attached to proteins by chemical modification with reagents such as fluorescein isothiocyanate (FITC).[30,51,52] In this approach, additional purification steps are necessary to separate labeled protein from the reagent and

[46] G. H. Patterson, S. M. Knobel, W. D. Sharif, S. R. Kain, and D. W. Piston, *Biophys. J.* **73,** 2782 (1997).
[47] R. Y. Tsien, *Annu. Rev. Biochem.* **67,** 509 (1998).
[48] A. B. Cubitt, L. A. Woollenweber, and R. Heim, *Methods Cell Biol.* **58,** 19 (1999).
[49] F. Yang, L. G. Moss, and G. N. Phillips, Jr., *Nature Biotechnol.* **14,** 1246 (1996).
[50] M. E. Jolley, *J. Biomol. Screening* **1,** 33 (1996).
[51] J. T. Radek, J. M. Jeong, J. Wilson, and L. Lorand, *Biochemistry* **32,** 3527 (1993).
[52] J. R. Lundblad, M. Laurance, and R. H. Goodman, *Mol. Endocrinol.* **10,** 607 (1996).

unlabeled protein. Further, labeling the protein at a single site can be difficult or impossible. In contrast, labeling a protein with GFP is complete and generates a single species. Purification of that species can be facilitated by the incorporation of an affinity tag such as His_6[53] or S · Tag.[54] The success of S65T GFP as the fluorophore in fluorescence gel retardation assays and fluorescence polarization assays arises largely from the altered spectral characteristics and increased fluorescence intensity of S65T GFP.[14] The availability of brighter S65T variants makes these assays more promising tools with which to investigate and analyze protein–protein interactions *in vitro*. We suggest that the role of fluorescein as a fluorescent label can be replaced by S65T GFP or its variants in many biochemical analyses.

Acknowledgments

This work was supported by Grant GM44783 (NIH). S.-H. P. was supported by a Korean Government Fellowship for Overseas Study.

[53] E. Hochuli, W. Bannwarth, H. Döbeli, R. Gentz, and D. Stüber, *Bio/Technology* **6,** 1321 (1988).

[54] R. T. Raines, M. McCormick, T. R. van Oosbree, and R. C. Mierendorf, *Methods Enzymol.* **326,** 362 (2000).

[17] Protein Fusions to Coiled-Coil Domains

By KRISTIAN M. MÜLLER, KATJA M. ARNDT, and TOM ALBER

Oligomerization plays key roles in protein function and regulation. As a consequence, powerful experimental tools have been created with defined, chimeric multimers made by genetic fusions to heterologous oligomerization domains. Coiled coils provide versatile fusion partners. They are particularly small domains with predictable quaternary structure and adjustable stability. Numerous coiled-coil fusions have been constructed to achieve diverse experimental aims. Here we review the principles of coiled-coil structure, highlight fusion domain sequences, and describe several applications.

Principles of Coiled-Coil Structure

The versatility of coiled coils for oligomerization derives from their diversity of oligomeric structures. Coiled coils are gently twisted, ropelike

0076-6879/00 $30.00

bundles containing two to five α helices in parallel or antiparallel orientation. The N and C termini of the helices are easily accessible, facilitating linkage to other proteins. Approximately 2–4% of amino acids in proteins are estimated to adopt coiled-coil folds.[1] This abundance offers many possible oligomerization domains with different properties.

The essential feature of coiled-coil sequences is a seven-residue repeat, $(\mathbf{abcdefg})_n$, with the first (**a**) and fourth (**d**) positions generally occupied by hydrophobic amino acids (Fig. 1A). The other amino acids in the repeat are mostly polar, and proline is largely excluded to preserve the helical architecture. The characteristic 3,4 hydrophobic repeat is strongly maintained for structural reasons. Each residue in a helix radially sweeps out ~100°, and seven residues lag two full turns by 20°. This lag generates a gentle, left-handed stripe of hydrophobic residues on an α helix. The coiled coil forms when such hydrophobic stripes associate. Deviations from the regular 3,4 spacing of nonpolar residues changes the angle of the hydrophobic stripe with respect to the helix axis, altering the crossing angle of the helices.[2–4]

Parallel dimers and trimers are by far the most commonly observed coiled coils. Coiled-coil oligomers can be distinguished on the basis of sequence patterns. Normalized sequence profiles for dimers and trimers revealed the relative frequencies of each amino acid at each heptad position.[5] As expected, strongly selective amino acids display skewed distributions. Covariation frequencies of selected residues provide a powerful basis (using the programs Paircoil[1] and Multicoil[6]) to identify coiled coils and distinguish dimers from trimers.[7]

The features that distinguish parallel dimers (Fig. 1B), trimers (Fig. 1C), and tetramers (Fig. 1D) are relatively well understood. Reviewing some of these ideas can facilitate the selection of oligomerization domains tailored to specific goals. In parallel coiled coils, the core (**a** and **d**) residues, which associate in register to form alternating layers, largely determine the oligomerization state.[8–10] Residues (**e** and **g**) on the edge of the helix

[1] B. Berger, D. B. Wilson, E. Wolf, T. Tonchev, M. Milla, and P. S. Kim, *Proc. Natl. Acad. Sci. U.S.A.* **92,** 8259 (1995).

[2] P. A. Bullough, F. M. Hughson, J. J. Skehel, and D. C. Wiley, *Nature* (*London*) **371,** 37 (1994).

[3] M. R. Hicks, D. V. Holberton, C. Kowalczyk, and D. N. Woolfson, *Fold. Des.* **2,** 149 (1997).

[4] R. B. Sutton, D. Fasshauer, R. Jahn, and A. T. Brunger, *Nature* (*London*) **395,** 347 (1998).

[5] D. N. Woolfson and T. Alber, *Protein Sci.* **4,** 1596 (1995).

[6] E. Wolf, P. S. Kim, and B. Berger, *Protein Sci.* **6,** 1179 (1997).

[7] A. Lupas, *Methods Enzymol.* **266,** 513 (1996).

[8] B. Y. Zhu, N. E. Zhou, C. M. Kay, and R. S. Hodges, *Protein Sci.* **2,** 383 (1993).

[9] P. B. Harbury, T. Zhang, P. S. Kim, and T. Alber, *Science* **262,** 1401 (1993).

[10] P. B. Harbury, P. S. Kim, and T. Alber, *Nature* (*London*) **371,** 80 (1994).

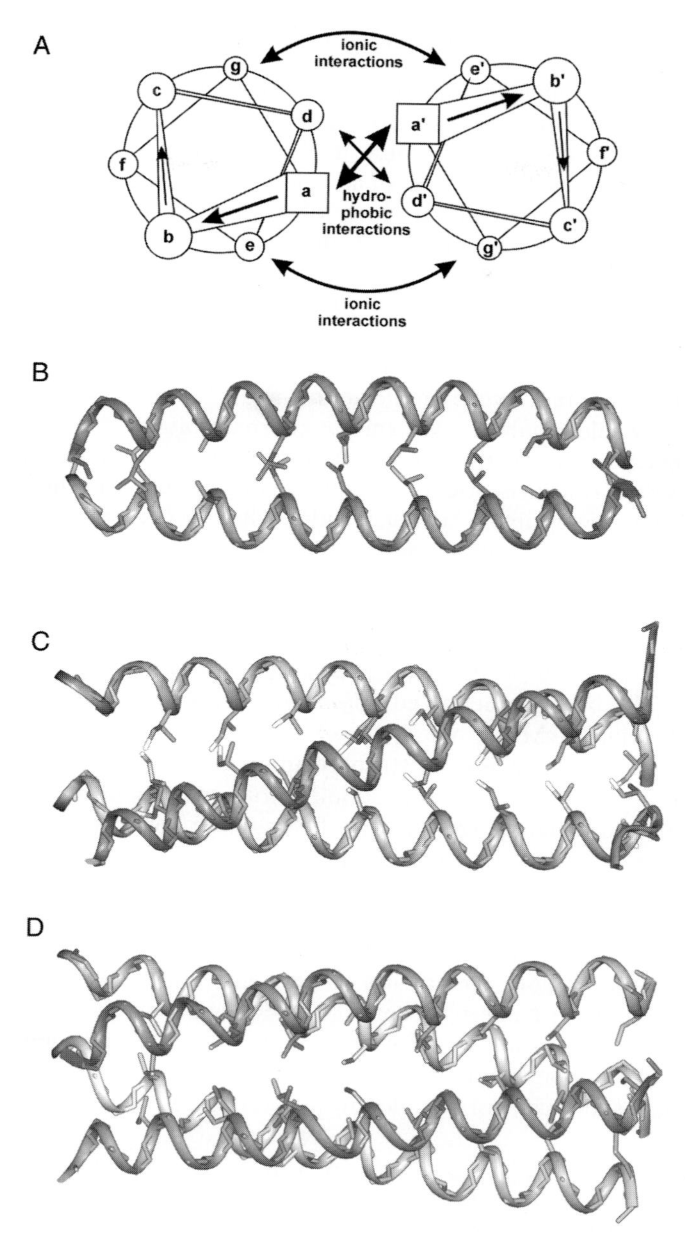

FIG. 1. Structures of parallel coiled coils. (A) Helical wheel diagram of a parallel, dimeric coiled coil illustrating the disposition of the characteristic heptad repeat. (B–D) X-ray crystal structures of dimeric, trimeric, and tetrameric GCN4 leucine zipper variants. The backbone and interface residues are shown superimposed on the ribbon representation of the helices. (B) Wild-type GCN4 leucine zipper (PDB code 2ZTA).[91] (C) GCN4-pII with isoleucines at positions **a** and **d** (PDB code 1GCM).[10] (D) GCN-pLI with leucine at the **a** positions and isoleucine at the **d** positions (PDB code 1GCL).[9]

interface also can play a secondary role in oligomer specification.[5,11,12] The dimer, trimer, and tetramer structures contain distinct packing spaces in the core. The fit of the core amino acids to the spaces intrinsic to each oligomer determines the number of associating helices.[9] Trimers appear to be the default oligomerization state for a random distribution of core hydrophobic amino acids. Dimers, tetramers, and pentamers require special distributions of core residues. This selectivity arises because β-branched amino acids, Ile, Val and Thr, cannot be accommodated in the **d** positions of dimers or the **a** positions of tetramers without significant structural distortions. Both core layers of trimers, however, generally permit β-branched and other hydrophobic amino acids. Core leucines are stabilizing, but they select only weakly among alternate oligomers.[8–10]

Distinct core sequences can be used to specify stable dimers, trimers, and tetramers in an otherwise identical sequence background.[9] In the context of the GCN4 leucine zipper sequence, ideal dimers contain Leu at the **d** positions, Ile at the **a** positions, and a single **a**-position Asn. Trimers contain Ile at all core positions, with or without a single **a**-position Gln.[5,13,14] The core polar residues in the dimer and trimer sequences confer structural specificity at the expense of stability. A GCN4 leucine zipper variant containing Leu at the **a** positions and Ile at the **d** positions forms a parallel tetramer. The best-characterized pentamer occurs in cartilage oligomeric matrix protein (COMP).[15]

Core polar amino acids can strongly influence oligomerization specificity.[16] Asn, Lys, and Arg are substantially more common at the **a** positions in dimers than trimers. Asn and Lys favor dimer formation by different mechanisms.[16] Asn forms a buried, interhelical hydrogen bond in dimers. In contrast, the Asn side chains are accommodated poorly in trimers, where they can make no direct, interhelical contacts. Rather than making direct contacts, Lys is solvated better in dimers than trimers. Gln is tolerated at the **a** positions of both dimers and trimers, and consequently Gln occurs more frequently in trimers. Core polar residues are destabilizing relative to hydrophobic substitutions. By disfavoring alternate arrangements that sequester polar atoms in nonpolar environments, however, core polar residues impart specificity for parallel, in-register structures.

[11] K. Beck, J. E. Gambee, A. Kamawal, and H. P. Bachinger, *EMBO J.* **16,** 3767 (1997).

[12] X. Zeng, H. Zhu, H. A. Lashuel, and J. C. Hu, *Protein Sci.* **6,** 2218 (1997).

[13] V. N. Malashkevich, B. J. Schneider, M. L. McNally, M. A. Milhollen, J. X. Pang, and P. S. Kim, *Proc. Natl. Acad. Sci. U.S.A.* **96,** 2662 (1999).

[14] S. Nautiyal and T. Alber, *Protein Sci.* **8,** 84 (1999).

[15] V. N. Malashkevich, R. A. Kammerer, V. P. Efimov, T. Schulthess, and J. Engel, *Science* **274,** 761 (1996).

[16] L. Gonzalez, Jr., D. N. Woolfson, and T. Alber, *Nature Struct. Biol.* **3,** 1011 (1996).

In parallel oligomers, oppositely charged residues at **g** and the succeeding **e′** position on neighboring helices can form hydrogen-bonded ion pairs (Fig. 1A). Although the stabilizing contribution of these ion pairs is controversial, equivalent charges at **g** and the succeeding **e′** positions are destabilizing.[17–20] These ionic interactions play essential roles in restricting the partners of coiled-coil sequences.

Fusions to heterotypic coiled coils have applications for constructing defined combinations of different proteins. Two main principles govern heterospecific associations of coiled coils. Complementary electrostatic interactions and complementary core polarity and packing impart heterospecificity.[21] Repulsive **g**-to-**e′** electrostatic interactions that are replaced by complementary charge pairs are sufficient to favor heterodimer and heterotrimer formation.[14,22–24] Complementary core residues also can mediate heterodimerization.[21] Heterologous polar interactions between core residues and neighboring **g** and **e** residues also favor hetero-oligomerization in the c-Myc/Max dimer,[25] the Fos/Jun Dimer,[26,27] and the SNARE complex.[4] An emerging trend in the structures of hetero-oligomers and antiparallel coiled coils is the tendency for one (or more) of the helices to be straighter than the others. This asymmetry coincides with distinct structural contexts for each helix.

Fusions to antiparallel coiled coils show promise for loop display[28] and protein multimerization. Antiparallel coiled coils, as exemplified by domains in seryl-tRNA synthetase,[29] hepatitis delta antigen (dimers),[30]

[17] K. J. Lumb and P. S. Kim, *Science* **268,** 436 (1995).

[18] D. Krylov, I. Mikhailenko, and C. Vinson, *EMBO J.* **13,** 2849 (1994).

[19] N. E. Zhou, C. M. Kay, and R. S. Hodges, *J. Mol. Biol.* **237,** 500 (1994).

[20] W. D. Kohn, C. M. Kay, and R. S. Hodges, *J. Mol. Biol.* **283,** 993 (1998).

[21] V. A. Sharma, J. Logan, D. S. King, R. White, and T. Alber, *Curr. Biol.* **8,** 823 (1998).

[22] E. K. O'Shea, K. J. Lumb, and P. S. Kim, *Curr. Biol.* **3,** 658 (1993).

[23] S. Nautiyal, D. N. Woolfson, D. S. King, and T. Alber, *Biochemistry* **34,** 11645 (1995).

[24] H. Chao, M. E. Houston, Jr., S. Grothe, C. M. Kay, M. O'Connor-McCourt, R. T. Irvin, and R. S. Hodges, *Biochemistry* **35,** 12175 (1996).

[25] P. Lavigne, M. P. Crump, S. M. Gange, R. S. Hodges, C. M. Kay, and B. D. Sykes, *J. Mol. Biol.* **281,** 165 (1998).

[26] E. K. O'Shea, R. Rutkowski, and P. S. Kim, *Cell* **68,** 699 (1992).

[27] J. N. Glover and S. C. Harrison, *Nature* **373,** 257 (1995).

[28] R. Miceli, D. Myszka, J. Mao, G. Sathe, and I. Chaiken, *Drug Des. Discov.* **13,** 95 (1996).

[29] S. Cusack, C. Berthet-Colominas, M. Hartlein, N. Nassar, and R. Leberman, *Nature (London)* **347,** 249 (1990).

[30] H. J. Zuccola, J. E. Rozzelle, S. M. Lemon, B. W. Erickson, and J. M. Hogle, *Structure* **6,** 821 (1998).

spectrin,[31] the designed peptide "Coil-Ser" (trimers),[32] and *lac* repressor (tetramer),[33] display distinct contacts compared with parallel structures. To match the spacing of hydrophobic residues, the **a** positions pack with **d′** positions of antiparallel helices (and the **d** residues pack with **a′** residues). Instead of **g**-to-**e′** contacts that characterize parallel coiled coils, **g** and **e** residues are sequestered in antiparallel oligomers. As a result, large or small core residues, the absence of core polar residues, and repulsive **g**-to-**e′** electrostatic interactions can favor antiparallel arrangements.[12,20,32,34,35] Introducing repulsive interactions at the **b** and **c** positions also allows the formation of antiparallel heterotetrameric coiled coils.[36]

Small size provides an important advantage of coiled coils for heterologous oligomerization. Dimers as short as 23 residues have been detected.[37] Concatemers of seven-residue repeats suggest that three heptads are the minimum required to form a stable coiled coil.[38] Hyperstable trimers and tetramers that remain folded even at boiling temperature have been constructed from GCN4 leucine zipper variants as short as 33 residues. Helical propensity[39] and core compatibility[40] also play large roles in oligomer stability. Systematically introducing helix-stabilizing residues at the surface positions (**b, c,** and **f**) may prove to be a simple approach to increasing the stability of coiled-coil oligomerization domains.

Short coiled-coil domains are sensitive to core substitutions that introduce destabilizing amino acids. The single Leu19Ile mutation in the dimeric, 33-amino acid GCN4 leucine zipper, for example, decreases T_m by ~20° and generates a sequence capable of trimerization.[41] Similarly, the single Asn16Ala replacement in the GCN4 leucine zipper results in a peptide that forms at least three states: parallel dimer, parallel trimer, and antiparallel trimer.[42,43] This conformational plasticity emphasizes that the 3,4 hydropho-

[31] Y. Yan, E. Winograd, A. Viel, T. Cronin, S. C. Harrison, and D. Branton, *Science* **262,** 2027 (1993).

[32] B. Lovejoy, S. Choe, D. Cascio, D. K. McRorie, W. F. DeGrado, and D. Eisenberg, *Science* **259,** 1288 (1993).

[33] M. Lewis, G. Chang, N. C. Horton, M. A. Kercher, H. C. Pace, M. A. Schumacher, R. G. Brennan, and P. Lu, *Science* **271,** 1247 (1996).

[34] K. J. Lumb and P. S. Kim, *Biochemistry* **34,** 8642 (1995).

[35] D. G. Myszka and I. M. Chaiken, *Biochemistry* **33,** 2363 (1994).

[36] R. Fairman, H. G. Chao, T. B. Lavoie, J. J. Villafranca, G. R. Matsueda, and J. Novotny, *Biochemistry* **35,** 2824 (1996).

[37] K. J. Lumb, C. M. Carr, and P. S. Kim, *Biochemistry* **33,** 7361 (1994).

[38] J. Y. Su, R. S. Hodges, and C. M. Kay, *Biochemistry* **33,** 15501 (1994).

[39] K. T. O'Neil and W. F. DeGrado, *Science* **250,** 646 (1990).

[40] J. Moitra, L. Szilak, D. Krylov, and C. Vinson, *Biochemistry* **36,** 12567 (1997).

[41] C. M. Huff, Master's thesis. University of California, Berkeley, 1999.

[42] L. Gonzalez, Jr., J. J. Plecs, and T. Alber, *Nature Struct. Biol.* **3,** 510 (1996).

[43] J. M. Holton, *et al.,* in preparation (2000).

bic repeat is compatible with multiple helical structures. Unique oligomers are distinguished by strong interactions, including steric repulsion and burial of polar groups.

The compatibility of heptad repeat sequences with different quaternary structures makes it essential to determine experimentally the oligomerization state of each expressed coiled-coil fusion protein. In one interesting case, for example, replacement of the trimerization domain of the yeast heat shock factor with the GCN4 leucine zipper yielded the expected dimeric fusion protein in solution.[44] Unexpectedly, the chimeric protein bound predominantly as a trimer to DNA containing three recognition sites. These results indicated that the specificity of coiled-coil oligomerization can be overwhelmed by sufficiently strong interactions with ligands.

The utility of coiled coils for oligomerization hinges in part on their stability. Unfolding of numerous short coiled coils is well described by a two-state transition between unfolded monomers and helical oligomers. Folded monomers are not populated significantly in aqueous buffers. Like any protein, the stabilities of coiled coils depend on solution conditions such as pH, salt concentration, and temperature. At low pH, for example, dimeric coiled coils are generally stabilized and trimers are generally destabilized. This difference in pH sensitivity suggests that ionized carboxylates generally stabilize trimers and destabilize dimers. Apparent stabilities of coiled coils vary widely. The GCN4 leucine zipper commonly used for dimerization displays an apparent ΔG of unfolding of 10.5 kcal/mol (corresponding to a K_D of 15 nM.[45] A hyperstable designed trimer with a completely hydrophobic core shows an apparent stability of 18.4 kcal/mol.[46] The E/K coil, a designed heterodimer, has a K_D in the 0.5–3.5 nM range.[24] Single substitutions of all 19 amino acids excluding proline shifted the stability of a designed coiled coil over the range of 11.3 kcal/mol (Gly) to 13.4 kcal/mol (Ala).[39] These results suggest that there is considerable scope for tuning the stability of heterologous oligomerization domains to suit diverse applications.

Examples of Coiled-Coil Fusions

Coiled-coil fusions have been used to achieve diverse experimental goals. The most straightforward application is replacement of natural oligomerization domains. For example, the dimerization domain of phage λ

[44] B. L. Drees, E. K. Grotkopp, and H. C. M. Nelson, *J. Mol. Biol.* **273,** 61 (1997).

[45] J. A. Zitzewitz, O. Bilsel, J. Luo, B. E. Jones, and C. R. Matthews, *Biochemistry* **34,** 12812 (1995).

[46] J. A. Boice, G. R. Dieckmann, W. F. Degrado, and R. Fairman, *Biochemistry* **35,** 14480 (1996).

repressor,[47] the constant region of antibodies,[48] and the trimerization domain of CD40 ligand[49] were replaced with coiled coils to generate active, chimeric proteins. Early applications aimed at the oligomerization of binding proteins in order to increase their avidity. The effect of apparent higher affinity by multivalent binding can be accommodated for by stepwise binding with a high local concentration after the first step.[50] Such increased avidity can have great significance. Low-affinity monomers can be preorganized to bind effectively to multimeric targets. This principle was used, for example, to produce oligomers of antibody fragments,[51,52] peptides,[53] and soluble trimeric CD40 ligand for signaling assays.[54,55] Another example is provided by fusion of the cytoplasmic domain of CD40 to a trimeric coiled coil to mimic the multimeric, activated state of the receptor.[56] Natural oligomers, such as human immunodeficiency virus glycoprotein 41 (HIV gp41) and Ebola virus Gp2, have been fused to coiled coils to augment stability and solubility.[57,58] Fusions of the intracellular domain of the aspartate receptor to different oligomeric coiled coils were constructed to analyze geometric requirements of signaling.[59] Acidic domains fused to natural leucine-zipper sequences have been created to inhibit bZIP transcription factors *in vitro* and *in vivo*.[60,61] Fusions to heterodimers can facilitate pro-

[47] J. C. Hu, E. K. O'Shea, P. S. Kim, and R. T. Sauer, *Science* **250,** 1400 (1990).

[48] K. M. Arndt, K. M. Müller, and A. Plückthun, in preparation (2000).

[49] A. E. Morris, R. L. Remmele, Jr., R. Klinke, B. M. Macduff, W. C. Fanslow, and R. J. Armitage, *J. Biol. Chem.* **274,** 418 (1999).

[50] K. M. Müller, K. M. Arndt, and A. Plückthun, *Anal. Biochem.* **261,** 149 (1998).

[51] P. Pack and A. Plückthun, *Biochemistry* **31,** 1579 (1992).

[52] P. Pack, K. Müller, R. Zahn, and A. Plückthun, *J. Mol. Biol.* **246,** 28 (1995).

[53] A. V. Terskikh, J. M. Le Doussal, R. Crameri, I. Fisch, J. P. Mach, and A. V. Kajava, *Proc. Natl. Acad. Sci. U.S.A.* **94,** 1663 (1997).

[54] J. F. McDyer, M. Dybul, T. J. Goletz, A. L. Kinter, E. K. Thomas, J. A. Berzofsky, A. S. Fauci, and R. A. Seder, *J. Immunol.* **162,** 3711 (1999).

[55] S. Gurunathan, K. R. Irvine, C. Y. Wu, J. I. Cohen, E. Thomas, C. Prussin, N. P. Restifo, and R. A. Seder, *J. Immunol.* **161,** 4563 (1998).

[56] S. S. Pullen, M. E. Labadia, R. H. Ingraham, S. M. McWhirter, D. S. Everdeen, T. Alber, J. J. Crute, and M. R. Kehry, *Biochemistry* **38,** 10168 (1999).

[57] W. Weissenhorn, L. J. Calder, A. Dessen, T. Laue, J. J. Skehel, and D. C. Wiley, *Proc. Natl. Acad. Sci. U.S.A.* **94,** 6065 (1997).

[58] W. Weissenhorn, L. J. Calder, S. A. Wharton, J. J. Skehel, and D. C. Wiley, *Proc. Natl. Acad. Sci. U.S.A.* **95,** 6032 (1998).

[59] A. G. Cochran and P. S. Kim, *Science* **271,** 1113 (1996).

[60] D. Krylov, D. R. Echlin, E. J. Taparowsky, and C. Vinson, *Curr. Top. Microbiol. Immunol.* **224,** 169 (1997).

[61] J. Moitra, M. M. Mason, M. Olive, D. Krylov, O. Gavrilova, B. Marcus-Samuels, L. Feigenbaum, E. Lee, T. Aoyama, M. Eckhaus, M. L. Reitman, and C. Vinson, *Genes Dev.* **12,** 3168 (1998).

tein purification[62] and assembly of enzymes[63,64] and bispecific minianti-bodies.[65]

The coiled coil can be genetically fused to the protein of interest via a flexible linker providing access to a large space. Flexible and semirigid attachments have been exploited to maintain the avidity of antibody fragment chimeras.[51,52] At the other extreme, direct fusions to a coiled coil in the protein of interest have been constructed to create rigid assemblies that facilitate crystallization.[66,67] In several instances, cysteines were included at the ends in a **d** position to enhance the stability of dimeric coiled coils.[53,68]

To be effective, coiled coils also must be stable in the expression host. In this regard, coiled coils have the advantage that disulfide bonds are not required for folding. On the other hand, it is our experience that *de novo* designed sequences can be sensitive to proteolysis. Genetic screens for stable oligomerization partners are just beginning.[69] The use of natural sequences or their variants can increase the chances that the oligomerization domain will express efficiently, fold properly, and resist degradation.

Table I[22,44,47–49,51–53,57–59,61–68,70–90] lists coiled-coil fusions, grouped by

[62] B. Tripet, L. Yu, D. L. Bautista, W. Y. Wong, R. T. Irvin, and R. S. Hodges, *Protein Eng.* **9,** 1029 (1996).

[63] J. N. Pelletier, F. X. Campbell-Valois, and S. W. Michnick, *Proc. Natl. Acad. Sci. U.S.A.* **95,** 12141 (1998).

[64] K. M. Arndt, J. N. Pelletier, K. M. Müller, T. Alber, S. W. Michnick, and A. Plückthun, *J. Mol. Biol.* **295,** 627 (2000).

[65] S. A. Kostelny, M. S. Cole, and J. Y. Tso, *J. Immunol.* **148,** 1547 (1992).

[66] W. Weissenhorn, A. Dessen, S. C. Harrison, J. J. Skehel, and D. C. Wiley, *Nature (London)* **387,** 426 (1997).

[67] W. Weissenhorn, A. Carfi, K. H. Lee, J. J. Skehel, and D. C. Wiley, *Mol. Cell* **2,** 605 (1998).

[68] R. Crameri and M. Suter, *Gene* **137,** 69 (1993).

[69] J. N. Pelletier, K. M. Arndt, A. Plückthun, and S. W. Michnick, *Nature Biotechnol.* **17,** 683 (1999).

[70] A. Blondel and H. Bedouelle, *Protein Eng.* **4,** 457 (1991).

[71] T. Schmidt-Dörr, P. Oertel-Buchheit, C. Pernelle, L. Bracco, M. Schnarr, and M. Granger-Schnarr, *Biochemistry* **30,** 9657 (1991).

[72] S. Alberti, S. Oehler, B. von Wilcken-Bergmann, and B. Müller-Hill, *EMBO J.* **12,** 3227 (1993).

[73] D. Porte, P. Oertel-Buchheit, M. Granger-Schnarr, and M. Schnarr, *J. Biol. Chem.* **270,** 22721 (1995).

[74] J. E. Lindsley, *Proc. Natl. Acad. Sci. U.S.A.* **93,** 2975 (1996).

[75] L. G. Riley, G. B. Ralston, and A. S. Weiss, *Protein Eng.* **9,** 223 (1996).

[76] M. K. Mohamed, L. Tung, G. S. Takimoto, and K. B. Horwitz, *J. Steroid Biochem. Mol. Biol.* **51,** 241 (1994).

[77] H. C. Chang, *et al., Proc. Natl. Acad. Sci. U.S.A.* **91,** 11408 (1994).

[78] K. Gramatikoff, O. Georgiev, and W. Schaffner, *Nucleic Acids Res.* **22,** 5761 (1994).

[79] J. de Kruif and T. Logtenberg, *J. Biol. Chem.* **271,** 7630 (1996).

TABLE I
FUSIONS TO COILED COILS

Aim	Coiled coil[a]	Fusion partner	Host, location	Comment	Ref.
		Homodimers			
In vivo assay for mutational effects on the GCN4 leucine zipper	GCN4[b] and mutated GCN4	N-terminal domain of the phage λ *c*I repressor	*E. coli*	GCN4 mutations at the **a** and **d** positions challenged with repression of a λP_R–*lacZ* fusion	47
Stabilize MalE dimer formation	GCN4	MalE	*E. coli,* periplasm		70
Study LexA homodimerization and Jun dimerization	Jun and mutated Jun	LexA DNA-binding domain	*E. coli*	*In vitro* studies of DNA binding and interactions of **e** and **g** positions	71
Multimerize scFv fragments for higher avidity	GCN4	scFv fragment	*E. coli,* periplasm	Leucine zipper with and without C-terminal disulfide bridge	51
Analyze the tetrameric leucine zipper of the Lac repressor	GCN4, modified GCN4	Lac repressor	*E. coli*	Hybrids of the coiled-coil domains of Lac repressor and GCN4	72
Define mutations that promote homodimerization of Fos	Fos variants	LexA	*E. coli*	LexA *in vivo* assay, randomizaton of **a** positions	73
Study activities of dimeric, disulfide-free topoisomerase II	GCN4	Yeast topoisomerase II variants	Yeast, strain BCY123	Linker contained 2 or 4 aa and a disulfide bridge	74
Test dimerization of Jun and Fos in a fusion protein	Jun or Fos	Glutathione *S*-transferase	*E. coli*	Thrombin cleavage site in 6-aa linker. Jun/GST fusion aggregated	75
Assess role of trimerization in DNA binding by heat shock factor	GCN4	Yeast heat shock factor DNA-binding domain	*E. coli* and *S. cerevisiae*	Fusion protein was predominantly dimeric in solution, but predominantly trimeric on DNA	44
Orient and dimerize a cytosolic chemotaxis signaling domain	GCN4-pR	Cytoplasmic domain of the *E. coli* aspartate receptor	*E. coli*	5- to 9-aa linker, N-terminal fusion in register with a putative coiled coil	59
Assemble active DHFR from fragments	GCN4	Murine dihydrofolate reductase fragments	*E. coli*	Established feasibility of a protein fragment complementation assay	63

		Heterodimers			
Produce a bispecific $F(ab')_2$	Jun + Fos	Fab antibody fragment	Mouse myeloma cells	Hinge with Cys-Gly-Gly linker	65
Create a phage display vector that allows N-terminal fusion for cDNA library screening	Jun + Fos	Jun fused to phase ml3 gIIIp, Fos fused to cDNA library	*E. coli,* periplasm	Dimers stabilized with disulfide bridge	68
Study DNA binding and activation of progesterone receptor (hPR) dimer	Jun + Fos	Jun and Fos fused to hPR_B or hPR_A	COS-1, HeLa, T47D-V22 cells	2-aa linker	76
Improve recombinant expression of soluble T cell receptor	Peptide "Velcro" (ACID-p1, BASE-p1)[22]	TCR α chain to BASE-p1, TCR β chain to ACID-p1	Insect cells (SF900II)	15-aa linker contained a thrombin cleavage site, S–S bridge in protein close to coiled-coil fusion	77
Select a cDNA library for proteins binding to c-Jun	Jun + cDNA library	Fragments of m13 gIIIp	*E. coli*	Jun binding selectively enhanced phage infectivity	78
Produce bispecific miniantibodies	Jun + Fos	scFv antibody fragments	*E. coli,* periplasm	10-aa linker + Gly-Gly-Cys tail for disulfide formation, homodimers obtained after separate expression were reduced and reoxidized to yield heterodimers	79
Inactivate bZIP proteins	C/EBP leucine zipper	Designed acidic peptide	*E. coli*	Acidic sequence complementary to the C/EBP basic region facilitated heterodimer formation and blocked DNA binding *in vitro* and *in vivo*	61, 80
Affinity purification or detection of coiled-coil partners	Designed K-coil[62]	17-aa PAK-pilin fused to E-coil[62]	*E. coli*	Gly_8 linker, K-coil was made synthetically and linked to biotin or a column matrix, respectively	62
Generation of an artificial metallorestriction enzyme	Jun + Fos	GlyLysHis sequence fused to Fos; Jun was expressed alone or fused to GST	*E. coli*	Cleavage of DNA containing an AP-1 DNA site was tested	81

(*continued*)

TABLE I (*continued*)

Aim	Coiled coil[a]	Fusion partner	Host, location	Comment	Ref.
Stabilize a T cell receptor	Peptide "Velcro"[22]	TCR α chain to ACID-p1, TCR β chain to BASE-p1	*E. coli,* secreted	Disulfide bridge in protein close to coiled-coil fusion	82
	Jun + Fos	Albumin-binding protein, alkaline phosphatase, ZZ IgG-binding domain	*E. coli,* periplasm	Two proteins fused to Fos, indirect immobilization via the protein fused to Jun	83
Target green fluorescent protein (GFP) for *in vivo* labeling by coiled coil	Designed peptides $R \cdot R_{34}$, $E \cdot E_{34}$	GFP to $R \cdot R_{34}$, α-actinin or vinculin to $E \cdot E_{34}$, FLAG tag	Foreskin fibroblasts	Proteins were fused in the order $R \cdot R_{34}$–GFP, and FLAG–$E \cdot E_{34}$–vinculin/actinin	84
Selection of heterodimeric coiled coils by a protein fragment complementation assay	Partially randomized coiled-coil library	Murine dihydrofolate reductase fragments	*E. coli*	14-aa and 15-aa linkers, metabolically stable heterodimers selected *in vivo*	64
Stabilize an Fv antibody fragment	WinZip-A2 and WinZip-B1[48]	Variable domain fragment V_H to WinZip-A2, V_L to WinZip-B1	*E. coli,* periplasm	14-aa linkers, the coiled-coil domain stabilizes V_H–V_L assembly	48
		Homotrimers			
Generate soluble II-2 receptor oligomers	Designed peptide[39]	IL-2R	Insect cells (Sf9)	Planned to make dimers, but fusion resulted in trimers	85
Assemble soluble gp41 for crystallography	GCN4-pII[9]	HIV-1 gp41 core	*E. coli*	The in-register, trimeric coiled coil increased solubility and facilitated crystallization	57, 66
Assemble a viral coat protein for crystallography	GCN4-pII[9]	Ebola membrane-fusion protein Gp2	*E. coli*	In-register fusion facilitated crystallization	58, 67

Assemble soluble gp41 core	GCN4-pI$_Q$I[86] with surface mutations	HIV-1 gp41 core	Chemically synthesized as L- and D-peptide	Used as target for mirror image phage display	87
Create active, soluble Cd40L	GCN4-pII[9]	CD40L extracellular domain	CHO cells	The coiled coil stabilizes the soluble, trimeric form	49
		Heterotrimer			
Mimic interleukin 2 cell surface receptor	LEALKEK- and LKALEKE- repeat[88]	IL-2R extracellular domain α chain to LEALKEK and β chain to LKALEKE	Insect cells (*Trichoplusia ni*)	Coexpression of fusions, trimers detected by ultracentrifugation	88
		Homotetramers			
Create high-avidity miniantibodies	GCN4-pLI[9]	scFv antibody fragment	*E. coli,* periplasm	Tetramerization increased apparent affinity	52
Substitute the tetramerization domain of p53	GCN4-pLI[9]	p53 aa 1–343	*In vitro* translation and Saos-2 cells	2-aa linker	89
		Homopentamer			
Increase avidity of peptide binding	Modified COMP pentamerization domain[53,90]	Peptide obtained from a phage display library	*E. coli*	24-aa hinge region from a camel antibody, used disulfide linkage	53

[a] Superscript numbers in this column indicate references.
[b] GCN4, Jun, and Fos refer to the coiled-coil regions of the proteins.

oligomerization state. Several specific applications are reviewed in Tables II[9,22,27,91] and III[48,51,92,93] to illustrate the versatility of the approach.

Constructing a Coiled-Coil Fusion

Genes encoding coiled-coil fusion domains are obtained generally by gene synthesis. Depending on the length of the synthetic gene, we favor three hybridization strategies utilizing 50- to 90-mer oligonucleotides. First, short gene fragments spanning a few heptads can be obtained directly by hybridizing two oligonucleotides encompassing the entire coding and noncoding strands. Second, when longer genes are desired, hybridizing oligonucleotides encoding each half of the sequence with a central overlap of about 15 bases and filling in the strands by primer extension is advisable. Third, long sequences of several hundred nucleotides have been constructed by polymerase chain reaction (PCR), using several oligonucleotides covering alternating coding and noncoding strands. For highly repetitive coiled-coil sequences, variations in codon usage help avoid incorrect hybridization and recombination.

As an example, we describe the construction of a heterodimeric ROP (repressor of primer)[94] variant, which was used to form bispecific minianti-bodies. Wild-type ROP is a homodimeric, antiparallel four-helix bundle with the knobs-into-holes packing typical of coiled coils. For the designed heterodimer, 10 synthetic oligonucleotides 69 to 84 bases in length were used to construct a dicistronic gene arrangement encoding two different

[80] D. Krylov, M. Olive, and C. Vinson, *EMBO J.* **14,** 5329 (1995).
[81] C. Harford, S. Narindrasorasak, and B. Sarkar, *Biochemistry* **35,** 4271 (1996).
[82] A. Golden, S. S. Khandekar, M. S. Osburne, E. Kawasaki, E. L. Reinherz, and T. H. Grossman, *J. Immunol. Methods* **206,** 163 (1997).
[83] P. Grob, S. Baumann, M. Ackermann, and M. Suter, *Immunotechnology* **4,** 155 (1998).
[84] B. Z. Katz, D. Krylov, S. Aota, M. Olive, C. Vinson, and K. M. Yamada, *BioTechniques* **25,** 298 (1998).
[85] Z. Wu, S. F. Eaton, T. M. Laue, K. W. Johnson, T. R. Sana, and T. L. Ciardelli, *Protein Eng.* **7,** 1137 (1994).
[86] D. M. Eckert, V. N. Malashkevich, and P. S. Kim, *J. Mol. Biol.* **284,** 859 (1998).
[87] D. M. Eckert, V. N. Malashkevich, L. H. Hong, P. A. Carr, and P. S. Kim, *Cell* **99,** 103 (1999).
[88] Z. Wu, K. W. Johnson, B. Goldstein, Y. Choi, S. F. Eaton, T. M. Laue, and T. L. Ciardelli, *J. Biol. Chem.* **270,** 16039 (1995).
[89] M. J. Waterman, J. L. Waterman, and T. D. Halazonetis, *Cancer Res.* **56,** 158 (1996).
[90] V. P. Efimov, A. Lustig, and J. Engel, *FEBS Lett.* **341,** 54 (1994).
[91] E. K. O'Shea, J. D. Klemm, P. S. Kim, and T. Alber, *Science* **254,** 539 (1991).
[92] J. L. Dangl, T. G. Wensel, S. L. Morrison, L. Stryer, L. A. Herzenberg, and V. T. Oi, *EMBO J.* **7,** 1989 (1988).
[93] K. M. Müller, K. M. Arndt, and A. Plückthun, *FEBS Lett.* **432,** 45 (1998).
[94] D. W. Banner, M. Kokkinidis, and D. Tsernoglou, *J. Mol. Biol.* **196,** 657 (1987).

TABLE II
EXAMPLES OF COILED-COIL SEQUENCES USED IN FUSION PROTEINS

Protein	Sequence	Ref.
GCN4-leucine zipper; homodimer	R MKQLEDK VEELLSK NYHLENE VARLKKL VGER	91
GCN4-pII,[a] trimer	R MKQIEDK IEEILSK IYHIENE IARIKKL IGER[b]	9
GCN4-pLI,[a], tetramer	R MKQIEDK LEEILSK LYHIENE LARIKKL LGER	9
Jun–Fos, heterodimer		
c-Jun	IARLEEK VKTLKAQ NSELAST ANMLREQ VAQL[c]	27
c-Fos	TDTLQAE TDQLEDE KSALQTE IANLLKE KEKL[c]	
"Velcro," heterodimer		
ACID-p1	AQLEKE LQALEKE NAQLEWE LQALEKE LAQ	22
BASE-p1	AQLKKK LQALKKK NAQLKWK LQALKKK LAQ	
WinZip-A2B1, heterodimer		
WinZip-A2	VAQL**R**E**R** VKTL**R**A**Q** **N**YEL**E**S**E** VQRL**R**E**Q** VAQL[d]	64, 69
WinZip-B1	VDEL**Q**A**E** VDQL**Q**D**E** **N**YAL**K**T**K** VAQL**R**K**K** VEKL[d]	

[a] The letter p refers to the peptide, the letters I and L refer to the amino acids occupying the **a** or **d** positions, respectively.
[b] An Ile16Gln preserves the trimer and reduces stability.[86]
[c] Serine at the **b** position in the third heptad is often exchanged to tyrosine in peptides.
[d] Positions shown in boldface were obtained by selection from a randomized library.

helical domains. The 10 oligonucleotides represented alternating coding and noncoding strands. Overlaps had estimated melting temperatures of about 56° (calculated by using 4° for G–C pairs and 2° for A–T pairs). The gene was assembled by PCR in two parts. Two gene fragments were constructed by mixing and extending nucleotides 1 to 4, corresponding to the first half of the sequence, and nucleotides 5 to 10, corresponding to the second half. These fragments were ligated after cutting with a shared, central restriction site. Assembling the whole gene in one step was not successful. For PCR amplification, the terminal oligonucleotides were used at a concentration of 0.8 pmol/μl and the inner oligonucleotides at 0.016 pmol/μl. A proof-reading polymerase (Vent; New England Biolabs, Beverly, MA) was used at a Mg^{2+} concentration of 5 m*M*. A 50-μl reaction volume in a 200-μl thin-walled reaction tube was cycled with the following program: 5 cycles of (92° for 1 min, cooling to 55° at 0.3°/sec, 55° for 1 min,

TABLE III
EXAMPLES OF LINKER SEQUENCES USED IN COILED-COIL FUSIONS

Protein	Sequence	Ref.
Murine IgG_3 upper hinge	PKPSTPPGSS	51, 92
Long flexible	SGGTSGSTSGTGST	48
Short flexible	GGSGGAP	93

72° for 1 min), 20 cycles of (92° for 1 min, fast cooling to 55° + 0.6° per cycle, 72° for 1 min), and finish with 72° for 3 min and cooling to 0°. Both PCR products were gel purified and cloned blunt end in the *Srf*I restriction site of the pCR-script(SK+) vector (Stratagene, La Jolla, CA). Inserts were identified by a standard blue–white plate screen and sequenced with standard primers. Equal expression of the two partners was achieved with a dicistronic expression vector in which the second Shine–Dalgarno sequence was located within the last codons of the first protein.[95]

Examples of Coiled-Coil Fusions

Increased Avidity from Multimerization: Fusions of Binding Domains to Dimeric, Trimeric, and Tetrameric Coiled Coils

The consummate binding proteins of mammals are antibodies. Complete antibodies cannot be produced in *E. coli,* however, because of the lack of glycosylation and the problem of multiple disulfide formation. Functional expression in *E. coli* is possible for Fv antibody fragments. The Fv fragment is a heterodimer of the heavy chain V_H and the light chain V_L that can be joined by a flexible linker in a single polypeptide chain to form a molecule named single-chain Fv (scFv). Antibodies are at least bivalent, as in the case of the IgG molecule, and in the early immune response, the decameric IgM molecules provide the primary defense. Besides cross-linking antigens, multivalent binding increases the avidity when the target is displayed in a geometrically similar array. To mimic the natural multivalency, coiled coils provide an effective small building block to generate miniantibodies.

Fusion of an scFv to the N terminus of a coiled coil leads to association of scFv fragments according to the oligomerization state of the coiled coil (Fig. 2A). To avoid steric clashes of scFv fragments and to allow simultaneous binding of separate ligands, a linker must be inserted between the scFv and the coiled coil. To dimerize and tetramerize an scFv fragment derived from the phosphorylcholine-binding antibody McPC603, for example, the upper hinge of a murine IgG_3 (Table III) was genetically interposed between the scFv and the wild-type GCN4 leucine zipper or the tetrameric variant GCN4-pLI (Table II, Fig. 2A).[51,52] Multimerization was verified by size-exclusion chromatography, and increased avidity was demonstrated by enzyme-linked immunosorbent assay (ELISA) and biosensor measurements.[52]

In the case of the dimeric scFv–GCN4–leucine zipper fusion, a variant with a C-terminal cysteine in the added sequence KANSRNC also was characterized.[51] This construct did not perform better in biological assays,

[95] K. M. Müller, K. M. Arndt, W. Strittmatter, and A. Plückthun, *FEBS Lett.* **422,** 259 (1998).

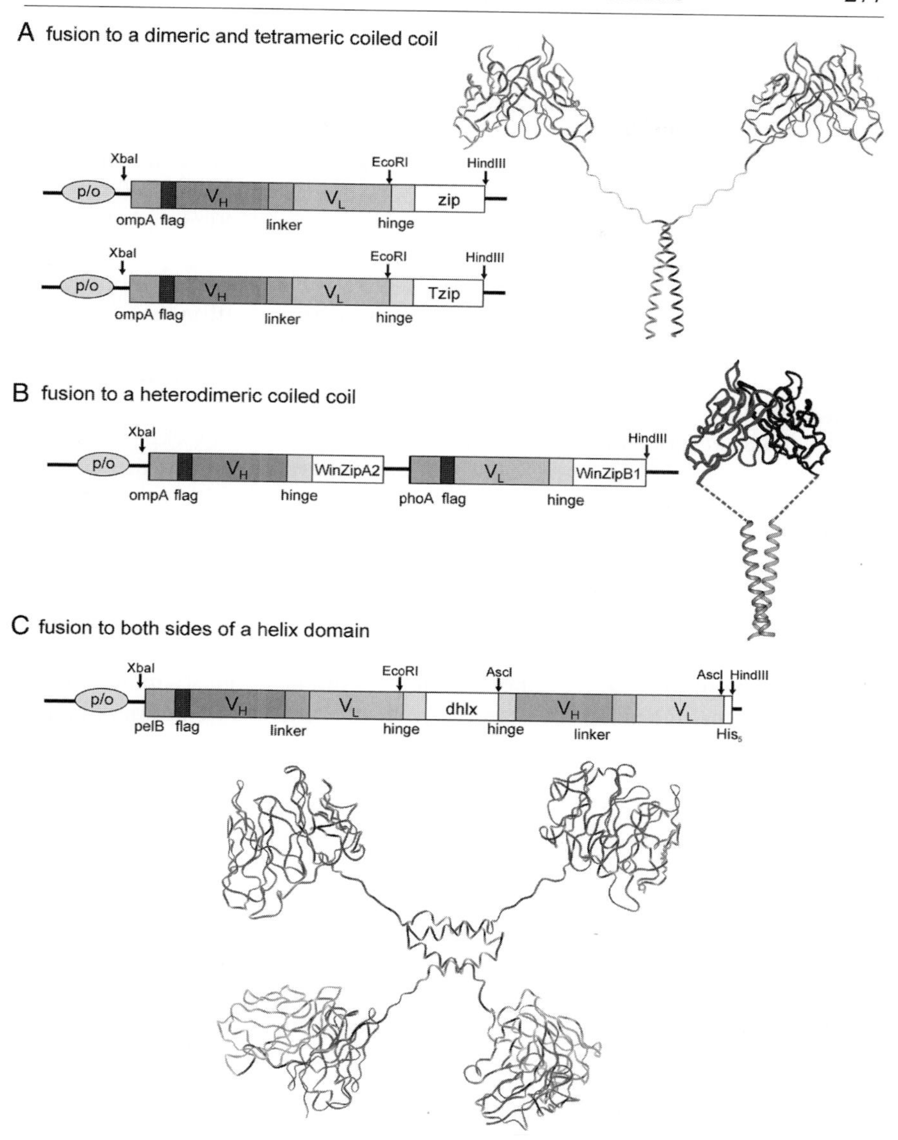

FIG. 2. Designs of coiled-coil fusions. (A) Fusions of dimeric or tetrameric coiled coils to the C terminus of single-chain Fv antibody fragments result in the formation of bivalent or tetravalent miniantibodies, respectively.[51,52] (B) Fusion of the heavy and light chains of an antibody Fv fragment to a heterodimeric coiled coil produced a stable, functional Fv fragment.[48] (C) Fusion of scFv antibody fragments to both sides of a homodimerizing helical motif yielded a dimeric, bispecific miniantibody.[93,98]

and the variant was more difficult to purify, probably because of the formation of nonnative disulfides. In general, additional disulfides may reduce miniantibody expression in the periplasm of *E. coli* and increase the heterogeneity of the products.

An increase in avidity by trimerization was shown for a fusion of the trimeric, GCN4-pII coiled coil to the cytoplasmic domain of CD40.[56] Compared with the apparent affinity of the monomeric CD40 cytoplasmic domain for the intracellular transducer, TRAF2, trimerization of CD40 enhanced TRAF2 binding by more than 10-fold.[56] This increase supports an avidity-based signaling model in which receptor trimerization by extracellular ligands promotes assembly of intracellular signaling complexes.[96] In general, oligomeric fusions to proteins aimed at multivalent targets may enhance binding sufficiently to support new assays.

Even larger molecular assemblies have been obtained by exploiting protein fusions to both ends of a multimerization domain, which are accessible in these simple helical motifs. For example, a bispecific miniantibody was created by following an scFv fragment by a dimerizing helical domain and a second scFv fragment (Fig. 2C).[93] Each scFv provided different binding specificity. The dimerization of the helical domain resulted in a dimeric, bispecific molecule with a total of four binding sites. In this case, the helical domain was not a typical coiled coil but a putative helix–loop–helix, four-helix bundle.[97,98] The murine IgG upper hinge sequence (Table III) was used as a spacer between the first scFv and the helix domain. The second scFv was connected by a short flexible linker with the amino acid sequence GGSGGAP. The chimera was shown by size-exclusion chromatography to be dimeric. Assembly of a ligand–miniantibody–ligand sandwich followed by surface plasmon resonance demonstrated the bispecificity and avidity of bivalent binding.[93]

Certain properties of the fusion proteins that can restrict the efficacy of such chimeras should be addressed at an early stage. For scFv fragments that tended to aggregate, bringing together two or more molecules resulted in poor to very poor yields. Solving this problem required improving the folding efficiency of the scFv and developing effective purification methods.[99] The linker joining the fused domain to the coiled coil is exposed and flexible, making it prone to proteolytic degradation. Examples given here, including the upper hinge of a murine IgG_3 (Table III), proved to be stable in *E. coli.* Other linkers, such as the flexible sequences based on repetitive

[96] S. M. McWhirter, S. S. Pullen, J. M. Holton, J. J. Crute, M. R. Kehry, and T. Alber, *Proc. Natl. Acad. Sci. U.S.A.* **96,** 8408 (1999).

[97] P. Pack, M. Kujau, V. Schroeckh, U. Knüpfer, R. Wenderoth, D. Riesenberg, and A. Plückthun, *Biotechnology* (*N.Y.*) **11,** 1271 (1993).

[98] J. J. Osterhout, *et al., J. Am. Chem. Soc.* **114,** 331 (1992).

[99] K. M. Müller, K. M. Arndt, K. Bauer, and A. Plückthun, *Anal. Biochem.* **259,** 54 (1998).

mixtures of Gly, Ser, and Thr, also have been used successfully. Some coiled coils, especially heterospecific partners expressed separately, are partially degraded during expression and purification when fused to scFv fragments, hinting at partial unfolding. In general, when the components of hetero-oligomers are unstable in isolation, coexpression is likely to reduce aggregation due to prolonged exposure of hydrophobic surfaces and degradation of partially folded domains.

Stabilization by Dimerization: An Antibody Fv Fragment Fused to a Heterodimeric Coiled Coil

An Fv fragment can be held together by adding a flexible linker from either V_H to V_L or from V_L to V_H, respectively. These approaches can have two disadvantages. First, fusion to the N terminus of either V_L or V_H located near the variable site may interfere with binding. Second, linking the domains might give rise to domain-swapped dimers and result in mixed species that are difficult to characterize.[100] To avoid these problems, the Fv heterodimer was stabilized by fusing heterodimerizing coiled coils to the C termini of the V_H and V_L domains (Fig. 2B).[48] Two 14-amino acid linkers, SGGTSGSTSGTGST and AGSSTGSSTGPGSTT, were used to span the distance between the two C termini. The domains were assembled by fusing them to the heterodimeric coiled coil WinZip-A2B1 (Table II) obtained by library-versus-library screening.[64] The partners were coexpressed to maximize yield.

Complementation by Dimerization: An Assay for Optimizing Coiled-Coil Interactions in Vivo

Fusions between coiled coils and other proteins also have been used to study the coiled-coil motif itself.[12,47] The origins of heterospecificity were investigated with a library-versus-library screen based on a protein fragment complementation assay *in vivo*.[63,64,69] The goal was to select for metabolically stable heterodimers with high specificity. Selection performed under different stringencies gave insights into which sequence features are the most crucial for heterodimerization and which impart more subtle optimization.

The two coiled-coil libraries were based on the c-Fos, c-Jun, and GCN4 leucine-zipper sequences. To achieve a high probability for complementarity, the **e** and **g** positions were randomized to Gln, Glu, Arg, or Lys. The core **a** position was randomized to Val or Asn, to investigate whether higher specificity or stability would be more advantageous. To produce an equimolar ratio of the desired amino acids, mixtures of trinucleotide build-

[100] K. M. Arndt, K. M. Müller, and A. Plückthun, *Biochemistry* **37,** 12918 (1998).

ing blocks encoding for the varied amino acids were used for the randomized positions in DNA synthesis. The libraries were genetically fused to two designed fragments of murine dihydrofolate reductase (mDHFR). Assembly of functional enzyme necessitated interactions between members of the two libraries. Selection for the coiled-coil interaction was achieved by making the mDHFR activity essential for cell growth.

Selection cycles with three different stringencies yielded sets of coiled-coil pairs, of which 80 clones were statistically analyzed. A strong bias toward an Asn pair in the core **a** position indicated selection for structural uniqueness, and reduction of repulsions in the **g**–**e′** positions indicated selection for stability. Increased stringency led to additional selection for heterospecificity by destabilizing the respective homodimers. Interestingly, the analysis revealed that the most successful variants do not consist simply of complementary charges in the interacting **g**–**e′** positions. Instead, the most successful charge distributions showed more complicated patterns that presumably fulfilled a variety of conflicting demands on the sequences. The absence of complete charge complementarity in the most successful sequences would have been extremely challenging to predict.

To validate the selection, the dominant pair from the most stringent selection was biophysically characterized.[64] The peptide pair proved to be stable, dimeric, and highly heterospecific. The heterodimer associated with a K_D of approximately 24 n*M*. Because this sequence pair was optimized *in vivo,* it should provide a valuable addition to the coiled coils serving as fusion partners.

Optimization by Elongation: Fusion of a Coiled Coil to a Coiled Coil

Coiled coils can be fused directly to existing coiled coils by maintaining the register of the heptad repeat. This approach was used to generate fragments of the HIV gp41 protein[57,66] (Fig. 3) and the Ebola virus mem-

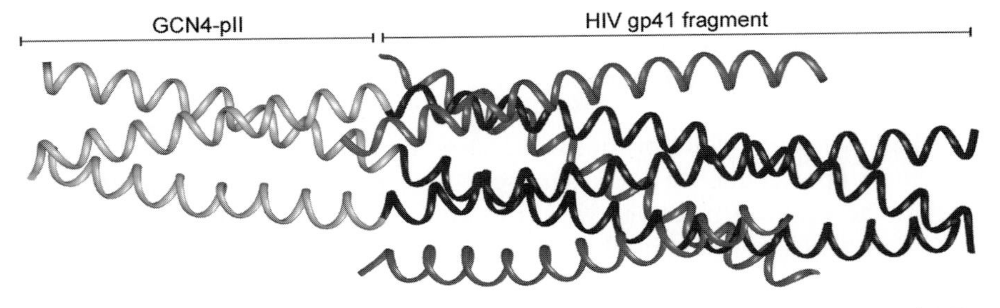

FIG. 3. Structure of the GCN4-pII peptide fused to the core domain of HIV gp41 (PDB code 1ENV).[66]

brane fusion subunit[58,67] for crystallization. In the case of the HIV gp41, a 49-amino acid core was predicted to form a triple-helical coiled coil, and this fragment was coexpressed with a 41-amino acid fragment predicted to associate with the core. To favor trimerization, the designed peptide GCN4-pII (Table II) was genetically fused in the coiled-coil heptad register to the N terminus of the core. The resulting protein was expressed in inclusion bodies in *E. coli* and refolded. Similarly, soluble fragments of the Ebola virus Gp2 ectodomain were constructed by replacing the N-terminal membrane fusion peptide with GCN4-pII. In these cases, one or two residues of GCN4-pII were mutated to cysteines to facilitate formation of heavy metal derivatives. The crystal structures of both chimeric proteins were similar to those of native fragments lacking heterologous sequences.[13,101] These similarities demonstrated that linked coiled coils can have minimal effects on the structure of the fusion partner.

Future Prospects

Oligomerization controls numerous cellular activities. The use of coiled coils to probe these functions will continue to grow with increasing knowledge of the principles of helical associations. *In vivo* applications that target natural coiled-coil domains offer particularly fruitful prospects. Expressing coiled-coil motifs alone or in combination with inactivating domains may be used increasingly to generate dominant negative proteins.[60,61] Fusions to fluorescent molecules such as green fluorescent protein can yield new methods to label coiled-coil structures.[84] Coiled-coil fusions to genetic systems for identifying protein–protein interactions may offer methods to screen genomic libraries for partner proteins.[47,68,69] To augment targeting of natural coiled-coil sequences, designed complementary coiled coils promise antibody-like affinity and specificity for numerous biochemical, genetic, diagnostic, and therapeutic applications.[21]

A variety of problems must be solved to increase the utility of coiled-coil motifs for heterologous oligomerization. Oligomers of six or more chains, for example, would offer even greater gains in binding avidity. Coiled coils whose oligomerization is sensitive to ligand binding[42] or phosphorylation[102] would allow exogenous control of multimerization. Coiled coils that function as heterologous protein receptors also may provide a novel approach to produce general binding proteins.[103] Overall, the importance of oligomerization in cellular physiology suggests that coiled-coil fusions will provide powerful tools for numerous future applications.

[101] D. C. Chan, D. Fass, J. M. Berger, and P. S. Kim, *Cell* **89,** 263 (1997).
[102] L. Szilak, J. Moitra, and C. Vinson, *Protein Sci.* **6,** 1273 (1997).
[103] G. Ghirlanda, J. D. Lear, A. Lombardi, and W. F. DeGrado, *J. Mol. Biol.* **281,** 379 (1998).

Acknowledgments

We thank A. Plückthun for helpful suggestions, for sharing of unpublished data, and for critical reading of the manuscript. This work was supported by NIH Grant GM48958 (T.A.), and by a fellowship of the Swiss National Science Foundation and the Human Frontier Science Program (K.M.).

[18] Molecular Applications of Fusions to Leucine Zippers

By JENNIFER D. RIEKER and JAMES C. HU

Fusion Proteins and Structure–Function

The use of chimeric proteins to study protein function dates back at least as far as the use of illegitimate recombination events among lambdoid phage to generate interspecies variants containing different amounts of polypeptide derived from each parent.[1] By examining functional and nonfunctional combinations of hybrids, subdomains involved in protein–protein interactions were identified for the interactions of the λ replication proteins O and P,[1] and for the terminase proteins A and Nu1.[2,3] Recombinant DNA methods made the use of fusion proteins more easily generalizable, and allowed fundamentally important insights to be derived for many biological systems. For example, our understanding of transcriptional activation was fundamentally altered by the experiments of Brent and Ptashne,[4] who showed that targeting of an activation domain to a particular chromosomal location through a fused bacterial LexA DNA-binding protein was sufficient to activate transcription.

Chimeric proteins involving known oligomerization motifs can be used to answer two simple, but important, classes of questions: Is bringing two proteins together sufficient to cause a change in their activities? Can we construct useful oligomeric proteins by forcing different kinds of subunits together? There are now many ways to use fused motifs to hold proteins together. Here, we will briefly discuss the use of natural and mutant leucine zippers as molecular clamps.

[1] M. E. Furth, C. McLeester, and W. F. Dove, *J. Mol. Biol.* **126,** 195 (1978).
[2] S. Frackman, D. A. Siegele, and M. Feiss, *J. Mol. Biol.* **180,** 283 (1984).
[3] S. Frackman, D. A. Siegele, and M. Feiss, *J. Mol. Biol.* **183,** 225 (1985).
[4] R. Brent and M. Ptashne, *Nature (London)* **312,** 612 (1984).

0076-6879/00 $30.00

Why Leucine Zippers?

Leucine zippers are short, parallel α-helical coiled coils found in a wide variety of proteins. The best studied leucine zippers are from the transcription factors GCN4, Fos, and Jun. Leucine zippers have a number of desirable properties for use as molecular clamps. Leucine zippers are small; in the case of the GCN4 leucine zipper, a 33-residue peptide is sufficient to form stable homodimers *in vitro* and *in vivo*. High-resolution X-ray crystallographic and nuclear magnetic resonance (NMR) structures are available for several leucine zippers.[5–11] As α-helical coiled coils, leucine zippers have simple secondary and tertiary structures. Leucine zipper dimerization can be stable, and can occur over a wide range of salt, temperature, and pH conditions. The large number of naturally occurring leucine zipper proteins and engineered leucine zipper mutants includes a wide variety of distinct and overlapping dimerization specificities.

At the sequence level, the eponymous leucines appear in every seventh position (**d**) over four or five heptad repeats, where the individual positions in the repeat are designated $(\mathbf{abcdefg})_n$. The hydrophobic core of the dimer interface is formed by residues at the **a** and **d** positions (Fig. 1). The **a** positions tend to be occupied by hydrophobic and branched amino acids. In most leucine zippers, one of the central **a** positions is occupied by an asparagine. In homodimers, the asparagines at these positions form an asymmetric hydrogen bond across the dimer interface. This buried asparagine is important for specifying that the helices form dimers; although replacing a buried polar group at **a** position Asn16 in the GCN4 leucine zipper dramatically increased thermal stability,[7,12] peptides lacking the asparagine pair tended to form trimers.

The solvent-accessible **e** and **g** positions are frequently occupied by charged amino acids.[13,14] Intersubunit salt bridges between oppositely charged amino acids at the **g** (*i*th heptad) and **e**′ (*i* + 1th heptad of the

[5] E. K. O'Shea, J. D. Klemm, P. S. Kim, and T. Alber, *Science* **254,** 539 (1991).

[6] T. E. Ellenberger, C. J. Brandl, K. Struhl, and S. C. Harrison, *Cell* **71,** 1223 (1992).

[7] P. B. Harbury, T. Zhang, P. S. Kim, and T. Alber, *Science* **262,** 1401 (1993).

[8] P. B. Harbury, P. S. Kim, and T. Alber, *Nature (London)* **371,** 80 (1994).

[9] J. N. Glover and S. C. Harrison, *Nature (London)* **373,** 257 (1995).

[10] V. Saudek, A. Pastore, M. A. Castiglione Morelli, R. Frank, H. Gausepohl, T. Gibson, F. Weih, and P. Roesch, *Protein Eng.* **4,** 3 (1990).

[11] F. K. Junius, J. P. Mackay, W. A. Bubb, S. A. Jensen, A. S. Weiss, and G. F. King, *Biochemistry* **34,** 6164 (1995).

[12] S. A. Potekhin, V. N. Medvedkin, I. A. Kashparov, and S. Y. Venyaminov, *Protein Eng.* **7,** 1097 (1994).

[13] J. C. Hu and R. T. Sauer, *Nucleic Acids Mol. Biol.* **6,** 82 (1992).

[14] H. C. Hurst, *Protein Profile* **2,** 101 (1994).

A

5 10 15 20 25 30

RMKQLEDKVEELLSKNYHLENEVARLKKLVGER

g a b c d e f g a b c d e f g a b c d e f g a b c d e f g a b c d

B C

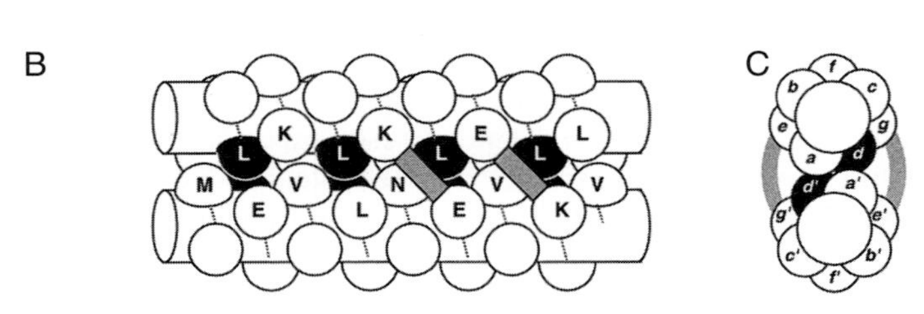

FIG. 1. Diagram of the leucine zipper of GCN4 arranged as a parallel coiled coil. (A) Amino acid sequence of the leucine zipper. Lower-case letters indicate positions in the heptad repeat. The **d** positions are highlighted. (B) Side view of the dimer. The amino acid backbones in a helical conformation are represented by cylinders, with the path of the polypeptide chain indicated by the stippled lines. Side chains are represented as knobs. The **d** positions are highlighted in black. (C) End view showing how different heptad positions are arranged in the dimer. The side chains at the **a** and **d** positions are buried in the dimer interface.

other monomer) positions[5,6,9,15] were observed in the crystal structures of leucine zippers, including GCN4 homodimers and Jun–Fos heterodimers. Interactions involving the **e** and **g** positions are clearly involved in dimerization specificity.[16–22]

Leucine zipper fusions have been used for a variety of experimental purposes. These can be divided into three classes of experiments: (1) those in which the fused domain provides an assay to allow the study of a leucine zipper or other protein–protein interaction, (2) those in which a leucine zipper was used to determine whether formation of homodimers or heterodimers was sufficient to replace the function of a naturally occurring protein domain, and (3) those in which proteins with novel and potentially useful properties were created by bringing new combinations of protein subunits together into a dimeric or multisubunit complex. Many systems that allow

[15] P. Konig and T. J. Richmond, *J. Mol. Biol.* **233,** 139 (1993).

[16] N. E. Zhou, C. M. Kay, and R. S. Hodges, *Protein Eng.* **7,** 1365 (1994).

[17] B.-Y. Zhu, N. E. Zhou, P. D. Semchuk, C. M. Kay, and R. S. Hodges, *Int. J. Peptide Protein Res.* **40,** 171 (1992).

[18] C. R. Vinson, T. Hai, and S. M. Boyd, *Genes Dev.* **7,** 1047 (1993).

[19] E. K. O'Shea, R. Rutkowski, and P. S. Kim, *Cell* **68,** 699 (1992).

[20] E. K. O'Shea, K. J. Lumb, and P. S. Kim, *Curr. Biol.* **3,** 658 (1993).

[21] D. Krylov, I. Mikhailenko, and C. Vinson, *EMBO J.* **13,** 2849 (1994).

[22] X. Zeng, H. Zhu, H. A. Lashuel, and J. C. Hu, *Protein Sci.* **6,** 2218 (1997).

detection of protein–protein interactions through protein fusions have used leucine zippers as test cases; the general features of these systems have been described and reviewed elsewhere.[23–26] Thus, this chapter focuses on the second and third applications of leucine zipper fusions.

Some Applications of Leucine Zipper Fusions

The activities of many regulatory proteins, ranging from transcription factors to membrane-bound receptors, are controlled by dimerization. These proteins often contain multiple domains, and one or more of the domains can be shown to account for the dimerization contacts in the regulator. In these cases, it is useful to determine whether homodimerization of the rest of the protein is sufficient to restore some or all of the activities of the normal protein.

The simplest examples of this approach are from the zipper swap experiments[27–29] that were done shortly after the leucine zipper motif was first described.[30] The ability of many of these chimeric proteins to activate transcription suggested that the primary function of the zipper motif was to hold dimers together. In contrast, fusion proteins in which the C-terminal domain of λ *c*I repressor was replaced by the leucine zipper of GCN4 restore high-affinity DNA binding to the N-terminal DNA-binding domain, but lack the abilities of the naturally occurring protein to bind cooperatively to multiple operator sites, or to be regulated by the SOS response. Fusion proteins of the GCN4 leucine zipper to the cytosolic domain of the *Vibrio cholerae* ToxR protein showed that dimerization was not sufficient to restore transcriptional activation; membrane localization appeared to be essential as well.[31]

The small size of the leucine zipper has been exploited to improve the expression and/or solubility of fusion proteins. For example, full-length antibodies are difficult to express in *Escherichia coli,* but single-chain Fv fragments can be overexpressed and secreted in high yields. Fusion proteins containing Fv fragments fused to leucine zippers or other small oligomeriza-

[23] J. C. Hu, M. G. Kornacker, and A. Hochschild, *Methods* **20,** 80 (2000).
[24] P. L. Bartel and S. Fields (eds.), "The Yeast Two Hybrid System." Oxford University Press, New York, 1997.
[25] R. Brent and R. L. Finley, Jr., *Annu. Rev. Genet.* **31,** 663 (1997).
[26] C. Bai and S. Elledge, *Methods Enzymol.* **273,** 331 (1996).
[27] P. Agre, P. F. Johnson, and S. L. McKnight, *Science* **246,** 922 (1989).
[28] D. R. Cohen and T. Curran, *Oncogene* **5,** 929 (1990).
[29] J. W. Sellers and K. Struhl, *Nature (London)* **341,** 74 (1989).
[30] W. H. Landschulz, P. F. Johnson, and S. L. McKnight, *Science* **240,** 1759 (1988).
[31] K. M. Otteman and J. J. Mekalanos, *Mol. Microbiol.* **15,** 719 (1995).

tion motifs can be efficiently secreted from *E. coli;* because cooperative binding can occur using two or more Fv domains, the fusion proteins may bind their antigens more tightly than the Fv fragments alone.

Altered Specificity Homodimers

Most artificial homodimers have been constructed with the GCN4 leucine zipper, although other leucine zippers have also been used. Construction of artificial systems involving more than one artificially dimerized protein will require a second dimerization motif. Although other dimerization motifs can be used, different combinations of homodimeric leucine zippers are available to allow expression of multiple homodimers. For this purpose, we have described a set of mutant forms of the GCN4 leucine zipper that have altered dimerization specificities.[32]

When a homodimeric leucine zipper assembles as a parallel coiled coil, residues at equivalent **a** positions are opposed to each other in the same layer of the interface (Fig. 1). To generate mutant leucine zippers with alterations in the dimer interface, residues in the **a** position of the hydrophobic core of the GCN4 leucine zipper were randomized between isoleucine and asparagine to generate different permutations of heptad repeats in which either isoleucine–isoleucine pairs or asparagine–asparagine pairs form in homodimers. Mutually primed DNA synthesis (see below) was used to generate a double-stranded DNA cassette from a pair of oligonucleotides with a random mixture of AAC and ATC codons at the four **a** positions in the GCN4 leucine zipper (Fig. 1). The cassette was cleaved at both ends and cloned into an expression vector to generate a library of candidate fusion proteins in which the leucine zippers would be fused downstream of residue 132 of λ repressor. From 64 candidates that were screened by sequencing, 40 mutant zippers had changes only at the **a** positions. Fourteen of the 16 possible sequences were recovered, and the remaining 2 sequences were constructed from the recovered sequences by exchanging DNA segments around an internal *Xho*I site.

Among the 16 possible sequences, Zeng *et al.*[32] described 7 different classes of dimerization specificities in which sequences within a class form homodimers and heterodimers, but fail to form heterodimers with members of other classes (Table I). The specificity differences among these mutant leucine zippers can be explained by the destabilizing effects of burying a polar asparagine in a nonpolar environment. In homodimers, the asparagine side chains at the **a** positions can form a hydrogen bond across the dimer interface, as is seen in the crystal structure.[5] In heterodimers, however, equivalent positions in the sequence may be occupied by different kinds

[32] X. Zeng, A. M. Herndon, and J. C. Hu, *Proc. Natl. Acad. Sci. U.S.A.* **94,** 3673 (1997).

TABLE I
ALTERED SPECIFICITY MUTANTS OF THE GCN4 LEUCINE ZIPPER

Specificity class[a]	**a** position sequences[b]
1	NNII
2	NINN
	NINI
3	NIIN
	NIII
4	INNN
	INNI
5	VNVV (wt GCN4)
	INII
	ININ
6	IINN
	IINI
7	IIIN
	IIII

[a] Mutants within a specificity class form homodimers and heterodimers with other members of the same class, but do not form heterodimers with members of other classes.

[b] Positions 9, 16, 23, and 30 in Fig. 1. Sequences of each mutant leucine zipper are as shown in Fig. 1, except at the indicated positions. wt, Wild type.

of amino acids. Whenever I or V pairs with N, the N will not be able to form intersubunit side-chain hydrogen bonds that it would form in homodimers. In this case, heterodimers will be less stable than either homodimer.

These altered specificity mutants could be useful even when only a single homodimeric protein is needed, but interaction with endogenous coiled-coil proteins could either lead to toxicity or interfere with the activity of the desired protein. This seems to be not much of a problem in expression of fusion proteins in *E. coli,* but many coiled-coil proteins exert "dominant negative" effects when expressed in eukaryotic cells.

Heterodimers

Amino acid residues required for function in an oligomeric protein are represented in each of the equivalent subunits of the oligomer. Thus, a mutation that changes an amino acid in the monomer can alter two or more residues in the protein complex. These changes may or may not contribute equally to any change in function observed for the mutant protein *in vivo* or *in vitro.* Coexpressing mutant and wild-type versions of the same subunit will generate mixed oligomers; whether these can be used to assign function to specific sites *in vivo* depends on whether the activities of the different oligomeric forms can be deconvoluted. For studies *in vitro,* mutant and wild-type proteins can be mixed and the desired mixed oligomer can be purified if (1) there is a basis for separation of the different forms and (2) reassortment by dissociation and reassociation is negligible.

Most naturally occurring leucine zippers are homodimeric; thus studies based on heterodimerization have used one of two systems: the natural leucine zippers Fos and Jun or the designed heterodimer ACID-p1 + BASE-p1[20] (Fig. 2). Although both pairs form predominantly heterodimers when mixed,[20,33] Jun is also able to form homodimers in the absence of Fos. In contrast, Fos, ACID, and BASE do not form detectable homodimers in isolation at neutral pH. Thus, ACID and BASE may be a better choice when one desires controls where the isolated fusion proteins are monomeric. Note, however, that because the homodimers are destabilized by repulsive electrostatic interactions involving the **e** and **g** positions, salt and pH can affect whether homodimers form. For example, at low pH, the glutamate side chains in ACID and Fos become protonated, and homodimers can be detected.

In ACID-p1 + BASE-p1, as in GCN4, a single asparagine pair at the central **a** position in each monomer helps to set registration between the helices, and prevents the formation of alternative oligomeric forms.[34] O'Shea *et al.*[20] designed these two peptides to contain leucines at both the **a** and **d** positions based on the model coiled coils previously described by the Hodges[35] and DeGrado[36] groups. Ironically, these earlier model peptides, which lack the central asparagine pair, form higher order oligomers[37,38]; variants of ACID-p1 + BASE-p1 in which the asparagines have been replaced by leucines form a variety of higher order oligomeric helical bundles.[34]

Kim and Hu used fusions to heterodimeric leucine zipper proteins to test the equivalence of sequences in the two subunits of a naturally homodimeric protein *in vivo.*[39] Bacteriophage λ repressor binds to its operator sites as a homodimer. Although the protein is presumably symmetric in solution,

[33] E. K. O'Shea, R. Rutkowski, W. Stafford III, and P. S. Kim, *Science* **245,** 646 (1989).
[34] K. J. Lumb and P. S. Kim, *Biochemistry* **34,** 8642 (1995).
[35] R. S. Hodges, N. E. Zhou, C. M. Kay, and P. D. Semchuk, *Peptide Res.* **3,** 123 (1990).
[36] K. T. O'Neil and W. F. DeGrado, *Science* **250,** 646 (1990).
[37] B.-Y. Zhu, N. E. Zhou, C. M. Kay, and R. S. Hodges, *Protein Sci.* **2,** 383 (1993).
[38] B. Lovejoy, S. Choe, D. Cascio, D. K. McRorie, W. F. DeGrado, and D. Eisenberg, *Science* **259,** 1288 (1993).
[39] Y.-I. Kim and J. C. Hu, *Proc. Natl. Acad. Sci. U.S.A.* **92,** 7510 (1995).

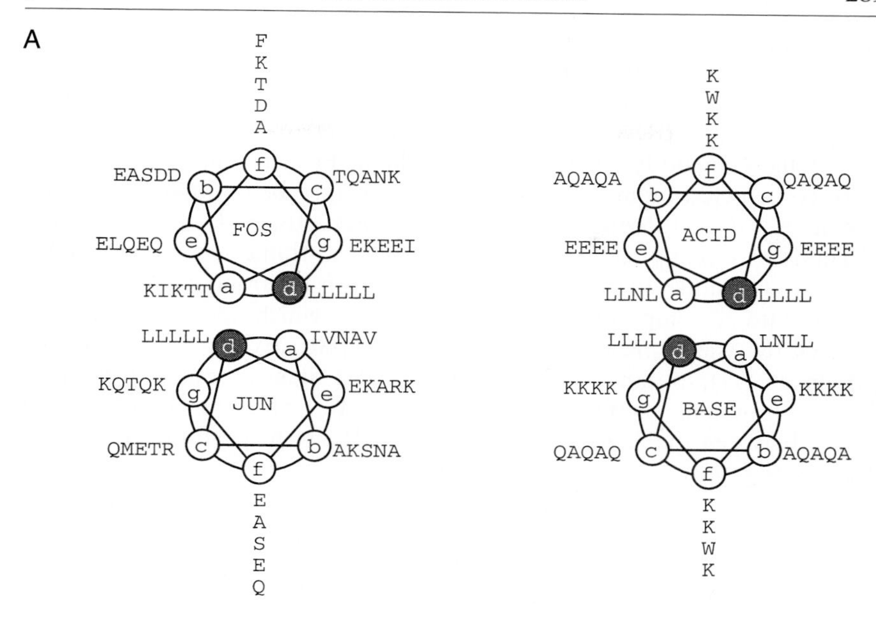

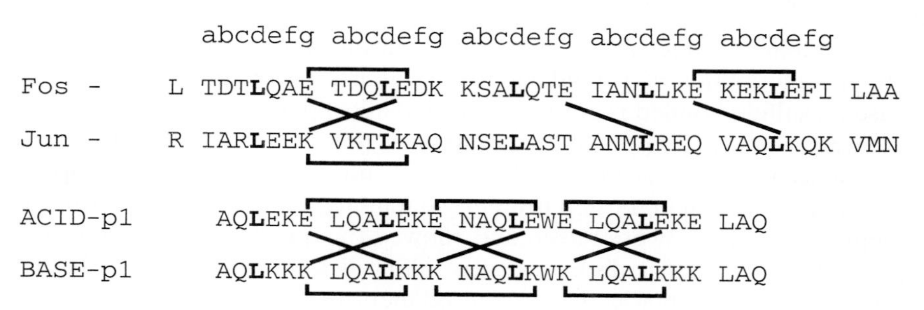

FIG. 2. Comparison of Fos–Jun with ACID-p1 + BASE-p1. (A) Helical wheels representing end views of the Fos–Jun (*left*) and ACID-p1 + BASE-p1 (*right*) heterodimers. (B) Amino acid sequences of the monomers. Diagonal lines indicate positions of putative intersubunit salt bridges that stabilize the heterodimers. Brackets above and below each sequence indicate repulsive electrostatic interactions that should destabilize the homodimers.

binding to the operator is asymmetric.[40,41] Repressor makes specific contacts with the operator by means of a helix–turn–helix motif and an N-terminal

[40] A. Sarai and Y. Takeda, *Proc. Natl. Acad. Sci. U.S.A.* **86,** 6513 (1989).
[41] L. J. Beamer and C. O. Pabo, *J. Mol. Biol.* **227,** 177 (1992).

"arm" composed of the first six amino acids of the DNA-binding domain. A wild-type protein fused to the Fos leucine zipper and an arm deletion variant fused to the Jun leucine zipper were coexpressed in *E. coli* from compatible plasmids to force the formation of heterodimers *in vivo* and *in vitro*. Oligonucleotide cassettes that encode the Fos and Jun leucine zippers were constructed by mutual primed synthesis (see below) and cloned between the *Sal*I and *Bam*HI sites of the *c*I fusion vector pJH391[42] to generate plasmids expressing ampicillin resistance and in-frame leucine zipper fusions after residue 132 of repressor. The fusion contructs were also subcloned into a compatible vector, pJH550, in which the *tet* gene from pACYC184 was replaced by the *tet* gene of pSELECT-1 from Promega (Madison, WI), to remove the *Sal*I and *Bam*HI sites and place an *Eco*RI–*Eco*RV fragment encoding residues 1–132 of *c*I and the GCN4 leucine zipper between the *Eco*RI site and the *Pvu*II site at position 515 on the pACYC184 map.

Heterodimeric leucine zippers have also been used to generate novel proteins. Fusions with heterodimeric leucine zippers have been used to drive the association of the extracellular segments of T cell receptors to generate soluble heterodimeric molecules for biochemical studies and crystallization[43–46] and to generate bispecific miniantibodies.[47]

Disulfide Bonded Versions

Cysteine residues at the ends of the leucine zippers can be introduced to allow formation of disulfide bonds to stabilize the dimers. Kim and Hu[48] used disulfide-bonded repressor fusions to allow purification of one-armed repressor–leucine zipper heterodimers. In this case, a Gly-Gly-Cys-His_6 sequence was introduced at the C-terminal end of the GCN4 leucine zipper in the fusion protein. Fusion proteins containing the wild-type and armless repressor domains were mixed and allowed to equilibrate under reducing conditions, and then disulfides were allowed to form by air oxidation. Although both homodimers and heterodimers were formed by this protocol, the different forms were separable by heparin chromatography.

[42] J. C. Hu, N. E. Newell, B. Tidor, and R. T. Sauer, *Protein Sci.* **2,** 1072 (1993).

[43] A. Kalandadze, M. Galleno, L. Foncerrada, J. L. Strominger, and K. W. Wucherpfennig, *J. Biol. Chem.* **271,** 20156 (1996).

[44] A. Golden, S. S. Khandekar, M. S. Osburne, E. Kawasaki, E. L. Reinherz, and T. H. Grossman, *J. Immunol. Methods* **206,** 163 (1997).

[45] H. C. Chang, Z. Bao, Y. Yao, A. G. Tse, E. C. Goyarts, M. Madsen, E. Kawasaki, P. P. Brauer, J. C. Sacchettini, S. G. Nathenson, and E. L. Reinherz, *Proc. Natl. Acad. Sci. U.S.A.* **91,** 11408 (1994).

[46] N. Patel, J. M. Herrman, J. C. Timans, and R. A. Kastelein, *J. Biol. Chem.* **271,** 30386 (1996).

[47] S. A. Kostelny, M. S. Cole, and J. Y. Tso, *J. Immunol.* **148,** 1547 (1992).

[48] Y.-I. Kim and J. C. Hu, *Mol. Microbiol.* **25,** 311 (1997).

Disulfide bonding should eliminate the potential for subunit exchange *in vitro*. However, it should be noted that many of the interior compartments of both eukaryotic and prokaryotic cells are highly reducing environments. Thus, it is unlikely that these disulfides will form *in vivo*.

Other Oligomeric Forms

The term *leucine zippers* was originally coined to describe part of the bZIP motif found in a variety of transcription factors.[30] Leucine zippers in this strict sense are short, autonomously folding, parallel, dimeric coiled coils. However, leucine zippers are often proposed to be present in other kinds of proteins on the basis of sequence analysis.[49–51] Some of these sequences form dimeric coiled coils, while others form other kinds of helical bundles, or are found as helical segments within globular domains. In addition, altering a small number of residues in a dimeric coiled coil can lead to different structures. Harbury *et al.*[7,8] have reengineered the hydrophobic core of the GCN4 leucine zipper to generate parallel three- and four-stranded coiled coils. These can also be used to drive the oligomerization of other proteins.[22,52] In addition, antiparallel leucine zippers have been designed by the Hodges[53,54] and Kim[55,56] groups. There are no reports yet of using these for driving oligomerization of fusion proteins, but they are likely to be useful when antiparallel orientation is needed (but see below).

Design Issues

In designing one or more fusion proteins, two issues must be addressed: (1) What leucine zipper should be used? and (2) where should it be fused to the protein of interest? As can be seen from the preceding discussion, there are many options for using leucine zippers to hold together pairs of fusion proteins. On the basis of the successes of other groups, GCN4 is a good choice for homodimers and either Fos–Jun or ACID-p1 + BASE-p1 can be used for heterodimers, depending on whether homodimerization of one of the proteins is desired.

The leucine zippers can be fused at either the N-terminal or C-terminal end of the fusion protein; in naturally occurring bZip proteins, the zippers

[49] A. Lupas, M. Van Dyke, and J. Stock, *Science* **252,** 1162 (1991).

[50] A. Lupas, *Curr. Opin. Struct. Biol.* **7,** 388 (1997).

[51] D. N. Woolfson and T. Alber, *Protein Sci.* **4,** 1596 (1995).

[52] X. Zeng and J. C. Hu, *Gene* **185,** 745 (1997).

[53] O. D. Monera, N. E. Zhou, C. M. Kay, and R. S. Hodges, *J. Biol. Chem.* **268,** 19218 (1993).

[54] O. D. Monera, N. E. Zhou, P. Lavigne, C. M. Kay, and R. S. Hodges, *J. Biol. Chem.* **271,** 3995 (1996).

[55] M. G. Oakley and P. S. Kim, *Biochemistry* **36,** 2544 (1997).

[56] M. G. Oakley and P. S. Kim, *Biochemistry* **37,** 12603 (1998).

are found at different positions in the primary sequence. Where the zipper is placed is more a function of the ability of the target protein to tolerate substitution at one or the other end than of constraints on the folding of the zipper.

Many of the applications of leucine zippers in fusion proteins involve holding together subunits that can fold on their own. However, leucine zippers have also been used to hold together protein fragments that would not autonomously fold. For example, Pelletier *et al.*[57] used GCN4 leucine zippers to reconstitute mouse dihydrofolate reductase (DHFR) from fragments. The two fragments of DHFR that they used corresponded to the N-terminal and C-terminal sides of a surface loop in the folded protein. However, these were not held together by fusing antiparallel leucine zippers at the site of the break; one leucine zipper monomer was fused at the N-terminal end of the protein, while the other was fused at the N-terminal end of the C-terminal fragment. Thus, flexible sequences at the ends of the zipper must have allowed the two fragments to come together.

This illustrates an important point about the design of fusion proteins containing leucine zippers. Although the parallel coiled-coil structure of the zippers places the ends of their monomers close in space, one does not need fusion end points from the target protein to be close to these distances in order to drive self-assembly. The target proteins can be tethered through relatively long flexible linkers; in the case of the fusions to the DNA-binding domain of λ repressor, the linker contains more amino acids than the leucine zipper.[58] Even in the presence of these flexible linkers, dimerization by the zipper increases the effective concentration of the fused proteins with respect to one another. Where shorter linkers are used,[59] fraying of the ends of the leucine zipper may allow for some flexibility.

Once design decisions have been made regarding which leucine zippers should be used, and where they should be placed, the specific sequences must still be designed. This is generally not a serious issue when using leucine zipper cassettes that have been previously used in other studies. However, it is important to note that codon usage should be considered when designing the zipper cassette. Cohen and Curran[28] showed that rare codons in the mammalian sequences for the Fos leucine zipper prevent its expression in *E. coli.*

Generating the Zipper Cassette

Generating leucine zipper fusion proteins is extremely straightforward and can be accomplished by several methods. Leucine zipper sequences

[57] J. N. Pelletier, F.-X. Campbell-Valois, and S. W. Michnick, *Proc. Natl. Acad. Sci. U.S.A.* **95,** 12141 (1998).

[58] J. C. Hu, E. K. O'Shea, P. S. Kim, and R. T. Sauer, *Science* **250,** 1400 (1990).

[59] M. D. Edgerton and A. M. Jones, *Plant Cell* **4,** 161 (1992).

can be swapped from existing vectors as restriction fragments; because this is straightforward subcloning, we do not describe it in detail here. However, because of their small size, it can be difficult to purify the zipper cassettes on gels. We often simply have the entire leucine zipper sequence synthesized by a commercial supplier of oligonucleotides as a pair of mutually priming sequences. Alternatively, we generate the zipper cassettes by polymerase chain reaction (PCR) amplification from existing plasmids. Our laboratory has most of the leucine zippers described above as C-terminal fusions to λ repressor, where the zipper parts are cloned between a *Sal*I site and a *Bam*HI site.

Construction of Zipper Cassettes by Mutually Primed Synthesis

A pair of oligonucleotides encoding the zipper sequences to be constructed can be ordered from a commercial source of synthetic oligonucleotides. One oligonucleotide encodes the N-terminal half of the leucine zipper and contains the appropriate restriction site at the 5′ end for cloning into the vector. The second oligonucleotide encodes the C-terminal half of the zipper and contains a different restriction site. Allow the N-terminal and C-terminal oligonucleotides to overlap by nine or more nucleotides so that they can be used as mutual primers. For our experiments we have used a variety of different oligonucleotide pairs, in which the overlapping region is in different positions within the leucine zipper in order to accommodate the incorporation of specific mutations or randomized bases at specific codons. We also design the oligonucleotides to contain on the order of 10 nucleotides beyond the restriction sites on either end. Although most restriction enzymes do not need this much DNA to cut, this allows us to monitor the digestion of the cassettes by gel electrophoresis.

We currently obtain our oligonucleotides from Sigma-Genosys (The Woodlands, TX). The oligonucleotides are ordered desalted on a 0.05-μmol scale and are directly resuspended in the tube supplied by the vendor in 300 μl of TE [10 m*M* Tris-HCl (pH 7.5), 1 m*M* EDTA] by vigorous shaking at room temperature for at least 2 hr.

Gel Purification of Oligonucleotides

The oligonucleotides are often supplied in sufficient purity to use directly in mutually primed synthesis. However, we routinely purify them on a 10% (w/v) acrylamide (19:1), 50% (w/v) urea, 1× Tris–borate–EDTA (TBE) gel. This seems to be especially important if the oligonucleotides are >50 nucleotides in length. The protocol below is a fairly standard protocol for a 14 × 16 cm gel, which is run in a standard large-format apparatus for protein gels. Volumes and gel conditions should be adjusted on the basis of the available gel apparatus. Any other purification protocol for oligonucleotides on this scale should also work.

1. Mix 150 μl of DNA with 50 μl of tracking dye [95% (v/v) formamide, 20 m*M* EDTA, 0.002% (w/v) bromphenol blue, 0.002% (w/v) xylene cylanol]. With our slot formers, this sample volume fills the slots almost completely. Because a large volume is being loaded on the gel, the concentrations of the tracking dyes are reduced relative to standard recipes in order to prevent the dyes from interfering with UV shadowing.
2. Load the DNA onto the gel and run at 25 W for 3 hr. When multiple oligonucleotides are being purified on the same gel, we usually skip a lane between each oligonucleotide.
3. Visualize the bands by UV shadowing. The gel is removed from the plates and placed on a piece of Saran wrap over an F_{254}-impregnated thin-layer chromatography (TLC) plate (EM Science, Cherry Hill, NJ). The gel is then illuminated from above with a hand-held UV lamp set on the long-wavelength setting, under which the TLC plate will give strong green fluorescence. The desired band will be the large band at the top of the smear.
4. Cut the desired band from the gel with a spatula (razor blades tend to slice all the way through to the TLC plate) and mash into small pieces with a glass rod or the end of a glass pipette in a 15-ml Falcon tube.
5. Elute the DNA by soaking the gel pieces in 0.5 ml of 0.3 *M* sodium acetate, 5 m*M* EDTA overnight in a roller drum in a 37° incubator.
6. Remove the pieces of the gel by filtering the solution through a plastic frit. We use empty Wizard (Promega) minipreparation columns. Wash the remaining gel pieces with 0.5 ml of 0.3 *M* sodium acetate and filter again. The overall yield is ~1 ml of 0.3 *M* sodium acetate containing the purified DNA.
7. Concentrate the DNA by ethanol precipitation and resuspend in 100 μl of TE. Use Beer's law to determine the concentration of the DNA by measuring the OD_{260}, and dilute the DNA to 10 μM stocks to be used in the annealing and extending reactions.

Mutually Primed Synthesis

The mutually primed synthesis is composed of two steps: annealing and extension. This protocol yields about 100 pmol of product.

1. Each of the annealing reactions contains a 1 μM concentration of each oligonucleotide, 1× Sequenase buffer [40 m*M* Tris-HCl (pH 7.5), 50 m*M* NaCl, 20 m*M* $MgCl_2$], 0.25 m*M* dNTPs, and 10 m*M* dithiothreitol (DTT) in a 100-μl reaction.
2. We perform the annealing reaction in a thermocycler to control the

rate of cooling; the cycler is programmed to heat the mixtures to 90° for 1 min, 70° for 1 min, and then cool at 1°/15 sec to 20°. However, it is not clear whether this is any better than simply immersing the mixture in boiling water for 1 min and then letting it cool to room temperature in a test tube rack.

3. After annealing the oligonucleotides, extend them by adding 13 units of Sequenase T7 DNA polymerase v2.0 (U.S. Biochemicals, Cleveland, OH) and incubating them for 30 min at room temperature.

We generally remove 10-μl samples from steps 1–3 to monitor annealing and extension reactions. For most of the leucine zipper cassettes, we run the samples on nondenaturing 12% (w/v) acrylamide (29:1) minigels and visualize the DNA by staining with ethidium bromide. The base pairing in the overlap region is not always strong enough to survive the gel, but when it is, the annealed but unextended oligonucleotides will run much slower than the expected size of the extension product, because of the largely single-stranded character of the hybrids. If the extension has worked well, a single prominent band will be seen at the expected size, although there is often a smear of staining around that size.

Construction of the Zipper Cassettes by Polymerase Chain Reaction

The zipper cassettes can be removed from an existing vector by PCR. PCR allows new restriction sites or mutations to be engineered in the zipper cassettes. Two primers, one for the 5′ end and one for the 3′ end of the zipper, need to be designed. The primers need to be approximately 20 nucleotides long and overlap the 5′ end and 3′ end of the zipper by 9 nucleotides. We order primers desalted on a 0.05-μmol synthesis scale from Sigma-Genosys. The oligonucleotides are resuspended in 300 μl of TE [10 m*M* Tris-HCl (pH 7.5), 1 m*M* EDTA] by shaking at room temperature for 2 hr. In our experience, it has not been necessary to purify the primers. Measure the OD_{260} and determine the concentration of the DNA by Beer's Law. Dilute the DNA to 10 μ*M* stocks to be used in the PCRs.

1. Each of the PCRs contains a 1 μ*M* concentration of each oligonucleotide, 1× buffer (based on the 10× stock provided by the supplier of the *Taq* polymerase), 0.25 m*M* dNTPs, 2 m*M* $MgCl_2$, and 5 units of *Taq* polymerase in a 100-μl reaction. For template, we typically use ~20 ng of plasmid preparation DNA. However, we have also had success by amplifying directly from colonies. In this case, a single colony of a strain that contains the desired plasmid is picked with a sterile toothpick and suspended in 100 μl of TE in a 1.5-ml microcentrifuge tube. The sample is boiled for 5 min and centrifuged in a microcentrifuge for 10 min. Five microliters of the supernatant is used as the template for PCR.

2. Perform the amplification. We use the standard amplification program on our thermal cycler: denaturation at 94° for 1 min followed by 30 cycles of denaturation at 92° for 40 sec, annealing at 60° for 40 sec, and extension at 75° for 90 sec. A 5-min extension at 75° is added at the end of the last cycle.

3. Clean the PCR DNA with a Qiagen (Chatsworth, CA) PCR purification kit. Elute the DNA with 50 μl of TE. The yield should be about 0.5 μg of DNA after PCR.

4. When the desired restriction sites are close to the ends of the cassette, it may be beneficial to treat the PCR product with proteinase K. We do this by adjusting the reaction to 10 m*M* Tris-HCl (pH 7.8), 5 m*M* EDTA, and 0.5% (w/v) sodium dodecyl sulfate (SDS). A 2.5-μl volume of a 2-mg/ml stock of proteinase K is added and the reaction is incubated at 56° for 1 hr. However, we have also had success in cloning cassettes without this proteinase treatment.

Cloning the Cassettes

The protocols described above yield 100–200 pmol of double-stranded leucine zipper cassette. This can be cleaved with the appropriate restriction enzymes and cloned into the appropriate vectors as with any other DNA. Two things should be kept in mind when cloning the cassettes: First, because the DNA is short, the number of restriction sites present will be higher than for an equivalent mass of plasmid or λ DNA, which is how the activity units for the restriction enzymes are typically defined. Thus, we typically digest 10% of the reaction with 10 units of enzyme for 2 hr. We run an aliquot of the digested DNA on a 12% (w/v) acrylamide gel to check whether the digestion has gone to completion and to estimate the concentration of DNA by comparison with the intensity of marker bands. The 50-bp ladder from BRL (Gaithersburg, MD) is convenient for this purpose, because bands of comparable size are easily found. Second, the amount of cassette produced is in vast excess over what is needed for cloning. A typical ligation reaction will contain 100 ng of vector, which is only 30–50 fmol. The digested cassette should be used at roughly equimolar amounts to the vector DNA, because an excess will lead to ligation of two cassettes to each digested vector; the resultant molecule will not be able to circularize and will not give a transformant.

Acknowledgments

We thank the members of the Hu lab and Debby Siegele for helpful discussions and suggestions. This work was supported by NSF grants MCB-9305403 and MCB-9808474 to JCH.

[19] Using the Yeast Three-Hybrid System to Detect and Analyze RNA–Protein Interactions

By BRIAN KRAEMER, BEILIN ZHANG, DHRUBA SENGUPTA, STANLEY FIELDS, and MARVIN WICKENS

Introduction

The interactions of RNAs and proteins are critical for a wide variety of cellular processes, ranging from translation to the proliferation of certain viruses. For this reason, several methods have been developed to facilitate the analysis of RNA–protein interactions by genetic means.[1–10] These complement an array of biochemical approaches, including RNA affinity chromatography.

We focus here on one such genetic method, the yeast three-hybrid system.[1] In this method, an RNA–protein interaction in yeast results in transcription of a reporter gene. The interaction can be monitored by cell growth, colony color, or the levels of a specific enzyme. Because the RNA–protein interaction of interest is analyzed independent of its normal biological function, a wide variety of interactions are accessible. The published applications of the method to date include the discovery of proteins that bind to known RNA sequences, confirmation of suspected interactions between an RNA and protein, and mutational analysis of interacting RNAs and proteins. In addition, one report suggests that it may be feasible to identify the previously unknown RNA partner of an RNA-binding protein,[11] and to find protein components of multiprotein–RNA complexes in which both protein–protein and RNA–protein interactions are important.[12]

[1] D. J. SenGupta, B. Zhang, B. Kraemer, P. Pochart, S. Fields, and M. Wickens, *Proc. Natl. Acad. Sci. U.S.A.* **93,** 8496 (1996).
[2] M. P. MacWilliams, D. W. Celander, and J. F. Gardner, *Nucleic Acids Res.* **21,** 5754 (1993).
[3] R. Stripecke, C. C. Oliveira, J. E. McCarthy, and M. W. Hentze, *Mol. Cell. Biol.* **14,** 5898 (1994).
[4] I. A. Laird-Offringa and J. G. Belasco, *Proc. Natl. Acad. Sci. U.S.A.* **92,** 11859 (1995).
[5] C. Jain and J. G. Belasco, *Cell* **87,** 115 (1996).
[6] K. Harada, S. S. Martin, and A. D. Frankel, *Nature (London)* **380,** 175 (1996).
[7] E. Paraskeva, A. Atzberger, and M. W. Hentze, *Proc. Natl. Acad. Sci. U.S.A.* **95,** 951 (1996).
[8] R. Tan and A. D. Frankel, *Proc. Natl. Acad. Sci. U.S.A.* **95,** 4247 (1998).
[9] U. Putz, P. Skehel, and D. Kuhl, *Nucleic Acids Res.* **24,** 4838 (1996).
[10] W. S. Blair, T. B. Parsley, H. P. Bogerd, J. S. Towner, B. L. Semler, and B. R. Cullen, *RNA* **4,** 215 (1998).
[11] D. SenGupta, M. Wickens, and S. Fields, *RNA* **5,** 596 (1999).
[12] J. Sonada and R. P. Wharton, *Genes Dev.* **13,** 2704 (1999).

0076-6879/00 $30.00

The three-hybrid system, like other genetic strategies, has the attractive feature that a clone encoding the protein of interest is obtained directly in the screen.

The various genetic methods developed to detect RNA–protein interactions[1–10] have distinct advantages. For example, several systems in eubacteria have been used to examine peptide–RNA interactions in detail.[5,6] Of the genetic methods presented so far, to our knowledge only the three-hybrid system has been used to identify new naturally occurring RNA-binding proteins of biological significance. We offer the protocols here as a starting point for investigators wanting to use the system as is, or to develop it further.

Principles of the Method

The general strategy of the three-hybrid system is diagrammed in Fig. 1A. DNA-binding sites are placed upstream of a reporter gene in the yeast chromosome. A first hybrid protein consists of a DNA-binding domain linked to an RNA-binding domain. The RNA-binding domain interacts with its RNA-binding site in a bifunctional ("hybrid") RNA molecule. The other part of the RNA molecule interacts with a second hybrid protein consisting of another RNA-binding domain linked to a transcription activation domain. When this tripartite complex forms at a promoter, even transiently, the reporter gene is turned on. Reporter expression can be detected by phenotype or simple biochemical assays.

The specific molecules most commonly used for three-hybrid screens at the time of writing are depicted in Fig. 1B. The DNA-binding site consists of a 17-nucleotide recognition site for the *Escherichia coli* LexA protein, and is present in multiple copies upstream of both the *HIS3* and *lacZ* genes. Hybrid protein 1 consists of LexA fused to bacteriophage MS2 coat protein, a small polypeptide that binds as a dimer to a short stem–loop sequence in its RNA genome. The hybrid RNA (depicted in more detail in Fig. 3) consists of two MS2 coat protein-binding sites linked to the RNA sequence of interest, X. Hybrid protein 2 consists of the transcription activation domain of the yeast Gal4 transcription factor linked to an RNA-binding protein, Y.

Although the components depicted in Fig. 1B are most commonly used, other RNAs and proteins can be substituted. For example, the MS2 components can be replaced by either the histone mRNA 3′ stem–loop and the protein to which it binds (SLBP), or by the NRE (Nanos response element) in the 3′ untranslated region (UTR) of *hunchback* mRNA and the protein to which it binds, Pumilio (B. Zhang and M. Wickens, unpublished, 1998; and Ref. 12).

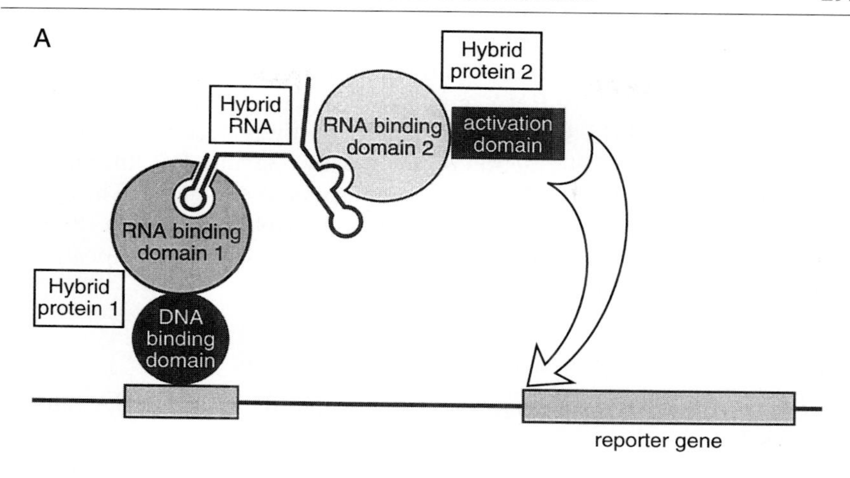

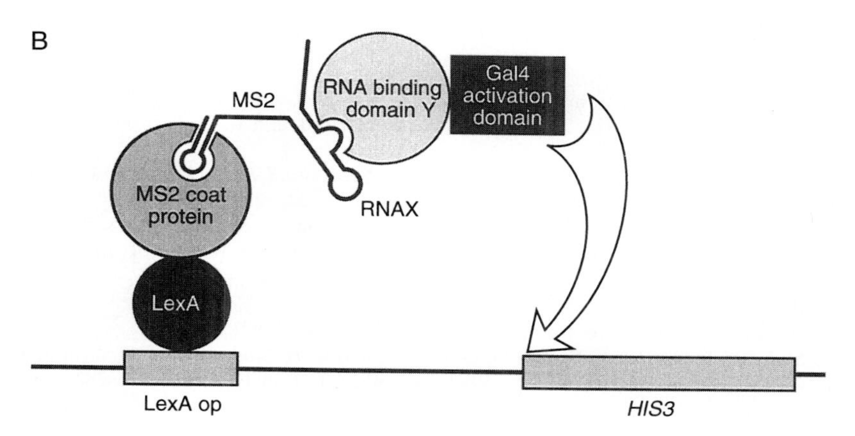

FIG. 1. The three-hybrid system to detect and analyze RNA–protein interactions. (A) General strategy of the system; (B) the specific protein and RNA components that typically have been used to date. Other RNA and protein components also can be used (see text). For simplicity, the following features are not indicated. Both *lacZ* and *HIS3* are present in strain L40 coat under the control of *lexA* operators, and so can be used as reporters. Multiple *lexA* operators are present (four in the *HIS3* promoters and eight in the *lacZ* promoter), and LexA protein binds as a dimer. The hybrid RNA contains two MS2-binding sites, and MS2 coat protein binds as a dimer to a single site. [Diagram adapted from Ref. 1.]

The three-hybrid approach has many of the same strengths and limitations as the two-hybrid system in detecting protein–protein interactions. By introducing libraries of RNA or protein, cognate partners can be identified. As in two-hybrid screens, the challenge then becomes to identify those molecules whose interaction is biologically relevant. For this purpose, mutations in the known RNA or protein component are useful. The system also makes it possible to identify regions of an RNA or protein required for a known interaction, and to test combinations of RNA and protein to determine whether they interact *in vivo*.

Perspective: Initial Considerations

The three-hybrid system has been used to identify specific RNA-binding proteins for a number of RNA targets. Many known or suspected combinations of RNA and protein partners have been tested, with a high frequency of success (Tables I[1,11–12] and II[19,20,22,35]). Several successful screens of cDNA libraries have been reported (Table III[21,24–32]) as well as one screen of a hybrid RNA library.[11] The system has enabled the analysis of protein–protein and RNA–protein interaction within a multiprotein–RNA complex,[12] the mutational analysis of RNA nucleotides important for protein

[13] A. B. Sachs, R. W. Davis, and R. D. Kornberg, *Mol. Cell. Biol.* **7,** 3268 (1987).
[14] D. J. Haile, M. W. Hentze, T. A. Rouault, J. B. Harford, and R. D. Klausner, *Mol. Cell. Biol.* **9,** 5055 (1989).
[15] H. A. Barton, R. S. Eisenstein, A. Bomford, and H. N. Munro, *J. Biol. Chem.* **265,** 7000 (1990).
[16] G. W. Witherell, A. Gil, and E. Wimmer, *Biochemistry* **32,** 8268 (1993).
[17] P. D. Zamore, D. P. Bartel, R. Lehmann, and J. R. Williamson, *Biochemistry* **38,** 596 (1999).
[18] L. W. Slice, E. Cadner, D. Antelman, M. Hooly, B. Wegrynski, J. Wang *et al., Biochemistry* **31,** 12062 (1992).
[19] E. Bacharach and S. P. Goff, *J. Virol.* **72,** 6944 (1998).
[20] S. W. Fewell and J. L. Woolford, *Mol. Cell. Biol.* **19,** 826 (1999).
[21] R. M. Long, P. Chartrand, W. Gu, G. Gonsalvez, and R. H. Singer, *J. Cell Biol.* (in press).
[22] E. G. Lee, A. Yeo, B. Kraemer, M. Wickens, and M. L. Linial, *J. Virol.* **73,** 6282 (1999).
[23] T. Boelens, D. Scherly, R. P. Beijer, E. J. Jansen, N. A. Dathan, I. W. Mattaj, and W. J. van Venrooij, *Nucleic Acids Res.* **19,** 455 (1991).
[24] F. Martin, A. Schaller, S. Eglite, D. Schumperli, and B. Muller, *EMBO J.* **16,** 769 (1997).
[25] Z. F. Wang, M. L. Whitfield, T. C. I. Ingledue, Z. Dominski, and W. F. Marzluff, *Genes Dev.* **10,** 3028 (1996).
[26] B. Zhang, M. Gallegos, A. Puoti, E. Durkin, S. Fields, J. Kimble, and M. P. Wickens, *Nature (London)* **390,** 477 (1997).
[27] E. Jan, C. K. Motzny, L. E. Graves, and E. B. Goodwin, *EMBO J.* **18,** 259 (1999).
[28] Z. F. Wang, T. C. Ingldue, R. Sanchez, and W. F. Marzluff, *Mol. Cell. Biol.* **19,** 835 (1999).
[29] Y. W. Park, J. Wilusz, and M. G. Katze, *Proc. Natl. Acad. Sci. U.S.A.* **96,** 6694 (1999).
[30] A. Dahanukar, J. A. Walker, and R. P. Wharton, *Mol. Cell* **4,** 1 (1999).
[31] R. M. Long, P. Chartrand, W. Gu, E. Lorimer, and R. H. Singer, submitted.
[32] H. G. Davies, F. Giorgini, M. A. Fajardo, and R. E. Braun, *Dev. Biol.* **221,** 87 (2000).

TABLE I
TESTS OF PREVIOUSLY KNOWN INTERACTIONS

RNA component		Protein component			
RNA	Length (nt)	Protein	Family	K_d[a]	Refs.[b]
TAR[c]	58	Tat[c]	Arg-rich	10^{-8}	1
Poly(A)[d]	30	PAB[d]	RRM	10^{-8}	*
IRE[e]	51	IRP-1[e]	Fe-S	10^{-10}–10^{-11}	1
IRES[f]	200	PTB[f]	RRM	10^{-7}–10^{-8}	*
U1 SL[g]	45	U1-70K[g]	RRM	—	11
NRE[h]	74	Pumilio[h]	Puf (PUM-HD)	10^{-9}	*, 12

[a] References for K_d values are as follows: TAR–Tat, Ref. 18; poly(A)–PAB, Ref. 13; IRE–IRP-1, Refs. 14 and 15; IRES–PTB, Ref. 16; NRE–Pumilio, Ref. 17.

[b] References for the use of the three-hybrid interaction are provided. An asterisk (*) indicates unpublished data from our laboratories. Analysis of the IRES–PTB interaction was performed in collaboration with the E. Wimmer laboratory.

[c] TAR and Tat from human immunodeficiency virus type 1 (HIV-1).

[d] Oligo$(A)_{30}$ and PAB (yeast Pab1p).

[e] IRE (iron response element from the rat ferritin gene) and IRP-1 (iron regulatory protein 1 from rabbit).

[f] IRES (internal ribosome entry site of EMCV virus) and human PTB (polypyrimidine tract-binding protein).

[g] U1 SL (the first stem–loop of yeast U1 snRNA) and the yeast U1-70K homolog (Snp1p).

[h] NRE (Nanos response element, from *D. melanogaster hunchback* mRNA) with *D. melanogaster* Pumilio.

TABLE II
TESTS OF SUSPECTED INTERACTIONS

RNA component		Protein component		
RNA	Length (nt)	Protein	Family	Refs.
RSV Mψ[a]	161	RSV Gag[a]	Retroviral CCHC[b]	22
HIV-1 ψ[c]	139	HIV-1 Gag[c]	Retroviral CCHC[b]	19
mTR[d]	393	hTP1[d]	WD40	35
18S rRNA SL[e]	72	S14[e]	—	20
*Rps14*B SL[f]	58	S14[f]	—	20

[a] RSV (Rous sarcoma virus) Mψ, the packaging site for this retrovirus, and RSV Gag.

[b] CCHC motifs characteristic of the NC region of retroviral capsid proteins.

[c] HIV-1 (human immunodeficiency virus 1) ψ and HIV-1 Gag.

[d] mTR (mouse telomerase RNA) and hTP1 (a human protein to which it binds).

[e] A portion of yeast 18S rRNA (nucleotides 876 to 948) and yeast ribosomal protein S14.

[f] A portion of yeast *rps14B* mRNA and yeast ribosomal protein S14.

TABLE III
THREE-HYBRID SCREENS OF cDNA LIBRARIES USING A KNOWN RNA SEQUENCE

RNA used as bait[a]		Protein recovered			
RNA	Length (nt)	Protein	Family	Specificity test	Refs.
fem-3 PME[b]	74	FBF-1, FBF-2	Puf (PUM-HD)	Point mutant	26
tra-2 TGE[c]	60	GLD-1	KH/STAR	Unrelated RNA	27
Histone SL[d]	35	SLBP1	—	Antisense	24
Histone SL[d]	35	SLBP1, SLBP2	—	Antisense	25, 28
Influenza NP[e]	50	GRSF-1	RRM	Deletion	29
nanos TCE A[f]	29	Smaug	—	Point mutants	30
ashl 3′ UTR[g]	126	Loc1p	—	Unrelated RNA	21
ashl 3′ UTR[g]	126	She3p	—	Unrelated RNA	31
prm-1 3′ UTR[h]	37	MSY4	Cold shock domain	Point mutant	32

[a] All RNAs were present as fusions with MS2-binding sites.
[b] *fem-3* PME, a portion of the 3′ UTR of *fem-3* from *C. elegans.* Hybrid RNA bait consisted of a tandem duplication of the 37-nucleotide site.
[c] *tra-2* TGE, a portion of the 3′ UTR of *tra-2* mRNA from *C. elegans.*
[d] Histone SL, a stem–loop structure at the 3′ end of histone mRNA. The same structure was used in screens by two different laboratories[24,25,28] and has been used to identify specific stem–loop binding proteins (SLBPs 1 and 2) in human,[24,25] mouse,[25] and *Xenopus*[25,28] libraries.
[e] Influenza NP, a sequence in the 5′ UTR of influenza virus.
[f] *nanos* TCE A, a regulatory element in the 3′ UTR of *nanos* mRNA from *D. melanogaster.*
[g] *ashl* 3′ UTR, from *Saccharomyces cerevisiae.*
[h] *prm-1* 3′ UTR, the first 27 nucleotides of the 3′ UTR of protamine mRNA.

binding (B. Zhang and M. B., unpublished, 1999; and, Refs: 22 and 30 and 32a below), and identification of novel RNA-binding domains in the protein component of an interaction.[25,26]

Several general properties of the system merit a brief discussion at the outset; in some cases, these points are considered in greater detail later in the text. Not surprisingly, both the sequence of the RNA and the affinity of the interaction impact the output of the reporter gene. The affinities (K_d) of RNA–protein interactions detected by the three-hybrid system range from subnanomolar to 0.1 μM (Table I). Although levels of expression of the reporter gene generally correlate with measured *in vitro* affinities, the precise relationship between expression and K_d has not yet been examined in detail. Reporter expression levels presumably vary with the abundance of the two partners and the influence of endogenous yeast proteins

[32a] F. Martin, F. Michel, D. Zenklusen, B. Muller, and D. Schumperli, *Nucleic Acids Res.* **28,** 1594 (2000).

and RNAs, as well as with the reporter gene used. (See Ref. 33 for a discussion of related points in the two-hybrid system.)

The RNase P RNA promoter has at least three desirable properties for the system: It is efficient, producing many copies of the hybrid RNA (up to 1300 copies per cell[33]; B. Kraemer and M. Wickens, unpublished, 1999); a large portion of the natural RNase P RNA molecule can be replaced with foreign sequence (Ref. 34; see Fig. 3); and the use of this promoter and the RNase P RNA leader and trailer may cause the RNA to be nuclear. The high level of RNA likely enhances the signal, because multicopy RNA plasmids yield higher reporter outputs than do centromeric plasmids (B. Zhang, B. Kraemer, and M. Wickens, unpublished, 1999). On the other hand, the use of this RNA polymerase III promoter limits the RNAs that can be analyzed, both in length and specific sequence (see Hybrid RNAs: Limitations, below). RNAs ideally should be 150 or fewer nucleotides long, and should lack oligouridylate stretches of four or more nucleotides. However, these limitations are not universal, and other promoters present different limitations. Interactions have been detected using both structured and ostensibly unstructured RNAs.

In screening cDNA or RNA libraries for interacting partners, secondary screens of the initial "positives" are needed to winnow down candidates for further analysis (see Finding Protein Partners of a Known RNA Sequence: Step 5, below). One crucial secondary screen identifies positives requiring the "bait" to activate the reporter gene. Another screen identifies positives exhibiting the correct sequence specificity. Subtle mutations that perturb the biological function of the bait are ideal. Subtly mutant baits will bind most biologically irrelevant, "nonspecific" interactors, which can then be discarded. If subtle mutations are not available, cruder specificity tests can be useful (Table III). Indeed, in the cDNA library screens carried out to date, virtually all positives that required the RNA to activate and possessed the correct sequence specificity proved to be the genuine interactor of interest (Table IV).

Plasmid Vectors

Several plasmids have been constructed to express hybrid RNA sequences, and are depicted in Fig. 2A and B. Each of these is a multicopy plasmid containing origins and selectable markers for propagation in yeast and bacteria. They are described below.

[33] J. Estojak, R. Brent, and E. Golemis, *Mol. Cell Biol.* **15,** 5820 (1995).
[34] P. D. Good and D. R. Engelke, *Gene* **151,** 209 (1994).

TABLE IV
cDNA SCREENS USING A KNOWN RNA SEQUENCE

	RNA used								
	Histone SL[a]	Histone SL[b]	*fem-3*[c]	*tra-2*[d]	Influenza NT[e]	*nanos* TCE[f]	*ash1* 3′ UTR[g]	*ash1* 3′ UTR[h]	*prm-1* 3′ UTR[i]
	Protein recovered								
	SLBP1	SLBP1	FBF-1, FBF-2	GLD-1	GRSF-1	Smaug	LOC1	SHE3	MSY4
Transformants	3×10^5	2×10^6	5×10^6	6×10^5	5×10^5	7.5×10^5	2×10^6	5×10^5	7.5×10^6
HIS3$^+$ ([3-AT])	4 (25 m*M*)	18 (5 m*M*)	5000 (5 m*M*)	87 (5 m*M*)	120 (3 m*M*)	77 (3 m*M*)	1500 (5 m*M*)	147 (1 m*M*)	113
White in *ADE2* screen	ND	ND	100	ND	ND	ND	66	49	ND
LacZ$^+$	4	18	60	ND	120	35	66	23	10
RNA dependent	1	7	12	20	10	9	11	13	5
Specificity test	1	7	3	1	2	1	4	ND	1
Genuine regulator	1	7	3	1	2	1	2	1	1

[a] See Ref. 24.
[b] See Ref. 25.
[c] See Ref. 26.
[d] See Ref. 27.
[e] See Ref. 29.
[f] See Ref. 30.
[g] See Ref. 21.
[h] See Ref. 31.
[i] See Ref. 32.

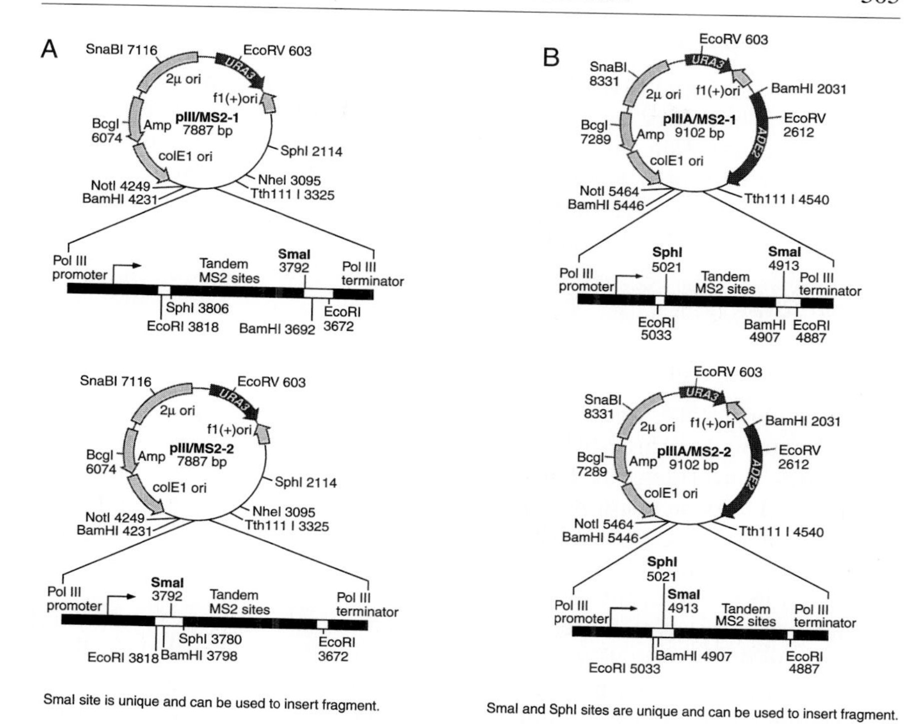

FIG. 2. Plasmid vectors used to express hybrid RNAs in the three-hybrid system. The restriction sites that are indicated either are unique or can be used to verify the presence of vaious elements on the plasmids. Sites suitable for insertion of sequences of interest are in boldface.

pIII/MS2-1 and pIII/MS2-2

pIII/MS2-1 and pIII/MS2-2 (Fig. 2A) are yeast shuttle vectors derived from pIIIEx426RPR, which was developed by Good and Engelke.[34] Sequences to be analyzed are inserted at the *Sma*I–*Xma*I site. pIII/MS2-1 and pIII/MS2-2 differ only in the relative position of the *Sma*I–*Xma*I site and the MS2-binding sites. Both plasmids carry a *URA3* marker and produce hybrid RNAs from the yeast RNase P RNA (*RPR1*) promoter, an RNA polymerase III promoter. The *RPR1* promoter was chosen for two reasons. First, it is efficient, directing the synthesis of more than 1000 molecules per cell. Second, the transcripts produced from this promoter presumably do not enter the pre-mRNA processing pathway, and may not leave the nucleus. An alternative method, in which an RNA polymerase II promoter is used to generate the hybrid RNA, has also been described.[9]

pIIIA/MS2-1 and pIIIA/MS2-2

pIIIA/MS2-1 and pIIIA/MS2-2 (Fig. 2B) are similar to pIII/MS2-1 and pIII/MS2-2, but carry the yeast *ADE2* gene (in addition to *URA3*). The *ADE2* gene is exploited in screening for RNA-binding proteins (see Finding Protein Partners of a Known RNA Sequence: Step 2, below). The two plasmids differ from one another only in the relative position of the *Sma*I–*Xma*I site and the MS2-binding sites.

Hybrid RNAs

General Considerations

The hybrid RNA, and the tripartite nature of the system, present unique considerations for three-hybrid versus two-hybrid screens. Among these are issues concerning the length of the RNA that can be analyzed, the ability of cDNA–activation domain proteins to activate transcription independent of binding to the RNA, and of RNAs to activate transcription independent of the cDNA–activation domain protein. These issues are discussed in the relevant sections.

RNA sequences to be tested are inserted into the unique *Sma*I–*Xma*I site of any of the four RNA vectors depicted in Fig. 2. The hybrid RNA molecule that is transcribed *in vivo* from one of these plasmids consists of the sequence of interest, X, linked to two MS2 coat protein-binding sites, and RNase P RNA 5′ leader and 3′ trailer sequences (Fig. 3). The MS2-binding sites are present in two copies because binding to coat protein is cooperative. Each site contains a single nucleotide change that enhances binding.[35]

In our experience, most RNA inserts of suitable size produce RNAs of comparable and high abundance. RNA abundance may be important in determining the level of signal produced from the reporter gene: transferring the hybrid RNA gene from the normal high-copy vectors to low-copy vectors substantially reduces levels of *lacZ* expression.

The relative order of the RNA sequence of interest and the MS2 sites can affect signal strength. In the few cases that have been tested systematically, both orientations yield activation and are specific. However, in the IRE–IRP interaction, placing the IRE upstream of the MS2 sites results in two- to threefold more transcription than does the alternative arrangement. Although RNA folding programs can be used to determine whether one arrangement is more likely to succeed, the accuracy of their predictions *in vivo* is problematic. In most cases tested we have placed the RNA sequence of interest upstream of the MS2 sites.

[35] P. T. Lowary and O. C. Uhlenbeck, *Nucleic Acids Res.* **15,** 10483 (1987).

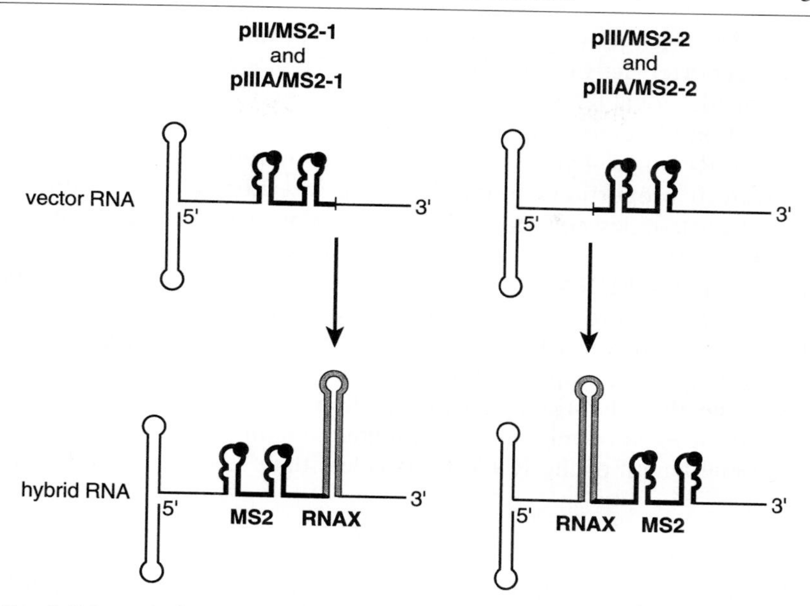

FIG. 3. Schematic diagram of the secondary structure of the hybrid RNA molecules. Thin lines, RNase P RNA 5′ leader and 3′ trailer sequences; thick lines, tandem MS2 recognition sites, including a point mutataion that increases affinity (black dot); gray lines, inserted RNA sequence, depicted as a stem–loop.

Limitations

RNA sequences to be analyzed are restricted in two respects at present. First, runs of four or more U's in succession can terminate transcription by RNA polymerase III. Second, typically, RNA inserts of lengths less than 150–200 nucleotides yield higher signals: longer inserts commonly reduce the level of activation of the reporter. In principle, both of these limitations might be overcome by using a different polymerase, such as a bacteriophage RNA polymerase.

Runs of U's. Four or more U's in succession can function as RNA polymerase III terminators, and so can prevent production of the desired hybrid RNA. The efficiency of termination at oligouridine tracts is context dependent; so it may be worth testing whether a suspect RNA sequence will function in the system. Northern blotting should always be performed to make certain that the hybrid RNAs are expressed at high levels. For long runs of U's, it may be necessary to eliminate the terminator by mutagenesis.

Size of the RNA Element. The size of the RNA insert appears to be an important determinant of three-hybrid activity. In reconstruction experiments using known RNA–protein partners, RNA sequences that are 30–150 nucleotides in length (e.g., TAR and IRE) typically yield substantial and specific reporter activation (Tables I and II). We have investigated the effects of additional RNA sequences flanking a known protein-binding site. The addition of heterologous sequences to the IRE caused a reduction in the IRE–IRP1 three-hybrid signal. The effect on reporter gene expression was inversely proportional to the size of the insertion, with the addition of 150–200 nucleotides commonly leading to almost complete abolition of the IRE–IRP1 three-hybrid signal. Similarly, we have tested the effect of additional natural RNA sequence on the ability to detect the interaction of the yeast Snp1 protein, a homolog of the mammalian U1-70K protein, with the loop I of yeast U1 RNA. Insertion of even the RNA sequences that normally flank loop I, but are not part of the Snp1-binding site, decreased the three-hybrid signal due to Snp1–U1 interaction. Taken together, these experiments suggest a minimal binding site is preferable, and the optimal length of the RNA insert is less than 150–200 nucleotides.

Yeast Strains

The yeast reporter strain L40coat is derived from L40-ura3 (a gift of T. Triolo and R. Sternglanz, Stony Brook, NY). The genotype of the strain is *MAT***a** *ura3-52 leu2-3,112 his3-200 trp1-1 ade2 LYS2*::(*lexA op*)-*HIS3 ura 3*::(*lexA-op*)-*lacZ LexA-MS2 coat* (*TRP1*). The strain is auxotrophic for uracil, leucine, adenine, and histidine. Each of these markers is exploited in the three-hybrid system. Both the *HIS3* and *lacZ* genes have been placed under the control of *lexA* operators, and hence are reporters in the three-hybrid system. A gene encoding the LexA–MS2 coat protein fusion has been integrated into the chromosome.

Strain R40coat is identical to L40coat, but of opposite mating type.

Testing Known or Suspected Interactors

The protocol to test a known or suspected RNA–protein interaction is straightforward. The RNA sequence of interest and the gene encoding the RNA-binding protein are cloned into an RNA plasmid and an activation domain vector, respectively. These are then introduced into yeast by transformation, and the level of expression of a reporter gene is determined. Table I indicates some known interactions detected by the three-hybrid system, and Fig. 4 depicts an analysis of the IRE–IRP1 interaction,[1] using the *lacZ* gene as a reporter. The results in Fig. 4 demonstrate that each segment of the system is required for activation of the reporter.

This approach also can be useful in testing candidate interactors obtained by other means (Fig. 1B). For example, several RNA–protein interactions that were suspected to occur on the basis of other criteria have

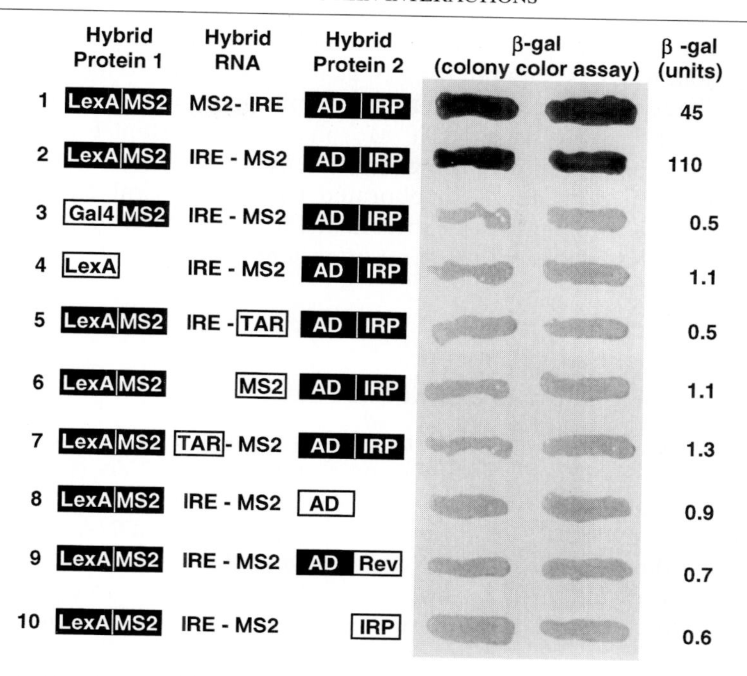

FIG. 4. The IRE–IRP1 interaction monitored by activation of the *lacZ* gene. Plasmids encoding the indicated hybrid RNAs and activation domain fusions were transformed into strain L40 coat. After selecting transformants for the presence of the plasmids, colonies were assayed for *lacZ* reporter activity. Only the two combinations that should lead to IRE–IRP1 interactions (lines 1 and 2) yield significant β-galactosidase activity.

been confirmed using the three-hybrid system, including some that were refractory to biochemical analysis. Table II provides a partial list of suspected interactions that have been detected by this system, including mammalian telomerase RNA (mTR) with newly cloned telomerase protein 1 (TP1)[36]; the Rous sarcoma virus (RSV) and human immunodeficiency virus type 1 (HIV-1) encapsidation signal RNAs with RSV and HIV-1 Gag[19,22]; and a helix in rRNA and a stem–loop in S14 pre-mRNA with ribosomal protein S14.[20] Similarly, binding of yeast ribosomal protein S9 with *CRY2* pre-mRNA, deduced by genetic experiments, has been demonstrated using the system (S. Fewell and J. Woolford, personal communication, 1999).

The protocol also provides a facile way to delineate the portions of the RNA or protein required for an interaction. In principle, it should be

[36] L. Harrington, T. McPhail, V. Mar, W. Zhou, R. Oulton, M. B. Bass, I. Arruda, and M. O. Robinson, *Science* **275,** 973 (1997).

possible to identify single amino acids and nucleotides important for the interaction. Regions required for RNA binding of FBF1 to its RNA target were inferred through such an analysis.[26] Similarly, regions of the packaging site of Rous sarcoma virus, RSV Mψ, that are important for RSV Gag binding have been identified using the three hybrid system and their importance for viral replication corroborated by *in vivo* viral encapsidation assays.[22] In such mutagenesis experiments, it is important to confirm that protein or RNA is present when reporter gene activation no longer is detected.

Finding Protein Partners of a Known RNA Sequence

The three-hybrid system can be used to identify a protein partner of a known RNA sequence. A library containing many cDNAs is introduced into a yeast strain carrying a plasmid that encodes an RNA sequence of interest as a hybrid RNA. From such screens emerge cDNAs capable of activating the reporter, some of which require the hybrid RNA to do so. Several successful three-hybrid screens have been reported, and have yielded proteins from diverse families of RNA-binding proteins (Tables III and IV).

If the RNA–protein interaction is particularly strong, as in the case of SLBP1 and its stem–loop target in histone mRNA, then the initial selection can demand high levels of expression of the reporter gene (Tables III and IV).[1,24,25] A stringent selection eliminates weak (and potentially nonspecific) activators. However, a less stringent selection is often preferable because the "strength" of the interaction being sought is not known. In turn, this reduced stringency leads to a higher background, and increased need for subsequent screens. While high levels of 3-aminotriazole (3-AT) may reduce background, they do so at the expense of the diversity of recovered cDNAs, as only the strongest interactors will be recovered.

Many RNA–protein complexes contain more than one protein molecule. In some cases, one protein can bind only in combination with another (e.g., U2B″).[23] The three-hybrid system can be used to detect and analyze such multiprotein complexes on an RNA. In one such case, a LexA–Pumilio fusion protein was used to tether a portion of the *hunchback* mRNA 3′ UTR to a promoter.[12] Nanos protein, present as an activation domain fusion, bound to the Pumilio–*hunchback* RNA complex, but not to either component individually. Clearly, variations on this theme could be used to screen for proteins that can bind only a specific RNA complex.

A protocol used to clone novel RNA-binding proteins is discussed below.

Step 1: Introduction of the RNA Plasmid and cDNA Library

The host strain, L40coat, is normally transformed with the RNA plasmid first. Cells containing the RNA plasmid are then transformed with a cDNA library fused with a transcription activation domain. (Hybrid RNAs are rarely toxic to the host cell, unlike some hybrid protein baits used in two-hybrid screens; as a result, there is no need to cotransform both plasmids.) The transformation mix is plated out on medium lacking both leucine (selecting for the cDNA plasmid) and histidine (selecting for *HIS3* gene expression). Maintenance of the RNA plasmid (i.e., selection for *ADE2* or *URA3*) is not demanded. This permits cells that can activate *HIS3* without the RNA to lose the RNA plasmid, permitting the colony color screen below (see Step 2, below).

Protein Libraries. Activation domain libraries prepared in *LEU2* vectors can be used with the hybrid RNA plasmids and yeast strains described above. Many such activation domain libraries have been prepared for use with the two-hybrid system. These may be obtained from individual laboratories and several commercial sources.

3-Aminotriazole in the Initial Selection. We typically add 3-aminotriazole (3-AT), a competitive inhibitor of the *HIS3* gene product, to select for stronger interactions. Some RNAs activate reporter genes weakly on their own, and many proteins appear to activate the reporter slightly, independent of the hybrid RNA; both situations yield "false positives." To eliminate weak activation by the RNA "bait," 3-AT should be titrated with a strain carrying only the RNA plasmid prior to undertaking the initial transformation. To diminish the number of false protein positives ("RNA-independent" positives; see below), concentrations of 3-AT in the range of 3 to 5 m*M* are a good starting point; they offer a reasonable balance between suppressing the background and permitting "real" positives to grow (see Tables III and IV). *In vitro* data on the affinity of the RNA–protein interaction may be valuable for determining the concentration of the 3-AT that should be used; stronger interactions allow more 3-AT to be used. In one of the screens yielding SLBP1, whose interaction with its RNA target is particularly robust *in vitro,* 25 m*M* 3-AT was included, reducing the background greatly.[24] In another screen for SLPB, however, "only" 5 m*M* 3-AT was used, and two related proteins were recovered (SLBP1 and SLBP2).[25,28]

Step 2: Elimination of RNA-Independent False Positives by Colony Color

Two classes of positives are obtained from the initial transformation. One class of transformants requires the hybrid RNA to activate *HIS3;* these are termed "RNA dependent." A second class of positives activates *HIS3* with or without the hybrid RNA; these are termed "RNA independent." The RNA-independent positives can carry proteins that may bind to the promoter regions of the reporter genes, or proteins that interact directly with the LexA–MS2 coat protein fusion. The RNA-independent class of transformants can be abundant, and can account for more than 95% of the total number of colonies.

To facilitate eliminating RNA-independent false positives, we exploit the *ADE2* gene on the RNA plasmid. The host strain is an *ade2* mutant. When the level of adenine in the medium becomes low, cells attempt to synthesize adenine *de novo* and accumulate a red purine metabolite due to lack of the *ADE2*-encoded enzyme. This accumulation renders the cell pink or red in color. In contrast, cells carrying the wild-type *ADE2* gene are white.

In the initial transformation, selection is imposed only for activation of *HIS3,* not for maintenance of the RNA plasmid. For RNA-dependent positives, selection for *HIS3* indirectly selects for the RNA plasmid, which carries the *ADE2* gene; thus these transformants are white and must remain so if they continue to grow in the absence of exogenous histidine. On the other hand, RNA-independent positives do not require the RNA plasmid to activate the *HIS3* gene and so can lose that plasmid, which they do with a frequency of a few percent per generation. These false positives therefore yield pink colonies or white colonies with pink sectors.

The initial transformation plates are usually incubated at 30° for 1 week. This duration allows positives to accumulate *HIS3* gene product and grow, and also provides enough time for the color to develop. If the pink color is not strong after 1 week, incubation at 4° overnight can help. Pink or pink-sectored colonies are discarded, and the uniformly white colonies are picked for further analysis. We usually pick all the white colonies (typically a few large, and many small, colonies) and patch onto medium, selecting again for both the cDNA plasmid and the RNA plasmid. Most of the small white colonies turn out to be RNA independent and fail to grow. White colonies that are able to grow on the selective medium are subject to further analysis.

The identification of RNA-dependent positives by colony color is not perfect; many, but not all, RNA-independent positives can be identified and discarded. A majority of the white colonies may still be RNA independent. It is important to rigorously eliminate the remaining RNA-independent activators from among the white colonies (Steps 4 and 5, below) before recovering plasmids in *E. coli.*

An additional and complementary means for eliminating false positives has been described.[37] In this protocol, $HIS3^{+}$ colonies from the initial plating

[37] Y. W. Park, S. L. Tan, and M. G. Katze, *BioTechniques* **26,** 1102 (1999).

are patched onto medium containing the drug 5-fluoroorotic acid (5-FOA). 5-FOA is toxic to yeast expressing the *URA3* gene. Cells requiring the hybrid RNA to activate the *HIS3* reporter will carry the hybrid RNA-encoding plasmid at high copy number and so will be 5-FOA sensitive. Cells not requiring the hybrid RNA to activate the *HIS3* reporter will carry the hybrid RNA plasmid at low copy number (or not at all) and so will be relatively 5-FOA insensitive.

Step 3: Assay of β-Galactosidase Activity

To corroborate that the *HIS3*$^+$ colonies contain cDNAs that activate through the three-hybrid system, the level of expression of the *lacZ* gene is monitored. In strain L40coat, the *lacZ* gene is integrated into the chromosome and placed under the control of LexA-binding sites. In our experience, most *HIS3*$^+$ transformants are also *lacZ*$^+$; thus this step sometimes is delayed until after the specificity test (Step 5, below).

β-Galactosidase can be assayed by measuring the conversion of a lactose analog to a chromogenic or luminescent product. This assay can be performed using either colonies permeabilized on a filter or a cell lysate. The filter assay yields qualitative results, while the liquid assay is more quantitative.

Qualitative (Filter) Assays. Perform the following steps.[38]

1. Restreak colonies from Step 2 onto the appropriate selective medium (SD –Leu –Ura) and grow overnight.
2. Replica colonies onto plate-sized nitrocellulose filters or filter papers (3MM; Whatman, Clifton, NJ).
3. Immerse each filter in liquid nitrogen for 20 sec.
4. Allow the filters to thaw on the benchtop (approximately 2 min).
5. Prepare petri dish-sized circles of 3MM Whatman paper, place in petri dishes, and saturate with Z buffer (60 m*M* Na_2HPO_4, 40 m*M* NaH_2PO_4, 10 m*M* KCl, 1 m*M* $MgSO_4$, 50 m*M*-mercaptoethanol, pH 7.0), supplemented with 5-bromo-4-chloro-3-indolyl-β-D-galactopyranoside (X-Gal, 300 μg/ml). The X-Gal should be added fresh. Remove excess buffer.
6. Overlay each filter onto Whatman paper and seal the dishes with Parafilm.
7. Incubate for 30 min to overnight at 30°. Examine the filters regularly.

A strong interaction (such as that between IRE and IRP) should turn blue within 30 min. With protracted incubation, weak interactions eventually yield a blue color. For this reason, it is important to examine the filters periodically to determine how long it takes for the color to develop.

[38] L. Breeden and K. Nasmyth, *Cold Spring Harbor Symp. Quant. Biol.* **50,** 643 (1985).

Quantitative (Liquid) Assays. The specific activity of β-galactosidase can be determined in yeast cell lysates with any of a variety of substrates. Colorimetric assays using O-nitrophenyl-D-galactoside (ONPG)[39] or chlorophenol red-beta-D-galactopyranoside (CPRG)[40] are common; CPRG is more sensitive, but also more expensive. Alternatively, luminescent substrates provide high sensitivity yet are relatively inexpensive. Below we provide a protocol using a luminescent substrate, Galacton-Plus (Tropix, Bedford, MA). The assay requires an instrument to detect luminescence. The protocol below was designed for a Monolight 2010 luminometer (Analytic Luminescent Laboratories, San Diego, CA). Certain details of the assay, such as sample volumes, will vary with the instrument used.

1. Inoculate 5-ml cultures of selective media in triplicate for each interaction to be tested. Grow overnight.
2. Inoculate fresh selective medium to an optical density (OD) of 0.1.
3. Grow to midlog phase (OD ~0.8)
4. Pellet ~1.0 OD unit of cells for each culture.
5. Resuspend the pellet in 100 μl of lysis buffer [100 m*M* potassium phosphate (pH 7.8), 0.2% (v/v) Triton X-100].
6. Lyse by freeze–thawing. This lysis requires three sequential cycles of freezing in liquid nitrogen for 10 sec, followed by incubation at 37° for 90 sec.
7. Vortex each tube briefly.
8. Pellet in a microcentrifuge at 12,000*g*, 2 minutes.
9. Collect the supernatant for luminometer assays. If necessary, samples may be frozen at −70°.
10. Add 10 μl of lysate to each luminometer tube. For strong interactions, it may be necessary to dilute the lysate to keep the assay in the linear range.
11. Dilute Galacton reagent 1 : 100 in reaction buffer (100 m*M* sodium phosphate, 1 m*M* magnesium chloride, pH 8.0). Add 100 μl to each luminometer tube.
12. Incubate at 25° for 60 min.
13. Measure luminescence as directed by the luminometer manufacturer (Analytic Luminescent Laboratories). Use Light Emission Accelerator II from Tropix.
14. Measure protein concentration in lysates by Bradford or equivalent assay.

[39] J. H. Miller, "Experiments in Molecular Genetics." Cold Spring Harbor Laboratory Press, Cold Spring Harbor, New York, 1972.

[40] K. Iwabuchi, B. Li, P. Bartel, and S. Fields, *Oncogene* **8,** 1693 (1993).

15. Normalize light emission by protein concentration, yielding a specific activity.

Step 4: Curing the RNA Plasmid and Testing Again

Most but not all of the RNA-independent false positives are eliminated by either the colony color or 5-FOA sensitivity assays in Step 2. To ensure that the positives are genuinely RNA dependent, the RNA plasmid is removed by counterselection against *URA3*. Expression of the reporter gene is then monitored. Candidates that fail to activate the reporter genes are analyzed further.

URA3 Counterselection. To select for cells that have lost the RNA plasmid, cells are plated on medium containing 5-fluoroorotic acid (5-FOA). 5-FOA is converted by the *URA3* gene product to 5-fluorouracil, which is toxic. Cells lacking the *URA3* gene product can grow in the presence of 5-FOA if uracil is provided, while cells containing the *URA3* gene product cannot.

1. Replica plate the positives from Step 2 to SD –Leu plates. Let the cells grow for 1 day, allowing the cells to lose the plasmid.
2. Replica plate onto SD –Leu + 0.1% (w/v) 5-FOA plates. Incubate at 30° for a few days.
3. Cells that grow can be streaked on an SD –Ura plate to confirm the loss of the RNA plasmid. A single pass through 5-FOA counterselection is usually sufficient.
4. Assay β-galactosidase activity.

The *ADE2* marker on the RNA plasmid can be useful in monitoring the loss of the RNA plasmid. Cells that lose the plasmid will turn pink in a few days. If the number of positives is small, then the use of 5-FOA, which is quite expensive, can be avoided. Cells can simply be grown in rich medium overnight, then spread onto SD –Leu plates. After a few days, some of the colonies become pink or show pink sectors. Uniformly pink colonies, which have lost the RNA plasmid, can be reassayed for β-galactosidase activity.

Step 5: Determining Binding Specificity with Mutant and Control RNAs

To test RNA-binding specificity, reintroduce plasmids encoding various hybrid RNAs into the strains that have been cured of their original RNA plasmid. If the number of positives is small, the various RNA plasmids can be introduced by transformation. Otherwise, mating is used to introduce the plasmids. Strain R40coat can be used for this purpose. A sample protocol for the mating assay follows.

1. Grow lawns of separate R40coat transformants carrying a specific hybrid RNA plasmid (e.g., (mutant versus wild type) on SD –Ura plates.
2. Replica plate the grid of Ura^- colonies from the 5-FOA plate to a YPD plate.
3. Replica plate the lawn from each R40coat strain to the same YPD plate.
4. Incubate the plates overnight at 30° to allow mating.
5. Replica plate to an SD –Leu –Ura plate to select for diploids.
6. Assay β-galactosidase activity.

Determining What RNAs Should Be Used in Specificity Tests. Ab initio, positives that survive Step 4 carry proteins that bind preferentially to the hybrid RNA relative to cellular RNAs. Although these positives recognize some features on the hybrid RNA, these need not be the ones recognized by the biologically relevant factor(s). Therefore the ideal controls are subtle (e.g., point) mutations that affect the biological functions or interaction in a sequence-specific manner. When point mutations are not available, RNAs carrying the antisense sequence, small deletions, or unrelated elements of similar predicted structure can serve as cruder specificity controls (Table III).

Determining What Fraction of All RNA-Dependent Positives Are "Correct." The fraction of physiologically relevant positives is unpredictable at the outset, as it is a function of the abundance of the protein (in a random library screen), strength of the interaction, and level of 3-AT used in the initial transformation, among other parameters. A synopsis of several published screens is shown in Table IV, which indicates the number of isolates obtained at each step of the screening protocol. For example, we identified the regulatory proteins FBF (*fem-3* binding factor)-1 and FBF-2 by screening a *Caenorhabditis elegans* cDNA–activation domain library, using a portion of the *fem-3* 3′ UTR as bait.[26] In this screen, 5 m*M* 3-AT was used in the initial transformation, selecting for reasonably strong interactions. In total, 5 million transformants were analyzed, yielding approximately 5000 His^+Leu^+ colonies. Of these, 100 were white. Sixty of these activated the *lacZ* gene. Twelve of the 60 proved to be genuinely RNA dependent, and 3 of the 12 displayed the appropriate binding specificity. Each of these was FBF, the genuine regulator. Thus, RNA-dependent positives were 0.2% of the total number of colonies obtained in the initial transformation. Importantly, 25% of all RNA-dependent positives were FBF. Moreover, in this screen and the others reported to date, all positives showing proper specificity were the genuine regulator.

Consequences if No Subtle RNA Mutations Are Available. Additional screens must be devised to identify the "correct" positives. Clearly, functional tests are ideal. The sequence of the cDNAs may be directly informative by comparison with known RNA-binding proteins, or by comparison with the predicted molecular weight of the expected protein (based on, e.g., UV cross-linking). Each case is idiosyncratic, and so no general discussion is offered here. However, we caution that such secondary screens are critical.

Step 6: Identifying the Positive cDNAs

cDNA plasmids that display the predicted RNA-binding specificity are recovered from the yeast cells and introduced into *E. coli* by transformation. The yeast cells can contain multiple cDNA plasmids, only one of which encodes the protein that binds to the RNA. Thus, plasmids should be isolated from multiple *E. coli* transformants, and reintroduced into yeast to ensure the correct plasmid has been obtained. The EZ yeast plasmid preparation kit (Genotech, St. Louis, MO) allows rapid isolation of plasmids from yeast. Alternatively, the following method of Robzyk and Kassir works well in our hands.[41]

1. Culture a yeast colony overnight in liquid SD –Leu mediun.
2. Pellet 1.5 ml of the overnight culture, and resuspend in 100 μl of lysis solution [5% (v/v) Triton X-100, 8% (w/v) sucrose, 100 m*M* NaCl, 50 m*M* Tris-HCl (pH 8.0), 50 m*M* EDTA].
3. Add about 0.2 g of acid-washed glass beads.
4. Vortex at high speed for 5 min. Add 100 μl of lysis solution. Mix.
5. Boil for 3 min. Chill on ice briefly.
6. Spin at high speed in a microcentrifuge for 10 min at 4°.
7. Transfer 100 μl of the supernatant to a clean tube containing 50 μl of 7.5 *M* ammonium acetate. Chill at −20° for 1 hr.
8. Spin at high speed in a microcentrifuge for 10 min at 4°.
9. Add 100 μl of supernatant to 200 μl of ice-cold ethanol to precipitate the DNA.
10. Spin at high speed in a microcentrifuge for 10 min at 4°.
11. Wash the pellet in 70% (v/v) ethanol.
12. Resuspend the pellet in 20 μl of H_2O. Use 10 μl to transform *E. coli.*

For convenience, to determine how many plasmids are present in each yeast transformant, we often perform PCR with lysed yeast colonies. To do so, we use primers that flank the inserts. The following is a protocol for yeast colony PCR.[42]

1. Touch the yeast colony with a sterile disposable pipette tip.
2. Rinse the tip in 10 μl of incubation buffer [1.2 *M* sorbitol, 100 m*M* sodium phosphate (pH 7.4), and Zymolyase (2.5 mg/ml; ICN Biomedicals,

[41] K. Robzyk and Y. Kassir, *Nucleic Acids Res.* **20,** 3790 (1992).
[42] M. Ling, F. Merante, and B. H. Robinson, *Nucleic Acids Res.* **23,** 4924 (1995).

Costa Mesa, CA)] by pipetting up and down several times. Incubate at 37° for 5 min.

3. Remove 1 μl for a 20-μl PCR. If desired, the PCR product can be purified by chromatography or by purification from a gel, and sequenced directly.

Step 7: Functional Tests or Additional Screens

Almost invariably, additional steps will be needed to identify those positives that are biologically meaningful. As stated earlier (see Step 5), those screens are idiosyncratic, depending on the interaction and organisms studied. In a screen with an RNA bait, it is not surprising that one might identify irrelevant RNA-binding proteins, as well as the legitimate interactor.

Finding an RNA Partner for a Known RNA-Binding Protein

The three-hybrid system can be used to identify a natural RNA ligand for a known RNA-binding protein by screening an RNA library with a protein–activation domain fusion as bait. Although only one such screen has been performed to date,[11] the objective is sufficiently general that we include discussion of our experience here.

A strain that carries a plasmid expressing an activation domain fusion with a known RNA-binding protein is transformed with a library of plasmids expressing hybrid RNAs, each composed of the MS2 coat protein-binding sites fused to an RNA element. In our initial experiments, the library consisted of fragments of yeast genomic DNA that are transcribed along with the coat protein-binding sites. We demonstrated that we could identify a fragment of the U1 RNA that binds to the yeast Snp1 protein in this type of search.[11] This hybrid RNA library should be useful in assigning RNA ligands to RNA-binding proteins of *Saccharomyces cerevisiae,* particularly those classified as RNA binding solely on the basis of primary sequence data. Indeed, the yeast RNA library has been used to identify yeast RNA sequences that bind specifically to different RNA-binding proteins (K. Evans, B. Zhang, and M. Wickens, unpublished, 1999).

A three-hybrid search to identify RNAs follows much the same logic as that described above to identify proteins. One particular class of false positives that must be eliminated consists of RNAs that are able to activate expression of the reporter genes on their own, without interacting with the protein fused to the activation domain (i.e., "protein-independent" positives).[11] A step must be included to classify RNAs as protein dependent, and thus worthy of additional analysis, or protein independent, and thus of no further interest in this context. In addition, the protein-dependent class of RNAs must be tested with the activation domain vector and other RNA-binding protein fusions to identify those RNAs specific to the protein of interest.

The strategy of using genomic DNA fragments to construct the hybrid RNA library can work with smaller genomes (as demonstrated for yeast in Ref. 11), but may not be practical with genomes that are much larger. A total of 200 base pairs is the typical upper limit for hybrid RNA vector inserts (see Hybrid RNAs: Limitations, above). Less than 1 million transformants of a genomic library containing 200-base pair inserts adequately represents the yeast genome. The number of transformants required to attain comparable coverage in other organisms is strictly proportional to the size of their genomes, and can become prohibitive. For example, approximately 300 million inserts would be required for comparable coverage of the human genome, using a 200-base pair insert size. In addition, increasing the number of nontarget sequences will elevate the background and so may complicate identifying the genuine target, and termination of polymerase III at oligouridylate tracts will discriminate against certain sequences in the library. For these reasons, it may be useful to construct RNA libraries that are preenriched for candidate interacting sequences, or to circumvent the length and sequence limitations of the RNA polymerase III vector.

Step 1: Introduction of the Activation Domain Plasmid and the Hybrid RNA Library

1. Transform L40coat with the plasmid expressing the RNA-binding protein as an activation domain hybrid, based on the pACTII vector.
2. Transform cells from a single colony carrying this plasmid with the RNA library, selecting on medium lacking tryptophan, leucine, uracil, and histidine, and containing 0.5 m*M* 3-AT. Without any 3-AT, most of the transformants will grow on a plate lacking histidine.

Construction of an RNA Library

The yeast RNA library used in our experiments was constructed as follows. Chromosomal DNA from *Sccharomyces cerevisiae* was partially digested with the following four enzymes, listed with their recognition sequences: *Mse*I (TTAA), *Tsp*509I (AATT), *Alu*I (AGCT), and RsaI (GTAC). The digests were pooled and fragments in the size range of 50 to 150 nucleotides were purified from a preparative agarose gel. The ends of the digested DNA were filled in with the Klenow fragment of DNA polymerase I where required. The plasmid pIII/MS2-2 was digested with

*Sma*I, treated with calf intestine phosphatase and ligated to the blunt-ended genomic DNA fragments. The ligations were used to transform electrocompetent HB101 *E. coli.* DNA fragments cloned at the *Sma*I site of the pIII/MS2-2 are expressed such that the RNA sequence corresponding to the yeast genomic fragment is positioned 5′ to the MS2 coat protein-binding sites within the hybrid RNA. More than 1.5 million *E. coli* transformants were obtained, pooled, and used to prepare plasmid DNA for the RNA library.

Step 2: Screening for Activation of the Second Reporter Gene

To ensure that activation of *HIS3* is not spurious, the level of expression of the *lacZ* gene is monitored.

1. Patch individual colonies that grew in the library transformation onto plates lacking leucine, tryptophan, and uracil.
2. Carry out filter β-galactosidase assays as described above.

Step 3: Elimination Protein-Independent False Positives

The identification of protein-independent false positives requires two successive steps. First, the activation domain plasmid must be removed from the strain. Then the level of *lacZ* expression must again be determined. The colonies that are $LacZ^+$ are cured of the activation domain plasmid. The plasmid is derived from pACTII. The plasmid is cured by growing a transformant overnight in YPD medium and then plating for single colonies on SD–Ura to select for the RNA plasmid. These colonies are then replica plated onto an SD–Leu plate to determine which of the colonies lack the *LEU2* marker on the activation domain plasmid. Assay cells cured of the activation domain plasmid for β-galactosidase activity by filter assays, using the protocols described above. Colonies that have lost activity on loss of the activation domain plasmid contain possible RNA ligands of interest.

Determining How Common Protein-Independent Activators Are. Protein-independent activators include hybrid RNAs that activate transcription when bound to a promoter. In our experiments to date, the frequency of such "activating RNAs" in a genomic library can sometimes be high. For example, 92% of all His^+LacZ^+ positives obtained after selection in 0.5 m*M* 3-AT were protein independent. However, at higher 3-AT concentrations only 10% were protein independent.

Step 4: Determining Binding Specificity

As in cDNA library screening, specificity tests are necessary in secondary screens to narrow down the number of candidates. Of course, the more subtle the mutation used, the better.

Transform cells from Step 3 that contain an RNA plasmid and appear to be protein dependent for three-hybrid activity with control activation domain plasmids such as pACTII and an IRP1–activation domain fusion in pACTII. An RNA ligand that is specific for the RNA-binding protein used in the library screen should not produce a three-hybrid signal with these control plasmids.

Step 5: Sequencing RNAs of Interest

Determine the sequence of the RNA ligand and test the binding by alternative methods such as *in vitro* binding.

Determining What Fraction of Protein-Dependent Activators Is Correct. In our experience with yeast Snp1 protein as bait, we screened 2.5×10^6 transformants, and obtained 13 that were protein dependent. Of these, the strongest by far was the appropriate segment of U1snRNA. Some of the other positives have weak sequence similarity to the relevant region of U1, and are now being analyzed further.

Prospects

In this review we have tried to explain how to use the yeast three-hybrid system both to identify partners in an RNA–protein interaction and to analyze either partner. With minimal modifications, the system may be altered to detect factors that enhance or prevent an RNA–protein interaction, to identify RNA ligands that enhance or prevent a protein–protein interaction, or to identify RNA–protein complexes consisting of a single RNA and multiple proteins and to detect RNA–RNA interactions.

With the proliferation of sequence databases and genomic information, a small amount of sequence information may be enough to help determine whether a given protein is a legitimate partner, or to shed light on its function. With the genome sequences of many organisms now available, it may be possible to create RNA–protein linkage maps connecting RNA and protein partners throughout a genome, or within families of RNA-binding proteins or RNA elements.

Acknowledgments

We are grateful to the Media Laboratory of the Biochemistry Department of the University of Wisconsin for help with the figures. We also appreciate the helpful comments and suggestions of members of both the Wickens and Fields laboratories, and communications of results from several laboratories prior to publication. Work in our laboratories is supported by grants from the NIH and NSF. B.K. was supported by an NIH Biotechnology Training Grant. S.F. is an investigator of the Howard Hughes Medical Institute.

[20] Principles and Applications of a Tat-Based Assay for Analyzing Specific RNA–Protein Interactions in Mammalian Cells

By BRYAN R. CULLEN

Introduction

Specific RNA–protein interactions are critical for the posttranscriptional regulation of cellular gene expression at the levels of mRNA splicing, nuclear export, translation, and stability and are key to the replication cycle of many pathogenic viruses, including human immunodeficiency virus type 1 (HIV-1), picornaviruses, and influenza viruses. Once a *cis*-acting RNA regulatory element has been defined, the next goal is frequently the identification and subsequent characterization of the cellular or viral protein that determines the phenotype exerted by that element. In addition to the conceptually simple but often technically challenging approach of biochemical purification based on RNA affinity, it is also possible to use an *in vivo,* genetic approach to this problem. Thus, bacterial assays have been described in which a specific RNA–protein interaction can be detected on the basis of its ability to inhibit translation or promote transcription antitermination.[1,2] The yeast three-hybrid RNA–protein interaction assay, in which a bivalent RNA bridges the interaction between a DNA-binding chimera and a chimera containing a transcription activation domain,[3] has proved to be a particularly useful way to either detect or characterize specific RNA–protein interactions. However, given that many of the most interesting RNA–protein interactions occur in higher eukaryotes and, more specifically, in human cells it is apparent that a simple and sensitive assay for RNA–protein binding in these same cells would be of considerable utility. Possible advantages of using human cells include the following:

1. An RNA–protein interaction that is dependent on a posttranslational modification that occurs only in higher eukaryotes could be detected only in higher eukaryotic cells.

2. If the RNA–protein interaction requires the formation of a protein heterodimer or heteromultimer, it might still be possible to identify or

[1] K. Harada, S. S. Martin, and A. D. Frankel, *Nature* (*London*) **380,** 175 (1996).

[2] C. Jain and J. G. Belasco, *Cell* **87,** 115 (1996).

[3] D. J. SenGupta, B. Zhang, B. Kraemer, P. Pochart, S. Fields, and M. Wickens, *Proc. Natl. Acad. Sci. U.S.A.* **93,** 8496 (1996).

0076-6879/00 $30.00

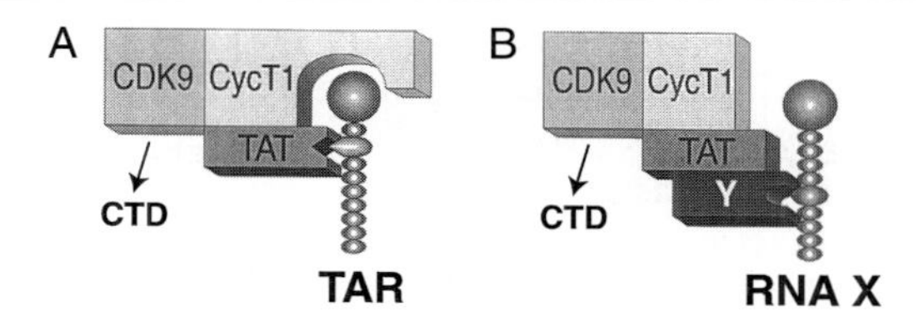

FIG. 1. Mechanism of action of Tat and of Tat–RNA-binding domain Y fusion proteins. See text for detailed discussion.

characterize one of these RNA-binding proteins in a higher eukaryotic cell that expresses the other subunits *in trans.*

3. If the RNA-binding protein of interest has been cloned, it may be possible to study the biological phenotype and RNA-binding properties of mutants in the same cellular context.

Principle of Method

The Tat-based assay for analyzing specific RNA–protein interactions relies on the unusual ability of the HIV-1 Tat protein to potently activate transcription from the viral long terminal repeat (LTR) promoter element when targeted to a promoter-proximal RNA target sequence (reviewed in Ref. 4). The HIV-1 LTR is an unusual promoter in that, in the absence of Tat, it efficiently and selectively initiates transcription by a poorly processive form of RNA polymerase II (PolII).[5] Transcriptional activation of the HIV-1 LTR promoter by Tat primarily reflects the modification of these initiated nonprocessive PolII complexes to processive forms, most probably by inducing the hyperphosphorylation of the PolII carboxy-terminal domain (CTD).[6] Proteins functionally analogous to HIV-1 Tat are also expressed by other primate lentiviruses as well as certain ungulate lentiviruses, such as equine infectious anemia virus (EIAV).[7]

The first step in the activation of the HIV-1 LTR by Tat is the direct interaction of Tat with a cellular cofactor called cyclin T1 (CycT1)[8] (Fig. 1A). CycT1 in turn forms part of a multicomponent cellular factor termed positive transcription elongation factor b (P-TEFb) and is known to bind the CDK9 cyclin-dependent kinase. Once formed, the Tat–CycT1/

[4] B. R. Cullen, *Cell* **93,** 685 (1998).
[5] M. B. Feinberg, D. Baltimore, and A. D. Frankel, *Proc. Natl. Acad. Sci. U.S.A.* **88,** 4045 (1991).
[6] C. H. Herrmann and A. P. Rice, *J. Virol.* **69,** 1612 (1995).
[7] S. J. Madore and B. R. Cullen, *J. Virol.* **67,** 3703 (1993).
[8] P. Wei, M. E. Garber, S.-M. Fang, W. H. Fischer, and K. A. Jones, *Cell* **92,** 451 (1998).

P-TEFb complex is efficiently recruited to a 59-nucleotide RNA stem–loop structure, termed TAR, that is encoded immediately 3′ to the transcription start site in the HIV-1 LTR promoter.[8] The binding of the Tat–CycT1/P-TEFb complex to the nascent TAR element is highly cooperative and involves direct interactions between Tat and the TAR U-rich bulge and between CycT1 and the TAR terminal hexanucleotide loop (Fig. 1A). After TAR recruitment, the CDK9 component of P-TEFb is believed to enhance elongation by phosphorylation of the PolII CTD and possibly also other target proteins.

As noted above, the Tat–CycT1 interaction occurs efficiently in the absence of TAR and is, in fact, a necessary prerequisite for TAR binding. Therefore, recruitment of Tat to an inserted, heterologous RNA target sequence X, substituted in place of TAR, by fusion of Tat to the cognate RNA binding protein Y, should also result in the recruitment of CycT1/P-TEFb to the RNA X target and thereby induce the activation of HIV-1 LTR-dependent transcription (Fig. 1B). Several groups have now demonstrated that this effect can indeed be readily observed with a range of different heterologous RNA targets and RNA-binding proteins (see below).[9–13]

Structure of Reporter Plasmids

Specific RNA binding by a Tat–protein Y chimera can in principle be readily detected by cotransfection of an appropriate indicator construct and chimeric Tat protein expression plasmid into a relevant cell line. The parental indicator construct, termed pTAR/CAT,[12] is schematically represented in Fig. 2A. Moving from left to right, the first functional element is an origin of replication from simian virus 40 (SV40). This origin induces a marked amplification of the signal observed on transfection of pTAR/CAT into an SV40 T antigen-expressing cell line, such as 293T, or in any primate cell line if a T antigen expression plasmid is cotransfected. The next functional element is an essentially intact HIV-1 LTR promoter element, including a complete copy of the HIV-1 TAR element. The TAR element contains unique *Bgl*II and *Sac*I restriction enzyme sites. Located immediately 3′ of TAR is an internal ribosome entry site (IRES), derived from poliovirus type II, that is designed to permit efficient translation of a downstream

[9] M. J. Selby and B. M. Peterlin, *Cell* **62,** 769 (1990).

[10] L. S. Tiley, S. J. Madore, M. H. Malim, and B. R. Cullen, *Genes Dev.* **6,** 2077 (1992).

[11] W. S. Blair, T. B. Parsley, H. P. Bogerd, J. S. Towner, B. L. Semler, and B. R. Cullen, *RNA* **4,** 215 (1998).

[12] Y. Kang and B. R. Cullen, *Genes Dev.* **13,** 1126 (1999).

[13] R. Tan and A. D. Frankel, *Proc. Natl. Acad. Sci. U.S.A.* **95,** 4247 (1998).

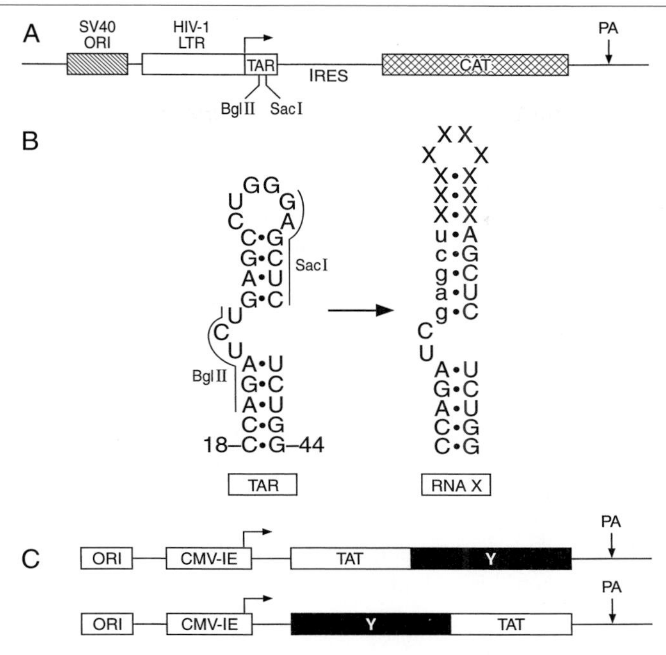

FIG. 2. Schematic representation of the structure of the indicator and effector plasmids used in the Tat-based RNA–protein interaction assay. See text for detailed discussion.

indicator gene even if an inserted RNA structure inhibits the ability of ribosomes to efficiently scan from the mRNA cap site. An indicator gene, such as chloramphenicol acetyltransferase (CAT), is inserted 3′ to the IRES. For use in screens, the green fluorescent protein (GFP) could be substituted in place of CAT (see below). Finally, pTAR/CAT also includes a functional polyadenylation site (PA) as well as sequences required for plasmid maintenance in bacteria.

Substitution of the "RNA X" target sequence in place of TAR can be performed as described in Fig. 2B. Cleavage of pTAR/CAT with *Bgl*II and *Sac*I removes part of the apex of TAR, including the essential terminal loop. The structured RNA X sequence can then be substituted in place of these TAR sequences by insertion of a DNA fragment, bearing the appropriate cohesive ends, that has been generated either by polymerase chain reaction (PCR) or by annealing complementary synthetic oligonucleotides. As shown in Fig. 2B, the integrity of the TAR stem should be restored by reinserting the 5′ nucleotides, indicated in lower-case letters, that constitute the apical part of the TAR stem. The true RNA X sequence, indicated by X in Fig. 2B, can then be formed as an apical extension of the TAR RNA stem.

The final required construct is an expression plasmid that encodes a TAT–protein Y chimera. We have previously described pcTat, which expresses a cDNA form of the HIV-1 Tat protein under the control of the powerful cytomegalovirus immediate early (CMV-IE) promoter element[10–12] and appropriate Tat–Y fusion protein expression plasmids can be derived from pcTat by standard cloning techniques (Fig. 2C). Importantly, the Tat protein is resistant to inactivation by fusion to a heterologous protein at either the amino or carboxy terminus. Nevertheless, both a Tat–Y and Y–Tat fusion protein expression plasmid should be prepared and tested, unless this is precluded by some biological property of the Y protein. In general, it is probably a good idea to insert two or three glycine residues at the fusion junction to generate a glycine hinge that should prevent the inappropriate extension of a protein structure across the fusion boundary. While the HIV-1 Tat protein has most commonly been used to generate Tat chimeras, the distinct EIAV Tat protein, which also directly binds to CycT1/P-TEFb, has been shown to function well in a fusion protein context[7] and may present technical advantages in certain instances.

Experimental Approach

Once a pTAR/CAT derivative bearing RNA X in place of the TAR apex has been generated (pX/CAT), and the relevant Tat fusion protein expression plasmids are available, biological activity can be tested by simple cotransfection into a relevant cell line. In our hands, 293T cells have proved to be the most sensitive system because of their high transfectability and the fact that they express SV40 T antigen. However, other human cell lines, such as HeLa, and murine cell lines, such as L cells, have also been shown to support this assay effectively.[7]

The parental plasmids pTAR/CAT and pcTat provide good specificity controls. Thus, pTAR/CAT is activated by 100- to 200-fold on cotransfection with pcTat into 293T cells and should also be efficiently activated by the Tat–Y and Y–Tat chimera. If this latter activation is not observed, then it is likely that the Tat chimera is defective due to a cloning error or simply due to misfolding. In the small number of cases that we have analyzed, both arrangements of the Tat fusion protein are generally active, although one usually gives a stronger signal than the other. Cotransfection of pcTat with the pX/CAT plasmid provides a useful negative control, as Tat should be unable to interact with the inserted X RNA target. If necessary, the expression and appropriate size of the Tat chimera can be further confirmed with anti-Tat antisera available commercially (e.g., Intracel,

Issaquah, WA) or from the NIH AIDS Research and Reference Reagent Program (Rockville, MD). Variants of Tat bearing a point mutation, such as K41A, that knocks out CycT1 recruitment[7] can also be used as negative controls, if desired. Of course, appropriate mutants of the RNA X target and of the protein Y RNA-binding domain are essential to fully prove the specificity of any observed *in vivo* interaction.

Table I lists the five published RNA–protein interactions that have thus far been analyzed by the Tat-based RNA–protein interaction assay. These interactions involve different types of RNA-binding motifs derived from a small but diverse group of viral and cellular proteins, including one prokaryotic example. Levels of activation of the CAT indicator enzyme, where reported, are in the range of 27- to 70-fold and are therefore readily detectable. The reported ability to detect the interaction of the Tat–Tap chimera with the Mason Pfizer monkey virus (MPMV) CTE[12] and, particularly, the interaction of the Tat–U1A chimera with the U1 snRNA hairpin II[13] is striking in that these interactions were detected in human cells in the face of competition for binding from a high level of endogenously expressed, unfused Tap and U1A proteins. This suggests that this assay may be sensitive enough to detect specific RNA–protein interactions of only moderate efficiency.

In designing plasmids that can be used in the Tat-based RNA–protein interaction assay, it is important to remember that transcription initiating from the HIV-1 LTR in the absence of Tat is, as noted above, only poorly processive.[1,5] In fact, transcription extending more than ~200 nucleotides

TABLE I
RNA–PROTEIN INTERACTIONS DETECTED IN HUMAN CELLS BY THE Tat-BASED RNA–PROTEIN INTERACTION ASSAY

RNA X[a]	Protein Y[b]	Reported activation[c]	Ref.
R17 operator	R17 coat protein	63×	9
Rev response element stem–loop IIB	HIV-1 Rev	70×	10
Poliovirus 5′ untranslated region stem–loop I	3C protease	60×	11
MPMV constitutive transport element (CTE)	Tap	27×	12
U1 hairpin II	U1A	+	13

[a] RNA substituted in place of TAR in pTAR/CAT or equivalent indicator plasmid.
[b] Protein expressed as a Tat fusion.
[c] Reported activation lists the CAT activity detected in transfected cells in the presence of the indicated Tat fusion protein divided by the CAT activity measured in its absence, as reported in the indicated reference. The U1 HPII interaction with the Tat:U1A chimera was reported as detectable in Ref. 13 but was not quantitated.

3′ to the LTR cap site is quite low in the absence of Tat and there is significant polarity even upstream of 200 nucleotides.[14] Therefore, the size of the RNA target to be tested in this assay should be as small as possible and certainly less than 200 nucleotides.

A second issue arises as to the nature of the RNA target being analyzed. Specifically, all five of the RNA targets analyzed in this assay thus far have been highly structured (Table I). Further, the described method of cloning into the pTAR/CAT indicator plasmid (Fig. 2B) is designed to place the inserted RNA target in place of the apical region of the TAR RNA stem–loop, a location that is comparable to that occupied by the normal Tat–CycT1-binding site (Fig. 1) and that may facilitate appropriate presentation of recruited protein complexes. Therefore, there is presently no experimental evidence relevant to the question of whether it might be possible to redesign this assay to detect specific interactions between proteins and unstructured RNA targets.

Potential Use of the Tat-Based RNA–Protein Interaction Assay in Screens

While published experiments that used the Tat-based assay for RNA–protein interactions have generally been focused on the *in vivo* characterization of the binding event, it is nevertheless in principle possible also to use this assay as a screening tool. The difficulty in screening for specific cDNAs by assaying for a phenotype in mammalian cells is generally twofold. First, it is difficult to efficiently introduce single plasmids into individual mammalian cells. Indeed, the fact that most transfection procedures introduce large numbers of plasmids into each mammalian cell is what makes cotransfection such a simple and reliable technique. Second, it is difficult to analyze enough mammalian cells at one time to screen a sufficiently large cDNA expression library to have a good statistical chance of obtaining the relevant clone.

In an effort to overcome these problems, Tan and Frankel generated stable HeLa cell lines containing an indicator construct structurally equivalent to pTAR/CAT but with the GFP indicator gene substituted in place of CAT. They then used the technique of bacterial protoplast fusion[15,16] to transfect these cells. Importantly, protoplast fusion, while relatively efficient, will nevertheless generally result in the fusion of only one protoplast, containing multiple copies of a single plasmid, with each mammalian

[14] M. J. Selby, E. S. Bain, P. A. Luciw, and B. M. Peterlin, *Genes Dev.* **3,** 547 (1989).

[15] B. Seed and A. Aruffo, *Proc. Natl. Acad. Sci. U.S.A.* **84,** 3365 (1987).

[16] R. M. Sandri-Goldin, A. L. Goldin, M. Levine, and J. C. Glorioso, *Mol. Cell. Biol.* **1,** 743 (1981).

cell.[13,15,6] To confirm the utility of this approach, Tan and Frankel[13] diluted bacteria containing a Tat expression plasmid with a 10^5-fold excess of bacteria containing an irrelevant plasmid. They then showed that a reiterative process consisting of (1) protoplast fusion with the target HeLa cells, (2) selection of GFP-positive cells by fluorescence-activated cell sorting (FACS) of $\sim 10^7$ transfected HeLa cells, (3) recovery of plasmid DNA from the sorted HeLa cells by an alkaline lysis procedure, and (4) retransformation of the recovered plasmids into bacteria could, after three rounds, lead to a 10^4-fold enrichment of the Tat expression plasmid.

To test whether the Tat-based RNA–protein interaction assay would indeed permit the selection of an unknown RNA-binding protein, Tan and Frankel[13] next generated an HeLa cell clone bearing an integrated copy of an HIV-1 LTR-based GFP indicator construct containing the RRE SLIIB RNA target substituted in place of TAR.[10] They then generated a library of fusion protein expression plasmids containing Tat linked to a Rev RNA-binding domain in which four amino acids had been randomized. Using this library, these workers were able to specifically select, after three rounds of protoplast fusion, FACS analysis and retransformation, several distinct variants of the Rev RNA-binding domain that displayed enhanced affinity for RRE SLIIB.

Although this result is encouraging, it remains to be seen whether the mammalian Tat-based RNA–protein interaction assay is indeed suitable for the recovery of novel cDNAs that encode RNA-binding proteins specific for an RNA target of interest or whether this approach is simply too unwieldy. What is clear, however, is that this assay can permit the detailed characterization of selected RNA–protein interactions in the cellular context in which they normally occur.

Section III

Phage Display and Its Applications

[21] Phage Display for Selection of Novel Binding Peptides

By SACHDEV S. SIDHU, HENRY B. LOWMAN, BRIAN C. CUNNINGHAM, and JAMES A. WELLS

Introduction

There are few ways, either natural or invented, that new binding entities can be discovered in a short period of time. The immune system is perhaps the best known natural example.[1] When challenged with a foreign antigen, high-affinity antibodies are generated in a matter of weeks. This fascinating process occurs in stages, with the first step being the selection of low-affinity antibodies. This is facilitated by two properties of the immune system. Initially, there is a resident library of naive antibodies (generally believed to be in the range of 10^7 separate molecules) that cover a wide range of diversity space. Second, these antibodies are presented in a decavalent, IgM format so that their overall binding affinity for the target surface is boosted significantly by multiple attachment to the antigen (which is often presented in a multimeric form). These avidity or chelate effects, coupled to the large number of naive antibodies in the starting pool, allow a small group of antibodies to be clonally selected. At this point, the intrinsic affinities are typically in the micromolar range.

The second stage in antibody selection involves an affinity maturation process. Mutations are introduced into the variable domains of selected antibodies by a process known as "somatic hypermutation." This process generates more diversity and permits improvements in the complementarity between the antibody and antigen. In addition, selected antibodies are switched in class from the decavalent IgM form to the bivalent IgG form; the lower valency format increases the stringency of selection for those antibodies with higher intrinsic affinity.

Phage display, as it is practiced today for selection of naive peptide binders, mirrors the natural immune system (Fig. 1). One begins with a large and diverse set of peptides that are presented in a polyvalent format. Peptides that are selected from this polyvalent format bind their target in the high micromolar range. Higher affinity peptides are generated from these leads by introducing additional mutations and transferring the pep-

[1] A. K. Abbas, A. H. Lichtman, and J. S. Pober, "Cellular and Molecular Immunology," 2nd Ed. W. B. Saunders, Philadelphia, 1994.

0076-6879/00 $30.00

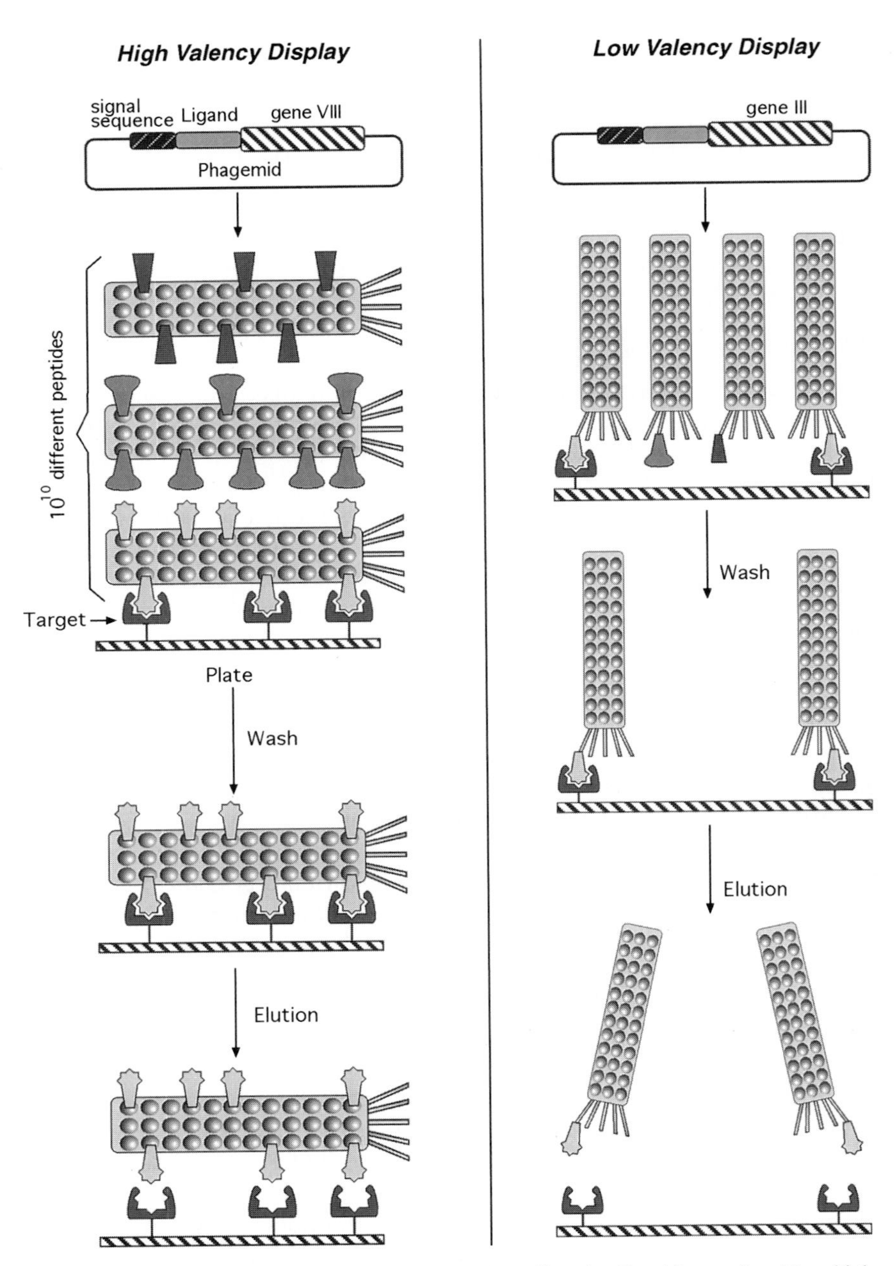

FIG. 1. Isolating specific ligands from peptide-phage libraries. Peptides are fused in a high-valency format to gene-8 on filamentous phage (*left*). Sequences that bind to the target are enriched and eluted. Low-affinity leads obtained from high-valency libraries are further randomized and displayed in a low-valency format on gene-3 (*right*). After several rounds of binding selection and elution, higher affinity binding peptides are obtained.

tides onto a lower valency format to allow for selection of peptides with affinities in the low micromolar to nanomolar range. Such peptides are useful reagents: they may serve as simple probes to understand molecular recognition, they may act as minimized surrogates for known or orphaned receptors, or they may even serve as lead molecules in drug design.

Origins and Development of Peptide Phage Display

The concept of phage display was first presented by G. Smith.[2] Gene fragments encoding small antigenic regions of a protein were fused to DNA encoding the gene 3 minor coat protein (protein 3, P3) which is present in five copies on the surface of the filamentous M13 phage. This established a physical linkage between phenotype and genotype: antigenic peptides were displayed on the surface of phage particles that also contained the coding DNA. Fusion phage displaying a particular antigenic peptide could be isolated from other phage by binding to antigen-specific antibodies, a process called "panning." Moreover, these phage could be amplified in *Escherichia coli* and subjected to additional rounds of panning to further enrich for the gene encoding the antigenic peptide. Thus, the gene for a binding entity could be selected *in vitro* and enriched away from nonbinding entities.

Later, several independent groups[3–5] showed that random peptides ($>10^8$ variants) could be displayed on P3. Weakly binding peptides (K_d values in the mid-micromolar range) were identified when these libraries were panned for binding to either an antibody that was known to bind a peptide, or to streptavidin, which binds the small molecule biotin. Thus, it was possible to identify novel binding peptides by phage display even for targets such as streptavidin that naturally bind small molecules.

At about the same time, it was shown that display of human growth hormone in a monovalent format allowed the identification of high-affinity binders in the nanomolar to picomolar range.[6,7] Monovalent display was accomplished by fusing the gene product onto gene 3 carried in a phagemid, and providing large amounts of the wild-type gene 3 product expressed from a helper phage (M13KO7).[8] A second development in display valency

[2] G. P. Smith, *Science* **228,** 1315 (1985).
[3] J. K. Scott and G. P. Smith, *Science* **249,** 386 (1990).
[4] J. J. Devlin, L. C. Panganiban, and P. E. Devlin, *Science* **249,** 404 (1990).
[5] S. E. Cwirla, E. A. Peters, R. W. Barrett, and W. J. Dower, *Proc. Natl. Acad. Sci. U.S.A.* **87,** 6378 (1990).
[6] S. Bass, R. Greene, and J. A. Wells, *Proteins Struct. Funct. Genet.* **8,** 309 (1990).
[7] H. B. Lowman, S. H. Bass, N. Simpson, and J. A. Wells, *Biochemistry* **30,** 10832 (1991).
[8] J. Vieira and J. Messing, *Methods Enzymol.* **153,** 3 (1987).

was the discovery that it was possible to display antigenic peptides at high levels by fusing onto the N terminus of the gene 8 major coat protein (protein 8, P8)[9–11] which is present at ~3000 copies per phage particle.

These technologies were assembled in a stepwise process showing that it was possible to develop peptides that bound to protein–protein interfaces with high affinity. A series of peptide libraries containing about 10^9 different peptide variants were constructed. These libraries were first panned in a high-valency format on P8 for binding to the immobilized receptors. Weakly binding peptides were transferred to a monovalent format on P3, where they were further randomized and panned. This resulted in the isolation of peptides that bound to the target receptors for interleukin-1 (IL-1),[12] thrombopoietin,[13] or erythropoietin (EPO)[14] with affinities in the low micromolar to nanomolar range. What was especially remarkable was that the peptides for the EPO and thrombopoietin receptors were agonists that functioned by dimerizing their respective cytokine receptor. When the structure of the complex between the EPO receptor and its agonistic peptide was solved, it revealed a simple, compact β-hairpin structure that dimerized with itself and two EPO receptors.[15]

This chapter is intended as an aid to those interested in using phage display to isolate novel binding peptides to their target of interest. We first outline the basic principles and requirements for the use of phage-displayed peptide libraries. Next, we describe detailed methods for library design, construction, and analysis. Finally, we present two examples from our group in which naive phage-displayed libraries were used to isolate highly specific peptide antagonists for biologically relevant protein–protein interactions. From these and other examples, it is apparent that selected peptides often have defined structures despite their small size, and that they fit into simple structural classes.

[9] A. A. Ilichev, O. O. Minenkova, S. L. Tat'kov, N. N. Karpyshev, A. M. Eroshkin, V. A. Petrenko, and L. S. Sandakhchiev, *Dokl. Akad. Nauk. S.S.S.R.* **307,** 481 (1989).

[10] J. Greenwood, A. E. Willis, and R. N. Perham, *J. Mol. Biol.* **220,** 821 (1991).

[11] F. Felici, L. Castagnoli, A. Musacchio, R. Jappelli, and G. Cesareni, *J. Mol. Biol.* **222,** 301 (1991).

[12] S. D. Yanofsky, D. N. Baldwin, J. H. Butler, F. R. Holden, J. W. Jacobs, P. Balasubramanian, J. P. Chinn, S. E. Cwirla, E. Peters-Bhatt, E. A. Whitehorn, E. H. Tate, A. Akeson, T. L. Bowlin, W. J. Dower, and R. W. Barrett, *Proc. Natl. Acad. Sci. U.S.A.* **93,** 7381 (1996).

[13] S. E. Cwirla, P. Balasubramanian, D. J. Duffin, C. R. Wagstrom, C. M. Gates, S. C. Singer, A. M. Davis, R. L. Tansik, L. C. Mattheakis, C. M. Boytos, P. J. Schatz, D. P. Baccanari, N. C. Wrighton, R. W. Barrett, and W. J. Dower, *Science* **276,** 1696 (1997).

[14] N. C. Wrighton, F. X. Farrell, R. Chang, A. K. Rashyap, F. P. Barone, L. S. Mulcahy, D. L. Johnson, R. W. Barrett, L. K. Jolliffe, and W. J. Dower, *Science* **273,** 458 (1996).

[15] O. Livnah, E. A. Stura, D. L. Johnson, S. A. Middleton, L. S. Mulcahy, N. J. Wrighton, W. J. Dower, L. K. Jolliffe, and I. A. Wilson, *Science* **273,** 464 (1996).

We illustrate further improvements in peptide phage display technology. Optimized methods provide much larger library sizes ($>10^{12}$), thereby facilitating the initial selection of weak binders. The valency switching process has become smoother and more tunable to the selection process, and the mutagenesis strategies allow for more rapid and complete affinity maturation. Overall, peptide phage display has become a dependable and general approach for identification of novel binding peptides.

Principles and Requirements

It is clear that the first requirement for peptide phage display is a suitable vector system designed to display fusion polypeptides on the surface of M13 phage. In addition, library construction requires efficient methods for the introduction of synthetic DNA into the phage display vector, and subsequently, for the introduction of library constructs into *E. coli.* Finally, effective affinity selection techniques are needed to isolate single clones with desired binding specificities, and simple methods for assaying the binding of phage-borne peptides are also useful.

Vector Design

A phagemid is a plasmid that contains sequences sufficient to allow packaging into phage particles.[16] We describe a phagemid that has been used to display peptides[17] and also large proteins.[18] In addition to origins of replication for both double- and single-stranded DNA synthesis, the vector contains a β-lactamase (*bla*) gene conferring resistance to ampicillin and carbenicillin. The final component is a fusion gene under the control of the alkaline phosphatase promoter (*PphoA*). The fusion gene encodes the stII signal sequence fused to either the mature P8 or to the C-terminal domain of P3. The signal sequence directs secretion of the coat protein and is subsequently cleaved by signal peptidase, leaving the coat protein embedded in the inner membrane.

In an *E. coli* host, the phagemid replicates as a double-stranded plasmid. On coinfection with a helper phage (such as M13KO7 or VCSM13), single-stranded DNA (ssDNA) replication is initiated, and the phagemid ssDNA is preferentially packaged into phage particles. The ssDNA can be easily

[16] J. Sambrook, E. F. Fritsch, and T. Maniatis, *in* "Molecular Cloning: A Laboratory Manual," 2nd Ed., Vol. 1, p. 4.17. Cold Spring Harbor Laboratory Press, Cold Spring Harbor, New York, 1989.

[17] H. B. Lowman, Y. C. Chen, N. J. Skelton, D. L. Mortensen, E. E. Tomlinson, M. D. Sadick, I. C. A. F. Robinson, and R. G. Clark, *Biochemistry* **37,** 8870 (1998).

[18] H. B. Lowman and J. A. Wells, *Methods: Comp. Methods Enzymol.* **3,** 205 (1991).

purified from phage and used for efficient sequencing, mutagenesis, or library construction. While the helper phage provides all the proteins necessary for phage assembly, copies of phagemid-encoded coat protein are also incorporated into the assembling virions. In this way, polypeptides fused to the N terminus of the phagemid-encoded coat protein are displayed in linkage to their encoding DNA.

Library Construction

With the phagemid system described above, peptide phage display can be achieved by inserting a peptide-encoding open reading frame (ORF) into the fusion gene, between the regions encoding the stII signal sequence and the coat protein. Insertion of degenerate DNA results in a diverse DNA population encoding many different peptides, in other words, a peptide library. While double-stranded cassette mutagenesis can be used to introduce synthetic, degenerate DNA into plasmids,[18] we describe a simple, robust method based on the method of Zoller and Smith,[19] with modifications introduced by Kunkel *et al.*[20]

Uracil-containing ssDNA (dU-ssDNA) is purified from an *E. coli* dut^-/ung^- strain such as CJ236[20] (Procedure 1, below). An appropriately designed synthetic oligonucleotide is first annealed to the dU-ssDNA, and then enzymatically extended and ligated to form covalently closed circular DNA (CCC-DNA) (Procedure 2, below). The oligonucleotide can be designed to anneal anywhere within the phagemid sequence; a heteroduplex results if the oligonucleotide is not perfectly complementary to the template. When the heteroduplex CCC-DNA is introduced into an *E. coli* ung^+ strain, the dU-ssDNA strand is preferentially destroyed and the *in vitro* synthesized strand is replicated. In this way, DNA mismatches (i.e., mutations) encoded by the synthetic oligonucleotide are incorporated into the phagemid. Efficient mutagenesis (>80%) is readily achieved, provided the template DNA is highly pure.

For library construction, the oligonucleotide is designed to insert a degenerate ORF into the phagemid-encoded fusion gene. The oligonucleotide begins with 18 base pairs complementary to the region preceding the insertion point, and it ends with 18 base pairs complementary to the region following the insertion point. Any desired sequences can be inserted between these complementary regions. The method is extremely versatile because the same template can be used with any oligonucleotide containing the required complementary regions, allowing for the construction of many different libraries from a single template.

[19] M. J. Zoller and M. Smith, *Methods Enzymol.* **154,** 329 (1987).
[20] T. A. Kunkel, J. D. Roberts, and R. A. Zakour, *Methods Enzymol.* **154,** 367 (1987).

In the final step of library construction, the heteroduplex CCC-DNA described above must be introduced into *E. coli* for amplification and phage propagation. The diversity of phage-displayed libraries is limited by the efficiency with which mutagenized DNA can be introduced into *E. coli.* Using high-voltage electroporation, the most efficient transformation method, it is commonly believed that the practical limit to phage-displayed diversity is on the order of 10^{10}.[21] Indeed, libraries of this size have been constructed, but only through the laborious combination of hundreds of separate electroporation reactions.[22] By optimizing the methods of Dower *et al.*,[23] we have greatly increased the phage-displayed diversities accessible through electroporation.

Electroporation is a two-component reaction in which the application of an electric field causes the uptake of DNA into *E. coli.* Increased transformant yields can be achieved by increasing the concentration of either *E. coli* or DNA. However, the reaction is extremely sensitive to salt concentrations, and the DNA concentration is limited by the need for low conductance. The standard method for producing low-conductance DNA solutions has been ethanol precipitation, but DNA prepared in this way can be used only at concentrations of about 10 μg/ml. We have found that, with affinity-purified DNA (Procedure 2, below), electroporation can be performed with DNA concentrations up to 500 μg/ml.

We also constructed a new *E. coli* strain for high-efficiency electroporation and phage production. A standard bacterial mating protocol[24] was used to transfer the F′ episome from *E. coli* XL1-Blue[25] to the high-efficiency electroporation strain *E. coli* MC1061.[26] Because *E. coli* MC1061 carries a streptomycin resistance chromosomal marker and the XL1-Blue F′ episome confers tetracycline resistance, the mating progeny were easily selected for resistance to both antibiotics. The new strain, *E. coli* SS320, retains the high electroporation efficiency of *E. coli* MC1061, and the introduced F′ episome allows for efficient phage infection. While standard electroporation procedures use *E. coli* at concentrations less than 5×10^{10}

[21] W. J. Dower and S. E. Cwirla, *in* "Guide to Electroporation and Electrofusion" (D. C. Chang, B. M. Chassy, J. A. Saunder, and A. E. Sowers, eds.), p. 291. Academic Press, San Diego, California, 1992.

[22] T. J. Vaughan, A. J. Williams, K. Pitchard, J. K. Osbourn, A. R. Pope, J. C. Earnshaw, J. McCafferty, R. A. Hodits, J. Wilton, and K. S. Johnson, *Nature Biotechnol.* **14,** 309 (1996).

[23] W. J. Dower, J. F. Miller, and C. W. Ragsdale, *Nucleic Acids Res.* **16,** 6127 (1988).

[24] J. H. Miller, *in* "Experiments in Molecular Biology," p. 190. Cold Spring Harbor Laboratory Press, Cold Spring Harbor, New York, 1972.

[25] W. O. Bullock, J. M. Fernandez, and J. M. Short, *BioTechniques* **5,** 376 (1987).

[26] M. J. Casadaban and Stanley Cohen, *J. Mol. Biol.* **138,** 179 (1980).

CFU/ml, *E. coli* SS320 can be concentrated to at least 3×10^{11} CFU/ml (Procedure 3, below).

Using highly concentrated *E. coli* and DNA, our electroporation protocol (Procedure 4, below) provides library diversities two orders of magnitude greater than those commonly obtained. A single electroporation reliably produces diversities of 2×10^{10}, and multiple electroporations can be combined to produce diversities greater than 10^{12}.

Naive Peptide Libraries

Naive peptide libraries are designed to be as unbiased as possible, and thus they can provide specific binders to many different targets. However, the binding interactions are often of low affinity ($K_d > 1\ \mu M$), and affinity maturation is usually necessary (see below).

In designing a library, peptide length is an important consideration. Naive libraries are usually "hard randomized," that is, degenerate positions are designed to encode all 20 natural amino acids. Because of the redundancy of the genetic code, the simplest way to encode the 20 natural amino acids is through the use of degenerate codons representing 32 unique codons (NNS or NNK; where N is any nucleotide, S is C or G, and K is G or T). Because the complexity of peptide libraries increases exponentially with length, even a library containing 10^{12} members cannot completely represent peptides containing more than 8 residues (Table I). On the other hand,

TABLE I
DIVERSITIES OF DNA-ENCODED PROTEIN LIBRARIES

Randomized positions (n)	Protein diversity (20^n)	DNA diversity[a] (32^n)	Transformants required for a complete library with:	
			90% confidence[b]	99% confidence[b]
1	20	32	74	149
2	400	1.0×10^3	2.4×10^3	4.8×10^3
3	8.0×10^3	3.3×10^4	7.6×10^4	1.5×10^5
4	1.6×10^5	1.1×10^6	2.4×10^6	4.9×10^6
5	3.2×10^6	3.4×10^7	7.7×10^7	1.6×10^8
6	6.4×10^7	1.1×10^9	2.5×10^9	5.0×10^9
7	1.3×10^9	3.4×10^{10}	7.9×10^{10}	1.6×10^{11}
8	2.6×10^{10}	1.1×10^{12}	2.5×10^{12}	5.1×10^{12}

[a] Because of the redundancy of the genetic code, representation of all 20 natural amino acids requires the use of degenerate codons containing at least 32 unique codons.

[b] A Poisson distribution was used to calculate the confidence that a library contains all possible amino acid sequences.

libraries of longer peptides contain more chemical diversity, and are thus more likely to contain binders with desired specificities and affinities. We find that libraries of 20-mer peptides work well. While such libraries are extremely incomplete ($32^{20} = 10^{30}$), they often provide specific binders with reasonable affinities ($K_d < 10\ \mu M$). In addition, peptides of this length can be readily synthesized by standard techniques.

A second consideration is the introduction of constraints that promote a discrete structure in solution. Structured peptides offer two important advantages. First, the loss in entropy on target binding is likely to be less than for an unstructured peptide, thus increasing the affinity of the interaction. Second, the determination of the free solution structure by nuclear magnetic resonance (NMR) analysis can provide insights into structure–function relationships (see Example 1, below).

A simple constraint is introduced by fixing two residues within the random peptide as cysteines. In an oxidizing environment, such as the *E. coli* periplasm, the cysteines form disulfide bonds. Because of proximity effects, intrachain disulfides are usually favored and the peptides are predominantly displayed as monomers containing a disulfide-constrained loop. A panel of libraries is generated by varying the length of the regions flanking the loop and also the size of the loop itself (Tables II and III). Fixing a Gly–Pro sequence within the disulfide loop may provide an additional constraint, because it is believed that this sequence promotes β turns.[14]

However, interchain disulfides can occur and may result in the display

TABLE II
PEPTIDE LIBRARIES SORTED FOR INSULIN-LIKE GROWTH FACTOR-BINDING PROTEIN 1

Library[b]	Design	Enrichment[a]		
		Round 2	Round 3	Round 4
A	$SGTAX_2GPX_4CSLAGSP$	1	3	16
B	$X_4CX_2GPX_4CX_4$	8	2000	680
C	X_{20}	9	23	70
D[c]	$X_7CX_4CX_7$	4	3500	3
	$X_7CX_5CX_6$			
	$X_6CX_6CX_6$			
	$X_6CX_7CX_5$			
	$X_5CX_8CX_5$			
	$X_5CX_9CX_4$			
	$X_4CX_{10}CX_4$			

[a] Enrichment = (phage eluted from target)/(phage eluted from BSA).
[b] Libraries were displayed on gene 8, under the control of *PphoA*.
[c] Seven designs were constructed separately and combined.

TABLE III
PEPTIDE LIBRARIES SORTED FOR BINDING TO INSULIN-LIKE GROWTH FACTOR-BINDING PROTEIN 1 OR BOVINE INSULIN-LIKE GROWTH FACTOR-BINDING PROTEIN 2

		Round 3 enrichment[b]	
Library[a]	Design	IGFbp-1	bIGFbp-2
A′	$X_4CX_2GPX_4CX_4$	1	1
B′	$X_7CX_4CX_7$	1	1
C′	$X_7CX_5CX_6$	1	1
D′	$X_6CX_6CX_6$	1	1
E′	$X_6CX_7CX_5$	1	1
F′	$X_5CX_8CX_5$	27	400
G′	$X_5CX_9CX_4$	1	90
H′	$X_4CX_{10}CX_4$	1	1
I′	X_8	1	1
J′	$X_2CX_2CX_2$	1	1
K′	$X_2CX_3CX_2$	1	2
L′	$X_2CX_4CX_2$	1	1
M′	$X_2CX_5CX_2$	1	1
N′	$X_2CX_6CX_2$	1	1
O′	$X_2CX_7CX_2$	1	1
P′	$X_2CX_8CX_2$	2300	110
Q′	$X_2CX_9CX_2$	1	900
R′	$X_2CX_{10}CX_2$	1	1

[a] Libraries were displayed on gene 8, under the control of the P_{tac} promoter.
[b] Enrichment = (phage eluted from target)/(phage eluted from BSA).

of homodimeric peptides. While usually rare, such homodimers may be preferentially selected during affinity sorting, because the affinity of even weak monomeric interactions can be greatly increased through avidity effects in the bivalent form. The presence of such homodimers is difficult to detect in phage selectants, and synthetic peptides are usually required to conclusively identify the active binding species.

Naive libraries are normally constructed as fusions with the major coat protein. Because each phage particle contains several thousand P8 molecules, the fused peptides are displayed in a polyvalent format. In a system in which the entire phage coat is formed by the P8 fusion, only peptides smaller than about eight residues can be displayed, because longer peptides disrupt the phage coat and result in inviable phage.[27] The phagemid system

[27] G. Iannolo, O. Minenkova, R. Petruzzelli, and G. Cesareni, *J. Mol. Biol.* **248,** 835 (1995).

circumvents this problem because additional wild-type P8 is supplied by the helper phage.[7] The wild-type P8 forms the majority of the phage coat, but multiple copies of the phagemid-encoded P8 fusion are also incorporated. In such a system, even large peptides can be displayed polyvalently without affecting phage viability.

Polyvalent display provides an avidity effect: When many displayed peptides simultaneously bind to an immobilized target, an extremely slow off-rate results.[5] This allows for the selection of weak interactions ($K_d > 1\ \mu M$) that could not be selected in a monovalent format. Because naive libraries usually contain only low-affinity binders, polyvalent avidity is often crucial for selection and amplification of specific clones.

Second-Generation Libraries

Second-generation libraries are designed to cover sequence space surrounding the sequence of a low-affinity binder. Because library members share significant homology with the low-affinity binder, such libraries usually contain many binders specific for a given target protein, and some binders may exhibit increased affinity. Thus, second-generation libraries can be used to affinity mature weakly binding lead sequences.

Second-generation library design depends on the nature of the low-affinity leads. If several lead sequences are available and a consenus is observed, a new library can be designed in which strongly consensed positions are fixed, while weakly consensed positions can be represented with "tailored codons" chosen to encode residues with a particular characteristic. For example, a position consensing to hydrophobic residues could be represented by an NTC codon, which encodes Phe, Leu, Ile, or Val. Remaining, nonconsensed positions can be hard randomized, as described above. Alternatively, if only a single lead is available, the sequence can be "soft randomized." In this case, a mutagenic oligonucleotide is synthesized such that each base is predominantly wild type, but small, equimolar amounts of the remaining nucleotides are added. Libraries constructed with such an oligonucleotide contain divergent sequences with homology to the original lead, and the mutation frequency depends on the percentage of wild-type base present in the oligonucleotide. In either case, sequence information from low-affinity leads is used to construct new libraries with some positions either fixed as, or biased toward, sequences likely to favor binding to a particular target. In this way, the search is focused on validated regions of sequence space.

While avidity is necessary for the isolation of low-affinity leads, it must be avoided for effective affinity maturation. In a polyvalent format, avidity effects may enhance apparent affinities to such a point that even intrinsic affinity differences of several orders of magnitude cannot be differentiated. Thus, second-generation libraries are usually displayed in a monovalent format on P3.

Selection and Analysis of Binding Peptides

Affinity sorting is the process whereby specific binders are isolated from phage libraries. A target protein is immobilized on a solid surface, which is often an immunosorbent plastic plate, or sometimes beads.[18] Library phage are added under conditions appropriate for binding, and after an incubation period, nonspecifically bound phage are removed by washing. The remaining phage are then eluted and amplified by passage through an *E. coli* host. The amplified phage pool is used in another round of the binding selection. In a successful sort, each round of the cycle enriches for clones that bind to the target protein, and the phage pool is eventually dominated by specific binders.

The progress of an affinity sort is conveniently monitored by calculating the enrichment ratio: the number of phage bound to a well coated with target protein divided by the number of phage bound to an uncoated well. An enrichment ratio is calculated for each round of the sort, and a ratio greater than one is usually indicative of phage binding specifically to the target protein. The enrichment ratio usually peaks at a particular round, and it may even decline in subsequent rounds, possibly because of a bias in the production of particular phage clones during amplification in *E. coli.* Sorting beyond the peak enrichment round generally favors phage clones that propogate well in *E. coli,* and these may not be the highest affinity binders. Thus, clones are typically analyzed when the enrichment ratio has reached a maximum.

DNA sequence analysis is used to derive the sequence of displayed peptides. In some cases, the pool may be dominated by identical DNA sequences (siblings) descended from a common progenitor. This situation usually arises under stringent sorting conditions, when the sort is carried on for many rounds, or when the initial library contains few binders specific for the target protein. In other cases, several unique DNA sequences may be obtained. If these sequences exhibit regions of consensus, the information can be used to design second-generation libraries, as described above.

Specific binding of peptide–phage to the target protein can be determined by a phage enzyme-linked immunosorbent assay (ELISA). Phage are incubated with immobilized target, and after a washing step, bound phage are detected with an M13-specific antibody conjugated with horseradish peroxidase (HRP). Detection involves spectrophotometric quantitation of an HRP-catalyzed colorimetric reaction. The signal strength depends on both the intrinsic affinity and the copy number of the displayed peptides; specific binding is evidenced by a greater signal from immunoplates coated with target protein in comparison with the signal from uncoated plates. Such assays cannot be used to determine intrinsic affinities, because the display level is an undefined variable.

Fortunately, phage ELISAs can be modified to measure intrinsic affinities, provided that polyvalent display is strictly avoided. For this purpose, monovalent phage particles at a fixed, nonsaturating concentration are allowed to bind to immobilized target in the presence of varying concentrations of soluble target. The soluble target inhibits binding to the immobilized target, and a median inhibitory concentration (IC_{50}) value can be determined (see Example 1, below). In the absence of avidity, this measurement provides an approximation of binding affinity (K_d), which often tracks closely with measurements of free peptide/protein binding.[28–30]

Phage ELISAs can also be used to determine whether peptide–phage binding is blocked by a known ligand of the target protein (see Example 1, below). In this application, phage particles are allowed to bind to immobilized target both in the presence and absence of the second ligand, and blocking is indicated by reduced phage binding in the presence of the second ligand. While such blocking suggests that free peptide alone may in turn antagonize the interaction between the target protein and the second ligand, this hypothesis can be proved only by direct assays using synthetic peptides corresponding to the phage-displayed sequence.

Indeed, the final test of any peptide phage sort is the efficacy of free peptides derived from selected sequences. To ensure that phage-displayed sequences are valid leads, and to minimize time wasted on phage-associated artifacts, peptides should be synthesized and tested as soon as possible. Synthetic peptides enable accurate measurements of intrinsic affinity, they can be used in biological assays for which phage-displayed peptides are unsuitable, and they allow for incorporation of nonnatural functionalities. Finally, and perhaps most importantly, synthetic peptides are amenable to detailed structure analysis using crystallographic and NMR techniques.

Materials and Methods

Enzymes are from New England BioLabs (Beverly, MA). Maxisorp immunoplates are from Nunc (Rosklide, Denmark). *Escherichia coli*

[28] B. C. Cunningham, D. G. Lowe, B. Li, B. D. Bennett, and J. A. Wells, *EMBO J.* **13,** 2508 (1994).

[29] Y. A. Muller, Y. Chen, H. W. Christinger, B. Li, B. C. Cunningham, H. B. Lowman, and A. M. de Vos, *Structure* **6,** 1153 (1998).

[30] Y. Dubaquié and H. B. Lowman, *Biochemistry* **38,** 6386 (1999).

MC1061 and *E. coli* CJ236 are from Bio-Rad (Hercules, CA). *Escherichia coli* XL1-Blue and VCSM13 are from Stratagene (La Jolla, CA). Bovine serum albumin (BSA), Tween 20, and *o*-phenylenediamine dihydrochloride (OPD) are from Sigma (St. Louis, MO). Anti-M13 antibody–horseradish peroxidase conjugate (anti-M13/HRP conjugate) and dNTPs and ATP from Pharmacia Biotech (Piscataway, NJ). Ultrapure glycerol is from Life Technologies (Gaithersburg, MD). Neutravidin, streptavidin, biotinylation reagents, and blocker casein buffer are from Pierce (Rockford, IL).

Solutions and Media

2YT: 10 g of Bacto-yeast extract, 16 g of Bacto-tryptone, 5 g of NaCl; add water to 1 liter and adjust to pH 7.0 with NaOH; autoclave
2YT/carb: 2YT, carbenicillin (50 μg/ml)
2YT/carb/VCS: 2YT/carb, VCSM13 (10^{10} PFU/ml)
2YT/tet: 2YT, tetracycline (5 μg/ml)
Glycerol (10%, v/v): 100 ml of ultrapure glycerol and 900 ml of H_2O; filter sterilize
TM buffer (10×): 500 m*M* Tris-HCl, 100 m*M* $MgCl_2$, pH 7.5
Coating buffer: 50 m*M* sodium carbonate, pH 9.6
OPD solution: 10 mg of OPD, 4 μl of 30% (v/v) H_2O_2, 12 ml of PBS
PBS: 137 m*M* NaCl, 3 m*M* KCl, 8 m*M* Na_2HPO_4, 1.5 m*M* KH_2PO_4; adjust to pH 7.2 with HCl; autoclave
PEG–NaCl solution: PEG 8000 (200 g/liter), NaCl(146 g/liter); autoclave
PT buffer: PBS, 0.05% (v/v) Tween 20
PBT buffer: PBS, 0.2% (w/v) BSA, 0.1% (v/v) Tween 20
SOC medium: 5 g of Bacto-yeast extract, 20 g of Bacto-tryptone, 0.5 g of NaCl, 0.2 g of KCl; add water to 1.0 liter and adjust to pH 7.0 with NaOH; autoclave; add 5 ml of 2.0 *M* $MgCl_2$ (autoclaved) and 20 ml of 1.0 *M* glucose (filter sterilized)
Superbroth: 24 g of Bacto-yeast extract, 12 g of Bacto-tryptone, 5 ml of glycerol; add water to 900 ml; autoclave; add 100 ml of 0.17 *M* KH_2PO_4, 0.72 *M* K_2HPO_4 (autoclaved)
dNTPs: 25 m*M* each of dATP, dCTP, dGTP, dTTP in H_2O

Procedure 1: Preparation of dU-ssDNA Template

1. Pick a single colony of *E. coli* CJ236 harboring the appropriate phagemid into 1 ml of 2YT/carb supplemented with chloramphenicol (10 μg/ml) for F′ episome selection. Incubate at 37° with shaking at 200 rpm for approximately 6 hr.

2. Add VCSM13 helper phage to a final concentration of 10^{10} PFU/ml. After 10 min, transfer the culture to 30 ml of 2YT supplemented with

carbenicillin (10 μg/ml) and uridine (0.25 μg/ml). Grow overnight at 37° with shaking at 200 rpm.

3. Centrifuge for 10 min at 15,000 rpm and 2° in a Sorvall SS-34 rotor (27,000*g*). Transfer the supernatant to a fresh tube and add a 1/5 volume of PEG–NaCl solution. Incubate for 5 min at room temperature. Centrifuge for 10 min at 10,000 rpm and 2° in an SS-34 rotor (12,000*g*).

4. Decant the supernatant. Respin briefly and remove the remaining supernatant with a pipette. Resuspend the phage pellet in 0.5 ml of PBS. Pellet insoluble matter by centrifuging for 5 min at 15,000 rpm and 2° in an SS-34 rotor. Transfer the supernatant to an Eppendorf tube.

5. Purify the single-stranded DNA using a Qiagen (Valencia, CA) QIAprep Spin M13 kit according to the manufacturer instructions, beginning with the addition of Qiagen MP buffer. Use one Qiaprep spin column; elute the DNA with 100 μl of 10 m*M* Tris-HCl, pH 8.0. Quantify the DNA spectrophotometrically (OD_{260} is 1.0 for ssDNA at 33 μg/ml). The yield should be approximately 20 μg.

Procedure 2: In Vitro Synthesis of Heteroduplex DNA

The following three-step procedure is an optimized, large-scale version of the method of Kunkel *et al.*[20] The oligonucleotide is first 5′-phosphorylated and then annealed to a dU-ssDNA phagemid template. Finally, the oligonucleotide is enzymatically extended and ligated to form CCC-DNA.

Step 1: Phosphorylation of the Oligonucleotide

1. Combine the following in an Eppendorf tube:

Oligonucleotide	0.6 μg
TM buffer (10×)	2 μl
ATP (10 m*M*)	2 μl
Dithiothreitol (DTT, 100 m*M*)	1 μl

Add water to a total volume of 20 μl.

2. Add 20 units of T4 polynucleotide kinase. Incubate for 1 hr at 37°.

Step 2: Annealing the Oligonucleotide to the Template

1. Combine the following in an Eppendorf tube:

dU-ssDNA template	20 μg
Phosphorylated oligonucleotide	0.6 μg
TM buffer (10×)	25 μl

Add water to a total volume of 250 μl. The DNA quantities provide an oligonucleotide : template molar ratio of 3 : 1, assuming that the oligonucleotide : template length ratio is 1 : 100.

2. Incubate at 90° for 2 min, 50° for 3 min, and 20° for 5 min.

Step 3: Enzymatic Synthesis of Covalently Closed Circular DNA

1. To the annealed oligonucleotide/template, add the following:

ATP (10 m*M*)	10 μl
dNTPs (25 m*M*)	10 μl
DTT (100 m*M*)	15 μl
T4 DNA ligase	30 weiss units
T7 DNA polymerase	30 units

2. Incubate at 20° for at least 3 hr.

3. Affinity purify and desalt the DNA with a Qiagen QIAquick DNA purification kit, according to the manufacturer instructions. Use one QIAquick column, and elute with 35 μl of ultrapure H_2O.

4. Electrophorese 1.0 μl of the reaction alongside the single-stranded template. Use a Tris–acetate electrophoresis buffer (TAE)–1.0% (w/v) agarose gel with ethidium bromide for DNA visualization. A successful reaction results in the complete conversion of single-stranded template to double-stranded DNA. Two product bands are usually visible. The lower band is correctly extended and ligated product (CCC-DNA) that transforms *E. coli* efficiently and provides a high mutation frequency (>80%). The upper band is an unwanted product resulting from an intrinsic strand-displacement activity of T7 DNA polymerase.[31] The strand-displaced product provides a low mutation frequency (<20%), but it also transforms *E. coli* at least 30-fold less efficiently than CCC-DNA. Thus, provided a significant proportion of the template is converted to CCC-DNA, a high mutation frequency will result. Occasionally, a third product band is visible. Migrating between the two bands described above, this band is correctly extended but unligated DNA, resulting either from insufficient T4 DNA ligase activity or from inefficient oligonucleotide phosphorylation. This product must be avoided, because it transforms *E. coli* efficiently but provides a low mutation frequency.

Procedure 3: Preparation of Electrocompetent Escherichia coli SS320

The method described below is a large-scale version of a previously described method.[23]

1. Pick a single colony of *E. coli* SS320 (from a fresh 2YT/tet plate) into 1 ml of 2YT/tet. Incubate at 37° with shaking at 200 rpm for about 8 hr. Transfer the culture to 50 ml of 2YT/tet in a 500-ml baffled flask and grow overnight.

[31] R. L. Lechner, M. J. Engler, and C. C. Richardson, *J. Biol. Chem.* **258,** 11174 (1983).

2. Inoculate 5 ml of the overnight culture into six 2-liter baffled flasks containing 900 ml of superbroth supplemented with tetracycline (5 μg/ml). Grow the cells to an OD_{600} of 0.6–0.8 (approximately 4 hr).

3. Chill three flasks on ice for 10 min with periodic shaking. All steps from here on should be done on ice and in a cold room where applicable.

4. Transfer the cultures to six 400-ml prechilled centrifuge tubes. Centrifuge for 5 min at 5000 rpm and 2° in a Sorvall GS-3 rotor (5000*g*). While the cultures are centrifuging. chill the remaining three flasks on ice. Decant the supernatant and add the cultures from the remaining three flasks to the same centrifuge tubes. Repeat the centrifugation and decant the supernatant.

5. Fill each tube with 1.0 m*M* HEPES, pH 7.0. Add a sterile, magnetic stir bar (the stir bars should be rinsed with sterile water before and after use, and they should be stored in ethanol). Use the stir bar to resuspend the pellet: Swirl briefly to dislodge the pellet from the tube wall and then stir at a moderate rate until the pellet is completely resuspended.

6. Centrifuge for 10 min at 5000 rpm and 2° in a GS-3 rotor. When removing the tubes from the rotor, be careful to maintain the angle so as not to disturb the pellet. Decant the supernatant, but do not remove the stir bars.

7. Repeat steps 5 and 6. Resuspend each pellet in 150 ml of 10% (v/v) glycerol. Do not combine the pellets at this point.

8. Centrifuge for 15 min at 5000 rpm and 2° in a GS-3 rotor. Decant the supernatant and remove the stir bars. Remove remaining traces of supernatant with a sterile pipette. Add 3.0 ml of 10% (v/v) glycerol to the first tube and resuspend the pellet by gently pipetting. Transfer the suspension to another tube and repeat until all the pellets are resuspended.

9. Aliquot 350 μl of cells into Eppendorf tubes, flash freeze on dry ice, and store at −70°. The procedure yields approximately 12 ml of cells at a concentration of 3×10^{11} CFU/ml.

Procedure 4: Escherichia coli Electroporation and Phage Production

1. Chill the purified DNA and a 0.2-cm gap electroporation cuvette on ice. Thaw a 350-μl aliquot of electrocompetent *E. coli* SS320 on ice. Add the cells to the DNA and mix by pipetting several times.

2. Transfer the mixture to the cuvette and electroporate. It is preferable to use a BTX (San Diego, CA) ECM-600 electroporation system with the following settings: 2.5 kV field strength, 129 Ω resistance, and 50 μF capacitance. Alternatively, a Bio-Rad Gene Pulser can be used with the following settings: 2.5 kV field strength, 200 Ω resistance, and 25 μF capacitance.

3. Immediately add 1 ml of SOC medium and transfer to a 250-ml baffled flask. Rinse the cuvette twice with 1 ml of SOC medium. Add SOC medium to a final volume of 25 ml and incubate for 30 min at 37° with shaking.

4. Plate serial dilutions on 2YT/carb plates to determine the library diversity. Transfer the culture to a 2-liter baffled flask containing 500 ml of 2YT/carb/VCS. Incubate overnight at 37° with shaking.

5. Centrifuge the culture for 10 min at 10,000 rpm and 2° in a Sorvall GSA rotor (16,000*g*). Transfer the supernatant to a fresh tube and add a 1/5 volume of PEG–NaCl solution to precipitate the phage. Incubate for 5 min at room temperature.

6. Centrifuge for 10 min at 10,000 rpm and 2° in a GSA rotor. Decant the supernatant. Respin briefly and remove the remaining supernatant with a pipette. Resuspend the phage pellet in a 1/20 volume of PBS or PBT buffer. Pellet insoluble matter by centrifuging for 5 min at 15,000 rpm and 2° in an SS-34 rotor. Transfer the supernatant to a clean tube.

7. Determine the phage concentration spectrophotometrically (OD_{268} of 1.0 for a solution containing 5×10^{12} phage/ml).

8. Use immediately, or flash freeze on dry ice and store at −70°.

Procedure 5: Affinity Sorting the Library

1. Coat Maxisorp immunoplate wells with 100 μl of target protein solution (2–5 μg/ml in coating buffer) for 2 hr at room temperature or overnight at 4°. The number of wells required depends on the diversity of the library. Ideally, the phage concentration should not exceed 10^{13} phage/ml and the total number of phage should exceed the library diversity by 1000-fold. Thus, for a diversity of 10^{10}, 10^{13} phage should be used and, using a concentration of 10^{13} phage/ml, 10 wells will be required.

2. Remove the coating solution and block for 1 hr with 200 μl of 0.2% (w/v) BSA in PBS. At the same time, block an equal number of uncoated wells as a negative control.

3. Remove the block solution and wash eight times with PT buffer.

4. Add 100 μl of library phage solution in PBT buffer to each of the coated and uncoated wells. Incubate at room temperature for 2 hr with gentle shaking.

5. Remove the phage solution and wash 10 times with PT buffer.

6. To elute bound phage, add 100 μl of 100 m*M* HC1. Incubate for 5 min at room temperature. Transfer the HCl solution to an Eppendorf tube. Neutralize with 1.0 *M* Tris-HCl, pH 8.0 (approximately 1/3 volume).

7. Add half the eluted phage solution to 10 volumes of actively growing *E. coli* SS320 or XL1-Blue ($OD_{600} < 1.0$). Incubate for 20 min at 37° with shaking.

8. Plate serial dilutions on 2YT/carb plates to determine the number of phage eluted. Determine the enrichment ratio: the number of phage eluted from a well coated with target protein divided by the number of phage eluted from an uncoated well.

9. Transfer the culture from the coated wells to 25 volumes of 2YT/carb/VCS and incubate overnight at 37° with shaking.

10. Isolate phage particles as described in Procedure 4.

11. Repeat the sorting cycle until the enrichment ratio has reached a maximum. Typically, enrichment is first observed in round 3 or 4, and sorting beyond round 6 is seldom necessary.

12. Pick individual clones for sequence analysis and phage ELISA.

Procedure 6: Phage Enzyme-Linked Immunosorbent Assay

This procedure is based on a previously described method.[32]

1. Isolate and quantitate phage as described above (Procedure 4).
2. Prepare threefold dilutions of phage stock, using PBT buffer.
3. Transfer 100 μl of phage solution to Maxisorp immunoplates coated and blocked as described above (Procedure 5). Incubate for 1 hr with gentle shaking.
4. Remove the phage solution and wash 10 times with PT buffer.
5. Add 100 μl of anti-M13/HRP conjugate (diluted 3000-fold in PBT buffer). Incubate for 30 min with gentle shaking.
6. Wash eight times with PT buffer and two times with PBS.
7. Add 100 μl of freshly prepared OPD solution. Allow color to develop for 5–10 min.
8. Stop the reaction with 50 μl of 2.5 M H_2SO_4 and read spectrophotometrically at 492 nm in a microtiter plate reader.

Example 1: Insulin-Like Growth Factor I-Binding Protein

Insulin-like growth factor (IGF-I) is a hormone with diverse metabolic and mitogenic activities.[33–35] Key to the activities of IGF-I are the insulin and type I IGF cell surface receptors to which IGF-I binds with high

[32] K. H. Pearce, Jr., B. J. Potts, L. G. Presta, L. N. Bald, B. M. Fendly, and J. A. Wells, *J. Biol. Chem.* **272,** 20595 (1997).

[33] M. Barinaga, *Science* **209,** 772 (1994).

[34] C. A. Bondy, L. E. Underwood, D. R. Clemmons, H.-P. Guler, M. A. Bach, and M. Skarulis, *Ann. Intern. Med.* **120,** 593 (1994).

[35] R. G. Clark and I. C. A. F. Robinson, *Cyto. Growth Factor Rev.* **7,** 65 (1996).

affinity.[36,37] However, the activity and pharmacokinetics of IGF-I are also mediated by a distinct class of soluble proteins known as the IGF-binding proteins (IGFbps), including six different IGFbps in humans.[38–40] By blocking the binding of IGF-I to its receptors, directing its tissue distribution, or directing its clearance from the circulation, the binding proteins may regulate IGF activity, and therefore may be useful targets for the modulation of IGF action. Indeed, it has been shown that mutants of IGF-I defective for binding receptor, but competent for binding IGFbps, can have significant IGF-like effects *in vivo*.[17,41] In this example, we discuss the search for novel peptides that bind to IGFbps and block IGF-I binding.

Selection of Peptides for Binding Insulin-Like Growth Factor-Binding Protein 1

This example illustrates selection using a panel of peptide libraries, with two forms of target presentation. Peptide libraries were fused to the N terminus of a linker peptide (GGGSGGG), followed by P8. Peptide–P8 expression and secretion were directed by the *PphoA* promoter and stII signal sequence, respectively.[17] Several naive peptide libraries (see Table II) were constructed to favor structured peptides, primarily through the formation of disulfide-constrained loops and in some cases by the introduction of Gly–Pro residues. Each library consisted of 10^8–10^9 independent transformants. For initial binding selections, libraries of the form $SGTACX_2GPX_4CSLAGSP$, $X_4CX_2GPX_4CX_4$, and an unconstrained X_{20} library were each sorted separately. Libraries of the form $X_iCX_jCX_k$ (where $j = 4$–10; $i + j + k = 18$) were pooled and sorted together.

For the first three rounds of selection, IGFbp-1 was covalently linked to biotin using EZ-Link NHS-SS-biotin (Pierce). This reagent is amine reactive and provides for release of phage bound to the labeled protein by reduction of the disulfide bound within the linker. To take advantage of surface (avidity) effects for identifying low- as well as high-affinity peptides, the biotin-SS–IGFbp-1 was first allowed to bind to immunosorbent plates coated with NeutrAvidin (Pierce). To minimize nonspecific binding effects, a varying set of blocking reagents (5.0 g/liter in each case) were added to

[36] J. Massagué and M. P. Czech, *J. Biol. Chem.* **257,** 5038 (1982).

[37] A. Ullrich, A. Gray, A. W. Tam, T. Yang-Feng, M. Tsubokawa, C. Collins, W. Henzel, T. Lebon, S. Kathuria, E. Chen, S. Jacobs, U. Francke, J. Ramachandran, and Y. Fujita-Yamaguchi, *EMBO J.* **5,** 2503 (1986).

[38] J. I. Jones and D. R. Clemmons, *Endocr. Rev.* **16,** 3 (1995).

[39] L. A. Bach and M. M. Rechler, *Diabetes Rev.* **3,** 38 (1995).

[40] R. C. Baxter and J. L. Martin, *Proc. Natl. Acad. Sci. U.S.A.* **86,** 6898 (1989).

[41] S. A. Loddick, X.-J. Liu, Z.-X. Lu, C. Liu, D. P. Behan, D. C. Chalmers, A. C. Foster, W. W. Vale, N. Ling, and E. B. De Souza, *Proc. Natl. Acad. Sci. U.S.A.* **95,** 1894 (1998).

the NeutrAvidin-coated plates prior to biotin–IGFbp-1: BSA in round 1, ovalbumin in round 2, and instant milk in round 3. Excess biotin was also added to minimize the capture of NeutrAvidin-binding peptide–phage. In spite of this precaution, some NeutrAvidin-binding phage were found; therefore, streptavidin (0.1 mg/ml) was included in the binding mix during the second and third rounds of selection. About 10^{10} phage/ml were added in PT buffer with 1.0 μM biotin and blocking agent (5.0 g/liter) to eight wells coated with biotin–IGFbp-1. Control wells were coated with NeutrAvidin or with albumin, but not with biotin–IGFbp-1. The phage were allowed to bind for 5–15 hr at room temperature, and plates were then washed 10 times with PT buffer. Phage remaining bound to the plates were eluted by incubating with 50 mM DTT for 1–2 hr at room temperature. The eluted phage were used to infect *E. coli* XL1-Blue cells directly for overnight amplification at 37°. Aliquots of eluted phage were titered after each sorting round to measure specific binding to IGFbp-1.

The fourth round of binding selection was carried out on immunosorbent plates directly coated with IGFbp-1 (2.0 μg/ml). This selection procedure was designed to eliminate any NeutrAvidin/streptavidin-binding peptide–phage, as well as any peptide–phage recognizing only biotinylated IGFbp-1. In this case, phage were eluted with 20 mM HCl and neutralized with 1.0 M Tris-HCl, pH 8.0, prior to colony counting and propagation. The results of phage-binding selections in cycles 2–4 are shown in Table II.

Survey of Peptide Libraries for Binding to Insulin-Like Growth Factor-Binding Proteins

In additional experiments, individual peptide–phage libraries with various arrangements of putative disulfide bonds were sorted for binding to either human IGFbp-1 or bovine IGF-binding protein 2 (bIGFbp-2). In this case, the target proteins were coated directly onto immunosorbant plates. Peptide–phage libraries were constructed as described above, except that peptide P8 expression was directed by the P_{tac} promoter,[42] with induction by 0.1 mM isopropyl-β-D-thiogalactopyranoside (IPTG).

Each library was subjected to three rounds of selection, using 100 mM HCl elution, followed by neutralization. Phage from round 1 were amplified in *E. coli* before use in round 2. The eluted phage from round 2 were not amplified before use in round 3. Table III shows the results of library selection in the form of enrichment ratios. For IGFbp-1, only libraries F′ and P′ showed significant enrichment, while for bIGFbp-2, these two libraries as well as G′ and Q′ showed significant enrichment. Common to both targets

[42] E. Amman and J. Brosius, *Gene* **40,** 183 (1985).

is the appearance of selectants containing CX_8C or CX_9C putative disulfide loops. The success of these libraries in yielding novel binding peptides is representative of our broader experience in that loop libraries of this size are often fruitful sources of binding and inhibitory peptides.

Screening of Polyvalent Phage Clones (Insulin-Like Growth Factor-Blocking Phage Assay)

Peptide–phage clones were isolated by mixing phage pools with *E. coli* cells, and plating onto antibiotic-containing medium. Colonies were isolated and grown with helper phage (as described above) to obtain single-stranded DNA for sequencing. Peptide sequences selected for binding IGFbp-1 were deduced from the DNA sequences of phagemid clones. A number of such clones are represented by the peptide sequences in Table IV. In general, few unique clones were found after three or four rounds of selection using either IGFbp-1 or bIGFbp-2; many clones were siblings (identical in DNA sequence).

Such peptide–phage clones could represent specific target-binding peptides that either do or do not block ligand binding (i.e., IGF-I binding to IGFbp), or nonbinding, background members of the selected pool. To distinguish among these possibilities, phage clones were tested for the ability to bind to IGFbp-1 in the presence and absence of IGF-I.

TABLE IV
BINDING PEPTIDES SPECIFIC FOR INSULIN-LIKE GROWTH FACTOR-BINDING PROTEIN 1 OR BOVINE INSULIN-LIKE GROWTH FACTOR-BINDING PROTEIN 2

Library[a]	Peptide sequence[b]
IGFbp-1-specific binders	
A	SEVGCRAGPLQLCEKYF
B	KDPVCGEGPLMRICERLF
C	EVDGRWWIVETFLAKWDHMA
F′	QGAMECEVVPRGVMCVLDSK
N′	EVCVQIEWWCSH
bIGFbp-2-specific binders	
F′	YMDSGCAQGWEEYCMWECS
G′	VTQESCWWEDMVGMECGRLE
K′	RDCDTSCGH
P′	SNCFWDGATVVCAQ
P′	TDCTGPWVWVWCPQV
Q′	DGCAWDGVQMVDCTG

[a] Library designs are shown in Tables II and III.

[b] Residues that are fixed in the library construction are underlined.

IGFbp-1 was coated directly onto Maxisorp plates as described above. Phage from clonal cultures were mixed with IGF-I (100 n*M* final concentration), and incubated with the immobilized IGFbp for 1 hr at room temperature. The plates were then washed 10 times, as described above, and a solution of rabbit anti-M13 antibody mixed with a goat anti-rabbit antibody–HRP conjugate was added. After an incubation of 1 hr at room temperature, the plates were washed 10 times and then developed with OPD solution. The reaction was stopped with addition of 1/2 volume of 2.5 *M* H_2SO_4. Optical density at 492 nm was measured on a spectrophotometric plate reader.

An assay of the dominant clone from library selections, bp102 (SEVGCRAGPLQWLCEKYF), is shown in Fig. 2A. The results of this assay demonstrate that even at high concentrations of polyvalent phage, IGF-I is able to effectively block the binding of phage to immobilized IGFbp-1. Such blocking could occur competitively, if the binding sites for IGF-I and peptide on IGFbp-1 overlap, or noncompetitively, if IGF-I alters the conformation of IGFbp-1 so as to alter the binding site for peptide. In either case, additional experiments are needed to test the converse: whether or not peptide binding can inhibit IGF-I binding to IGFbp-1.

Note that this experiment does not provide binding affinity information. To illustrate this, the 50% binding saturation point in Fig. 2A occurs at $<10^{10}$ phage/ml. If we assume that 100 copies of peptide are displayed per phage particle, then the total concentration of peptide in this solution is at most 10^{-9} *M* (i.e., 10^{12} molecules/ml) at the midpoint of this phage dilution curve. On the other hand, as shown below and in peptide solution,[17] the binding affinity (K_d) of this peptide is on the order of 10^{-7} *M*. A likely explanation for this discrepancy is the avidity effect: polyvalent phage are more efficiently captured by an immobilized target than would be predicted on the basis of the intrinsic affinity of the displayed molecule for binding the target in solution.

Affinity Measurement by Monovalent Phage Enzyme-Linked Immunosorbent Assay

While polyvalent phage display enables the detection and selection of even low-affinity peptides, monovalent phage display allows for quantitative binding affinity measurements by phage ELISA.[28,43] For monovalent phage display of the IGFbp-1-binding peptide bp102, the peptide-coding DNA sequence was inserted in place of the human growth hormone (hGH)-coding gene of the phagemid phGHam-g3.[18] The resulting construct displays

[43] H. B. Lowman, *in* "Methods in Molecular Biology" (S. Cabilly, ed.), Vol. 87, p. 249. Humana Press, Totowa, New Jersey, 1998.

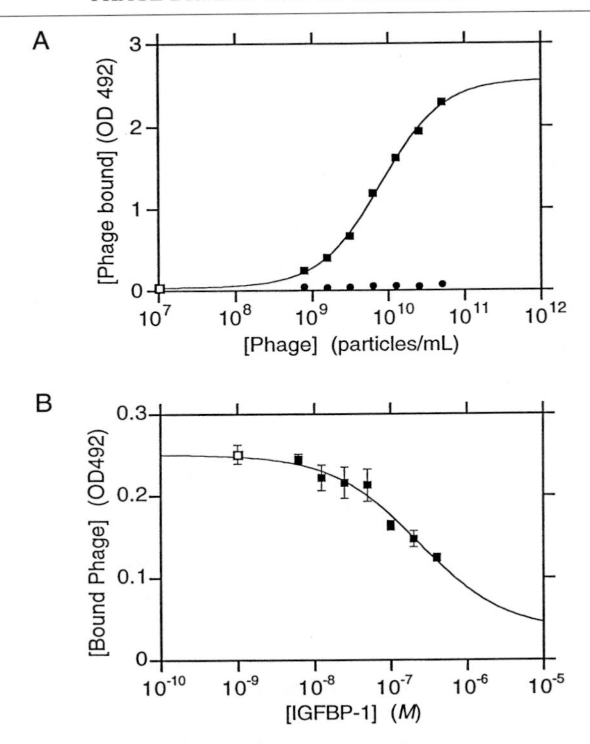

FIG. 2. ELISAs with phage displaying the IGFbp-1-binding peptide bp102. (A) Inhibition assay showing the ability of IGF-I to block the binding of polyvalent bp102-phage to an IGFbp-1-coated plate. A phage solution (5.0×10^{10} phage/ml) displaying bp102 as a P8 fusion was serially diluted and mixed with IGF-I (circles; 1 μM final concentration) or PBT buffer (filled squares), before being added to an IGFbp-1-coated plate. A no-phage control is also shown (open square). (B) Competitive phage ELISA showing the relative amount of monovalent bp102-phage bound to an IGFbp-1-coated plate in the presence of increasing concentrations of IGFbp-1 in solution. A phage solution (1.2×10^{11} phage/ml) displaying bp102 as a fusion with the P3 carboxy-terminal domain was premixed with serial dilutions of IGFbp-1 (filled squares). A no-competition control is shown (open square). The curve shows an IC_{50} of 230 ± 30 n*M*.

bp102 fused to the C-terminal domain of P3. Phage ELISAs were carried out as described above, using immunosorbent plates coated with IGFbp-1 at 2.0 μg/ml. bp102–P3 phage were serially diluted, and binding was measured to determine a phage concentration giving <50% of the ELISA signal at saturation. Thereafter, a fixed concentration of phage was premixed with serial dilutions of IGFbp-1, then added to an IGFbp-1-coated plate. The results (Fig. 2B) show that soluble IGFbp-1 was able to bind bp102–P3 phage, inhibiting phage binding to immobilized IGFbp-1 with an IC_{50} of 230 n*M*.

A Structured Peptide for Binding Insulin-Like Growth Factor-Binding Protein 1

Two forms of the IGFbp-1-binding peptide were synthesized for structure and activity testing: bp101, CRAGPLQWLCEKYFG(cyclic), and bp102, SEVGCRAGPLQWLCEKYFG(cyclic). Both showed activity for blocking IGF-I binding to IGFbp-1, with submicromolar activities: an IC_{50} of 50–190 n*M* for bp102 and an IC_{50} of 180–400 n*M* for bp101.[17] These activities compare favorably with the apparent affinity (230 ± 30 n*M*) of peptide bp102 for IGFbp-1 as measured in a monovalent phage ELISA (Fig. 2B). The structure of bp101 was determined by NMR analysis (Fig. 3A; see color insert).[17]

As noted above, relatively few unique clones were identified for binding each of the IGFbp's tested. A possible explanation for this is that the peptide epitopes for binding IGFbp's are small, perhaps consisting of only four to six residues. In that case, even an incomplete peptide–phage library might be expected to yield a near-optimal peptide binder, assuming that the remaining residues, although perhaps necessary for overall peptide structure, are more tolerant to variation. Consistent with this, extensive randomization and reselection within the 14-mer bp101 framework have shown some variability at many positions, but little improvement in binding affinity compared with the initial polyvalent phage clone (Y. Chen and H. Lowman, unpublished observations, 1999).

Ongoing efforts are directed at understanding how bp101 interacts with IGFbp-1. In contrast to IGF-I itself, which binds at least six distinct IGFbps as well as cell surface receptors, bp101 is highly specific for IGFbp-1. Work has shown that the binding epitopes for IGFbp-1 and IGFbp-3 on IGF-I are overlapping but have distinct features.[30] Small peptides identified from phage-displayed libraries for binding IGFbps may therefore mimic discrete patches of the IGF-I surface and, in so doing, provide extraordinarily specific inhibitory reagents for probing the biological role of IGFbps and IGF-I.

Example 2: Vascular Endothelial Growth Factor

Vascular endothelial growth factor (VEGF) is a hormone involved in modulating the development and permeability of blood vessels (for review see Dvorak *et al.*[44]). Antagonists that block the action of VEGF are of

[44] H. F. Dvorak, L. F. Brown, M. Detmar, and A. M. Dvorak, *Am. J. Pathol.* **146,** 1029 (1995).

major biomedical interest because of the role that the hormone may play in the angiogenesis of solid tumors.[45] Thus, it was of interest to develop high-affinity peptides that block binding of VEGF to its receptors.[46]

Identification of Weak Binding Peptides on Gene 8

For identifying peptides that bind to VEGF, polyvalent phage libraries were used as described above. The gene 8 peptide libraries B and D (Table II) were sorted separately for two rounds as follows. A form of VEGF containing only residues 8 through 109 was treated with 1.5 equivalents of EZ-Link NHS-SS-biotin. The lysine-modified VEGF was allowed to bind to NeutrAvidin-coated Maxisorp plates in PBS, and the plates were incubated with blocker casein buffer (BCB). Plates were washed briefly with BCB and phage libraries were added ($\sim 10^{12}$ phage particles per well) and shaken for 1–2 hr at room temperature. Unbound phage were rinsed away and phage were eluted by a 5-min treatment with 75 m*M* DTT, which reduces the disulfide that links VEGF to the plate. Phage were propagated in *E. coli* XL1-Blue and prepared for another round of selection.

For the third and fourth rounds of selection, VEGF was coated directly onto plates and blocked with 0.5% (w/v) BSA in PBS to avoid selection for NeutrAvidin binders. Bound phage were eluted with 10 m*M* HCl [in 100 m*M* NaCl and 0.3% (w/v) BSA], neutralized with 1.0 *M* Tris, pH 8.0, and propagated for the next round.

After four rounds of selection, we titered the number of phage from the selected pools from libraries B and D that bound to plates that were coated with VEGF versus BSA. We found that for library B or D there was a 19,000 or 6000-fold difference, respectively, suggesting significant enrichment for VEGF-binding peptides. Twenty clones from each of the libraries were sequenced to reveal a single consensus sequence from library B of GERWCFDGPLTWVCGEES (called class 1; fixed residues are underlined) and two consensus sequences from library D of RGWVEICVADDNGMCVTEAQ and GWDECDVARMWEWECFAGV (called class 2 or 3, respectively). By the phage-blocking assay described above, it was possible to displace phage containing each of these sequences with a natural VEGF ligand (the kinase insert domain-containing receptor, KDR), indicating that they bound at or near the KDR-binding site on VEGF.

[45] J. Folkman, *Nature Med.* **1,** 27 (1995).

[46] W. J. Fairbrother, H. W. Christinger, A. G. Cochran, G. Fuh, C. J. Keenan, C. Quan, S. K. Shriver, J. Y. K. Tom, J. A. Wells, and B. C. Cunningham, *Biochemistry* **37,** 17754 (1998).

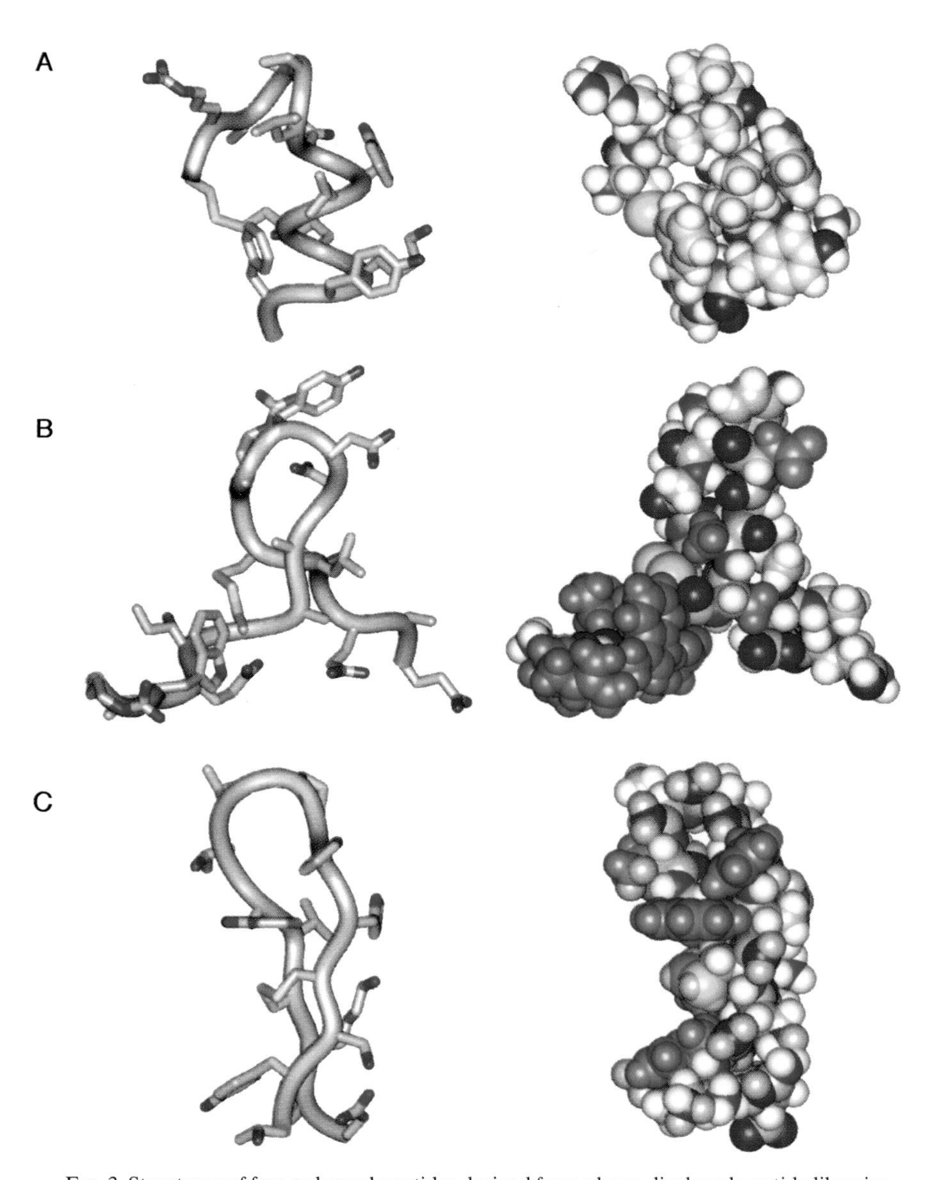

FIG. 3. Structures of free or bound peptides derived from phage-displayed peptide libraries. In each panel, a model of the structure at left illustrates side chains, with oxygen, nitrogen, or sulfur atoms shown in red, blue, or yellow, respectively. At right are shown space-filling models to illustrate the peptide surfaces. When the bound structure is known, the atoms making contact with the binding partner (protein) are shown in green. (A) The structure of the bp101 peptide in solution as determined by NMR[17] shows a turn–helix conformation. (B) The structure of a bound form of the v108 peptide as determined by X-ray crystallography of the peptide: VEGF.[47] (C) The EMP1 peptide is a β hairpin that forms a symmetric, noncovalent dimer in complex with two molecules of EPO receptor.[15] The monomer structure is shown.

Initial Optimization of Peptides by Soft Randomization on Gene 3

Given the limited library size ($\sim 10^{10}$), it was impossible to sample all the diversity space in the original 20 residue peptides (20^{16} or 20^{18} possible sequences for library B or D, respectively). Using the gene 8-selected sequences as leads, we then wanted to search nearby sequences for improved variants. To do this, we employed soft randomization, in which the sequence is mildly randomized so that, on average, each residue can mutate to a non-wild-type residue with a frequency of 40%. When this is carried over the 20-residue sequence, there is only a small fraction of the original sequence left (<0.01%). Such multiply mutated sequences allow for selection of highly cooperative effects. However, the fact that the library is large ($>10^9$ variants) still allows for rescue of sequences closely related to the original. To achieve the soft randomization conditions, the mutagenic DNA was synthesized so that each base was 80% wild type and 7% for each of the other nucleotides.

To increase the stringency of selection, we transferred the three sequences described above into a monovalent format by fusing to gene 3 (as described above). Phage libraries were sorted for six rounds for binding to VEGF directly adsorbed to Maxisorp plates. Clones from each library were sequenced and those derived from class 1 are shown in Fig. 4A. A clear consensus developed having W–C–F around the first Cys, and W-V-C-G around the second Cys. The Gly–Pro turn was also completely conserved in all the selectants. Many of the other residues were observed to occur predominantly as one type or another. Some of these selectants were analyzed by phage ELISA using KDR to displace the phage and found to have IC_{50} values in the range of 0.6 to 6.0 μM.

Final Optimization by Tailored Randomization on Gene 3

On the basis of the conservation or predominance of particular residues in the soft randomization experiment for the class 1 peptide, it was possible to fix 8 of the 18 positions, and to partially fix many of the others. Because the diversity field had been substantially narrowed, it was now possible to focus the mutagenesis on those residues that had shown only partial consensus. To do this, a tailored randomization approach was used. A library was constructed with fixed codons at eight positions and partially randomized codons at other positions. At each varied position, a tailored codon was chosen to include primarily those residues that were selected from the soft randomization. For example, Arg, Leu, or Val was selected at position 10, so this position was represented by an SKT codon (where S is G or C and K is G or T), which encodes Arg, Leu, Val, or Gly. The total number of possible variants allowed was limited to $\sim 5 \times 10^5$ possible sequences to

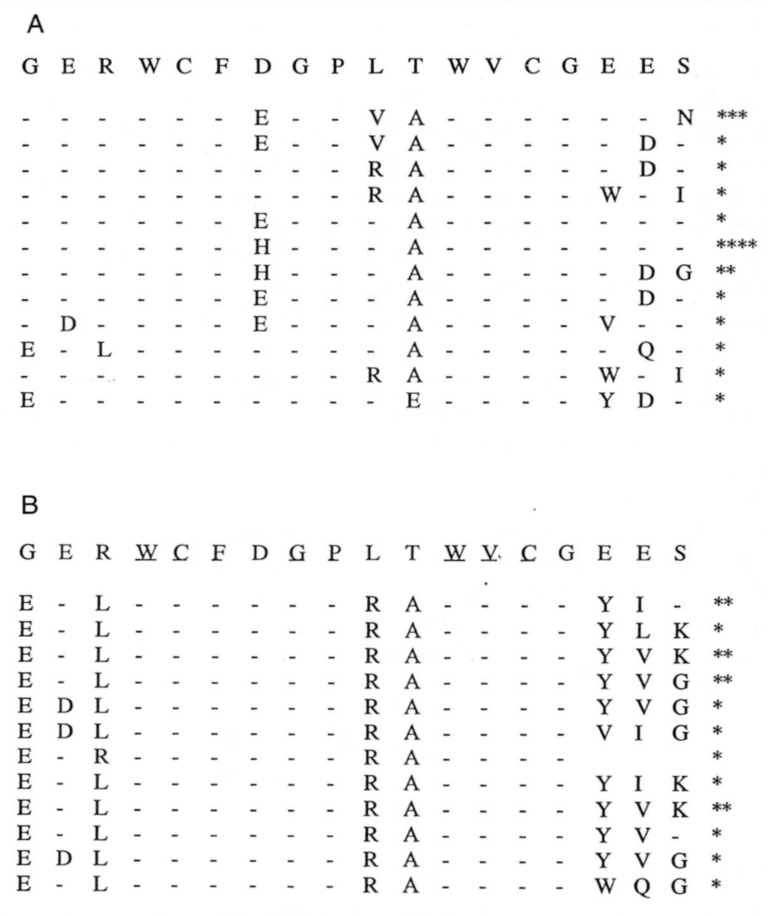

FIG. 4. Affinity-optimized VEGF binding sequences from class 1. (A) Sequences from the soft randomization selection. (B) Sequences from the tailored randomization selection. Dashes indicate identity with the parent sequence (shown at the top of each alignment with fixed residues underlined). Asterisks (*) represent the number of individual clones identified for each amino acid sequence. See Example 2 for details.

ensure good representation in the new libraries ($\sim 10^8$ independent clones). To increase the stringency of selection during this phase, we progressively lowered the concentration of the VEGF used to coat from 50 to 1.0 n*M* over the six rounds of selection. Clones were sequenced after selection, revealing a similar kind of consensus as seen in the soft randomization for class 1 (Fig. 4B). In general, the tailored randomization experiments sharpened the consensus residues that were observed for all three classes

TABLE V
IC_{50} VALUES FOR VASCULAR ENDOTHELIAL GROWTH FACTOR-BINDING SYNTHETIC PEPTIDES

Peptide	Sequence	IC_{50} (μM)[a]
Class 1		
v106	GERWCFDGPRAWVCGWEI-NH_2[b]	7.0 ± 0.9[c]
V127	EELWCFDGPRAWVCGYVK-NH_2[d]	10.0
Class 2		
v108	RGWVEICAADDYGRCLTEAQ-NH_2[b]	2.2 ± 0.1
v113	RGWVEICAADDYGRCL-NH_2[b]	4.0 ± 0.1
v128a	RGWVEICESDVWGRCL-NH_2[d]	1.1
Class 3		
v107	GGNECIARMWEWECFERL-NH_2[b]	0.70 ± 0.06
v114	VEPNCDIHVMWEWECFERL-NH_2[d]	0.22

[a] The IC_{50} values are the concentrations of peptide that resulted in 50% inhibition of VEGF (residues 8–109) binding to KDR (domains 1–3), using a BIAcore-based assay.
[b] Peptides derived from sequences identified in the soft randomization selection.
[c] Values with uncertainties reported result from duplicate measurements.
[d] Peptides derived from sequences identified in the tailored randomization selection (Fig. 4).

of peptides, and led to isolation of peptides that were about 5- to 10-fold improved in their affinities when compared with the best isolates after soft randomization.

Characterization of Binding Properties and Structures for Selected Peptides

To identify the best peptide sequences for synthesis and further characterization, we routinely used a phage ELISA with soluble VEGF as competitor (as described above). The sequences for a number of the best selectants were chemically synthesized. Their binding constants were estimated from their ability to block the binding of VEGF to immobilized KDR in a BIAcore (Biacore, Uppsala, Sweden) assay.[46] The affinities for these peptides ranged from 10 to 0.2 μM, depending on the class of peptides (Table V).

We also analyzed their binding sites to VEGF either by HSQC measurements[46] or by X-ray crystallography of the complex.[47] Each peptide was

[47] C. Weismann, H. W. Christinger, A. G. Cochran, B. C. Cunningham, W. J. Fairbrother, C. J. Keenan, G. Meng, and A. M. de Vos, *Biochemistry* **38,** 17765 (1998).

shown to bind to an overlapping but nonidentical site on VEGF. The structure of the best Class 2 peptide is shown in Fig. 3B (see color insert) (taken from Weismann *et al.*[47]).

Structures of Peptides from Phage Displayed Libraries

Several high-resolution structures of phage-derived peptides have been obtained by NMR (Fig. 3A; see color insert) or X-ray crystallography (Fig. 3B and 3C; see color insert). The structures of these peptides allow for the design of recombinant or synthetic analogs to probe their binding interactions with their respective targets.

In the case of the bp101 peptide, attempts to cocrystallize peptide with IGFbp-1 were not successful. However, alone in solution, the 15-residue peptide still yielded well-defined NMR spectra, allowing for high-resolution structure determination of the unbound form. Unlike the remaining examples shown here, no structure of the bound form has been determined. However, it has been argued that the bound form likely resembles the free peptide structure.[17] The bp101 structure (Fig. 3A; see color insert) reveals a turn–helix motif, with hydrophobic side chains Leu-6, Leu-9, Trp-8, Tyr-13, and Phe-14 forming a cluster on one side of the helix. The packing of these hydrophobic side chains, in addition to the disulfide bond, appears to stabilize the turn–helix structure.

Among VEGF-binding peptides, the best class 2 peptide, v108, which also derives from a CX_8C library, was crystallized in complex with VEGF[47] and found to have a different structure (Fig. 3B; see color insert). Unlike the bp101 peptide, v108 was unstructured in solution. In the bound state, this peptide has a loop structure, but with an extended N terminus that makes extended main-chain hydrogen-bonding interactions with VEGF.

The EMP1 peptide, identified for binding to and activating the EPO receptor,[14] was found by crystallography to form a β-hairpin structure when bound to a dimer of EPO receptor.[15] In this case, the peptide actually undergoes noncovalent dimerization so that the complex consists of two molecules of peptide bound to two molecules of receptor. While a structure of the free peptide has not been reported, the bound structure has been used in extensive structure–activity analyses of this peptide.[48]

The structural and binding determinants of peptides are closely intertwined. For example, the bp101 peptide shares the sequence motif CxxGPLxWxC with the EMP1 peptide, GGTYSCHFGPLTWVCKP-

[48] D. L. Johnson, F. X. Farrell, F. P. Barbone, F. J. McMahon, J. Tullai, K. Hoey, O. Livnah, N. C. Wrighton, S. A. Middleton, D. A. Loughney, E. A. Stura, W. J. Dower, L. S. Mulcahy, I. A. Wilson, and L. K. Jolliffe, *Biochemistry* **37,** 3699 (1998).

QGG.[14] The Cys and Gly–Pro residues were in fact invariant in the library from which bp101 was derived,[17] yet these peptides have quite different structures, in one case a turn–helix and in the other a β hairpin. Thus, a variety of different side-chain conformations can be presented even within the context of the CX_8C family. Clearly peptides of this size provide an abundance of different binding epitopes for protein targets.

Conclusions

The technology of peptide phage display has exploded in recent years. It clearly has the ability to generate small peptides (10–20 residues long) that will bind to almost any protein with moderate to high affinity. The technology effectively mimicks the immune system and even offers some advantages. Selections are performed *in vitro* so that buffer conditions, temperatures, and other aspects of the process can be adjusted to benefit the selection process. Library sizes can now be made much larger ($>10^{12}$) than the initial antibody library, and the mutagenesis for affinity maturation can be focused on discrete areas of design. The selected peptides are functionally robust and in many cases highly structured. These seem to fall into a small number of structural classes, which may not be surprising given their small size. It is also remarkable that they often seek ligand-binding sites even though the selection process exerts no pressure for this to happen. Finally, the product of selection is a peptide that can be tinkered with and further improved by synthetic chemistry.

Acknowledgments

We thank Yvonne Chen, Gerald Nakamura, and Mark Dennis for phage constructs and assistance with phage libraries, Nicholas Skelton and Christian Wiesmann for preparing figures, and David Reifsnyder and David Clemmons for providing IGFbp-1 and bIGFbp-2.

[22] Selectively Infective Phage Technology

By KATJA M. ARNDT, SABINE JUNG, CLAUS KREBBER, and ANDREAS PLÜCKTHUN

Introduction

In this chapter we describe a selection technology related to phage display, the selectively infective phage technology (SIP). In both SIP and phage display, libraries of proteins or peptides of interest are displayed on the tip of a filamentous phage fused to the gene 3 protein (g3p). However, in contrast to conventional phage display, in which separation of binders with an immobilized ligand is required, in SIP technology the selection step is carried out in solution and directly coupled to the infection event. Therefore, no elution of binders is ever necessary, and incomplete elution of the most tightly binding molecules is not a limitation.

This is achieved by taking advantage of the modular structure of g3p, which occurs probably in five copies on the tip of the filamentous phage and consists of the N-terminal domains N1 [68 amino acids (aa)], N2 (131 aa), and the C-terminal domain CT (150 aa), connected by the glycine-rich linkers G1 (18 aa) and G2 (39 aa) (Fig. 1A and Fig. 2A).[1,2] In a SIP phage the N-terminal domains required for infection (N1 or N1–N2) are replaced by a protein or peptide of interest in all copies of g3p. This leads to a noninfective phage that displays the protein or peptide to be selected as a fusion to the C-terminal domain of g3p. The "adapter," consisting of either N1 or N1–N2, is fused to the second protein or peptide of interest or chemically coupled to a nonpeptidic ligand (Fig. 1B and C). No wild-type g3p must be present on a SIP phage. Therefore, in its simplest form, the CT-fusion protein directly replaces the g3p in the phage genome. However, we discuss below other strategies for carrying out SIP with phagemid/helper phage systems. To simplify the nomenclature, we refer to the protein or peptide library fused to the CT domain of g3p, and thus displayed on the phage, as "A" and denote the target fused to or chemically coupled to the N-terminal part of g3p, the soluble adapter, as "B." Cognate interaction between these two molecules (A and B) restores the gene 3 protein in a noncovalent form and thus allows for phage infection. There are two variants of this selection system, termed *in vitro* SIP and *in vivo* SIP.

In the *in vitro* SIP procedure only the fusion of library A to the C-

[1] I. Stengele, P. Bross, X. Garces, J. Giray, and I. Rasched, *J. Mol. Biol.* **212,** 143 (1990).
[2] J. Armstrong, R. N. Perham, and J. E. Walker, *FEBS Lett.* **135,** 167 (1981).

0076-6879/00 $30.00

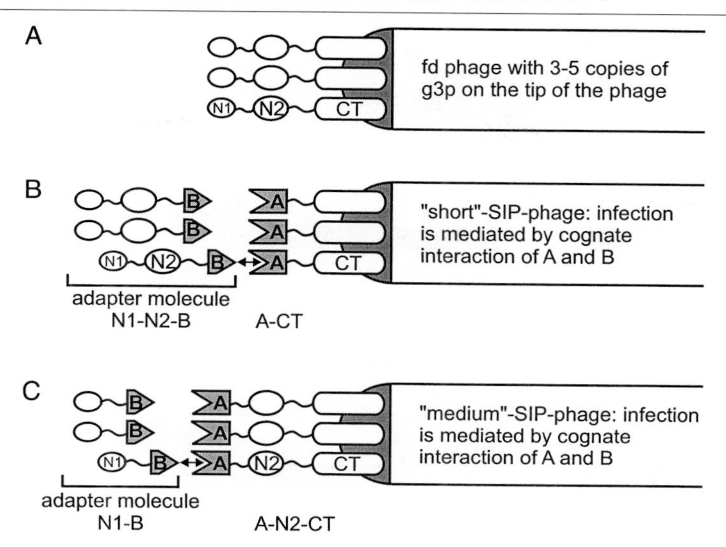

FIG. 1. Overview of the arrangement of the gene 3 protein in various phages. (A) Fd phages carry on their tip three to five copies of the gene 3 protein, which consists of three domains, the two N-terminal domains N1 and N2, required for infection and binding to the F pilus, respectively, and the C-terminal domain CT, which anchors the N-terminal domains in the phage particle. (B) SIP phage as used for *in vivo* and *in vitro* SIP. One protein of interest is displayed as a fusion to CT on the phage ("short" phage), whereas N1–N2 are expressed as a fusion to or chemically coupled to the other protein/molecule of interest. In *in vitro* SIP, the adapter molecule is expressed and purified separately and added later for infection, whereas in *in vivo* SIP both proteins are expressed simultaneously on the phage genome. (C) Alternative SIP phage used in *in vitro* SIP. In this construct, the "medium" phage displays the protein of interest fused to N2–CT, whereas the adapter consists only of N1 and the second protein of interest.

terminal part of g3p is encoded by the phage or phagemid vector (Fig. 2C and D). The adapter molecule with the N-terminal part of the gene 3 protein is expressed and purified separately and can then be chemically coupled to a great variety of targets, including nonpeptidic ones. The separate expression of the adapter molecule allows for better control of the infection experiments by enabling quality control of the adapter molecule, titration of adapter, and measurement of background infectivity of purified phage particles. For *in vivo* SIP, both parts of the gene 3 protein are expressed simultaneously *in vivo,* and thus the genetic information of both parts remains linked to the phage phenotype. This means that, in principle, a library-versus-library approach should be feasible, in which one library is fused to the CT domain (library A) and the other library is fused to the N-terminal (adapter) part of the gene 3 protein (library B) (Fig. 2E).

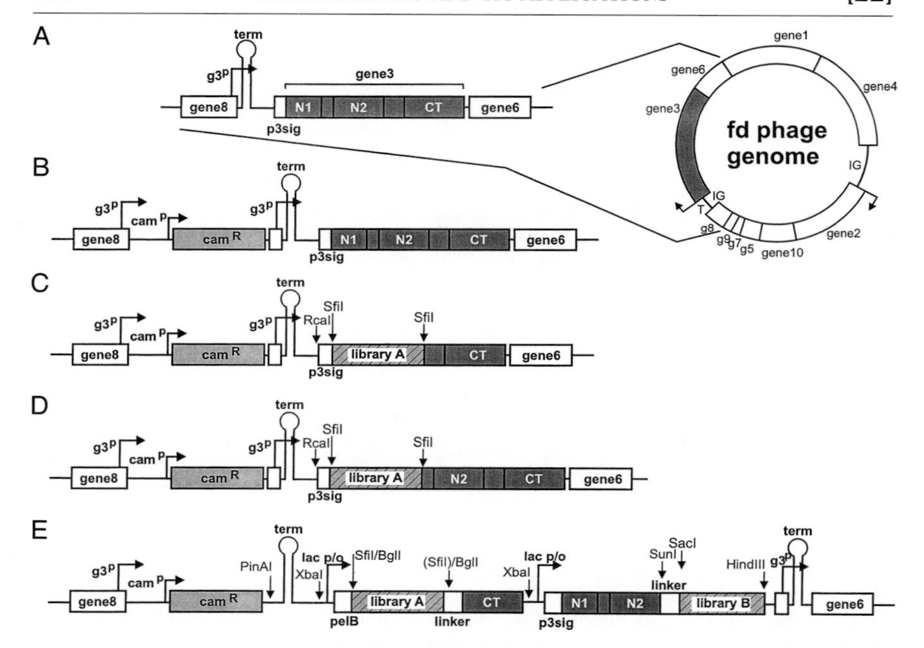

FIG. 2. (A) Organization of the fd phage genome. The genome of the wild-type phage is subdivided by two intergenic regions (IG) into a region of strongly expressed genes (genes 2, 5, 7, 8, 9, and 10) and a region of weakly expressed genes (genes 1, 3, 4, and 6). A terminator (*term*) separates the weakly expressed genes from the strongly expressed ones. The organization of the gene segment of the fd phage showing gene 8, gene 3 promotor ($g3^p$), gene 3 terminator (*term*), gene 3 (dark gray) encoding the three domains N1, N2, and CT of the wild-type gene 3 protein, and gene 6 is shown in more detail. *p3sig* indicates the signal sequence for the gene 3 protein, necessary for transport into the periplasm. (B) In the derivative of the fd phage fCKC, which served as a starting construct for all other constructs, a chloramphenicol resistance gene (Cam^R, light gray) is inserted downstream of gene 8 and upstream of the natural terminator. The end of gene 8 is repeated, as it carries the promotor for g3p.[7] (C) Gene arrangement of the "short" phage (library A-CT) and (D) "medium" phage (library A-N2-CT) as used for *in vitro* SIP. The fusion is expressed under the natural gene 3 promotor ($g3^p$), and library A can be exchanged by two *Sfi*I sites with different cutting sequences. (E) Gene arrangement of the two parts of the gene 3 protein used in *in vivo* SIP. Libraries A and B mark the places where the proteins/peptides of interest are encoded, which can be a single sequence or a library. In this case, both fusions are under the control of the *lac* promotor/operator system (*lac* p/o), and a second terminator is inserted in front of gene 6. Both libraries can be cloned by unique restriction sites as indicated. Restriction enzymes in parentheses indicate sites that are not present in all constructs.

However, this system, which is more powerful but also less controllable, can be used only with proteins and peptides for both partners.

In the following section we provide several general protocols, which apply for *in vitro* as well as *in vivo* SIP. In the second and third part, we focus on *in vitro* and *in vivo* SIP, and compare in more detail the use of both methods by means of a few examples. Two other names, DIRE (direct rescue interaction)[3] and SAP (selection and amplification of phage),[4] have also been used for this technology.

General Procedures

Bacterial Strains

XL1-Blue (Stratagene, La Jolla, CA) (*recA1 endA1 gyrA96 thi-1 hsdR17 supE44 relA1 lac* [*F′ proAB lacI*^{q}*Z*Δ*M15* Tn*10* (Tet^{R})]) is the preferred host for electroporation and for infection experiments because of its *recA1* genotype. For infection, it is possible to switch resistances by alternatingly using the tetracycline- or the kanamycin-resistant strain XL1-Blue MRF′ Kan (Stratagene; Δ(*mcrA*) *183* Δ(*mcrCB-hsdSMR-mrr*) *173 endA1 supE44 thi-1 recA1 gyrA96 relA1 lac* [F′ *proAB lacI*^{q} *Z*Δ*M15* Tn5 (Kan^{R})]). This ensures that during phage infection only phages but no cells are carried over. An additional precaution is to filter the phage-containing supernatant directly after phage production and prior to polyethylene glycol (PEG) precipitation (see Phage Production, below). BL21 (DE3) (Novagen, Madison, WI) [F^{-} *ompT hsdS*$_{B}$ (r_B^- m_B^-) *gal dcm,* harboring DE3, a λ prophage carrying the T7 RNA polymerase gene] is used for the cytoplasmic production of N1–N2 adapter molecules for *in vitro* SIP. BL21 (DE3)pLysS (Novagen) [same as BL21 (DE3), but harboring the plasmid pLysS Cam^{R}, which encodes T7 lysozyme, a natural inhibitor of T7 RNA polymerase; resistant to chloramphenicol at 40 μg/ml] is used for the cytoplasmic production of N1 adapter molecules for *in vitro* SIP.

Vector Constructs: Phage versus Phagemid System

In SIP, the infection is strictly dependent on the reconstitution of the gene 3 protein. Therefore, when using a phagemid/helper phage system, it is absolutely necessary that the helper phage does not deliver any wild-type g3p to the phage particle. For this purpose, Δg3 helper phages were constructed.[5,6] These phages, while not carrying the gene, do need to carry

[3] K. Gramatikoff, O. Georgiev, and W. Schaffner, *Nucleic Acids Res.* **22,** 5761 (1994).
[4] M. Dueñas and C. A. Borrebaeck, *Bio/Technology* **12,** 999 (1994).
[5] M. Dueñas and C. A. Borrebaeck, *FEMS Microbiol. Lett.* **125,** 317 (1995).
[6] F. K. Nelson, S. M. Friedman, and G. P. Smith, *Virology* **108,** 338 (1981).

functional g3p on their coats, because the helper phage must be able to infect the bacteria harboring the phagemid library for which it needs the g3 protein. There are two ways to produce such a phage. First, the helper phage can be produced in cells harboring an unpackageable g3p expression plasmid, which does not contain an F1 origin.[4,5] Nonetheless, we still observed some packaging of the g3p expression plasmid, which increased the background in subsequent infection experiments, as the population will produce phages carrying wild-type g3p. This problem has more recently been solved by engineering an *Escherichia coli* strain with a g3p expression cassette integrated in the chromosome (S. Rondot and F. Breitling, personal communication, 1999). An alternative way is to transform the phagemid library into cells that already contain Δg3 helper phage.[3] However, at least in our experience, the helper phage Δg3 M13KO7 showed genetic instabilities and lowered the viability and transformation frequency of bacteria.[7]

A third approach, favored by the authors, is to directly use a phage genome, in which the complete genetic information of the phage is encoded. The circular genome of the filamentous phage consists of two main transcriptional units that are separated by a central terminator on one side and the origin of replication on the other (Fig. 2A).[8] The wild-type gene 3 is modified accordingly for *in vitro* and *in vivo* SIP, and the resulting gene 3 cassettes are described further below. In addition, an antibiotic resistance gene was inserted into the fd genome, immediately downstream of gene 8 and upstream of the original gene 3, to enable detection of infection events by screening for resistance (Fig. 2B). Several antibiotic resistances were tested, and resistance against chloramphenicol was chosen, because it gave, along with ampicillin resistance, the highest level of infectivity, and it had the advantage that it does not allow growth of plasmid-free cells due to enzyme leakage as observed for ampicillin. This resulted in the phage vector fCKC,[7] which is the basis for all other constructs.

Vector Constructs: Different Gene 3 Fusions

Similar to phage display, a peptide or protein library of interest can be fused to different parts of the gene 3 protein of the phage. In the "short" fusion, the protein of interest is expressed as a fusion to the CT domain, while in the "medium" fusion, the protein or peptide library of interest is fused to the N terminus of the N2–CT fragment. Conversely, the adapter can consist either of only the N1 "bait" fusion or an N1–N2 "bait" fusion (Fig. 1B and C). In one combination of phage and adapter, no N2 domain at all is present, while in another N2 is present both on the adapter and

[7] C. Krebber, S. Spada, D. Desplancq, and A. Plückthun, *FEBS Lett.* **377,** 227 (1995).
[8] E. Beck and B. Zink, *Gene* **16,** 35 (1981).

the phage. All four combinations lead to infectivity as tested for *in vitro* SIP, albeit at different levels and with a different concentration dependence of the adapter.[9] Only the "short" fusion has been tested for *in vivo* SIP so far.

Library Construction

The library will of course depend on the particular question to be addressed with SIP. A number of standard libraries is available nowadays, including peptide libraries, antibody libraries, or cDNA libraries, some of which are even commercially available. In addition, several techniques have been developed for generating genetic diversity using, e.g., PCR (polymerase chain reaction) techniques,[10–13] recombination sites,[14] mutator strains or degenerated codons,[15,16] and trinucleotides[17] for randomization of synthetic genes. These methods are numerous and well established, and therefore are not detailed here. However, we discuss the issue of library complexity and quality in connection with the *in vivo* SIP procedure.

Phage Production

Phage expression in XL1-Blue cells carrying the phage DNA is performed in shake flasks in 2× YT medium[18] supplemented with 1% (w/v) glucose, chloramphenicol (Cam; 30 μg/ml), and tetracycline (Tet; 5 μg/ml). Overnight cultures are grown at 37° (or 25° when there is reason to believe that the *in vivo* folding yield of the ligands might be increased on lowering the temperature). For *in vitro* SIP with the constructs shown, in which the gene 3 fusion protein is expressed under the natural gene 3 promotor, phages are obtained directly from the supernatant of the overnight culture. For *in vivo* SIP, where both parts of gene 3 are under *lac* promoter/operator (*lac* p/o) control, a main culture containing Cam (30

[9] C. Krebber, S. Spada, D. Desplancq, A. Krebber, L. Ge, and A. Plückthun, *J. Mol. Biol.* **268,** 607 (1997).
[10] R. C. Cadwell and G. F. Joyce, *Genome Res.* **3,** S136 (1994).
[11] A. Crameri, S. A. Raillard, E. Bermudez, and W. P. C. Stemmer, *Nature (London)* **391,** 288 (1998).
[12] W. P. C. Stemmer, *Nature (London)* **370,** 389 (1994).
[13] H. Zhao, L. Giver, Z. Shao, J. A. Affholter, and F. H. Arnold, *Nature Biotechnol.* **16,** 258 (1998).
[14] N. Tsurushita, H. Fu, and C. Warren, *Gene* **172,** 59 (1996).
[15] S. Kamtekar, J. M. Schiffer, H. Xiong, J. M. Babik, and M. H. Hecht, *Science* **262,** 1680 (1993).
[16] E. Wolf and P. S. Kim, *Protein Sci.* **8,** 680 (1999).
[17] B. Virnekäs, L. Ge, A. Plückthun, K. C. Schneider, G. Wellnhofer, and S. E. Moroney, *Nucleic Acids Res.* **22,** 5600 (1994).
[18] J. Sambrook, E. F. Fritsch, and T. Maniatis, "Molecular Cloning: A Laboratory Manual," 2nd Ed. Cold Spring Harbor Laboratory Press, Cold Spring Harbor, New York, 1989.

μg/ml), Tet (5 μg/ml), and glucose (1%, w/v) is inoculated from the overnight culture to an initial OD_{550} of 0.15 (typical dilution of 1:25–1:30) and grown at 37° until an OD_{550} of 0.5 to 0.6 is reached. The cells are then pelleted by centrifugation (3000*g*, 10 min, 4–10°) and resuspended in fresh 2× YT medium without glucose containing the appropriate antibiotics and 0.5 m*M* isopropyl-β-D-thiogalactopyranoside (IPTG) to induce the expression of the two fusion proteins, protein A–CT and N1–N2–protein B, which results in phage packaging. Expression is continued for at least 6 hr at 37°, or in the case of less stable proteins at 24–26° overnight. In both cases, phages are separated from the cells by centrifugation for 10–15 min at 5000*g* and 4°. The supernatant containing the phages can then be optionally filtered (0.22-μm pore size filters) to remove any remaining cells. The yield is usually in the range of 5×10^{11} to 2×10^{12} phages per milliliter of culture supernatant.

Phage Purification

For the infection experiments, phage particles are enriched by two precipitations with polyethylene glycol (PEG). The most convenient way is to prepare a five-times concentrated stock solution with 16% (w/v) PEG 6000 and 3.3 *M* NaCl, passed through a 0.22-μm pore size filter. For the PEG precipitation, one part of this solution is added directly to four parts of the supernatant obtained after phage production and cell precipitation, and incubated on ice or at 4° for at least 1 hr or overnight. Phages are pelleted by centrifugation at least at 5000*g* for 30 min at 4° and redissolved in about 1 ml of TBS buffer [Tris-buffered saline, 50 m*M* Tris-HCl (pH 8.0), 150 m*M* NaCl]. If higher purity or concentrations are required, a second PEG precipitation can be performed similar to the first one. For this purpose, PEG–NaCl solution is added in the same ratio as for the first PEG precipitation, and after 1 hr of incubation at 4° phages are pelleted in a 2-ml tube by centrifuging at maximal speed for 20 min at 4°. Phages are then dissolved in 50–200 μl of TBS buffer. They can be stored at 4° for several months, depending mainly on the stability of the fusion protein.

However, if phages of greater purity are required as, e.g., for a more accurate quantification or electron microscopy, phages are best purified by ultracentrifugation in a CsCl gradient.[19] Phages are first enriched 80-fold by PEG precipitation, and then added to a solution of 1.6 g of CsCl in TBS, filled to a final volume of 4 ml. The CsCl solution is transferred to a 13 × 38 mm polyallomer tube (Quick-Seal; Beckman, Fullerton, CA) and centrifuged at 100,000 rpm (541,000*g*) for 4 hr in a TLN-100 rotor (Beckman) at 4°. After ultracentrifugation the phages become visible as a white

[19] G. P. Smith and J. K. Scott, *Methods Enzymol.* **217,** 228 (1993).

band through light scattering when shining light from the top to the bottom of the tube and by looking through the wall of the tube at a right angle to the light beam. Phages are removed by puncturing the side of the tube with a hypodermic needle and gently removing the banded phages. To remove the remaining CsCl, phages are then pelleted by ultracentrifugation (50,000 rpm/135,000*g*, 1 hr, 4°, TLA-100.3 rotor), redissolved in 3 ml of TBS buffer, pelleted again (same conditions), and redissolved in 50 μl of TBS buffer.

Determination of the Phage Concentration

Phage particles are quantified spectrophotometrically by measuring the absorption between 200 and 350 nm. A broad peak between 260 and 280 nm is obtained due to the presence of both DNA and proteins, and the absorption at 269 nm is used to estimate the phage concentration[20] as follows:

$$\frac{\text{Phage particles}}{\text{ml}} = \left(\frac{\text{OD}_{269}}{\text{number of base pairs of ssDNA}}\right)(5.98 \times 10^{16})$$

This method, however, gives only a crude estimate and upper limit of phage concentration for phages that are prepared merely by PEG precipitation, because proteins from the culture supernatant can be coprecipitated by PEG. Especially in cases of low phage production, coprecipitated proteins can influence the absorption significantly, as seen by a red shift of the absorption maximum. This can be avoided by using phages purified by CsCl gradient ultracentrifugation. However, in our experience, "average" phage titers (about 10^{12} phages per milliliter of supernatant), prepared by two subsequent PEG precipitations, have a "normal" absorption spectrum with no red shift (comparable to purified phages), and usually give reliable results in concentration determinations, when compared with phages purified by ultracentrifugation.

Another possibility for phage quantification is by using an enzyme-linked immunosorbent assay (ELISA), by immobilizing the phage and detecting with an antibody against gene 8 protein. However, this method does not yield absolute phage concentrations, but relative numbers, and is therefore best suited for comparative infection experiments. In this case, phages prepared by PEG precipitation are coated overnight at 4° on an ELISA plate in a twofold dilution series. The ELISA plate is washed three times with TBS and blocked with 4–5% (w/v) milk powder in TBS for 1 hr at room temperature. After washing twice with TBS, phages are detected with a 1:5000 dilution of anti-M13–peroxidase conjugate (Pharmacia, Pis-

[20] L. A. Day, *J. Mol. Biol.* **39,** 265 (1969).

cataway, NJ) in TBST [TBS with 0.05% (v/v) Tween 20]. The amount of bound anti-M13–peroxidase is measured with peroxidase (POD) soluble substrate (Boehringer Mannheim, Mannheim, Germany) at 405 nm after various incubation times. To obtain absolute concentrations, a control phage dilution needs to be included with known concentration. Preferably, the concentration of this calibration standard should have been determined by UV absorption after purification by CsCl gradient ultracentrifugation.

In Vitro Selectively Infective Phage Procedures

In the *in vitro* SIP procedure, the adapter molecule is purified separately, and later added to phages displaying the protein or peptide library of interest, which is fused to the C-terminal part of the gene 3 protein. Thus, nonpeptidic "baits" can also be used, which can be chemically coupled to N1 or N1–N2, respectively. In addition, the concentration of the adapter molecule can be varied.

Design of Vectors Suitable for in Vitro Selectively Infective Phage

The phage vector used in these examples is based on fCKC[7] and encodes the protein or library of interest with the gene 3 leader sequence as fusion to either N2–CT ("medium" phage) or CT ("short" phage) under the control of the natural gene 3 promotor. The phage production is carried out in XL1-Blue as described in the general procedures.

Native, periplasmic expression of the N-terminal domains of gene 3 was found to be difficult, as the native protein always led to complete lysis of the culture and relatively low yields requiring a tedious concentration step. On the other hand, the N1 domain alone, lacking the glycine-rich linker, can be obtained by secretion.[21] However, we favor cytoplasmic expression of the gene 3 N-terminal domains under the control of the strong T7 promoter, using the vector pTFT74,[22] as this appears to be more reliable and simpler for downstream processing. The deletion of the signal sequence and the T7 expression system leads to cytoplasmic inclusion bodies, which are easily refolded. Interestingly, the toxicity effects of secreted proteins vary significantly for the various constructs. The highest toxicity is observed for the N1 domain with a C-terminal glycine-rich linker, whereas the expression of the N1–N2 complex caused significantly fewer problems. However, N1–N2 with a C-terminal glycine-rich linker showed again more problems with expression (lower yield and higher toxicity) than N1–N2 without the

[21] P. Holliger and L. Riechmann, *Structure* **5,** 265 (1997).

[22] C. Freund, A. Ross, B. Guth, A. Plückthun, and T. A. Holak, *FEBS Lett.* **320,** 97 (1993).

A

```
  1  AETVESCLAK SHTENSFTNV WKDDKTLDRY ANYEGCLWNA TGVVVCTGDE
 51  TQCYGTWVPI GLAIPENEGG GSEGGGSEGG GSEGGGTKPP EYGDTPIPGY
101  TYINPLDGTY PPGTEQNPAN PNPSLEESQP LNTFMFQNNR FRNRQGALTV
151  YTGTVTQGTD PVKTYYQYTP VSSKAMYDAY WNGKFRDCAF HSGFNEDLFV
201  CEYQGQSSDL PQPPVNAPSG CPHHHHHH*
```

B

```
  1  AETVESCLAK SHTENSFTNV WKDDKTLDRY ANYEGCLWNA TGVVVCTGDE
 51  TQCYGTWVPI GLAIPENEGG GSEGGGSEGG GSGCPHHHHH H*
```

FIG. 3. Amino acid sequence of (A) N1–N2 and (B) N1 as used for *in vitro* SIP. The initiator methionine, which is needed for expression in the T7 system and that becomes cleaved afterward, is not shown. The engineered cysteine is underlined and the hexahistidine (His_6) tag is shown in boldface. (A) N1–N2 domain: N1 (amino acids 1–68) followed by G1 (amino acids 69–86), N2 (amino acids 87–217), and the Cys–His_6 tag (228 residues; 24,854.9 Da, p*I* 4.58; $\varepsilon = 41{,}100\ M^{-1}\ cm^{-1}$). (B) N1 domain: N1 (amino acids 1–68) followed by the first 14 amino acids of G1 and the Cys–His_6 tag for purification and coupling (91 residues; 9,658.37 Da; p*I* 4.87; $\varepsilon = 21{,}210\ M^{-1}\ cm^{-1}$).

long glycine-rich linker. To obtain cytoplasmic expression, the gene 3 signal sequence (amino acids 1–18) is deleted, and a methionine is added as a start codon, which becomes cleaved, as detected by mass spectrometry and N-terminal sequencing. In addition, a C-terminal Ser-Gly-Cys-Pro-His_6 tag is introduced to allow chemical coupling (Cys) and easy purification (hexahistidine, His_6) (Fig. 3A and B). We use the N1–N2 domain with only three additional residues before the engineered cysteine (Fig. 3A) and the N1 domain alone followed by the first 14 amino acids from the linker G1 before the cysteine (Fig. 3B) for *in vitro* SIP experiments. In addition, we could produce in the same way the N1–N2 complex followed by a glycine-rich linker and the N2 domain alone followed by its glycine-rich linker. Similar expression constructs have been used to produce N1–N2 for X-ray crystallography,[23] and N1–TolA for X-ray crystallography.[24] As mentioned above, N1 lacking the glycine-rich linker and Cys-His_6 tag could be purified from culture supernatant for nuclear magnetic resonance (NMR).[21]

[23] J. Lubkowski, F. Hennecke, A. Plückthun, and A. Wlodawer, *Nature Struct. Biol.* **5,** 140 (1998).

[24] J. Lubkowski, F. Hennecke, A. Plückthun, and A. Wlodawer, *Structure Fold. Des.* **7,** 711 (1999).

Expression of the N1–N2 Domain of the Gene 3 Protein

Plasmids carrying the gene for cytoplasmic expression of N1–N2 (Fig. 3A) are transformed into BL21 (DE3), plated onto 2× YT agar plates,[18] containing Amp (200 μg/ml) and 1% (w/v) glucose, and grown overnight at room temperature. A single colony of each host/plasmid is grown in 2× YT [containing Amp at 200 μg/ml, 1% (w/v) glucose] for 8 hr. Subsequently, a two-thirds volume of 50% (v/v) glycerol is added and the cells are frozen as glycerol stocks at −80°. For the expression culture, 50 μl of the glycerol stock is used to inoculate 50 ml of SB medium [2% (w/v) tryptone, 1% (w/v) yeast extract, 0.5% (w/v) NaCl], containing Amp (200 μg/ml), 1% (w/v) glucose, 2% (w/v) glycerol, 50 m*M* K_2HPO_4, and 10 m*M* $MgCl_2$, and the culture is grown overnight at 37°. The overnight culture is used to inoculate the main culture (2 liters of SB medium with the same additives). The main culture is grown at 37° to an OD_{550} of 0.9 to 1.2, and then IPTG is added to a final concentration of 1 m*M*, and the cells are grown for another 3–4 hr.

Expression of the N1 Domain of the Gene 3 Protein

Because of the high toxicity and thus instability of the expression culture for the N1 domain alone (Fig. 3B), it is produced in BL21(DE3) harboring the plasmid pLysS, which encodes T7 lysozyme, a natural inhibitor of T7 RNA polymerase.[25] This plasmid also confers resistance to chloramphenicol (up to 40 μg/ml). After transformation, cells are plated onto 2× YT agar plates, containing Amp at 200 μg/ml, Cam at 30 μg/ml, and 1% (w/v) glucose, and grown overnight at room temperature. For glycerol stocks (see preceding section), 2× YT medium supplemented with Amp (200 μg/ml) and 1% (w/v) glucose is inoculated with a single colony and grown for 8 hr. For production of N1 alone a more elaborate protocol must be used than in the case of N1–N2 production. A dilution of 50 μl of glycerol stock with 250 μl of SB medium is plated on three 100-mm SB–agar plates, containing Amp (200 μg/ml), Cam (30 μg/ml), 1% (w/v) glucose, and 5% (v/v) glycerol. After incubation overnight at 37°, the colonies (near confluence) are scraped from the plates with 20 ml of SB medium, which is then used to inoculate 2 liters of the main culture. The main culture [2 liters of SB medium containing Amp (200 μg/ml), Cam (30 μg/ml), 1% (w/v) glucose, 2% (v/v) glycerol, 50 m*M* K_2HPO_4, 10 m*M* $MgCl_2$] is grown at 37° to an OD_{550} of 0.9 to 1.2, IPTG is added to a final concentration of 1 m*M*, and the cells are grown for another 3–4 hr.

As the expression of the N1 domain with the single cysteine is not

[25] F. W. Studier, *J. Mol. Biol.* **219,** 37 (1991).

totally reliable even in the described experimental setup, two or three main cultures (2 liters each) are started separately and small aliquots of each are analyzed. In contrast, fusion proteins of N1 with other proteins give reliable inclusion bodies. Cells from 50 ml of culture are centrifuged and resuspended in 1 ml of 25 m*M* Tris-HCl, pH 7.5. Four milliliters of 6 *M* guanidine hydrochloride, 25 m*M* Tris-HCl, pH 7.4, is added to lyse the cells. Because of the liberated nucleic acids the lysate becomes viscous and must be sonicated. After centrifugation (SS34, 20,000 rpm/48,000*g*, 10 min, 4°) the supernatant is filtered through a 0.45-μm pore size filter. The filtrate is analyzed by coupled IMAC-AIEX as described in detail below. Small aliquots of the peak fractions can be loaded onto a sodium dodecyl sulfate (SDS)–polyacrylamide gel, as the sample is eluted from the AIEX in a nonionic urea buffer.

Cell Rupture and Enrichment of the Inclusion Bodies

Cells are harvested by centrifugation (GS3, 5000 rpm/4000*g*, 10 min, 4°). The cell pellet of a 2-liter culture is resuspended in 25 ml of 50 m*M* Tris-HCl, 1 m*M* $MgCl_2$ (pH 8.0), and 10 mg of RNase A, 10 mg of DNase I, and 25 mg of hen egg white lysozyme are added. The cell suspension is ruptured in a French press (Aminco). After centrifugation (SS34, 20,000 rpm/48,000*g*, 30 min, 4°) the pellet containing the inclusion bodies is resuspended in 25 ml of 20 m*M* Tris-HCl, 23% (w/v) sucrose, 0.5% (v/v) Triton X-100, 1 m*M* EDTA, pH 8.0. The suspension is stirred by a magnetic stir bar at 4° for 30 min, pelleted again, and the wash step is repeated. During the washing procedure the inclusion body pellet should change its color from brownish-yellow to white. To the final washed inclusion body pellet, 20 ml of 5.5 *M* guanidine hydrochloride, 25 m*M* Tris-HCl, pH 7.5, is added and the pellet is solubilized by stirring at room temperature. Insoluble material is removed by centrifugation (SS34, 20,000 rpm/48,000*g*, 30 min), and the supernatant is filtered through a 0.45-μm pore size filter.

Protein Purification

Two routes can be chosen to obtain purified gene 3 N-terminal domains from the inclusion bodies. The first procedure is the more classic way: refolding followed by purification. For the second, the protein is first purified under denaturing conditions. Both procedures are equivalent in yield. However, the second procedure may be advantageous especially when proteases are of concern and when a more defined and pure denatured sample is necessary (e.g., to test refolding parameters for more complicated fusion proteins). Using the second route, we were able to refold the protein by dialysis and did not have to refold by dilution, thus saving an additional concentration step. Overall, the second method is faster and more straightforward. Chromatography can be performed with any equipment; however, the automated two-column format simplifies the procedure significantly.[26] Immobilized metal-ion affinity chromatography (IMAC) is usually able to enrich the His-tagged adapter protein to a high extent, although some contaminants are still present. Anion- or cation-exchange chromatography is a good second step after IMAC, as the relatively pure IMAC-purified protein can often be baseline separated. In both cases the chromatography, performed with a BioCAD 60 system (PerSeptive Biosystems, Framingham, MA) takes about 30 min to obtain pure protein.

Route 1: Refolding and Subsequent Purification of the Native Protein

Step A: Refolding by Dilution

Twenty milliliters of the solubilized inclusion bodies (in 5.5 *M* guanidine hydrochloride, 25 m*M* Tris-HCl, pH 7.5) is reduced by addition of dithiothreitol (DTT) to 100 m*M* and EDTA to 10 m*M* final concentration. Most conveniently, this is done overnight at 4°, but reduction can also be performed for 4 hr at room temperature. As the DTT interferes with the formation of the disulfide bonds in the gene 3 N-terminal domains, the DTT concentration is reduced to 0.1 m*M* by three dialysis steps for 2 hr each at room temperature in 10 volumes (200 ml) of 5.5 *M* guanidine hydrochloride, 25 m*M* Tris-HCl, 10 m*M* EDTA, pH 6.0. The reduced, denatured protein is then slowly added dropwise to 1 liter of refolding buffer [0.4 *M* L-arginine, 0.2 *M* Tris-HCl, 0.2 *M* guanidine hydrochloride 0.1 *M* $(NH_4)_2SO_4$, 2 m*M* EDTA, pH 8.5] which contains 1 m*M* oxidized and 0.2 m*M* reduced glutathione. After overnight stirring at 10°, the refolding mixture is concentrated in an RA2000 concentrator (Amicon, Danvers, MA) to 120–150 ml, and further concentrated to a volume of 10–15 ml with a Centriprep YM10 (Amicon). Precipitates (mainly made up of contaminating proteins) are removed by centrifugation. Finally, the protein is dialyzed against 25 m*M* HEPES, 900 m*M* NaCl, pH 7.5, and filtered through a 0.45-μm pore size filter.

Step B: Coupled Immobilized Metal-Ion Affinity–Anion Exchange Chromatography under Native Conditions

To purify the refolded N-terminal domains of gene 3 protein, the protein is first bound to an IMAC column, using the C-terminal His_6 tag. After

[26] A. Plückthun, A. Krebber, C. Krebber, U. Horn, U. Knüpfer, R. Wenderoth, L. Nieba, K. Proba, and D. Riesenberg, *in* "Antibody Engineering: A Practical Approach" (J. McCafferty and H. R. Hoogenboom, eds.), p. 203. IRL Press, Oxford, 1996.

the protein is eluted directly onto the anion-exchange column (AIEX) under such conditions that it binds there. On the AIEX, the protein can be purified to a high degree by an NaCl gradient. The whole procedure is automated on a BioCAD 60 (PerSeptive Biosystems): First, the 1.66-ml HQ/M AIEX column (PerSeptive Biosystems) is equilibrated with 16 ml of 50 m*M* Tris-HCl, pH 7.5. The switches are then set such that the column is taken out of the flow and the 1.66-ml MC/M IMAC column (PerSeptive Biosystems) is switched in-line, washed with 20 ml of water, loaded with 2 ml of 100 m*M* $NiCl_2$ through one of the water-washed sample lines, and subsequently washed with 20 ml of water. After Ni^{2+} charging the MC/M column, it is equilibrated with 16 ml of 25 m*M* HEPES, 900 m*M* NaCl, pH 7.5. Five milliliters of sample is loaded onto the MC/M column. It is then washed first with 12 ml of 25 m*M* HEPES, 150 m*M* NaCl, 1 m*M* imidazole, pH 7.5, and then with 24 ml of 25 m*M* HEPES, 12 m*M* imidazole, pH 7.5. After the wash steps, the HQ/M column is set in-line such that the flow is directed from the MC/M to the HQ/M column. To transfer the protein from the MC/M onto the HQ/M column it is directly eluted from the MC/M onto the HQ/M column by 12 ml of 120 m*M* imidazole, pH 7.5, which results in strong binding of the protein to the HQ/M column. The MC/M column is then set off-line and the HQ/M column is washed with 16 ml of 50 m*M* Tris-HCl, pH 7.5. A 20-ml gradient of 0 to 500 m*M* NaCl in 50 m*M* Tris-HCl, pH 7.5, is employed to elute the N-terminal domain protein. As part of the protein dimerizes in the refolding reaction due to the free cysteine in the tag (to be used for chemical coupling), two major gene 3 N-terminal protein domain peaks are observed. The monomer and dimer peak fractions are combined and DTT is added to 2 m*M* and EDTA to 10 m*M* final concentration, respectively. The protein solution is passed through a 0.22-μm pore size filter and stored at 4°. If necessary, the protein solution can be concentrated with a Centriprep YM10 (Amicon).

Route 2: Purification and Subsequent Refolding

Step A: Coupled Immobilized Metal-Ion Affinity–Anion Exchange Chromatography under Denaturing Conditions

The purification under denaturing conditions is basically the same as under native conditions. However, because the purification starts with freshly solubilized inclusion bodies, the reduction with DTT and the subsequent dialysis to remove it are not necessary. Oxidized protein is reduced on the column and only the monomeric species must be separated from contaminants. As the pure protein can be refolded at higher concentrations, the protein is refolded by dialysis and not by dilution, which further simplifies downstream processing.

First, the 8-ml HQ/M AIEX column (PerSeptive Biosystems) is equilibrated with 60 ml of 6 *M* urea, 50 m*M* Tris-HCl, pH 7.5. The column is switched out of the flow and the 4-ml MC/M IMAC column (PerSeptive Biosystems) is set in-line, washed with 40 ml of water, loaded with 5 ml of 100 m*M* $NiCl_2$ through one of the water-washed sample lines, and subsequently washed with 80 ml of water. After charging the MC/M column it is equilibrated with 30 ml of 8 *M* urea, 25 m*M* HEPES, 1.5 *M* NaCl, pH 7.5. Five milliliters of solubilized inclusion body sample is loaded onto the MC/M column. It is then washed first with 20 ml of 8 *M* urea, 25 m*M* HEPES, 1.5 *M* NaCl, 4 m*M* imidazole, pH 7.5, and then with 60 ml of 8 *M* urea, 25 m*M* HEPES, 15 m*M* imidazole, pH 7.5. After the wash step the HQ/M column is set in-line such that the flow would be directed from the MC/M to the HQ/M column. To transfer the protein from the MC/M onto the HQ/M column it is eluted from the MC/M onto the HQ/M column by 40 ml of 8 *M* urea, 100 m*M* imidazole, pH 7.5, which results in strong binding of the protein to the HQ/M column. The MC/M column is then set off-line and the HQ/M column is washed, and at the same time the protein is reduced on the column with 20 ml of 8 *M* urea, 100 m*M* Tris-HCl, 100 m*M* DTT, pH 9.0, at a low flow rate of 2 ml/min (10 min). Subsequently, the column is washed with 60 ml of 6 *M* urea, 25 m*M* Tris-HCl, pH 7.5. The reduced, denatured protein is eluted by a 60-ml gradient from 0 to 250 m*M* NaCl in 6 *M* urea, 50 m*M* Tris-HCl, pH 7.5.

Step B: Refolding by Dialysis

The purified, reduced and denatured protein (10–20 ml) is immediately dialyzed against 500 ml refolding buffer [0.4 *M* L-arginine, 0.2 *M* Tris-HCl, 0.2 *M* guanidine hydrochloride, 0.1 *M* $(NH_4)_2SO_4$, 2 m*M* EDTA, pH 8.5], which contains 1 m*M* oxidized and 0.2 m*M* reduced glutathione, overnight at 10°. The buffer is exchanged by two dialysis steps, each against 500 ml of 50 m*M* Tris-HCl, 50 m*M*, NaCl, pH 7.5. DTT is added to 2 m*M* and EDTA to 10 m*M* final concentration, respectively. The protein solution is passed through a 0.22-μm pore size filter and stored at 4°, and if necessary concentrated on a Centriprep YM10 (Amicon). However, in case of direct peptide or protein fusions to the C terminus of N1 or N1–N2, the Cys residue in the C-terminal tag is replaced by the interaction partner B, and the refolding and purification procedure might have to be varied accordingly.

Chemical Coupling of Nonpeptidic Ligands

Coupling is carried out via the free sulfhydryl group, introduced by the Cys residue in the C-terminal tag (Fig. 3). Coupling chemistry and purification may vary depending on the bait. We will give an example for coupling fluorescein,[9] which is already coupled to a Lys residue to give the compound FluCad (5-[(5-aminopentyl)thioureidyl]fluorescein). Coupling is achieved by using the heterobifunctional cross-linker *N*-succinimidyl-6-maleimidocaproate (Fluka, Ronkonkoma, NY). The coupling of FluCad to the cross-linker is carried out in dimethylformamide (DMF) for 1 hr at 30° in the dark, at a ratio of FluCad to cross-linker of 3:2. Completion of the reaction is controlled by thin-layer chromatography in 80% (v/v) ethyl acetate–10% (v/v) methanol–10% (v/v) acetic acid. At the same time the DTT in the N1 or N1–N2 stock is removed by gel filtration on a short Sephadex G-25 column (Pharmacia) in 25 m*M* sodium phosphate, pH 6.8. This leads to N1 or N1–N2 molecules that contain a free sulfhydryl group in the cysteine of the tag. The amount of protein is quantified by spectrophotometry, using the calculated extinction coefficient.[27] Three molar equivalents of FluCad are then reacted with one molar equivalent of the free sulfhydryl group of N1 or N1–N2 in 25 m*M* sodium phosphate, pH 6.8, for 1 hr at 25° and then at 4° overnight. The pH of the reaction buffer is critical, as at higher pH the maleimide would start to react with primary amines in the N1 or N1–N2 domain. The resulting adapter molecules are gel filtrated on a Sephadex G-25 column (Pharmacia) in 50 m*M* Tris-HCl, pH 7.5. Because fluorescein tends to noncovalently interact with the adapter protein (N1 or N1–N2), another purification step is needed. Therefore, the fluorescein-coupled adapter molecules are separated from the uncoupled N1 or N1–N2 by anion-exchange chromatography on a perfusion chromatography HQ column NaCl gradient (0 to 600 m*M* NaCl in 50 m*M* Tris-HCl, pH 7.5), using a BioCAD 60 system (PerSeptive Biosystems). Mass spectrometry should be used to verify the success of the coupling reaction and purification. In general, the coupling method and the heterobifunctional linker can be used for the coupling of any primary amine compound that does not contain a free sulfhydryl group. Using an analogous strategy every compound activated by a maleimide should be linkable to the free cysteine containing N1 or N1–N2. The disulfide bonds of N1 or N1–N2 were not affected, neither by the DTT nor the maleimide under the conditions employed.

Phage Infection Experiment

In vitro SIP experiments are performed by incubating 1 μl of SIP-phage supernatant, concentrated 500 times by two PEG precipitations (as described under General Procedures), with 350 n*M* adapter N1–N2 bait

[27] S. C. Gill and P. H. von Hippel, *Anal. Biochem.* **182,** 319 (1989).

at 4°. For infection, 100 μl of an exponentially growing XL1-Blue culture is added to the mixture and incubated for 1 hr at 37° with shaking, and plated subsequently on 2× YT–agar plates containing Cam (25–30 μg/ml) and Tet (15 μg/ml). Addition of 50 m*M* $MgCl_2$ to the cells prior to and during infection increases infectivity four- to sixfold.[9] The number of phages can be varied according to library size and expected infection rate. The concentration of adapter molecule, especially for N1–N2 bait, can influence the infectivity, and thus, when maximal infectivity is needed, a titration of adapter molecules might be advisable (see the next section).

Dependence of Infectivity on Concentration of Adapter Molecule

The concentration of adapter molecules influences the phage infectivity. Because of the law of mass action, higher concentrations of adapter will shift the equilibrium to the bound and thus infective state. However, differences between N1 bait and N1–N2 bait are apparent, presumably because of the ability of N2 to bind to the pili and thus block infection at high concentration. Both variants have previously been tested with fluorescein as bait, fused to N1 or N1–N2, respectively, and a fluorescein-binding single-chain variable fragment (scFv), fused to N2–CT ("medium" phage) or CT ("short" phage), respectively (Fig. 4).[9] The infectivity of 10^{10} medium

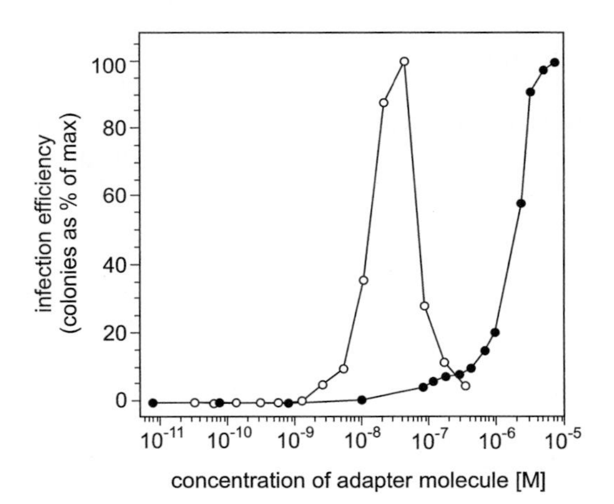

FIG. 4. Dependence of infectivity on concentration of adapter molecules and g3p arrangement in *in vitro* SIP. Open circles represent the situation depicted in Fig. 1C with the "short" phage and N1–N2 bait as adapter molecule. In this case, the same infectivity was observed when using "medium" phages instead (data not shown). Filled circles represent the infectivities obtained with medium phage and N1 bait as adapter molecules (Fig. 1D). [Adapted from Krebber *et al.*[9]]

phages increases as a function of N1–fluorescein up to the highest concentration tested (10^{-5} *M*), at which point it levels off, yielding 1.2×10^6 colonies. In contrast, using the adapter N1–N2–fluorescein with short or medium phages, a maximum of infectivity is reached at 3×10^{-8} *M*, yielding 5×10^4 colonies from 10^{10} input phages. In summary, using N1 bait as adapter and medium phages, the maximal infectivity is about 20-fold higher than with the N1–N2 bait, but a higher adapter concentration is needed. Consequently, the N1–N2 bait is recommended for selection of high-affinity systems, as the adapter concentrations can be kept low, whereas the N1 bait is more suitable for lower affinities, as the adapter concentration can be increased without inhibitory effects.

Examples from the Literature using in Vitro Selectively Infective Phage Selection

In vitro SIP was used to select for novel, nonrepetitive linkers for antibody scFv fragments.[28] In this case, medium phages and adapters with N1 bait were used, and phage production was performed at 30 and 37°. After one round of SIP selection, 22 of 22 clones gave positive signals in ELISA, whereas before selection, only 1 of 23 clones was positive. Nine clones were characterized further, and shown to be comparable to the parental sequence, which had $(Gly_4Ser)_3$ as the linker, in terms of binding, folding, expression, and solubility.

A selection from an antibody Fab library was carried out and compared with selection results from conventional phage display.[4] In this case, a phagemid system was used, and the adapter, consisting of the first 98 amino acids of the mature gene 3 protein fused to the antigen, was secreted from a separate *E. coli* culture, and the supernatant was used directly for infection. The N2 domain was missing completely in this experimental setup. Infection was carried out by adding 25% (v/v) adapter-containing supernatant to the supernatant from the phage-producing cells and incubating the mixture overnight at 4° before adding XL1-Blue recipient cells. From this library, half the phages gave positive ELISA signals after three rounds of SIP selection, while no antibodies against the antigen could be obtained from phage display after three rounds of phage panning. Thus, SIP is faster than conventional phage display, using the identical library.

In two further model systems, the properties of selection were investigated in more detail by using a small set of molecules with defined thermodynamic and kinetic parameters. The phage infectivity of various point mutants with slightly different K_D values and activities was compared and a

[28] F. Hennecke, C. Krebber, and A. Plückthun, *Protein Eng.* **11,** 405 (1998).

clear correlation was observed.[29] In these experiments, short phages and N1–N2 bait adapter molecules were used. In a competitive SIP experiment with all mutants, only the tightest binders remained after three rounds of selection. In another model selection with six different Fab fragments, the ability to select for affinity or even kinetic constants was investigated.[30] In individual experiments using the phagemid system (mentioned above), phages expressing higher affinity clones were enriched preferentially with low concentrations of antigen. However, it was also possible to select clones with lower affinity by increasing antigen concentration. In another experiment the incubation time was varied. Using short incubation times (30 min), clones with high association rate constants were preferred, whereas after long incubation (16 hr), clones with the lowest dissociation rate constant were preferred.

Two further examples use techniques closely related to *in vitro* SIP. In the first example, linkage of covalent catalysis to infectivity was demonstrated by fusing a catalytic antibody fragment to the CT domain and the substrate was coupled to N1–N2. Only covalent catalysis was able to produce infective phages.[31] In the second example, selection for protease resistance was carried out by fusing the proteins of interest between N1–N2 and CT and performing several rounds of *in vitro* proteolysis, infection, and propagation.[32,33]

In Vivo Selectively Infective Phage Procedures

In the *in vivo* SIP procedure, all steps are carried out with crude *E. coli* supernatant, without the need for purification of any compound or any *in vitro* panning steps. In particular, the target for the library does not have to be expressed and purified. Both interacting partners (protein A–CT and N1–N2–protein B) are encoded on the same phage vector, as described below. Alternatively, a phagemid system[3,4] (see General Procedures) or a combination of phage and phagemid vectors that are copackaged[34] can be used. Although two-vector systems provide for more convenient cloning,

[29] G. Pedrazzi, F. Schwesinger, A. Honegger, C. Krebber, and A. Plückthun, *FEBS Lett.* **415,** 289 (1997).

[30] M. Dueñas, A. C. Malmborg, R. Casalvilla, M. Ohlin, and C. A. Borrebaeck, *Mol. Immunol.* **33,** 279 (1996).

[31] C. Gao, C. H. Lin, C. H. L. Lo, S. Mao, P. Wirsching, R. A. Lerner, and K. D. Janda, *Proc. Natl. Acad. Sci. U.S.A.* **94,** 11777 (1997).

[32] V. Sieber, A. Plückthun, and F. X. Schmid, *Nature Biotechnol.* **16,** 955 (1998).

[33] P. Kristensen and G. Winter, *Fold. Des.* **3,** 321 (1998).

[34] F. Rudert, C. Woltering, C. Frisch, C. Rottenberger, and L. L. Ilag, *FEBS Lett.* **440,** 135 (1998).

it is important to ensure the quality of both vectors and the complete transformation or transfection of one library to the other, which is usually verifiable only by sequencing cotransformants.

Cloning of Two Libraries in One Vector

If a library-versus-library experiment is to be carried out, two libraries must be cloned (library A–CT and N1–N2–library B). If the phage system is used, both libraries must be cloned in one vector with the highest efficiency possible. Two strategies are available. First, to avoid multiple transformations, both libraries can be combined with the interconnecting vector fragment in an assembly PCR, and then cloned in one step into the vector. Thereby, only one transformation is required. Alternatively, both libraries can be cloned one after the other, which then requires two ligations and two transformations, however. We tested both possibilities by sequencing individual clones obtained after cloning. Using the assembly PCR approach, fewer clones analyzed were generally correct, and the remainder were mainly frameshift mutants with deletions of 1 base pair up to a stretch of several base pairs. Also in the region between both libraries, deletions were occasionally observed. However, using the second approach with two subsequent ligations and transformations, the result was significantly better. It appears that the assembly PCR, even though performed with a proofreading polymerase, introduces more errors, and because the overall transformation rates are comparable for both approaches, the sequential cloning appears to be the method of choice for library cloning in one-vector systems.

Design of Vectors Suitable for in Vivo Selectively Infective Phage to Avoid Genetic Instabilities

In the case of *in vivo* SIP, where both parts of the gene 3 protein are expressed in the same cell, whether on one or two vectors, great care must be taken to prevent recombination. In initial constructs with the arrangement N1–N2–protein B and protein A–CT, homologous recombination between duplicated glycine-rich regions was observed, leading to restoration of a wild type-like gene 3 protein at high frequency. Therefore, in more recent vectors[7] the order of the two fusion proteins has been switched to protein A–CT being upstream of N1–N2–protein B, and the glycine-rich linkers between protein A–CT and N2–protein B were shortened such that there were no long identical sequence stretches present in both fusion proteins. By this measure, recombination could be prevented successfully in most cases.

However, the occurrence of recombination of course still depends on the nature of the two interacting proteins or peptides A and B. In the case that both parts have similar sequence stretches, either in constant parts or in a randomized region, the risk of recombination increases significantly. In an example of a library-versus-library experiment using two semirandomized libraries, recombinations were found only in clones in which an identical stretch of 8 consecutive base pairs (bp) happened to exist in libraries A and B. In these cases, the sequence stretch between those 8 bp, which served as recombination sites, was doubled, which led to a wild type-like arrangement of gene 3. Such clones can therefore easily be recognized by wild type-like infectivity.

However, such recombined clones can be eliminated in a simple and fast way. Phage DNA of the entire library after the appropriate SIP round, where recombined clones are found, is prepared, and the gene 3 cassette is excised by restriction digest (in the vector described here, *Pin*AI/*Hin*dIII; Fig. 2E). The correctly sized band of the gene 3 cassette (1963 bp in our example) can then be easily separated by gel electrophoresis from the larger band created by the recombination (3379 bp). The purified, correctly sized band is cloned back into fresh vector and then used for further SIP selection. With this approach, another round of SIP selection yields only clones with correctly sized gene 3 cassette, as judged by analytical restriction digest, and significant recombination usually does not occur for another two rounds. It is advisable to perform an analytical restriction digest as a control for recombination from the entire library or single clones obtained after selection.

Selective Infection Experiments

For infection, Tet^R or Kan^R XL1-Blue cells are grown in 2× YT medium, supplemented with 1% (w/v) glucose, 50 m*M* $MgCl_2$, and Tet (5 μg/ml) or Kan (50 μg/ml), respectively, to an OD_{550} of 0.5 to 0.8. It has been shown previously that the addition of 50 m*M* $MgCl_2$ increases the infectivity four- to sixfold.[9] The number of phages used for infection should be on the order of 10^8 to 10^{11} phages, and is varied depending on the size of the libraries and the expected infection rate. The appropriate amount of phages is added to 0.5 ml of cells and shaken for 1 hr at 37°. Bacteria are plated on 2× YT–agar containing 1% (w/v) glucose, Cam (30 μg/ml), and Tet (5 μg/ml) or Kan (50 μg/ml), depending on the host strain. For library experiments 245 × 245 mm plates are usually most appropriate. The infectivity is determined by plating a series of 10-fold dilutions on small plates. Plates are incubated overnight at 37 or 25° for more unstable proteins. With this step, one round of SIP selection is finished. To proceed with the library for further selection rounds, the cells are pooled from the plates into 5–10 ml of rich medium, shaken for 5 min at 37° to allow for good mixing and cell separation, and used for inoculation of a new overnight culture. At the same time, glycerol stocks are prepared as a backup (stored at −80°), by adding glycerol to a final concentration of 30% and making aliquots of 500–1000 μl. Individual clones can be analyzed by restriction digest or sequencing of minipreparation DNA.

Occurence of Cysteine Mutations and Effects of Dithiothreitol

Libraries created by using synthetic oligonucleotides, even those of high quality, almost always have a low percentage of frameshift mutants due to imperfect oligonucleotide synthesis. Similarly, cysteines are introduced by random mutagenesis techniques such as DNA shuffling or error-prone PCR[12,13,35] and will inevitably be present in cDNA libraries. Most of the frameshift mutants will not be selected, because they lack the ability to fold and bind their partner, but some might still show up in selection: Provided a cysteine is introduced both at the end of the adapter and the beginning of the CT domain, a covalently linked "wild type-like" gene 3 protein is formed. In the following section, we highlight this problem and point out general solutions.

In an example of a library-versus-library selection (library A–CT and N1–N2–library B) for noncovalently interacting pairs, strong selection for a single cysteine in each library peptide was observed, even though the designed sequences contained no cysteines at any place. The cysteines were found to be caused by either point mutations (in library A) or by frameshifts (in library B). The cysteine pairs caused a strong increase in infectivity of four orders of magnitude in one SIP round. The same phenomenon was observed in a one-library approach (library A–CT and N1–N2–peptide B), in which also a point mutation in library A and a frameshift in peptide B led to single cysteines (for details see Ref. 36). These covalent interactions are mostly unspecific but nonetheless provide a strong selective advantage. Consequently, the high infectivity of phages in the absence of DTT was significantly reduced after DTT treatment. Control experiments showed that incubation with DTT has only a minor effect on phage infectivity itself, consistent with previous experiments, where it was reported that only phage production but not phage infection and phage DNA replication is prevented by 5 m*M* DTT.[37] Furthermore, all four disulfide bridges in the native g3p

[35] D. W. Leung, E. Chen, and D. V. Goeddel, *Technique* **1,** 11 (1989).

[36] S. Jung, K. M. Arndt, K. M. Müller, and A. Plückthun, *J. Immunol. Methods* **231,** 93 (1999).

[37] M. Vaccaro, B. Boehler-Kohler, W. Müller, and I. Rasched, *Biochim. Biophys. Acta* **923,** 29 (1987).

are inaccessible to the alkylating agent vinylpyridine after treatment with DTT (50°, 50 min, 100-fold molar excess of DTT over cysteines).[38]

This stability of phages against reducing agents can be extremely useful when working with complex libraries, where the occurrence of single cysteines cannot be excluded. In our experience, the purified wild-type phage tolerates a 3- to 5-hr incubation at 37° with 5 m*M* DTT, pH 8.0, prior to infection very well, with only a slight loss of infectivity. We usually dilute the phages after DTT incubation and prior to infection to obtain a final DTT concentration of about 10 μM during infection, but the infection worked equally well at a 100-fold higher DTT concentration (1 m*M*). However, the exact amount of DTT, the incubation time, and temperature depend not only on the stability of the phage but also on those of the proteins of interest. Nonetheless, DTT incubation can significantly reduce but not fully eliminate the problem as some disulfide bridges might survive the DTT treatment or reform during the infection process.

Detection of Translational Frameshifts by the Appearance of Polyphages

In a further SIP experiment using the same library-versus-library approach previously mentioned, two clones were selected with a −1 frameshift in the library peptide before the CT domain. Extensive studies revealed that neither the frameshift peptide, which would be produced instead of the CT–library A fusion, nor the peptide from library B fused to N1–N2, could replace the function of CT. Most likely the observed −1 frameshift is accompanied by a second, translational +1 frameshift caused by two rare arginine codons (AGG) that were present in both clones due to the −1 frameshift.[39] This was further confirmed by the observation that these clones formed polyphages, most likely due to insufficient capping of the nascent phages caused by the limited amounts of CT domain, created by a translational +1 frameshift (for details refer to Ref. 36).

It can therefore be concluded that the CT domain, and probably the transmembrane helix, is not only needed for infection, but can definitely not be replaced by "sticky peptides," which would nonspecifically attach N1–N2 to the phage. Therefore, the SIP procedure must clearly bring together N1–N2 and CT on the phage.

Examples from the Literature using in Vivo Selectively Infective Phage Selection

In an example of *in vivo* SIP selection, a conserved region in the immunoglobulin variable domain was investigated by using a small synthetic

[38] A. Kremser and I. Rasched, *Biochemistry* **33,** 13954 (1994).

[39] R. A. Spanjaard and J. van Duin, *Proc. Natl. Acad. Sci. U.S.A.* **85,** 7967 (1988).

library.[40] The library was fused to the CT domain, and the peptidic antigen was fused to N1–N2. After only three rounds of SIP, a strong enrichment was found of clones similar to the most abundant sequences in the Kabat database. In another example, the coiled-coil domain of c-Jun with an engineered free cysteine was fused to the CT domain of gene 3 protein and used as bait to select against a human cDNA library fused to the soluble N-terminal domains.[3,41] Several clones were isolated that were predicted to encode potential coiled-coil structures. However, in our hands, constructs with free cysteines, even though they show higher infectivity, showed less specificity in their interaction. In addition, in the one-library approach we prefer to fuse the library to the CT domain rather than to N1–N2. To keep phenotype and genotype connected, it is advantageous to have the library tightly connected to the phage, in case the ligand dissociates and exchanges to a different phage during the infection in the library pool.

A further model experiment showed the feasibility of a two-vector packaging of adapter and CT fusion, which would be useful for library-versus-library selection.[34] In this case, protein A–CT was displayed on phage and encoded on the phage genome and the N1–N2–bait was encoded on a phagemid. After cotransformation both vectors were copackaged in polyphages, which were screened on special filter plates.

Discussion

SIP is a powerful strategy to select for protein–ligand interactions as well as for other desired features such as protein folding and stability, provided the binding function of the protein is limited to the native state. The selection is carried out completely in solution without any need for solid-phase panning steps. Furthermore, the enrichment cycle is fast and thus less time consuming. Most importantly, the selection power is extremely high and enrichment factors of 10^5 to 10^6 have been observed,[42] while in phage display enrichment factors of 10 to 10^3 are normal.[43] Therefore, in most applications SIP requires only one selection round to separate binders from nonbinders,[28,29] whereas phage display usually needs three or four rounds. In three or four SIP rounds, it is moreover possible even to discriminate between more subtle affinity, folding, and stability differences.[29,40]

[40] S. Spada, A. Honegger, and A. Plückthun, *J. Mol. Biol.* **283,** 395 (1998).

[41] K. Gramatikoff, W. Schaffner, and O. Georgiev, *Biol. Chem.* **376,** 321 (1995).

[42] M. Dueñas, L. T. Chin, A. C. Malmborg, R. Casalvilla, M. Ohlin, and C. A. Borrebaeck, *Immunology* **89,** 1 (1996).

[43] G. Winter, A. D. Griffith, R. E. Hawkins, and H. R. Hoogenboom, *Annu. Rev. Immunol.* **12,** 433 (1994).

However, be aware that this extremely strong selection ability can sometimes lead to unwanted solutions, by covalently linking the N terminal to the C-terminal part of g3p. This is only a problem in the *in vivo* SIP procedure, where both partners can potentially acquire mutations, because in *in vitro* SIP the infection-mediating particle is produced separately. With highly complex libraries, products of random mutagenesis, and cDNA libraries, a selection for disulfide bridges and enrichment of recombined clones is possible. Genetic recombination is a rare event, easy to check and to remedy by cutting out and recloning restriction fragments. The selection for disulfide bonds from spurious cysteines can be most effectively controlled by working with high-quality libraries in the beginning. Reducing agents also help, but do not eliminate the problem. Currently, the work with cDNA is still a challenge.

For a library-versus-library selection, only *in vivo* SIP is applicable, but needs to be strictly controlled. It also remains to be determined what the maximum size of the fusion protein can be, whether there are geometric restrictions imposed by the infection process, and what the minimum affinity for the interaction is. In an *in vivo* SIP library-versus-library approach, a weak interaction might lead to the exchange of adapter molecules between different clones and thus destroy the genotype–phenotype correlation. Another limitation for complex libraries is the moderate infection efficiency. The highest observed was 1 infection per 10^4 SIP phages,[9] but much lower infectivities have also been reported.

Taking everything together, SIP is at this point most useful in the fast screening of less complex libraries and in molecular improvement. We generally recommend the use of the *in vitro* SIP procedure over *in vivo* SIP whenever possible, because it is more robust and easier to control. Furthermore, by choosing the right adapter and titrating its concentration, it allows for selection of medium as well as high-affinity interactions. Furthermore, the *in vivo* SIP complements very well the other two techniques available to directly select two libraries simultaneous against each other, the yeast two-hybrid system[44] and the protein fragment complementation assay,[45] which select in the nucleus of yeast or the cytosol of *E. coli,* respectively. In *in vivo* SIP proteins are folded in the oxidative environment of the periplasm, which is essential for selection from libraries containing proteins that require the formation of disulfide bridges.

[44] S. Fields and O. Song, *Nature* (*London*) **340,** 245 (1989).

[45] J. N. Pelletier, K. M. Arndt, A. Plückthun, and S. W. Michnick, *Nature Biotechnol.* **17,** 683 (1999).

[23] Use of Phage Display and Transition-State Analogs to Select Enzyme Variants with Altered Catalytic Properties: Glutathione Transferase as an Example

By MIKAEL WIDERSTEN, LARS O. HANSSON, LISA TRONSTAD, and BENGT MANNERVIK

Introduction

Available evidence suggests that naturally occurring proteins adopt a limited number of folds, whereas the variety of primary structures is essentially unlimited. It can therefore be concluded that most naturally evolved proteins have arisen by redesign of previously existing protein scaffolds. This approach to protein design appears particularly obvious in protein families where structural similarities are pronounced even though functional properties are highly divergent.

The soluble glutathione transferases (GSTs) are a family of detoxication enzymes with diverse substrate specificities.[1] The multiple forms of GSTs have been divided into classes, and as a first approximation, functional properties were found to be correlated with the amino acid sequences.[2] Even though the overall identities in the amino acid sequences are limited among the known GSTs, the protein folds of the different classes are similar.[3] This finding agrees with the proposal that the GSTs have undergone divergent evolution from a common ancestral protein.[4] The evolutionary perspective thus suggests that GSTs have stable structures, which by combinatorial rearrangements can be redesigned for novel functions.

An important paradigm in enzyme catalysis is stabilization of transition states of the reactions catalyzed.[5] It has therefore been proposed that antibodies with affinity for a transition-state analog should be able to catalyze the corresponding reaction.[6] This postulate has been put into practice by the development of catalytic antibodies.[7] However, this principle for

[1] B. Mannervik and U. H. Danielson, *CRC Crit. Rev. Biochem.* **23,** 283 (1988).
[2] B. Mannervik, P. Ålin, C. Guthenberg, H. Jensson, M. K. Tahir, M. Warholm, and H. Jörnvall, *Proc. Natl. Acad. Sci. U.S.A.* **82,** 7202 (1985).
[3] R. N. Armstrong, *Chem. Res. Toxicol.* **10,** 2 (1997).
[4] B. Mannervik, *Adv. Enzymol. Relat. Areas Mol. Biol.* **57,** 357 (1985).
[5] L. Pauling, *Chem. Eng. News* **24,** 1375 (1946).
[6] W. P. Jencks, "Catalysis in Chemistry and Enzymology," p. 288. McGraw-Hill, New York, 1969.
[7] P. G. Schultz and R. A. Lerner, *Science* **269,** 1835 (1995).

0076-6879/00 $30.00

creating new biocatalysts should not be restricted to antibodies, but be applicable to any protein (or nucleic acid) that can evolve binding affinities that favor the transition state in comparison with the ground-state structure in a chemical reaction.

This chapter describes the use of transition-state analogs for selection of GSTs with novel catalytic properties from libraries of active site mutants. The GST variants are expressed in functional form on filamentous phage and variant GSTs with proper affinities are isolated by affinity adsorption, using transition-state analogs as ligands.

Transition-State Stabilization in Design of Catalysts

A well-evolved enzyme (efficient catalyst) is expected to have higher affinity for the transition state than for substrate and product. However, a transition state cannot be studied directly owing to its extremely short life span. The Hammond postulate states that "if there is an unstable intermediate on the reaction pathway, the transition state for the reaction will resemble the structure of this intermediate."[8] Intermediates occur between two transition states on the reaction coordinate, and for many reactions such an intermediate is useful for visualizing the structure of the transition state.

S_NAr Reactions. The nucleophilic aromatic substitution reaction (S_NAr) is an addition–elimination reaction that involves an unstable σ-complex intermediate (Meisenheimer complex; Fig. 1a). Uncatalyzed S_NAr reactions in polar solvents are proposed to be rate-limited by the addition of a nucleophile to the aromatic nucleus.[9] The reaction of the thiol group of glutathione (GSH) and 1-chloro-2,4-dinitrobenzene (CDNB) is catalyzed by GSTs and is assumed to proceed via the negatively charged σ-complex 1-chloro-2,4-dinitro-1-(*S*-glutathionyl)hexadienate intermediate (Fig. 1a). It contains an sp^3-hybridized carbon, a nonplanar ring, and the negative charge delocalized over the conjugated π system. The compound 1,3,5-trinitrobenzene (TNB), structurally similar to CDNB, does not provide a good leaving group, and an elimination step therefore does not occur on addition of GSH (Fig. 1b). A semistable σ complex is formed as an equilibrium product, which can reasonably be assumed to structurally mimic the transition state of the CDNB S_NAr reaction.

Ester Hydrolysis. Hydrolysis of carboxylic acid esters involves a nucleophilic attack by water in the form of a hydroxide ion on the electrophilic carbonyl carbon followed by a rearrangement of the electrons shared be-

[8] G. S. Hammond, *J. Am. Chem. Soc.* **77,** 334 (1955).

[9] F. Terrier, Nucleophilic aromatic displacement: The influence of the nitro group. *In* "Organic Nitro Chemistry Series." VCH, New York, 1991.

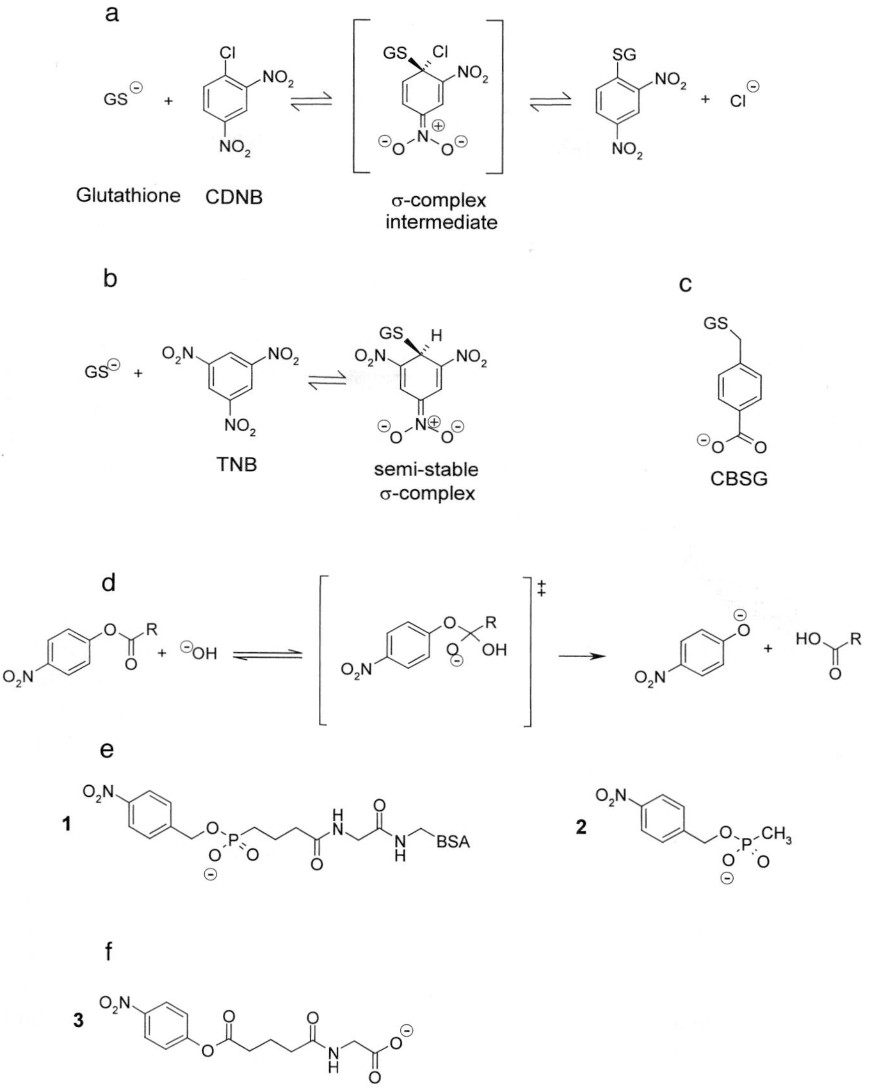

FIG. 1. Substrates and transition-state analogs. (a) S_NAr reaction of glutathione and 1-chloro-2,4-dinitrobenzene (CDNB). (b) Formation of a semistable σ complex as the equilibrium product of glutathione and 1,3,5-trinitrobenzene (TNB). (c) *S*-(*p*-Carboxy)-benzylglutathione (CBSG). (d) Ester hydrolysis. (e) Phosphonate transition-state analogs used in the phage dislay selection of GST A1-1 mutants. Phosphonate **1** was immobilized by conjugation to BSA. Phage bound to **1** during the selection were eluted by addition of 0.5 m*M* phosphonate **2**. (f) Ester substrate **3** used to assay for mutants isolated by phage display toward phosphonates **1** and **2**. The synthesis of phosphonates **1** and **2**, and of ester **3**, has been reported earlier [D. S. Tawfik, B. S. Green, R. Chap, M. Sela, and Z. Eshhar, *Proc. Natl. Acad. Sci. U.S.A.* **90,** 373 (1993)].

tween the carbon and the oxygen in the carbonyl to form the transition state (Fig. 1d). Decomposition of the transition state and release of the alkoxide group complete the reaction. The reaction rate is dependent on stabilization of the negative charge developed in the transition state as well as on the acidity of the conjugate acid of the alkoxide leaving group. Enzymes catalyzing this reaction provide an electrostatic environment, e.g., an "oxyanion hole," stabilizing the negative charge and, in some cases, an acidic group protonating the leaving group, thus facilitating its release.

Phosphonate derivatives have proved to be good mimics of transition states in ester hydrolysis. The similarities comprise both charge distribution and geometric configuration, including a negatively charged oxyanion intermediate, sp^3 hybridization of the central carbon atom, and increased bond lengths similar to those in the transition state. Phosphonates can hence be used as probe ligands for selection of proteins with esterase activity (Fig. 1e).

Affinity Selection by Phage Display

In the functional affinity selection afforded by a phage display system, the protein (the phenotype) is linked to its corresponding genotype, the DNA encoding the selected protein (Fig. 2). When a mutant protein is isolated from a large library of variant proteins by any means of selection, its primary structure can thus be deduced from the accompanying DNA.

The displayed protein is exposed on the surface of a filamentous bacteriophage by fusion with a coat protein of the phage. Two variant fusion constructs have been frequently employed.[10] In one of them, the displayed protein is fused with the gene 3 protein (g3p, three to five copies on the tip of the phage participating in infection of bacteria). In the second, the protein is fused with the gene 8 protein (g8p; more than 2000 copies forming the cylindrical wall of the phage particle). In polyvalent display, all copies of either g3p or g8p are replaced by fusion proteins. Monovalent display (one fusion protein per phage) is generally preferred, such that one-to-one binding is achieved in the affinity selection. To accomplish monovalent display of g3p fusion proteins, an excess of unmodified g3p (encoded by a DNA of a helper phage) is expressed in parallel. Consequently, only a small fraction of all phage particles will present a fusion protein product (0.1–10%),[11,12] predominantly one per phage.

[10] G. P. Smith, *Science* **228,** 1315 (1985).

[11] S. Bass, R. Greene, and J. A. Wells, *Proteins* **8,** 309 (1990).

[12] S. Demartis, A. Huber, F. Viti, L. Lozzi, L. Giovannoni, P. Neri, G. Winter, and D. Neri, *J. Mol. Biol.* **286,** 617 (1999).

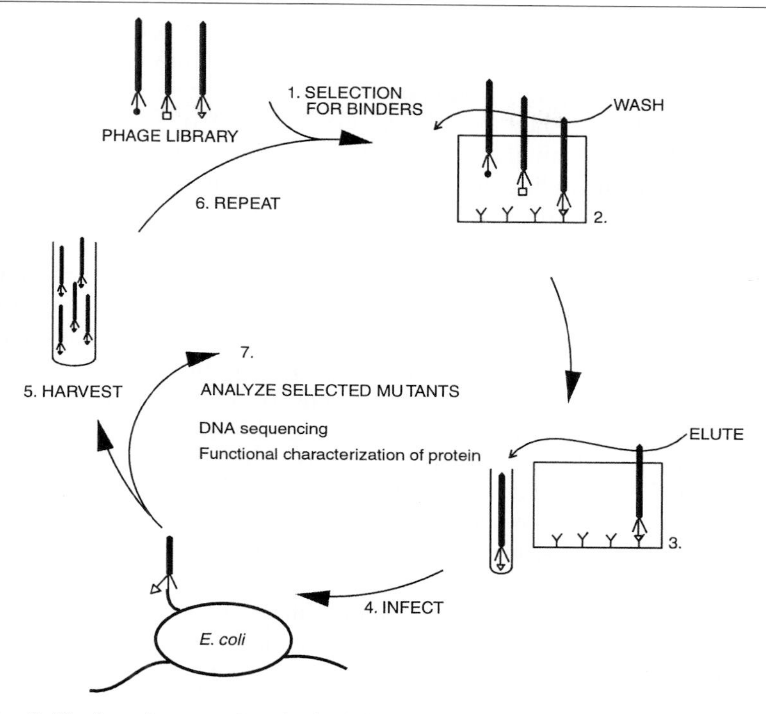

FIG. 2. The iterative procedure for isolation of proteins (or peptides) with affinity for an immobilized ligand. A library of mutant proteins displayed on phage is incubated with an immobilized ligand (1), nonbinding proteins are washed off (2), and mutant proteins with binding affinity are eluted (3). The DNA encoding binding proteins is amplified by infecting bacteria with the eluted phage particles (4). A new set of phages is harvested from the bacterial culture (5) and the affinity selection can be repeated (6), if necessary, with higher stringency. Finally, the structure and function of the selected clones are characterized (7).

Methods

Active Site of GST A1-1 and Rationale for Construction of Mutant Libraries

The enzyme chosen for phage display was the human GST A1-1. Its crystal structure has been determined and its active site topography has been deduced from the bound *S*-benzylglutathione.[13] For the present application, the prime focus is the subsite binding the electrophilic substrate

[13] I. Sinning, G. J. Kleywegt, S. W. Cowan, P. Reinemer, H. W. Dirr, R. Huber, G. L. Gilliland, R. N. Armstrong, X. Ji, P. G. Board, B. Olin, B. Mannervik, and T. A. Jones, *J. Mol. Biol.* **232,** 192 (1993).

that will be conjugated with glutathione. This largely hydrophobic subsite (called the H-site) is formed by three modules structurally converging to form the binding site (Fig. 3). Module 1 is the loop connecting the $\beta1$ strand and the $\alpha1$ helix; module 2 is the C-terminal part of the $\alpha4$ helix; and module 3 is the C-terminal segment of the protein consisting of amino acid residues 208–222 including the $\alpha9$ helix (Fig. 3). As judged from the crystal structure, the walls of the H-site are formed by the side chains of residues 9, 10, 12, and 15 from module 1, residues 104, 107, 108, 110, and 111 from module 2, and residues 208, 213, 216, 219, 220, and 222 from module 3. The number of possible combinations of amino acid residues in n positions is 20^n, and for an active site with 10 or more residues all combinations cannot be generated in a common mutant library. In the work described in this chapter two experimental approaches have been taken. First, 10 amino acids were randomly mutagenized with the expectation that even if only a fraction of the 10^{13} possible variants could be tested, this limited sample would still provide a reasonable representation of the variations theoretically possible. The second approach was to limit the number of randomized residues to four or five, corresponding to 10^5 to 10^6 possible mutants.

The rationale for generating the mutant libraries was to create modifications in the active site topography and thereby alter the substate selectivity of the enzyme. This was achieved by making synthetic oligonucleotides

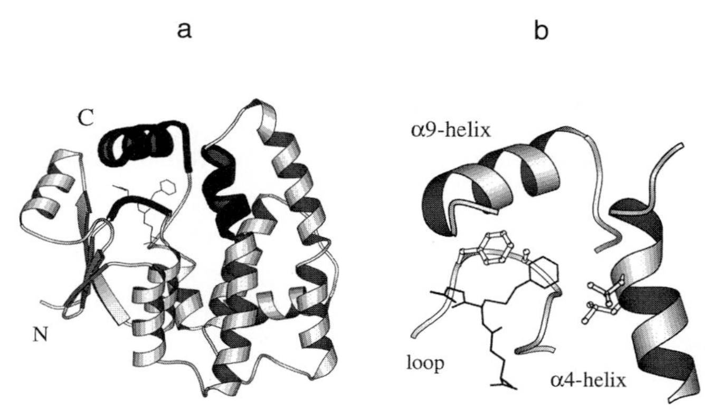

FIG. 3. Three-dimensional structure of a human GST A1-1 subunit with the ligand *S*-benzylglutathione bound to the active site [I. Sinning, G. J. Kleywegt, S. W. Cowan, P. Reinemer, H. W. Dirr, R. Huber, G. L. Gilliland, R. N. Armstrong, X. Ji, P. G. Board, B. Olin, B. Mannervik, and T. A. Jones, *J. Mol. Biol.* **232,** 192 (1993)]. (a) A monomer with the three modules of the H-site highlighted. (b) The H-site with active site residues Phe-10 and Ala-12 in module 1, as well as Leu-107 and Leu-108 in module 2, shown in ball-and-stick representation; these residues were randomly substituted in mutant library B.

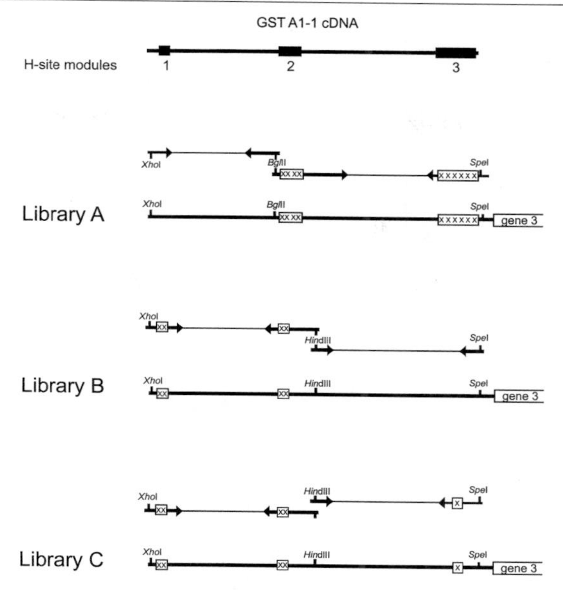

FIG. 4. Construction of libraries of random active site GST A1-1 mutants expressed in fusion with gene 3 protein of phage M13. The targeted H-site modules forming the substrate-binding site are indicated as black boxes in the sequence at the top. The assembly of the different libraries from segments amplified with mutagenic PCR primers (arrows with randomized codons marked by "X") using suitable restriction sites is indicated below. The digested PCR fragments were ligated to generate full-length cDNA molecules. The mutagenized cDNAs were ligated into pC3ΔNX [M. Widersten and B. Mannervik, *J. Mol. Biol.* **250,** 115 (1995)] and transferred into *E. coli* XL1-Blue by electroporation. After a recovery period, transformed bacteria were superinfected with helper phage and incubated overnight for phage production [C. F. Barbas III, A. S. Kang, R. A. Lerner, and S. J. Benkovic, *Proc. Natl. Acad. Sci. U.S.A.* **88,** 7978 (1991); M. Widersten and B. Mannervik, *J. Mol. Biol.* **250,** 115 (1995)].

with NNN or NNS (where N is any deoxyribonucleotide, A, C, G or T; and S is either G or C) replacing the wild-type codon for the targeted amino acid residue in the protein sequence. A schematic presentation of the assembly of the recombinant DNA encoding three different mutant libraries (A–C) used is given in Fig. 4. Table I shows the positions of randomized residues in the different libraries.

It is clear that structural components remote from the active site may contribute to catalytic efficiency,[14–17] but it is a reasonable proposition that

[14] L. Hedstrom, L. Szilagyi, and W. J. Rutter, *Science* **255,** 1249 (1992).

[15] P. A. Patten, N. S. Gray, P. L. Yang, C. B. Marks, G. J. Wedemayer, J. J. Boniface, R. C. Stevens, and P. G. Schultz, *Science* **271,** 1086 (1996).

[16] S. Oue, A. Okamoto, T. Yano, and H. Kagamiyama, *J. Biol. Chem.* **274,** 2344 (1999).

TABLE I
ACTIVE-SITE RESIDUES MUTATED IN GST A1-1 TO PRODUCE LIBRARIES OF VARIANT ENZYME FORMS

	Active site module															
	Module 1: loop connecting $\beta 1$ strand and $\alpha 1$ helix					Module 2: C-terminal portion of $\alpha 4$ helix					Module 3: C-terminal $\alpha 9$ helix with surrounding loops					
Residue:	9	10	12	14	15	104	107	108	110	111	208	213	216	219	220	222
Wild type:	Y	F	A	G	R	E	L	L	P	V	M	L	A	I	F	F
Library A	—	—	—	—	—	—	X	X	X	X	X	X	X	X	X	X
Library B	—	X	X	—	—	—	X	X	—	—	—	—	—	—	—	—
Library C	—	X	X	—	—	—	X	X	—	—	—	—	X	—	—	—

the determinants for substrate specificity reside primarily in the amino acid side chains forming the active site. In the case of GSTs, this assumption has support from the finding that the closely related isoenzymes GST A1-1 and GST A2-2 differ more than 100-fold in 3-oxosteroid isomerase activity. The primary structures of these two enzymes differ in only 11 of 221 amino acids, and 4 of the variant residues are located in the H-site.

Generation of Phage Libraries Expressing Mutant GST A1-1

The libraries of mutant GST A1–g3p fusion proteins were constructed from DNA fragments generated by polymerase chain reaction (PCR), using the human GST subunit A1 cDNA as template (Fig. 4), and inserted into an expression vector allowing for monovalent phage display expression of the encoded mutant protein. Monovalent expression was chosen in order to increase the stringency in the affinity selection by allowing for only a one-phage-to-one-ligand binding stoichiometry. The total numbers of individual clones in the phage libraries containing mutated GST A1 subunits were estimated from titrations to be 5×10^6 (libraries A and B) and 1×10^7 (library C).

Design of Transition-State Analogs

S_NAr Reactions. Different ligands featuring targeted properties of the proposed transition state of an S_NAr reaction were tested in affinity selection experiments by phage display (Fig. 5). The application of the ligands found most useful is described in further detail below.

1. *CBSG:* The *p*-carboxy-substituted *S*-benzylglutathione (CBSG; Fig. 1c) can be viewed as a crude analog for the transition state of the reaction between GSH and a *p*-nitro-substituted phenylhalide, which is considered to involve a negatively charged σ complex (Fig. 1a). This glutathione conjugate was immobilized through a bifunctional linker with thiolated bovine serum albumin (BSA) and used in affinity selection.[18]
2. *TNB:* The formation of an immobilized σ complex of GSH and TNB (Fig. 1b) was applied as the selection basis.[19] GSH was immobilized to a Sepharose matrix via the amino group of the γ-glutamyl residue and a σ complex was formed on addition of TNB. Phage-displayed active site mutants of GST A1-1 were thus selected by means of a combination of affinity for preformed σ complex and the ability to promote its formation in the

[17] L. O. Hansson, R. Bolton-Grob, M. Widersten, and B. Mannervik, *Protein Sci.* **8,** 2742 (1999).
[18] M. Widersten and B. Mannervik, *J. Mol. Biol.* **250,** 115 (1995).
[19] L. O. Hansson, M. Widersten, and B. Mannervik, *Biochemistry* **36,** 11252 (1997).

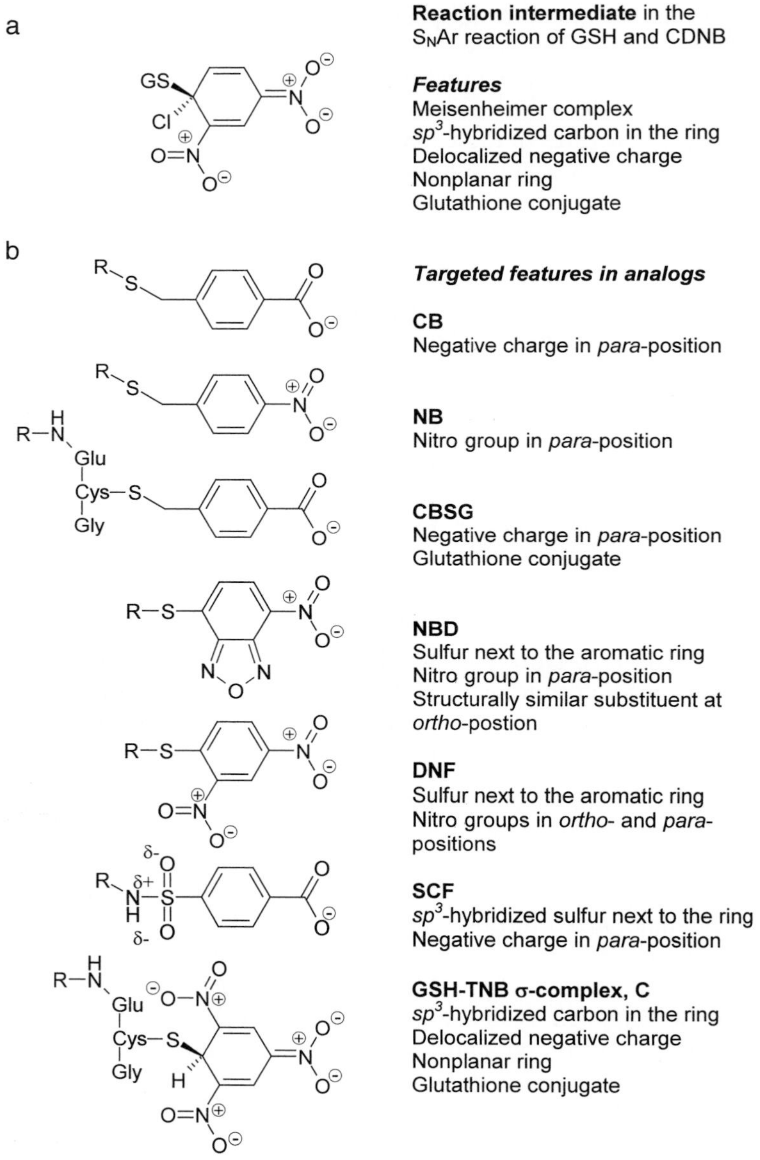

a

Reaction intermediate in the S_NAr reaction of GSH and CDNB

Features
Meisenheimer complex
sp^3-hybridized carbon in the ring
Delocalized negative charge
Nonplanar ring
Glutathione conjugate

b

Targeted features in analogs

CB
Negative charge in *para*-position

NB
Nitro group in *para*-position

CBSG
Negative charge in *para*-position
Glutathione conjugate

NBD
Sulfur next to the aromatic ring
Nitro group in *para*-position
Structurally similar substituent at *ortho*-postion

DNF
Sulfur next to the aromatic ring
Nitro groups in *ortho*- and *para*-positions

SCF
sp^3-hybridized sulfur next to the ring
Negative charge in *para*-position

GSH-TNB σ-complex, C
sp^3-hybridized carbon in the ring
Delocalized negative charge
Nonplanar ring
Glutathione conjugate

FIG. 5. Comparison of transition-state analogs for the S_NAr reaction of GSH and CDNB. (a) The proposed intermediate 1-chloro-2,4-dinitro-1-(*S*-glutathionyl)hexadienate in the reaction of GSH and CDNB (Fig. 1a). (b) A set of immobilized transition-state analogs used for phage display selection of novel GST A1-1 variants.

active site, i.e., to mimic two essential features of the reaction. Adequate binding affinity for GSH was assumed to be retained by the mutants, because the residues contributing to the binding of GSH were not altered.

Ester Hydrolysis. For the isolation of novel GST variants with esterase activity, a phosphonate transition-state analog, immobilized by conjugation to BSA, was used as affinity ligand. These compounds have previously been used for the isolation of antibodies catalyzing the hydrolysis of *p*-nitro-substituted aromatic esters such as ester **3** (Fig. 1f).[20]

Phage-Display Selection of Mutant GST A1-1 Proteins

GST A1-1 mutants on phage, monovalently displayed, in fusion with the gene 3 protein (g3p), were allowed to interact with the corresponding immobilized transition-state analog. After washing, adsorbed phage were eluted specifically by addition of unconjugated ligand (CBSG or phosphonate **2**, Fig. 1, depending on the selection targeted) or unspecifically by addition of acid. After neutralization, eluted phage were added to *E. coli* XL1-Blue in the exponential growth phase for amplification. The affinity selection was repeated for four or five rounds.

Sequence Analysis of Isolated Clones

S_NAr Reaction. The deduced amino acid sequences of mutants isolated from library A after five rounds of selection toward the CB–BSA, NB–BSA, and CBSG–BSA ligands, and from library B after four rounds of selection against the GSH–TNB ligand, are given in Table II. An examination of the sequences from the library A selection does not provide any unequivocal explanation as to why these particular protein variants were selected. It is clear, though, that the hydrophilicity of the H-site has been increased by a number of amino acid replacements to residues with polar side chains, i.e., residues capable of stabilizing a negative charge on the nitro groups during the reaction.

The deduced primary structures of mutants isolated after the GSH–TNB selection, however, shared a marked preponderance of aromatic residues in the randomized positions. This feature is also observed in a number of antibodies raised against immobilized nitrobenzenes.[21,22]

Ester Hydrolysis. Sequences of selected clones are presented in Ta-

[20] D. S. Tawfik, B. S. Green, R. Chap, M. Sela, and Z. Eshhar, *Proc. Natl. Acad. Sci. U.S.A.* **90,** 373 (1993).

[21] P. Gettins, D. Givol, and R. A. Dwek, *Biochem. J.* **173,** 713 (1978).

[22] J. Anglister, M. W. Bond, T. Frey, D. Leahy, M. Levitt, H. M. McConnell, G. S. Rule, J. Tomasello, and M. Whittaker, *Biochemistry* **26,** 6058 (1987).

TABLE II
DEDUCED AMINO ACID SEQUENCES IN GST A1-1 VARIANTS ISOLATED AFTER PHAGE DISPLAY SELECTION TOWARD TRANSITION-STATE ANALOGS

		Module 1		Module 2				Module 3					
	Residue:	10	12	107	108	110	111	208	213	216	219	220	222
Clone ID[a]	Wild type:	F	A	L	L	P	V	M	L	A	I	F	F
				Library A									
CBSG:165, CB:9, 22, NB:26		—	—	K	V	H	R	A	G	H	—	N	G
CBSG:197, CB:18, 32, 38, NB:4, 18, 19		—	—	T	T	G	G	L	T	S	T	T	P
CBSG:265, CB:21		—	—	P	G	R	A	S	P	T	H	R	?
CBSG:167		—	—	A	S	D	S	C	S	N	E	L	—
CBSG:169		—	—	R	Q	R	R	L	M	S	G	C	R
CBSG:191		—	—	F	M	A	R	T	V	R	S	P	P
CBSG:1912		—	—	S	F	I	A	P	—	T	Q	E	—
CBSG:134		—	—	M	P	I	H	E	F	P	?	I	?
CBSG:138		—	—	S	—	V	M	P	C	S	P	I	W
				Library B									
C:6		L	R	P	W	—	—		—	—	—	—	—
C:7		V	Y	T	—	—	—	—	—	—	—	—	—
C:8		W	D	—	T	—	—	—	—	—	—	—	—
C:9		A	W	T	H	—	—	—	—	—	—	—	—
C:10		D	Q	R	V	—	—	—	—	—	—	—	—
C:36		P	W	F	R	—	—	—	—	—	—	—	—
				Library C									
P:19, 34		V	S	H	S	—	—	—	—	D	—	—	—
P:5		I	S	H	R	—	—	—	—	N	—	—	—
P:8		S	Y	H	A	—	—	—	—	C	—	—	—
P:13		D	G	H	K	—	—	—	—	R	—	—	—
P:2		H	G	R	I	—	—	—	—	Q	—	—	—

[a] Clone ID is defined according to ligand used in the phage display selection. A dash (—) indicates identity to wild-type amino acid residue. See Figs. 1 and 5 for detailed structures of ligands.

ble II. A clear preference for His was observed in residue 107. From modeling studies a His residue in this position of the active site structure would be able to interact with the negatively charged oxygen of a bound phosphonate ligand, hence taking on the function of an oxyanion hole in catalysis. The majority of the mutated amino acid residues had been changed from residues of hydrophobic character to residues with charged or polar side chains. Two of the selected mutant proteins, P:19 and P:34, were found to have the same DNA sequence and were scored as siblings of the same mutant cDNA clone. Residues (His, Ser, Cys, Arg, Tyr) known to be present in the active sites of naturally occurring hydrolases as well as in catalytic antibodies were identified in the selected mutants.

Functional Analysis of Isolated Mutant Glutathione Transferases

The isolated mutant GST cDNAs were excised from the phage display vector and introduced into plasmid expression systems, which are better suited for production of soluble proteins. Selected mutants were purified according to standard procedures.[18,19]

S_NAr Reaction. Mutants isolated after affinity selection toward the transition-state analogs for the S_NAr reaction were tested for enzymatic activity with GSH and CDNB (Table III, Fig. 1). In general, the mutants isolated with the less elaborated analog CBSG were less efficient in catalyzing the conjugation reaction, even though mutants CBSG:265 and C:36, isolated in the selection toward different ligands, displayed comparable rate accelerations. Mutant C:36 was also shown to promote the formation of the σ complex between GSH and TNB at its active site, when these compounds were presented in solution to this GST A1-1 variant.

Ester Hydrolysis. Hydrolytic activity of mutant and wild-type GST A1-1 proteins was measured with the *p*-nitro-substituted ester **3** (Fig. 1f). The mutants tested displayed insignificant levels of esterase activity with the tested substrates (Table III). Mutant P:8, however, was inhibited by addition of the transition-state analog used as eluent during the phage display selection, an indication of improved binding affinity between this protein and the phosphonate ligand.

Discussion

Comparison with the Design of Catalytic Antibodies

Catalytic antibodies have successfully been designed by challenging the immune system to produce immunoglobulins with affinity for transition-

TABLE III
KINETIC PROPERTIES[a] OF PURIFIED GST A1-1 MUTANTS

	Substrate						
	GSH + CDNB				Ester **3**		
Enzyme	k_{cat} (sec^{-1})	K_m (mM)	k_{cat}/K_m (sec^{-1} M^{-1})	Rate enhancement (-fold)	I_{50} (mM)	k_{cat}/K_m (sec^{-1} M^{-1})	Rate enhancement (-fold)
Wild type	88	0.56	160,000	1.1×10^7	—	2	170
CBSG:165	0.097	1.89	51	3.5×10^3	—	—	—
CBSG:197	0.78	12	67	4.6×10^3	—	—	—
CBSG:265	—	—	2,200	1.5×10^5	—	—	—
C:36	—	—	6,700	4.6×10^5	—	—	—
P:8	—	—	7	4.6×10^2	0.22	5	500
P:34	—	—	1	50	—	2	170

[a] Steady state kinetics were studied in the presence of 0.05–1.8 mM CDNB and 5 mM GSH in 0.1 M sodium phosphate, pH 6.5. The kinetic parameters k_{cat}, K_m, and k_{cat}/k_m were obtained by nonlinear regression analysis of the experimental data, using the SIMFIT program package [W. G. Bardsley, P. B. McGinlay, and M. G. Roig, *J. Theor. Biol.* **139,** 85 (1989)]. The reaction rate enhancement was determined from the ratio of the second-order rate constants for the catalyzed (k_{cat}/K_m) and uncatalyzed reactions (k_2). The value of k_2 was determined as 1.45×10^{-2} sec^{-1} M^{-1} for the reaction of GSH with CDNB. Values for k_{cat} and K_m could not be determined separately for mutants CBSG:265 and C:36, because the catalyzed reaction did not show rate saturation with increasing substrate concentration. All assays with ester **3** were carried out in 0.1 M sodium phosphate at pH 7.6 at 30°. Ester **3** was dissolved in DMSO and the steady state rate of formed product was measured spectrophotometrically at 405 nm in the presence of 0.25 to 2 mM ester **3**. The kinetic parameter k^3_{cat}/K^3_M was obtained by nonlinear regression analysis of the steady state initial rates, $\Delta\varepsilon$ = 14,200 M^{-1} cm^{-1}. The value of the rate constant of the uncatalyzed reaction k_2 ($v = k_2[OH^-][\mathbf{3}]$) was extrapolated to zero buffer concentration, derived from measurements of the uncatalyzed reaction rate at different buffer concentrations. Inhibition studies of mutant P:8 was carried out in 1 mM ester **3** in the presence of increasing concentrations of phosphonate **2**. The concentration of phosphonate **2** resulting in 50% enzymatic activity was denoted as the IC_{50} value.

state analogs of the targeted chemical reaction.[7] In some cases, catalysts have been produced for which a counterpart is unknown in the repertoire of naturally occurring enzymes. However, it may be argued that the immunoglobulin fold of antibodies has its biological function optimized for binding, whereas the essential feature of a catalyst is to facilitate the chemical transformation of reactants into products. Protein scaffolds distinct from the immunoglobulin fold may be more suited to guide the reacting molecular species along the favorable reaction pathway. For example, the release of products is often rate limiting in antibody-catalyzed reactions. The choice of a protein structure naturally evolved to serve as a catalyst may therefore have advantages over immunoglobulins. In biological systems, the TIM barrel has proven to be a particularly useful motif for the evolution of diverse catalysts.[23]

In the present case, we have adopted the GST structure for redesign of its catalytic properties. The GST molecule is stably folded and large quantities of soluble protein can be produced by heterologous expression in *Escherichia coli.* The versatility of GSTs as catalysts is evidenced by the occurrence of multiple forms displaying diverse substrate specificities.[4,24]

The use of phage display for selection of GSTs with novel catalytic properties has obvious similarities with the generation of catalytic antibodies. In both cases the affinity for a transition-state analog may be used to mold a catalyst for a targeted reaction. For the phage display of GSTs, libraries of active site mutants are generated from which individuals with affinity for a transition-state mimic are selected *in vitro* with the entire mutant population presented at the same time. In contrast, antibodies with affinity for a transition-state analog are produced *in vivo* by the continuous evolution of binding specificity in the immune system, such that novel binding specificities continuously emerge with time. Both strategies have advantages and limitations. For phage display, the construction of the mutant library can be made by a rational choice of regions to be mutated but the significance of other parts of the structure may be neglected. The production of antibodies is governed by clonal selection of cells in the immune system and restructuring of sequences in unpredictable combinations. Thus, the latter method may afford a more general stochastic exploration of the available "sequence space," but leaves less opportunity for rational design of protein structure. The phage display method provides the option of targeting an active site cavity for random mutations as well as making it possible to complement cassette mutagenesis with random recombinations such as those obtainable by DNA shuffling.[25]

Active site mutants selected by phage display may be further improved by additional amino acid substitutions, either of second-sphere residues or of other residues contributing to the active site. In the case of the S_NAr reaction, the catalytic efficiency of one mutant (C:36) selected from library B was further improved by a fifth substitution in the active site. The targeted residue 208 was chosen so that all three modules contributing to the active

[23] P. C. Babbitt and J. A. Gerlt, *J. Biol. Chem.* **272,** 30591 (1997).

[24] B. Mannervik and M. Widersten, *in* "Advances in Drug Metabolism in Man" (G. M. Pacifici and G. N. Fracchia, eds.), p. 407. European Commission, Luxembourg, 1995.

[25] W. P. C. Stemmer, *Nature* (*London*) **370,** 389 (1994).

site were explored (see Table II) and an isolated Cys-208 variant exhibited a twofold improvement in catalytic efficiency.[26]

Conclusions

Phage display is a powerful method for linking phenotype to genotype. In the application presented in this chapter binding affinity for an active site ligand was used as the basis for selecting GST variants with altered catalytic activities from libraries of mutants.

Acknowledgments

The authors thank Drs. Carlos F. Barbas III and Richard A. Lerner (The Scripps Research Institute) for the generous gift of the pComb3 vector, and Drs. Daniel S. Tawfik (Cambridge, UK) and Zelig Eshhar (Weizmann Institute of Science, Rehovot, Israel) for providing the phosphonate and ester compounds used in this study. The work from the authors' laboratory has been supported by the Swedish Research Council for Engineering Sciences, the Swedish Natural Science Research Council, and the Carl Trygger Foundation.

[26] L. O. Hansson, M. Widersten, and B. Mannervik, *Biochem. J.* **344,** 93 (1999).

[24] Selecting and Evolving Functional Proteins *in Vitro* by Ribosome Display

By JOZEF HANES, LUTZ JERMUTUS, and ANDREAS PLÜCKTHUN

Introduction

Most techniques used for the screening and selection of protein libraries are based on *in vivo* systems. They either use living cells directly, such as cell surface display,[1,2] the yeast two-hybrid system,[3] or the protein–fragment complementation assay (see [14] in this volume[4]), or they use cells indirectly for the production of phages or viruses in techniques such as phage display[5,6]

[1] E. T. Boder and K. D. Wittrup, *Nature Biotechnol.* **15,** 553 (1997).
[2] S. Stahl and M. Uhlen, *Trends Biotechnol.* **15,** 185 (1997).
[3] C. Bai and S. J. Elledge, *Methods Enzymol.* **283,** 141 (1997).
[4] S. W. Michnick, I. Remy, F. X. Campbell-Valois, A. Vallée-Belisle, and J. N. Pelletier, *Methods Enzymol.* **328,** 208 (2000).
[5] G. P. Smith and J. K. Scott, *Methods Enzymol.* **217,** 228 (1993).
[6] G. P. Smith, *Science* **228,** 1315 (1985).

0076-6879/00 $30.00

or the selectively infective phage (SIP) technology.[7] However, *in vivo* systems have a number of limitations. One important restriction is that the library size is limited by the transformation efficiency,[8] a step that all of these techniques have in common. Usually, libraries of not more than 10^9 to 10^{10} independent members can be prepared in *Escherichia coli,* and yeast libraries are still smaller by several orders of magnitude because transformation is less efficient. Another limitation of *in vivo*-based selection methods becomes apparent if a diversification step must be included. It requires either repeatedly switching between *in vivo* selection and *in vitro* diversification[9–11] or the use of mutator strains.[12] The former approach is quite laborious, as it makes a new library generation and large-scale transformation necessary for each cycle of sequence diversification and selection. The latter approach may have the disadvantage that possible candidate molecules may be removed from the library by mutations generated either in the host genome or in the plasmid regions important for expression or replication.

All these limitations can be simultaneously overcome in ribosome display. Ribosome display is an *in vitro* technology for the simultaneous selection and evolution of proteins from diverse libraries (reviewed in Refs. 13 and 14). Because no transformation is necessary, large libraries can be prepared and applied for selection. Furthermore, diversification is conveniently introduced in this method, making evolutionary approaches easily accessible. Ribosome display was first applied to short peptides[15,16] and has subsequently been improved to work with folded proteins.[17–19] In this

[7] S. Spada, C. Krebber, and A. Plückthun, *Biol. Chem.* **378,** 445 (1997).
[8] W. J. Dower and S. E. Cwirla, *in* "Guide to Electroporation and Electrofusion" (D. C. Chang, B. M. Chassy, J. A. Saunders, and A. E. Sowers, eds.), p. 291. Academic Press, San Diego, California, 1992.
[9] A. C. Braisted and J. A. Wells, *Proc. Natl. Acad. Sci. U.S.A.* **93,** 5688 (1996).
[10] R. Schier, A. McCall, G. P. Adams, K. W. Marshall, H. Merritt, M. Yim, R. S. Crawford, L. M. Weiner, C. Marks, and J. D. Marks, *J. Mol. Biol.* **263,** 551 (1996).
[11] W. P. Yang, K. Green, S. Pinz-Sweeney, A. T. Briones, D. R. Burton, and C. F. Barbas III, *J. Mol. Biol.* **254,** 392 (1995).
[12] N. M. Low, P. H. Holliger, and G. Winter, *J. Mol. Biol.* **260,** 359 (1996).
[13] J. Hanes and A. Plückthun, *in* "Combinatorial Chemistry in Biology" (M. Famulok and E. L. Winnacker, eds.). *Curr. Top. Microbiol. Immunol.* **243,** 107 (1999).
[14] L. Jermutus, L. A. Ryabova, and A. Plückthun, *Curr. Opin. Biotechnol.* **9,** 534 (1998).
[15] L. C. Mattheakis, R. R. Bhatt, and W. J. Dower, *Proc. Natl. Acad. Sci. U.S.A.* **91,** 9022 (1994).
[16] L. C. Mattheakis, J. M. Dias, and W. J. Dower, *Methods Enzymol.* **267,** 195 (1996).
[17] J. Hanes and A. Plückthun, *Proc. Natl. Acad. Sci. U.S.A.* **94,** 4937 (1997).
[18] J. Hanes, J. Jermutus, S. Weber-Bornhauser, H. R. Bosshard, and A. Plückthun, *Proc. Natl. Acad. Sci. U.S.A.* **95,** 14130 (1998).
[19] J. Hanes, L. Jermutus, C. Schaffitzel, and A. Plückthun, *FEBS Lett.* **450,** 105 (1999).

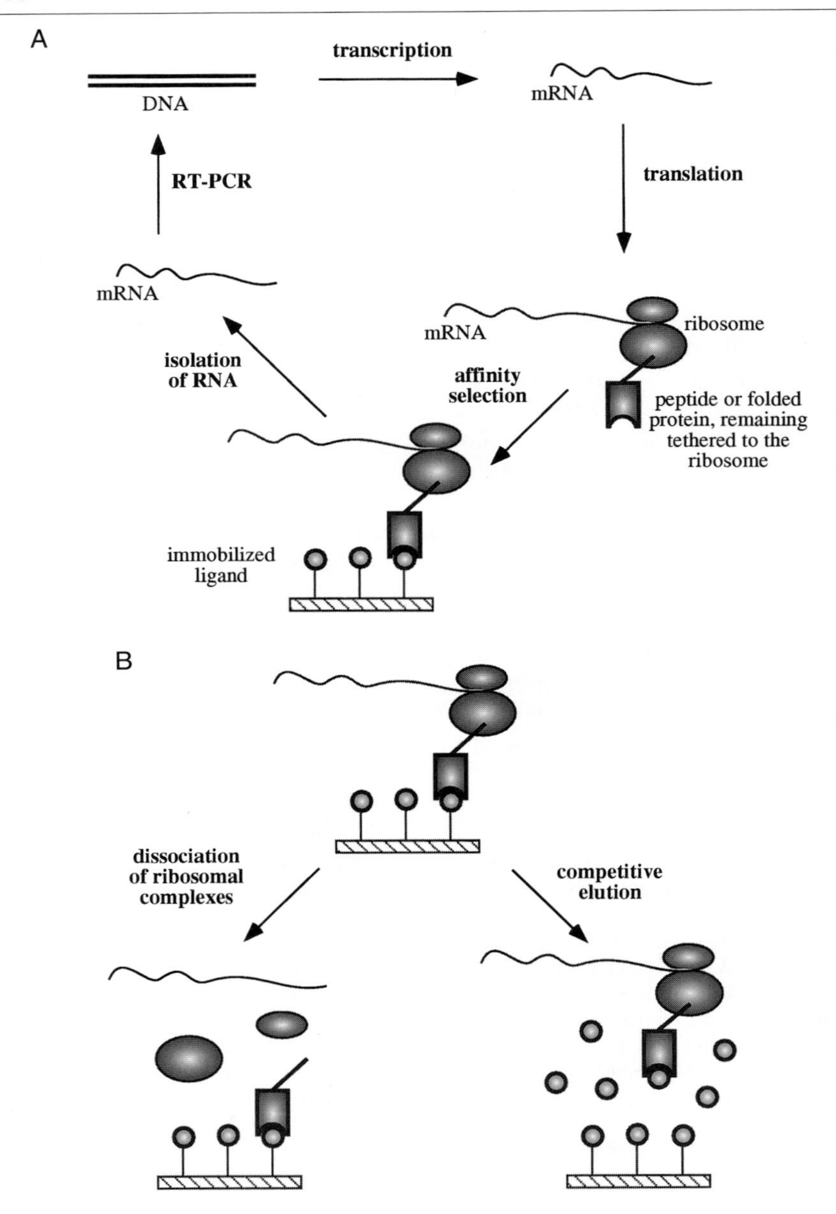
A
transcription
DNA
mRNA
translation
RT-PCR
mRNA
mRNA
ribosome
isolation
of RNA
affinity
selection
peptide or folded
protein, remaining
tethered to the
ribosome
immobilized
ligand
B
dissociation
of ribosomal
complexes
competitive
elution

chapter we describe in detail the methodology and the nature of the improvements.

The principle of ribosome display is shown in Fig. 1A. DNA, usually a polymerase chain reaction (PCR) product, which encodes a protein library in a special ribosome display cassette (discussed below), is transcribed *in vitro* to mRNA. The mRNA is purified and used for *in vitro* translation. This *in vitro* translation is performed under such conditions that ternary ribosomal complexes are formed, consisting of mRNA, ribosome, and translated polypeptide, which do not dissociate. These complexes can then be used for affinity selection. Complexes that do not encode and produce a polypeptide that specifically recognizes the target are removed by intensive washing. The mRNA of the selected ribosomal complexes, which encode a polypeptide cognate for the target, is isolated and used for reverse transcription and PCR. This amplified DNA can then be used for another round of ribosome display or for analysis. The mRNA can be isolated from bound ribosomal complexes either directly by removing Mg^{2+} with an excess of EDTA, and thus causing dissociation of all bound complexes, or by competitive elution of ribosomal complexes with free ligand, followed by RNA isolation from the eluted complexes (Fig. 1B). In this chapter we provide the information necessary to perform ribosome display for the selection and evolution of functional proteins, using either an *E. coli* or a rabbit reticulocyte translation system.

The *Escherichia coli* Ribosome Display System

Construction of a DNA Library and Its Transcription to mRNA

The ribosome display construct (Fig. 2A) contains, on the DNA level, a T7 promoter for efficient transcription to mRNA. On the RNA level,

---

FIG. 1. (A) Principle of ribosome display for screening protein libraries for ligand binding. A DNA library containing all important features necessary for ribosome display (for details see text) is first transcribed to mRNA and after its purification, mRNA is translated *in vitro*. Translation is stopped by cooling on ice, and the ribosome complexes are stabilized by increasing the magnesium concentration. Ribosomal complexes are affinity selected from the translation mixture by the native, newly synthesized protein binding to immobilized ligand. Nonspecific ribosome complexes are removed by intensive washing, and mRNA is isolated from the bound ribosome complexes, reverse transcribed to cDNA, and cDNA is then amplified by PCR. This DNA is then used for the next cycle of enrichment, and a portion can be analyzed by cloning and sequencing and/or by ELISA or RIA. (B) Two methods for mRNA isolation from bound ribosomal complxes. The bound ribosomal complexes can either be dissociated by an excess of EDTA and then RNA is isolated, or they can first be eluted specifically with free ligand followed by RNA isolation.

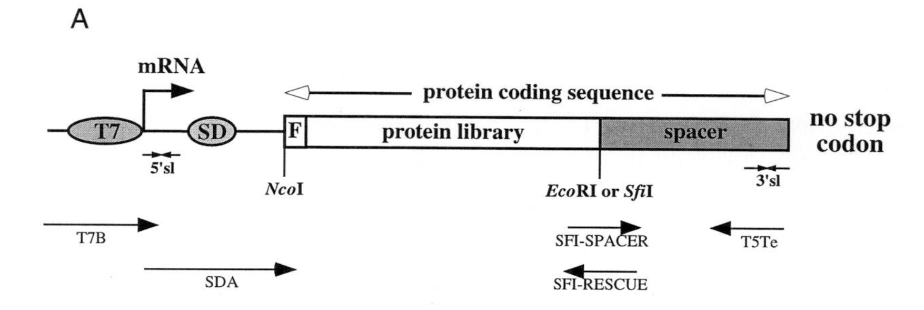

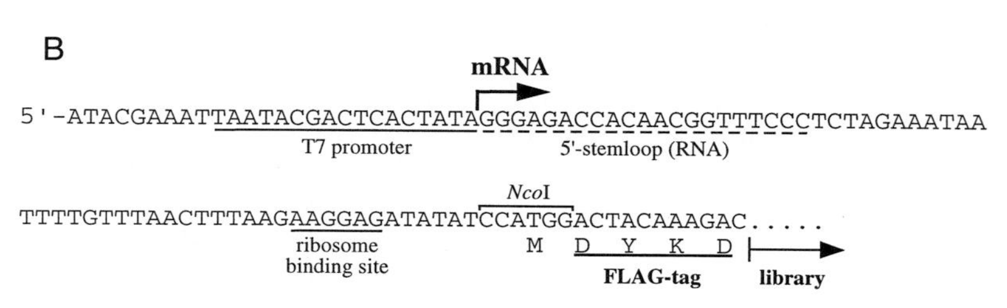

FIG. 2. (A) Schematic drawing of the DNA construct used for *E. coli* ribosome display system. *T7* denotes the T7 promoter, *SD* the ribosome-binding site, *spacer* the part of protein construct connecting the folded protein to the ribosome, *F* the FLAG tag sequence and *5′sl* and *3′sl* the stem–loops on the 5′ and 3′ end of the mRNA. The filled arrow indicates the transcriptional start and open arrows the protein coding sequence. The primers used for the PCR amplification are shown below the scheme. If the construct is prepared by ligation of the library coding sequence to the spacer, in the first PCR this ligation mixture is amplified with primers SDA and T5Te, followed by a second amplification with primers T7B and T5Te. For the preparation of the construct by assembly PCR, the library sequence and the spacer are amplified separately with primers SDA and SFI-RESCUE for library or SFI-SPACER and T5Te for spacer. Purified PCR products are then assembled by PCR without primers for several cycles after amplification with primers SDA and T5Te in the first step. The second amplification is carried out with primers T7B and T5Te. (B) DNA upstream sequence used for the *E. coli* ribosome display constructs.

the construct contains a prokaryotic ribosome binding site (Shine–Dalgarno sequence), followed by the open reading frame encoding the protein library without any stop codon, to avoid release of synthesized peptide, and its mRNA from the ribosome. The mRNA also contains 5′ and 3′ stem–loops, which are known to stabilize mRNA against RNases and therefore increase its half-life *in vivo* as well as *in vitro*.[20] In our most frequently used constructs

[20] J. G. Belasco and G. Brawerman, "Control of Messenger RNA Stability." Academic Press, San Diego, California, 1993.

the 5′-untranslated region of the mRNA, including the 5′ stem–loop and the ribosome-binding site, is derived from gene 10 of phage T7[21] (Fig. 2B), and the 3′ stem–loop is derived from the early terminator of phage T3[22] (Fig. 3) that has been slightly modified to give an open reading frame.

The protein-coding sequence comprises two portions: the N-terminal part, which encodes the polypeptide to be selected (the library), and the C-terminal part, which is constant and serves as a spacer (Fig. 2A). The spacer has two functions: (1) it tethers the synthesized protein to the ribosome by maintaining the covalent bond to the tRNA, which is bound at the P-site of the ribosome, and (2) it provides an unstructured portion at the C terminus of the protein to be folded, such that the ribosomal tunnel can cover at least 20–30 amino acid residues[23–25] of the emerging polypeptide without interfering with the folding of the main portion of the protein. We currently use two different spacers, named here spacer A and spacer B, both derived from gene III of filamentous phage M13mp19[26] (Swissprot: P03662), covering amino acids (aa) 211–318 or 249–318, respectively. The use of the longer spacer A (121 aa; Fig. 3A) results in an approximately twofold higher efficiency of ribosome display than that of spacer B (82 aa; Fig. 3B). However, it is somewhat more difficult to introduce the longer spacer by PCR amplification with oligonucleotides that anneal to the ends of the cDNA, usually resulting in lower amount and quality of the PCR products than the shorter one.

The ribosome display library can be constructed *in vitro,* without any *E. coli* transformation, either by ligating the DNA encoding the library to the spacer, or by assembly PCR. If the library is prepared by ligation, the spacer is most conveniently isolated from a restriction fragment of a phage or phagemid. To keep restriction sites convenient, either *Eco*RI/*Hin*dIII or *Sfi*I/*Hin*dIII fragments are used (Fig. 2A). Thus, spacer A can be isolated from an fd phage (f17/9),[27] which displays a single-chain Fv fragment and had been engineered to contain appropriate restriction sites in gene III, by *Eco*RI/*Hin*dIII digestion. Similarly, spacer B can be isolated from the vector pAK200[28] by cutting with *Sfi*I/*Hin*dIII. The DNA library must be

[21] F. W. Studier, A. H. Rosenberg, J. J. Dunn, and J. W. Dubendorff, *Methods Enzymol.* **185,** 60 (1990).

[22] R. Reynolds, R. M. Bermudez-Cruz, and M. J. Chamberlin, *J. Mol. Biol.* **224,** 31 (1992).

[23] W. P. Smith, P. C. Tai, and B. D. Davis, *Proc. Natl. Acad. Sci. U.S.A.* **75,** 5922 (1978).

[24] R. Jackson and T. Hunter, *Nature (London)* **227,** 672 (1970).

[25] L. I. Malkin and A. Rich, *J. Mol. Biol.* **26,** 329 (1967).

[26] P. M. van Wezenbeek, T. J. Hulsebos, and J. G. Schoenmakers, *Gene* **11,** 129 (1980).

[27] C. Krebber, S. Spada, D. Desplancq, and A. Plückthun, *FEBS Lett.* **377,** 227 (1995).

[28] A. Krebber, S. Bornhauser, J. Burmester, A. Honegger, J. Willuda, H. R. Bosshard, and A. Plückthun, *J. Immunol. Methods* **201,** 35 (1997).

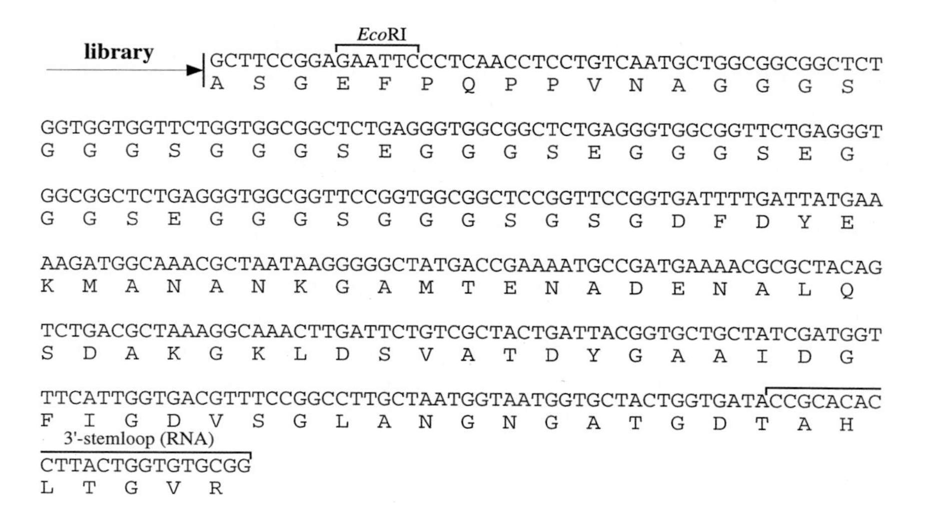

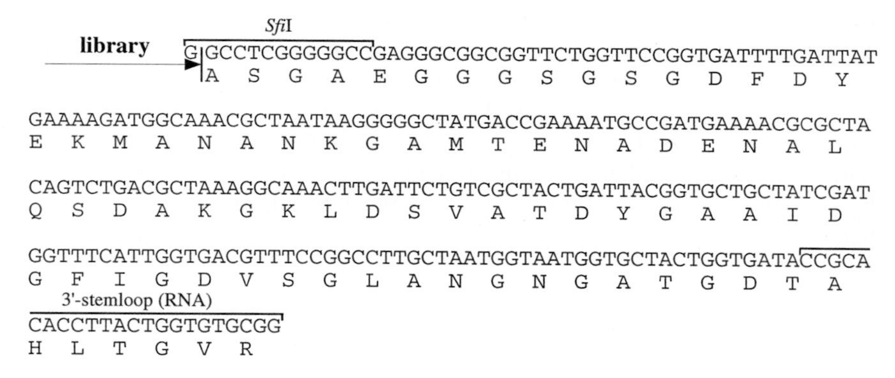

FIG. 3. DNA and deduced protein sequence of spacer A (A) and spacer B (B) used for ribosome display. The arrow indicates the 3′ end of the protein library coding sequence.

prepared in such a way (usually by PCR) that it contains *Sfi*I or *Eco*RI flanking sequences at its 3′ end and carries at least 12 nucleotides of a constant 5′-flanking sequence. The 5′-flanking sequence is necessary for the annealing of PCR primers, with which the entire DNA library is amplified to generate the full expression cassette (Fig. 2). In this way, the required upstream elements (T7 promoter, Shine–Dalgarno sequence) are introduced. For this constant sequence, either a part of the 5′-untranslated mRNA sequence can be used, or a protein tag can be added to the translated region that can later also be used for protein detection. For actually assembling the cassette (Fig. 2) by ligation, the DNA library encoding the polypeptide of interest is cut with an appropriate restriction enzyme (*Sfi*I or *Eco*RI) at the 3′ end of the polypeptide region and ligated to the selected spacer. The ligation mixture is subsequently amplified in two PCRs with two pairs of oligonucleotides (SDA and T5Te, followed by T7B and T5Te; Fig. 2A), which introduce all above-mentioned features important for ribosome display.[17]

If the whole ribosome display library construct (Fig. 2A) is prepared by assembly PCR instead of ligation, the actual protein library DNA and the constant spacer are first amplified separately. Primers are designed in such a way that the DNA library at its 3′ end and the spacer DNA at its 5′ end each contain an identical sequence of 18 nucleotides. Both PCR products are purified and used for assembly PCR (SDA and SFI-RESCUE, and SFI-SPACER and T5Te; Fig. 2A). First, during a few cycles of PCR, both fragments anneal together and are completed with *Taq* polymerase in the absence of any primers, and then this assembled template is further amplified in the presence of the "outer" primers (T7B and T5Te; Fig. 2A).

Library diversity can be increased before starting affinity selection or in between cycles of ribosome display by, e.g., DNA shuffling,[29] error-prone PCR,[30] or the staggered extension process.[31] In this case, only the DNA encoding the protein library is diversified by these PCR-based techniques, amplified with primers specific for this polypeptide cassette (e.g., SDA and SFI-RESCUE; Fig. 2A), and then either ligated to the spacer and then subjected to PCR amplification or, alternatively, directly used for assembly PCR to regenerate the whole cassette.

PCR products encompassing the whole construct with promoter, Shine–Dalgarno sequence, protein library and spacer, prepared either from liga-

[29] W. P. Stemmer, *Nature (London)* **370,** 389 (1994).

[30] R. C. Cadwell and G. F. Joyce, *Genome Res.* **3,** S136 (1994).

[31] H. Zhao, L. Giver, Z. Shao, J. A. Affholter, and F. H. Arnold, *Nature Biotechnol.* **16,** 258 (1998).

tion or from assembly PCR, are directly used for *in vitro* transcription with T7 RNA polymerase,[32] and RNA is purified by LiCl precipitation.

As an example, we provide here a procedure for the preparation of an antibody single-chain Fv fragment (scFv) library from immunized mice. Briefly, light (V_L) and heavy (V_H) chains are separately amplified by PCR from the natural mRNAs. After their purification they are assembled by PCR to create a DNA library *in vitro,* without *E. coli* transformation. The resulting PCR product encodes V_L linked to V_H via a flexible linker $(Gly_4Ser)_4$. This procedure, which is identical for assembling an scFv library for either phage display or ribosome display, has been described elsewhere.[28] After digestion with *Sfi*I (which flanks the scFv portion at its 3′ end), this library is used for the preparation of the ribosome display construct. The coding sequence starts with a FLAG tag (sequence MDYKD) and, of course, carries no signal sequence. To this FLAG sequence the primer SDA anneals, which is used for the PCR amplification of the whole library and that introduces the ribosome-binding site.

Materials

PCR primers:
- SDA: 5′-AGA CCA CAA CGG TTT CCC TCT AGA AAT AAT TTT GTT TAA CTT TAA GAA GGA GAT ATA TCC ATG GAC TAC AAA GA-3′; the ribosome-binding site is underlined
- T7B: 5′-ATA CGA AAT TAA TAC GAC TCA CTA TAG GGA GAC CAC AAC GG-3′; the T7 promoter is underlined
- T5Te: 5′-CCG CAC ACC AGT AAG GTG TGC GGT ATC ACC AGT AGC ACC-3′; the sequence creating a 3′ stem–loop at the RNA level is underlined

*Sfi*I fragment of mouse scFv DNA library (150 ng, ca. 2×10^{11} molecules)
QIAquick gel extraction kit (Qiagen, Hilden, Germany)
T4 DNA ligase (Roche Diagnostics, Mannheim, Germany)
Taq DNA polymerase (GIBCO-BRL, Gaithersburg, MD)
dNTPs, 20 m*M* each (Roche Diagnostics)
Dimethyl sulfoxide (DMSO); Fluka, (Buchs, Switzerland)
10 m*M* Tris-HCl, pH 8.5
T7 RNA polymerase (New England BioLabs, Beverly, MA)
T7 RNA polymerase buffer (5×): 1 *M* HEPES-KOH (pH 7.6), 150 m*M* magnesium acetate, 10 m*M* spermidine, 0.2 m*M* dithiothreitol (DTT)
RNasin (Promega, Madison, WI)

[32] I. D. Pokrovskaya and V. V. Gurevich, *Anal. Biochem.* **220,** 420 (1994).

LiCl, 6*M*
RNase-free water (Milli-Q-purified)

Additional Material for DNA Library Construction by Assembly Polymerase Chain Reaction

Primers:
SFI-RESCUE: 5′-GCC CTC GGC CCC CGA GGC-3′
SFI-SPACER: 5′-GCC TCG GGG GCC GAG GGC GGC GGT T-3′

Protocol for DNA Library Construction by Ligation or Assembly Polymerase Chain Reaction and Its Transcription to mRNA. The vector pAK200[28] is cut by *Sfi*I/*Hin*dIII and the resulting 481-bp fragment, encoding part of the C-terminal domain of the gene III protein, which serves as the spacer (Fig. 2A), is purified with the QIAquick gel extraction kit. The *Sfi*I fragment of an scFv library (150 ng), prepared by assembly PCR as described in Ref. 28, is ligated to a threefold molar excess of the purified 481-bp *Sfi*I/*Hin*dIII fragment (300 ng) in a 30-μl reaction overnight at 16° with 10 units (U) of T4 DNA ligase. The PCR amplification is performed in 50 μl, containing 5 μl of ligation mixture, 5 μl of 10× PCR buffer, 1.5 μl of 50 m*M* $MgCl_2$ (the ligation mixture also contains Mg^{2+}), 0.5 μl of 20 m*M* dNTPs, 2.5 μl of DMSO, 0.5 μl each of 100 μM primers SDA and T5Te, and 0.5 μl (2.5 U) of *Taq* DNA polymerase (4 min at 94°, and then 5 cycles of 30 sec at 94°, 30 sec at 37°, 2.5 min at 72°, followed by 15 to 20 similar cycles at 50 instead of 37°, and finished by 10 min at 72°). PCR products are separated by agarose gel electrophoresis, and the DNA band corresponding to the library is purified with the QIAquick gel extraction kit and eluted in 30 to 50 μl of 10 m*M* Tris-HCl, pH 8.5 or water. This DNA is subsequently used for the second PCR amplification, using the same conditions as for the first one, except that primers T7B and T5Te and 10 to 20 μl of purified PCR products are used as template for the PCR (usually 12 to 16 cycles).

To prepare a DNA library by assembly PCR instead of ligation to the 3′-tether cassette, first the vector pAK200[28] is amplified with primers SFI-SPACER and T5Te, and the *Sfi*I fragment of the mouse scFv DNA library is amplified with primers SDA and SFI-RESCUE, using the same conditions as described above. After purification of both fragments with the QIAquick gel extraction kit, the library and the spacer fragment are applied for assembly PCR. Fifteen nanograms of the library fragment and 30 ng of spacer fragment (about a threefold molar excess) are mixed and amplified by PCR, first without primers (4 min at 94°, then 5 cycles of 30 sec at 94°, 30 sec at 50°, 2.5 min at 72°), followed by amplification with primers SDA and T5Te (12 cycles of 30 sec at 94°, 30 sec at 50°, 2.5 min at 72°, and finished by 10 min at 72°), under similar conditions as described above for the amplification from the ligation mixture. The second PCR with primers T7B and T5Te is performed as described above.

It is essential for obtaining the best ribosome display performance that the quality of the PCR product used for transcription be extremely high: the amplified sample should contain a single strong band and the DNA should be at least 10 to 20 ng/μl in concentration, without by-products or smears, when analyzed by agarose gel electrophoresis. The mRNA from a ribosome display library is prepared by *in vitro* transcription in a 150-μl reaction containing 30 μl of 5× T7 RNA polymerase buffer, 21 μl of NTPs (50 m*M* each), 3 μl (120 U) of RNasin, 6 μl (300 U) of T7 RNA polymerase, and 30 μl of the unpurified PCR products from the second PCR amplification (see above). After incubating the mixture for 2 to 3 hr at 37°, the RNA is purified as follows. The reaction is mixed with the same volume of 6 *M* LiCl, incubated on ice for 30 min, and subsequently centrifuged 20 min at 4° and 16,000*g*. The RNA pellet is washed once with 500 μl of 70% ethanol, dissolved (without prior drying) in 100 μl of RNase-free water, and precipitated a second time by adding a one-tenth volume of 1 *M* NaCl and 3 volumes of ethanol and incubating on ice for 30 min. The mixture is centrifuged as for the first precipitation, and the pellet is washed with 70% ethanol, dried, and dissolved in RNase-free water. The concentration of RNA is estimated by optical density (OD) measurement at 260 nm (1 OD_{260} = 40 μg/ml). For successful ribosome display, the mRNA quality is important. This can be controlled by agarose gel electrophoresis.[33] The sample should contain at least 90% full-length mRNA.

Preparation of S-30 Extract

In vitro translation is performed by using an S-30 *E. coli* extract, which is prepared from the *E. coli* strain MRE600 by a slightly modified procedure of Lesley,[34] based on the procedure of Chen and Zubay.[35] Our protocol for S-30 ribosome display extract preparation differs from the published method[34] only by the fact that DTT, 2-mercaptoethanol, and *p*-toluene sulfonyl fluoride are omitted in all solutions. However, if a library of proteins is to be screened that does not contain disulfide bonds, the extract can be prepared with DTT and 2-mercaptoethanol. The quality of the S-30 extract

[33] S. K. Goda and N. P. Minton, *Nucleic Acids Res.* **23,** 3357 (1995).

[34] S. A. Lesley, *Methods Mol. Biol.* **37,** 265 (1995).

[35] H. Z. Chen and G. Zubay, *Methods Enzymol.* **101,** 674 (1983).

is important for the efficiency of ribosome display and also for the *in vitro* translation itself. The cells must be harvested in early exponential growth phase, when their translational machinery is highly active. According to our experience, the cells should not be harvested at OD_{550} higher than 0.5, when 5-liter baffled flasks are used for culturing. However, cells can be harvested at a higher OD_{550} if a fermenter is used.

Material

Escherichia coli strain MRE600
Growth medium (per liter of distilled water): 5.6 g of KH_2PO_4 (anhydrous), 28.9 g of K_2HPO_4 (anhydrous), 10 g of yeast extract, 15 mg of thiamin, and 25 ml of 40% (w/v) glucose (added after separate sterilization)
S-30 buffer: 10 m*M* Tris–acetate (pH 7.5), 14 m*M* magnesium acetate, and 60 m*M* potassium acetate. The solution must be chilled on ice before use
French pressure cell (SLM-Aminco, Rochester, NY)
Dialysis tubing (6000–8000 molecular weight cutoff), prewashed in S-30 buffer
Preincubation mix (per 10 ml): 3.75 ml of 2 *M* Tris–acetate (pH 7.5), 71 μl of 3 *M* magnesium acetate, 75 μl of a solution containing the 20 amino acids (10 m*M* each), 0.3 ml of 0.2 *M* ATP, 0.2 g of phosphoenolpyruvate (trisodium salt), and 50 U of pyruvate kinase. The preincubation mix must be prepared immediately before use

Protocol for S-30 Extract Preparation. A starter culture of *E. coli* MRE600 is grown overnight at 37° in 100 ml of yeast extract containing growth medium with shaking. On the following day, 1-liter aliquots of medium in 5-liter baffled flasks are inoculated with 10 ml of the overnight culture and growth is allowed to continue at 37° up to an OD_{550} of 0.5. Then, approximately the same volume of ice is added to the *E. coli* culture, and cells are harvested by centrifugation at 3500 *g* for 15 min at 4°. The supernatant is discarded, and another 1-liter aliquot of the remaining culture can be added to the centrifuge bottles and centrifuged. After centrifugation, pellets are thoroughly resuspended in 200 ml (per liter of culture) of ice-cold S-30 buffer and centrifuged at 3500*g* for 15 min at 4°. This washing step is repeated once more, and after centrifugation, pellets are combined and weighted. At this point, pellets can be stored at −80° or immediately used for extract preparation.

The washed cells are thoroughly resuspended in ice-cold S-30 buffer at a ratio of 4 ml of buffer per gram of wet cells. The cells are lysed by a passage through a chilled French pressure cell at 6000 psi. Repeated passing of the cell suspension through French pressure cell will result in a decrease of translational activity of the extract. The lysed cells are immediately centrifuged at 30,000*g* for 30 min at 4°. The supernatant is transferred to a clean centrifuge tube and centrifuged again at 30,000 *g* for 30 min at 4°. The supernatant of the second centrifugation is transferred again to a clean flask, and for each 6.5 ml of S-30 extract, 1 ml of preincubation mix is added, and this solution is slowly shaken for 1 hr at 25° (there should be no foaming). By this so-called "run-off" procedure the ribosomes finish translation of endogenous mRNA, and endogenous mRNA and DNA are degraded by nucleases present in the extract. This step is necessary to make the system dependent only on exogenous template. Afterward, the S-30 extract is transferred to dialysis tubing and dialyzed in the cold room three times against chilled S-30 buffer (500 ml of buffer per aliquot of extract prepared from 1 liter of culture). Each dialysis solution is replaced after 4 hr (one step can also be overnight). The extract is divided into aliquots of 100 to 500 μl and frozen in liquid nitrogen. The frozen extract can be stored for several months at −80° without loss of activity. After thawing, the extract can be frozen once more for later use. It is not recommended to thaw and freeze the extract more than twice, because it will begin losing activity.

In Vitro Translation for Ribosome Display

The *in vitro* translation for ribosome display is usually performed at 37° for a relatively short time of only about 6 to 8 min. This short time has been optimized to minimize decay of the ternary complexes.[17] Despite the general tendency of proteins to fold with higher efficiency at lower temperature *in vitro,* the yield of functional molecules from *in vitro* translation was experimentally found to be higher at 37° at least for the scFv fragments examined, which are not unusually stable molecules. This higher yield may be due to the action of molecular chaperones in the extract, and the optimal temperature is probably a complicated function of the temperature dependence of translation, folding, RNases, and perhaps proteases. The translation mixture contains the whole translational machinery present in the S-30 extract and also all low molecular weight compounds necessary for *in vitro* translation (ATP, GTP, amino acids, tRNA, the energy regeneration system, etc.). The efficiency of forming correctly folded disulfide-containing proteins such as antibody scFv fragments in the *E. coli* ribosome display system was found to be increased if eukaryotic protein disulfide isomerase (PDI), which catalyzes disulfide bond formation and rearrangement,[36,37] was used during the reaction.[17] Furthermore, the efficiency of the *E. coli* system was found to be higher if the 10Sa-RNA (the product of the *ssrA* gene), which is responsible for tagging and releasing truncated peptides from *E. coli* ribosomes,[38–40] was eliminated. This was achieved by including an antisense oligonucleotide in the reaction mixture.[17]

To achieve the maximal efficiency of the *E. coli* ribosome display system, the *in vitro* translation must be optimized for each batch of extract regarding the concentration of magnesium and potassium ions, the concentration of the extract itself, and the translation time under ribosome display conditions (see Optimization of the *Escherichia coli* Ribosome Display System, below).

Here, we provide an example of *in vitro* translation, using optimized conditions for one particular batch of S-30 extract and using mRNA encoding an scFv library, prepared according to the protocol of the section Construction of a DNA Library and Its Transcription to mRNA, above).

Material

- S-30 *E. coli* extract
- Premix Z: 250 m*M* Tris–acetate (pH 7.5), a 1.75 m*M* concentration of each amino acid, except methionine, 10 m*M* ATP, 2.5 m*M* GTP, 5 m*M* cAMP, 150 m*M* acetyl phosphate, *E. coli* tRNA (2.5 mg/ml), folinic acid (0.1 mg/ml), 7.5% (w/v) polyethylene glycol (PEG) 8000 (Sigma, St. Louis, MO)
- Methionine, 200 m*M*
- Magnesium acetate, 100 m*M*
- Potassium glutamate, 2 *M*
- Anti-*ssrA* oligonucleotide, 200 μM (5′-TTA AGC TGC TAA AGC GTA GTT TTC GTC GTT TGC GAC TA -3′)
- Protein disulfide isomerase (PDI), 44 μM (Sigma)
- Library mRNA, 1 μg/μl
- Washing buffer WBTH: 50 m*M* Tris–acetate (pH 7.5), 150 m*M* NaCl, 50 m*M* magnesium acetate, 0.1% (v/v) Tween 20, heparin (2.5 mg/ml)

Protocol for Ribosome Display in Vitro Translation. A tube containing 440 μl of washing buffer WBTH is prepared in an ice–water bath. On ice,

[36] L. A. Ryabova, D. Desplancq, A. S. Spirin, and A. Plückthun, *Nature Biotechnol.* **15,** 79 (1997).

[37] R. B. Freedman, T. R. Hirst, and M. F. Tuite, *Trends Biochem. Sci.* **19,** 331 (1994).

[38] K. C. Keiler, P. R. Waller, and R. T. Sauer, *Science* **271,** 990 (1996).

[39] Y. Komine, M. Kitabatake, T. Yokogawa, K. Nishikawa, and H. Inokuchi, *Proc. Natl. Acad. Sci. U.S.A.* **91,** 9223 (1994).

[40] B. K. Ray and D. Apirion, *Mol. Gen. Genet.* **174,** 25 (1979).

the following ice-cold solutions are combined: 22 μl of premix Z, 1.1 μl of 200 m*M* methionine, 11 μl of 2 *M* potassium glutamate, 7.6 μl of 100 m*M* magnesium acetate, 2 μl of 200 μ*M* anti-*ssrA* oligonucleotide, 1.5 μl of 44 μ*M* PDI, 40 μl of S-30 *E. coli* extract, and sterile double-distilled water is added up to 100 μl. The aliquot of the library mRNA to be used should be thawed just before use, and the unused part should be immediately snap-frozen. Ice-cold library mRNA (10 μg or approximately 2×10^{13} molecules) in 10 μl of RNase-free water is added to the mixture, gently vortexed, and immediately placed in a 37° water bath. After 7 min of *in vitro* translation, the reaction mixture is immediately transferred to the chilled tube with 440 μl of WBTH in an ice–water bath, briefly and gently vortexed, and placed back in the ice–water bath.

Affinity Selection of Ribosomal Complexes

The ribosomal complexes are stabilized by both chilling the translation mixture and by increasing the Mg^{2+} concentration from about 14 m*M* during the reaction to 50 m*M*,[17] by diluting the mixture with ice-cold Mg^{2+}-containing washing buffer. The diluted translation mixture is applied for affinity selection in the presence of 2% (w/v) sterilized milk (from milk powder) and heparin (2 mg/ml),[18] which together significantly reduce nonspecific binding of ribosomal complexes to the surface of microtiter wells, streptavidin magnetic beads, or avidin immobilized on agarose beads. Milk is sterilized to eliminate its intrinsic RNase activity. Another way to reduce nonspecific binding is to apply the diluted translation mixture for a so-called "prebinding" step prior to affinity selection. For this purpose, the translation mixture is first incubated in milk-coated microtiter wells or immunotubes, where "sticky" ribosomal complexes can bind to the milk components on the surface. The supernatant of this preselected solution is then used for affinity selection, which can be carried out either with ligand immobilized on a polystyrene surface or in solution by using biotinylated ligand and subsequent capture by streptavidin magnetic beads or avidin immobilized on agarose beads. After affinity selection, either Mg^{2+} ions are complexed with an excess of EDTA, resulting in the dissociation of ribosomal complexes and elution of mRNA, or whole ribosomal complexes can be eluted with an excess of free ligand. Although EDTA elution is generally preferred as tight binders can be selected, mRNA from nonspecifically bound ribosomal complexes will also be eluted by the EDTA treatment.

For affinity selection with a ligand immobilized on a polystyrene surface, microtiter strips or plates are usually used. However, for large-scale experiments, panning tubes can also be used. If the ligand is a small molecule, e.g., a hapten or peptide, it must be coupled to the carrier protein, e.g., bovine serum albumin (BSA) or transferrin, or covalently coupled to activated surfaces (e.g., CovaLink NH_2; Nunc, Roskilde, Denmark).

Affinity selection in solution requires the ligand to be biotinylated, and to be captured by streptavidin magnetic particles or avidin immobilized on agarose beads. The ligand should contain a spacer arm (according to our experience at least 30 Å long) between the biotin moiety and the ligand itself, as this must be long enough to present the ligand outside of the deep streptavidin binding pocket.[41] First, the biotinylated ligand is added to the ribosomal complexes and incubated on ice with shaking for at least 1 hr, to allow the ribosomal complexes to bind to the ligand. Then, unbound biotinylated ligand and also the bound ribosomal complexes are captured by streptavidin magnetic particles, which must therefore be present in sufficient excess.

One of the main keys to success in ribosome display system is to carry out the entire ribosome display procedure, after *in vitro* translation up to the RNA isolation, on ice and to keep all material necessary for this part of the experiment (e.g., pipette tips, tubes, microtiter plates) ice-cold.

Material

Washing buffer WBT: 50 m*M* Tris–acetate (pH 7.5), 150 m*M* NaCl, 50 m*M* magnesium acetate, 0.1% (v/v) Tween 20
Washing buffer WB (10×): 0.5 *M* Tris–acetate (pH 7.5), 1.5 *M* NaCl, 0.5 m*M* magnesium acetate
PBS: 137 m*M* NaCl, 2.7 m*M* KCl, 10 m*M* Na_2HPO_4, 1.8 m*M* KH_2PO_4
Microtiter plate strips or plates (Nunc)
Panning tubes (Nunc)
Ligand
Streptavidin magnetic particles (Roche Diagnostics)
Avidin immobilized on agarose beads (Sigma)
Biotinylated ligand
Sterilized 12% (w/v) milk powder (in water)
Elution buffer EB20: 50 m*M* Tris–acetate (pH 7.5), 150 m*M* NaCl, 20 m*M* EDTA, *Saccharomyces cerevisiae* RNA (50 μg/ml; Sigma)

Protocol for Affinity Selection of Ribosomal Complexes with Ligand Immobilized on a Polystyrene Surface. Sterilized 12% (w/v) milk powder is combined with 10× washing buffer WB to give 10% (w/v) milk powder in 1× WB, and this can be stored on ice for several weeks. Microtiter strips are coated overnight at 4° with 100 μl of PBS solution, containing at least 4 mg of ligand per milliliter. The next day, the coated strips are washed

[41] W. A. Hendrickson, A. Pahler, J. L. Smith, Y. Satow, E. A. Merritt, and R. P. Phizackerley, *Proc. Natl. Acad. Sci. U.S.A.* **86,** 2190 (1989).

with PBS and, together with the same number of noncoated strips, blocked with 4% (w/v) milk powder in PBS for 1 hr at room temperature with shaking. After washing the strips three times with PBS and two times with washing buffer, the strips are filled with 250 μl of ice-cold washing buffer, placed on ice in a cold room, and incubated for at least 20 min prior to affinity selection. This procedure is usually performed before the ribosome display *in vitro* translation is started. It is important that the strips be ice-cold, and the pipette tips temperature equilibrated in the cold room before use, otherwise it will result in a failure of ribosome display.

The washing buffer from the ice-cold strips is removed just before use, and the strips are placed on ice in an appropriate shaker in the cold room. The diluted ice-cold translation mixture (see preceding section) is centrifuged for 5 min at 4° and 16,000*g* to remove insoluble components. The supernatant is mixed with one-fifth volume of ice-cold 10% (w/v) milk powder in 1× WB, pipetted into milk-blocked microtiter strips (which do not contain ligand) for prebinding (200 μl per well), and the strips are gently shaken for 20 min on ice in the cold room. The content of the wells is then transferred to the ligand-coated and milk-blocked strips, and the strips are gently shaken for 1 hr in the cold room on ice. After five washes with ice-cold washing buffer WBT, the retained ribosomal complexes are dissociated with 200 μl of ice-cold elution buffer EB20 for 5 min on ice by gentle shaking, resulting in mRNA elution. The eluted mRNA can be frozen (e.g., in liquid nitrogen), or is immediately used for isolation.

Protocol for Affinity Selection of Ribosomal Complexes in Solution. Because the protocol for affinity selection in solution is similar to the protocol described above for affinity selection on a surface, we focus only on the differences. For the affinity selection a sufficient amount (to allow the capturing of all biotinylated ligand) of streptavidin magnetic particles is washed four times with ice-cold washing buffer WBT, then resuspended in their original volume in ice-cold WBT and aliquoted (if necessary) in ice-cold tubes. Panning tubes (5-ml volume) are blocked with 4% (w/v) milk powder for 1 hr by end-over-end rotation at room temperature, and then washed three times with PBS and three times with washing buffer WBT, filled with WBT, and placed on ice. Biotin-free milk is prepared by end-over-end rotation of 1 ml of sterilized 12% (w/v) milk powder with 100 μl of streptavidin magnetic particles for 1 hr at room temperature and then, after removing the streptavidin magnetic particles, stored on ice. The diluted and centrifuged translation mixture, containing 2% biotin-free milk, is supplemented with biotinylated ligand to the required concentration, and the mixture is transferred to the prepared panning tubes. The tubes are placed inside of an appropriate larger tube or flask filled with ice and rotated end over end for 1 hr in the cold room. In this way, affinity selection affinity selection (with biotinylated ligand in solution) and prebinding (to the tube surface) are performed simultaneously.

After affinity selection, the solution from the panning tubes is added to streptavidin magnetic particles and rotated end over end on ice for 10 to 15 min in the cold room. Washing and elution of mRNA are performed as described in the previous section, except that a magnet is used for the capturing of streptavidin magnetic particles. The capacity of streptavidin magnetic particles for binding of biotinylated ligand is dependent on the ligand size. Because it is important to capture all ligand-bound ribosomal complexes, especially if libraries of high diversity are used for selection, we do not use more than 10 pmol of biotinylated ligand for 100 μl of streptavidin magnetic particles (the particles can bind 100 times more free biotin). To eliminate the selection of streptavidin binders from the library, we recommend switching between streptavidin magnetic particles and avidin agarose beads for the capturing of the biotinylated ligand after each or each second ribosome display cycle. For the products of the suppliers indicated above, 1 volume of avidin immobilized on agarose has the same binding capacity as 5 volumes of streptavidin magnetic particles.

mRNA Purification, Reverse Transcription, and Polymerase Chain Reaction

mRNA is isolated by using a commercial RNA isolation kit. In some cases, eluted mRNA samples may contain residual DNA. Therefore, we recommend the use of an RNA isolation kit that includes a DNase treatment of the sample. Purified RNA is subsequently used for reverse transcription and PCR, and the resulting DNA is used for another round of ribosome display, for radioimmunoassay (RIA) analysis, or for cloning and subsequent analysis of single clones.

PCR amplification is usually performed in two steps in a similar way as described in the section Construction of a DNA Library and Its Transcription to mRNA (above), by using the SDA and T5Te primers (Fig. 2A) in the first step, subsequent purification of the resulting PCR product by agarose gel electrophoresis, and amplification of the purified template with primers T7B and T5Te in the second PCR. In principle, it is possible to carry out only one PCR with primers T7B and T5Te; however, the quality of the PCR products and also of the mRNA prepared from them is quite often not sufficient to perform a ribosome display experiment for the subsequent round.

Material

Primers T5Te, SDA, and T7B (see above)
High Pure RNA isolation kit (Roche Diagnostics)

Superscript reverse transcriptase (GIBCO-BRL)
RNasin (Promega)
Taq DNA polymerase (GIBCO-BRL)
Solution of dNTPs (20 m*M* each)
DMSO (Fluka)
QIAquick gel extraction kit (Qiagen)
Tris-HCl, pH 8.5 (10 m*M*)

Protocol for mRNA Purification, Reverse Transcription, and Polymerase Chain Reaction. RNA is isolated according to the manufacturer's instructions, using the High Pure RNA isolation kit. The DNase treatment, which is carried out after loading the columns with the RNA sample, is performed for 5 to 10 min at room temperature. Purified RNA is eluted in 35 μl of RNase-free water, and immediately placed in a 70° water bath for 10 min. During the incubation the premix for the reverse transcription reaction is prepared. The following solutions are combined on ice for each reaction: 0.5 μl of T5Te primer (100 μM), 1.25 μl of each dNTP (20 m*M* each), 1.25 μl of RNasin (50 U), 10 μl of 5× first-strand synthesis buffer, 5 μl of 0.1 *M* DTT, and 1.25 μl of Superscript reverse transcriptase (250 units) (the last three components are from the Superscript reverse transcriptase kit). After the 70° incubation, the samples are chilled on ice for 1 to 2 min, centrifuged at 4°, and the whole RNA preparation of approximately 30 μl is combined with the prepared reverse transcription premix and incubated for 1 hr at 50°.

Because it is difficult to predict the number of cycles necessary for PCR amplification at this point, as this depends on the amount of isolated mRNA, a test PCR should first be performed in a 50-μl reaction, containing 5 μl of reverse transcript, 5 μl of 10× PCR buffer, 1.7 μl of 50 m*M* $MgCl_2$, 0.5 μl of dNTPs (20 m*M* each), 2.5 μl of DMSO, 0.5 μl of 100 μM of each of the primers SDA and T5Te, and 0.5 μl (2.5 U) of *Taq* DNA polymerase (4 min at 94°, followed by 20 to 30 cycles of 30 sec at 94°, 30 sec at 50°, 2.5 min at 72°, and finished by 10 min at 72°). Usually, the more ribosome display cycles have already been performed with a diverse library the more mRNA encoding a cognate product to the antigen is isolated and the fewer PCR cycles are necessary. It is better to perform the test PCR first for fewer cycles and to add further cycles, if necessary, after analysis of PCR products by agarose gel electrophoresis. For this purpose, usually no additional components need to be added even after overnight PCR. It is not recommended to overamplify DNA. If PCR products are too smeary, usually resulting from overamplification, the RNA quality will not be sufficient for successful ribosome display. The remaining reverse transcript (about 45 μl) or a portion of it is used for a large-scale PCR, under the optimal PCR conditions established from the small aliquot of reverse transcript. The whole reverse transcript should be used for amplification mainly in cycles of the ribosome display experiment to quantitatively amplify all recovered binders from the library.

PCR products are purified by agarose gel electrophoresis and used for the second amplification, using primers T7B and T5Te as described for the second PCR in the section Construction of a DNA Library and Its Transcription to mRNA, above.

Optimization of the Escherichia coli Ribosome Display System

To achieve maximal efficiency, the *E. coli* ribosome display system should be optimized. For this purpose, a model protein is required, which gives a clear enrichment on a surface coated with a cognate ligand, compared to a control surface in a ribosome display experiment (e.g., for an scFv library, the antibody scFvhag can be used[17]). The system should be optimized for the concentration of magnesium and potassium ions, the concentration of the extract itself and the translation time for each batch of *E. coli* extract, using ribosome display conditions. The optimization of the system is performed in such a way that the mRNA ribosome display construct, encoding a model protein, is used for several ribosome display *in vitro* translation reactions, which differ either in the concentration of Mg^{2+}, potassium glutamate, or extract. After various incubation times at 37° aliquots of the reactions are stopped and applied for affinity selection. The affinity selection is carried out either with the appropriate ligand, immobilized on the surface, or in solution, using biotinylated ligand. After affinity selection either the mRNA is isolated, reverse transcribed to cDNA, amplified by PCR, and the PCR products analyzed by agarose gel electrophoresis, or the isolated mRNA is analyzed by Northern blot hybridization with a probe specific for the mRNA construct. We prefer the latter approach, because it is possible to directly quantify the amount of isolated mRNA, and this technique is more sensitive in the detection of differences among the amounts of recovered mRNA from different samples. In the optimization of the system, first the magnesium concentration is optimized, and this optimal Mg^{2+} concentration is then used for the further optimization of potassium glutamate and extract concentrations. For the optimization of the Mg^{2+} concentration it should be kept in mind that the extract itself already contains 14 m*M* Mg^{2+}.

We provide here an example for the optimization of the *E. coli* ribosome display system using a ribosome display construct, in which a model protein is fused to spacer B.

Material

High Pure RNA isolation kit (Roche Diagnostics) (optional, see protocol)

Washing buffer WBTH
Sterilized milk, 4% (w/v) milk powder in WBT
Elution buffer EB5: 50 m*M* Tris–acetate (pH 7.5), 150 m*M* NaCl, 5 m*M* EDTA, *S. cerevisiae* RNA (50 μg/ml; Sigma)
RNA denaturation buffer (prepared fresh before use): 10 μl of formamide, 3.5 μl of formaldehyde, 2 μl of 10× morpholinepropanesulfonic acid (MOPS) buffer [0.2 *M* MOPS (pH 7.0), 80 m*M* sodium diacetate, 10 m*M* EDTA]
Gel loading buffer: 50% (w/v) glycerol, 1 m*M* EDTA, 0.25% (v/v) bromphenol blue
Turboblotter with Nytran Nylon membrane (Schleicher & Schuell, Dussel, Germany)
Oligonucleotide SFI-LINK: 5′-CCT TTA AGC AGC TCA TCA AAA TCA CCG GAA CCA GAA CCG CCG CCC TCG GCC CCC GAG GCC G-3′. This oligonucleotide anneals to the *Sfi*I site and a part of the spacer B of ribosome display construct mRNA
Digoxigenin (DIG) oligonucleotide tailing kit (Roche Diagnostics)
Chemiluminescent substrate disodium 3-(4-methoxyspiro[1,2-dioxetane-3,2′-tricyclo[3.3.1.1^{3,7}]decan]4-yl)phenyl phosphate (CSPD; Roche Diagnostics)

Protocol for the Optimization of the Escherichia coli Ribosome Display System by Northern Blot Hybridization. First, the mRNA of a ribosome display construct of a model protein, containing the spacer B, is prepared as described above for the library construction. To optimize the system, *in vitro* translations and affinity selections of the ribosome display system are carried out with a single model mRNA under similar conditions as described in the section *In Vitro* Translation for Ribosome Display (above), with the following modifications. From our experience, the optimal conditions for ribosome display *in vitro* translation can vary in the following ranges: 11 to 15 m*M* magnesium acetate, 180 to 220 m*M* potassium glutamate, 30 to 50 μl of S-30 extract for a 110-μl reaction, and 6 to 9 min for the time of translation. Longer translation times might be necessary if a longer library mRNA is used for ribosome display.

In vitro translation reactions, differing in the concentration of the compound to be tested (magnesium acetate, potassium glutamate, or amount of extract), each prepared in a 110-μl volume, are incubated at 37°. After selected time points (e.g., for an antibody scFv library, after 5, 7, 9, and 11 min) aliquots of 25 μl are taken and immediately added to tubes containing 100 μl of ice-cold washing buffer WBTH (prepared in advance in a water–ice bath), gently vortexed, and stored in the ice bath until the experiment is finished. Samples are then centrifuged, 110 μl of supernatant is mixed with the same volume of 4% (w/v) sterilized milk in WBT, and two 100-

μl volumes of each sample are applied for affinity selection. After washing, mRNA is eluted with 200 μl of elution buffer EB5, and RNA is immediately precipitated by adding 600 μl of ice-cold ethanol and incubating the samples for 30 min on ice. After 30 min of centrifugation at 4° at 16,000*g*, the supernatants are removed and RNA pellets are dried at room temperature for 10 to 15 min, dissolved in 10 μl of RNA denaturation buffer on ice, and incubated for 10 min at 70°. For the estimation of the amount of recovered mRNA several other control samples containing between 0.2 and 10 ng of the original model mRNA can be prepared in a similar way. RNA samples of up to 5 μl can be directly mixed with 10 μl of RNA denaturation buffer. After the 70° incubation, samples are chilled on ice, mixed with 1 μl of gel loading buffer, and separated by 1.5% (w/v) agarose gel electrophoresis in the presence of TBE and 20 m*M* guanidinium isothiocyanate.[33] The use of elution buffer containing higher EDTA concentrations and/or RNA precipitation at lower temperatures may lead to the coprecipitation of EDTA, resulting in a failure of the Northern analysis. Alternatively, RNA may be eluted with buffer EB20 and subsequently purified by the High Pure RNA isolation kit. This procedure, however, requires an additional concentration step of the RNA samples by precipitation.

RNA samples separated by agarose gel electrophoresis are blotted to Nytran Nylon membrane, using a Turboblotter according to the manufacturer recommendations. Hybridization is carried out at least for 4 hr at 60° as described[17] with the oligonucleotide SFI-LINK, labeled by 3′ tailing with digoxigenin–11-dUTP/dATP using the DIG oligonucleotide tailing kit. The hybridized oligonucleotide probe is detected with the DIG DNA labeling and detection kit with the chemiluminescent substrate CSPD and exposure to X-ray film.

Monitoring the Enrichment of a Ribosome Display Library for Specific Binders and Analysis of Single Clones by Radioimmunoassay

After each round of ribosome display, isolated mRNA from affinity-selected ribosomal complexes is reverse transcribed to cDNA and amplified by PCR. The PCR products are used for mRNA preparation, which is used for the next round of ribosome display. This mRNA can also be used for RIA analysis of the ribosome display library to check for the presence of specific binders. Although this mRNA contains no stop codon, translation for more than 30 min in the absence of anti-*ssrA* oligonucleotide and the use of buffers without magnesium for affinity selection lead to protein release from ribosomal complexes, and a RIA can be performed. After *in vitro* translation, which is carried out in the presence of [^{35}S]methionine, samples are tested for ligand binding, with ligand immobilized on a polystyrene surface. The RIA is performed both in the presence and absence of free ligand to verify specific binding, as true specific binding always must be inhibitable by free ligand. After affinity selection, bound radioactive protein is eluted with sodium dodecyl sulfate (SDS) and quantified by scintillation counting. The whole procedure after *in vitro* translation can be performed at room temperature. If the analyzed proteins are not stable under such conditions, however, lower temperatures (e.g., 4°) can also be used. The elution must be performed at room temperature, because SDS precipitates at 4°.

If the translation product of the mRNA pool binds to the ligand and this binding can be clearly inhibited with free ligand, the library is already enriched for specific binders. The library-encoded PCR product, from which the mRNA was transcribed for RIA analysis, is cloned, and plasmids are isolated from single clones and used for *in vitro* transcription. After purification, these plasmid mRNAs are used for *in vitro* translation in the presence of [^{35}S]methionine and the translation mixtures are applied for RIA analysis as described above. If the enriched DNA pool is cloned into a plasmid that adds a peptide detection tag to the protein, binders can also be analyzed by enzyme-linked immunosorbent assay (ELISA).

Material

S-30 *E. coli* extract
Premix Z
[^{35}S]Methionine (10 mCi/ml, 1175 Ci/mmol; New England Nuclear, Boston, MA)
Magnesium acetate, 100 m*M*
Potassium glutamate, 2 *M*
PBST: PBS containing 0.5% (v/v) Tween 20
SDS elution buffer: 4% SDS in PBS

Protocol for Radioimmunoassay Analysis of the DNA Pool Enriched for Specific Binders or for Radioimmunoassay of Single Clones. In vitro translation is performed as described in the section *In Vitro* Translation for Ribosome Display (above) with the following modifications. In a 110-μl *in vitro* translation reaction 1.1 μl of cold methionine is replaced with 2 μl of [^{35}S]methionine, anti-*ssrA* oligonucleotide is omitted, and the translation is performed with either 10 μg of enriched library mRNA or 10 μg of mRNA, obtained by transcription from a plasmid, for 30 to 40 min at 37°. After translation, samples are diluted fourfold or more, if required, in PBST, and centrifuged for 5 min to remove any insoluble components. The ligand is diluted in 4% (w/v) milk in PBST to a twofold higher concentration than used in the inhibition reaction. This ligand solution is then mixed with an equal volume of the supernatant from the centrifuged translation mixture applied for binding to immobilized ligand on the surface of microtiter wells. After washing five times with PBST, bound radioactive protein is eluted with SDS elution buffer for 15 min with shaking, and the eluted protein is quantified by liquid scintillation counting.

The Rabbit Reticulocyte Ribosome Display System

As has been shown,[19] the use of a rabbit reticulocyte lysate as a ribosome source does not offer any clear advantage over the *E. coli*-based display system. In a direct comparison with antibody scFv fragments, which are of course derived from eukaryotic proteins, the rabbit reticulocyte lysate system gave rise to lower amounts of functional complexes, at least for those tested, lower enrichment factors, and higher costs. However, because the lysates are commercially available and contain a lower intrinsic RNase activity, the use of this system might be perceived as being more convenient, and it is also possible that different proteins might be expressed with different efficiencies in the two translation systems.

Construction of a DNA Library and Its Transcription to mRNA

The ribosome display construct for the eukaryotic translation system differs from the construct for the *E. coli*-based system in two respects. First, because of the lower nuclease activity in the reticulocyte translation system, the 5′ and 3′ stem–loops may be omitted, although their presence has no negative influence on the efficiency of the rabbit reticulocyte ribosome display system. Second, to allow efficient translation of any sequence, the gene 10 region of phage T7 is replaced with the translational enhancer sequence of the *Xenopus laevis* β-globin gene, together with an optimal Kozak sequence,[42] the eukaryotic ribosome-binding sequence (Fig. 4). In our construct, the 5′ stem–loop is omitted, while the entire 3′ part of the construct is kept the same as in the *E. coli* ribosome display construct. Therefore, the primer SDA is replaced by the primer SDA-RRL and the primer T7B by T7RR-EN (Fig. 4A).

The preparation of the construct is carried out as described for the *E. coli*-based system. For transcription, we tested the use of a cap analog, which has been reported to enhance the efficiency of *in vitro* translation in some cases.[43,44] However, for maximal enhancement probably both a poly(A) tail and a 5′ cap may be required as the interaction between these

[42] D. Falcone and D. W. Andrews, *Mol. Cell. Biol.* **11,** 2656 (1991).
[43] D. R. Gallie, *Gene* **216,** 1 (1998).
[44] H. J. Song, D. R. Gallie, and R. F. Duncan, *Eur. J. Biochem.* **232,** 778 (1995).

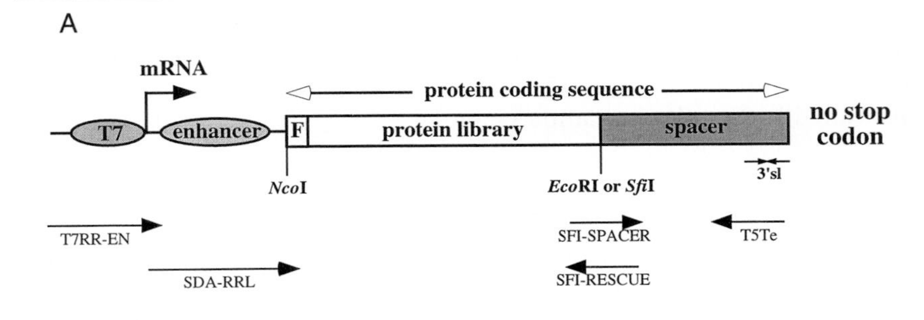

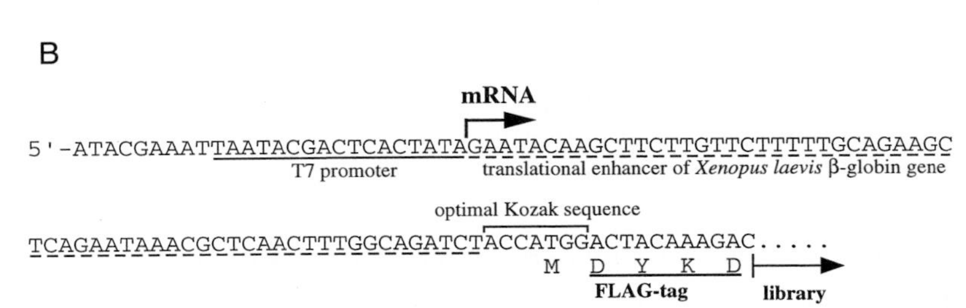

FIG. 4. (A) Schematic drawing of the DNA construct used for rabbit reticulocyte ribosome display system. *T7* denotes the T7 promoter, *enhancer* the translational enhancer of *Xenopus laevis* β-globin gene, *spacer* the part of protein construct connecting the folded protein to the ribosome, *F* the FLAG tag sequence, and *3'sl* the stem–loop at the 3' end of the mRNA. The filled arrow indicates the transcriptional start and open arrows the protein coding sequence. The primers used for the PCR amplification are shown below the scheme. If the construct is prepared by ligation of the library coding sequence to the spacer, in the first PCR this ligation mixture is amplified with primers SDA-RRL and T5Te, followed by a second amplification with primers T7RR-EN and T5Te. For the preparation of the construct by assembly PCR, the library sequence and the spacer are amplified separately with primers SDA-RRL and SFI-RESCUE for library or SFI-SPACER and T5Te for spacer. Purified PCR products are then assembled by PCR without primers for several cycles after amplification with primers SDA-RRL and T5Te in the first step. The second amplification is carried out with primers T7RR-EN and T5Te. (B) DNA upstream sequence used for the rabbit reticulocyte ribosome display constructs.

termini has been shown to allow for efficient translation initiation.[45] In ribosome display, we could observe only a marginal effect of about a twofold improvement. However, the yield of RNA synthesis drops dramatically, if the cap analogs are used.

[45] M. Piron, P. Vende, J. Cohen, and D. Poncet, *EMBO J.* **17,** 5811 (1998).

The procedures for the construction of a DNA library and its transcription to mRNA remain unchanged relative to the *E. coli* system.

Materials

PCR primers:
SDA-RRL: 5-TTT GCA GAA GCT CAG AAT AAA CGC TCA ACT TTG GCA GAT CTA CCA TGG ACT ACA AAG A-3
T7RR-EN: 5-ATA CGA AAT TAA TAC GAC TCA CTA TAG AAT ACA AGC TTC TTG TTC TTT TTG CAG AAG CTC-3
Cap analog ^{7m}GpppG (Stratagene La Jolla, CA)

Ribosome Display In Vitro Translation and Affinity Selection

The best results in using the rabbit reticulocyte system in ribosome display were obtained with a commercial kit, in which DTT can be omitted and concentrations of Mg^{2+} and K^+ can be optimized.[19] We found that the efficiency of the rabbit reticulocyte ribosome display system is related to the amount of protein produced in the *in vitro* translation reaction. Therefore, potassium and magnesium concentrations are optimized by carrying out *in vitro* translation, which is performed for 20 min at 30° in the presence of radioactive methionine, followed by SDS–polyacrylamide gel electrophoresis (SDS–PAGE) of the translation products and autoradiography. We did not find it necessary to adjust the optimal conditions for each new batch from the commercial supplier. The efficiency for disulfide containing proteins, such as antibody single-chain Fv fragments, can be increased if eukaryotic protein disulfide isomerase (PDI) is used cotranslationally. The translation is stopped in a similar way as for the *E. coli* system. However, we observed that 50 m*M* Mg^{2+} present during affinity selection may occasionally lead to artifacts, when affinity selection is carried out on polystyrene surfaces.[19] Therefore, the magnesium concentration is reduced to 5 m*M* in the eukaryotic system, without significantly altering the efficiency of this system. Because the preferred anion in the rabbit reticulocyte translation system is chloride, the Tris–acetate-based buffer of the *E. coli* system is replaced with a PBS-based buffer.

Materials

Flexi rabbit reticulocyte lysate kit (Promega)
[^{35}S]Methionine
SDS–PAGE equipment
Washing buffer PBSM (10×): 1.37 *M* NaCl, 0.027 *M* KCl, 0.1 *M* Na_2HPO_4, 0.018 *M* KH_2PO_4, 0.05 *M* $MgCl_2$
Washing buffer PBSTM: PBS with 0.05% (v/v) Tween 20, 5 m*M* $MgCl_2$

Washing buffer PBSTMH: PBSTM with heparin at 2.5 mg/ml
Elution buffer PEB20: PBS with 20 m*M* EDTA

Protocol for Ribosome Display in Vitro Translation and Affinity Selection. Sterilized 12% (v/v) milk is combined with 10× PBSM to obtain 10% (v/v) milk in 1× PBSM, and this solution can be stored on ice for several weeks. In an ice–water bath, a tube containing 200 μl of PBSTMH buffer and 62.5 μl of 10% (v/v) sterilized milk in PBSM is prepared. On ice, the following ice-cold solutions are combined: 0.8 μl of 2.5 *M* KCl, 0.2 μl of 200 m*M* methionine, 1 μl of 1 m*M* each amino acid except methionine, 33 μl of Flexi rabbit reticulocyte lysate, and sterile, double-distrilled water is added up to 40 μl. The amino acid mix (1 m*M* each) without methionine is included in the translation kit, as is KCl. The library mRNA should be thawed only directly before use, and the remainder should be immediately frozen. Five micrograms (approximately 1×10^{13} molecules) of ice-cold library mRNA, either capped or uncapped, in 10 μl is added to the mixture, gently vortexed, and immediately placed in a 30° water bath. After 20 min of *in vitro* translation, the mixture is pipetted out of the reaction tube, immediately added to the tube containing buffer PBSTMH and milk, briefly and gently vortexed, and placed in the ice–water bath. We did not find it necessary to centrifuge the translation mixture prior to selection. Selection is carried out as described for the *E. coli* system, except that buffers WBT and EB20 are replaced by buffers PBSTM and PEB20, respectively. mRNA purification, reverse transcription, and PCR are performed as described for the *E. coli* system, except that primers SDA and T7B are replaced by primers SDA-RRL and T7RR-EN.

[25] Yeast Surface Display for Directed Evolution of Protein Expression, Affinity, and Stability

By ERIC T. BODER and K. DANE WITTRUP

Background

Many platforms are available for the construction of peptide and polypeptide libraries, allowing directed evolution or functional genomics studies.[1] Currently, the two most widely used polypeptide library methods are phage display and the yeast two-hybrid method. However, neither of these methods is effective for complex extracellular eukaryotic proteins, because

[1] E. V. Shusta, J. J. Van Antwerp, and K. D. Wittrup, *Curr. Opin. Biotechnol.* **10,** 117 (1999).

0076-6879/00 $30.00

of the absence of such posttranslational modifications as glycosylation and efficient disulfide isomerization. We have developed a yeast surface display method that addresses this deficiency by utilizing the yeast secretory apparatus to process cell wall protein fusions.[2] Yeast surface display is well suited to engineer extracellular eukaryotic proteins such as antibody fragments, cytokines, and receptor ectodomains.

A further advantageous characteristic of yeast surface display is that soluble ligand-binding kinetics and equilibria may be measured in the display format, and as a result quantitatively optimized screening protocols may be designed.[3] Using such optimal screening conditions, numerous mutants with small improvements may be finely discriminated with high statistical certainty, and further recombination may be used to achieve greater improvements.[4]

To date, we have applied yeast display in the following studies: affinity maturation of the 4-4-20 anti-fluorescein single-chain antibody (scFv) to femtomolar affinity[5]; affinity maturation of the KJ16 anti-T cell receptor scFv[6]; affinity maturation of the D1.3 anti-lysozyme scFv[7]; display of a single-chain T cell receptor (scTCR)[8]; stabilization and increased secretion of an scTCR[9]; affinity maturation of an scTCR against a superantigen[10]; and activation of T cells by contact with yeast-displayed KJ16 scFv.[11]

Yeast Display of a Protein of Interest

A given protein may be displayed on the surface of yeast by expression as a protein fusion to the Aga2p mating agglutinin protein. We have constructed the pCT302 plasmid (Fig. 1) for expression of such fusions under control of the GAL1,10 galactose-inducible promoter. Because the Aga2p protein is tethered in the cell wall via disulfide bridges to the Aga1p protein, we have constructed yeast strain EBY100 (**a** *GAL1-AGA1*::*URA3 ura3-52 trp1 leu2Δ1 his3Δ200 pep4*::*HIS2 prb1Δ1.6R can1 GAL*), in which Aga1p

[2] E. T. Boder and K. D. Wittrup, *Nature Biotechnol.* **15,** 553 (1997).

[3] E. T. Boder and K. D. Wittrup, *Biotechnol. Prog.* **14,** 55 (1998).

[4] W. P. Stemmer, *Nature* (*London*) **370,** 389 (1994).

[5] E. T. Boder and K. D. Wittrup, *Proc. Natl. Acad. Sci. U.S.A.,* in press (2000).

[6] M. C. Kieke, B. K. Cho, E. T. Boder, D. M. Kranz, and K. D. Wittrup, *Protein Eng.* **10,** 1303 (1997).

[7] J. J. Van Antwerp and K. D. Wittrup, unpublished data (1999).

[8] M. C. Kieke, E. V. Shusta, E. T. Boder, L. Teyton, K. D. Wittrup, and D. M. Kranz, *Proc. Natl. Acad. Sci. U.S.A.* **96,** 5651 (1999).

[9] E. V. Shusta, M. C. Kieke, E. Parke, D. M. Kranz, and K. D. Wittrup, *Nature Biotechnol.* **18,** 754 (2000).

[10] M. C. Kieke, K. D. Wittrup, and D. M. Kranz, unpublished data (1999).

[11] B. K. Cho, M. C. Kieke, E. T. Boder, K. D. Wittrup, and D. M. Kranz, *J. Immunol. Methods* **220,** 179 (1998).

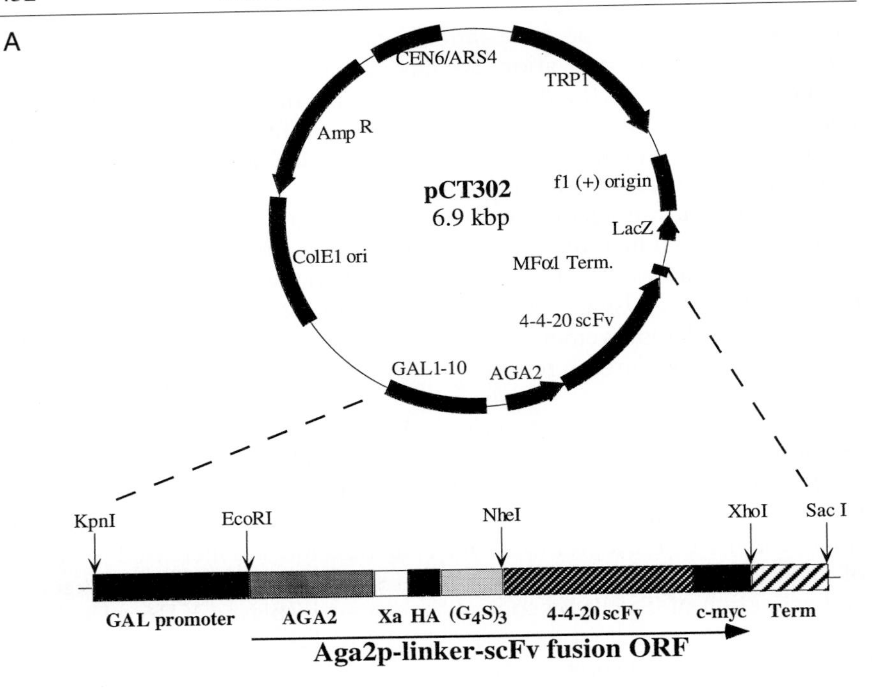

N-terminal flanking sequence:

```
ATAAACACACAGTATGTTTTTAAGGACAATAGCTCGACGATTGAAGGTAGATACCCATAC
I  N  T  Q  Y  V  F  K  D  N  S  S  T  I  E  G  R  Y  P  Y   -
      Aga2p  <-|-> Linker  <- |Factor Xa^|-> HA

               PstI/
GACGTTCCAGACTACGCTCTGCAGGCTAGTGGTGGTGGTGGTTCTGGTGGTGGTGGTTCT
D  V  P  D  Y  A  L  Q  A  S  G  G  G  G  S  G  G  G  G  S   -
epitope tag <-|->          Linker

          NheI/   AatII/
GGTGGTGGTGGTTCTGCTAGCGACGTCGTTATGACTCAAACACCACTATCACTTCCTGTT
G  G  G  G  S  A  S  D  V  V  M  T  Q  T  P  L  S  L  P  V   -
                <-|-> VL of 4-4-20 scFv
```

C-terminal flanking sequence:

```
                                          XhoI/   /BglII
TCCTCAGAACAAAAGCTTATTTCTGAAGAAGACTTGTAATAGCTCGAGATC
S  S  E  Q  K  L  I  S  E  E  D  L  *                  -
<-|  |->  c-myc epitope tag <-|
```

FIG. 1

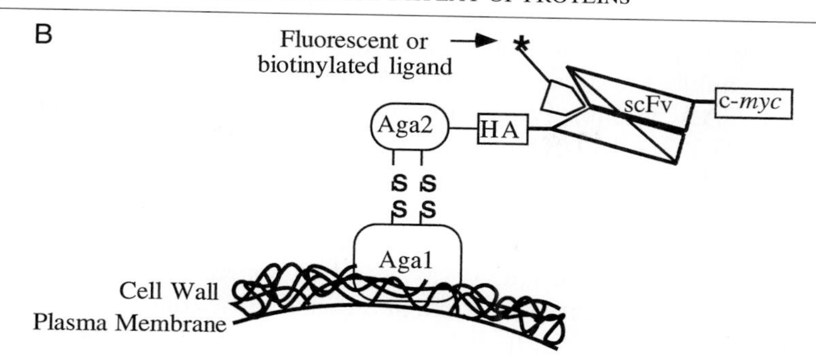

FIG. 1. (A) Map of plasmid pCT302 for expression of Aga2p fusion proteins. Flanking sequences are shown. We have generally inserted new ORFs as *Nhe*I–*Xho*I fragments. (B) Schematic of fusion topology on the cell surface. Aga1p is covalently attached to the cell wall via a phosphatidylinositol glycan tail, and Aga2p is disulfide bonded to Aga1p. The N-terminal HA and C-terminal c-Myc epitope tags enable detection of a fusion protein independent of its ligand-binding characteristics. For purposes of illustration, an scFv antibody fragment is shown as the protein of interest. Objects are not drawn to actual proportionate scale. Steric hindrance by cell wall components is not generally an issue, because an Aga2p–scFv fusion is accessible for labeling with 2×10^6 kDa fluorescein dextran.[2]

expression is inducible from a single, integrated open reading frame (ORF) downstream of the *GAL1,10* promoter.

The presence of the fusion protein on the cell surface may be detected independently of its ligand-binding activity by immunofluorescent labeling of epitope tags. Quantitative detection of fusion polypeptide surface levels via epitope tag labeling is important to normalize for random variation in surface expression levels when measuring ligand-binding kinetics and equilibria. Two epitope tags are used in the fusion, one N terminal (hemagglutinin, HA) and one C terminal (c-Myc) to the protein of interest. If the 3′ *Xho*I site is used to subclone the ORF into pCT302, the c-Myc coding sequence must be reconstructed. Both tags are of value; immunofluorescent detection of the c-Myc epitope on the cell surface confirms the presence of the full-length polypeptide, and the quality control apparatus of the endoplasmic reticulum (ER) provides high confidence that the protein on the surface is correctly folded.[12] The HA tag serves as a useful internal control should the fusion be displayed in a partially proteolyzed form.[9] The availability of two epitope tags also allows one to alternate their use, thereby circumventing artifactual isolation of epitope tag mutations during library screening.[6]

Once the ORF of interest is subcloned in-frame with Aga2p in pCT302,

[12] C. Hammond and A. Helenius, *Curr. Opin. Cell Biol.* **7,** 523 (1995).

the expression plasmid may be transformed into the yeast strain EBY100 by any of a number of standard yeast transformation methods, with selection on SD + CAA medium [yeast nitrogen base (6.7 g/liter), Na_2HPO_4 (5.4 g/liter), $NaH_2PO_4 \cdot H_2O$ (8.56 g/liter), dextrose (20 g/liter), Bacto Casamino acids (CAA, 5 g/liter; Difco, Detroit, MI)]. This allows selection of EBY100 and pCT302 under Trp– and Ura– conditions, and yeast generally grow faster and to higher density in this semidefined medium. The transformants should be maintained on glucose as a carbon source in order to repress Aga1p and Aga2p fusion expression, and thus repress consequent counter-selection against potentially toxic heterologous expression effects. The ORF of interest should be examined for Lys–Arg sequences; if the expressed protein contains this sequence, an alternative *Saccharomyces cerevisiae* strain containing a deletion of the chromosomal Kex2 protease ORF may be required to prevent endoproteolytic processing within the *trans*-Golgi network.[13,14]

The Aga2p protein fusion is expressed on the cell surface by induction in galactose-containing medium. From plates (SD + CAA containing 1 *M* sorbitol), 3- to 5-ml liquid test tube cultures (SD + CAA) are inoculated and grown at 30° for 12–48 hr. Length of the lag phase is highly variable and depends on the age of the colony and the size of the inoculum. The time of growth on glucose is not directly significant; however, best surface expression usually results when exponentially growing cells are switched to galactose. For EBY100 transformants grown under these conditions, this corresponds to an OD_{600} of 2.0–5.0; generally, overnight growth from a fresh colony will yield a culture in this density range in approximately 12–20 hr. Cultures grown to stationary phase ($OD_{600} > 10$) may be induced for surface expression; however, a smaller fraction of cells display the protein of interest and lower expression levels per cell are typical. Stationary-phase liquid cultures may be reprepared for optimal surface display by pelleting cells, resuspending in fresh SD medium, and grown for ~6–10 hr at 30°. To induce fusion expression, the culture is centrifuged for 10–20 sec at maximum speed in a microcentrifuge, and resuspended in SG + CAA medium, which is identical to SD medium except that glucose is replaced by galactose as the carbon source. Induction of surface expression must be performed at 20° for best results for single-chain antibodies; at >25° surface expression levels vary substantially with induction time, and lower surface levels are generally found. However, other mammalian proteins are better displayed by induction at 30 or 37°. Cells should be induced in SG at starting densities of 0.2–2.0 OD_{600}. Cultures may be induced

[13] N. C. Rockwell and R. S. Fuller, *Biochemistry* **37,** 3386 (1998).
[14] A. Bevan, C. Brenner, and R. S. Fuller, *Proc. Natl. Acad. Sci. U.S.A.* **95,** 10384 (1998).

from 16 to 48 hr, most typically 20–24 hr. The culture usually undergoes approximately one doubling during induction at 20°, resulting in a final culture density of approximately 0.3–3.0 OD_{600}. The additional stability afforded by expression at 20° may be a potential advantage for unstable proteins such as single-chain antibodies, by comparison with bacterial or mammalian expression platforms.

Yeast cultures may be fluorescently labeled for examination by flow cytometry, by the following procedures.

1. Centrifuge a volume of induced culture equivalent to 0.2 OD_{600} ml (e.g., 0.2 ml of a culture with OD_{600} of 1.0; OD_{600} of 1.0 corresponds to $\sim 10^7$ cells/ml for EBY100) in a microcentrifuge at maximum speed for ~10–20 sec. Remove and discard the supernatant.

2. Resuspend the cell pellet by vortexing in ice-cold buffered saline solution (BSS) containing bovine serum albumin (BSA, 1 mg/ml; Sigma). BSA is present to serve as a carrier protein for ligands added in subsequent labeling steps, to minimize adsorptive losses on plasticware. Either PBS [NaCl (8 g/liter), KCl (0.2 g/liter), Na_2HPO_4 (1.44 g/liter), KH_2PO_4 (0.24 g/liter), pH 7.4] or TBS (50 m*M* Tris-HCl, 100 m*M* NaCl, pH 8.0) can be used. Pellet the culture again by microcentrifuge.

3. Labeling of the c-Myc and HA epitope tags may be performed after or during certain steps of the more specific ligand-binding protocols described in the next section. For instance, we have commonly combined ligand and epitope tag primary antibody labeling in a single step. Likewise, we have combined secondary epitope tag labeling with ligand labeling in one step. The 12CA5 monoclonal antibody (MAb) labeling step has also been performed during 25° kinetic library screening steps (see below).

4. Resuspend the cell pellet in 100 μl of BSS containing either primary antibody against the epitope tags, or biotinylated or fluorescently labeled ligand for the protein of interest. Specific conditions of interest follow:

   HA epitope tag: 12CA5 monoclonal antibody (Boehringer Mannheim, Indianapolis, IN) at 10 μg/ml final concentration. Incubate 30 min on ice

   c-Myc epitope tag: 1 : 100 dilution of the 9E10 monoclonal antibody ascites fluid (Berkeley Antibody Company, Richmond, CA) for 30 min on ice

   Biotinylated ligand detection: 1 : 100 streptavidin–phycoerythrin (PharMingen, San Diego, CA) 30 min on ice

5. After incubation with the primary antibody, pellet the cells, remove and discard the supernatant, and wash once with 0.5 ml of ice-cold BSS.

6. Label the cells with fluorescently labeled ligand or fluorescently labeled secondary antibody [e.g., fluorescein isothiocynate (FITC)- or phy-

coerythrin-conjugated goat anti-mouse IgG (Sigma, St. Louis, MO) at titers of ~1:25, for 12CA5 or 9E10 MAb]. Incubate on ice for 30–45 min.

7. Pellet the cells and wash at least with 0.1 ml of ice-cold BSS. Resuspend the cells in ice-cold BSS at a density suitable for microscopy (~10^7 cell/ml) or flow cytometry (~$1–5 \times 10^6$/ml). Cultures should be kept on ice after all primary and secondary antibody labeling steps, prior to and during flow cytometry analysis.

8. Immunofluorescently labeled yeast may be examined by flow cytometry to obtain data of the type shown in Fig. 2. Setting of the instrument's gains should be chosen to allow detection of the background fluorescent peak.

One feature worthy of note in the flow cytometry data is the presence of a peak with background fluorescence equivalent in intensity to that obtained with negative controls. This negative peak is present for all proteins examined to date, and represents at least 15% of the cells examined in every case. Nonfluorescent cells appear to contain pCT302 vector and intracellular HA-tagged proteins[15]; the nonfluorescent fraction of the population is a strong function of the stability of the protein fusion expressed,[8] as well as of the temperature of induction.[16] This leads us to speculate that the nonfluorescent cells represent those cells for which the secretory apparatus has become saturated and blocked with misfolded protein,[17] a phenomenon that has now been well documented for solubly expressed and secreted proteins.[17–19] In any case, in more than 20 library screens to date we have found that the presence of the unlabeled cell population does not interfere with library screening for improved protein properties.

Construction of Yeast Display Libraries

If the protein of interest is displayed on the cell surface in active form, one may proceed directly to library construction in order to identify mutants with improved properties. We have also found that if the original protein is not displayed, it is possible to isolate "displayable" mutants of the protein[8] to enable further directed evolution.

The full range of available mutagenesis methods is consistent with yeast surface display library construction. We have had particular success with

[15] E. T. Boder, S. M. Elliott, and K. D. Wittrup, unpublished data (1999).

[16] M. C. Kieke, E. T. Boder, D. M. Kranz, and K. D. Wittrup, unpublished data (1999).

[17] A. S. Robinson and K. D. Wittrup, *Biotechnol. Prog.* **11,** 171 (1995).

[18] R. N. Parekh and K. D. Wittrup, *Biotechnol. Prog.* **13,** 117 (1997).

[19] E. V. Shusta, R. T. Raines, A. Pluckthun, and K. D. Wittrup, *Nature Biotechnol.* **16,** 773 (1998).

error-prone polymerase chain reaction (PCR)[20] of the entire ORF of the protein of interest, as well as mutagenesis of the entire plasmid by an *Escherichia coli* mutator strain (e.g., XL1-Red; Stratagene, La Jolla, CA). We provide an example of error-prone PCR mutagenesis of an scFv ORF. Primers were first designed to amplify the scFv gene with >100bp of Aga2p sequence 5′ of the *Nhe*I cloning site and 50 bp of sequence 3′ of the *Xho*I cloning site (see Fig. 1). Primer sequences are 5′-GGCAGCCCCATAAA-CACACAGTAT-3′ and 5′-GTTACATCTACACTGTTGTTAT-3′. Alternatively, standard T7 and T3 primers may be used to amplify the entire expression cassette from outside the promoter and transcriptional terminator. These large overhangs improve the efficiency of restriction digestion and dramatically increase the number of recombinants generated when subcloning the PCR products. The PCR is made mutagenic by the presence of 0.3–0.375 m*M* manganese chloride along with 2.25 m*M* magnesium chloride.[20,21] Under these conditions we obtain error rates of from 0.5 to 1.0%. Four 75-μl PCR are pooled and purified electrophoretically on a 1% (w/v) low melting agarose gel. The product band is excised from the gel and DNA is eluted in TAE buffer with a Bio-Rad (Hercules, CA) Electroeluter model 422, following the manufacturer recommended protocol. To reduce the EDTA concentration before restriction digestion, eluted products are diluted to 2 ml in doubly distilled H_2O, concentrated to ~60 μl in a Centricon-30 cartridge (Amicon, Danvers, MA), diluted to 0.5 ml in doubly distilled H_2O, and concentrated to ~10–20 μl in a Microcon-50 cartridge (Amicon). Final products are digested with *Nhe*I and *Xho*I and again gel purified with low melting point agarose. The appropriate band is cut out with a razor blade under illumination from a hand-held high-UV lamp, and the DNA is recovered from the gel slices with a Wizard PCR Prep kit (Promega, Madison, WI). This mutagenized insert is then ligated into a similarly gel-purified pCT302 backbone at an insert-to-vector ratio of ~2 : 1. Multiple 40-μl ligation reactions are performed to make use of as much mutagenized DNA as possible. One key to large library size is maximization of amount of DNA in each step (>1 μg).

We have also used DNA shuffling.[4,22] Key variables for optimization of that protocol are choice of DNase concentration and/or incubation time, and use of primers for final PCR amplification that are nested within the amplified region of those used for the original template amplification.

The ligation mixture is transformed into maximum-competency *E. coli* cells (e.g., XL10-Gold; Stratagene). Alternatively, to achieve maximal li-

[20] D. W. Leung, E. Chen, and D. V. Goeddel, *Technique* **1,** 11 (1989).
[21] R. C. Cadwell and G. F. Joyce, *PCR Methods Appl.* **3,** S136 (1994).
[22] H. Zhao and F. H. Arnold, *Nucleic Acids Res.* **25,** 1307 (1997).

brary size, multiple ligation reactions may be combined, diluted to 0.5 ml in doubly distilled H_2O to reduce salt concentration, concentrated to 10 μl in a Microcon-50 cartridge (Amicon), and transformed into electroporation-competent *E. coli* in multiple parallel transformation (e.g., DH10B Electo-MAX cells; Life Technologies, Bethesda, MD). This method has yielded DNA libraries containing $>10^7$ recombinants. The *E. coli* culture is maintained in liquid LB medium with carbenicillin (100 μg/ml) and ampicillin (50 μg/ml). Aliquots are plated to determine transformation efficiency. The liquid culture is then inoculated into 200-ml cultures and grown for 16–20 hr at 30°. Plasmid DNA is purified from the 200-ml culture by Qiagen Maxiprep kit or similar method.

Finally, the mutagenized plasmid pool is transformed into yeast by the high-efficiency protocol essentially as described by Gietz and Schiestl.[23] Multiple parallel transformations are performed; after gentle resuspension of cells in doubly distilled H_2O, cells are pelleted at 6000 rpm, resuspended in 100 μl of doubly distilled H_2O, and pooled. The transformed culture is amplified directly in SD + CAA liquid culture without plating, after plating of aliquots to determine yeast library size. Particularly important variables are the time of heat shock, quality of the single-stranded DNA (ssDNA) preparation, and gentle resuspension of cells after pelleting. For EBY100, a heat shock time of 30–35 min is optimal.[24]

Screening of Yeast Display Libraries

Successful engineering of proteins by directed evolution depends not only on a suitably diverse library from which to select altered phenotypes, but also critically on a quantitatively designed screening and isolation methodology. Specific parameters important in screening and sorting of yeast displayed libraries are ligand concentration, kinetic competition time, thermal denaturation time, and fluorescence-activated cell sorting (FACS) stringency. Mathematical estimation of these parameters enhances the utility of the surface display approach.[3]

Two screening approaches exist for identifying desirable affinity mutants within a surface displayed library. The most suitable method depends on the values of the affinity and kinetic constants of the wild-type protein–ligand binding interaction. Mutants may be distinguished by equilibrated binding with low concentrations of fluorescently labeled ligand in cases of fairly low affinity interactions ($K_d > 1$ n*M*, or no affinity if the library is being screened to isolate a novel binding specificity). However, for applications

[23] R. D. Gietz and R. H. Schiestl, *Methods Mol. Cell. Biol.* **5,** 255 (1995).
[24] E. T. Boder, B. G. Goekner, and K. D. Wittrup, unpublished data (1999).

designed to evolve tight-binding proteins, excessively large volumes of dilute ligand solutions are necessary to maintain molar ligand excess, complicating handling of samples. In such cases, improvements in binding affinity may be approximated by changes in dissociation kinetics. Kinetic competition for a stoichiometrically limiting ligand can be used to identify improved clones within the population[25]; however, this method eliminates the quantitative predictability of the screening approach and is not recommended in general. General strategies for equilibrium or kinetic screening of yeast displayed libraries are outlined below.

Quantitative Equilibrium Binding Screen

To verify the protein–ligand dissociation constant K_d within the surface display context a titration of the wild-type protein is performed by flow cytometric analysis.[6] A useful procedure for this analysis is as follows.

1. Grow and induce yeast cultures as described above, and harvest multiple samples containing $\sim 2 \times 10^6$ cells (i.e., ~ 0.2 OD_{600}-ml). If necessary the number of cells may be reduced to 1×10^6.
2. Label samples with 12CA5 MAb and biotinylated or fluorescently labeled ligand as described above. Use 10 or more dilutions of ligand such that the expected K_d of the interaction is effectively spanned. For example, for an expected K_d of 100 n*M*, ligand concentrations from ~ 10 n*M* to 1 μM should yield adequate results. Importantly, a 10-fold or greater molar excess of ligand must be maintained at all dilutions. A conservative estimate of the displayed protein concentration can be made by assuming $\sim 10^5$ copies/cell. Thus, 2×10^6 cells per sample yields ~ 0.33 pmol of displayed protein per sample, and incubation volumes should be adjusted to ensure >3 pmol of total ligand at the desired concentration. Volumes up to 50 ml have been used successfully. Note also that lower ligand concentrations may require longer incubations to ensure equilibrium.
3. Label with secondary antibodies and/or streptavidin–phycoerythrin as described previously.
4. Analyze cell populations by flow cytometry. Gate on only the displaying fraction of the population (observed by 12CA5 labeling). Determine the mean fluorescence intensity (i.e., the arithmetic mean) due to ligand binding of the displaying population. Note that geometric mean fluorescence (i.e., mean logarithmic histogram channel of fluorescence) is not useful for equilibrium or kinetic analysis. Therefore, alternative statistics such as peak or median fluorescence should be used if the instrument reports only geometric mean.

[25] R. E. Hawkins, S. J. Russell, and G. Winter, *J. Mol. Biol.* **226,** 889 (1992).

5. Plot (fluorescence intensity)/(ligand concentration) versus fluorescence intensity and apply Scatchard analysis to determine K_d. Deviations from linearity at higher ligand concentrations reflect saturation binding of surface protein, and data points beyond the saturating concentration should be ignored. An alternative and more rigorous procedure is to use a nonlinear least-squares routine to fit the binding equilibrium equation.

Once the K_d of the wild-type interaction has been measured, the optimum ligand concentration for discriminating mutants improved by a defined increment may be calculated from the following equation[3]:

$$\frac{[L]_{opt}}{K_d^{wt}} = \frac{1}{(S_r K_r)^{1/2}}$$

where $[L]_{opt}$ is the concentration of ligand yielding the maximum ratio of mutant to wild-type fluorescence, S_r is the maximum signal-to-background ratio for yeast saturated with fluorescent ligand, and K_r is the minimum affinity improvement desired (e.g., $K_r = 5$ if mutants improved fivefold in affinity are desired). S_r is the ratio of fluorescence of yeast saturated with fluorescent ligand over autofluorescence of unlabeled yeast, and is dependent on the particular flow cytometer and efficiency of protein expression. Our experience suggests a ligand concentration of $\sim 0.05–0.1 \times K_d$ of the wild-type interaction should generally yield adequate discrimination of mutants improved 3- to 10-fold.

Quantitative Kinetic Binding Screen

Screening by dissociation rate is achieved by labeling yeast to saturation with fluorescently labeled ligand followed by incubation in the presence of excess nonfluorescent ligand competitor. Prior to screening a displayed library for improved dissociation kinetics, the K_{off} of the wild-type protein–ligand reaction must be obtained. A protocol for determining K_{off} by flow cytometry of yeast displaying the protein of interest follows.

1. Grow and induce yeast cells as described above. Harvest $\sim 2 \times 10^7$ cells (2 OD_{600}-ml) and label for two-color fluorescence with anti-HA peptide MAb and fluorescently labeled ligand. Label the cells with a saturating amount of fluorescent ligand for a sufficient time to saturate labeling.
2. Pellet, remove, and discard the supernatant, wash with ice-cold BSS, pellet by centrifugation, and keep on ice until ready to begin flow cytometric analysis.
3. Add 2 ml of nonfluorescent ligand preequilibrated to room temperature (or other temperature of interest). The concentration of nonfluorescent ligand should be adjusted to yield a 10 to 100-fold excess over saturated

displayed protein complexes. A conservative estimate of this value may be calculated by assuming $\sim 10^5$ receptors per cell.

4. Analyze the fluorescence of the displaying population (i.e., gated by anti-HA epitope labeling) as a function of time. This may be performed by analysis of aliquots taken at time points and quenched on ice, or kinetic data may be taken on-line with some flow cytometers. Arithmetic mean, median, or peak fluorescence values may be used to extract K_{off}.

After determination of wild-type dissociation rate, time of competition with nonfluorescent ligand yielding the maximal fluorescence discrimination of mutants improved by a defined increment can be calculated from the following equation[3]:

$$k_{off,wt} t_{opt} = 0.293 + 2.05 \log k_r + \left(2.30 - 0.759 \frac{1}{k_r}\right) \log S_r$$

where t_{opt} is the optimal duration of competition, S_r is the signal-to-background ratio of flow cytometrically analyzed yeast, and k_r is the minimum fold improvement desired in k_{off}. S_r is best calculated as the ratio of fluorescence of displaying yeast saturated with fluorescent ligand over that of displaying yeast following competition to complete dissociation. Alternatively, mathematical analysis and experience suggest competition times of $\sim 5/k_{off}$ of the wild-type interaction should allow discrimination of mutants improved threefold under most experimental conditions.

Stability Screen by Thermal Denaturation Kinetics

A convenient method for evolving improved stability in a protein makes use of the protein denaturation rate at temperatures up to 50°. Viability of yeast may be maintained at these temperatures by pretreatment at 37° to induce stress response proteins. Prior to screening for improved denaturation kinetics, the wild-type denaturation rate should be measured by the following or similar methods.

1. Grow and induce yeast as described above. Harvest $\sim 2 \times 10^6$ cells per sample in six samples of 100 μl each.
2. Heat shock the samples for 50 min at 37°.
3. Incubate the cells at 50° for various times up to complete denaturation of the protein of interest, and then quench by adding 1 ml of ice-cold BSS.
4. Label the cells for two-color fluorescence as described with 12CA5 MAb and fluorescent ligand or conformation-sensitive antibody.
5. Analyze ligand-associated fluorescence intensity of the displaying fraction (as observed by 12CA5 labeling) as a function of time. Arithmetic mean, peak, or median fluorescence (but not geometric mean) may be fit

as a first-order exponential decay to determine k_{den}, the rate for constant for denaturation.

After determination of k_{den} of the wild-type protein, the optimal duration of 50° incubation for screening libraries can be calculated by using the t_{opt} kinetic equation given above, substituting k_{den} for k_{off}.

Library Sorting

The yeast display library should be oversampled by at least 10-fold to improve the probability of isolating rare clones (e.g., analyze $\sim 10^8$ cell from a yeast library with 10^7 clones). At typical flow cytometry sorting rates of 10^3–10^4 cells/sec, screening of 10^8 yeast may be performed in a full work day.

Diagonal sorting windows as shown in Fig. 2A should be drawn to take advantage of quantitative normalization by surface expression level. Trial windows should be drawn until the desired fraction of the population falls within the sort window. Labeled wild-type control cultures should be prepared each day for assistance in setting sort windows and confirmation of progress in library enrichment. In the first sort of a library, it is best to isolate the top 5% of the population in high-recovery mode (enrichment), to ensure retention of rare clones. Ensuing rounds of screening should use windows set to the top 0.1–1% of the population in purifying mode, with stringency increasing each sorting round. Sorted yeast remain viable in buffered sheath fluid for the duration of the sort. Sorted cells should be inoculated into SD + CAA, containing kanamycin (25 mg/ml) and adjusted to pH 4.5 with citrate buffer [sodium citrate (14.7 g/liter), citric acid monohydrate (4.29 g/liter)], to discourage growth of bacterial contaminants. The sorted cells may be passaged in liquid culture directly to another round of induction, labeling, and sorting. If necessary, these SD + CAA liquid cultures may be stored in the refrigerator for several weeks prior to revival at 30° and subsequent induction. We have found substantial enrichment of clones within the sorting window as early as the second screen, consistent with a frequency of approximately 1% of improved clones in the library. More typically, substantial enrichment (i.e., appearance of a minor, flow cytometrically observable population) is obtained by the third screen of a given library. If no enrichment is evident in the sort window by the fourth sort, there would be little justification for progressing to a fifth screen, as single clones from the original library should be enriched by that point. This has not been an issue to date, as we have isolated improved clones from each library screened.

Once the analyzed population exhibits a substantial fraction (>10%)

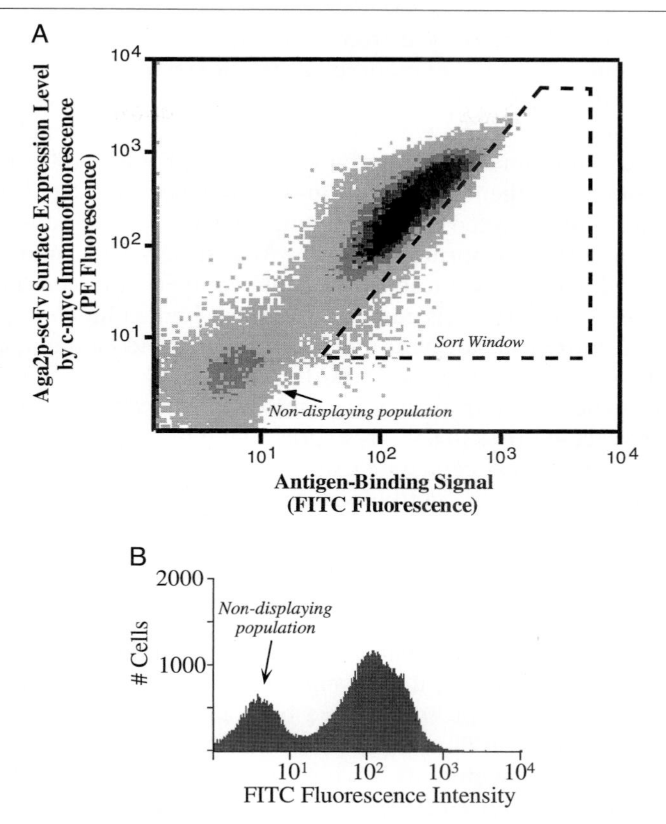

FIG. 2. (A) Flow cytometric analysis of a culture expressing an Aga2p-4-4-20 scFv fusion protein. In the dot plot, each dot represents a single analyzed cell. On the y axis, cell surface levels of the HA epitope are shown, while levels of binding to a fluorescein–dextran conjugate are shown on the x axis. The ratio of these two signals allows detection of ligand-binding activity normalized by the number of fusion proteins on the cell, enabling sort windows to be set as shown for isolation of improved mutants. (A) More common fluorophore pair is streptavidin–phycoerythrin to detect biotinylated ligand, and FITC-labeled secondary antibody to detect the 12CA5 or 9E10 MAb. (B) The projection of the single-cell fluorescence intensity histogram on of the FITC axis is shown.

within the sort window, the sorted culture may be plated to isolate individual clones. For simplicity, these monoclonal cultures can be analyzed individually by flow cytometry for improved affinity, dissociation rate, or stability. In our experience, the precision of equilibrium constant measurements by flow cytometry is ±40%, and ±10% for the dissociation rate. It is convenient to perform this screening process without the necessity of subcloning, expressing, and purifying the mutant proteins.

Plasmids may be recovered from yeast by rescue to *E. coli* either by use of a commercial kit (Zymoprep; Zymo Research, Orange, CA), or essentially as described previously.[26] Briefly, sorted cells are inoculated into SD–CAA medium and grown overnight at 30°. Cells are pelleted and resuspended in lithium chloride buffer containing Triton X-100, mixed with an equal volume of phenol–chloroform–isoamyl alcohol (25 : 24 : 1, v/v/v), and mechanically disrupted with zirconium oxide beads. The aqueous phase is collected and further purified using the Wizard DNA Cleanup kit (Promega). Eluted plasmids are transformed into competent *E. coli.* For recovery of a sorted library, it is important to use high-competency *E. coli,* maintain the entire transformed *E. coli* culture in liquid medium, and plate an aliquot to determine the total transformants.

Mutant genes of interest may be subcloned into expression/secretion vectors for yeast in order to solubly express the mutant proteins for further analysis. *Saccharomyces cerevisiae* expression systems for secretion of single-chain antibodies at 20 mg/liter in shake flask culture have been developed.[19]

Summary

The described protocols enable thorough screening of polypeptide libraries with high confidence in the isolation of improved clones. It should be emphasized that the protocols have been fashioned for thoroughness, rather than speed. With library plasmid DNA in hand, the time to plated candidate yeast display mutants is typically 2–3 weeks. Each of the experimental approaches required for this method is fairly standard: yeast culture, immunofluorescent labeling, flow cytometry. Protocols that are more rapid could conceivably be developed by using solid substrate separations with magnetic beads, for instance. However, loss of the two-color normalization possible with flow cytometry would remove the quantitative advantage of the method.

Yeast display complements existing polypeptide library methods and opens the possibility of examining extracellular eukaryotic proteins, an important class of proteins not generally amenable to yeast two-hybrid or phage display methodologies.

Acknowledgments

Helpful comments on the manuscript were provided by C. Graff, M. Kieke, E. Shusta, and J. VanAntwerp.

[26] A. C. Ward, *Nucleic Acids Res.* **18,** 5319 (1990).

Section IV

Construction of Hybrid Molecules by DNA Shuffling and Other Methods

[26] Methods for *in Vitro* DNA Recombination and Random Chimeragenesis

By ALEXANDER A. VOLKOV and FRANCES H. ARNOLD

Introduction

In vitro polymerase chain reaction (PCR)-based methods for recombining homologous DNA sequences are capable of creating highly mosaic chimeric sequences. Several different methods have been reported for *in vitro* recombination or "DNA shuffling": the original Stemmer method of DNase I fragmentation and reassembly,[1] the staggered extension process (StEP),[2,3] and random priming recombination.[4] We have found that slight variations in the shuffling protocols can affect the outcome of the experiment. Furthermore, different genes sequences recombine most efficiently under different conditions. Here we provide protocols that are designed to give a high likelihood of success. The protocols presented here are known to work for recombining sequences of ~>85% identity.

Methods

Stemmer Method

The Stemmer method involves digestion of parental DNA molecules into small fragments, using DNase I, and reassembly by overlap extension PCR. In the presence of Mg^{2+} or Mn^{2+} DNase I exhibits no obvious sequence specificity and generates DNA fragments randomly distributed over the gene length. Fragments are assembled in a cyclic PCR-like reaction with denaturation, annealing, and extension steps. During the annealing step single-stranded molecules associate with complementary molecules from any of the parent DNAs present to create novel sequence combinations.

Stemmer Protocol Using DNase I Fragmentation

1. Prepare DNA templates. Templates can be plasmids carrying target sequences, sequences excised by restriction endonucleases or amplified by

[1] W. P. C. Stemmer, *Nature (London)* **370,** 389 (1994).
[2] H. Zhao, L. Giver, Z. Shao, J. A. Affholter, and F. H. Arnold, *Nature Biotechnol.* **16,** 258 (1998).
[3] M. S. B. Judo, A. B. Wedel, and C. Wilson, *Nucleic Acids Res.* **26,** 1819 (1998).
[4] Z. Shao, H. Zhao, L. Giver, and F. H. Arnold, *Nucleic Acids Res.* **26,** 681 (1998).

0076-6879/00 $30.00

PCR. Prepare 2–5 μg of DNA templates mixed in equal proportions in a volume not exceeding 44 μl.

2. Add 2.5 μl of 1 *M* Tris-HCl (pH 7.5) and 2.5 μl of 200 m*M* $MnCl_2$, and bring the volume to 49 μl with deionized water. Equilibrate the mixture at 15° for 5 min.

3. Add 1 μl of DNase I (10 U/μl; Boehringer Mannheim, Indianapolis, IN) freshly diluted 1:100 in deionized water and perform digestion at 15°. Take 10-μl aliquots after 30 sec and after 1, 2, 3, and 5 min of incubation and immediately mix them with 5 μl of ice-cold stop buffer containing 50 m*M* EDTA and 30% (v/v) glycerol.

4. Separate the fragments by electrophoresis in a 2% (w/v) agarose gel. Figure 1 shows an example of a DNase I digest separated on a gel. It is important not to use standard loading buffers with their high concentration of tracking dyes, as dyes can mask fluorescence from DNA fragments. The stop buffer already contains glycerol and does not require additional loading buffer; however, a small amount of bromphenol blue can be added to it to simplify loading samples on the gel.

5. Cut DNA fragments in the desired size range from the gel and extract by an appropriate elution protocol.

6. Combine 10 μl of purified fragments, 5 μl of 10× *Pfu* buffer, 5 μl of 10× dNTP mix (each dNTP 2 m*M*), and 0.5 μl of *Pfu* polymerase (Stratagene, La Jolla, CA) in a total volume of 50 μl.

7. Run the assembly reaction, using the following thermocycler program: 3 min at 94° followed by 40 cycles of 30 sec at 94°, 1 min at 55°, and 1 min + 5 sec/cycle at 72°. The number of cycles depends on the fragment size. Assembly from small fragments may require more cycles than assembly from large fragments. Small fragments also may require a lower annealing temperature, at least during the initial cycles. Extension time depends on

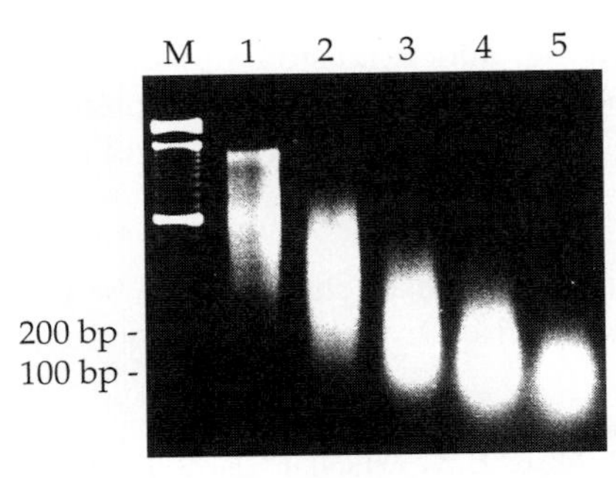

FIG. 1. Gel electrophoresis of DNase I digestion products. Lanes 1–5 are aliquots taken 30 second 1, 2, 3, and 5 min after the beginning of the reaction. M, Molecular weight marker, 100-bp ladder (Life Technologies, Bethesda, MD).

the gene size and should be adjusted accordingly for genes larger than 1–2 kb.

8. Amplify recombinant genes in a standard PCR, using serial dilutions of the assembly reaction (1 μl each of undiluted reaction, 1:10 dilution, and 1:50 dilution).

9. Run a small aliquot of the amplified products on an agarose gel to determine the yield and quality of amplification. If amplification produces a smear with low yield of full-length sequence reamplify these products with nested primers separated from the previously used primers by 50–100 bp. Run a small aliquot on an agarose gel.

10. Select the reaction with high yield and low amount of nonspecific products. Purify the reaction products, digest with appropriate restriction endonucleases, and ligate into the cloning vector.

Notes

1. DNA fragmentation with DNase I requires the presence of either Mg^{2+} or Mn^{2+} ions. The original version of the DNA shuffling protocol[1] recommended using $MgCl_2$. Experience indicates that $MnCl_2$ may be more suitable to maintain low point mutagenesis rates.[5,6] Mutation rate can also be reduced by using high-fidelity polymerases, *Pfu* or Vent, instead of *Taq* polymerase.[6]

2. Manganese and magnesium ions affect DNase I digestion differently: Mg^{2+} stimulates formation of single-stranded cuts, while Mn^{2+} stimulates cleavage of both strands.[7,8] Both metal ions promote random fragmentation, but some Mg^{2+}-generated nicks may remain undetected by agarose gel electrophoresis in nondenaturing conditions and lead to overestimation of the real fragment sizes.

3. Another important feature of this protocol is the use of EDTA rather than thermal inactivation to stop the DNase digestion. The presence of either Mg^{2+} or Mn^{2+} makes DNase I thermostable.[8–10] A high temperature does eventually inactivate the enzyme, but it remains active up to about 60°, continuing digestion with increasing speed. This high thermostability of DNase I is not by itself a problem, as long as the same inactivation protocol is used in all experiments. Switching from one inactivation method to another may require adjustment of DNase concentration or incubation

[5] I. A. J. Lorimer and I. Pastan, *Nucleic Acids Res.* **23,** 3067 (1995).

[6] H. Zhao and F. H. Arnold, *Nucleic Acids Res.* **25,** 1307 (1997).

[7] V. W. Campbell and D. A. Jackson, *J. Biol. Chem.* **255,** 3726 (1980).

[8] H.-M. Eun, "Enzymology Primer for Recombinant DNA Technology." Academic Press, San Diego, California, 1996.

[9] S. W. Bickler, M. C. Heinrich, and G. C. Bagby, *BioTechniques* **13,** 64 (1992).

[10] F. M. Pohl, R. Thomae, and A. Karst, *Eur. J. Biochem.* **123,** 141 (1982).

time. EDTA inactivation is recommended because it is technically simpler and more reproducible.

4. Gel electrophoresis of DNase products is not absolutely necessary for successful shuffling. It is possible to adjust reaction conditions and stop the reaction at a selected time. This approach may work well for some templates, but in some cases, even after digestion for extended periods of time, there is a significant amount of full-length template remaining in the reaction. Using the unfractionated mixture for the assembly reaction would generate an unacceptably large fraction of parental, nonrecombinant molecules. Other advantages of gel separation are visual control of the reaction and the ability to collect only the fragments in the desired size range.

Staggered Extension Process

Staggered extension process (StEP) recombination is based on template switching during polymerase-catalyzed primer extension. The abbreviated denaturation and annealing cycles limit the primer extension in a single cycle. Extension interrupted by denaturation resumes during the next annealing step, where the partially extended primers can anneal to different parent sequences present in the reaction. Multiple cycles of partial extension then create a library of chimeric sequences.[2] With no template digestion or fragment reassembly, this protocol is simple.

Staggered Extension Process Protocol

1. Prepare DNA templates. Templates can be plasmids carrying target sequences, sequences excised by restriction endonucleases or amplified by PCR.

2. Combine 5 μl of 10$\times$ *Taq* buffer, 5 μl of 10$\times$ dNTP mix (each dNTP 2 m*M*), 1–20 ng of each template DNA, 30–50 pmol of each primer, and 0.5 μl of *Taq* polymerase in a total volume of 50 μl.

3. Run 80–100 extension cycles: 94° for 30 sec and 55° for 5–15 sec.

4. Run a small aliquot of the reaction on an agarose gel. Possible reaction products are full-length amplified sequence, a smear, or a combination of both.

5. If plasmids were purified from a *dam*-methylation positive strain (DH5α, XL1-Blue) the extension reaction can be incubated with *Dpn*I endonuclease to remove parent DNA and decrease the background of nonrecombinat clones. Combine 2 μl of the extension reaction, 1 μl of DpnI reaction buffer, 6 μl of H_2O, and 1 μl of *Dpn*I restriction endonuclease (5–10 U/μl). Incubate at 37° for 1 hr.

6. Amplify the target sequence in a standard PCR, using serial dilutions of the previous reaction (1 μl each of undiluted reaction, 1:10 dilution, and 1:50 dilution).

7. Run a small aliquot of the amplified products on an agarose gel to determine the yield and quality of amplification. If amplification produces a smear with a low yield of full-length sequence, reamplify these products with nested primers separated from the previously used primers by 50–100 bp. Run a small aliquot on an agarose gel.

8. Select the reaction with high yield and a low amount of nonspecific products. Purify the reaction products, digest with appropriate restriction endonucleases, and ligate into the cloning vector.

Notes

1. Appearance of the extension products in step 4 may depend on the specific sequences recombined or the type of template used. Using whole plasmids in StEP recombination may result in nonspecific annealing of primers and their extension products all over the vector sequence, which would appear as a smear on the gel. As the number of cycles increases these products may be extended further, and the smear would shift up on the gel. A similar effect may be observed for large targets, even in the absence of any vector sequences. Small genes prepared by PCR amplification or endonuclease digestion are most likely to show gradual accumulation of the full-length product with increasing number of cycles.

2. DNA polymerases currently used in DNA amplification are fast enzymes. Even brief cycles of denaturation and annealing provide time for these enzymes to extend primers for hundreds of nucleotides. Therefore, it is not unusual for the full-length product to appear after only 10–15 cycles.

3. The faster the full-length product appears in the extension reaction, the fewer the template switches that have occurred and the lower the recombination frequency. Everything possible should be done to minizize time spent in each cycle: selecting a faster thermocycler, using smaller test tubes with thin walls, and, if necessary, reducing the reaction volume. Polymerases are not all equally fast. The proofreading activity of *Pfu* and Vent polymerases slows them down, offering another way to increase recombination frequency.[3] Polymerases with proofreading activity are also recommended during the amplification step to keep the mutagenic rate to a minimum.

4. As a general rule, annealing temperature should be decreased when higher recombination frequency is required or when templates have low GC content. Genes with GC pairs unevenly distributed along the gene length may present significant problems due to nonspecific annealing. These sequences should be amplified by PCR or excised from their cloning vectors prior to recombination to minimize the amount of nonspecific DNA (vector) present in the reaction.

Random-Priming Recombination

Random-priming recombination uses extension of random primers to generate fragments for reassembly.[4] Random primers are annealed to the template DNAs and are extended by a DNA polymerase at room temperature or below. The low temperature provides enough stabilization of the annealed primers to allow the use of random hexamers. Hexanucleotides are long enough to form stable duplexes with template DNA and short enough to ensure random annealing. Although longer primers can also be used, annealing may not remain random for short genes.

Random-Priming Protocol

1. Prepare DNA templates. Templates can be plasmids carrying target sequences, sequences excised by restriction endonucleases or amplified by PCR.
2. Combine 0.2–0.5 pmol of each template DNA and 7 nmol of $dp(N)_6$ random primers (Pharmacia Biotech, Piscataway, NJ) in a total volume of 65 μl.
3. Incubate for 5 min at 100° and transfer on ice.
4. Add 10 μl of 20 m*M* dithiothreitol (DTT), 10 μl of 10× buffer [0.9 *M* HEPES (pH 6.6), 0.1 *M* $MgCl_2$], 10 μl of dNTP mix (5 m*M* each), and 5 μl of Klenow fragment (2 U/μl).
5. Incubate for 3–6 hr at 22°.
6. Run a small aliquot of the reaction on an agarose gel. A faint, low molecular weight smear should be visible.
7. Add 100 μl of deionized water and purify extension products from the template by passing the reaction mixture through a Microcon-100 filter (Amicon, Beverly, MA) at 500*g* for 10–15 min at 25°. Do not exceed the recommended centrifugation speed, as a significant amount of template may pass through the filter.
8. Concentrate the flowthrough fraction on a Microcon-3 or -10 filter at 14,000*g* for 30 min at 25° to remove primers and small fragments.
9. Recover the retentate fraction and continue with the assembly reaction (steps 6–10 of the Stemmer protocol).

Notes

1. Random primer extension is a versatile method for generating random fragments for recombination. Unlike DNase I fragmentation in Stemmer shuffling, this method does not require double-stranded DNA and can be used on both double-stranded and single-stranded substrates. Moreover, random primers can also be used with RNA substrates. In fact, because the stability of RNA–DNA duplexes is higher than that of DNA–DNA

duplexes, RNA templates have an advantage over DNA templates that may allow extension at higher temperatures.

2. Another important feature of this method is the ability to use any DNA polymerase. Because the primer extension step does not involve thermal cycling, polymerase choice is not limited to thermophilic enzymes. This feature gives more options for selecting the best polymerase.

Applications of *in Vitro* Recombination

The original Stemmer protocol, first introduced in 1994, has been successfully applied to the directed evolution of a large number of proteins. For example, protein folding and solubility were improved for the green fluorescent protein from *Aequorea victoria*[11] and single-chain antibody fragments (scFv) produced in *Escherichra coli.*[12] Enzyme substrate specificity has been changed: for example, a galactosidase was converted to a fucosidase,[13] while the substrate specificity of a biphenyl dioxygenase was modified and extended.[14] Enzyme thermostability has been improved.[15,16] Enzymatic activity of cephalosporinases was dramatically increased by shuffling homologous naturally occurring genes.[17] Sequencing of the evolved genes has proved the ability of this method to recombine closely spaced mutations and create highly mosaic genes.

A relatively new DNA shuffling method, StEP recombination, has been used to improve the catalytic activity and thermostability of subtilisin.[2,18] In these studies, mutations as close as 34 bp were recombined. The random-priming method has been successfully used to recombine mutants of subtilisin E.[18] Mutations separated by as few as 12 bp were recombined. An efficient, hybrid *in vitro–in vivo* recombination method is described in the next chapter.[19]

[11] A. C. Crameri, E. A. Whitehorn, E. Tate, and W. P. C. Stemmer, *Nature Biotechnol.* **15,** 315 (1996).

[12] K. Proba, A. Worn, A. Honegger, and A. Pluckthun, *J. Mol. Biol.* **275,** 245 (1998).

[13] J.-H. Zhang, G. Dawes, and W. P. C. Stemmer, *Proc. Natl. Acad. Sci. U.S.A.* **94,** 4504 (1997).

[14] T. Kumamaru, H. Suenaga, M. Mitsuoka, T. Watanabe, and K. Furukawa, *Nature Biotechnol.* **16,** 663 (1998).

[15] F. Buchholz, P.-O. Angrand, and A. F. Steward, *Nature Biotechnol.* **16,** 657 (1998).

[16] L. Giver, A. Gershenson, P.-O. Freskgard, and F. H. Arnold, *Proc. Natl. Acad. Sci. U.S.A.* **95,** 12809 (1998).

[17] A. Crameri, S.-A. Raillard, E. Bermudez, and W. P. C. Stemmer, *Nature (London)* **391,** 288 (1998).

[18] H. Zhao and F. H. Arnold, *Protein Eng.* **12,** 47 (1999).

[19] A. A. Volkov, Z. Shao, and F. H. Arnold, *Methods Enzymol.* **328,** Chap. 27, 2000 (this volume).

Recombination Results

All three PCR-based methods presented can create libraries of recombined sequences. The different methods each have their own advantages and disadvantages, and their relative performance will probably differ for different templates. We compared the three methods for their ability to recombine the truncated green fluorescent protein (GFP) genes in the recombination test system described in the next chapter.[19] Table I presents the results of that comparison. In these experiments, the Stemmer protocol using DNase I fragmentation and StEP show the highest recombination efficiency. With DNase fragmentation, using smaller fragments (<100 bp) generates a slightly higher efficiency than larger fragments (100–200 bp).

DNA recombination may differ from sequence to sequence and be affected by base composition or other specific features of the sequence. For example, secondary structure formation in single-stranded DNA may adversely affect the performance of all the methods. However, the extent of this effect is probably not equal for all of them. The assembly step in the Stemmer protocol and random-priming recombination can be affected by secondary structure. StEP recombination, with its short annealing times, may be even more sensitive to secondary structure. Random priming can also be affected by secondary structure at the most crucial step of the reaction, annealing and extension of the random primers. Small primers

TABLE I

FLUORESCENT *Escherichia coli* COLONIES OBTAINED BY RECOMBINING TWO GREEN FLUORESCENT PROTEIN TEMPLATES CONTAINING MUTATIONS AT THE INDICATED SITES[a]

Distance between mutations (bp)	Fraction of fluorescent colonies (%)			
	Stemmer protocol			
	<100-bp fragments	100 to 200-bp fragments	StEP recombination	Random-printing recombination
423	20.5	19.2	18.5	5.0
315	14.5	9.7	13.1	4.8
207	11.5	8.3	9.8	3.1
99	9.6	8.4	8.2	1.2
24	5.8	5.1	4.8	0.9
99 + 99	6.1	3.3	1.8	0.2

[a] See [27] in this volume[19] for detailed explanation of the templates. Generation of fluorescence requires recombination between the sites to restore the wild-type sequence. The last row shows the results of recombining one single and one double mutant with 99-bp distances between each mutation.

will be more sensitive to secondary structure than longer primers at this annealing step. The low temperature used in the extension of the random primers also stabilizes secondary structure. Increasing extension temperature would force the use of longer primers, which would probably lead to less efficient recombination, at least for short genes. Termination of elongation of both short and long primers is sensitive to secondary structure. Formation of stable stem–loop structures may be the most important factor causing nonrandom distribution of extended fragments in random-priming recombination.

An important parameter for determining the utility of any given method is the average number of recombination events per gene (crossover frequency) that can achieved. The Stemmer protocol probably offers the best possible cross-over frequency. The number of recombination events per gene is inversely proportional to the fragment size. Preparing shorter fragments increases recombination frequency, as demonstrated in Table I, although the effect is not large. Small genes, however, will have limited capacity for such an improvement, because small fragments will be inefficiently reassembled.

In random priming, the fragment sizes can be controlled by adjusting the primer-to-template molar ratio. High primer concentration limits exten-

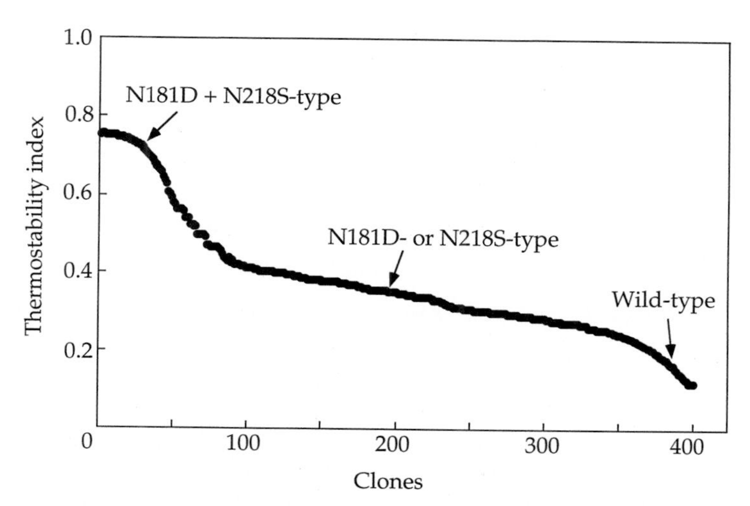

FIG. 2. Results of screening a library of recombined thermostable subtilisin E mutants.[21] Data are sorted and plotted in descending order of thermostability (residual/initial activity). Arrows indicate thermostability of wild type, parents N181D and N218S, and recombinant N181D + N218S. The plateau regions corresponding to the parent and recombined sequences are characteristic of successful recombination of a limited number of improved mutants.

sion of all primers.[20] There is one more yet-unexplored opportunity for size control during primer extension: shortening the incubation time may significantly decrease fragment size. Optimal incubation times will have to be determined experimentally, and they will vary from gene to gene.

DNA is not the only nucleic acid that can be used in recombination experiments. Certainly cDNA copies can always be synthesized from RNA templates, but the random-priming and StEP recombination protocols can potentially use RNA directly.

Detailed information on recombinants is, of course, provided by DNA sequencing. Restriction digests are also useful, provided restriction sites change on recombination. Screening the functional properties of some number of clones may also give useful information about recombination and also about associated point mutation rates. Figure 2 shows the results of a thermostability assay carried out on a library of recombined thermostable subtilisin E mutants.[21] The data are sorted and plotted in descending order. When the mutations are additive, one can clearly distinguish the recombinant clones from the parental ones. The relative fraction of highly thermostable recombinants indicates the efficiency of the recombination process (in this example, perfect recombination with no point mutation would yield 25% highly thermostable clones). Point mutation rates can be deduced from the fraction of inactive clones.[6]

[20] C. P. Hodson and R. Z. Fisk, *Nucleic Acids Res.* **15,** 6295 (1987).
[21] H. M. Zhao and F. H. Arnold, *Proc. Natl. Acad. Sci. U.S.A.* **94,** 7997 (1997).

[27] Random Chimeragenesis by Heteroduplex Recombination

By ALEXANDER A. VOLKOV, ZHIXIN SHAO, and FRANCES H. ARNOLD

Introduction

DNA recombination is an important tool for directed evolution of proteins and nucleic acids. Genetic variations existing in nature or created in the laboratory can be recombined to generate libraries of molecules containing novel combinations of sequence information from any or all of the parent sequences. By combining beneficial mutations and removing deleterious ones, recombination may help to accelerate the evolution of single molecules toward a specified function. Novel chimeric sequences

0076-6879/00 $30.00

generated by recombination of homologous, naturally occurring genes also provide an extremely rich source of genetic diversity for directed evolution.

Various methods by which DNA sequences can be recombined ("shuffled") in the laboratory to create chimeric sequence libraries have been described. Detailed protocols for *in vitro* methods, including the DNA shuffling method first described by Stemmer,[1] the staggered extension process (StEP),[2] and random-priming recombination,[3] are presented in [26] in this volume.[4] All these polymerase chain reaction (PCR)-based methods require synthesis of significant amounts of DNA during the assembly/recombination step and subsequent amplification of the final products. PCR-based methods work well with sequences of moderate size. Recombination of long sequences by these methods may prove problematic because of inefficient DNA amplification and excessive accumulation of unwanted mutations.

In vivo approaches to random chimeragenesis include tandem cloning recombination[5–7] and recombination in yeast cells.[8,9] The tandem cloning approaches recombine two parental genes cloned in tandem on the same plasmid and utilize homologous recombination in bacterial cells[5] or bacterial enzymes responsible for double-strand break repair[6,7] to generate the desired chimeric genes. Although PCR amplification is not required, tandem cloning recombination cannot generate more than one cross-over in a single experiment.

Yeast cells can efficiently join linear, partially overlapping double-stranded DNA fragments introduced by transformation, and restore a functional, covalently closed plasmid.[8,9] This approach is also free from the size limitation associated with the PCR-based approaches. The number of cross-overs introduced in one recombination experiment may be higher than with tandem cloning recombination, but is still likely to be low compared with *in vitro* recombination methods.

We described a simple method for creating libraries of chimeric DNA sequences, derived from homologous parental sequences, that uses *in vivo*

[1] W. P. C. Stemmer, *Nature* (*London*) **370,** 389 (1994).

[2] H. Zhao, L. Giver, Z. Shao, J. A. Affholter, and F. H. Arnold, *Nature Biotechnol.* **16,** 258 (1998).

[3] Z. Shao, H. Zhao, L. Giver, and F. H. Arnold, *Nucleic Acids Res.* **26,** 681 (1998).

[4] A. A. Volkov and F. H. Arnold, *Methods Enzymol.* **328,** Chap. 26, 2000 (this volume).

[5] G. L. Gray, U.S. Patent 5,093,257 (1992).

[6] H. Weber and C. Weissmann, *Nucleic Acids Res.* **11,** 5661 (1983).

[7] M. D. van Kampen, N. Dekker, M. R. Egmond, and H. M. Verheij, *Biochemistry* **37,** 3459 (1998).

[8] D. Pompon and A. Nicolas, *Gene* **83,** 15 (1989).

[9] J. S. Okkels, PCT application WO 97/07205 (1997).

repair of heteroduplexes for recombination.[10] This "heteroduplex recombination" approach relies on the mismatch repair system of the host cells to repair regions of nonidentity in the heteroduplex and creates a library of new sequences composed of elements from each parent. This method combines the advantages of *in vitro* and *in vivo* recombination methods and avoids some of their drawbacks. Heteroduplex recombination does not require PCR amplification and should be useful for recombination of large DNA sequences.

Methods

Heteroduplexes of two homologous sequences are often formed by whole-plasmid methods.[11,12] Each plasmid carrying a homologous gene is cut with a different restriction endonuclease with a unique recognition site in the vector sequence. The digested plasmids are mixed and annealed together. Homologous genes form circular heteroduplex plasmids containing one single-stranded break in each strand. Although this protocol has worked well for studying the mechanisms of mismatch repair, it is not always efficient in recombination. A modified protocol with significantly improved efficiency of recombination is presented here (Fig. 1). Experiments were performed on a recombination test system especially designed to evaluate different recombination methods.[4,10]

Recombination Templates

The DNA templates of this test system are truncated genes for the green fluorescent protein (GFP) from *Aequorea victoria.* The templates are created by introducing the sequence TAAT, containing a stop codon, at selected positions along the gene by site-directed mutagenesis of plasmid pGFP (Clontech, Palo Alto, CA; GenBank accession no. U17997). The stop codons interrupt translation and result in the synthesis of truncated products that are not fluorescent. Table I lists the mutagenic primers, mutation positions, and restriction endonucleases associated with each mutant. Templates 205, 313, 421, 529, 604, and 637 contain a single stop codon at the designated position, while template 421–637 contains two stop codons approximately 200 bp apart.

Recombination between truncated variants of GFP generates the full-length wild-type gene and restores fluorescence. (The multiply mutated

[10] A. A. Volkov, Z. Shao, and F. H. Arnold, *Nucleic Acids Res.* **27,** e18 (1999).

[11] J. P. Abastado, B. Cami, T. H. Dinh, J. Igolen, and P. Kourilsky, *Proc. Natl. Acad. Sci. U.S.A.* **81,** 5792 (1984).

[12] J. Westmoreland, G. Porter, M. Radman, and M. A. Resnik, *Genetics* **145,** 29 (1997).

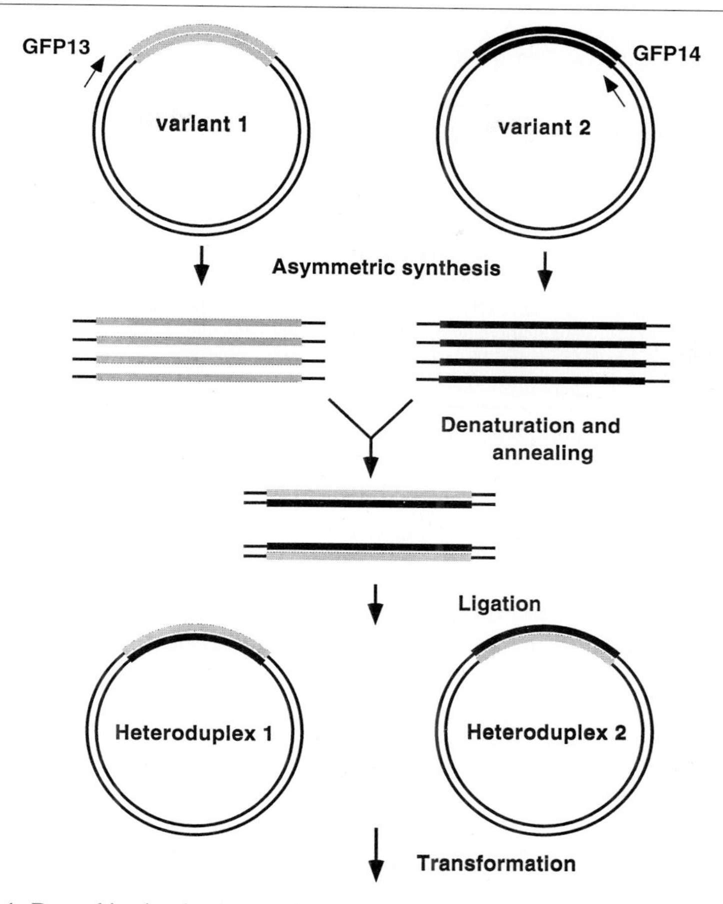

FIG. 1. Recombination by *in vitro* heteroduplex formation followed by *in vivo* repair: outline of optimized protocol.

gene is also made, presumably in equal proportions.) Mutations distributed over the gene length provide the opportunity to evaluate how the efficiency of recombination depends on the distance between the nonidentities. For example, template 637 recombined pairwise with each of the other single-stop codon templates restores fluorescence on a single recombination event between the mismatches that are 24, 99, 207, 315, and 423 bp apart. When double mutant 421–637 is recombined with single-stop codon template 529, two recombination events are required to restore the wild-type sequence, each one to occur within 99 bp.

A second pair of plasmid recombination templates was constructed to

TABLE I

GREEN FLUORESCENT PROTEIN RECOMBINATION TEMPLATES AND MUTAGENIC PRIMERS USED TO CONSTRUCT THEM[a]

GFP recombination template (position of stop codon)	Mutagenic primer sequence	Restriction endonuclease site
205	ctttctcttatggtgtt**TAA**tgctagctcaagatacccagatc	*Nhe*I
313	caaagatgacggg**TAA**tagatctacacgtgctgaagtc	*Bgl*II
421	gaaacattcttggacacaaa**TAA**tatgcataactataactcacacaatg	*Nsi*I
529	gaagatggaagcgtt**TAA**tggatccgaccattatcaacaaaatactc	*Bam*HI
604	gacaaccattacctg**TAA**tggtacccctgccctttcg	*Kpn*I
637	cgaaagatcccaac**TAA**ttctagagaccacatggtcc	*Xba*I

[a] Stop codon TAA is indicated by boldface capital letters, and restriction endonuclease sites are underlined. GFP variants are identified by the positions of the stop codons (base pairs from start of gene).

test the ability of this method to recombine large, nonhomologous sequences. One was plasmid pGFP encoding the wild-type GFP gene. The second plasmid, pKan, was constructed by inserting the aminoglycoside 3′-phosphotransferase gene from pUC4K (Pharmacia, Piscataway, NJ) into the *Aat*II site of pGFP and deleting an *Spe*I–*Xba*I fragment containing the GFP gene. Figure 2 illustrates the heteroduplex formed by these two plasmids. The GFP and aminoglycoside 3′-phosphotransferase genes do not have complementary sequences on the opposite strands and therefore form large single-stranded loops (826 and 1435 bp, respectively). The loops are stabilized by the 336-bp double-stranded region common to both plasmids and located between the *Spe*I and *Aat*II sites.

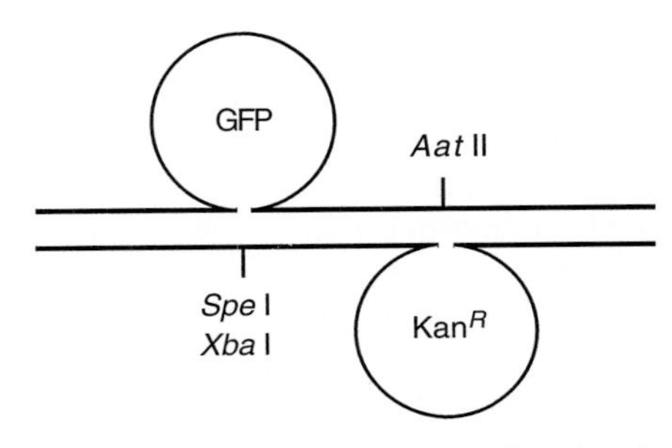

FIG. 2. Heteroduplex formed by plasmids carrying GFP and aminoglycoside 3′-phosphotransferase (Kan^R) genes. Recombination will generate plasmids containing both genes.

Heteroduplex Recombination Protocol

1. Prepare DNA templates with a Qiagen (Valencia, CA) minipreparation purification kit or equivalent.
2. Combine 5 μl of 10× *Taq* buffer, 5 μl of 10× dNTP mix (each dNTP 2 m*M*), 30–50 pmol of one primer, 5 ng of template, and 0.5 μl of *Taq* polymerase (10 U/μl; Boehringer Mannheim, Indianapolis, IN) in a final volume of 50 μl. One DNA template is combined with forward primer GFP 13 (5′-CCGACTGGAAAGCGGGCAGTG-3′), another with reverse primer GFP14 (5′-CCGCATAGTTAAGCCAGCCCCG-3′). Both primers are directed to the pGFP vector upstream or downstream of the GFP gene.
3. Synthesize single-stranded DNA templates in a PCR-like reaction for 100 cycles of 30 sec at 94°, 30 sec at 56°, and 1 min at 72°.
4. Combine the reaction products and purify with a Qiagen PCR purification kit or equivalent to remove primers, dNTPs, and *Taq* polymerase. Elute DNA in 50 μl of 10 m*M* Tris-HCl, pH 8.0.
5. Combine 47.5 μl of purified products and 2.5 μl of 20× SSPE buffer (180 m*M* NaCl, 1 m*M* EDTA, 10 m*M* NaH_2PO_4, pH 7.4).
6. Incubate the reaction at 96° for 4 min, immediately place on ice for 4 min, and continue the incubation for 2 hr at 68°.
7. Purify the annealed products, digest with appropriate restriction endonucleases, and ligate into the cloning vector.
8. Transform *Escherichia coli* competent cells and grow on agar plates supplemented with ampicillin.

Results and Discussion

We found the following features to be important for optimizing recombination efficiency in heteroduplex recombination: repairing unsealed nicks in the heteroduplex plasmid, small vector size, and eliminating nonrecombinant parent clones.[10] The initial events in DNA mismatch repair are the recognition of a mismatch by MutS and MutL, activation of MutH endonuclease, and formation of a nick on one of the strands.[13] Plasmid heteroduplexes already have two nicks and present a perfect substrate for the remaining part of the repair mechanism. The preexisting nicks may serve as primary initiation points for repair or as a shunt, suppressing initiation at other regions. The location of the nicks outside the target sequence means that productive recombination can occur only when repair polymerization terminates within the target sequence. Sealing these nicks by DNA ligase

[13] R. S. Lahue, K. G. Au, and P. Modrich, *Science* **245,** 160 (1989).

treatment increases the efficiency of recombination severalfold (cf. the third and fourth columns [Heteroduplex plasmid] in Table II).

Plasmid heteroduplex formation results in a mixture of products, but only circular, heteroduplex plasmids efficiently transform bacterial cells. The efficiency of the circularization reaction decreases rapidly with vector size. For large vectors, the amount of the circular form may be so low that the background from uncut molecules becomes significant. Annealing the insert rather than the entire plasmid provides a simple solution to this problem. Furthermore, the cloning vector, prepared once, can last for hundreds of experiments, substituting for the lengthy procedure of digesting each parent. The fifth column in Table II [Insert heteroduplex (%)] shows the further improvement in recombination efficiency achieved by annealing only the insert and ligating the insert heteroduplexes into a cloning vector.

While solving one problem, insert annealing creates another: It reintroduces a large nonrecombinant background from parent sequences. (With whole-plasmid heteroduplex formation, parent homoduplexes remain linear and do not efficiently transform.) Asymmetric, or single-primer target sequence, synthesis was therefore used to suppress formation of parental duplexes. By amplifying each parent template in a PCR with a single primer, each parent contributes only one strand to the resulting heteroduplex, which greatly limits the possibility of homoduplex formation. Although asymmetric synthesis cannot completely prevent homoduplex formation, running the synthesis for many cycles ensures that the asymmetrically synthesized strands are the predominant species. The last column in Table II

TABLE II
PERCENTAGE OF FLUORESCENT COLONIES OBTAINED AFTER HETERODUPLEX RECOMBINATION BY DIFFERENT PROTOCOLS[a]

Recombination templates	Distance between mutations (bp)	Heteroduplex plasmid: With single-stranded breaks (%)	Heteroduplex plasmid: Treated with DNA ligase (%)	Insert heteroduplex (%)	Insert heteroduplex, asymmetric strand synthesis (%)
205 and 637	423	3	10	17	29
313 and 637	315	3	9	13	25
421 and 637	207	1.3	7	9	18
529 and 637	99	1.1	8	11	16
604 and 637	24	0	<1	0.9	1.2
421–637 and 529	99 + 99	—	—	1.8	2.8

[a] Recombination efficiencies are higher than the fluorescent fractions, because recombination generates the double mutants as well as functional GFP.

[Insert heteroduplex, asymmetric strand synthesis (%)] presents the cumulative results of sealing nicks, insert hybridization, and asymmetric synthesis, all of which generate a more than 10-fold improvement in recombination efficiency. Heteroduplex recombination performs well on the small GFP gene. However, its most useful application may well be for recombination of much larger DNA sequences.

We also applied this method to recombine two plasmids carrying nonhomologous genes to create a plasmid carrying both genes in tandem orientation. Parental plasmid pGFP was linearized by digestion with *Xmn*I; plasmid pKan was linearized with *Sca*I. The heteroduplex was prepared as described in steps 5 and 6 of the heteroduplex recombination protocol (above). Transformed bacterial cells were plated on LB–agar containing ampicillin and kanamycin. Successful recombinants were expected to form kanamycin-resistant and fluorescent colonies. Approximately 1% of the kanamycin-resistant colonies were fluorescent. Plasmid DNA from 10 fluorescent and kanamycin-resistant colonies was analyzed by endonuclease digestion. All 10 samples had digestion patterns that indicated the presence of both genes. This experiment demonstrates that heteroduplex recombination can be used to create chimeric constructs between sequences with large regions of nonidentity. This feature may be useful for engineering entire operons and metabolic pathways.

[28] Use of Chimeras Generated by DNA Shuffling: Probing Structure–Function Relationships among Glutathione Transferases

By LARS O. HANSSON and BENGT MANNERVIK

Introduction

Site-directed mutagenesis is a tool frequently used to study the functional importance of one or several amino acid residues in a protein. In a similar manner, modules of the polypeptide chain can be exchanged between homologous proteins in order to analyze contributions of the selected regions to the functional properties. In many studies of this kind the foundation for the chimera construction is structural information, for example, exon borders, protein domains, or structural motifs derived from a three-dimensional structure. This chapter presents a procedure for structure–function analysis based on chimeric enzymes without necessarily knowing more than their primary structures. The approach is to create a library of

METHODS IN ENZYMOLOGY, VOL. 328

0076-6879/00 $30.00

```
           1
GST M2-2   PMTLGYWNIR GLAHSIRLLL EYTDSSYEEK KYTMGDAPDY DRSQWLNEKF
GST M1-1   ..I....D.. ...#A..... ......#... .......... ..........

           51
GST M2-2   KLGLDFPNLP YLIDGTHKIT QSNAILRYIA RKHNLCGESE KEQIREDILE
GST M1-1   .......... .....A.... ..#..#C... .....#.#T. E.K.#V....
           101
GST M2-2   NQFMDSRMQL AKLCYDPDFE KLKPEYLQAL PEMLKLYSQF LGKQPWFLGD
GST M1-1   ..T..NH... GMI.#N.E.. ...#K.#EE. ..K#....E. ...R...A#N

           151
GST M2-2   KITFVDFIAY DVLERNQVFE PSCLDAFPNL KDFISRFEGL EKISAYMKSS
GST M1-1   ..#.#.#LV. ...DLHRI.. .N.#.#...# ......#... ..........

           201
GST M2-2   RFLPRPVFTK MAVWGNK
GST M1-1   ........S. .......
```

FIG. 1. Protein sequence alignment of human GST M1-1 and GST M2-2. A hatch mark (#) denotes a difference in the codon for a given amino acid; a dot (.) denotes identical codons.

randomly composed chimeras of two homologous enzymes. This is achieved by constructing chimeric cDNA by random recombination of DNA fragments, DNA shuffling.[1,2] The chimeric enzymes are expressed in *Escherichia coli* and isolated clones are assayed for catalytic activity in crude cell lysates. The primary structures, derived from the DNA sequences, are determined for a subset of the analyzed clones. The selection for the sequence analysis is based on sampling of chimeras that display divergent activities. Finally, the differences in activity among the chimeric variant enzymes can be correlated to specific alterations in the primary structure of the chimeric enzymes.

The study of two glutathione transferase (GST) isoenzymes will exemplify structure–function analysis by random chimerization.[3] The GSTs catalyze two-substrate reactions with the tripeptide glutathione (GSH) and a large variety of electrophilic compounds. The two human isoenzymes being studied, GST M1-1 and GST M2-2, belong to a common structural class of GSTs (the Mu class) and they exhibit 84% sequence identity at the protein level (183 of 217 residues are identical; Fig. 1) and 89% identity at the level of coding DNA sequence. The residues differing between the two enzymes give rise to a>100-fold higher specific activity for GST M2-2 with

[1] W. P. C. Stemmer, *Nature (London)* **370,** 389 (1994).

[2] W. P. C. Stemmer, *Proc. Natl. Acad. Sci. U.S.A.* **91,** 10747 (1994).

[3] L. O. Hansson, R. Bolton-Grob, T. Massoud, and B. Mannervik, *J. Mol. Biol.* **287,** 265 (1999).

some electrophilic compounds, e.g., aminochrome, undergoing an additional reaction, and 2-cyano-1,3-dimethyl-1-nitrosoguanidine (cyano-DMNG), undergoing a substitution reaction (Fig. 2). However, not all activities known are as different and GST M2-2 exhibits only a twofold higher activity than GST M1-1 with 1-chloro-2,4-dinitrobenzene (CDNB; Fig. 2). A relevant problem is therefore the structural basis for the diverse activities. Comparison of the GST M1-1 and GST M2-2 structures, both at the primary and the three-dimensional levels, does not give an obvious explanation for the differences in catalytic properties. To elucidate structure–activity relationships a library of randomly composed chimeras was constructed from the homologous enzymes GST M1-1 and GST M2-2 by DNA shuffling of the corresponding cDNA. From activity screening of 125 of the corresponding variant enzymes in combination with primary sequence data from 18 of the clones, it was concluded that a number of residues situated in α helices 4–6 are responsible for the 100-fold difference in activity.

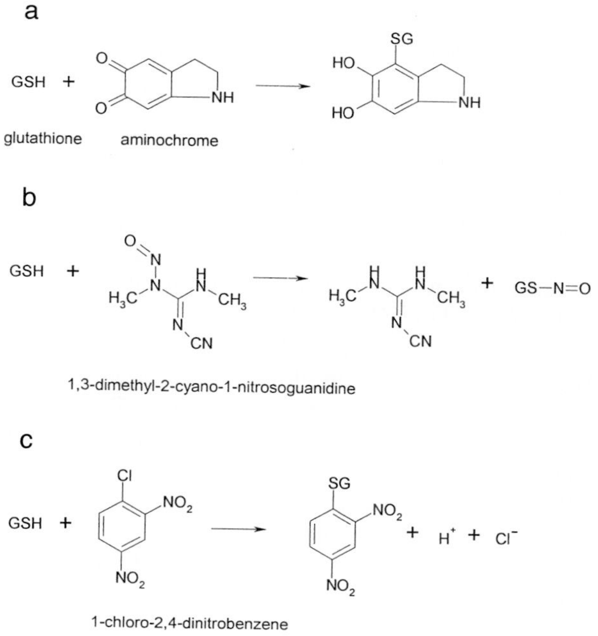

FIG. 2. Enzyme-catalyzed reactions studied with cell lysates containing chimeras of GST M1-1 and GST M2-2. (a) Addition of GSH to the *o*-quinone aminochrome. (b) Denitrosation of 2-cyano-1,3-dimethyl-1-nitrosoguanidine (cyano-DMNG) by GSH. (c) Substitution of chlorine in 1-chloro-2,4-dinitrobenzene (CDNB) by GSH.

Method

DNA shuffling[1,2] is used to create a library of randomly composed chimeras of GST M1-1 and GST M2-2. This method consists basically of three steps: fragmentation of template DNA by DNase I, random reassembly of isolated fragments by primerless polymerase chain reaction (PCR), and amplification of recombined products by a PCR with primers flanking the targeted DNA segment. To succeed in creating a diverse library of chimeras the template cDNA sequences of the homologous enzymes need to exhibit a reasonably high identity and preferably with the mismatches well distributed over the sequences. A lower limit of segment similarity for successful recombination cannot be rigorously defined, but the known cases have >50% identity. Also, the two wild-type enzymes should exhibit a major difference in activity or substrate specificity in order to make possible the identification of significant differences among the chimeras being screened. A large number of library clones need to be functionally analyzed to obtain a representative sampling. An activity assay suited for a reasonably high throughput is thus desirable.

Preparation of DNA Template for DNA Shuffling

1. The cDNAs of human GST M1-1 and GST M2-2 are amplified by PCR, using pGΔETacM1b[4] and pGEM-3Zf(+)M2[5] as templates in the reaction, respectively. The reaction mixture (10 tubes, each containing 100 μl) is composed of template (10 ng/ml), 1.5 mM $MgCl_2$, a 0.2 mM concentration of each of the four deoxyribonucleotide triphosphates (dNTPs), *Taq* DNA polymerase (10 units/ml; Boehringer Mannheim, Mannheim, Germany), and a 1 μM concentration of the vector primers M13 forward (5′-AAT TGT GAG CGG ATA ACA AT) and M13 reverse (5′-AGC GGA TAA CAA TTT CAC ACA GGA). Forty cycles of PCR (1 min at 94°, 2 min at 55°, 2 min at 72°) are carried out with a hot start for 3 min at 95° and a final extension for 10 min at 72°. The PCR products are purified with Wizard PCR Preps DNA purification resin (Promega, Madison, WI).

2. The GST M2-2 PCR fragment is digested with *Sac*II in order to eliminate 32 nucleotides from the coding sequence and ensure that all chimeric clones produced contain the 5′ end of the high-level expression GST M1-1 clone.[4]

[4] M. Widersten, M. Huang, and B. Mannervik, *Protein Expression Purif.* **7,** 367 (1996).

[5] S. Baez, J. Segura-Aguilar, M. Widersten, A.-S. Johansson, and B. Mannervik, *Biochem. J.* **324,** 25 (1997).

3. The PCR products of GST M1-1 and GST M2-2 (digested by *Sac*II) are subjected to agarose gel (1%) electrophoresis. DNA fragments (1100 bp for GST M1-1 and 720 bp for GST M2-2) are purified with the GeneClean II silica matrix (Bio 101, La Jolla, CA), with elution into water.

Fragmentation by Digestion with DNase I

1. Equimolar amounts of the prepared DNA encoding GST M1-1 (9 μg) and GST M2-2 (7 μg) are degraded separately. The DNA (130 μl) is first preincubated at 15° for 15 min. Digestion buffer (15 μl, 10 times concentrated) is added to a final concentration of 10 m*M* $MnCl_2$ in 50 m*M* Tris-HCl, pH 7.5, and preincubation at 15° continues for a further 5 min.
2. The digestion is initiated by the addition of DNase I (Boehringer Mannheim) to a final concentration of 0.4 U/ml (6 μl of a 1:1000 dilution). The reaction is allowed to proceed for 3.5 min at 15°.
3. The DNA degradation is interrupted by incubating the reaction mixture for 10 min at 85°.
4. The DNA fragments obtained are size fractionated by agarose gel (2%) electrophoresis. Fragments of approximately 50 to 120 bp in length are purified by inserting a strip of DEAE membrane (Schleicher & Schuell, Keene, NH) into the gel at the position of 50 bp and running the DNA onto the strip by further electrophoresis until DNA fragments of 120 bp have reached the DEAE membrane.
5. The membrane is removed and the DNA fragments are eluted by incubating the membrane in 20 m*M* Tris-HCl, pH 8.0, containing 1 *M* NaCl and 0.1 m*M* EDTA for 30 min at 65°.
6. The DNA fragments are precipitated with ethanol and redissolved in water.

Reassembly of Fragmented DNA

1. The purified DNA fragments of GST M1-1 and GST M2-2 are pooled and subjected to PCR in the presence of 1.5 m*M* $MgCl_2$, 0.2 m*M* dNTPs, and *Taq* DNA polymerase (10 U/ml), thus without any added primers. PCR (1 min at 94°, 2 min at 50°, 2 min at 72°) is repeated 40 times with a final elongation reaction for 10 min at 72°.
2. Agarose gel (2%) electrophoresis of an aliquot of the product shows a smear of DNA fragments with a higher intensity at the size of the original GST M1-1 DNA subjected to digestion.

Amplification of DNA Shuffled Product

1. The product from the reassembly PCR (diluted 40 times) is used as template in a new PCR to amplify DNA of the correct size. The reaction

conditions and reaction constituents are identical to those described above for the cDNA amplification, except that 1 μM Lac Z propr (5′-AAT TGT GAG CGG ATA ACA AT) is used as forward primer, the annealing temperature is 52°, and 20 cycles of PCR are carried out.

2. The PCR product is subjected to agarose gel (1%) electrophoresis and amplified DNA of the expected size is excised from the gel and purified with GeneClean II.

Subcloning

1. The purified DNA is digested with the restriction enzymes *Eco*RI and *Hin*dIII. Chloroform extraction of the digestion mixture is undertaken and the DNA is precipitated with ethanol and eluted in water.

2. The digested DNA is subjected to agarose gel (1%) electrophoresis and purified with GeneClean II.

3. The randomly recombined GST M1-1/M2-2 DNA is subcloned into the vector fragment of *Eco*RI/*Hin*dIII-digested pGΔETacM1b with T4 DNA ligase (Boehringer Mannheim), with a molar ratio of fragment to vector of 10 : 1. The site of insertion is under the control of the *tac* promoter.

4. Transformation of electrocompetent *E. coli* XL-1 Blue cells (Stratagene, La Jolla, CA) is carried out by electroporation with the resulting plasmid construct. Recovery of transformants in 2 ml of 2TY medium [1.6% (w/v) tryptone, 1% (w/v) yeast extract, and 0.5% (w/v) NaCl] at 37° for 1 hr is followed by titration of the library by plating aliquots onto LB–ampicillin plates [1% (w/v) tryptone, 0.5% (w/v) yeast extract, 1% (w/v) NaCl, 1.5% (w/v) Bacto-agar, and ampicillin at 25 μg/ml] in order to estimate the number of the individual DNA clones in the constructed library of shuffled GSTs.

5. A stock of vectors containing DNA-shuffled GSTs is prepared with Wizard Midipreps (Promega) from bacteria grown overnight.

Preparation of Crude Cell Lysate for Activity Screening

1. Randomly picked colonies of transformed *E. coli* XL-1 Blue cells grown on LB–ampicillin plates are inoculated in 2 ml of 2TY medium supplemented with ampicillin (50 μg/ml) and grown overnight at 37°.

2. Samples of 0.1 ml of culture are diluted 200-fold into 20 ml of 2TY (ampicillin at 50 μg/ml), incubated until log phase is reached, and GST expression is induced with 0.3 mM isopropyl-β-D-thiogalactopyranoside (IPTG). Protein expression is continued overnight.

3. Cells are harvested by centrifugation at 10,000g for 5 min at 4°, resuspended in 1.0 ml of 0.1 M sodium phosphate, pH 6.5, and lysed by ultrasonication. Cell debris is removed by centrifugation at 14,000g for

20 min at 4° and 0.02% (w/v) sodium azide is added to prevent microbial growth.

4. As controls, crude cell lysates containing the wild-type GST M1-1 and GST M2-2 are prepared from cultures of cells carrying pGΔETacMlb and pGΔETacM2.[6]

Glutathione Transferase Activity Assays

GST-catalyzed reactions are monitored spectrophotometrically at wavelengths in the UV region. The assays of enzyme activity with the substrates CDNB,[7] aminochrome,[5] and cyano-DMNG[8] are described elsewhere. The cyano-DMNG assay is modified to use 0.5 m*M* substrate instead of 1 m*M* in order to avoid a high initial absorbance, because crude cell lysate absorbs at the wavelength used. For the activity assays, the lysates are diluted at least 20 times in the cuvette. The nonenzymatic reaction rate between GSH and the nucleophilic substrate is subtracted from the rate determined in the presence of cell lysate.

Comments on the Method

Template DNA for Fragmentation

PCR was used to prepare template DNA in the presented study of GST isoenzymes. The frequency of spurious point mutations should be kept low in all steps of the library construction, and it is therefore recommended to prepare DNA template by plasmid digestion rather than by PCR. Alternatively, a DNA polymerase with proofreading activity should be used (e.g., *Pwo* and *Pfu* polymerases) instead of the *Taq* polymerase.[9]

The template DNA used exceeded the size of the coding region for the enzymes, thus containing vector DNA in both the 5′ end (approximately 300 bp) and in the 3′ end (85 bp). This is advantageous for two reasons. First, vector primers can be used to amplify the reassembled DNA. If instead internal primers are used, these need to anneal equally well to both variants of template cDNA. Second, in the subcloning step it is advantageous to be able to verify that the restriction enzyme digestion has been efficient, i.e., after digestion to identify a significant change in agarose gel mobility of the digested DNA.

[6] A.-S. Johansson, R. Bolton-Grob, and B. Mannervik, *Protein Expression Purif.* **17,** 105 (1999).

[7] W. Habig, M. J. Pabst, and W. B. Jakoby, *J. Biol. Chem.* **249,** 7130 (1974).

[8] D. E. Jensen and G. J. Stelman, *Carcinoginesis* **8,** 1791 (1987).

[9] H. Zhao and F. H. Arnold, *Nucleic Acids Res.* **6,** 1307 (1997).

In the present study conservation of the 5′ end from the GST M1-1 DNA was achieved by specifically removing the 5′ end of the GST M2-2 template DNA. This operation was performed to improve the expression level of the final chimeras. The 5′ end of the GST M1-1 cDNA has previously been optimized for high-level expression in *E. coli* by silent mutations,[4] while the GST M2-2 was not yet optimized and only a low yield could be obtained in *E. coli*. By removing the 5′ end of GST M2-2 all constructed chimeras should contain the first 11 codons of GST M1-1 and thereby have a high probability of providing good yields when expressed.

Fragmentation

In the fragmentation of DNA by DNase I $MnCl_2$ was used in the buffer rather than $MgCl_2$, which is most commonly used in DNA shuffling experiments. The manganese ion has been shown to improve the fidelity of the DNA shuffling,[9,10] thus reducing the frequency of spurious mutations in the final product. The choice of Mn^{2+} also slows down the further hydrolysis of small DNA fragments.[10] This prevents excessive degradation of DNA in the fragmentation step.

Screening of Chimeric Clones

Bacteria were transformed with chimeric clones and 1-ml samples of crude cell lysates were prepared from 20-ml cell cultures. The volume of 20 ml appeared to be suitable for clear detection of GST activity in lysates also from clones with poor activity, due to either a low specific activity or a low expression level. The activities of the wild-type enzymes were measured in several cultures by the same procedure as the chimeras, and confidence limits of the values were calculated. These limits were used as references to identify chimeras that deviate from the parental enzymes (see Fig. 3).

The amount of GST protein in bacterial lysates for different chimeras could not be determined accurately, and therefore the specific activities were not obtained for unpurified enzymes. In principle, an immunoassay could have been used in crude samples, provided that an antibody recognizing all chimeric variants of GST M1-1 and GST M2-2 was available. Instead the ratio of activities (substrate preference) was applied as a parameter independent of protein concentration. The wild-type enzymes differ only 2-fold in the specific activity with CDNB, in contrast to the 100-fold difference with aminochrome, and their substrate preferences of aminochrome over CDNB therefore differ by approximately 50-fold.

[10] I. A. J. Lorimer and I. Pastan, *Nucleic Acids Res.* **23,** 3067 (1995).

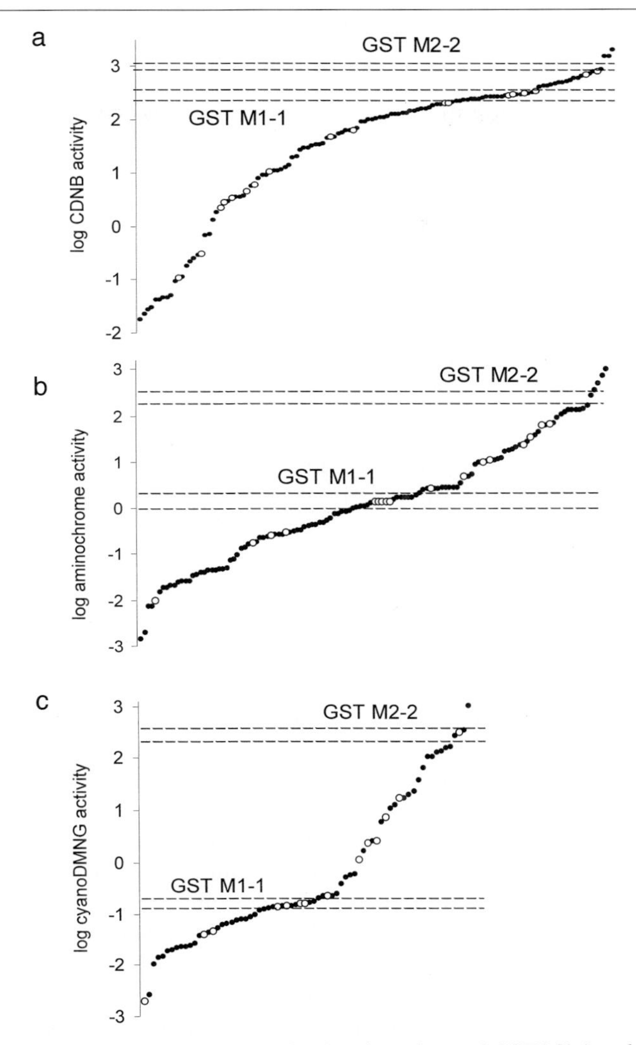

FIG. 3. Distribution of GST activities of chimeric variants of GST M1-1 and M2-2 determined in bacterial cell lysates. The data are ordered according to increasing values and the dotted lines indicate the confidence limits (95%) calculated from measurements on several wild-type enzyme lysates obtained under identical conditions. The wild-type GST M2-2 used as reference has separately been optimized for high-level expression in *E. coli* [A.-S. Johansson, R. Bolton-Grob, and B. Mannervik, *Protein Expression Purif.* **17,** 105 (1999)]. Open circles indicate the values of the clones that were DNA sequenced as presented in Fig. 5b. (a) Activity with CDNB. (b) Activity with aminochrome. (c) Activity with cyano-DMNG.

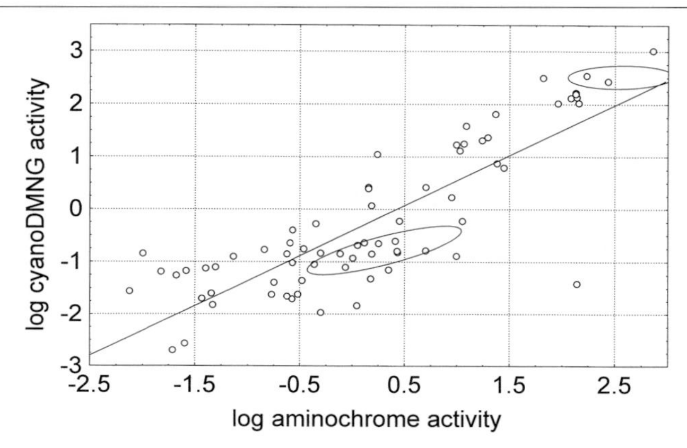

FIG. 4. Correlation of activities determined with aminochrome and cyano-DMNG for 72 GST variants. The straight line was obtained by linear regression analysis of the data set. Confidence contours (95%) for the wild-type enzymes GST M1-1 (*bottom*) and GST M2-2 (*top*) were calculated on the basis of the assumption of a bivariate normal distribution.

Results and Discussion

Characterization of the Library of Chimeric Glutathione Transferases

The library of chimeric GST M1-1/M2-2 DNA sequences in transformed *E. coli* cells was estimated to contain 5×10^5 independent clones. A number of clones were first randomly chosen and screened by activity assays. At a second stage DNA sequences were determined. The majority of the sampled clones (89% of 140 clones) were found to be catalytically active, based on detectable activity with aminochrome. The 125 active clones are represented in Figs. 3, 4, and 5.

The first 32 base pairs of GST M2-2 were removed prior to the DNA shuffling in order to obtain a high expression level for the chimeras. The 5′ end of GST M1-1 has previously been optimized by silent mutations for expression in *E. coli*,[4] and this optimized sequence was accordingly introduced into all clones. It was found that a large fraction of the sampled library clones exhibited as high activity with CDNB as was obtained with wild-type GST M1-1 (Fig. 3a). The two wild-type enzymes show similar specific activity with CDNB (GST M1-1, 170 μmol min^{-1} mg^{-1} [11] and GST M2-2, 460 μmol min^{-1} mg^{-1} [6]) but differ by more than 100-fold with the alternative substrates aminochrome and cyano-DMNG (Fig. 3). The distri-

[11] M. Widersten, W. R. Pearson, Å. Engström, and B. Mannervik, *Biochem. J.* **276,** 519 (1991).

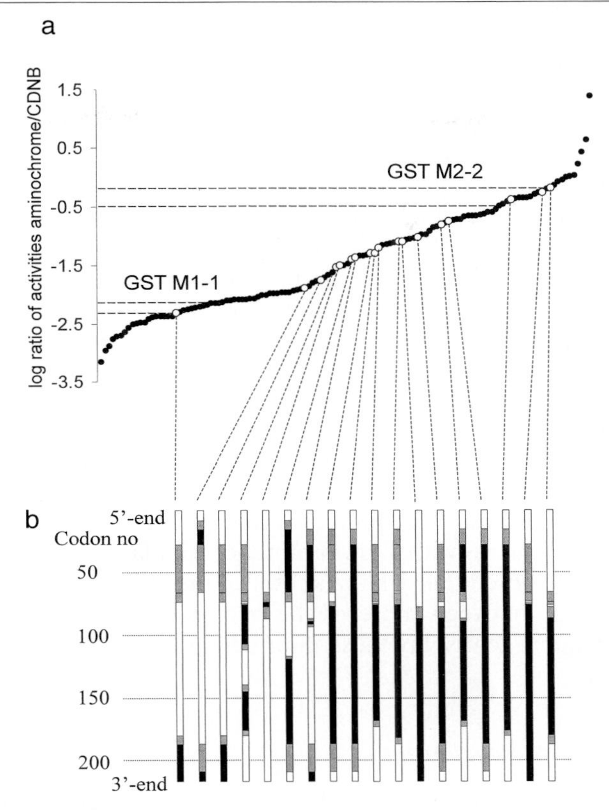

FIG. 5. Combined functional and structural analysis of variant GSTs constructed by DNA shuffling. (a) Distribution of substrate preferences (ratio of activities with aminochrome and CDNB) of 125 clones analyzed in cell lysates. Dotted horizontal lines indicate the confidence limits (95%) calculated from measurements on wild-type enzyme lysates expressed under identical conditions. (b) DNA sequences determined for 18 variant GSTs with varying preferences for aminochrome over CDNB. The individual clones are identified in (a) by dotted lines. In addition to substitutions due to the chimerization the sequenced clones contain on average one spurious mutation (specified in Ref. 3). White segment, GST M1-1 sequence; black segment, GST M2-2 sequence; gray segment, sequence in which recombination has occurred.

butions of the latter activities determined for the chimeras are presented in Fig. 3b and c. In both cases the sampled variant enzymes show a wide distribution of activity values ranging above and below the activities of the parental enzymes. Because the wild-type enzymes differ in activity with more than one substrate, the variant enzymes can be analyzed in several dimensions. Figure 4 presents a two-dimensional activity plot based on the substrates aminochrome and cyano-DMNG. The values for the parental enzymes are indicated with two-dimensional 95% confidence contours and the sampled chimeras appear to be distributed mainly between the values of the parental GSTs, rather than being distributed evenly over the plot. No clone is represented in the upper left or the lower right quadrants of the plot (Fig. 4), except for a single clone. The activities with aminochrome and cyano-DMNG seem to be correlated and the same segments of primary structure can thus be assumed to be essential to the two types of activities within the structural resolution of the chimera population.

The substrate preference for aminochrome over CDNB (the ratio of the activities) was used as a parameter independent of protein concentration. Figure 5a presents the ratio values obtained for the 125 chimeras analyzed and shows that a large subset of the library has a parameter value similar to that of GST M1-1. It is reasonable to assume that the probability is high for the offspring to have reduced rather than raised catalytic activity when segments of an efficient and a poor enzyme are randomly recombined. Thus, it is not surprising that the number of chimeras exhibiting a high preference for aminochrome, similar to that of GST M2-2, is smaller than the number of chimeras with a low preference, similar to that of GST M1-1. It should also be noted that a large subset of the chimeras exhibits activity ratios in a range between those of the parental enzymes. The structural basis for a chimera to display these intermediate activities is then an obvious question.

Structure–Activity Relationships

The DNA sequence was determined for 18 of the DNA shuffled clones (Fig. 5b). These individuals were not picked randomly but rather were selected so that the complete range of preferences for aminochrome over CDNB would be represented. In this manner it may be possible to correlate changes in substrate preference to alterations in primary structure of the individual chimeras. All sequenced clones showed chimeric structures as seen in Fig. 5b. In summary, 15 unique recombination sites were identified at the DNA level. Because of the degeneracy of the genetic code, some recombinations are silent, but 10 unique cross-over points in the primary structures of the proteins were obtained.

Figure 5 combines both functional and structural information about the chimeric enzyme variants. For clones exhibiting a high preference for aminochrome over CDNB the central part of the DNA (codons 89–168) is consistently derived from the cDNA of GST M2-2. In contrast, the DNA segments at the 5′ end (codons 1–88) and the segments at the 3′ end (codons 169–217) are not always obtained from GST M2-2. From this structure–function analysis it can be concluded that the residues specific to GST M2-2 in the segment 89–168 are essential for the catalytic feature investigated. These residues correspond to α helices 4, 5, and 6 in the three-dimensional structure. From the crystal structure of GST M2-2 in complex with glutathione,[12] it was found that residues situated in α helices 4 and 6 are indeed contributing to the lining of the active site cavity, probably interacting with the substrates studied. A structure of the enzyme in complex with any of the electrophilic substrates studied is not available.

Functional Studies of Chimeric Enzymes

An increasing number of enzymes subjected to chimeric redesign are being described in the literature. In particular, isoenzymes forming a common structural class as well as homologous enzymes from different species are suitable for modular exchange (see Ref. 13). In the field of GSTs a number of studies have applied the approach of chimera construction for elucidation of the basis of the diverse substrate specificities of isoenzymes. Hybrids of human GST A1-1 and the homologous rat GST,[14,15] as well as rat isoenzymes 1-1 and 8-8 (now known as GSTs A1-1 and A4-4, both belonging to the structural Alpha class), have been studied.[16] In the latter case an identification was made of a residue that was essential for the high activity with 4-hydroxyalkenals. In another study the difference in substrate stereoselectivities of the rat isoenzymes GST M1-1 and GST M2-2 was probed by a number of chimeric variants.[17] In this study the fusion points were selected to correspond to exon borders, because the sequence variability is most pronounced in certain exons.

The procedure described in this chapter deals with randomly assembled chimeras of homologous enzymes, a stochastic process for which no information other than the primary structure is required. The degree of sequence identity of their cDNA is the principal limitation. There are few examples of randomly generated chimeras, because most hybrid enzymes investigated so far have been rationally designed, based on structural knowledge or

[12] S. Raghunathan, R. J. Chandross, R. H. Kretsinger, T. J. Allison, C. J. Penington, and G. S. Rule, *J. Mol. Biol.* **238,** 815 (1994).

[13] R. N. Armstrong, *Chem. Rev.* **90,** 1309 (1990).

[14] B. Mannervik, P. G. Board, K. Berhane, R. Björnestedt, V. M. Castro, U. H. Danielson, X.-Y. Hao, R. Kolm, B. Olin, G. B. Principato, M. Ridderström, G. Stenberg, and M. Widersten, *in* "Glutathione *S*-Transferases and Drug Resistance" (J. D. Hayes, C. B. Pickett, and T. J. Mantle, eds.), p. 35. Taylor & Francis, London, 1990.

[15] R. Björnestedt, M. Widersten, P. G. Board, and B. Mannervik, *Biochem. J.* **282,** 505 (1992).

[16] R. Björnestedt, S. Tardioli, and B. Mannervik, *J. Biol. Chem.* **270,** 29705 (1995).

[17] P. Zhang, S. Liu, S. Shan, X. Ji, G. L. Gilliland, and R. N. Armstrong, *Biochemistry* **31,** 10185 (1992).

assumptions about functionally important residues, van Kampen *et al.*[18] have provided an example of random chimerization *in vivo* of two homologous lipases. The recombination system of *E. coli* that was used restricted the chimera formation to one recombination event per chimera. In the present GST study the analyzed clones exhibited from one to six recombination sites per clone (see Fig. 5b). The *in vitro* recombination technique has the advantage of providing a higher diversity among the chimeras. Furthermore, the random chimerization approach generates variant enzymes that probably would not have been constructed by rational design. From a set of chimeric structures, such as that represented in Fig. 5b, it may be possible to isolate individual clones with interesting structures and activities for further investigation. A number of the clones described in Fig. 5b were indeed analyzed in greater detail by Hansson *et al.*[19]

Evolution of Enzyme Function

Recombination of DNA segments is an important principle in the natural evolution of proteins, and exon shuffling[20] has been proposed as a mechanism for recombining genetic material *in vivo*. Regarding the multiple forms of naturally occurring GSTs, it has been suggested that the diversity of isoenzymes in the enzyme family has arisen from gene fusion and DNA recombinations.[14,21] In particular the Mu-class GSTs have been studied from an evolutionary point of view. Taylor *et al.*[22] have proposed that the high sequence identity of these genes is evidence of a recent gene conversion event within the Mu-class gene cluster.

DNA shuffling *in vitro* is a way of mimicking natural molecular evolution in biological systems. Originally, this approach was applied to a single DNA template, using the combinatorial power to accumulate favorable spurious mutations,[1,2,23–25] but lately shuffling of DNA sequences encoding distinct but homologous enzymes has achieved attention. This type of hybridization of enzymes has been described as "family shuffling" or "breeding of en-

[18] M. D. van Kampen, N. Dekker, M. R. Egmond, and H. M. Verhij, *Biochemistry* **37,** 3459 (1998).
[19] L. O. Hansson, R. Bolton-Grob, M. Widersten, and B. Mannervik, *Protein Sci.* **8,** 2742 (1999).
[20] W. Gilbert, *Nature* (*London*) **271,** 501 (1978).
[21] B. Mannervik, *Adv. Enzymol. Related Areas Mol. Biol.* **57,** 357 (1985).
[22] J. B. Taylor, J. Oliver, R. Sherrington, and S. E. Pemble, *Biochem. J.* **274,** 587 (1991).
[23] D. R. Liu, T. J. Magliery, M. Pasternak, and P. G. Schultz, *Proc. Natl. Acad. Sci. U.S.A.* **94,** 10092 (1997).
[24] T. Yano, S. Oue, and H. Kagamiyama, *Proc. Natl. Acad. Sci. U.S.A.* **95,** 5511 (1998).
[25] F. Buchholz, P.-O. Angrand, and A. F. Stewart, *Nature Biotechnol.* **16,** 657 (1998).

zymes."[26] A mixture of such variant enzymes has proved to be an excellent basis for functional selection of novel enzyme functions.[26,27]

Eigen[28] has proposed that molecular evolution *in vivo* operates on a repertoire of mutants rather than on individual molecular species. The population of GST chimeras obtained by the presented method exhibits a broad spectrum of activities, as displayed in Fig. 3. The repertoire of variant GSTs, which is described by both random recombination and spurious point mutations, is thus well suited for further molecular evolution *in vitro.* It is noteworthy that the activities of the chimeric structures in the first-generation mutant library span four to five orders of magnitude (Figs. 3–5). In particular, clones with properties outside the range of the parental enzymes may be of interest for discovering enzyme variants with novel activities. Such molecular breeding of GSTs from libraries of random structures have great potential in view of the catalytic versatility of the proteins.

In contrast to other family shuffling experiments, the study of GSTs described in this chapter deals with functional screening rather than selection. The combination of functional and structural studies has proved useful in structure–function analysis of two functionally distinct GST isoenzymes.[3,19] The presented procedure of random chimerization is thus applicable to the analysis of other pairs of homologous enzymes.

Acknowledgments

This work was supported by grants to B.M. from the Swedish Research Council for Engineering Sciences and the Swedish Natural Science Research Council. We thank Dr. Robyn Bolton-Grob (Department of Physiology and Pharmacology, University of Queensland) and Tahereh Massoud for essential contributions to the original work. Dr. Mikael Widersten and Per Jemth of our department have given valuable advice, and Dr. David E. Jensen (Thomas Jefferson University, Philadelphia) generously provided cyano-DMNG.

[26] A. Crameri, S.-A. Raillard, E. Bermudez, and W. P. C. Stemmer, *Nature (London)* **391,** 288 (1998).

[27] T. Kumamaru, H. Suenaga, M. Mitsuoka, T. Watanabe, and K. Furukawa, *Nature Biotechnol.* **16,** 663 (1998).

[28] M. Eigen, "Steps towards Life: A Perspective on Evolution." Oxford University Press, Oxford, 1992.

[29] Protein Engineering by Expressed Protein Ligation

By ULRICH K. BLASCHKE, JONATHAN SILBERSTEIN, and TOM W. MUIR

Introduction

The power of organic chemistry for studying structure–activity relationships in small bioactive peptides is now well established, and indeed represents a pillar of medicinal chemistry.[1] Consequently, it is reasonable to expect that the application of these same chemical engineering strategies to proteins will have no less an impact in understanding the molecular details of how they work. Going beyond the 20 amino acids dictated by the genetic code permits all manner of modifications to be made in the backbone and amino acid side chains of a protein, allowing, for example, the precise introduction into proteins of fluorophores, isotopic labels, photoactivable cross-linkers, electron paramagnetic resonance (EPR) spin labels, posttranslational modifications, and countless other unnatural amino acids. Given the possibilities in protein engineering, it is not surprising that an enormous amount of effort has gone into the development of approaches for the site-specific introduction of unnatural amino acids into proteins. Indeed, several chemical[2–4] and biosynthetic[5–7] strategies are now available for this purpose that, to varying degrees, allow most if not all of the preceding modifications to be incorporated into proteins. This chapter focuses on one such approach, expressed protein ligation (EPL), which allows a target protein to be assembled from a series of recombinant and synthetic polypeptide building blocks.[8–10]

EPL is an extension of the previously described native chemical ligation approach,[11] originally developed for the total chemical synthesis of small

[1] V. J. Hruby and F. Al-Obeidi, *J. Biochem.* **268,** 249 (1990).
[2] T. W. Muir and S. B. H. Kent, *Curr. Opin. Biotech.* **4,** 420 (1993).
[3] J. Wilken and S. B. H. Kent, *Curr. Opin. Biotech.* **9,** 412 (1998).
[4] G. J. Cotton and T. W. Muir, *Chem. Biol.* **6,** R247 (1999).
[5] V. W. Cornish, D. Mendel, and P. G. Schultz, *Angew. Chem. Int. Ed. Engl.* **34,** 621 (1995).
[6] C. J. A. Wallace, *Curr. Opin. Biotech.* **6,** 403 (1995).
[7] D. Y. Jackson, J. Burnier, C. Quan, M. Stanley, J. Tom, and J. A. Wells, *Science* **266,** 243 (1994).
[8] K. Severinov and T. W. Muir, *J. Biol. Chem.* **273,** 16205 (1998).
[9] T. W. Muir, D. Sondhi, and P. A. Cole, *Proc. Natl. Acad. Sci. U.S.A.,* **95,** 6705 (1998).
[10] T. C. Evans, Jr., J. Benner, and M.-Q. Xu, *Protein Sci.* **7,** 2256 (1998).
[11] P. E. Dawson, T. W. Muir, I. Clark-Lewis, and S. B. H. Kent, *Science* **266,** 776 (1994).

METHODS IN ENZYMOLOGY, VOL. 328
0076-6879/00 $30.00

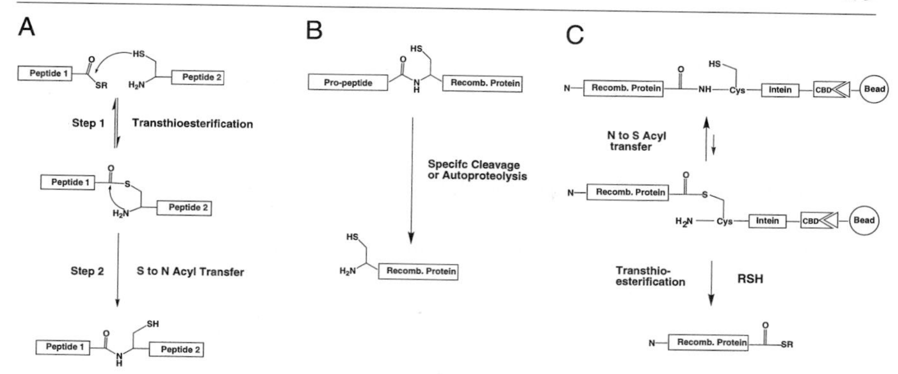

FIG. 1. Components of expressed protein ligation. (A) Mechanism of the native chemical ligation approach used to chemoselectively link unprotected polypeptides via a normal peptide bond.[11] (B) Fusion protein strategy used to generate recombinant proteins possessing an N-terminal cysteine residue. The propeptide sequence can contain the recognition motif for a specific protease,[18] or can itself be an autoprocessing protein domain such as an engineered intein.[24–26] (C) Intein fusion protein strategy used to generate recombinant protein α-thioesters. CBD, Chitin-binding domain affinity handle.[30,33]

proteins.[12] Native chemical ligation is arguably the most powerful of the so-called peptide ligation approaches[4] to protein synthesis and allows fully unprotected synthetic peptide building blocks to be regioselectively linked together through a normal peptide bond. As illustrated in Fig. 1A, the approach is based on the chemoselective reaction of a polypeptide containing a C-terminal thioester with a second polypeptide containing an N-terminal cysteine residue.[11] The first step in this process is the transthioesterification of the thioester with the cysteine thiol group, a chemoselective process at pH ~7. The resulting thioester-linked intermediate undergoes a rapid and irreversible S → N-acyl shift to form a native peptide bond at the ligation site. It is noteworthy that additional cysteine residues are permitted in one or both peptide segments because of the reversible nature of the initial transthioesterification step.[11,13,14] Native chemical ligation has been widely used in the synthesis of small proteins and protein domains and is emerging as the method of choice for the rapid generation of molecules of this type.[3] Moreover, the development of sequential ligation[14,15] and solid-

[12] T. W. Muir, *Structure* **3,** 649 (1995).

[13] T. M. Hackeng, C. M. Mounier, C. Bon, P. E. Dawson, J. H. Griffin, and S. B. H. Kent, *Proc. Natl. Acad. Sci. U.S.A.* **94,** 7845 (1997).

[14] T. M. Hackeng, J. H. Griffin, and P. E. Dawson, *Proc. Natl. Acad. Sci. U.S.A.* **96,** 10068 (1999).

[15] T. W. Muir, P. E. Dawson, and S. B. H. Kent, *Methods Enzymol.* **289,** 266 (1997).

phase sequential ligation[16,17] approaches will extend the scope of native chemical ligation to the assembly of somewhat larger protein targets.

Expressed Protein Ligation

Expressed protein ligation is a protein semisynthesis technique that allows recombinant proteins and synthetic peptides to be joined together under native chemical ligation conditions.[8–10] The approach combines the structural flexibility associated with synthetic peptides with the extended size range of recombinant polypeptides, thereby allowing "chemical mutagenesis" to be performed on extremely large protein systems. EPL is made possible as a result of advances in protein engineering that allow the generation of recombinant proteins possessing N-terminal cysteine residues and/or C-terminal thioester groups.

Methods for generating recombinant proteins possessing amino-terminal cysteine residues all involve specific removal of an N-terminal leader sequence from a precursor fusion protein (Fig. 1B). In the proteolytic approach developed by Verdine and co-workers, a factor Xa recognition sequence is appended immediately in front of the cryptic N-terminal cysteine in the protein of interest.[18] Treatment of this recombinant fusion protein with the protease gives the requisite N-terminal cysteine protein directly, which can then be used in subsequent ligation reactions.[18–22] Other protease that cleave on the C-terminal side of their recognition site, such as enterokinase or ubiquitin C-terminal hydrolase,[23] should also be compatible with such an approach. An alternative strategy has been described that does not require a separate proteolysis step but instead makes use of an autoprocessing fusion protein system.[24–26] This approach was developed from studies of protein splicing (see below) and utilizes an engineered

[16] J. A. Camarero, G. J. Cotton, A. Adeva, and T. W. Muir, *J. Peptide Res.* **51,** 303 (1998).

[17] L. E. Canne, P. Botti, R. J. Simon, Y. Chen, E. A. Dennis, and S. B. H. Kent, *J. Am. Chem. Soc.* **121,** 8720 (1999).

[18] D. A. Erlanson, M. Chytil, and G. L. Verdine, *Chem. Biol.* **3,** 981 (1996).

[19] M. Chytil, B. R. Peterson, D. A. Erlanson, and G. L. Verdine, *Proc. Natl. Acad. Sci. U.S.A.* **95,** 14076 (1998).

[20] R. Xu, B. Ayers, D. Cowburn, and T. W. Muir, *Proc. Natl. Acad. Sci. U.S.A.* **96,** 388 (1999).

[21] G. Cotton, B. Ayers, R. Xu, and T. W. *Muir, J. Am. Chem. Soc.* **121,** 1100 (1999).

[22] J. A. Camarero and T. W. Muir, *J. Am. Chem. Soc.* **121,** 5597 (1999).

[23] R. T. Baker, *Curr. Opin. Biotechnol.* **7,** 541 (1996).

[24] T. C. Evans, J. Benner, and M. Q. Xu, *J. Biol. Chem.* **274,** 3923 (1999).

[25] S. Mathys, T. C. Evans, Jr., C. I. Chute, H. Wu, S. Chong, J. Benner, X.-Q. Liu, and M.-Q. Xu, *Gene* **231,** 1 (1999).

[26] M. W. Southworth, K. Amaya, T. C. Evans, M.-Q. Xu, and F. B. Perler, *BioTechniques* **27,** 110 (1999).

intein splicing domain, which spontaneoulsy cleaves itself off the precursor fusion protein, again to give the requisite N-terminal cysteine protein.

Reactive recombinant protein α-thioesters can be prepared by exploiting protein splicing, a naturally occurring process that is known to involve thioester intermediates.[27,28] Protein splicing, the protein equivalent of RNA splicing, is a posttranslational process in which a proprotein undergoes a series of intramolecular rearrangements resulting in the precise excision of an internal domain (termed an intein) and the ligation of the flanking regions (termed the *N*- and *C*-exteins). This remarkable process requires no additional proteins or other cofactors, and indeed can take place *in vitro* with purified proproteins, a feature that has allowed many aspects of the mechanism to be elucidated.[29,30] Accordingly, the first step in protein splicing involves an N $\rightarrow$ S (or N $\rightarrow$ O)-acyl shift in which the *N*-extein polypeptide segment is transferred to the side-chain group of a conserved Cys/Ser/Thr residue, always located at the immediate N terminus of the intein (the structure of the intein domain most likely catalyzes this unfavorable rearrangement by twisting the scissile amide bond into a higher energy conformation).[31,32] There then follows a transesterification step in which *N*-extein acyl unit is transferred to a second conserved Cys/Ser/Thr side chain, this time located at the intein–*C*-extein junction (+1 position). In the final step, this "branched intermediate" is resolved through a cyclization reaction involving a conserved asparagine residue at the C terminus of the intein. This results in cleavage of the amide bond between the intein and the *C*-extein (the intein element is excised as a C-terminal succinimide derivative) and the formation of a new amide bond between the two exteins following a spontaneous S(O) $\rightarrow$ N shift (identical to native chemical ligation).

These mechanistic insights have led to the generation of mutant inteins containing an Asn $\rightarrow$ Ala substitution and still capable of promoting the initial N $\rightarrow$ S-acyl shift, but unable to proceed any further in the splicing process.[24–26,30,33] Because there are no sequence requirements in the *N*-

[27] F. B. Perler, *Cell* **92,** 1 (1998).
[28] H. Paulus, *Chem. Soc. Rev.* **27,** 375 (1998).
[29] M.-Q. Xu and F. B. Perler, *EMBO J.* **15,** 5146 (1996).
[30] S. Chong, Y. Shao, H. Paulus, J. Benner, F. B. Perler, and M.-Q. Xu, *J. Biol. Chem.* **271,** 22159 (1996).
[31] T. Klabunde, S. Sharma, A. Telenti, W. R. Jacobs, Jr., and J. C. Sacchettini, *Nature Struct. Biol.* **5,** 31 (1998).
[32] Q. Xu, D. Buckley, C. Guan, and H.-C. Guo, *Cell* **98,** 651 (1999).
[33] S. Chong, F. B. Mersha, D. G. Comb, M. E. Scott, D. Landry, L. M. Vence, F. B. Perler, J. Benner, R. B. Kucera, C. A. Hirvonen, J. J. Pelletier, H. Paulus, and M.-Q. Xu, *Gene* **192,** 271 (1997).

extein, any polypeptide expressed as an in-frame fusion to one of these mutant inteins can be cleaved by thiols (or other nucleophiles)[33] via intermolecular transthioesterification to give the corresponding recombinant polypeptide α-thioester derivative (Fig. 1C). Several protein expression vectors are now commercially available that allow a gene or gene fragment, P, to be expressed as a fusion protein of the type P–intein–CBD, where CBD is a chitin-binding domain affinity handle.

With the availability of recombinant polypeptide building blocks for use in native chemical ligation, it is now possible to chemoselectively attach a synthetic peptide to either the N terminus,[18,19,21] or C-terminus blocks[8–10] of a recombinant protein. Moreover, we have developed an extension of expressed protein ligation that allows a synthetic peptide to be inserted into the interior of a recombinant protein framework, or vice versa.[21] Thus, expressed protein ligation offers a powerful route to the chemical modification of proteins and, as summarized in Table I,[8–10,18–22,24,33–39] the technique has already been applied to a wide range of different protein systems allowing a variety of noncoded groups to be site-specifically introduced.[4]

Methods

General Materials and Methods

Analytical gradient high-performance liquid chromatography (HPLC) is performed on a Hewlett-Packard (Palo Alto, CA) 1100 series instrument with 214- and 280-nm detection, using a Vydac (Hesperia, CA) C_{18} column (5 μm, 4.6 $\times$ 150 mm) at a flow rate of 1 ml/min. Preparative HPLC is performed on a Waters (Milford, MA) DeltaPrep 4000 system fitted with a Waters 486 tunable absorbance detector, using a Vydac C_{18} column (15–20 μm, 50 $\times$ 250 mm) at a flow rate of 30 ml/min. All runs use linear gradients of 0.1% (w/v) aqueous trifluoroacetic acid (TFA; solvent A) versus 90% (v/v) acetonitrile plus 0.1% (v/v) TFA (solvent B). Electrospray mass spectrometric (ESMS) analysis is routinely applied to all synthetic peptides and components of reaction mixtures. ESMS is performed on a SCIEX API-100 single quadrupole electrospray mass spectrometer (PE Biosystems, Foster City, CA). Calculated masses are obtained with

[34] G. J. Cotton and T. W. Muir, *Chem. Biol.* **7,** 253 (2000).

[35] J. A. Porter, S. C. Ekker, W.-J. Park, D. P. von Kessler, K. E. Young, C.-H. Chen, Y. Ma, A. S. Woods, R. J. Cotter, E. V. Koonin, and P. A. Beachy, *Cell* **86,** 21 (1996).

[36] T. W. Muir, unpublished (1999).

[37] J. Lykke-Andersen and J. Christiansen, *Nucleic Acids Res.* **26,** 5631 (1998).

[38] T. C. Evans, Jr., J. Brenner, and M.-Q. Xu, *J. Biol. Chem.* **274,** 18359 (1999).

[39] C. Kinsland, S. V. Taylor, N. L. Kelleher, F. W. McLafferty, and T. P. Begley, *Protein Sci.* **7,** 1839 (1998).

TABLE I
PROTEINS STUDIED BY EXPRESSED PROTEIN LIGATION

System	Size[a]	Region modified	Comments	Refs.
AP-1	70	N terminus	Fe-EDTA probe introduced, allowing protein–DNA interactions to be studied	18, 19
c-Abl-SH(32)	160	N and C terminus	Two recombinant domains ligated together, permitting segmental isotopic labeling of product	20
c-Abl-SH(32)	160	Middle	Sequence ligation approach used to introduce dansyl fluorescence probe in middle of protein, allowing conformational dynamics to be studied	21
c-Crk	304	N and C terminus	Dual labeling with tetramethylrhodamine and fluorescein probes, permitting protein–protein interactions to be studied	34
c-Crk-N-SH3	56	N and C terminus	Structure–function studies using head-to-tail backbone-cyclized version of protein	22
Csk	450	C terminus	Phosphotyrosine and fluorescein probes introduced to induce and sense, respectively, conformational changes	9
Hedgehog (Hh-Np)	~170	C terminus	Semisynthesis used to elucidate the chemical mechanism of hedgehog autoprocessing	35
*Hpa*I	254	C terminus	Semisynthesis facilitated generation of potentially cytotoxic protein	10
MBP	395	N and C terminus	Variety of C-terminal modifications have been introduced into this model system, including thioesters and hydroxamic acids	33
RNAP β'	1407	C terminus	Fluorescein probe introduced into ~500-kDa *E. coli* RNA polymerase holoenzyme	36
RNase A	124	C terminus	Semisynthesis facilitated generation of potentially cytotoxic protein	10
σ^{70}	613	C terminus	Synthetic DNA-binding domain introduced into protein	8
T4 DNA ligase	487	N terminus	EPL reactions performed using N-terminal Cys derivative prepared with novel intein-based N-terminal cleavage system	24
T7 RNAP	882	C terminus	Mg^{2+}-dependent catalysis probed with a series of semisynthetic amide and ester-containing analogs	37
Thioredoxin	135		Backbone cyclized and polymeric versions of protein prepared	38
ThiS	66	C terminus	α-Thioacid group introduced to study biosynthesis of thiamin	39

[a] Number of amino acids.

the program MacProMass (S. Vemuri and T. Lee, City of Hope, Duarte, CA). Expressed proteins are routinely analyzed by sodium dodecyl sulfate–polyacrylamide gel electrophoresis (SDS–PAGE), using the standard procedures.

Synthesis of H-Cys-Gly-Gly-Gly-Gly-Gly-Lys [N^{ε}-levulinic acid]-Arg-NH_2 and H-Cys-Gly-Gly-Gly-Gly-Gly-Lys [N^{ε}-D-biotin]-Arg-NH_2

The peptide H-Cys-Gly-Gly-Gly-Gly-Gly-Lys-Arg-NH_2 is synthesized (0.93-mmol scale) manually on a 4-methylbenzhydrylamine resin according to the *in situ* neutralization/*o*-benzotriazol-1-yl *N,N,N′,N*-tetramethyluronium hexafluorophosphate (HBTU) activation protocol for *t*-butyloxycarbonyl/benzyl (Boc/Bzl) solid-phase peptide synthesis (SPPS).[40] Orthogonal protection of the lysine ε-NH_2 group with the base-labile 9-fluorenylmethoxycarbonyl (Fmoc) group allows directed attachment of the levulinic acid and biotin groups (activated with HBTU) before the final cleavage step. After removal of the N^{α}-Boc protecting group from the peptide, global deprotection and cleavage from the support are achieved by treatment with HF containing 4%(v/v) *p*-cresol, for 1 hr at 4°. After evaporation of the HF, crude peptide products are precipitated and washed with anhydrous cold diethyl ether (Et_2O) before being dissolved in 50% (v/v) buffer A/B and lyophilized. Polypeptides are purified by preparative HPLC and in all cases, polypeptide composition and purity are confirmed by ESMS and analytical HPLC. H-Cys-Gly-Gly-Gly-Gly-Gly-Lys(N^{ε}-levulinic acid)-Arg-NH_2: 268 mg (yield 63%), observed mass, 788.9 ± 1.0 Da, calculated mass for $C_{30}H_{53}N_{13}O_{10}S$ (average isotope composition), 788.8 Da. H-Cys-Gly-Gly-Gly-Gly-Gly-Lys(N^{ε}-biotin)-Arg-NH_2: 232 mg (overall yield 63%), observed mass, 915.9 ± 1.0 Da, calculated mass for $C_{35}H_{62}N_{15}O_{10}S_2$ (average isotope composition), 915.9 Da.

Construction of Plasmids

The DNA encoding the SH2 domain (residues Met-1 to Gly-124) of murine c-Crk is amplified by polymerase chain reaction (PCR) from a full-length clone of m-Crk, using the oligonucleotide primers mCrk#1 (5′- ATG AAA GCG CGG AGG CTC ATA TGG CGG GCA ACT TCG ACT CG-3′) and mCrk#2 (5′-CTG CCT GAG AGC TCT TCC GCA ACC CTG CCT TGA TCT GGC-3′). The primers are designed to introduce an *Nde*I site (mCrk#1) on the 5′-end and a *Sap*I site (mCrk#2) on the 3′ end of the gene fragment. The PCR product is purified and digested with *Nde*I

[40] M. Schnölzer, P. Alewood, A. Jones, D. Alewood, and S. B. H. Kent, *Int. J. Peptide Protein Res.* **40,** 180 (1992).

and *Sap*I and then recloned into the *Nde*I- and *Sap*I-treated pTXB1 (GyrA intein) or pTYB1 (VMA intein) plasmids (both from New England BioLabs, Beverly, MA). The resulting T7-driven expression plasmids, $pTXB1_{Crk\text{-}SH2}$ and $pTYB1_{Crk\text{-}SH2}$, have been shown to be free of mutations by automated DNA sequencing.

Expression and Purification of mCrk SH2–Intein–CBD Fusion Proteins

Escherichia coli BL21 cells, transformed with plasmids $pTXB1_{Crk\text{-}SH2}$ or $pTYB1_{Crk\text{-}SH2}$, are grown at 37° in Luria Bertani (LB) medium containing ampicillin (100 μg/ml) to an OD_{590} of 0.3–0.4 ($pTXB1_{Crk\text{-}SH2}$) or 0.6–0.7 ($pTYB1_{Crk\text{-}SH2}$). For each of the two plasmids, induction with 0.4 m*M* isopropylthiogalactoside (IPTG) at 15° for 10 hr has been found to give optimal soluble expression of the desired intein fusion ($pTXB1_{Crk\text{-}SH2}$, ~1.0 mg/liter of cells; $pTYB1_{Crk\text{-}SH2}$, 0.6–0.8 mg/liter of cells). Cells are harvested by centrifugation (8500*g*, 4°, 10 min) and stored as a pellet at −80°. After thawing, cells are resuspended in 25 ml of lysis buffer [20 m*M* phosphate (pH 7.0), 500 m*M* NaCl, 1 m*M* EDTA, 5% (v/v) glycerol] and disrupted with a French press. After centrifugation (31,000*g*, 4°, 30 min), the supernatants are diluted with 75 ml of column buffer [20 m*M* HEPES (pH 7.0), 250 m*M* NaCl, 1 m*M* EDTA, 0.1% (v/v) Triton X-100] and then mixed with chitin beads (~2 ml/liter of cultured cells) preequilibrated in the same column buffer. These suspensions are incubated at 4° overnight and the beads carrying the SH2–intein–CDB fusion proteins are filtered off with a disposable column, washed extensively with column buffer, and stored at 4° until further use.

Cleavage of the Crk–SH2–Intein–CBD Fusion Proteins with Ethanethiol

An aliquot of chitin beads (0.05–4.0 ml), loaded with either the Crk–SH2–GyrA–CBD fusion or the Crk–SH2–VMA–CBD fusion, is transferred into a siliconized tube and suspended in an equal volume of cleavage buffer [0.2 *M* phosphate (pH 7.2), 0.2 *M* NaCl] containing 2% (v/v) ethanethiol. The thiolysis reaction is allowed to proceed at room temperature for 24 hr with gentle agitation. [In the kinetic analyses, the amount of the SH2 ethyl α-thioester derivative present at different time points is determined by analyzing aliquots of the reaction supernatant by analytical HPLC and ESMS; observed mass = 13,681.7 ± 3.3 Da, calculated mass for $C_{611}H_{949}N_{175}O_{186}S_2$. (*Note:* N-Terminal methionine residue is removed) = 13,679.1 Da (average isotope composition).] The reaction supernatant is then removed by centrifugation and the chitin beads are washed with cleavage buffer. The Crk–SH2 ethyl α-thioester is then purified from the combined supernatant and washes by semipreparative HPLC (30–65%

buffer B over 45 min). The yield of purified Crk–SH2 ethyl α-thioester is ~0.25 mg/liter using the GyrA intein and ~0.1 mg/liter using the VMA intein.

Expressed Protein Ligation Reactions

Chitin beads (0.05–1.0 ml) carrying the immobilized SH2–intein–CBD fusion protein are equilibrated with ligation buffer [0.2 *M* sodium phosphate (pH 7.2), 0.2 *M* NaCl, 0.1% (v/v) Triton X-100]. Typically, a 3- to 10-fold molar excess of the synthetic peptide is dissolved in ligation buffer containing 2% (w/v) 2-mercaptoethanesulfonic acid (MESNA) and mixed with the beads. The ligation reaction is allowed to proceed for 24 hr at room temperature with gentle mixing, at which point the beads are filtered off and washed with 50% (v/v) buffer A/B. (The progress of the ligation reactions is followed by periodically removing aliquots of the reaction supenatants and analyzing by analytical HPLC and ESMS.) In each case, the semisynthetic SH2 domain is purified from the combined supernatant and washes by semipreparative HPLC [30–65% (v/v) buffer B over 45 min] and characterized by ESMS. Crk–SH2–Cys-Gly-Gly-Gly-Gly-Gly-Lys[N^{ε}-biotin]-Arg-NH_2 (yield, >90%): mass found, 14,535.5 ± 4.0 Da; calculated mass for $C_{641}H_{995}N_{188}O_{195}S_2$ (average isotope composition), 14,534.0 Da. Crk–SH2–Cys-Gly-Gly-Gly-Gly-Gly-Lys[N^{ε}-levulinic acid]-Arg-NH_2 (yield, >90%): mass found, 14,406.9 ± 2.6 Da; calculated mass for $C_{636}H_{987}N_{186}O_{195}S$ (average isotope composition), 14,405.9 Da.

Results and Discussion

In the following sections, the expressed protein ligation approach is illustrated with the Src homology 2 (SH2) domain of the adapter protein c-Crk (residues Met-1 to Gly-124 of murine c-Crk) as an example, with particular emphasis placed on some of the experimental design and practical aspects of the approach. Note that for a rigorous practical guide to native chemical ligation using synthetic and/or recombinant peptides, we also recommend Ref. 41.

Choosing the Ligation Site

The first step in the generation of a semisynthetic protein by EPL is to decide on the location of the ligation site. This depends on several factors.

[41] J. A. Camarero and T. W. Muir, "Current Protocols in Protein Science," Chap. 18.4, pp. 1–21. John Wiley & Sons, New York, 1999.

1. The position(s) in the protein primary sequence at which the non-coded probe is to be introduced: If this region is within ~50 residues of the N or C terminus of the protein, then it should be possible to generate the target molecule from two fragments, one synthetic and one recombinant. *Note:* Polypeptides of up to ~50 residues can be chemically synthesized with reasonable confidence by optimized SPPS protocols.[40] If the region of interest is located further from the N or C terminus then it may be necessary to assemble the target molecule from three polypeptides: two flanking recombinant fragments and a synthetic peptide cassette. This scenario requires that a sequential expressed protein ligation approach be employed.[21]

2. Where possible, naturally occurring X-Cys motifs in the native sequence should be chosen as the ligation sites, where X refers to the amino acid residue N-terminal to the Cys. If there is no cysteine at the appropriate position in the sequence, then it will be necessary to mutate a wild-type residue in order to facilitate the ligation reaction. The effect of such a mutation on the structure and function of a protein can often be minimized by following a few simple rules: (1) If the three-dimensional structure of the protein is known (or can be modeled with reasonable confidence), choose a region remote from the active site of the protein, preferably in a flexible surface loop or in an interdomain linker region; (2) choose the mutation to be as conservative as possible, e.g., Ser → Cys or Ala → Cys; (3) try to avoid mutating residues that are highly conserved across a protein family; and (4) take advantage of any known mutational data about the system because the effect on structure–function of mutating a particular residue may already be known.

3. It is also important to consider the amino acid identity of X at the ligation site, because this will become the C-terminal residue in the thioester-containing peptide segment. Model studies indicate that all 20 naturally occurring amino acids are able to support native chemical ligation when placed at the C terminus of an α-thioester peptide.[14] However, there are dramatic differences in the kinetics of ligation depending on the chemical properties of X, for example, His Cys, and Gly are associated with rapid ligation reactions whereas Ile, Val, and Pro give extermely slow ligation rates. The nature of X also appears to influence the yields of recombinant polypeptide α-thioesters obtained with intein fusion systems (here X refers to the -1 position relative to the intein N terminus). Certain amino acids are associated with high levels of premature *in vivo* cleavage of the fusion protein (in particular Asp), whereas other residues are linked to poor yields in the *in vitro* thiolysis reaction (e.g., Pro).[26,42] The ranking of amino acids

[42] S. Chong, K. S. Williams, C. Wotkowicz, and M.-Q. Xu, *J. Biol. Chem.* **273,** 10567 (1998).

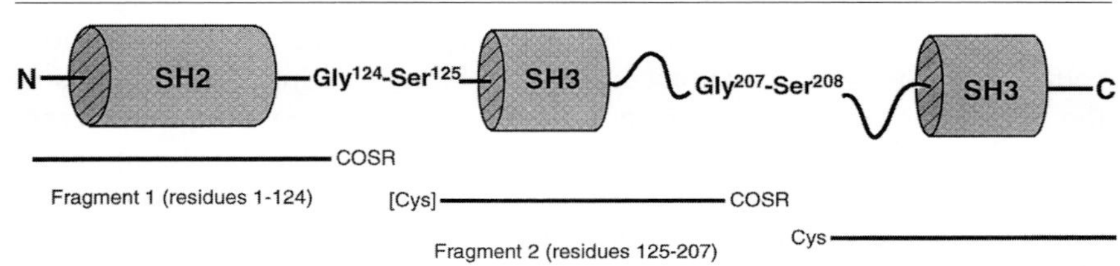

FIG. 2. Schematic representation of the domain structure of the 304-residue c-Crk adapter protein. The protein contains three Src homology domains (SH2, residues 13–118; N-SH3, residues 132–193; C-SH3, residues 256–304) connected by linkers of different length. As part of ongoing NMR solution structure studies, we propose to assemble the full-length protein from three recombinant polypeptide fragments, using a sequential ligation strategy.[4] The ligation sites, $G^{124}S^{125}$ and $G^{207}S^{208}$, are both located in the interdomain linkers, and in each case a Ser → Cys mutation is required to facilitate the chemistry.

with respect to these two phenomena seems to vary to some degree depending on the intein being used.[26,42] Moreover, it cannot be ruled out (and indeed seems likely[22]) that the relative levels of *in vivo* versus *in vitro* cleavage are influenced by additional structural features in the *N*-extein.

Example. We described an extension of EPL that allows the segmental isotopic labeling of proteins for solution structure analysis by nuclear magnetic resonance (NMR) spectroscopy.[20] This allows discrete regions of a protein to be labeled with NMR-sensitive nuclei (^{13}C or ^{15}N) and, in combination with new methods in spectral observation,[43] has the potential to allow NMR structure analysis to be applied to larger proteins than is currently possible. As part of a general research program in this area, we are interested in isotopically labeling each of the structural domans within the adapter prtein, c-Crk, including internal domains of the protein. The examples described in this chapter are taken from this ongoing work.

Full-length mouse c-Crk contains 304 amino acids (33 kDa) and is composed of a series of Src homology domains linked by polypeptide spacers of different length. As illustrated in Fig. 2, our objective is to assemble the full-length protein from three recombinant polypeptide fragments (each encompassing one of the SH domains), thereby allowing the systematic segmental isotopic labeling of the protein. The two ligation sites $G^{124}S^{125}$ and $G^{207}S^{208}$ are both located in the linker regions and in each case a conservative Ser → Cys mutation is required to facilitate the ligation reaction. Because these mutations are in the linker regions, they are unlikely to affect the folding properties of the assembled protein. It is also important to note that the N-terminal and middle fragments will have glycine residue

[43] V. Dötsch and G. Wagner, *Curr. Opin. Struct. Biol.* **8,** 619 (1998).

at their C termini, a feature that is predicted to result in minimal *in vivo* cleavage of the corresponding intein fusion protein and, in addition, to promote high yields of thiol-induced cleavage and expressed protein ligation. In this chapter we focus exclusively on the N-terminal fragment of c-Crk containing the SH2 domain and part of the first linker region (residues Met-1 to Gly-124, 13.8 kDa).

Choosing the Intein

Approximately 100 different intein domains have been identified to date, most of which originate from Archaea and Eubacteria, although a small number of have been identified in Eucarya (for a comprehensive listing of intein sequences, see *www.neb.com/neb/inteins.html*). All these inteins are characterized by a series of short sequence motifs, termed blocks A, B, F, and G, which contain conserved residues important for splicing activity.[44] Broadly speaking, inteins can be divided into two groups: those that contain a core homing endonuclease domain (which is not involved in protein splicing) inserted between conserved blocks B and F, and those that do not. The latter are in the minority and are often referred to as mini-inteins.[45]

Escherichia coli expression vectors are now commercially available that allow the generation of protein fusions to two different engineered inteins, the *Saccharomyces cerevisiae* VMA intein (454 residues and containing an endonuclease domain) and the *Mycobacterium xenopi* GyrA miniintein (198 residues). Depending on the nature of the polypeptide fusion (see above), these two inteins may result in variations in the relative levels of *in vivo* versus *in vitro* cleavage.[26,42] Moreover, the considerable difference in size between these two inteins may well affect soluble expression levels of the corresponding polypeptide fusion. Note that in our experience soluble expression of the intein fusion is desirable because of the poor yields associated with reconstituting a functional intein from bacterial inclusion bodies.[46] In practical terms it is difficult to know, a priori, which intein expression system will work best for a given polypeptide and thus it is advisable to try both in parallel.

Example. The commercial expression plasmids pTYB1 (VMA intein) and pTXB1 (GyrA intein) contain identical multiple cloning sites, thereby making it possible to prepare both Crk–SH2 constructs by the same PCR

[44] S. Pietrokovski, *Protein Sci.* **7,** 64 (1998).

[45] A. Telenti, M. Southworth, F. Alcaide, S. Daugelat, W. R. Jacobs, Jr., and F. B. Perler, *J. Bacteriol.* **179,** 6378 (1997).

[46] B. Ayers, K. U. Blaschke, J. A. Camarero, G. J. Cotton, M. Holford, and T. W. Muir, *Biopolymers* **51,** 343 (2000).

subcloning strategy. Primers were designed to introduce the appropriate restriction sites (5′-*Nde*I, 3′-*Sap*I) on either side of the DNA encoding the Crk–SH2 domain, which was amplified by PCR using the full-length mouse c-Crk gene as a template. Note, use of a 3′-*Sap*I restriction site avoids the addition of any extra residues between the C terminus of the polypeptide of interest (the c-Crk fragment) and the beginning of the intein sequence. Ligation of the digested and purified insert into the predigested pTYB1 and pTXB1 vectors gave the required IPTG-inducible expression plasmids $pTYB1_{Crk\text{-}SH2}$ and $pTXB1_{Crk\text{-}SH2}$, respectively.

To optimize the expression of soluble fusion proteins in transformed BL21 cells, a series of preliminary studies was undertaken in which the cell density (OD_{590} = 0.3–0.7), induction temperature (15 and 25°), and induction time (2, 5, and 16 hr) were systematically varied. Even under optimal conditions (see Methods), the yield of soluble fusion protein was low (~1 mg/liter for the GyrA fusion and 0.6–0.8 mg/liter for the VMA fusion), although significant amounts of both fusion proteins were found in the pellet (presumably in inclusion bodies). The presence of the C-terminal chitin-binding domain allowed the desired fusion proteins to be readily purified from cell lysate supernatants by affinity chromatography with chitin beads. This purification system proved to be extremely effective as no major contaminating proteins were detected by SDS–PAGE analysis of the loaded chitin beads. In addition, little or no *in vivo* cleavage was observed for either intein fusion protein, for example, analysis of the SH2–GyrA–CBD chitin beads gave a major band on the gel with an apparent molecular mass of 40 kDa, consistent with the intact fusion protein (Fig. 3, lane 1).

Our experience suggests that prolonged storage (weeks) of immobilized intein fusion proteins can lead to the accumulation of immobilized cleavage products, i.e., the labile thioester bond between the polypeptide of interest and the intein is slowly hydrolyzed, thereby reducing the effective loading of the beads.[46] To avoid this problem, it is advisable either to perform the ligation reaction immediately or to release the polypeptide from the immobilized fusion protein in the form of a stable ethyl α-thioester derivative; alkyl thioesters of this type are relatively unreactive as acyl donors, making them ideal for storage and purification purposes. However, the reactivity of the polypetide α-thioester can be increased during a ligation reaction through transthioesterification with, for example, 2-mercaptoethanesulfonic acid or thiophenol.[8–10] An additional advantage of isolating the recombinant polypeptide α-thioester is that it allows the subsequent ligation reactions to be performed in the presence of high concentrations of solublizing agents such as 6 *M* guanidinium hydrochloride or 8 *M* urea.

To compare the sensitivity of the GyrA and VMA intein fusions to

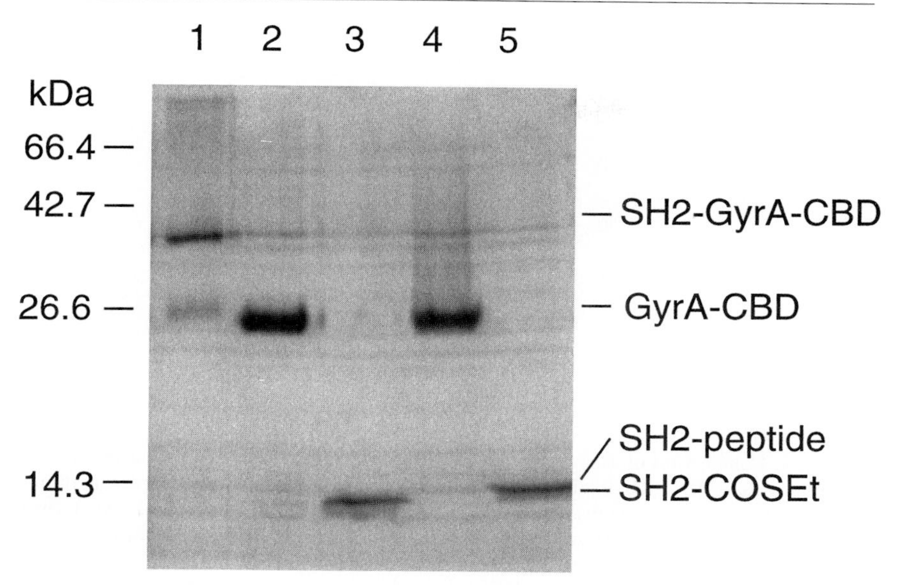

FIG. 3. Bacterial expression, thiolysis, and ligation of the Crk-SH2-GyrA-CBD fusion protein. Coomassie-stained SDS–15% (w/v) polyacrylamide gel analysis of: Lane 1, the affinity-purified recombinant fusion protein; lane 2, the chitin beads after exposure to ethanethiol for 24 hr (thiolysis); lane 3, the thiolysis reaction supernatant at 24 hr; lane 4, the chitin beads after treatment with a biotinyated peptide for 24 hr (EPL); lane 5, the EPL reaction supernatant at 24 hr. The efficiency of the two reactions is illustrated by the apparent absence of starting material in lanes 2 and 4.

thiolytic cleavage, the immobilized Crk–SH2–GyrA–CBD and the Crk–SH2–VMA–CBD fusion proteins were exposed to ethanethiol, and the release of the Crk–SH2 ethyl α-thioester derivative into solution was monitored by analyzing aliquots of the reaction supernatant by HPLC and electrospray mass spectrometry (ESMS) at different time points (Fig. 4). This kinetic study revealed essentially no difference between the two inteins; in both cases ~80% of the SH2 domain was released from the beads after 4 hr and the reaction was >90% complete after 24 hr. HPLC and ESMS analysis also indicated that in both cases the cleavage reaction was extremely clean (Fig. 5), the reaction supernatant contained a single component whose mass was consistent with the ethyl α-thioester derivative of the c-Crk–SH2 domain (note that the N-terminal Met residue had been completely removed). Under optimized conditions, the Crk–SH2–GyrA–CBD fusion system afforded 0.25 mg of HPLC-purified Crk-SH2 ethyl α-thioester derivative per liter of cell culture, compared with 0.1 mg/liter

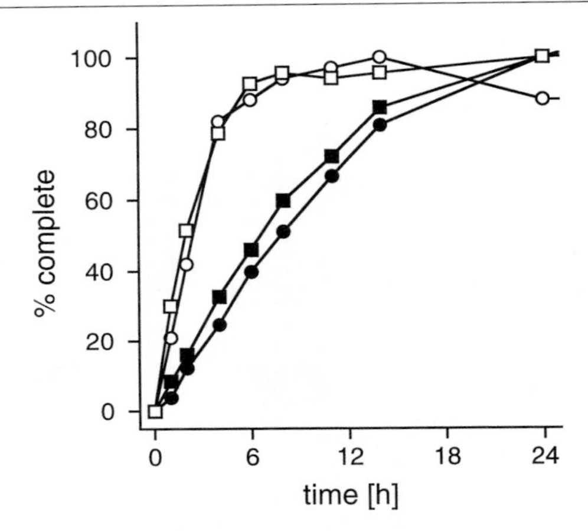

FIG. 4. Comparison of kinetic properties of the GyrA and VMA intein fusion proteins. The immobilized Crk-SH2-intein-CBD fusion protein was treated with either ethanethiol or a biotinylated peptide in the presence of MESNA. Aliquots of the reaction supernatants were taken at different time points and analyzed by analytical HPLC. For both inteins [VMA (□, ■); GyrA (○, ●)], the cleavage of the fusion protein with ethanethiol (□, ○) was almost complete after 6 hr of reaction. The MESNA-catalyzed expressed protein ligation is slightly slower (■, ●), but again showed a similar pattern for both inteins.

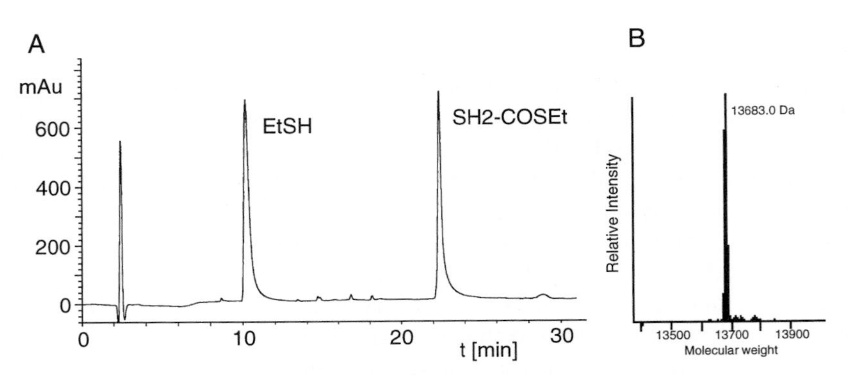

FIG. 5. Cleavage of Crk-SH2-GyrA intein fusion protein with ethanethiol. (A) Analytical HPLC analysis (linear gradient of 0–73% B over 30 min, with 214-nm detection) of the reaction supernatant after 24 hr. The ethyl α-thioester derivative of the Crk-SH2 domain elutes as a single peak at 22.3 min. Importantly, no hydrolysis product was observed. (B) Reconstructed electrospray mass spectrum of the 22.3-min peak. Expected mass for the Crk-SH2 ethyl α-thioester is 13,679.1 Da.

using the VMA intein fusion system. Therefore, the GyrA intein expression system was deemed the superior of the two for this polypeptide example.

More generally, it is worth pointing out that thioester formation offers a relatively mild way of activating carboxylates to attack by O-, N-, S-, and H-nucleophiles. Indeed, a variety of C-terminal modifications (e.g., esters, thio acids, amides, hydroxamic acids, alcohols, aldehydes, and hydrazides) can be introduced into synthetic peptides by using α-thioester intermediates.[47] By analogy, it is also possible to cleave polypeptide–intein fusions with different nucleophiles to give a number of C-terminal derivatives of recombinant polypeptides. For example, we and others have shown that polypeptide–intein fusions can be cleaved with various alkyl and aryl thiols to give the corresponding α-thioester derivatives.[8,10,33,48] These recombinant polypeptide α-thioesters can then be isolated and subsequently used in expressed protein ligation reactions. Other studies have shown that α-hydroxamic acid[33] and α-thio acid[39] derivatives of recombinant proteins can be prepared by treating polypeptide–intein fusions with hydroxylamine and ammonium sulfide, respectively. Thus, the intein fusion system provides a versatile synthetic hub from which to introduce a number of biochemically interesting and synthetically useful chemical groups into recombinant proteins.

Expressed Protein Ligation

As summarized in Table I, EPL allows the site-specific incorporation of a wide variety of biochemical and biophysical probes into recombinant proteins. In principle, this can be achieved either through the direct introduction of such probes into the synthetic peptide fragment, or through the incorporation of a unique reactive handle into the synthetic peptide, which can then be chemically elaborated as necessary after generation of the semisynthetic protein. In this section we illustrate each of these two labeling strategies, using the c-Crk–SH2–intein system.

Example. Two related synthetic peptides were used in EPL reactions with the recombinant c-Crk–SH2 domain, namely H-Cys-Gly-Gly-Gly-Gly-Gly-Lys[N^{ε}-D-biotin]-Arg-NH_2 and H-Cys-Gly-Gly-Gly-Gly-Gly-Lys[N^{ε}-levulinic acid]-Arg-NH_2. Biotin is widely used in biochemistry as a way of affinity-immobilizing peptides and proteins (on avidin beads) and is thus representative of the type of probe one might wish to site specifically incorporate into semisynthetic proteins. The ketone group within the levu-

[47] J. A. Camarero, A. Adeva. and T. W. Muir, *Lett. Peptide Sci.* **7,** 17 (2000).

[48] E. Welker and H. A. Scheraga, *Biochem. Biophys. Res. Commun.* **254,** 147 (1999).

linic acid (oxopentanoic acid) moiety represents a versatile synthetic handle that can be chemoselectively derivatized with a large number of commercially available probes, using established hydrazone and oxime-forming chemistries.[49–51] This two-step labeling strategy is especially useful for the introduction of acid- or thiol-sensitive probes incompatible with either SPPS or EPL.

Both synthetic peptides were synthesized by the *in situ* neutralization/HBTU activation protocol for Boc/Bzl solid-phase peptide synthesis.[40] Use of an orthogonal side-chain protection strategy allowed straightforward attachment of the levulinic acid and D-biotin moieties to the N^{ε}-amino group of Lys during the solid-phase chain assembly (in principle, a wide variety of labels can be introduced into synthetic peptides by this general procedure). Both modifications were stable to the liquid HF deprotection/cleavage procedure and the peptides were subsequently purified in satisfactory yields (~60%) by preparative HPLC, using individually optimized gradients.

Expressed protein ligation is typically carried out in aqueous buffers (pH 7.0–8.0) at temperatures between 4 and 40°, and in the presence of a thiol cofactor such as thiophenol or 2-mercaptoethanesulfonic acid (MESNA). Ligation of the synthetic and recombinant polypeptides can be carried out using either the affinity-immobilized intein fusion protein directly (i.e., one-pot transthioesterification and peptide ligation) or a purified recombinant polypeptide α-thioester (essentially native chemical ligation). The former approach has the advantage that it involves fewer steps; however, it is necessary to use near physiological conditions in order to retain a functional intein—moderate concentrations of chemical denaturants (up to 2 *M* guanidinum hydrochloride), detergents [0.1% (v/v) Triton X-100], and organic solvents [10% (v/v) dimethylsulfoxide (DMSO)] are permitted.[46] In contrast, prior isolation of the recombinant polypeptide α-thioester allows the ligation reaction to be performed in the presence of a wide range of chemical denaturants, detergents, and organic solvents.[41] In either scenario, it is desirable to have the synthetic peptide in large molar excess and at high concentration, typically 1 m*M* or higher. Both the biotin- and levulinic acid-containing peptides were soluble at millimolar concentrations in our standard ligation buffer [0.2 *M* sodium phosphate (pH 7.2), 0.2 *M* NaCl containing 0.1% (v/v) Triton X-100 and 2% (w/v) MESNA]. Studies

[49] H. F. Gaertner, K. Rose, R. Cotton, D. Timms, R. Camble, and R. E. Offord, *Bioconj. Chem.* **3,** 262 (1992).

[50] H. F. Gaertner, R. E. Offord, R. Cotton, D. Timms, R. Camble, and K. Rose, *J. Biol. Chem.* **269,** 7224 (1994).

[51] K. Rose, *J. Am. Chem. Soc.* **116,** 30 (1994).

indicate that MESNA catalyzes EPL reactions to almost the same extent as thiophenol but is significantly more soluble and is completely odorless.[10]

A series of model EPL reactions were undertaken with the biotinylated synthetic peptide and the immobilized GyrA and VMA fusion proteins. These studies were designed to explore whether the nature of the intein has any effect on the kinetics and/or efficiency of ligation. Aliquots of the reaction supernatants were periodically removed and analyzed by HPLC and ESMS, which revealed an extremely clean reaction, regardless of the nature intein in the original fusion protein (Fig. 6). As illustrated in Fig. 4, the rate of ligation was essentially the same using the two intein fusions, in both cases greater than 90% of the SH2 domain was ligated to the synthetic peptide after 24 hr, as judged by SDS–PAGE analysis of the beads and reaction supernatants (Fig. 3). The near-quantitative yields of the reactions allowed the semisynthetic ligation product to be easily purified from the excess synthetic peptide by semipreparative HPLC (in principle, purification under more physiological conditions should be possible by size-exclusion chromatography). Interestingly, the rates of the two ligation reactions were slightly slower than for the corresponding ethanethiol cleavage reactions (Fig. 4). This observation combined with the fact that at no stage during either ligation reaction was the thioester derivative of the SH2 domain detected by HPLC/ESMS, suggests that the slowest step in the process is the initial transthioesterification reaction, followed by rapid ligation of the SH2 α-thioester intermediate and the synthetic peptide.

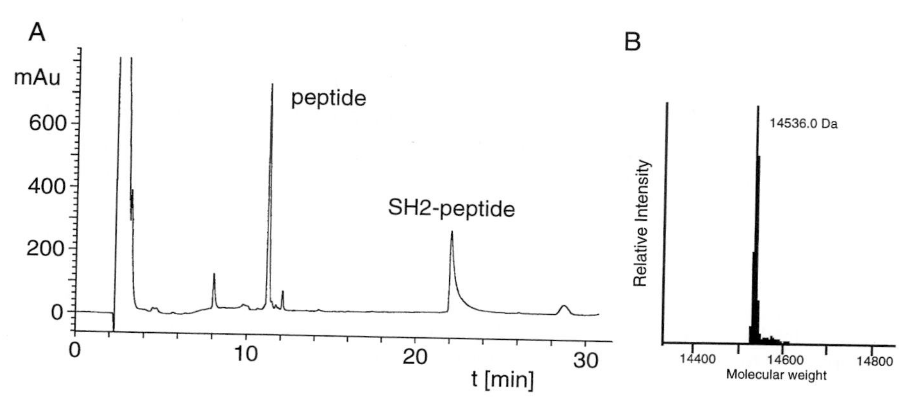

FIG. 6. Expressed protein ligation of H-Cys-Gly-Gly-Gly-Gly-Gly-Lys[N^{ε}-D-biotin]-Arg-NH$_2$ and the Crk-SH2-VMA intein fusion protein. (A) Analytical HPLC analysis (linear gradient of 0–73% B over 30 min, with 214 nm detection) of the reaction supernatant after 24 hr. The ligation product, Crk-SH2-Cys-Gly-Gly-Gly-Gly-Gly-Lys[N^{ε}-biotin]-Arg-NH$_2$, elutes as single peak at 22.0 min. (B) Reconstructed electrospray mass spectrum of the 22.0-min peak. Expected mass for the ligation product is 14,534.0 Da.

The levulinic acid-containing peptide was also ligated to the Crk–SH2 domain, using the immobilized GyrA and VMA intein fusion proteins (comparative data not shown). Again, little or no difference was observed between the two intein fusion proteins, and in both cases the desired ligation product was obtained in excellent yield (>90%) after 24 hr of reaction (data not shown).

Summary

By allowing the controlled assembly of synthetic peptides and recombinant polypeptides, expressed protein ligation permits unnatural amino acids, biochemical probes, and biophysical probes to be specifically incorporated into semisynthetic proteins.[4] A powerful feature of the method is its modularity; once the reactive recombinant pieces are in hand and the optimal ligation conditions have been developed, it is possible to quickly generate an array of semisynthetic analogs by simply attaching different synthetic peptide cassettes—in most cases the synthetic peptides will be small and easy to make. From a practical perspective, the rate-determining step in the process is usually not the ligation step (it is based on a simple and efficient chemical reaction[52]), but rather the generation of the reactive polypeptide building blocks. In particular, optimizing the yields of recombinant polypeptide building blocks can require some initial effort. However, it should be noted that the initial investment in time required to optimize the production of the recombinant fragment is offset by the ease and speed with which one can produce the material thereafter. In the example described in this chapter, the yield of soluble intein fusion protein was slightly better using the GyrA intein than for the VMA intein, although in both cases significant amounts of fusion protein were present in the cell pellet. Studies are currently underway to identify optimal refolding conditions for GyrA fusion proteins solubilized from inclusion bodies.

Acknowledgments

This research was supported by a Pew Scholarship in the Biomedical Sciences (T.W.M.), the National Institutes of Health (GM55843-01; T.W.M.), a Burroughs-Wellcome New Investigator Award (T.W.M.), and a Merck Postdoctoral Fellowship (U.K.B.).

[52] T. Wieland, *in* "The Roots of Modern Biochemistry" (v. D. Kleinkauf Jaeniche, ed.), pp. 213–221. Walter de Gruyter, Berlin, 1988.

Section V

Applications of Fusions in Functional Genomics

[30] A Systematic and General Proteolytic Method for Defining Structural and Functional Domains of Proteins

By JANNETTE CAREY

Introduction

Creation of chimeras implies dissection of existing proteins at appropriate boundaries as a basis for cloning corresponding gene fragments. A structural model of the protein, or homologies to known structures, can guide choices of boundaries for domains, subdomains, or secondary structural elements. However, the sometimes disproportionate effect of terminal sequences on stability of protein fragments, and even on intact proteins or domains, suggests that an empirical definition of segment boundaries may be preferable even when structural models are available. Such dissection studies can be carried out at the gene level or at the protein level. Proteolysis is a well-established technology, and for the goal of domain definition it brings an essential advantage over genetic dissection: labile segments of proteins, and the complementary structured regions, are identified by a probe that is directly sensitive to the structural parameters of interest. Thus the resulting proteolytic data often contain exactly the kind of structural information that is sought, at the resolution of the individual peptide bonds at the sites of cleavage. In contrast, to acquire this information by genetic dissection as a first step requires a trial-and-error approach. Proteolysis can be applied to proteins that are available only in relatively low yield and purity, as all analysis can be conducted by sodium dodecyl sulfate–polyacrylamide gel electrophoresis (SDS–PAGE). As is discussed here, the proteolytic fragmentation pattern itself can be used to identify candidate structured and nonstructured parts of the molecule as a guide to cloning.

Limited proteolysis has historically been an important method for dissecting proteins into structural and functional domains and subdomains (for reviews, see Refs. 1–3), and is finding increasing application in generating targets suitable for structural analysis by nuclear magnetic resonance

[1] A. Fontana, P. Polverino deLaureto, and V. DeFilippis, *in* "Protein Stability and Stabilization" (W. Van den Tweel, A. Harder, and M. Buitelaar, eds.), p. 101. Elsevier Science, Amsterdam, 1993.

[2] S. J. Hubbard, *Biochim. Biophys. Acta* **1382,** 191 (1998).

[3] Z. Peng and L. C. Wu, *Adv. Protein Chem.* **53,** 1 (2000).

0076-6879/00 $30.00

(NMR) spectroscopy and X-ray diffraction as well. From many years of using partial proteolysis extensively to identify and characterize folded substructures of proteins, certain general features have emerged. The most important of these is the observation that, even when using enzymes of relatively low sequence specificity, typically only a small number of surprisingly discrete fragments is produced during the course of a digestion, and some of these fragments persist even after long times of digestion. These observations led to two empirical criteria that we use in our laboratory to judge whether fragments resulting from partial proteolysis are candidates to represent folded substructures of the native protein. First, proteases with different sequence preferences should cleave in similar regions of the protein. This result indicates the likelihood that the cleavage sites lie in a loop or linker, or in an otherwise incompletely ordered part of the structure (which may sometimes be an active site, however, as in the tryptophan repressor[4]). Second, such fragments should persist during digestion for a time significantly longer than the time in which they were generated. This result indicates that the fragments contain no bonds as labile to proteolysis as those that were cut in the intact protein to generate the fragments. Provided the fragments contain additional potential proteolytic sites that are not cleaved, this result supports the likelihood of folded structure in the fragments themselves or in a nicked fragment complex. Careful analysis of the appearance and disappearance of starting material and fragments can often permit the use of these criteria to identify candidate structured domains of a given protein.

The implementation of the proteolytic technique has typically been protein specific, in the sense that conditions for generating desirable fragments have been quite specific for a given protein. Such conditions are then often borrowed for other proteins, for which they may or may not work well and may be adapted for the new substrate by exploring the nearby condition space. Yet it is quite simple to implement the proteolytic procedure in a systematic, general empirical approach that quickly identifies the requirements for dissection of any new protein under conditions that are optimal for that protein, or, for that matter, under virtually any desired conditions. The purpose of this chapter is to lay out such an empirical procedure. Its basic features are a range-finding experiment to calibrate enzyme activity under the chosen conditions, and a time-course experiment during which the complete progress of digestion can be observed.

[4] M. L. Tasayco and J. Carey, *Science* **255,** 594 (1992).

Procedure

Preparation of Protein Substrate

The protein should be as pure and as concentrated as is practical. Impurities may be tolerated in the analysis depending on their behavior during proteolysis (see below). The amount of intact protein required is the amount that will provide adequate staining intensity after SDS–PAGE. The relevant consideration is that proteolysis divides the total staining intensity of the intact protein band into multiple fragment bands. Because proteins (and fragments) differ in their staining intensity, and the number of fragments is not known in advance, standard amounts cannot be given and the amounts must be determined empirically. If staining intensity were linearly related to size then all fragments would stain proportionally to their length, but anomalous intensities are often observed, presumably due to amino acid composition effects, and some empirical adjustment of the amount of starting material may be necessary. A reasonable starting point is to use an amount of intact protein that is about 10 times larger than the amount needed to detect a well-stained band. This is the amount of protein required per lane for the SDS gel analysis.

The concentration of the protein should be high enough so that its volume can comprise ~50% of volume loaded onto the SDS gel, but more dilute samples can be concentrated prior to SDS–PAGE analysis by using trichloroacetic acid (see Secondary Structural Analysis, below). The protein should be in as weak a buffer as is practical, containing as few additional components as possible. If little or nothing is known about its requirements for stability and solubility, a good starting point is phosphate-buffered saline (PBS, pH 7.0), containing 10 to 20 m*M* sodium phosphate and 0.1 to 0.2 *M* sodium chloride. If the protein is in a much stronger buffer and/or at a pH very different from the enzyme reaction conditions, it may be necessary to use a 2× or higher stock of protease reaction buffer (see Range-Finding Experiment, below) in order to ensure achieving the required final reaction conditions on mixing of the protein and the enzyme. Stability of the protein at room temperature under the chosen buffer conditions is convenient, but if this is not known in advance, protease reactions should be conducted on ice, where the enzyme activity will be reduced by severalfold.

Preparation of Proteases

Solid proteases stored at lower than room temperature should be permitted to warm to room temperature before opening the bottle, to prevent condensation of water vapor on the surface of the cold solid. Weigh out approximately 1 to 5 mg into a tared microcentrifuge tube. Add 1 ml (ice-

TABLE I
PROTEASE CHARACTERISTICS

Enzyme	Suggested quality[a]	Storage and dilution buffer	Reaction pH optimum	Inhibitors	Cleavage preferences
Bromelain	Sigma B-5144	50 m*M* Tris, pH 8	5–7	Ag^{+2}; Cu^{+2}; tosyl chloromethyl ketones (T[L/P]CK)	Nonspecific
Chymotrypsin	Sigma C-3142	1 m*M* HCl	7–9	Phenylmethylsulfonyl fluoride; T[L/P]CK	Aromatics, Leu, Glu
Elastase	Sigma E-0258	1 m*M* HCl	8–9	PMSF	Uncharged nonaromatics
Proteinase K[b]	Roche 745-723	50 m*M* Tris, pH 8, 5 m*M* $CaCl_2$	7.5–12	PMSF	Hydrophobic aliphatics and aromatics
Subtilisin	Sigma P-5380	50 m*M* Tris, pH 8	6–11	PMSF	Most
Thermolysin[c]	Sigma P-1512	20 m*M* $CaCl_2$	7–9	EDTA; Hg^{+2}; Ag^{+2}	Bulky hydrophobics
Trypsin	Sigma T-4665	1 m*M* HCl	~8	TLCK; PMSF	Lys, Arg
Endoproteinase Glu-C (V8)	Roche 91-156	H_2O	7.8; 4.0	Dichloroisocoumarin	Glu, Asp (pH 7.8); Glu (pH 4.0)

[a] Sigma (St. Louis, MO); Roche (Nutley, NJ).
[b] Stimulated by SDS, urea.
[c] Reaction temperature 4–80°.

cold) of the manufacturer-recommended storage buffer (see Table I and Ref. 5) to form the enzyme stock solution. Mix thoroughly but avoid foaming. Place the tube of enzyme stock solution on ice. Aliquot the stock into small (~ 20 μl) and large aliquots (~ 200 μl). The size of the small aliquots depends on the volume of enzyme stock required for each day's experiments as described below. Quickly freeze aliquots on dry ice or liquid nitrogen, and then transfer (without thawing) to storage at −70 or −80°.

Choice of Gel

The SDS–PAGE system used must be capable of resolving fragments of the sizes that will be generated by proteolysis; this cannot be determined in advance, but a reasonable starting point is to choose a gel system in which the intact protein is well resolved near the center of the gel when

[5] A. J. Barrett, N. D. Rawlings, and J. F. Woessner, eds., "Handbook of Proteolytic Enzymes." Academic Press, New York, 1998.

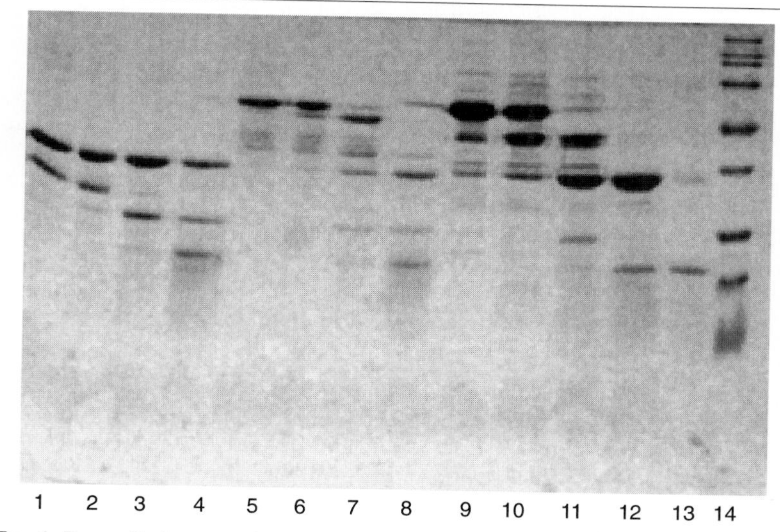

FIG. 1. Range-finding experiment using chymotrypsin (S. Osterdahl-Brijker, M. Sunnerhagen, and J. Carey, unpublished, 1999). Two different GST fusion proteins were used as substrates: GST–A (lanes 1–8) and GST–B (lanes 9–12). In lanes 1–4 the GST–A fusion protein had been previously cleaved at the factor X cleavage site engineered between the GST and A portions of the fusion; both portions of the substrate were thus subjected to chymotryptic cleavage simultaneously. Lanes 1, 5, and 9 contain only the substrate proteins and no enzyme; lanes 2, 6, and 10 contain chymotrypsin at 1:1000 dilution, lanes 3, 7, and 11 at 1:100 dilution, and lanes 4, 8, and 12 at 1:10 dilution. Lane 13 contains only chymotrypsin at 1:10 dilution in an amount equal to that in lanes 4, 8, and 12. Lane 14 contains molecular mass markers, top to bottom (kDa): 200, 116, 97, 66, 45, 31, 21.5, 14.5, and 6.5. Each reaction contained 15 μl of substrate protein in 50 m*M* Tris-HCl (pH 7.5), 150 m*M* NaCl, 1 m*M* $CaCl_2$ (lanes 1–8), or 50 m*M* Tris-HCl (pH 8.0), 10 m*M* glutathione-SH (lanes 9–12), and 5 μl of chymotrypsin diluted in 50 m*M* Tris-HCl, pH 8.0, from a stock of ~1 mg/ml in that buffer. Reactions were stopped after 30 min at room temperature by adding 6 μl of a 4× Laemmli sample buffer and boiling for 3 min. SDS–PAGE was carried out with the standard discontinuous buffer system[5a] with a 15% (w/v) acrylamide resolving gel.

the dye front is run to the bottom. Run appropriate molecular weight standards on a test gel along with the intact protein, to estimate the sizes of fragments that can be resolved in the chosen system. Many domains of proteins are in the molecular mass range of 8–10 kDa. Fragments of this size are easily resolved on standard Laemmli-type Tris–glycine SDS–PAGE minigels with an acrylamide concentration of between 15 and 17% (w/v) (see Figs. 1[5a] and 2). As is shown by example below, information can be extracted from the pattern of proteolytic consumption of the starting material with relatively little reference to the banding pattern. Thus, highest

[5a] U. Laemmli, *Nature* (*London*) **227,** 680 (1970).

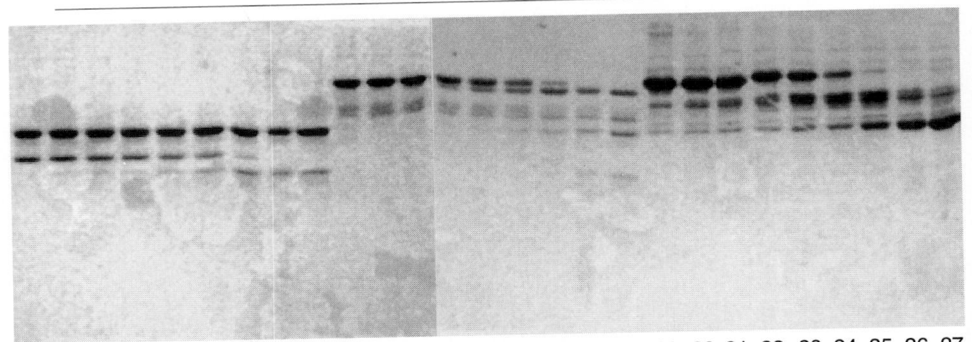

FIG. 2. Time course of chymotrypsin digestion (S. Osterdahl-Brijker, M. Sunnerhagen, and J. Carey, unpublished, 1999). Lanes 1–9, GST and A proteins as in Fig. 1; lanes 10–18, GST–A fusion protein; lanes 19–27, GST–B fusion protein. Each group of nine lanes represents a time course containing time points of 0, 1, 2, 5, 10, 20, 50, 100, and 200 min, respectively. The reaction mix contained 88 μl of substrate protein and 32 μl of chymotrypsin diluted 1:100. At each time point a 15-μl aliquot was removed and added to 4 μl of 6× Laemmli sample buffer followed by boiling; all other conditions were as in Fig. 1.

resolution of products is not required in the earliest stages of the analysis (e.g., Fig. 3).

Once the pattern of proteolysis is determined with standard gels, higher resolution precast gels employing gradients or alternative buffers can be selected with improved resolution in the molecular weight range of the fragments, according to the manufacturer specifications (e.g., Figs. 4 and 5). As well, it can sometimes be informative to run protease digests on native polyacrylamide gels, and to correlate the results with the SDS gel patterns. Such analyses can be especially useful in probing multisubunit proteins or interactions with other proteins or nucleic acids. The choice of native gel conditions is unique to each system. The major relevant consideration is that both charge and size/shape affect mobility in such gels. As a consequence, separations are not based on size alone, but both the pH and the pore size of the gel exert significant influence, and can be manipulated experimentally to optimize the separation. These considerations have been discussed with respect to gel mobility shift assays of DNA binding ("gel retardation") in this series,[6] and an example of their application to proteolysis is given in Ref. 7. A general overview of protein electrophoretic methods is given in Ref. 8.

[6] J. Carey, *Methods Enzymol.* **208,** 103 (1991).

[7] J. Carey, *J. Biol. Chem.* **264,** 1941 (1989).

[8] T. E. Zewert and M. G. Harrington, *Curr. Opin. Biotechnol.* **4,** 3 (1993).

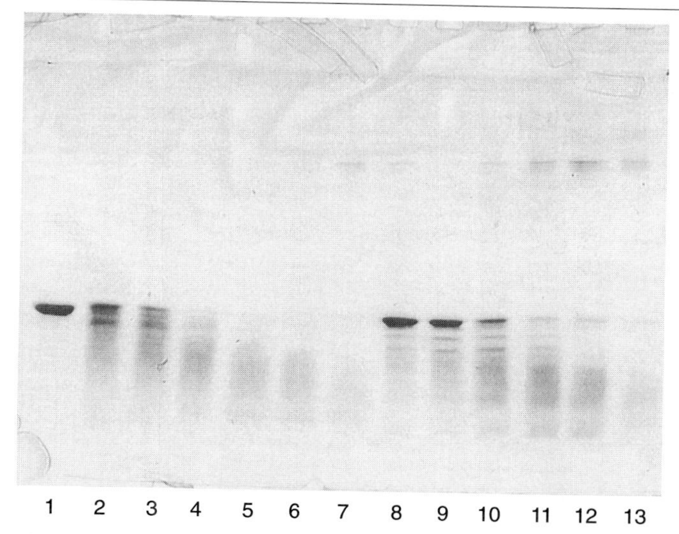

FIG. 3. Time course digestions of a small protein substrate resolved on a 17% (w/v) acrylamide Laemmli gel (A. Kesarwala and J. Carey, unpublished, 1999). Lane 1, starting material with molecular mass ~12 kDa. Lanes 2–7, digestion with proteinase K at 1:100 dilution; lanes 8–13, digestion with thermolysin at 1:1000 dilution. Each time course includes time points of 1, 2, 5, 10, 20, and 50 min; performed at room temperature.

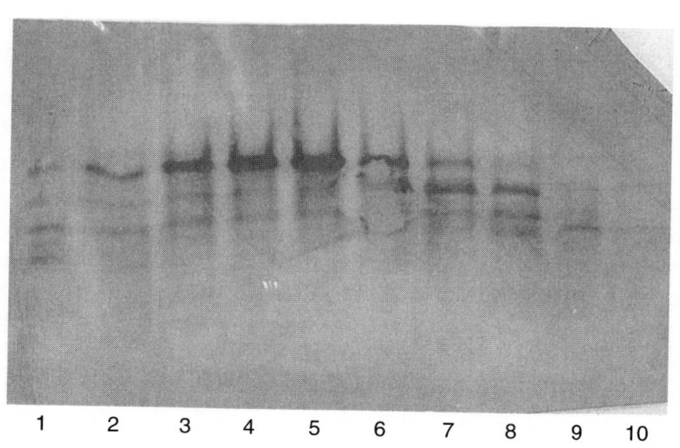

FIG. 4. Higher resolution gel of time course digests (A. Kesarwala and J. Carey, unpublished, 1999). As in Fig. 3 but samples were resolved on a 16.5% (w/v) acrylamide, Tris–Tricine gel system (Bio-Rad, Hercules, CA). The Coomassie blue-stained Western blot of the gel is shown. Lanes 1–5, proteinase K digest at 20, 10, 5, 2, and 1 min, respectively; lanes 6–10, thermolysin digest at 1, 2, 5, 10, and 20 min, respectively.

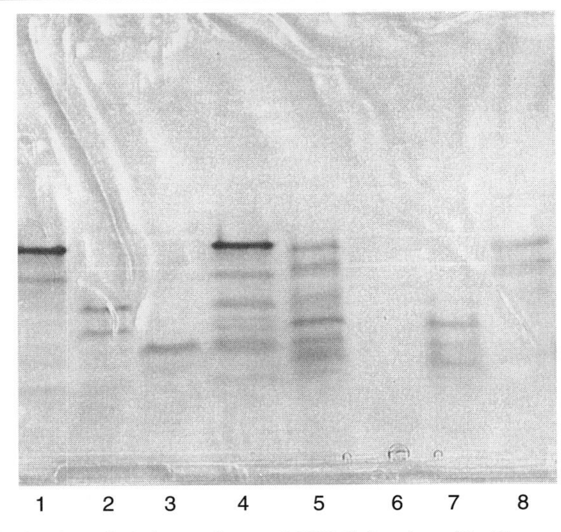

FIG. 5. Analysis of pooled time points and HPLC fractions (A. Kesarwala and J. Carey, unpublished, 1999). Lanes 1–4, thermolysin; lanes 5–8, proteinase K. Lanes 4 and 5 contain the pool of reserved time point samples from the digests shown in Figs. 3 and 4. Lanes 1–3 and 6–8 contain pools of all peaks collected within 1-min intervals from a C_{18} reversed-phase HPLC column eluted with a gradient of acetonitrile in 0.1% (w/v) trifluoroacetic acid.

Range-Finding Experiment

For dilution of the enzyme, prepare five tubes each containing 45 μl of the protease reaction buffer. Most proteases have relatively broad reaction pH optima (Table I), and few have specific additional requirements for activity that are absolute.[5] If the protein substrate is already in a buffer of suitable pH for the enzyme reaction (Table I), this buffer can also be used as the enzyme diluent. If not, choose a reaction buffer with a p*K* near the pH optimum for the enzyme if possible, and use this as the enzyme diluent. Adjust its concentration if necessary to provide adequate buffering capacity to achieve to desired final conditions. Even if the final conditions are not ideal for the enzyme, the range-finding experiment will determine its activity under those conditions. Most enzymes have a wide tolerance of conditions, with concomitant changes in activity, but only under quite extreme conditions does activity approach zero. Label the five enzyme dilution tubes 1, 2, 3, 4, and 5. Set aside on ice while thawing one 20-μl aliquot of enzyme stock on ice.

Prepare seven reaction tubes containing the amount of protein determined above (see Preparation of Protein Substrate), and an eighth tube containing only the protease reaction buffer in a volume equal to that of the protein added to each of the first seven tubes. The substrate should be aliquoted before diluting the enzyme. Into enzyme tube 1, aliquot 5 μl of the protease stock, rinsing the pipette tip well in the reaction buffer. Mix the tube well by gentle but thorough finger tapping, and return to ice. With a fresh pipette tip remove 5 μl from tube 1 and add it to tube 2, rinsing and mixing as before. Working quickly because the enzymes are activated for autodigestion once they are diluted into the reaction buffer, continue to make serial 10-fold dilutions in this manner in all five enzyme tubes, placing each back on ice. To the first reaction tube, add an equal volume of protease reaction buffer. To the second tube, add an equal volume of enzyme dilution number 5, mix well, and start timing the reactions, noting the approximate interval between enzyme additions to successive reaction tubes. To the third tube, add an equal volume of enzyme dilution 4, to the fourth tube dilution 3, etc., until all the enzyme dilutions have been added. To the seventh tube add an equal volume of the undiluted enzyme stock. To the eighth tube also add this same volume of the undiluted enzyme stock.

After 10 min at room temperature or 30 min on ice, begin to stop the reactions by adding an appropriate volume of SDS loading dye mix and immediate boiling of the samples for 3 min. Stagger the additions of SDS loading dye at approximately the intervals at which the successive reactions were initiated. With some enzymes, digestion does not stop on adding SDS and is stopped only during boiling. Samples should be loaded onto SDS gels as soon as possible. If absolutely necessary they can be stored (after boiling) at $-20°$ overnight but no longer, and even then further digestion can occur with some enzymes, so storage is not recommended.

The reproducibility of the range-finding experiment is paramount. Discard all tubes of enzyme dilutions and the remaining enzyme stock. Under no circumstance should any enzyme solutions be refrozen and used again, as the activity will be different from that established in the initial experiment. If any other conditions of the reactions are changed, such as use of a new protein stock or change in the protein buffer, change of reaction temperature, or addition of ligands, new buffer components, other potential effectors, etc., the range-finding experiment must be repeated to reestablish the protease activity under the new conditions. It should also be repeated if the initial small-volume enzyme aliquots are used up and the larger aliquots must be used. These should be thawed on ice, aliquoted ($\sim$20 μl), and calibrated prior to continuing a series of protease experiments.

Analysis. This experiment is used to identify an enzyme concentration that can be expected to give a well-developed time course of digestion, in which it will be easier to identify stable fragments that are candidates to be structured domains. Because the only purpose of the range-finding experiment is to calibrate enzyme activity, the volumes of enzyme and substrate and the length and temperature of the fixed-time assay are arbitrary and need only be reproducible in the time-course digest; thus, the volumes, times, and temperatures recommended here are only a guideline. For this same reason, the fragment pattern obtained in the range-finding experiment is not important; rather, focus on the disappearance of the intact protein. The time course will be conducted over a long period, tens to hundreds of times longer than the range-finding experiment. Therefore, look for an enzyme concentration that digests a small fraction of the intact protein in 10 min at room temperature (or in 30 min on ice). The next higher enzyme concentration than this gives an indication of the approximate extent of digestion that can be expected after 10 times as long a digestion period, and the next higher concentration indicates the extent of digestion after 100 times as long a digestion, etc. Choose an enzyme concentration in which digestion at room temperature can be expected to require a period of about 200 to 1000 min to proceed to completion, i.e., to digest all or most of the intact protein. The time course on ice may need to be much longer, say, 20 to 50 hr. To a first approximation, the product of enzyme concentration and time yields a constant extent of digestion at fixed temperature. In other words, if enzyme concentration is lowered by 10-fold, the time course will look similar over a 10-fold longer period, although some additional differences can occur because of the change in enzyme-to-substrate ratio.

An example of a range-finding experiment using chymotrypsin is shown in Fig. 1. Three glutathione *S*-transferase (GST) fusion proteins were digested in PBS buffer with chymotrypsin at enzyme dilutions of 1 : 10, 1 : 100, and 1 : 1000 from an ~1-mg/ml stock. Fixed-time assays were conducted for 30 min at room temperature and analyzed on 15% (w/v) polyacrylamide–SDS Laemmli gels. For each protein, little or no digestion occurs at the lowest enzyme concentration, and nearly complete digestion of all but the most resistant species is observed at the highest enzyme concentration. [In the example shown in Fig. 1, the resistant species is the GST portion of the fusion protein at ~30 kDa, represented by the top band in lane 1, which is common to all lanes. In lanes 1–4 (Fig. 1) the GST portion had previously been detached from the fusion by cleavage with factor X at an engineered site. Interestingly, chymotrypsin appears to cleave both GST-fused proteins (lanes 5–8 and 9–12, Fig. 1) at a site near the factor X site. In all lanes, the GST protein can be considered a contaminant.] Molecular weight markers are run in the last lane, and chymotrypsin at the highest concentration used in the reactions is run in the adjacent lane to permit identification of bands arising from the enzyme itself. From the results of this experiment, the 1 : 100 dilution of chymotrypsin was selected for a 200-min time course (Fig. 2).

Time-Course Experiment

Prepare 12 stop tubes containing the required amount of concentrated SDS loading dye. Starting from a new, freshly thawed (on ice) tube of enzyme stock, and using 10-fold serial dilutions of no less than 50-μl total volume each, prepare the protease at the required dilution chosen from the range-finding experiment; the minimum final volume of protease dilution required is equal to the volume of 10 protein samples. Be sure to prepare the required dilution serially from the stock rather than in one step; the best reproducibility of the range-finding results will be obtained by following the identical dilution steps. Prepare 1 reaction tube containing enough protein for 11 gel lanes. Remove one-eleventh of this sample into the first stop tube and add an equal volume of enzyme reaction buffer; this is the "zero time" aliquot. To the remaining ten-elevenths of the protein in the reaction tube add an equal volume of the chosen protease dilution, mix thoroughly but gently, and begin timing the reaction. After 1 min, remove one-tenth of the remaining sample volume, add it to the second stop tube, and place in a boiling water bath; after 2 min remove another volume equal to the first and add it to the third stop tube, etc. After boiling, each reaction can be quickly frozen on dry ice or liquid nitrogen until the time course is complete. Proceed similarly for all time points, stopping each one by boiling as soon as it is added to the stop tube, until the time points acquired correspond to 1, 2, 5, 10, 20, 50, 100, 200, 500, and 1000 min. These time points are approximately evenly spaced on a log axis, and because such a long time period is covered, it is convenient to compress the time axis onto a log scale. Some adjustment may be required in the time points for reactions conducted on ice, but the principle of the log time scale with evenly spaced points should still be used as it is convenient for later analysis. For example, 0.5, 1, 2, 5, 10, 20, and 50 hr might be chosen, depending on the results of the range-finding experiment. Run the samples on the SDS gel. Discard all enzyme solutions. The time course should be highly reproducible when using fresh aliquots of the enzyme, and should also reproduce closely the corresponding time point in the range-finding experiment. Pipetting errors, which can be substantial for concentrated protein solutions and/or small volumes, are the principal factor leading to irreproducibility.

Analysis. At this stage, high resolution of the product bands (as, e.g., in Fig. 2) becomes essential in order to evaluate precursor–product relationships and to permit further analytical characterization of the products. However, even a low-resolution gel analysis can provide important information about the fragmentation pattern that can be useful to justify the sometimes difficult search for suitable high-resolution gel conditions. An example of such a case is presented in Figs. 3 and 4. The poor resolution obtained in the standard Laemmli gel (Fig. 3) is typical for small fragments such as these, which are in the range of 4 to 9 kDa. The gel reveals a smeary zone of products representing most of the original staining intensity; the entire zone gradually shifts to higher mobility with time. The smearing could be due to heterogeneous cleavage, poor resolution, or both, and these possibilities can be distinguished only with a higher resolution separation. However, the time dependence of the average size of fragments suggests the presence of residual structure in a substantial fraction of the population. This inference is directly supported when the time points are run on a tricine 16.5% minigel (Fig. 4), which resolves the smeared zone into several discrete bands. A similar resolution improvement, achieved with a urea–SDS gel system, was used to identify an early cleavage close to the N terminus of the tryptophan repressor.[7]

Precursor–product relationships among fragment bands can often be inferred directly from the resolved gel patterns when they are sufficiently simple. However, even in such cases, and certainly whenever the patterns are more complex, it is wise to use caution in inferring such relationships between fragment bands; the only time a precursor is certain is when it is the only band present. As well, caution is needed in inferring the number of cuts required to generate a given fragment band; sometimes this is reasonably clear, as, for example, when one precursor gives rise to two products summing to the mass of the precursor, but often more analytical data are required to establish these relationships. In general, establishing precursor–product relationships is not as important as identifying fragment bands that are relatively resistant to further proteolysis, in order to apply the pair of empirical criteria cited in the introduction to identify fragments that are candidates to be structured domains. To apply these two criteria requires a high-resolution separation of products, and additional analytical data to determine the end points of each fragment within the protein sequence (see N-Terminal Sequencing and Mass Spectrometry, below). The production of similarly sized fragments by different proteases is often an indication of residual structure, but by itself cannot be used to conclude that the fragments are the same. Similarly, it can be useful to indicate the potential cleavage sites for each chosen protease on a linear map of the protein sequence, but identification of the cleavage sites actually used requires additional analytical data.

Most enzymes have even broader cleavage preferences than are generally reported, and sequence or structural context dependencies have not been well studied (but see Refs. 2 and 9). Thus, an empirical determination

[9] V. Schellenberger, K. Braune, H. J. Hofman, and H. D. Jakubke, *Eur. J. Biochem.* **199,** 623 (1991).

of the sequence preferences for cleavage is as reliable as anything. For example, chymotrypsin is widely regarded as having a preference for the aromatic amino acids, yet it is often observed to cleave at leucine residues,[9] which are nearly as bulky or hydrophobic as the aromatics but more often accessible in protein structures. The preferences listed in Table I should be used only as a guide, and reliable evidence supporting cleavage at other amino acid residues should be respected. The other side of this coin is that, if cleavage at, say, leucine is observed at one site in the protein, then the other leucine residues of the protein can be considered as potential substrate sites, and failure to cleave them can be taken as evidence of the presence of protecting structure. However, because the C terminus of each fragment is not directly analyzed (see Mass Spectrometry, below), the assignment of precise cleavage sites is not always secure; this complication should be borne in mind when inferring sites of protection.

N-Terminal Sequencing

Once a promising cleavage pattern is obtained reproducibly with good resolution of the fragments, a scaled-up time-course experiment should be conducted with enough protein to provide in each time point approximately 0.1 pmol for one gel and the usual amount for a second gel. After stopping the time points, load the two portions onto duplicate SDS gels. Stain one gel as usual and blot the other to polyvinylidene difluoride (PVDF) membrane according to standard procedures for Western blotting.[10] Stain the blot briefly and excise relevant fragment bands for automated Edman degradation.[10] Such analyses are now routine, and many institutions have in-house facilities; publicly available facilities can be found at *www.abrf.org/*. An alternative to blotting that may be more sensitive is a so-called negative or reversible gel stain using zinc (e.g., GelCode; Pierce, Rockford, IL). In this case the fragment bands are eluted directly from the gel for Edman degradation. It is also useful to bear in mind that Edman degradations can be performed on unresolved mixtures of fragments, provided that the number of components is limited to, say, three or four. One example of the utility of this approach is in identifying a cleavage product that may be too short to be resolved or retained on the SDS gel. With adequate amounts of protein (about 0.1 pmol per species), Edman degradation can easily be carried out for five cycles, although three is often adequate for unambiguous sequence determination from either a blot or a mixture. If the sequence results reveal a high level of heterogeneity in the cleavage sites for one enzyme, the same may not be true with other proteases.

[10] P. Matsudaira, *Methods Enzymol.* **182,** 602 (1990).

Mass Spectrometry

The mass of the fragments must be measured because there is no simple and reliable method for direct C-terminal sequence determination that is in wide use. Thus, to define both ends of a fragment, N-terminal sequence and mass data are combined. Current mass spectrometry (MS) methods have high resolution (±1 mass unit in 10,000) but are sensitive to SDS, salts, and many buffer components.[11] The usual approach is to purify fragments by high-performance liquid chromatography (HPLC),[12] which can remove most contaminants, but SDS must still be avoided as its interaction with proteins is difficult to reverse. Thus, an alternative approach is required for stopping the time points. Many proteases are readily inhibited by commonly available small molecules, e.g., phenylmethylsulfonyl fluoride (PMSF), tosylsulfonyl phenylalanyl chloromethyl ketone (TPCK), EDTA, certain metal ions, or reducing agents (see Table I and Ref. 5). Because these agents generally bind to or react with the enzyme directly, their effective concentrations can be quite low, comparable to the enzyme concentration, although they are normally used in vast excess. As well, many proteases have relatively narrow pH optima. To stop reactions destined for MS, we use both a specific chemical inhibitor chosen for the particular enzyme, at a final concentration of 10 m*M*, and acidification to about pH 4 with acetic acid [one-tenth volume from a 25% (v/v) stock], followed by thorough mixing and immediate freezing on dry ice. This degree of acidification is not adequate for the V8 enzyme, which is not inactivated at pH 4 (see Table I).

To minimize the number of HPLC separations, it can be convenient to have one sample containing all the proteolytic fragments produced during an entire time course, provided the resolution of the HPLC method is adequate to separate them, which is often the case. Thus, at the end of the time course, quickly thaw all time points on ice, and reserve a portion of each for an SDS gel to check the reproducibility of the time course and adequacy of the stopping conditions. Combine the remainder of all time points into one tube, mix well, divide into ~80 and 20% portions, and refreeze followed by lyophilization of both portions in a SpeedVac (Savant, Hicksville, NY). The major (80%) sample can now be fractionated by HPLC prior to MS. A small portion of each HPLC peak should be lyophilized for follow-up SDS gel analysis to permit correlation of the MS data with the Edman data. The 20% portion of the combined time points should be run on the gel alongside the HPLC peaks. An example of a gel analysis of

[11] For general review of protein MS, see M. Wilm, *Adv. Protein Chem.* **54,** 1 (2000).
[12] For many useful references on HPLC methods, see *Methods Enzymol.* **104** (1984).

HPLC peaks and reserved, pooled time points is shown in Fig. 5. If buffer components interfere with the separation or analysis, it may be useful to employ a volatile buffer such as ammonium acetate or ammonium bicarbonate, but we have not found this to be necessary.

Secondary Structural Analysis

It is relatively easy to acquire low-resolution biophysical data that can provide supporting evidence for the presence of structure in the fragment products. The methods used typically require larger amounts of material than the SDS gel analyses, in the range of a few nanomoles. Two complementary approaches can be taken using circular dichroism spectropolarimetry, which requires approximately 1 to 10 nmol of protein in 0.1 to 1.0 ml. Either fragments from the digest can be isolated by traditional biochemical methods and subjected to standard circular dichroic (CD) analysis alone or in various combinations, or the time course can be conducted in the CD with continuous monitoring of CD signal changes. An example of the former type of analysis can be found in Ref. 4. The latter type of experiment is set up similarly as a standard time course, but with the volume, protein concentration, and enzyme dilution adjusted appropriately. The relevant considerations are as follows: volume large enough to permit removal of all time points from the cuvette without the meniscus falling into the light beam; protein concentration adequate for a strong CD signal without severely reducing transmitted light; and enzyme concentration low enough to be transparent, or nearly so, relative to the protein. The protein sample is prepared in the cuvette and placed in the spectropolarimeter to record a starting value. Enzyme is added, mixed well, and timing begun. Aliquots are removed for SDS gel analysis at each time point. The CD signal can be monitored either continuously at a single wavelength or by taking spectra at discrete time points. In this way secondary structural changes can be correlated with the fragment pattern on the gels. An example of the application of this method can be found in Ref. 7; a general guide to the interpretation of CD data can be found in Ref. 13.

If the protein concentration in the cuvette is too low to show detectable bands on the SDS gel, larger aliquots can be taken at each time point and precipitated with ice-cold trichloroacetic acid (TCA) at a final concentration of 10%. Such samples must be carefully aspirated to remove TCA, but residual acid entrained in the pellet may yet turn the SDS sample buffer from blue to yellow; it should be back-titrated with 0.5-μl aliquots of 1.5 *M* Tris-HCl, pH 8.8.

[13] W. C. Johnson, Jr., *Proteins Struct. Funct. Genet.* **7,** 205 (1990).

Functional Analysis

There are three general ways of proceeding: classic biochemical fractionation of proteolytic fragments, and assay of the purified products individually or in various combinations; resolution of fragments by gel electrophoresis and assay of the fragments *in situ* in the gel or in a blot made from it; and real-time assay during proteolysis. Details of the first two approaches are system specific, and a general treatment of such methods is beyond the scope of this chapter. There are two alternative approaches to real-time assay during proteolysis: either the sample can be assayed continuously during the course of proteolysis with occasional sampling for gel analysis of the digestion progress, or individual time points can be taken from the digestion and assayed. In both cases, the goal is to correlate the proteolytic fragment pattern with changes in activity(ies). The feasibility of the former approach of course hinges on compatibility of the assay and proteolysis conditions. However, bear in mind that the proteases generally have a wide range of tolerance toward solution conditions, and their activities are easily calibrated under the conditions required for functional assay by following the steps outlined in this chapter. With a wide range of proteases to choose from, it is likely to find one or more that can be used in the required assay conditions. In the latter approach, the method of stopping the protease must be compatible with the activity measurement, or else the assay time should be short relative to the digestion time. Often this can be achieved simply by reducing the protease concentration, thus extending the digestion time course.

Acknowledgments

I am indebted to Susanne Osterdahl-Brijker and Maria Sunnerhagen for kindly providing Figs. 1 and 2, and to both of them as well as Prof. Gottfried Otting and the Wenner-Gren Foundation for enabling my visit to the Karolinska Institute, which inspired this chapter. I am also grateful for discussions with and the contribution of Figs. 3–5 by Aparna Kesarwala, and for discussions with Prof. Fred Hughson and Dr. Saw Kyin. I also thank Ingrid Hughes for expert assistance with the manuscript.

[31] High-Throughput Expression of Fusion Proteins

By MARC NASOFF, MARK BERGSEID, JAMES P. HOEFFLER, and JOHN A. HEYMAN

Genome-scale DNA sequencing efforts have identified a large number of putative open reading frames (ORFs). Methods to rapidly create plasmids that express these ORFs are needed to determine gene function.

A critical requirement for the high-throughput generation of expression plasmids is a means to identify plasmids that direct synthesis of the desired ORFs. If the cloning process includes a method [such as the polymerase chain reaction (PCR)[1]] to define the termini of the full-length open reading frames, then it is possible to clone each ORF into an expression vector so that it is in-frame with an epitope tag. This facilitates the use of Western blotting as the means to identify plasmids that direct synthesis of the desired fusion protein.

Western blotting for epitope-tagged proteins is a convenient means to test for plasmid function for several reasons: (1) no expensive equipment, such as mass spectrometers, is required; (2) the "read-out" from a Western blot (predicted band size vs. actual band size) is easy to interpret and can be done visually and rapidly; (3) production of ORFs fused to an epitope tag allows the use of one antibody for the detection of all expressed recombinant proteins; and (4) epitope tags will be useful in subsequent experiments such as immunolocalization, immunoprecipitation, and protein purification

For high-throughput applications, it is preferable to encode ORFs with C-terminal, rather than N-terminal, tags. A protein tagged at its N terminus can be expressed, appear by Western blot analysis to run at approximately the correct molecular weight, but in fact contain a stop codon that results in an incorrectly truncated protein. In contrast, a C terminal-tagged protein containing a frame shift or premature stop codon will prevent translation of the tag sequence and will not be detected by Western blot analysis. Plasmids that fail to produce the correct size protein product will be judged to be incorrect and not be used in subsequent experiments.

Detailed protocols for the testing and use of plasmids that direct synthesis of epitope-tagged proteins are presented, including (1) methods for the rapid introduction of plasmids into bacterial, mammalian, and yeast cells, (2) streamlined methods for Western blot detection of epitope-tagged re-

[1] R. K. Saiki, S. Scharf, F. Faloona, K. B. Mullis, G. T. Horn, H. A. Erlich, and N. Arnheim, *Science* **230,** 1350 (1985).

0076-6879/00 $30.00

combinant proteins, and (3) a robust method for bacterial gene expression and subsequent purification of recombinant proteins.

All the protocols assume that diagnostic tests (restriction analysis, diagnostic PCR, and/or sequencing) have demonstrated that each plasmid to be expression tested contains the desired insert in the correct orientation. In addition, a spreadsheet-driven method (e.g., Filemaker) should be used to track plasmid location and results.

Materials

The procedures described are essentially standard molecular biology protocols that have been adapted to a 96-well format. Thus, the required equipment is found in most laboratories. In addition, for laboratory techniques that are now standard (preparation of competent *Escherichia coli* and *Saccharomyces cerevisiae,* Western blotting incubation and washing protocols, etc.) detailed protocols are not given. Rather, we point out those reagents that might not be in all laboratories but will greatly facilitate the protocols. In addition, formulas for buffers that are used in several protocols are listed in this section, whereas formulas for buffers that are mentioned only once will be given in the protocol. *Note:* There is always some reagent lost during multichannel pipetting, and one should prepare 15% extra for all solutions that will be multichannel distributed into several wells.

Multichannel pipettors: Eight-channel pipettors with a volume range from 2.5 to 150 μl are required. Those made by Costar (Corning Costar, Cambridge, MA) and BrandTech (BrandTech Scientific, Essex, CT) are reliable

Eight-channel repeater pipette: Useful for distributing large (>300-μl) volumes from pipetting reservoirs to 96-deep-well blocks (Eppendorf, Madison, WI)

Disposable pipetting reservoirs: Corning Costar

Centrifuge with adaptors for 96-well plates: (Sorvall RT 6000; Sorvall Newton, CT)

Jitterbug shaker (Boekel Industries, Feasterville, PA)

Nine-well polyacrylamide gels: Must have lane spacing suitable for loading with eight-channel pipettors [Bio-Rad (Hercules, CA) 12% Tris–glycine gels, for example]

Eight-channel glass syringe: Crucial for easy multichannel loading of protein gels (Hamilton, Reno, NV)

See Blue protein standards (Novex, San Diego, CA): Or use equivalent prestained protein standards

Deep-well 96-well blocks: (Corning Costar): These blocks can hold up to a 1.7-ml volume/well. However, if the block is to be shaken, the liquid in each well should not exceed 1.3 ml, to prevent well-to-well contamination

AirPore tape (Qiagen, Valencia, CA): This tape allows aerobic and clean culturing of cells in 96-well blocks

Buffers

DNase buffer: 1× DNase buffer [50 m*M* Tris-HCl (pH 7.4), 5 m*M* $MgCl_2$, DNase I (grade II, 0.1 mg/ml; Boehringer Mannheim, Chicago, IL), 1 m*M* phenylmethylsulfonyl fluoride (PMSF), and 5% (v/v) glycerol]

Protein loading sample buffer: 2× sample buffer [0.5 *M* Tris-HCl (pH 6.8), 20% (v/v) glycerol, 10% (w/v) sodium dodecyl sulfate (SDS), 0.1% (w/v) bromphenol blue, 700 m*M* 2-mercaptoethanol]

Methods

Transfection in 96-Well Plates and Analysis of Expression in Mammalian Cells

The following protocols have been developed for use with pcDNA3.1/GS-based plasmids (Invitrogen, Carlsbad, CA). This vector directs constitutive gene expression [from the cytomegalovirus (CMV) promoter[2,3]] in mammalian cells, and encodes a C-terminal V5–hexahistidine (His_6)[4] epitope tag. For Western blotting, a horseradish peroxidase (HRP)-conjugated monoclonal antibody directed against the V5 epitope is commercially available (Invitrogen). These protocols are suitable for use with any mammalian expression vector that directs constitutive synthesis of epitope-tagged proteins, provided that an effective anti-epitope antibody is available.

The protocols are all developed for use in a 96-well format, and manipulations can be performed with either 12- or 8-channel pipettors. However, it should be noted that precast SDS–polyacrylamide gels (for protein analysis) compatible with 8-channel pipettors are commercially available, but gels compatible with 12-channel pipettors are not. Thus, prior to transfection, DNA samples should be arranged by column in 96-well plates so that samples 1–8 are in wells 1A–1H, samples 9–16 are in wells 2A–2H, etc. After transfection and lysate preparation, each column of eight samples

[2] M. Boshart, F. Weber, G. Jahn, K. Dorsch-Hasler, B. Fleckenstein, and W. Schaffner, *Cell* **41,** 521 (1985).

[3] J. A. Nelson, C. Reynolds-Kohler, and B. A. Smith, *Mol. Cell. Biol.* **7,** 4125 (1987).

[4] E. Hochuli, *In* "Genetic Engineering, Principles and Practice" (J. Setlow, ed.), Vol. 12, pp. 87–98. Plenum, New York, 1990.

can then be loaded by an eight-channel pipette onto one acrylamide gel. When working with fewer than 96 plasmids, it is still best to work in the 96-well format, to take full advantage of multichannel pipetting.

Transfection efficiencies are affected by DNA quality. We have found that both the Qiagen Qiaprep Turbo and Invitrogen SNAP kits yield DNA of sufficient quality and quantity (~0.2 μg/μl) to achieve high-efficiency transfection of mammalian cells.

Growth of CHO Suspension Cells prior to Transfection. Chinese hamster ovary (CHO) suspension cells are an excellent mammalian expression host. They are easy to transfect with lipid reagents[5] and can be grown in suspension, which allows the transfection of a relatively large (3×10^5) number of cells in a small, multichannel-compatible format (i.e., 96-deep-well blocks).

At times, CHO cells in spinner flasks tend to clump. This can be alleviated by the addition of heparin (Elkin-Sinn, Cherry Hill, NJ) to a final concentration of 30 units/ml.

PROTOCOL

1. CHO cells [CCL 61; American Type Culture Collection (ATCC), Manassas, VA] are cultured in CHO-S-SFM II serum-free growth medium (GIBCO-BRL, Gaithersburg, MD) in spinner flasks at 37° with 8% CO_2. CHO cells cultured under these conditions will grow as suspension cells. Care is taken to dilute cells to 3×10^5 cells/ml into fresh medium when a density of 1.2×10^6 cells/ml is reached (roughly every 48 hr).

2. The day prior to transfection, dilute cells to a cell density of $3–4 \times 10^5$ cells/ml. CHO cells double every ~24 hr under these conditions, so 1×10^6 cells/ml can be expected on the day of transfection.

Transfection and Lysate Preparation. Lipid-mediated transfection[5] is well suited to high-throughput protocols. In general, a 96-well transfection is performed as follows: First, a master mixture of growth medium and lipid transfection reagent is prepared and aliquoted into a 96-deep-well block. After the addition of plasmid DNA, CHO cells are then added to the DNA/medium/lipid mixtures and transfection is carried out for 4 hr at 37°. The cells are then pelleted, growth medium is added, and the cells are allowed to grow for an additional 40–48 hr. After centrifugation, cells are resuspended in lysis buffer, and protein gel loading buffer is added.

The protocol assumes that DNA plasmids have been arranged into eight-channel pipetting format.

[5] P. L. Felgner and G. M. Ringold, *Nature* (*London*) **337,** 387 (1989).

PROTOCOL

1. Determine the amount of CHO cells needed for transfection. (Although the cells will not be used until step 4 below, cells are counted at the beginning of the protocol so that steps 3 and 4 below are less hurried.)
   a. Number of transfections multiplied by 3×10^5 cells/transfection equals the CHO cells required. (Remember that some reagent is lost during multichannel pipetting. Assume that 15% more cells than calculated will be required.)
   b. CHO cells required divided by CHO cell density equals total volume of CHO suspension culture. CHO cells will be concentrated to the required density immediately after the medium/lipid/DNA transfection mix is prepared—see step 3 below.
2. Prepare and aliquot a master mix of Opti-MEM I reduced serum medium (GIBCO-BRL) and PerFect Lipids (pFx-6) (Invitrogen).
   a. Determine the amount of mixture that will be used. Each transfection (each well of a 96-well block) will require 24 μg (12 μl) of pFx-6 lipids and 488 μl of Opti-MEM I.
   b. Combine the above components into a reservoir and mix gently. Use an eight-channel repeating pipette to distribute 500 μl of the mixture to each well of a 96-deep-well block.
3. Pipette 4–6 μg of miniprepped plasmid DNA to each well of the 96-well block prepared in step 2. Cover the block with AirPore tape and shake on a Jitterbug shaker at speed 5 for 5 min at room temperature.
4. Remove the appropriate volume of CHO cells (determined in step 1) from the spinner flask and transfer into a 50-ml conical tube. Centrifuge the cells at 670*g* (2000 rpm in a Sorvall RT 6000 equipped with rotor H-1000B microtiter plate holders) for 5 min at room temperature.
   a. Remove the medium by aspiration and resuspend the cell pellet in an appropriate amount of Opti-MEM I. Avoid creating air bubbles at this stage.
   b. Decant the cell suspension into a pipetting reservoir, and use a multichannel pipette to distribute 500 μl to each well of the 96-deep-well block that contains a lipid/medium/DNA mixture. Take care to avoid touching the lipid/medium/DNA mixture with the pipette tips. Cover the wells with AirPore tape.
5. Place the deep-well block on a Jitterbug shaker and shake for 5 min at setting 5. Place in a 37° humidified incubator for 4 hr.
6. Centrifuge the block at 2000 rpm, at room temperature for 5 min, and resuspend the pelleted cells in 1.5 ml of CHO-S-SFM medium

(GIBCO-Life Technologies, Bethesda, MD). Incubate the cells for 42–48 hr at 37°. Add heparin to 5 units/μl final concentration as needed to prevent cell clumping.

7. Pellet the cells (2000 rpm for 5 min) and decant the supernatant. At this point, lysates can be prepared (go on to step 8) or the block can be frozen for future use.
8. Prepare lysates for gel loading.
   a. Multichannel pipette 15 μl of 1× DNase buffer (see Materials) to each cell pellet in the 96-well block. Vortex the blocks for 3 min to fully lyse pellets. The use of the DNase buffer ensures that chromosomal DNA will not interfere with subsequent gel loading.
   b. Add 15 μl of 2× loading dye (see Materials) to each well. Boil the block for 5 min, place on ice for 5 min, or freeze for later use.

Western Blot Analysis of Transfected Cells. Techniques for Western blot analysis are well known.[6,7] In addition, some antibodies require particular blocking, incubation, and washing conditions. For these reasons, we point out the aspects of our protocol that are of general use for high-throughput Western blot detection of epitope-tagged proteins.

PROTOCOL

1. Loading lysates.
   a. Use an eight-channel glass syringe (Hamilton) to load lysates onto nine-well 12% Tris–glycine polyacrylamide gels (Bio-Rad). One column of wells is loaded onto lanes 2–9 of the acrylamide gel. The syringe is essential for loading because it enables the loading of a large (25-μl) volume without lane spillover (the pins of the syringe fit into the gel wells).
   b. Five microliters of See Blue marker (Novex), along with 50 ng of a control protein for the anti-V5/HRP-conjugated antibody (Invitrogen), are loaded onto lane 1 of each gel. It is helpful to use prestained protein standards because several gels will be transferred onto the same nitrocellulose sheet, and it is essential to have some landmarks so that the filters can be labeled correctly.
2. Proteins are transferred to Schleicher & Schuell (Keene, NH) Optitran membrane filters by use of a Hoefer (San Francisco, CA) TE42

[6] J. Sambrook, E. F. Fritsch, and T. Maniatis, "Molecular Cloning: A Laboratory Manual," 2nd Ed. Cold Spring Harbor Laboratory Press, Cold Spring Harbor, New York, 1989.

[7] F. M. Ausubel, R. Brent, R. E. Kingston, D. D. Moore, J. G. Seidman, J. A. Smith, and K. Struhl, "Current Protocols in Molecular Biology." Greene Publishing Associates and Wiley-Interscience, New York, 1994.

transfer apparatus according to the manufacturer instructions. The large size of this apparatus allows one to place six gels (each containing eight transfection samples) on one large filter. Thus, a 96-well transfection can be transferred onto just two 17.5 × 24.5 cm filters. These filters are of the correct size to fit in standard film cassettes, so that 96 samples can be analyzed on two standard size (8 × 10 in.) films, and the nitrocellulose sheets do not need to be cut to size after transfer. This greatly simplifies the Western blotting manipulations.

3. After transfer, stain filters with Ponceau S (Sigma-Aldrich, St. Louis, MO) according to the manufacturer instructions. A permanent record of the protein stain can be obtained by placing a stained filter between two sheets of acetate and photocopying. This, coupled with the prestained protein standards, should enable accurate labeling of the filters.
4. Wash Ponceau stain from the filters and probe with the HRP-conjugated anti-V5 antibody (Invitrogen). Use of a conjugated primary antibody saves time (no incubations with a secondary antibody) and results in clean Western blots. Immunolocalized antibody is detected by incubation with SuperSignal Ultra (Pierce, Rockford, IL) chemiluminescence reagent and subsequent exposure to film. Transfected plasmids that direct synthesis of the correctly sized fusion protein are marked as Western positive. Figure 1 is a typical result obtained by the above procedure.

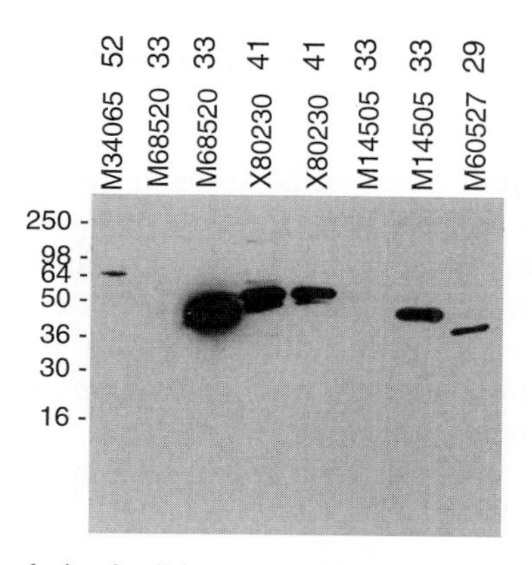

FIG. 1. Western blot analysis of cell lysates from CHO cells transfected with pcDNA3.1/GS-based plasmids. Plasmids contain the full-length human gene indicated by the GenBank accession number. Predicted size of the recombinant protein is given in kilodaltons above the accession number.

Transformation and Expression Testing of Plasmids Designed for Inducible Expression in Yeast Cells

The following protocol has been developed for use with the vector pYES2 (Invitrogen). This vector features the *GAL1*[8–10] promoter for galactose-inducible expression in appropriate *S. cerevisiae* strains, the 2 μm origin of replication for maintenance of high copy number, and the *URA3* gene for stable selection of transformants in *ura3* strains.[11] This general protocol can be adapted for use with any yeast expression vector, provided that the strains and vectors have compatible markers, and that the correct selective growth medium is provided. It should be noted that our protocol is designed for the rapid identification of functional expression plasmids, not for the generation of transformed yeast cell lines. Thus, for time-saving purposes, we do not streak our transformations out to single colonies. Rather, cells are transformed, cultured in selective medium, and then transferred to induction medium. The protocols can be easily adapted to include a step to streak for single colonies.

Preparation of Competent Yeast Cells. Protocols have been developed for use with yeast cells made competent with the Invitrogen S.c. EasyComp transformation kit. However, any method for preparation of competent yeast is acceptable, provided that it gives cells with transformation efficiency of $>10^3$ colony-forming units (CFU)/μg DNA (after freezing and thawing once). We recommend that cells be frozen in aliquots of volumes for 96-well transformations, that is, if 25 μl of cells will be used for each individual transformation, freeze cells in 3.5-ml aliquots (25 μl/well × 96 wells + some extra for loss in multichannel pipetting).

Yeast Transformation, Induction, and Lysate Preparation

PROTOCOL

1. Thaw competent yeast cells, transfer to pipetting reservoir, and pipette 25 μl of cells to each well of a 96-deep-well block on ice.
2. Add 1–3 μg of each plasmid DNA to competent INVSc1 *S. cerevisiae* cells (*his3Δ1 leu2 trp1-289 ura 3-52*) in a 96-deep-well block.
3. Add 250 μl of solution III (Invitrogen EasyComp kit) to each DNA/yeast cell mixture and mix by pipetting up and down several times. Cover the block with AirPore tape.
4. Incubate the block for 1 hr at 250 rpm in a 30° shaker/incubator.

[8] E. Giniger, S. M. Barnum, and M. Ptashne, *Cell* **40,** 767 (1985).
[9] R. W. J. West, R. R. Yocum, and M. Ptashne, *Mol. Cell. Biol.* **4,** 2467 (1984).
[10] M. Johnston and R. W. Davis, *Mol. Cell. Biol.* **4,** 1440 (1984).
[11] P. Philippsen, A. Stotz, and C. Schert, *Methods Enzymol.* **194,** 199 (1991).

5. Pellet the cells by centrifugation at 2000*g* for 5 min at room temperature. Discard the supernatant.

6. Resuspend the cells in 1.3 ml of selective growth medium [1.3% (w/v) yeast nitrogen base, 2% (w/v) glucose, histidine (20 μg/ml), tryptophan (20 μg/ml), and leucine (30 μg/ml)] per well and cover the block with AirPore tape. Incubate for 3–4 days at 30° on a shaker at 250 rpm.

7. Pellet the cells at 3500 rpm for 5 min, and replace the medium with 1.3 ml of induction medium [1.1% (w/v) yeast nitrogen base, 2% (w/v) galactose, 1% (w/v) raffinose, histidine (20 μg/ml), tryptophan (20 μg/ml), and leucine (30 μg/ml)] per well. This block will be referred to as the induction block.

8. After 3 hr of induction, remove 800 μl of medium from each well and transfer to corresponding wells of a fresh 96-well block (storage block). Pellet these cells, decant the supernatant, and store at −20°. (These cells will later be pooled with cells that are induced overnight, and lysates from the pooled cells will then be tested for protein expression. In this way, lysates from short and long induction time points can be loaded in the same wells on the same gel.) Add 800 μl of fresh induction medium to each well of the induction block and culture overnight.

9. Transfer cultures from the induction block to the corresponding wells of the storage block. Centrifuge the storage block at 2000*g* for 5 min, and resuspend each pellet in 20 μl of 1× DNase buffer.

10. Add 20 μl of 2× sample buffer to each well, place the entire 96-well block in boiling water for 5 min, and then place on ice.

11. Samples are analyzed by Western blot as described above for the CHO cell lysates. Load 20 μl of sample per lane and save the remaining lysates at −20°. If subsequent Ponceau staining and Western blotting reveal the lanes to be overloaded, it is possible to go back and load a fraction of the stored lysates.

Rapid Transformation and Expression in Bacterial Cells

The bacterial expression vector pBAD/Thio-TOPO (Fig. 2[12,13,14]; Invitrogen) directs synthesis of fusion proteins that contain a His-Patch

[12] J. A. Heyman, J. Cornthwaite, L. Foncerrada, J. R. Gilmore, E. Gontang, K. J. Hartman, C. L. Hernandez, R. Hood, H. M. Hull, W. Y. Lee, R. Marcil, E. J. Marsh, K. M. Mudd, M. J. Patino, T. J. Purcell, J. J. Rowland, M. L. Sindici, and J. P. Hoeffler, *Genome Res.* **9,** 383 (1999).

[13] S. Shuman, *J. Biol. Chem.* **269,** 32678 (1994).

[14] S. Shuman, *J. Biol. Chem.* **266,** 1796 (1991).

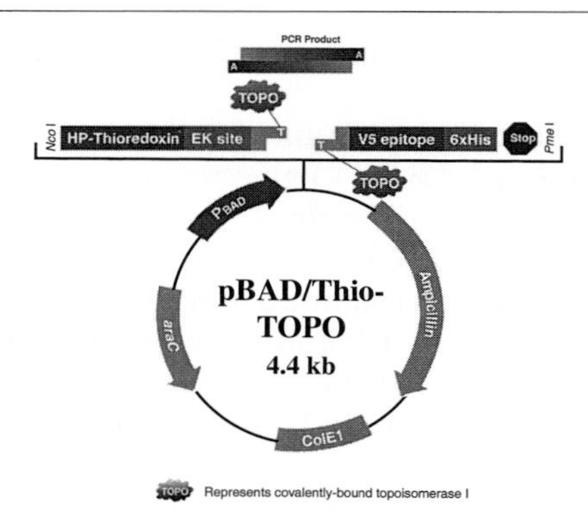

FIG. 2. Several aspects of the pBAD/Thio-TOPO expression vector are depicted in this map. (1) Transcriptional regulation. AraC, the product of the *araC* operon, functions both as a negative and as a positive regulator of the P_{BAD} promoter (in the absence and presence of L(+)-arabinose, respectively). (2) Thioredoxin leader sequence. His–Patch thioredoxin (HP–thioredoxin) has been shown to enhance expression of peptides/proteins that are cloned in-frame with this leader sequence. HP–thioredoxin can also be used as a purification epitope in metal affinity chromatography performed under nondenaturing conditions.[4,15] An enterokinase cleavage site (EK site) is encoded after HP–thioredoxin. (3) The TOPO cloning site. PCR products with single 3′ A-base overhangs are cloned into the T-base overhang site shown. This vector is depicted as adapted for topoisomerase I-mediated cloning (TOPO cloning; Invitrogen, Carlsbad, CA). In this cloning method,[12,13,14] topoisomerase I, rather than DNA ligase, is used to join PCR products to an appropriately prepared vector. High-throughput TOPO cloning of full-length open reading frames into expression vectors has been described in detail.[12] (4) The optional V5–His_6 epitope. This encodes a 24-amino acid domain composed of the V5 epitope fused to six tandem histidine residues.

(HP)[15]–thioredoxin[16] domain fused to the N terminus of the protein of interest. In addition, the gene can be cloned to encode an optional C-terminal V5–His_6 tag. This vector is extremely well suited to the high-throughput synthesis of proteins in *E. coli.* The strongly inducible P_{BAD} promoter[17–21] can be tightly regulated, the N-terminal thioredoxin leader sequence aids protein translation (and in some cases, protein solubility[22]),

[15] Z. Lu, E. A. DiBlasio-Smith, K. L. Grant, N. W. Warne, E. R. LaVallie, L. A. Collins-Racie, M. T. Follettie, M. J. Williamson, and J. M. McCoy, *J. Biol. Chem.* **271,** 5059 (1996).
[16] A. Holmgren, *Annu. Rev. Biochem.* **54,** 237 (1985).
[17] N. Lee, C. Francklyn, and E. P. Hamilton, *Proc. Natl. Acad. Sci. U.S.A.* **84,** 8814 (1987).
[18] C. G. Miyada, L. Stoltzfus, and G. Wilcox, *Proc. Natl. Acad. Sci. U.S.A.* **81,** 4120 (1984).
[19] S. Ogden, D. Haggerty, C. M. Stoner, D. Kolodrubetz, and R. Schleif, *Proc. Natl. Acad. Sci. U.S.A.* **77,** 3346 (1980).
[20] R. S. Schleif, *Annu. Rev. Biochem.* **61,** 199 (1992).
[21] L. M. Guzman, D. Belin, M. J. Carson, and J. Beckwith, *J. Bacteriol.* **177,** 4121 (1995).
[22] E. R. LaVallie, E. A. DiBlasio, S. Kovacic, K. L. Grant, P. F. Schendel, and J. M. McCoy, *Bio/Technology* **11,** 187 (1993).

and the C-terminal V5–His_6 domain provides an excellent tool for Western blot detection and protein purification.[23–25] One advantage of the pBAD/Thio-TOPO vector is that unlike T7 polymerase-based expression systems[26,27] it does not require a heterologous RNA polymerase. Thus, virtually any *E. coli* strain can be used as a host. In addition, the P_{BAD} promoter system is forgiving. For each vector tested, there is a wide range of induction conditions (arabinose concentrations, cell density, induction temperature, length of induction) that will give good results.

The following protocol is designed for use with this vector system, but it can be adapted to any vector that directs expression of polyhistidine-tagged recombinant protein.

Preparation of Competent Cells. The transformation will involve transferring miniprepped plasmids into cells for expression. These plasmids have been determined (through diagnostic PCR, minipreparation analysis, or sequencing) to contain the appropriate insert in the correct orientation. Thus, only a few transformants are required and any protocol that yields cells with efficiencies of $>10^5$ colonies/μg plasmid DNA will suffice. Cells should be frozen in convenient aliquots for 96-well transformations. Aliquots of 5.5 ml are handy if the transformation protocol requires 50 μl of cells per reaction.

Bacterial Transformation and Plating in 96-Well Plates

1. Arrange plasmids to be transformed in the eight-channel format and dilute to 50 pg/μl.
2. Thaw on ice a 5.5-ml aliquot of competent *E. coli* and transfer to a multichannel pipetting reservoir on ice.
3. Aliquot 50 μl of competent cells to each well of a 96-well microtiter dish on ice.
4. Multichannel pipette 3 μl (150 pg) of each plasmid into the wells of the microtiter dish.
5. Incubate on ice for 20 min, heat shock by placing the tray on a 42° 96-well heat block for 1 min, and transfer the tray to ice for 1 min (see Materials for a description of the heat block). Add 150 μl of SOC [2% (w/v) tryptone, 0.5% (w/v) yeast extract, 10 m*M* NaCl, 2.5 m*M* KCl, 10

[23] E. Hochuli, W. Bannwarth, H. Dobeli, R. Gentz, and D. Stuber, *Bio/Technology* **6,** 1321 (1988).

[24] D. Stuber, H. Matile, and G. Garotta, *Immunol. Methods* **4,** 121 (1990).

[25] A. Hoffman and R. Roeder, *Nucleic Acids Res.* **19,** 6337 (1991).

[26] S. Tabor and C. C. Richardson, *Proc. Natl. Acad. Sci. U.S.A.* **82,** 1074 (1985).

[27] F. W. Studier, A. H. Rosenberg, J. J. Dunn, and J. W. Dubendorff, *Methods Enzymol.* **185,** 60 (1990).

m*M* $MgCl_2$, 10 m*M* $MgSO_4$, and 20 m*M* glucose] to each well, cover the plate with foil tape, and incubate for 45 min at 37°.

6. Plate 50 μl from each well onto a selective plate. *Note:* Because the P_{BAD} promoter of pBAD/Thio-TOPO-based plasmids will express well under a variety of induction conditions, it is possible to omit the plating step. One can simply inoculate each transformation into a well of selective medium in a 96-deep-well block and incubate with shaking to stationary phase (225 rpm, 37°) overnight. One can then proceed to step 3 in the next section (Rapid Identification of Highly Expressing Clones).

Rapid Identification of Highly Expressing Clones

1. Pick cells from each transformation plate and inoculate into the corresponding well of a 96-well block that contains 1.2 ml of LB plus ampicillin (50 μg/ml) (LB AMP) per well. *Note:* Because each transformation is performed with a single plasmid species, it is not necessary to pick a single colony for the inoculation.
2. Grow overnight. Cells will reach stationary phase (OD_{600} of ~3 for TOP10 cells in deep-well blocks).
3. Pipette 50 μl of each stationary culture into the corresponding wells of a second LB AMP 96-well block. These cultures will have a starting OD_{600} of ~0.1.
4. Grow to OD_{600} 0.4 (1.5–3 hr) and induce with L(+)-arabinose [0.02% (w/v) final concentration].
5. Grow for 3 hr, pellet the cells by centrifugation, and decant the supernatant.
6. Resuspend the cells with 200 μl of 2× sample buffer [0.5 *M* Tris-HCl (pH 6.8), 20% (v/v) glycerol, 10% (w/v) SDS, 0.1% (w/v) bromphenol blue, 700 m*M* 2-mercaptoethanol] and lyse by boiling the block for 15 min.
7. Load 15 μl of lysate onto SDS–polyacrylamide gels using an eight-channel glass syringe, and run the gels.
8. Stain the gels with Sypro Orange (Molecular Probes, Eugene, OR) or Coomassie stain[6] and analyze the gels for the presence of induced bands.
9. Record the plasmids (strains) that give induced band as "high-expression" clones. In Fig. 3, crude lysates from arabinose-induced cells bearing pBAD/Thio-TOPO-based plasmids have been separated by SDS–PAGE and Coomassie stained.

Metal Affinity Purification of Recombinant Fusion Proteins

The identification of clones that express high levels of polyhistidine-tagged recombinant protein allows the use of a streamlined metal affinity protein purification[22–25] protocol. This protocol is adapted from Shi *et al.*[28]

[28] P. Y. Shi, N. Maizels, and A. M. Weiner, *BioTechniques* **23,** 1036 (1997).

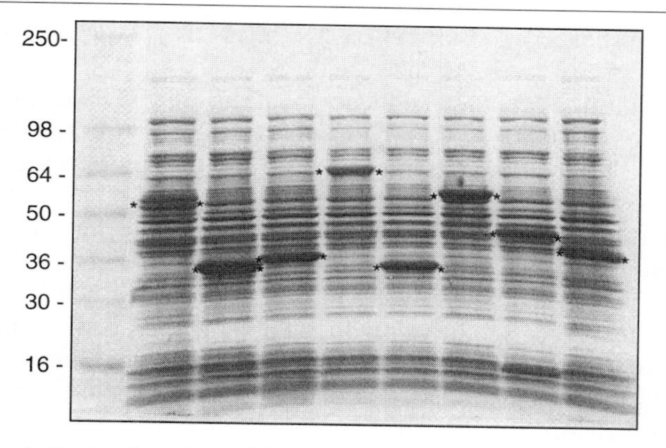

FIG. 3. Crude lysates from bacterial cells bearing pBAD/Thio-TOPO-based plasmids. Lysates were separated by SDS–PAGE and Coomassie stained. Results are representative of pBAD/Thio-TOPO-based plasmids used according to the methods described in text. Induced recombinant protein bands are bracketed with asterisks (*). Molecular mass standards (See Blue; Novex) were loaded in lane 1. Masses of protein standards are given in kilodaltons to the left of the lane.

and takes advantage of the fact that recombinant proteins that are highly expressed in bacterial cells almost always form inclusion bodies. In the protocol, cells are initially lysed by sonication and then centrifugation is used to separate inclusion bodies from soluble proteins. These inclusion bodies are then solubilized and the desired recombinant protein is purified by immobilized metal affinity chromatography (IMAC) using Probond (Invitrogen), a nickel-chelating agarose resin that binds to proteins containing a His_6 sequence. After extensive washing, bound proteins are refolded on the resin by changing to aqueous buffer conditions. The aqueous, soluble proteins are eluted by the addition of imidazole. In some cases[28] proteins purified with this type of protocol will be enzymatically active.

PROTOCOL

1. LB + ampicillin (100 μg/ml) (125 ml) is prepared in 250-ml baffled culture flasks. Each flask is labeled to correspond to 1 well of a 96-deep-well block.

2. One milliliter of LB + ampicillin is added to each well of the 96-well plate, and then each well is inoculated with single colonies from clones confirmed to be expression positive by SDS–PAGE separation followed by Coomassie blue staining[6] (described above). The 96-well plate is covered with AirPore tape and shaken overnight at 37°.

3. One milliliter of the overnight culture from each well of the plate is inoculated into the corresponding flask containing the LB + ampicillin, and

the flasks are shaken at 250 rpm at 37° to final OD_{600} of 0.5 (approximately 3 hr).

4. The cultures are then induced by addition of 125 μl of 20% (w/v) L(+)-arabinose to each flask, to a final arabinose concentration of 0.02% (w/v). After 3 hr, the flasks are place on ice, and the contents are transferred to 150-ml centrifuge tubes.

5. The tubes are centrifuged at 4500 rpm for 20 min at 4°, and the supernatants are decanted. At this point, the pelleted cells may be frozen on dry ice. This freezing step will improve subsequent cell lysis.

6. To lyse the cells, the centrifuge tubes are placed on ice, and the pellets in each tube are resuspended in 5 ml of Probond native lysis buffer (PNLB) at pH 7.8 [final concentrations of 50 mM K_2HPO_4, 400 mM NaCl, 100 mM KCl, 10% (v/v) glycerol, 0.5% (v/v) Triton X-100, 10 mM imidazole] and 50 μl of 100 mM phenylmethylsulfonyl fluoride (PMSF).

7. The resuspended pellets are sonicated (Vibra Cell; Sonics & Materials, Danbury, CT) for 30 sec at full power with a 3-mm tip, transferred to 30-ml Oakridge tubes, and centrifuged at 12,000 rpm for 15 min at 4°. The supernatants are decanted, and then 5 ml of guanidine lysis buffer at pH 8.0 (6 M guanidine hydrochloride, 0.1 M K_2HPO_4, 0.01 M Tris-HCl) is added to each Oakridge tube, and the pellets are sonicated for 20 sec and centrifuged at 12,000 rpm for 15 min at 4°. The supernatant from each tube is poured onto a 0.5-ml Probond resin column (previously prepared and equilibrated with 5 ml of guanidine lysis buffer), the lysates and the

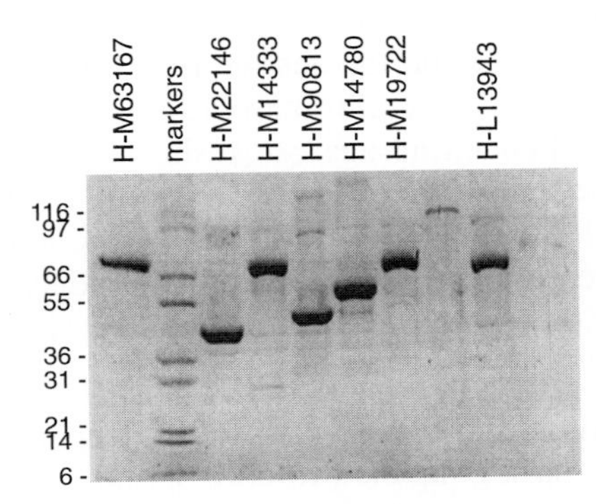

FIG. 4. SDS–polyacrylamide gel loaded with purified human recombinant proteins and stained with Sypro Orange. Full-length human genes (GenBank accession number given above lanes) were cloned into the pBAD/Thio-TOPO vector so that each protein would be synthesized with HP–thioredoxin at its N terminus and the V5–His_6 epitope tag at its C terminus. Proteins were purified according to the procedure in text. Peak elution fractions were pooled and approximately 0.5% of the pooled fraction for each protein was loaded in the indicated lane. Novex Mark12 markers (1 μg/band) were loaded in markers lane. Unlabeled lanes were loaded with eluate, but the protein purification was unsuccessful.

resin are mixed, and, after a 30-min equilibration, the lysate is drained, leaving the protein of interest bound to the resin. The columns are washed successively with 10 ml of guanidine lysis buffer at pH 8.0, 5 ml of urea buffer at pH 8.0 (8 *M* urea-HCl, 0.1 *M* K_2HPO_4, 0.01 *M* Tris-HCl), 5 ml of urea buffer at pH 6.3 (8 *M* urea-HCl, 0.1 *M* K_2HPO_4, 0.01 *M* Tris-HCl), and twice with 5 ml of Probond Native buffer (PNB) at pH 7.8 [50 m*M* K_2HPO_4, 400 m*M* NaCl, 100 m*M* KCl, 10% (v/v) glycerol, 10 m*M* imidazole]. The columns can be stored at 4° overnight. Each column is eluted successively with 50 m*M*, 100 m*M*, 200 m*M*, 500 m*M*, and 1 *M* imidazole in PNB buffer at pH 7.8 (2 ml of each) into separate tube, and the fractions are separated by SDS–PAGE and analyzed by Sypro Orange or Coomassie staining.[6] In Fig. 4, proteins purified by the above procedure were run on an SDS–polyacrylamide gel and stained with Sypro Orange.

Summary

The expression of putative ORFs as fusion proteins can accelerate research greatly. The availability of an epitope tag allows the use of Western blotting as an efficient means to identify useful recombinant plasmids, which can then be used to study protein function. In addition, the epitope tag can be extremely useful in downstream applications such as protein purification, immunolocalization, and immunoprecipitation experiments.

The preceding protocols should be applicable to a variety of expression vectors, and should be useful in the identification of functional plasmids. The protocols require no exotic equipment and can be adapted for use in high-throughput transfection assays, protein purification protocols, and immunolocalization studies.

Acknowledgment

This work was supported by National Institutes of Health Grant 1 R01 CA80224-01.

[32] Rapid Construction of Recombinant DNA by the Univector Plasmid-Fusion System

By QINGHUA LIU, MAMIE Z. LI, DOU LIU, and STEPHEN J. ELLEDGE

Introduction

Since the advent of recombinant DNA technology in 1970s,[1] a wide variety of molecular techniques have evolved that allow for the cloning and manipulation of gene sequences for genetic, biochemical, and cell biological purposes. However, the functional analysis of a single gene often requires many cloning events of the same gene into various vectors for different purposes. Each of these manipulations consumes significant amounts of time and energy for the following reasons. First, each gene must be individually tailored for each vector. This is because not only is the sequence of every gene different, but also the majority of existing vectors were developed independently by different scientists, and thus contain different sequences and restrictions sites for insertions of genes. Second, the conventional "cut and paste" cloning strategy is time-consuming and requires many *in vitro* manipulations including restriction endonuclease digestion, agrose gel electrophoresis, DNA fragment isolation, and ligation. Finally, a rational cloning strategy must be designed prior to each cloning event by identifying a compatible vector and suitable restriction enzymes. This normally requires detailed knowledge of the gene sequence as well as that of the recipient vector. The advent of the polymerase chain reaction (PCR) and site-directed mutagenesis has greatly facilitated the alteration of gene sequences and the creation of compatible restriction sites for cloning purpose.[2,3] The high error rate of thermostable polymerases, however, requires each PCRderived DNA fragment to be verified by DNA sequencing, which is clearly another time-consuming process.

The Univector Plasmid-Fusion System

To facilitate the rapid, efficient, and uniform construction of recombinant DNA molecules, a novel cloning strategy, the univector plasmid-

[1] S. N. Cohen, A. C. Chang, H. W. Boyer, and R. B. Helling, *Proc. Natl. Acad. Sci. U.S.A.* **70,** 3240 (1973).

[2] K. Mullis, F. Faloona, S. Scharf, R. Saiki, G. Horn, and H. Erlich, *Cold Spring Harbor Symp. Quant. Biol.* **51,** 263 (1986).

[3] G. Winter, A. R. Fersht, A. J. Wilkinson, M. Zoller, and M. Smith, *Nature (London)* **299,** 756 (1982).

0076-6879/00 $30.00

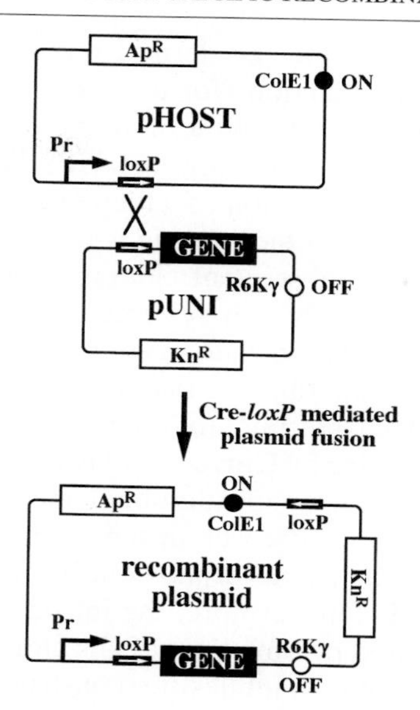

FIG. 1. A schematic representation of UPS. The Cre–*loxP*-mediated site-specific recombination results in fusion of pUNI and pHOST, thus generating a recombinant plasmid in which the gene of interest from pUNI is placed under the control of pHOST regulatory elements.

fusion system (UPS), was developed based on the Cre–*loxP* site-specific recombination system of bacteriophage P1.[4–6] *loxP* is a 34-bp DNA sequence that consists of two 13-bp inverted repeats flanking an 8-bp central sequence. Cre is the site-specific recombinase that catalyzes recombination between two *loxP* sites. The scheme for the UPS is shown in Fig. 1. The Cre–*loxP*-mediated site-specific recombination catalyzes plasmid fusion between the univector (pUNI) containing the gene of interest and the recipient vector (pHOST) that carries the regulatory information. Fusion events can be genetically selected and generate a recombinant plasmid in which the gene is placed under the control of novel regulatory elements. The UPS eliminates the need for restriction enzymes and DNA ligases; instead, both functions are carried out simultaneously by a single enzyme, Cre. Cloning by the UPS is rapid because it eliminates the many *in vitro* manipulations required for the restriction enzyme-mediated cloning. The UPS is also efficient because up to 17% of pUNI and pHOST plasmids can be fused

[4] N. Sternberg, D. Hamilton, S. Austin, M. Yarmolinsky, and R. Hoess, *Cold Spring Harbor Symp. Quant. Biol.* **45,** 297 (1981).

[5] K. Abremski, R. Hoess, and N. Sternberg, *Cell* **32,** 1301 (1983).

[6] Q. Liu, M. Z. Li, D. Leibham, D. Cortez, and S. J. Elledge, *Curr. Biol.* **8,** 1300 (1998).

in a single reaction to generate the desired recombinant plasmids.[6] And through a simple genetic selection (for details see Cloning by Univector Plasmid-Fusion System, below), 100% of transformants will contain the correct recombinant plasmids.[6] Finally, the UPS is uniform because the Cre–*lox* site-specific recombination is independent of plasmid size, sequences, and restriction sites. Once a gene is properly introduced into the univector, the majority of subsequent cloning events for the same gene can be accomplished uniformly and systematically simply by fusing with different recipient vectors. There is no need to worry about cloning strategies, reading frames, or compatible restriction sites, because all recombinant clones are constructed by plasmid fusion and will be automatically inframe (for details see Cloning by Univector Plasmid-Fusion System, below). Furthermore, we have developed additional methods to facilitate gene manipulation by using the UPS. For instance, the precise open reading frame (ORF) transfer (POT) method allows only the ORF to be transferred from pUNI to a pHOST. Furthermore, the introduction of genes into the univector can be facilitated by directional PCR cloning by *lacO* reconstitution. While the UPS greatly simplifies the construction of protein fusions at the amino (N) terminus of the protein of interest, the recombination-based "C-terminal tagging" method allows for the rapid and systematic generation of protein fusions at the carboxyl (C) terminus of the target protein.

The Univector

The univector is the plasmid into which the gene of interest should be cloned before using the UPS for the construction of recombinant DNA. In general, the univector is composed of a *loxP* site for plasmid fusion, a polylinker for insertion of genes, a conditional origin of replication from plasmid R6Kγ, and the *neo* gene encoding kanamycin resistance (Kn^R) from transposon Tn*5* for propagation and selection in bacteria. The *loxP* sequence is placed immediately upstream of the polylinker and its open reading frame is in-frame with the ATG of the *Nde*I and *Nco*I sites within the polylinker. This was designed to facilitate the subsequent generation of N-terminal protein fusions by the UPS. After the polylinker are eukaryotic and bacterial transcriptional terminators for the formation of 3′ ends of transcripts. The R6Kγ origin of replication is conditional because it is active in bacterial hosts expressing the *pir* gene but becomes completely silent in strains lacking *pir*. The *pir* gene, derived from plasmid R6Kγ, encodes a replication protein π that is essential for the function of the R6Kγ origin. Furthermore, depending on the *pir* allele present in the bacterial host, the R6Kγ origin can exist at two copy numbers: 15 copies/cell in the BW23473 strain (wild-type *pir*$^+$ allele) or 250 copies/cell in the BW23474 strain (mutant *pir-116* allele).[7] This property will become useful when potentially toxic genes are manipulated. The maps for several versions of the univector are shown in Fig. 2: pUNI10 (AF143506), pUNI50 (AF149258), pUNI60 (AF149259), and pUNI70 (AF149260), with the GenBank accession numbers in parentheses. The pUNI50, pUNI60, and pUNI70 univectors also include the T3 promoter placed immediately upstream of the *loxP* site for *in vitro* transcription and translation, and additional site-specific recombination sites, such as *RS,* to facilitate the precise ORF transfer (POT). In addition, the pUNI50 univector contains a *lacO* site to facilitate the C-terminal tagging, whereas the pUNI60 univector contains only half of a *lacO* site for directional PCR cloning by *lacO* reconstitution.

Recipient Vectors

Common features for the recipient vectors include the ColE1 origin of replication, the *bla* gene encoding ampicillin resistance (Ap^R) for propagation and selection in bacteria, a specific promoter, which will eventually control the expression of the gene of interest, immediately followed by a *lox* site for plasmid fusion. Recipient vectors may also contain specific sequences responsible for propagation, selection and maintenance in organisms other than *Escherichia coli.* In general, the pHOST vectors can be subdivided into two categories, those intended for transcriptional fusions (between a coding sequence and a promoter) and those intended for translational fusions (between two coding sequences). For translational fusions, recipient vectors must also contain additional coding sequences and the appropriate translational initiation sequences for the designated host.

Transcriptional Fusions

When transcriptional fusions are made using the UPS, a *loxP* site will be introduced between the promoter and the gene of interest, and subsequently become part of the 5′ untranslated region (UTR) of future transcripts. Because *loxP* contains two 13-bp perfect inverted repeates, it can form a stem–loop structure and interfere with the initiation of translation when the UPS-derived recombinant plasmids are expressed in eukaryotic cells.[6,8] However, we have found this is not a problem in the case of translational fusions made by the UPS because the *loxP* site is present within the coding regions. To circumvent the transcriptional fusion problem, we have de-

[7] W. W. Metcalf, W. Jiang, and B. L. Wanner, *Gene* **67,** 31 (1994).

[8] M. Kozak, *Mol. Cell. Biol.* **9,** 5134 (1989)

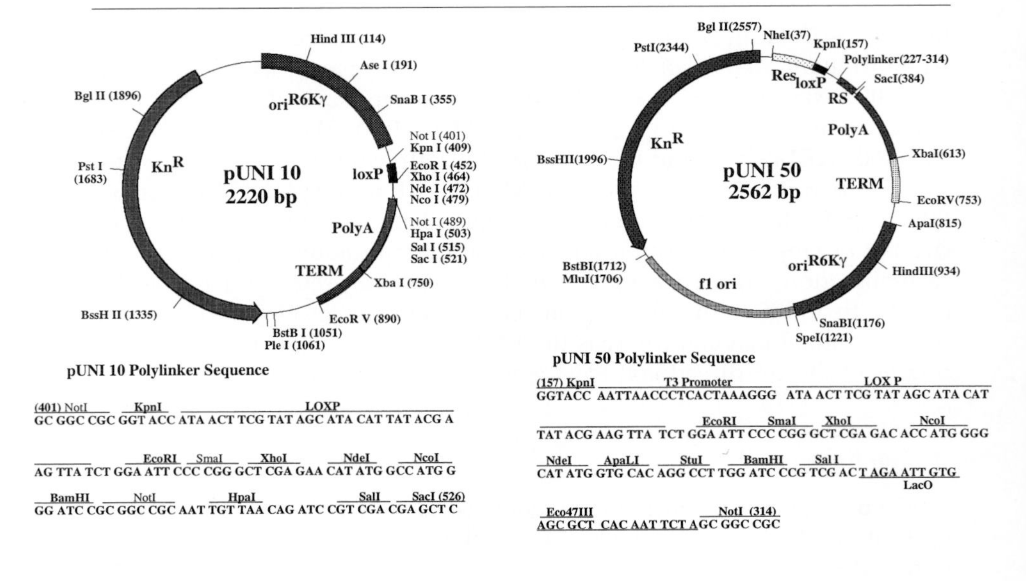

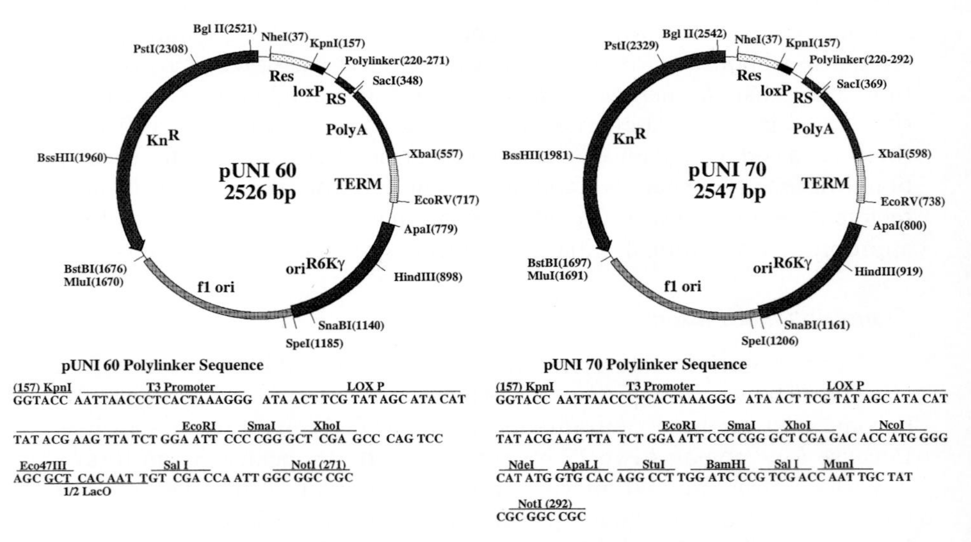

FIG. 2. Maps of pUNI10, pUNI50, pUNI60, and pUNI70. Nucleotide positions are shown with unique sites in boldface. Below each map is the polylinker sequence displayed as triplet codons according to the *loxP* ORF.

signed a mutant *loxP* site called *loxH* (ATtACcTCaTATAGCATACAT-TATACGAAGTTAT, mutations in lower case) by introducing mutations at nucleotides 3, 6, and 9 of the *loxP* sequence. Thus *loxH* will have three evenly dispersed mismatches between the two inverted repeats, diminishing its ability to form a secondary structure. This *loxH* site, when placed within the 5′ UTR of a reporter gene, failed to display any inhibitory effects on the expression of the reporter gene, while in the same context *loxP* reduced the expression by fourfold.[6] Furthermore, when pairing with *loxP, loxH* maintains 25% of the recombination efficiency of *loxP,* which is well within the range useful for recombinant plasmid construction by the UPS. Therefore, we recommend that *loxH* be used in recipient vectors for transcriptional fusions to maximize expression, whereas *loxP* should be used for other applications due to its higher recombination efficiency.

Translational Fusions

All translational fusions constructed by UPS will be automatically in-frame due to the precise nature of the Cre–*lox*-mediated site-specific recombination. This can be achieved when all recipient vectors are correctly constructed and the gene is properly introduced into the univector. Because *loxP* contains only one ORF, it should be placed immediately downstream and in-frame with the coding sequences carried by the recipient vectors. On the other hand, the gene or ORF should be cloned into the univector according to the *loxP* ORF, thus it will be eventually in-frame with the coding sequences of the recipient vectors. Although several amino acids encoded by *loxP* will be added at the fusion junction, they do not appear to affect either the folding or function of the fusion proteins generated by the UPS (our unpublished observations, 1998).

Construction of pHOST Vectors

The construction of pHOST vectors is simple, requiring only the insertion of a *lox*-containing oligonucleotide linker into the polylinker of a preexisting vector. Therefore, any vector can be easily adapted for UPS use. We are in the process of generating prototype recipient vectors for all general expression needs, including bacteria, yeast, and mammalian and insect cells. Many of these have been successfully tested by ourselves and others. In each case, expression was equivalent to the levels achieved by conventional cloning. Furthermore, we have made a series of base vectors to facilitate the construction of recipient vectors used for translational fusions (Fig. 3A). Each base vector carries a specific epitope tag coding sequence immediately followed by an in-frame *loxP*–*neo*(Kn^R)–*loxP* se-

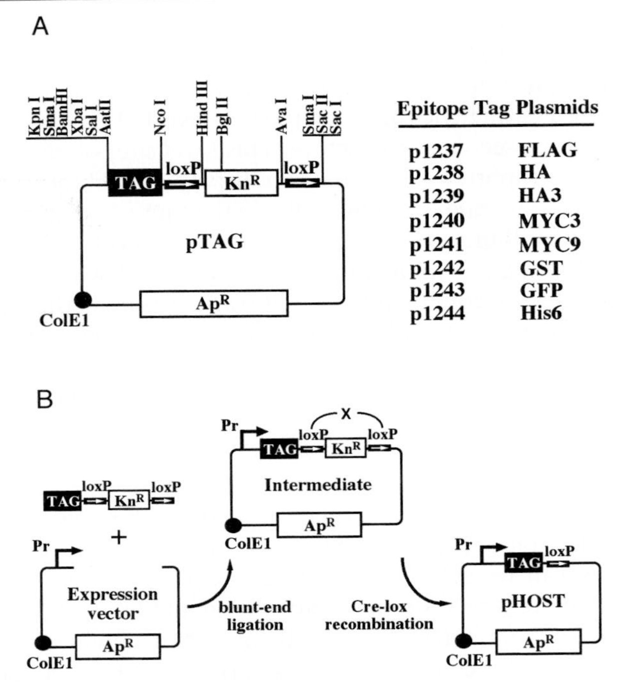

FIG. 3. (A) A schematic diagram of the pTAG vectors. The epitope tag-coding sequence is followed immediately by a *loxP–neo–loxP* cassette of which the first *loxP* site is placed in-frame with the epitope tag sequence. The unique restriction sites are shown without considering the internal restriction sites of specific epitope tags. (B) A universal scheme of constructing pHOST vectors from the pTAG vectors. The Tag–*loxP–neo–loxP* cassette excised from pTAG by *Sma*I is subcloned into a preexisting expression vector by blunt-end ligation to create a precursor plasmid. The pHOST plasmid is then made from the precursor by the Cre–*lox*-mediated excision of the *neo* gene.

quence. The Tag–*loxP–neo–loxP* cassette can be excised by *Sma*I cleavage and subcloned into preexisting vectors by blunt-end ligation and selection for kanamycin resistance. Those subclones containing inserts in the correct orientation can be identified by restriction analysis. The recipient vectors are then made by the Cre–*loxP* mediated excision of *neo* (Fig. 3B). This can be accomplished *in vivo* by directly transforming into a Cre-expressing strain, such as BUN13, and screening for kanamycin-sensitive colonies or *in vitro* by briefly incubating the precursor plasmid with purified glutathione *S*-transferase (GST)–Cre enzyme followed by bacterial transformation. We recommend the *in vivo* method because it is efficient.

GST–Cre

For the *in vitro* UPS reaction, GST–Cre is used, which can be easily produced in large quantities in *E. coli* and purified to near homogeneity by a single-step purification.[9] The recombinase activity of purified GST–Cre is equivalent to that of native Cre enzyme.[6]

Protocol 1: Preparation of Active GST–Cre Enzyme

Reagents

BL21 strain containing pQL123 (GST–Cre)
LB/ampicillin (100 μg/ml) medium
0.4 *M* Isopropyl-β-D-thiogalactoside (IPTG) in H_2O
Glutathione–Sepharose beads (Pharmacia, Piscataway, NJ)
Lysozyme (10 mg/ml in TE)
Leupeptin and antipain (both 5 mg/ml in H_2O)
NETN [0.5% (v/v) Nonidet P-40 (NP-40), 1 m*M* EDTA, 20 m*M* Tris-HCl (pH 8), 100 m*M* NaCl]
NT [100 m*M* Tris-HCl (pH 8), 120 m*M* NaCl]
NTG [100 m*M* Tris-HCl (pH 8), 120 m*M* NaCl, 20 m*M* glutathione]

Methods

1. From 0.5 ml of fresh overnight culture, grow a 50-ml culture of BL21 (pQL123) at 37° in LB/ampicillin medium until OD_{600} of 0.6 is reached. Occasionally, repeated passage of expression clones in BL21 shows reduced expression. If this occurs, using a fresh transformant of BL21 with the GST–Cre-expressing plasmid pQL123 should yield more reproducible results.
2. Add IPTG to a final concentration of 0.4 m*M*. Induce overnight at room temperature.
3. Pellet the cells at 4,500 rpm for 10 min at 4° in a 50-ml conical tube. Discard the supernatant. The cell pellet may be frozen at −80° at this point.
4. Resuspend the cell pellet in 2.8 ml of ice-cold NETN that is freshly supplemented with 1 m*M* phenylmethylsulfonyl fluoride (PMSF), leupeptin (5 μg/ml), and antipain (5 μg/ml).
5. Add 200 μl of lysozyme (10 mg/ml in TE) and incubate on ice for 15 min. If using the BL21 (pLysS) strain expressing lysozyme, this step can be skipped.
6. Sonicate the cell suspension five times [15 sec on setting 2 of a Branson (Danbury, CT) sonifier 450] in a cold room with a 30-sec interval

[9] D. B. Smith and K. S. Johnson, *Gene* **67,** 31 (1988).

of incubation on ice between sonications. Avoid foaming, which denatures proteins, by keeping the sonicator tip well below the aqueous surface.

7. Transfer the cell lysate from step 6 into three clean Eppendorf tubes (1 ml/tube). Clear the lysates by spinning (Eppendorf) at 14,000 rpm for 15 min at 4° to remove cell debris.

8. Prepare glutathione–Sepharose beads in a 100-μl bed volume per 1.5-ml Eppendorf tube. Wash three times with 1 ml of ice-cold NETN. Mix by inverting the tubes. Pellet the beads gently between washes by spinning at 3000 rpm for 15 sec at room temperature. Remove the liquid after each wash. After the last wash, add the clear supernatant from step 7 to the beads and mix by inverting the tubes. Rotate the tubes at 4° for 1 hr on a rotating rod.

9. Pellet beads as in step 8 and wash four times with 1 ml of ice-cold NETN that is freshly supplemented with protease inhibitors, as in step 4. For convenience, these washes can be performed at room temperature.

10. Wash twice with 1 ml of NT to replace NETN. After the last wash, remove excess liquid with thin-tip pipette tips (PHENIX, Hayward, CA) that will not take up the beads.

11. Elute the GST–Cre protein by adding 100 μl of freshly prepared NTG. Mix by inverting the tubes and rotate the tubes at 4° for 1 hr. Pellet beads as in step 8 and collect the supernatant. Repeat the elution once and combine the two elutions. Typically 0.1–0.4 μg of GST–Cre per microliter is obtained.

12. Distribute the GST–Cre elution into 20-μl aliquots and store at −20 or −80°. Once an aliquot is thawed, it can be stored at 4° for 1 month without significant loss of enzymatic activity. Measure the quantity of the GST–Cre preparation by running 5 μl of elution on a sodium dodecyl sulfate (SDS)–10% (w/v) polyacrylamide gel and judge the quality by performing a standard *in vitro* UPS reaction.

Cloning of Genes into the Univector

When cloning a gene into the univector, it is best to include only its coding sequence. The entire 5′ UTR should be excluded because it may complicate the subsequent generation of N-terminal protein fusions, whereas deletion of 3′ UTR will facilitate the generation of C-terminal protein fusions. Furthermore, the gene of interest should be cloned in-frame with the *loxP* ORF on the univector; only then can it be used to seamlessly create translational fusion expression constructs by fusing with different recipient vectors. All translational fusions made by the UPS will be automatically in-frame due to the precise nature of the Cre–*lox* site-specific recombination. The polylinker of the univector contains many useful restriction sites for insertion of genes, including *Nde*I and *Nco*I sites, of which the ATG is in-frame with the *loxP* ORF. Thus, it is desirable to engineer a restriction site, such as *Nde*I or *Nco*I, at the translational start codon of the target gene to facilitate its cloning into the univector. When cloning PCR-derived fragments, the following sequencing primers can be used to verify the DNA sequences introduced into the univector: the 5′ primer, 5′-ATAACTTCGTATAGCATACAT-3′, located at the *loxP* site and the 3′ primer, 5′-ATTTTATTAGGAAAGGACAG-3′, located within the mammalian polyadenylation sequence.

Strains

BW23473: Δ*lac-169 rpoS*(Am) *robA1 creC510 hsdR514* ***endA recA1*** *uidA* (Δ*Mlu*I)::***pir***$^{+}$

BW23474: Δ*lac-169 rpoS*(Am) *robA1 creC510 hsdR514* ***endA recA1*** *uidA* (Δ*Mlu*I)::***pir-116***

The relevant genotypes are in boldface type.

The propagation of univector-based plasmids requires transformation into bacterial strains expressing the *pir* gene, such as the BW strains listed above. There are a number of protocols available for making heat-shock competent cells. Protocol 2, described below, works well for the BW23473, BW23474, and other typical *E. coli* strains.

Protocol 2: Preparation of Heat-Shock Competent Cells

Reagents

TfbI: 30 m*M* Potassium acetate, 100 m*M* RbCl, 10 m*M* $CaCl_2$, 50 m*M* $MnCl_2$, 15% (v/v) glycerol; adjust to pH 5.8 with 1:10 diluted acetic acid. Sterilize by filtration

TfbII: 10 m*M* MOPS, 75 m*M* $CaCl_2$, 10 m*M* RbCl, 15% (v/v) glycerol; adjust to pH 6.5 with 1 *N* KOH. Sterilize by filtration

LB: Yeast extract (5 g/liter), tryptone (10 g/liter), NaCl (10 g/liter); agar at 20 g/liter for plates.

Methods

1. Grow a 5-ml overnight starter culture in LB medium from a fresh single colony. The next day, inoculate 500 ml of LB medium with the starter culture and grow at 37° until an OD_{550} of ~0.45 (normally takes 2–3 hr).

2. Chill the culture on ice for 5 min. Pellet the cells at 6000 rpm for 5 min at 4° [Sorvall (Newtown, CT) GS3 rotor]. All further steps should be performed in a cold room with chilled solutions and pipettes. Once cells

are treated with salts, they must be handled as gently as possible if high efficiency is to be obtained. Resuspend the cell pellets in 200 ml of TfbI buffer and leave on ice for 5 min.

3. Pellet the cells as in step 2. Resuspend the cell pellets in 20 ml of TfbII buffer and leave on ice for 15 min.

4. Aliquot and freeze the cells immediately in liquid nitrogen. Store at $-80°$ (cells maintain excellent efficiency for 2 years). The transformation efficiency is expected to be at least 5×10^6 colony forming unit (cfu)/μg supercoiled plasmid DNA.

Directional PCR Cloning by lacO Reconstitution. Because the introduction of genes or ORFs into the univector often utilizes PCR-amplified materials, we have developed a simple method for directional cloning of PCR fragments into the univector or other plasmids. This method relies on the reconstitution of a *lac* operator (*lacO*) site on DNA ligation. When present on a phage or high copy number plasmid, a functional *lacO* site can induce the expression of the endogenous *lacZ* gene by titrating out a limiting number of *lac* repressors (lacR).[10–12] Thus, plasmids carrying an intact *lacO* can be distinguished from those lacking it by their ability to induce endogenous *lacZ* and the formation of blue colonies on a 5-bromo-4-chloro-3-indolyl-β-D-galactopyranoside (X-Gal) plate (Fig. 4A and B). On the basis of these observations, we utilized a 20-bp symmetrical *lacO* derivative site, AATTGTGAGCGCTCACAATT, with an *Eco*47III restriction site located at the center.[13] We then placed the 3′ half-sequence of this *lacO* site into the polylinker of the pUNI60 univector and incorporated the 5′ half-sequence of *lacO* into the 3′ PCR primer used for amplifying the ORF of interest (Fig. 4A). Thus, the PCR-derived DNA fragment will contain the ORF of interest followed by the 5′ half of *lacO* at its carboxyl terminus. A functional *lacO* site will be reconstituted if the PCR fragment is ligated to the *Eco*47III-cleaved pUNI60 DNA in the correct orientation. When transformed into the *pir-116 lac*$^+$ BUN10 strain (for genotype see C-Terminal Tagging, below), these *lacO*-containing clones will generate dark blue color colonies on LB/kanamycin plates containing X-Gal at 80 μg/ml. However, if the PCR fragment is inserted in the opposite direction, it will be flanked by two half-*lacO* sites and only white or pale

[10] J. H. Miller and W. S. Reznikoff, "The Operon." Cold Spring Harbor Laboratory Press, New York, 1978.

[11] K. J. Marians, R. Wu, J. Stawinski, T. Hozumi, and S. A. Narang, *Nature* (*London*) **263,** 748 (1976).

[12] H. L. Heyneker, J. Shine, H. M. Goodman, H. W. Boyer, J. Rosenberg, R. E. Dickerson, S. A. Narang, K. Itakura, S. Lin, and A. D. Riggs, *Nature* (*London*) **263,** 748 (1976).

[13] J. R. Sadler, H. Sasmor, and B. L. Betz, *Proc. Natl. Acad. Sci. U.S.A.* **80,** 6785 (1983).

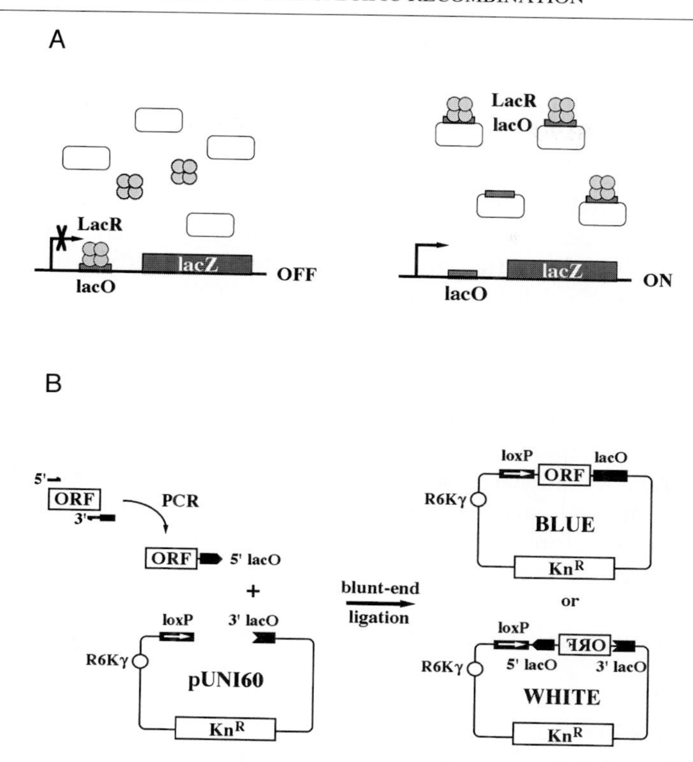

FIG. 4. Directional cloning of PCR fragments into the univector by reconstitution of a functional *lacO* site on ligation. (A) In the presence of plasmids lacking a *lacO* site, the endogenous *lacZ* gene is normally repressed by the lacR repressors binding to its *lacO* site. In the presence of a *lacO*-carrying high copy number plasmid, the limited number of LacR repressors is titrated out by the plasmid *lacO* sites, and thereby the expression of *lacZ* gene is derepressed. (B) A schematic representation of directional PCR cloning by *lacO* reconstitution. PCR is used to amplify the ORF of interest. Because the 3′ PCR primer contains on its 5′ end the 5′ half sequence of a *lacO* site (see text), the PCR-amplified ORF will be immediately followed by the 5′ half of a *lacO* site. On ligation to the *Eco*47III-linearized pUNI60 DNA that carries the 3′ half of *lacO,* an intact *lacO* site is reconstituted. When transformed into the BUN10 strain, this *lacO*-positive pUNI60 clone will activate the transcription of endogenous *lacZ* and result in the formation of blue colonies on plates containing X-Gal. In contrast, if the ligation occurs in the opposite direction, a *lacO*-negative clone is created and only white colonies will be formed.

blue colonies will be formed. This blue–white selection is efficient because approximately 90% of the dark blue colonies host plasmid clones with correctly oriented inserts. This method works best when the PCR primers used are phosphorylated. If *Taq* polymerase is used, it is necessary to briefly treat the material with T4 DNA polymerase and dNTPs to remove the 3′ overhangs.

Cloning by Univector Plasmid-Fusion System

Once the ORF of interest is properly introduced into the univector, dozens of constructs can be made simultaneously by the UPS in one afternoon simply by using different recipient vectors. The UPS reaction is performed as described in protocol 3. After the UPS reaction, the reaction mixture is transformed directly into a *pir*⁻ bacterial strain and selected for kanamycin resistance [typical *E. coli* strains are *pir*⁻, such as DH5α, JM107, one-shot cells from InVitrogen (Carlsbad, CA)]. Because of a lack of either a functional replication origin or the *neo* gene, neither pUNI nor pHOST can survive the kanamycin selection. The recombinant plasmids, however, can generate kanamycin-resistant colonies because the fusion between pUNI and pHOST covalently links *neo* to a functional ColE1 origin of replication. Therefore, in principle, 100% of Kn^R transformants will contain the correct recombinant plasmids.

Protocol 3: In Vitro Univector Plasmid-Fusion System Reaction

Reagents

Buffer (10×): 0.5 *M* Tris-HCl (pH 7.5), 0.1 *M* $MgCl_2$, 0.3 *M* NaCl, bovine serum albumin (BSA, 1 mg/ml)
GST–Cre (0.1–0.4 μg/μl)
pUNI plasmid DNA
pHOST plasmid DNA
pir⁻ *E. coli* competent cells
LB medium
LB/kanamycin (50 μg/ml) plates

Methods

1. Assemble the UPS mixture on ice as described below. Plasmid DNA prepared by the CsCl gradient, QIAGEN, RPM, and alkaline lysis methods can all be used for the UPS, although in general cleaner DNA gives better results. The relative ratio of pUNI and pHOST is not critical, but we frequently use an equal molar ratio of the two plasmids.

pUNI	0.2–0.4 μg
pHOST	0.2–0.4 μg
Buffer (10×)	2.0 μl
H_2O	X μl
	19 μl

2. Add 1 μl of GST–Cre. Mix well and immediately incubate at 37° for 20 min followed by 65° for 5 min to inactivate GST–Cre.

3. Thaw 100 μl of heat-shock competent cells on ice. Add 5–10 μl of the UPS mixture to the cells and mix by swirling with a pipette tip. Do not pipette cells up and down because it will decrease the transformation efficiency. Store on ice for 15 min followed by heat shock at 42° for 90 sec. Alternatively, use 1 μl of the UPS mixture for electroporation to avoid arcing.

4. Immediately after electroporation or after returning heat-shocked cells to ice for 1–2 min, add 1 ml of LB and recover at 37° on a rotating wheel or in a shaking water bath for 1 hr. Plate the transformants on LB/kanamycin plates and incubate at 37° overnight. Typically 100–10,000 colonies will be obtained. The number of Kn^R transformants depends on the amount of substrates, the quality of GST–Cre, and the efficiency of the competent cells used.

5. If converting libraries, the UPS mixture should be purified to remove salts and concentrated by phenol–chloroform extraction followed by ethanol precipitation or by using a GeneClean kit (BIO 101, Vista, CA), and then transformed by electroporation to maximize the transformation efficiency.

Restriction Analysis of Recombinant Plasmids. Although 100% of Kn^R transformants will contain the correct fusion products of pUNI and pHOST and most are dimeric plasmids, occasionally trimers are observed. These trimers function as well as dimers for most needs and are generated primarily by single fusion events between preexisting monomeric and dimeric plasmids rather than sequential fusion events of monomeric plasmids (our unpublished observations, 1997). Because the UPS results in a fusion between pUNI and pHOST at the *lox* sites, on restriction analysis the recombinant plasmids will generate unique patterns rather than a simple combination of the parental bands. For instance, dimers will have one or two parental bands missing and the same number of new bands appearing at new positions. Trimers will maintain one of the missing parental bands in addition to the dimeric bands. To keep the digestion pattern simple and informative, it is advisable to choose restriction enzymes having fewer sites and to avoid enzymes that cut near the *loxP* sites on both pUNI and pHOST. It is also helpful if simple drawings of both parental and recombinant plasmids are used to predict their patterns prior to the restriction analysis.

Precise Open Reading Frame Transfer

The precise ORF transfer (POT) utilizes both the R–*RS* and Cre–*lox* site-specific recombination to precisely transfer only the ORF of interest

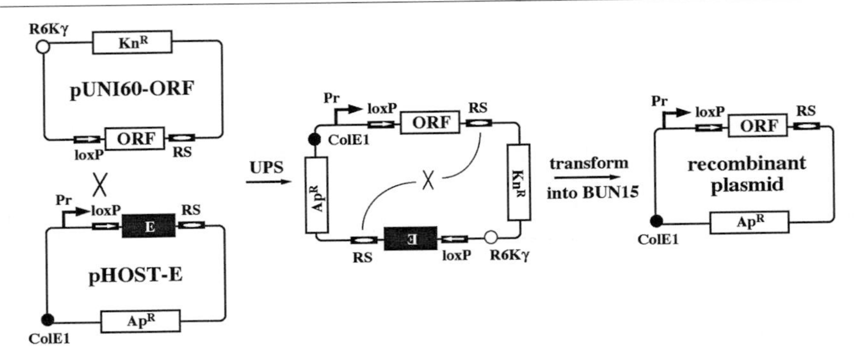

FIG. 5. A schematic representation of POT. The Cre–*loxP*-mediated site-specific recombination promotes the initial fusion between pUNI60-ORF and pHOST-E. The UPS mixture is transformed directly into BUN15 cells, which transiently express the R recombinase from a Ts plasmid. The R–*RS*-mediated site-specific recombination will eliminate both the lethal *E* gene and the univector backbone, resulting in the precise transfer of the ORF from pUNI60 to pHOST.

from the univector into a recipient vector. As illustrated in Fig. 5, the ORF of interest is cloned into the univector, such as pUNI60, and is flanked by the *loxP* and *RS* sites. The recipient vector, pHOST-E, contains the *loxP* and *RS* sites flanking a lethal gene *E* from phage ϕX174 under the control of the *tac* promoter.[14,15] After the UPS reaction the reaction mixture is transformed directly into the BUN15 strain and selected for ampicillin resistance at 42° in the presence of IPTG. BUN15 is a *pir*$^-$ strain, which can transiently express the R recombinase from a temperature-sensitive (Ts) plasmid. This plasmid carries a Ts origin of replication and will be lost at restrictive temperature, 42°. Under these selection conditions, the pUNI60-ORF plasmid fails to propagate because it lacks either a functional origin of replication or an ampicillin-resistant marker. Because of the induction of the lethal gene *E* by IPTG, neither the pHOST plasmid nor the pUNI60-pHOST cointegrant can produce viable transformants. Thus, only the recombinant plasmids that have completed POT will propagate and form ampicillin-resistant colonies. The toxic *E* gene, together with the univector backbone, is eliminated from the pUNI-pHOST cointegrant by the second R–*RS*-mediated site-specific recombination, resulting in the precise transfer of the ORF of interest into the recipient vector.

[14] D. Maratea, K. Young, and R. Young, *Gene* **40,** 39 (1985).
[15] W. D. Roof, H. Q. Fang, K, Young, J. Sun, and R. Young, *Mol. Microbiol.* **25,** 1031 (1997).

Strain

BUN15: XL1 Blue strain [F′::Tn*10 proA⁺B⁺ lacI*^{Q} Δ(*lacZ*)*M15*/*recA1 endA1 gyrA96* (*Nalr*) *thi hsdR17* (*rk-mk⁺*) *supE44 relA1 lac*] containing plasmid pML66, which carries a Ts replication origin, a spectinomycin-resistant marker, and a *tac:R* expressing cassette. To prepare competent cells, BUN15 cells are cultured at 30° in LB/spectinomycin (50 μg/ml) medium containing 0.4 m*M* IPTG to induce to expression of the R recombinase.

Protocol 4: Precise Open Reading Frame Transfer Reaction

Reagents

Phenol–chloroform (1 : 1, v/v)
Sodium acetate, pH 5.3 (3 *M*)
Ethanol
BUN15 electrocompetent cells
LB/ampicillin (100 μg/ml)–IPTG (0.1 m*M*) plates

Methods

1. Perform UPS reaction as described in protocol 3.
2. Add 80 μl of H_2O and 100 μl of phenol–chloroform (1 : 1, v/v) to the 20-μl UPS mixture. Mix well by vortexing. Spin at 14,000 rpm for 5 min at room temperature in an Eppendorf centrifuge. Transfer the supernatant to a new Eppendorf tube and add 10 μl 3 *M* sodium acetate, pH 5.3, and 200 μl ethanol. Mix well by vortexing and incubate at −80° for 30 min. Pellet the DNA at 14,000 rpm for 10 min at room temperature. Wash with 70% (v/v) ethanol, dry briefly, and resuspend in 10 μl of sterile water.
3. Mix 2 μl of DNA from step 2 with 40 μl of electrocompetent BUN15 cells thawed on ice. Perform the electroporation, add 1 ml of LB, and allow recovery at 30° on a rotating wheel or in a shaking water bath for 45 min. Plate the transformants on LB/ampicillin–IPTG plates and incubate at 42° overnight.

C-Terminal Tagging

It was previously shown that the *E. coli recBCsbcA* or *recD* mutants could take up linear DNA and recombine it onto the chromosome or resident plasmids.[16,17] Unlike wild-type *E. coli* strains, these mutant strains

[16] S. C. Winans, S. J. Elledge, B. B. Mitchell, L. Marsh, and G. C. Walker, *J. Bacteriol.* **161,** 1219 (1985).

[17] C. B. Russell, D. S. Thaler, and F. W. Dahlquist, *J. Bacteriol.* **171,** 2609 (1989).

not only lack the exonuclease activity that rapidly degrades linear DNA fragments, but also is able to efficiently produce near its ends single-stranded DNA required for *recA*-mediated homologous recombination. On the basis of these observations, we designed a simple and versatile method for generating protein fusions at the carboxyl (C) terminus of the gene of interest on the univector as well as other plasmids (Fig. 6). First, the epitope tag coding sequence is amplified by PCR from the pUNI(Ap^R)-TAG plasmid, a univector derivative containing *bla* (Ap^R) instead of *neo* (Kn^R). We have made a series of pUNI(Ap^R)-TAG plasmids, for which the GenBank accession numbers are as follows: pUNI(Ap^R) (AF149261), pUNI(Ap^R)-MYC3 (AF149262), pUNI(Ap^R)-HA3 (AF149263), pUNI(Ap^R)-GFP (AF149264), pUNI(Ap^R)-GST (AF149265), pUNI(Ap^R)-DBD (AF149266), and pUNI(Ap^R)-AD (AF149267). All these epitope tag sequences contained in the pUNI(Ap^R)-TAG plasmids are preceded in-frame by a common anchor sequence, 5′-GGATCCCCGGGTTAATTAA-3′. This was designed such that a single primer pair can be used to modify the C terminus of a given gene by any of the epitope tags listed above. As illustrated in Fig. 6A, the 5′ primer A (70 nucleotides) contains the last 50 nucleotides (excluding the stop codon) of the yeast *SKP1* ORF immediately followed in-frame by the anchor sequence. The 3′ primer B (22 nucleotides), 5′-GGATATAGTTCCTCCTTTCAGC-3′, recognizes a sequence on the univector 367 bp downstream of the epitope tag. Thus, the PCR-derived tag sequence will be flanked by 50 bp homologous to the C terminus of the *SKP1* ORF and 367 bp homologous to the univector. The PCR fragment, together with the linearized pUNI-SKP1 DNA that has been cleaved by unique restriction enzymes, such as the rare cutter *Not*I, located immediately 3′ to the C terminus of *SKP1,* are then cotransformed into the *recBCsbcA* BUN10 strain and selected for kanamycin resistance. Homologous recombination between the two linear DNA fragments will generate a circular plasmid in which the C terminus of *SKP1* ORF is fused in-frame with the epitope tag coding sequence. Because homologous recombination occurs much more efficiently than the religation of the linear pUNI-SKP1 plasmid, up to 95% of the Kn^R transformants contained the correct recombinant plasmids. The recombinant clones are usually identified by restriction analysis or by PCR, using the same primers as previously used for epitope tag amplification. It can also be facilitated by blue–white selection when using the pUNI50 univector, a *lacO*-containing plasmid that can activate the endogenous *lacZ* gene and give rise to blue colonies even in the absence of IPTG (for details see Directional PCR Cloning by *lacO* Reconstitution, above). Plasmids having undergone 3′ end recombination events will generate white instead of blue colonies because the *lacO* site is removed by

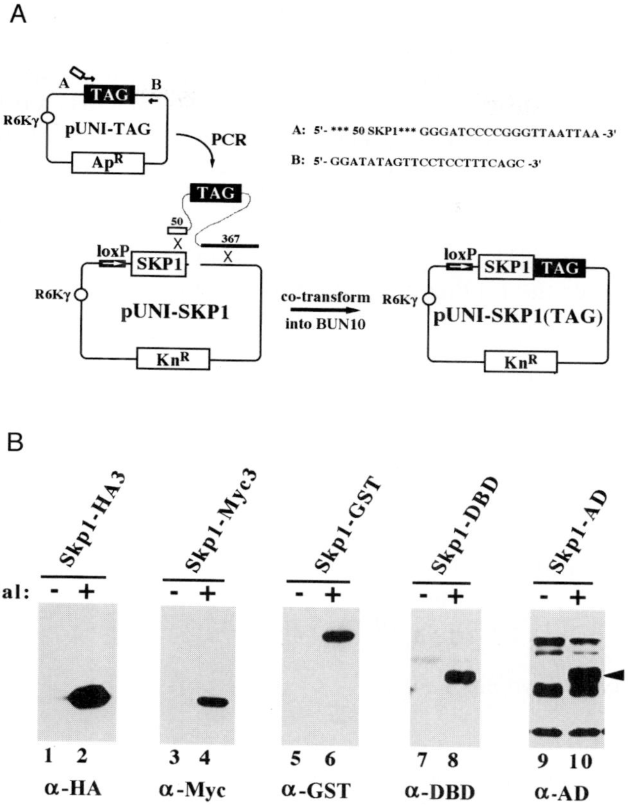

FIG. 6. C-terminal tagging by homologous recombination in *E. coli.* (A) A schematic representation of C-terminal tagging. The epitope tag sequence is amplified from the pUNI(Ap^R)-TAG plasmid with primers A and B, such that it will eventually be flanked by 50 bp homologous to the C terminus of the *SKP1* ORF and 367 bp homologous to the univector (for details see text). The PCR fragment is then cotransformed into the BUN10 strain along with pUNI-SKP1 DNA that has been linearized immediately 3′ to the *SKP1* ORF. Homologous recombination between the two linear DNA fragments results in a circular pUNI-SKP1 (TAG) plasmid in which the *SKP1* ORF is correctly tagged at its C terminus. (B) Expression of C-terminally epitope-tagged Skp1 proteins in *S. cerevisiae.* A series of pUNI-Skp1(TAG) clones made as described in (A) were fused to the pHY326 (2 μm *URA3 GAL*) recipient vector by UPS to create yeast expression constructs. Whole-cell lysates are subjected to SDS–PAGE followed by immunoblotting with relevant antibodies that recognize the individual tags. Shown here are Skp1–HA3 (lanes 1 and 2), Skp1–Myc3 (lanes 3 and 4), Skp1–GST (lanes 5 and 6), Skp1–DBD) (lanes 7 and 8), and Skp1–AD (lanes 9 and 10) either in the absence (odd lanes) or presence (even lanes) of galactose induction. The arrowhead indicates the position of the Skp1–AD fusion protein.

homologous recombination. However, the recombination reaction is usually so efficient that the blue/white screening is unnecessary.

We have successfully tested several genes with multiple epitope tags by this recombination method, including HA3, Myc3, GST, GFP, and the DNA binding domain (DBD) and activation domain (AD) of the yeast Gal4 protein (our unpublished observations, 1999). Shown in Fig. 6B are the Skp1–HA3, Skp1–Myc3, Skp1–GST, Skp1–DBD, and Skp1–AD fusion proteins expressed from the yeast *GAL1* promoter in *Saccharomyces cerevisiae.* In all cases, these epitope tags function normally and do not seem to affect either the folding or function of the fusion proteins.

Strain

BUN10: ***recB21 recC22 sbcA23 lac⁺ hsdR::cat-pir-116*** (Cm^R) *hisG4 thr-1 leuB6 thi-1 lacY1 galK2 ara-14 xyl-5 mtl-1 argE3 rfbD1 mgl-51 kdgK51* Δ(*gpt-proA*)*62 rpsL31 tsx33 supE44*

The mutant *pir-116* allele allows the propagation of univector clones at 250 copies/cell. The wild-type *lacZ* gene confers the blue–white selection for the C-terminal tagging and the directional PCR cloning by *lacO* reconstitution. The *recBCsbcA* mutations are required for the C-terminal tagging, and so is the *hsdR* mutation, which prevents the restriction of unmethylated PCR-amplified DNA by *E. coli* host restriction enzymes.

Protocol 5: C-Terminal Tagging

Reagents

PCR primers and reagents
Restriction enzymes
BUN10 electrocompetent cells
LB/kanamycin (50 μg/ml) plates
Low melting point (LMP) agarose (GIBCO-BRL, Gaithersburg, MD)
GeneClean kit (BIO 101, Vista, CA)

Methods

1. Amplify the epitope tag sequence by PCR, using pUNI(Ap^R)-TAG DNA as template. Isolate the PCR fragment on a 1% (w/v) LMP agarose gel and purify it from the agarose with a GeneClean kit or equivalent method.
2. Linearize the pUNI-ORF DNA with a unique restriction enzyme, such as *Not*I, that cuts immediately 3′ to the C terminus of the ORF of interest. Isolate and purify the linearized univector DNA by LMP agarose gel electrophoresis and GeneClean.
3. Use 0.1 μg each of the PCR fragment and the linear pUNI-ORF DNA to cotransform BUN10 by electroporation.

4. Add 1 ml of LB and allow recovery at 37° for 45 min. Plate the transformants onto LB/kanamycin plates and incubate at 37° overnight.

5. The correct recombinant plasmids can be identified by restriction analysis or by PCR, using the same primers previously used for amplifying the epitope tag.

Concluding Remarks

Genome projects will have an enormous impact on modern biological research. The access to whole genome sequences allows gene analysis to be performed on unprecedentedly large scales. However, the inherent defects of restriction enzyme-mediated cloning will be exponentially amplified when large sets of genes need to be analyzed in parallel. Therefore, modern biological research is in great need of new cloning technologies to facilitate the advancement of genome projects into functional proteome projects. Here we have described a novel cloning strategy, the univector plasmid-fusion system (UPS), that facilitates the rapid, efficient, and uniform construction of recombinant DNA molecules without the use of restriction enzymes and DNA ligases. This system, together with several adaptations described herein, including the precise ORF transfer, directional PCR cloning by *lacO* reconstitution, and the C-terminal tagging by homologous recombination, provides a comprehensive new approach for the manipulation of DNA sequences. The simplicity and uniformity of UPS make it possible to manipulate large sets of genes with minimal effort, a feature that is required for postgenome research. Furthermore, for each model organism whose genome is completely sequenced, all the identified ORFs could be cloned into the univector to make a "unigenome" array or library. It is highly feasible that, using the rapid and uniform UPS method, these arrayed clones could be systematically converted by automation into various recipient vectors. The highly efficient UPS method, together with the extremely efficient electroportion method of bacterial transformation, also makes it possible to convert the whole "unigenome" library in a single reaction into other specialized vectors, turning one library into many libraries.[6] Finally, this "unigenome" approach will eliminate many of the redundant cloning efforts that are currently a common practice in molecular biological research.

[33] High-Throughput Methods for the Large-Scale Analysis of Gene Function by Transposon Tagging

By ANUJ KUMAR, SHELLEY ANN DES ETAGES, PAULO S. R. COELHO, G. SHIRLEEN ROEDER, and MICHAEL SNYDER

Introduction

As a discipline, molecular biology has never been static: Newly emerging technologies have always fueled fundamental shifts in the paradigms of biological studies. For example, advances in DNA sequencing technology during the mid-to-late 1980s catalyzed efforts to sequence entire genomes during the early 1990s. These large-scale sequencing efforts enabled the release in 1995 of the complete genome sequence of the bacterium *Haemophilus influenzae*[1]; further large-scale projects have since resulted in the determination of well over 40 complete genome sequences,[2,3] with ever-increasing quantities of genomic sequence information expected within the immediate future. This unprecedented volume of raw sequence data presents an imposing challenge: To effectively utilize this resource, researchers must develop innovative experimental approaches to define and assess gene function on a global, genome-wide scale.

A number of such large-scale functional studies are already in place, most notably as a means of studying the budding yeast, *Saccharomyces cerevisiae.* With its fully sequenced genome[4] and straightforward genetics, *Saccharomyces* has provided the template for large-scale projects analyzing gene expression and protein function. For example, a variety of techniques have been used to obtain genome-wide expression data in yeast. The serial analysis of gene expression (SAGE) offers a means of measuring mRNA abundance by sequencing chimeric cDNAs consisting of concatenated tags uniquely identifying each expressed gene; transcript abundance can be determined from the frequency with which each unique tag is present within this concatenated product.[5] Advances in biochemistry and chemical engineering have enabled the development of high-density microscopic

[1] R. D. Fleischmann, M. D. Adams, O. White, R. A. Clayton, E. F. Kirkness, A. R. Kerlavage, C. J. Bult, J. F. Tomb, B. A. Dougherty, J. M. Merrick, *et al., Science* **269,** 496 (1995).

[2] G. J. Olsen and C. R. Woese, *Cell* **89,** 991 (1997).

[3] E. V. Koonin, *Genome Res.* **7,** 418 (1997).

[4] H. W. Mewes, K. Albermann, M. Bahr, D. Frishman, A. Gleissner, J. Hani, K. Heumann, K. Kleine, A. Maierl, S. G. Oliver, F. Pfeiffer, and A. Zollner, *Nature* (*London*) **387,** 7 (1997).

[5] V. E. Velculescu, L. Zhang, B. Vogelstein, and K. W. Kinzler, *Science* **270,** 484 (1995).

METHODS IN ENZYMOLOGY, VOL. 328
0076-6879/00 $30.00

arrays containing miniature ordered collections of oligonucleotide[6] or cDNA[7] samples. These microarrays serve as hybridization targets for fluorescently labeled cellular mRNA probes; after hybridization, quantification of fluorescence intensity can provide a measure of expression levels for tens of thousands of genes in parallel.[8] In addition, gene function can be investigated on a large scale by a method termed genomic footprinting,[9] in which overexpression of the *Ty1* transposon is used to generate large numbers of mutagenic insertion events throughout the yeast genome. Subsequent growth of these *Ty1*-mutagenized strains under various selective conditions can be used to identify genes contributing to growth under those particular conditions, as strains carrying insertions within such genes will be underrepresented in comparison with the starting population. Standard methods of deletion analysis are also now being applied on a genomic scale. Polymerase chain reaction (PCR)-based gene deletions are being systematically employed to generate a comprehensive set of tagged deletion strains, with each strain deleted for 1 of the nearly 6200 predicted genes in the *Saccharomyces* genome.[10]

Functional studies of the *Saccharomyces* genome have also encompassed efforts to analyze directly all predicted protein products. Such studies have primarily utilized methodologies to generate and analyze chimeric protein products. For example, the classic two-hybrid method[11] is currently being modified[12] for use in high-throughput, systematic screens as a means of constructing a complete protein interaction map for yeast. Translational fusions of *lacZ* to randomly tagged yeast genes have been generated on a large scale by transposon-based mutagenesis techniques.[13] Yeast strains carrying these fusion proteins have been examined by immunofluorescence microscopy with anti-β-galactosidase (β-Gal) antibodies in order to determine the subcellular localization of each β-Gal chimera; these β-Gal fusions have also been useful in generating gene expression data on a large scale.[13,14]

[6] D. Lockhart, H. Dong, M. Byrne, K. Follettie, M. Gallo, M. Chee, M. Mittmann, C. Wang, M. Kobayashi, H. Horton, and E. Brown, *Nature Biotechnol.* **14,** 1675 (1996).

[7] J. DeRisi, L. Penland, P. O. Brown, M. L. Bittner, P. S. Meltzer, M. Ray, Y. Chen, Y. A. Su, and J. M. Trent, *Nature Genet.* **14,** 457 (1996).

[8] M. Schena, D. Shalon, R. Heller, A. Chai, P. O. Brown, and R. W. Davis, *Proc. Natl. Acad. Sci. U.S.A.* **93,** 10614 (1996).

[9] V. Smith, K. N. Chou, D. Lashkari, D. Botstein, and P. O. Brown, *Science* **274,** 2069 (1996).

[10] R. Rothstein, *Methods Enzymol.* **194,** 281 (1991).

[11] S. Fields and O. Song, *Nature* (*London*) **340,** 245 (1989).

[12] M. Fromont-Racine, J. C. Rain, and P. Legrain, *Nature Genet.* **16,** 277 (1997).

[13] N. Burns, B. Grimwade, P. B. Ross-Macdonald, E. Y. Choi, K. Finberg, G. S. Roeder, and M. Snyder, *Genes Dev.* **8,** 1087 (1994).

[14] P. Ross-Macdonald, N. Burns, M. Malcynski, A. Sheehan, G. S. Roeder, and M. Snyder, *Methods Mol. Cell Biol.* **5,** 298 (1995).

Similar gene expression and protein localization data may be obtained from a genomic study designed to generate gene fusions to the green fluorescent protein (GFP) in budding yeast.[15]

While these approaches are all positive steps toward developing efficient functional genomic methodologies, each technique in itself is limited. Most of these approaches rely on the prior identification of genes presumed to be open reading frames and are, therefore, ineffective means of finding new genes. In addition, the majority of these procedures are too labor intensive to be performed within a variety of genetic backgrounds, potentially limiting the utility of the data collected. Even more importantly, each approach provides only a specific type of biological information; to more fully understand gene function, the application of multiple approaches would be necessary. To address these drawbacks, we have developed a multifunctional transposon-based mutagenesis system for the large-scale accumulation of expression, phenotypic, and localization data. Our approach is not biased toward previously annotated genes and offers the additional advantage of generating plasmid-borne alleles of mutagenized genes for convenient analysis within any desired genetic background. Furthermore, our system offers the potential to determine gene functions associated with essential genes—a target population previously refractory to large-scale study.

Principle

The workhorse of our mutagenesis system is a multifunctional minitransposon (mTn) derived from the bacterial transposable element Tn*3*.[16,17] The minitransposon mTn-3xHA/*lacZ* (GenBank accession number U54828) is shown in Fig. 1; this construct has been used for the bulk of our genomic studies and is the focus of this discussion. In itself, mTn-3xHA/*lacZ* is transposition defective, as it lacks the transposase gene *tnpA*. The *tnpA* gene product transposase directs transposition of Tn*3*-type transposons through recognition of two *cis*-acting 38-bp sequences corresponding to the Tn*3* termini.[16] These Tn*3* terminal repeats flank the mTn-3xHA/*lacZ* construct as indicated. Supplied *in trans,* TnpA initiates mTn transposition, forming a complex cointegrate structure consisting of fused donor and target molecules with duplicated mTn sequences at the junction.[18] This

[15] R. K. Niedenthal, L. Riles, M. Johnston, and J. H. Hegemann, *Yeast* **12,** 773 (1996).

[16] M. F. Hoekstra, D. Burbee, J. Singer, E. Mull, E. Chiao, and F. Heffron, *Proc. Natl. Acad. Sci. U.S.A.* **88,** 5457 (1991).

[17] P. Ross-Macdonald, A. Sheehan, G. S. Roeder, and M. Snyder, *Proc. Natl. Acad. Sci. U.S.A.* **94,** 190 (1997).

[18] H. S. Seifert, E. Y. Chen, M. So, and F. Heffron, *Proc. Natl. Acad. Sci. U.S.A.* **83,** 735 (1986).

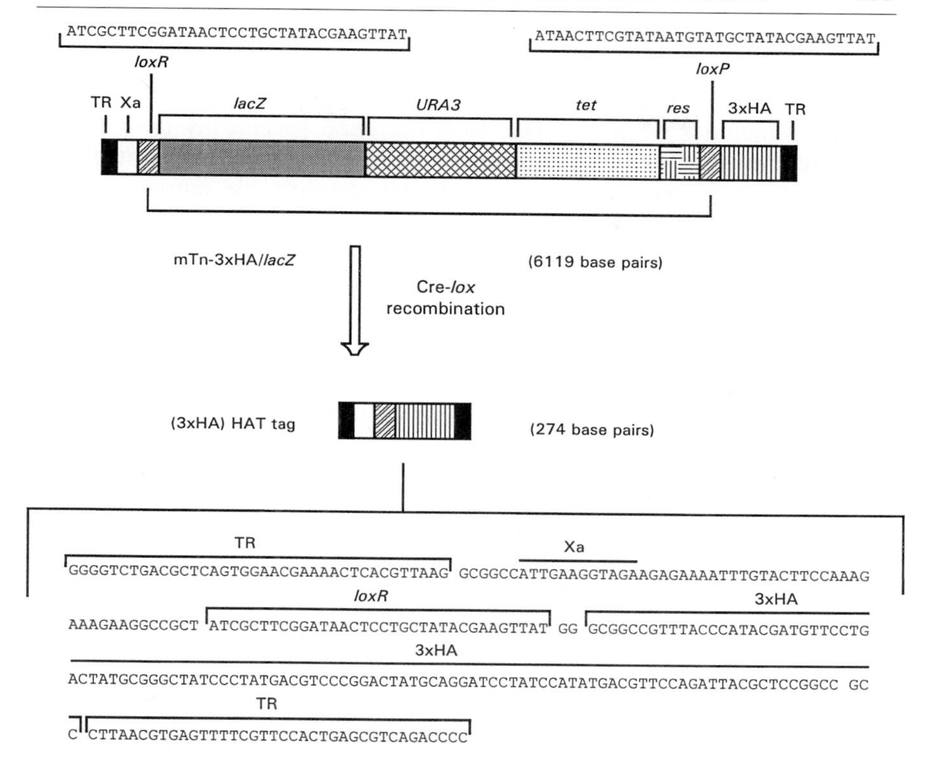

FIG. 1. Schematic representation of mTn-3xHA/*lacZ* and its HAT tag derivative. Constructs are annotated as follows: TR, Tn*3* terminal repeats; *loxR, loxP,* target sites for the Cre recombinase; 3xHA, sequence encoding three tandem copies of the HA epitope; *res,* Tn*3 res* site for resolution of transposition cointegrates; *lacZ,* a modified form of the *lacZ* gene lacking an initiator methionine codon as well as upstream promoter sequnce; *tet,* sequence encoding the tetracycline efflux protein; *URA3,* sequence encoding the Ura3 protein for selection in yeast. The nucleotide sequence of *loxR* and *loxP* is shown above each respective site; these target sites are divergent from one another to ensure low levels of spontaneous recombination in the absence of Cre. By Cre–*lox* recombination, the full-length mTn-3xHA/*lacZ* construct may be reduced to an epitope insertion element consisting primarily of three tandemly arranged copies of the HA epitope. Corresponding nucleotide sequence is indicated below the HAT tag element; complete DNA sequence is accessible from GenBank (accession number U54828). Schematics are not drawn to scale.

cointegrate form is subsequently resolved through the action of the *tnpR* gene product, which encodes a site-specific recombinase recognizing the *res* site[16] carried within mTn-3xHA/*lacZ*.[17] The TnpR resolvase, also provided *in trans,* catalyzes cointegrate resolution, yielding distinct donor and target molecules each containing a copy of the mTn construct. Therefore, the 38-bp Tn*3* terminal repeats and *res* site incorporated into mTn-3xHA/*lacZ* are sufficient to direct transposition in *Escherichia coli* through the *trans*-acting TnpA and TnpR gene products.

The remainder of mTn-3xHA/*lacZ* has been specifically designed to facilitate a variety of functional studies. mTn-3xHA/*lacZ* carries the bacterial selectable marker *tet* as well as the yeast selectable marker *URA3*, enabling insertional mutagenesis of target genes in *E. coli* and subsequent functional analysis in *Saccharomyces*. This shuttle mutagenesis scheme offers the advantage of generating insertion alleles carried on plasmids for potential analysis in a variety of *ura3* yeast strain backgrounds. To measure gene expression, mTn-3xHA/*lacZ* carries a modified form of the *lacZ* gene lacking both an ATG initiation codon and upstream promoter sequences. Production of β-Gal in yeast strains carrying mTn3x-HA/*lacZ* should result from mTn insertion within a transcribed and translated region of the genome, typically corresponding to an in-frame fusion of *lacZ* to yeast protein coding sequence. Because the mTn is not targeted to a subpopulation of previously annotated yeast open reading frames, this system is capable of uncovering previously unidentified coding sequences within the *Saccharomyces* genome. Productive *lacZ* fusions can be used to generate expression data by quantitative measurement of β-Gal activity; such measurements have previously been shown to provide a reliable index of gene expression.[19] In addition, mTn insertion creates a truncation of the mutagenized gene, thereby generating disruption alleles for subsequent phenotypic analysis.

mTn-3xHA/*lacZ* carries two *lox* elements, one located near each Tn*3* terminal repeat (Fig. 1). Specifically, a *loxP* element is located internal to sequence encoding three copies of an epitope from the influenza virus hemagglutinin protein (the HA epitope).[20] Toward the opposite end of mTn-3xHA/*lacZ* a *loxR* element lies adjacent to coding sequences specifying the cleavage site of factor Xa protease (Ile-Glu-Gly-Arg).[21] These *lox* sequences are target sites for the Cre recombinase, which catalyzes site-specific recombination between the *lox* sites.[22] Expression of the Cre recombinase, therefore, results in excision of the central body of the transposon, reducing the mTn construct to a 274-bp sequence consisting of a single *loxR* site flanked by the HA epitope tag, factor Xa protease site,

[19] P. Ross-Macdonald, A. Sheehan, C. Friddle, G. S. Roeder, and M. Snyder, *Methods Enzymol.* **303,** 512 (1999).

[20] I. A. Wilson, H. L. Niman, R. A. Houghten, A. R. Cherenson, M. L. Connolly, and R. A. Lerner, *Cell* **37,** 767 (1984).

[21] S. Magnusson, T. E. Peterson, L. Sottrup-Jensen, and H. Claeys, *in* "Proteases and Biological Control" (E. Reich, D. B. Rifkin, and E. Shaw, eds.), p. 123. Cold Spring Harbor Laboratory Press, Cold Spring Harbor, New York, 1975.

[22] B. Sauer, *Mol. Cell. Biol.* **7,** 2087 (1987).

and Tn*3* terminal repeats (Fig. 1). Because of a 5-bp target site duplication associated with Tn*3* transposition, this reduced construct actually corresponds to a small insertion element encoding precisely 93 amino acids. As this final product encodes no translational stops, a yeast strain bearing an in-frame insertion of mTn-3xHA/*lacZ* may be used to derive a corresponding strain containing an in-frame 93-amino acid HA epitope insertion element (or HAT tag).

HAT tags are useful in a variety of functional studies. Of foremost relevance to our work, full-length HAT-tagged proteins can be effectively immunolocalized with antibodies directed against the HA epitope. Furthermore, this HAT tag element offers the benefit of potentially generating conditional phenotypes, as demonstrated from a pilot mutagenesis study of the yeast *SER1* gene.[17] *SER1* encodes 3-phosphoserine aminotransferase, an enzyme required for serine biosynthesis; accordingly, *ser1* mutant strains are auxotrophic for serine.[23] Random insertion of the HAT tag element within *SERI* generated a series of mutant strains exhibiting temperature-sensitive serine auxotrophic phenotypes, confirming, the utility of mTn-3xHA/*lacZ* as a means of generating conditional alleles. Similarly, HAT tag insertions may provide a means of studying cellular functions associated with essential genes on a large scale. The HAT insertion element provides the potential to create hypomorphic mutants exhibiting partial gene function; HAT tag insertion within an essential gene may generate viable mutant alleles for subsequent phenotypic analysis under desired test conditions. Previous small-scale studies have supported the feasibility of this approach to the genome-wide study of essential genes.[17]

In summary, mTn-3xHA/*lacZ* can be used to generate a wide variety of mutant alleles. Null mutations, reporter fusions, epitope-tagged alleles, and hypomorphic mutants can all be obtained from a single mutagenesis, thereby facilitating the rapid acquisition of sufficient data to effectively characterize gene function on a large scale. As such, the protocols described in this text, although specifically tailored to the study of *Saccharomyces,* represent a general blueprint for the functional analysis of numerous genomes.

Method

An outline of the transposon tagging strategy is presented in Fig. 2. As indicated, we utilize a shuttle mutagenesis approach: mTn-3xHA/*lacZ* insertions are generated in a plasmid library of yeast genomic DNA in *E. coli* for subsequent analysis in *S. cerevisiae.* This initial mutagenesis is

[23] P. Belhumeur, N. Fortin, and M. W. Clark, *Yeast* **10,** 385 (1994).

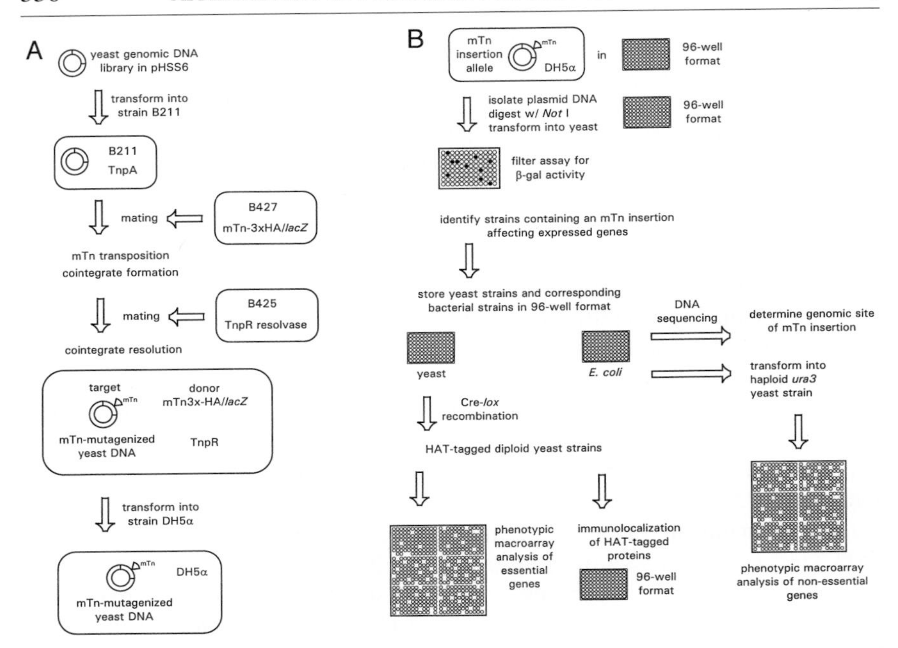

FIG. 2. The mTn insertion project. (A) Through the bacterial mating steps indicated, a plasmid-based yeast genomic DNA library may be mutagenized in *E. coli* with mTn-3xHA/*lacZ*. (B). High-throughput functional analysis of mTn-mutagenized genes. After transposition, mTn-insertion alleles can be recovered by isolation of plasmid DNA from individual bacterial transformants. After digestion with an appropriate restriction enzyme, mutagenized yeast DNA fragments may be transformed into a desired genetic background; those strains carrying productive *lacZ* fusions can be used to analyze gene function. Using mTn-encoded *lacZ* as a reporter, gene expression can be monitored under a variety of growth conditions. In addition, yeast DNA fragments carrying mTn-3xHA/*lacZ* may be transformed into a haploid strain for phenotypic macroarray analysis of nonessential genes. Essential genes can be studied using mTn-3xHA/*lacZ* in a diploid background; Cre-mediated recombination can be used to derive strains carrying HAT-tagged essential proteins, potentially generating viable hypomorphic mutants for phenotypic analysis. HAT-tagged proteins may also be immunolocalized using antibodies directed against the HA epitope. As indicated, all techniques have been adapted for performance using a 96-well format for high-throughput.

accomplished primarily through a series of bacterial matings (Fig. 2A). Plasmids carrying inserts of yeast genomic DNA are transformed into a bacterial strain producing the Tn*3* transposase, TnpA. An F plasmid bearing mTn-3xHA/*lacZ* is then transferred into these cells by mating. Resulting cointegrate structrues (containing both plasmids) are resolved through a second mating with bacterial cells expressing *tnpR*. The mutagenized yeast genomic library is subsequently recovered and transformed into a standard laboratory *E. coli* strain.

Individual transformants exhibiting mTn-mediated tetracycline resistance are selected for further analysis (Fig. 2B); in particular, plasmid DNA isolated from these transformants is useful for a variety of purposes. To identify productive *lacZ* fusion events, this plasmid DNA is digested with an appropriate restriction enzyme to liberate yeast genomic DNA. Mutagenized yeast DNA is subsequently transformed into an appropriate diploid yeast strain. Transformants carrying the mTn-borne *URA3* marker are selected and assayed for β-Gal activity after incubation under desired growth conditions. Yeast strains producing β-Gal are subsequently selected and maintained in a permanent collection. Plasmid DNA prepared from corresponding bacterial clones can be sequenced in order to determine the precise site of mTn insertion within the yeast genome. Yeast strains containing in-frame fusions to *lacZ* may then be treated to induce the Cre recombinase, present on a plasmid under transcriptional control of the *GAL* promoter. Ura^- colonies, which have presumably undergone the *lox* excision event, are selected and maintained within a collection used for immunolocalization of epitope-tagged proteins with anti-HA antibodies. In addition, these epitope-tagged strains may be used to determine mutant phenotypes resulting from HAT insertion in essential genes. Disruption phenotypes within nonessential genes are analyzed after transformation of a haploid yeast strain with plasmid-derived insertion alleles; haploid yeast transformants carrying full-length mTn insertions are subsequently transferred to appropriate test media to characterize mutant phenotypes. These phenotypic studies, when considered in conjunction with the expression data and protein localization studies outlined above, offer a thorough means of characterizing gene function on a genomic scale, applicable to *Saccharomyces* as well as other organisms possessing tractable genetic systems.

Protocols

Strains recommended for use in this study are indicated in Table I; all bacterial and yeast strains are cultured according to established protocols described elsewhere.[24] Standard experimental manipulations are performed as described in Sambrook *et al.*[25] Unless otherwise noted, all antibiotics are to be used at the following concentrations: kanamycin (40 μg/ml), chloramphenicol (34 μg/ml), tetracycline (3 μg/ml), streptomycin (100 μg/ml).

[24] C. Guthrie and G. R. Fink, *Methods Enzymol.* **194,** 933 (1991).

[25] J. Sambrook, E. F. Fritsch, and T. Maniatis, "Molecular Cloning: A Laboratory Manual." Cold Spring Harbor Laboratory Press, Plainview, New York, 1989.

TABLE I
STRAINS USED IN THIS STUDY

Strain	Description	Ref.
Escherichia coli		
R1123	Strain XL1-Blue (Stratagene) carrying pHSS6 (kanamycin resistance)	*a*
B211	Strain RDP146 (F^- *recA'* (Δ*lac-pro*) *rpsE*; spectinomycin resistant) with plasmid pLB101 (pACYC184 with *tnpA*; active transposase, chloramphenicol resistance)	*a*
B425	Strain NG135 (F^- K12 *recA56 gal-ΔS165 strA*; streptomycin resistant) with plasmid pNG54 (pACYC184 with *tnpR*; active resolvase, chloramphenicol resistance)	*b*
B427	Strain RDP146 with pOX38::mTn-3xHA/*lacZ* (F factor derivative carrying mTn3 derivative; tetracycline resistance)	*b*
B227	Strain DH5-α carrying p*GAL-cre* (*amp, ori, CEN, LEU2*)	*a*
Saccharomyces cerevisiae		
Y800	*MAT***a** *leu2-Δ98* $cry1^R$/*MATα leu2-Δ98 CRY1 ade2-101 HIS3/ade2-101 his3-Δ200 ura3-52*$can1^R$/*ura3-52 CAN1 lys2-801/lys2-801 CYH2*/$cyh2^R$ *trp1-1/TRP1* Cir^0	*b*
Y800-T1a	*MAT***a** *ade2-1 leu2-Δ98 ura3-52 lys2-801 trp1-1 his3-Δ200* Cir^0	*b*
BY4741	*MAT***a**/*MATα leu2Δ0/leu2Δ0 his3Δ0/his3Δ0 met15Δ0/MET15 ura3Δ0/ura3Δ0 LYS2/lys2*	
Y2279	*MAT***a** *ura3-52 trp1Δ1 ade2-101 lys2-801 leu2Δ98*	*b*

[a] M. F. Hoekstra, H. S. Seifert, J. Nickoloff, and F. Heffron, *Methods Enzymol.* **194,** 329 (1991).
[b] P. Ross-Macdonald, A. Sheehan, G. S. Roeder, and M. Snyder, *Proc. Natl. Acad. Sci. U.S.A.* **94,** 190 (1997).

Transposon Mutagenesis

To use Tn*3*-derived minitransposons for insertional mutagenesis, target DNA must be cloned into a vector free of any Tn*3* terminal repeats: vectors containing a Tn*3* repeat (e.g., all common *E. coli* ampicillin-resistant plasmids) are immune to Tn*3* transposition. This immunity operates only *in cis,* affecting any DNA element carrying at least one terminal Tn*3* repeat while leaving other molecules in the same cell accessible to transposition.[26] As the *E. coli* chromosome is immune to Tn*3* transposition, this property

[26] M. K. Robinson, P. M. Bennet, and M. H. Richmond, *J. Bacteriol.* **129,** 407 (1977).

can be used to direct mTn insertion into appropriate vectors carrying target DNA. For this purpose, we utilize derivatives of the pHSS series described by Hoekstra *et al.*[16] These Tn*3*-free constructs possess a small vector backbone consisting primarily of sequences essential for replication functions and antibiotic resistance; therefore, recovered mTn insertions fall preferentially within cloned DNA inserted into the vector. Transposon insertions occur once per plasmid (again due to transposition immunity) with little or no site preference within most genomic inserts.[27] In our hands, the procedure listed below can be reliably used to generate 10^5 independent insertions from a single mutagenesis.

1. A library of genomic DNA must be constructed (or obtained) in an appropriate vector as discussed. Plasmid DNA prepared from this library is then transformed into competent cells of strain B211, selecting on plates containing kanamycin (if using pHSS6 vectors) and chloramphenicol.
2. B211 transformant colonies are recovered for use in subsequent mating steps; an aliquot of this cell suspension may be stored in 15% (v/v) glycerol at −70° if desired.
3. To introduce mTn-3xHA/*lacZ,* strain B427 (carrying mTn-3xHA/*lacZ* on the F plasmid derivative pOX38[28] is grown overnight at 37° with aeration in LB medium supplemented with tetracycline. Strain B427 is then subcultured at 1 : 100 dilution into fresh LB medium (without antibiotics); the cell suspension obtained in step 2 is also diluted to the same density. Each strain is grown at 37° with aeration for 2–3 hr (early log phase), or until a cell suspension is first visible. Care should be taken at this stage not to vortex strain B427, as bacterial pili can be easily broken.
4. To initiate mating, 500 μl of each culture is mixed in a microcentrifuge tube; this mixture is subsequently incubated at 37° for 20 min to 1 hr without agitation. The mating process is highly sensitive to agitation and detergent. Also, a shorter mating time (such as 20 min) has been found to minimize bacterial replication and, therefore, increase the probability that the insertions obtained will be independent.
5. After mating, aliquots of 200 μl are plated onto LB medium supplemented with tetracycline, kanamycin, and chloramphenicol. As a control, a 20-μl aliquot from each starting culture should also be plated on this medium. Plates are subsequently incubated at 30° for 1–2 days to allow cointegrate formation; after this incubation, thousands of colonies should be visible (control plates should be free of growth).

[27] P. Ross-Macdonald, A. Sheehan, C. Friddle, G. S. Roeder, and M. Snyder, *Methods Microbiol.* **26,** 161 (1998).

[28] M. F. Hoekstra, H. S. Seifert, J. Nickoloff, and F. Heffron, *Methods Enzymol.* **194,** 329 (1991).

6. To resolve cointegrates, cells are mated to strain B425. In preparation, B425 should be grown overnight at 37° with aeration in LB medium supplemented with chloramphenicol. Prior to use, strain B425 is subcultured at 1:100 dilution in LB medium (without antibiotic); colonies containing cointegrates are washed off plates and subsequently diluted (in LB without antibiotic) to approximately the same cell density. Strains are grown and mated as in steps 3 and 4 above. After mating, 200-μl aliquots are plated onto LB medium supplemented with tetracycline, kanamycin, streptomycin, and chloramphenicol; controls should be included as described previously. After incubation overnight at 37°, plates should contain thousands of colonies (control plates should again be free of growth).

7. To recover DNA from this strain, colonies from the mating are eluted in LB medium; an aliquot of this suspension is diluted in LB medium supplemented with tetracycline and kanamycin to give a culture of almost saturated density. This culture is grown at 37° with aeration for 1–2 hr. The remaining suspension may be stored as stock in 15% (v/v) glycerol at −70°. Plasmid DNA is subsequently prepared from this culture by alkaline lysis; roughly one-tenth of this plasmid DNA can be used to transform a standard *recA endA E. coli* strain (e.g., DH5α) with selection on LB medium supplemented with tetracycline and kanamycin. Many thousands of transformants should be recovered.

8. This transformant pool may be stored as stock in 15% (v/v) glycerol at −70°.

High-Throughput Preparation of Plasmid DNA and Yeast Transformations

Individual colonies from the transformant pool generated above may be selected for further analysis. To isolate plasmid DNA from these colonies, we utilize a standard alkaline lysis protocol[25] adapted for use with samples in a 96-well format. Prepared plasmid DNA is digested to release mTn-mutagenized yeast genomic DNA for subsequent transformation into an appropriate yeast strain (e.g., any diploid or haploid *ura3* strain, although a diploid *ura3 leu2* strain would be required for the analysis of HAT tag insertions within essential genes as discussed below). By homologous recombination, the mutagenized genomic DNA fragment should replace its chromosomal locus, generating a yeast strain now carrying a chromosomal mTn insertion. Transformations are performed on a large scale by using a modified form of the lithium acetate/single-stranded DNA/polyethylene glycol (PEG) method described by Schiestl and Gietz.[29] These

[29] R. H. Schiestl and R. D. Gietz, *Curr. Genet.* **16,** 339 (1989).

high-throughput applications can be performed most efficiently with a multichannel pipetter or programmable 96-channel microdispenser (available from Robbins Scientific, Sunnyvale, CA). Using these methods, a single skilled technician can comfortably complete 768 plasmid DNA preparations and initiate corresponding yeast transformations in a 2-day period. Furthermore, plasmid DNA prepared according to this protocol is suitable for most standard laboratory manipulations, including DNA sequencing.

1. Bacterial transformants are selected after growth on solid LB medium supplemented with tetracycline and kanamycin. Subsequently, individual colonies are each transferred into 1.75 ml of TB medium[25] (maintaining antibiotic selection) in 96-well microplates (Whatman Polyfiltronics, Rockland, MA). Cultures are incubated overnight at 37° with aeration on either a rotary or platform shaker.
2. Bacterial cultures are harvested by centrifugation at 2000 rpm for 7 min in a Sorvall (Newtown, CT) H-1000B rotor. Pelleted cells are resuspended in 110 μl of solution I (50 mM Tris, 10 mM EDTA, pH 7.5); these bacterial suspensions are then transferred into 96-well 0.65 μm-pore filtration plates (Millipore, Bedford, MA) containing 100 μl of solution II per well [200 mM NaOH, 1% (w/v) sodium dodecyl sulfate (SDS)]. Samples are mixed by repeated pipetting prior to the addition of 100 μl of solution III (3 M potassium acetate, 5 M glacial acetic acid). Samples are mixed again and incubated immediately at −20° for at least 10 min.
3. Filtration plates are subsequently thawed at room temperature. Cell debris and genomic DNA are cleared from plasmid preparations by filtration with a vacuum manifold; filtrate containing plasmid DNA is collected in 96-well titer plates (1-ml Deep-Well plates available from Beckman Coulter, Fullerton, CA).
4. Plasmid DNA is precipitated with 2 volumes of ethanol; samples are mixed by inversion prior to incubation at −20° for a minimum of 5 min. Precipitated DNA is collected by centrifugation at 3100 rpm (2190g) for 25 min in a Sorvall H-1000B rotor. Recovered pellets are washed in 80% (v/v) ethanol and allowed to air dry before final resuspension in TE; DNA resuspension is facilitated by incubation at 37° for 30 min.
5. Yeast DNA is liberated from plasmid vector by digestion with a suitable enzyme. Our yeast genomic DNA libraries were constructed in pHSS6-based vectors using convenient *Not*I restriction sites; we, therefore, digest our prepared plasmid DNA with *Not*I overnight according to the manufacturer suggested protocols (New England BioLabs, Beverly, MA). After heat inactivation, this restriction digest can be used to transform competent yeast cells without further purification.

6. To prepare competent yeast, the strain to be transformed is inoculated into 50 ml of 2× SC medium[24] lacking leucine (if plasmid retention is required in this given strain) in a 250-ml flask; this culture is incubated overnight at 30° with shaking to yield roughly 10^7 cells/ml (OD_{600} of approximately 0.1). A 40-ml volume of YPAD[24] is subsequently added, and incubation at 30° with aeration is continued for an additional 2 hr. Cells are then harvested and washed in 100 ml of sterile water. Washed cells are resuspended in a 90-ml mixture of 1× TE (pH 7.5)–100 m*M* lithium acetate. This cell suspension is incubated at 30° with shaking for 30 min; cells are subsequently treated with 600 μl of 2-mercaptoethanol prior to an additional incubation at 30° with shaking for 30 min. Denatured carrier DNA (17 mg total salmon sperm DNA in a volume of 1.7 ml) is then added to the mixture.

7. Yeast cells are now competent to be transformed with the digested DNA generated in step 5 above. An aliquot of competent cells (100 μl) is dispensed into each well of a Deep-Well plate containing digested plasmid DNA (the entire 30-μl digest volume from step 5). The plate is then capped tightly and incubated on its side at 30° for 30 min; the plate is placed on its side in order to maximize surface contact between yeast cells and digested DNA. Subsequently, a 250-μl volume of polyethylene glycol solution [1× TE, 100 m*M* lithium acetate, 40% (w/v) PEG] is added to each well; samples are again incubated on their sides for an additional 15–30 min at 30°. Cells are then incubated at 45° for 15 min. This heat shock treatment can be performed in a water bath: The wells of Deep-Well plates are conical in shape, allowing each well adequate contact with the surrounding water.

8. After heat-shock treatment, samples are centrifuged at 3100 rpm (2190*g*) for 15 min at 4°. The PEG mixture is removed simply by decanting; cell pellets are resuspended in 2× SC dropout medium by gentle vortexing. Deep-Well plates containing yeast transformants are incubated on a rotary shaker at 30° for 3–5 days (medium may be refreshed after 4 days of growth). Transformants can now be screened for β-Gal activity.

Assay of β-Galactosidase Production in Yeast Transformants

To identify strains carrying productive *lacZ* fusions, Ura^+ transformants are screened for β-Gal activity using a filter-based chloroform lysis procedure.[14] After transformation, liquid cultures are transferred onto filter paper with a 96-channel microdispenser; in this way, an ordered array corresponding to each microplate is replicated on each filter. Filters are then incubated on appropriate SC dropout medium (e.g., 2× SC-ura) prior to cell lysis by chloroform vapor. Colonies producing β-Gal are subsequently identified by incubating filters on growth medium supplemented with X-Gal. β-Gal levels can be estimated by eye, or more quantitatively if desired. As mTn-3xHA/*lacZ* carries a *lacZ* reporter gene lacking an initiator methionine codon, only one in six insertion events within yeast coding sequence is predicted to yield an in-frame fusion to *lacZ*. Accordingly, we have observed β-Gal production in roughly 10% of all transformants (a reasonable finding in light of the large fraction of the yeast genome expected to encode proteins[4]). The precise genomic site of mTn insertion within these strains can be determined by directly sequencing corresponding plasmid-borne insertion alleles.

1. After growth in appropriate 2× SC dropout medium, liquid cultures of transformants can be replicated from Deep-Well plates onto Whatman 3MM filter paper (Clifton, NJ), using either a 96-prong replicator or a multichannel microdispenser. If using a programmable microdispenser, a small aliquot (e.g., 6 μl) of culture may be withdrawn from each well of the Deep-Well plates and subsequently dispensed onto filter paper overlaid on solid 2× SC dropout medium.

2. Filters are incubated on growth medium for approximately 20 hr prior to treatment with chloroform. To permeabilize cells, filters are placed inside a closed chamber containing chloroform vapor for 15–30 min; during this incubation, filters should not actually be immersed in chloroform. Subsequently, filters are transferred face up onto fresh X-Gal plates [5-bromo-4-chloro-3-indolyl-β-D-galactopyranoside (x-Gal, 120 μg/ml), 0.1 *M* $NaPO_4$ (pH 7.0), 1 m*M* $MgSO_4$ in 1.6% (w/v) agar] to be incubated at 30° for up to 3 days. After several days of growth, β-Gal levels can be reliably estimated from the observed intensity of blue staining.

3. Once transformants of interest have been identified, several methods may be used to determine the genomic site of mTn insertion within these clones. Most simply, corresponding plasmid-borne insertion alleles can be sequenced with the following primer (complementary to sequence near the 5′ end of mTn-3xHA/*lacZ*): GGCCTTCTTTCTTTGGAAGTAC. Alternatively, a number of plasmid rescue protocols and PCR-based approaches are useful for this purpose: these approaches have been outlined previously.[19]

Phenotypic Analysis of Strains Carrying an mTn Insertion within a Nonessential Gene

Yeast strains carrying a productive mTn insertion event may be analyzed further as a means of identifying functions associated with these transposon-tagged genes. Specifically, insertion of full-length mTn-3xHA/*lacZ* results in a truncation of its host gene, thereby generating disruption mutants for phenotypic analysis. This analysis is again amenable to high-throughput methods: specific insertion alleles can be transformed on a large scale into any desired yeast background for subsequent analysis under a variety of test conditions. We typically utilize a haploid *ura3* strain to analyze insertions within nonessential genes (the phenotypic analysis of mTn insertions within essential genes requires a different approach discussed separately in this chapter). Large-scale phenotypic assays are easily accomplished with a 96-pronged replicator (Fig. 3A). Similar replicators are commercially available; however, they may need to be custom modified in order to ensure prongs of sufficient length (i.e., to contact the bottom of a 2-ml microplate). Using this transfer device, transformants can be plated in an ordered fashion onto appropriate selective growth medium, where they form large arrays ("macroarrays") (Fig. 3B). In this manner, at least 576 strains can be simultaneously screened for a given mutant phenotype in a single macroarray, as detailed below.

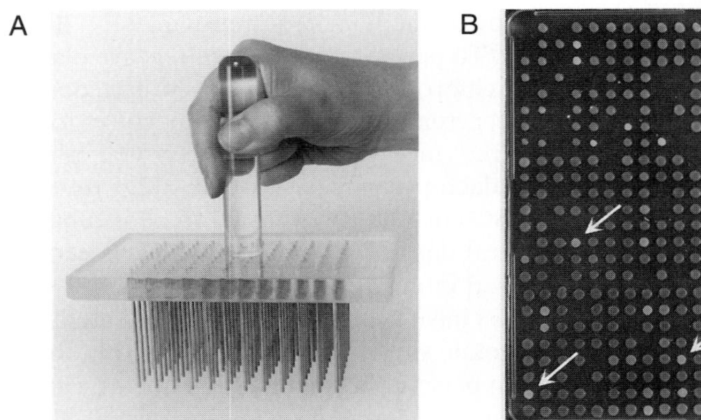

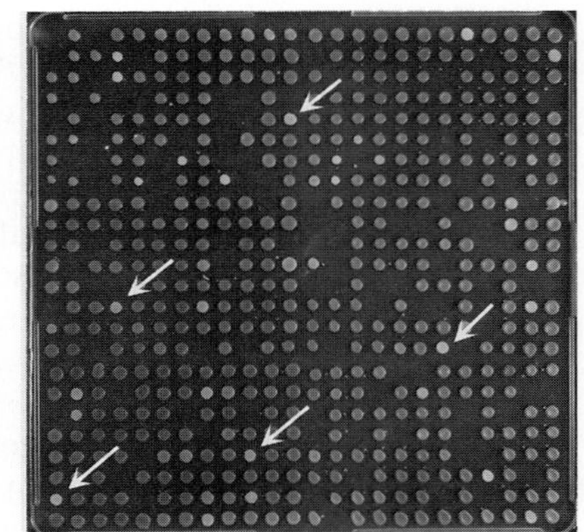

FIG. 3. Phenotypic macroarray analysis. (A) Haploid transformants carrying an mTn insertion may be quickly transferred onto test growth medium using a 96-pronged replicator, such as the one shown here. Note that this replicator contains prongs of sufficient length to contact the bottom of a 2-ml well within a 96-sample microplate. (B) Phenotypic macroarrays. Transfer tools can be used to generate ordered arrays of transformants on solid medium (macroarrays), enabling the simultaneous screening of hundreds of transformants for a given phenotype. Specifically, we routinely analyze 576 strains in a single array generated on a 21 × 21 cm plate. One such array is shown here; this particular macroarray can be used to identify yeast mutants deficient in oxidative phosphorylation. Mitochondrial mutants unable to carry out oxidative phosphorylation will lack red pigmentation when grown on YPD medium; red pigment formation on YPD within *ade2* mutant strains requires oxidative phosphorylation. A sampling of white transformants is indicated with arrows. Similar assays may be designed to screen for any number of additional mutant phenotypes.

1. Preparations of plasmid DNA containing insertion alleles of interest are transformed into a chosen haploid yeast strain; this procedure is performed as described above, with the following modifications. When preparing competent cells, the strain to be transformed is inoculated into 250 ml 2× SC-Ura dropout medium. This culture is grown overnight at 30° with shaking. A 200-ml volume of YPAD is subsequently added, and incubation is continued for an additional 2–3 hr. All subsequent steps are as described previously. Haploid transformants may be stored as stock in 15% (v/v) glycerol; glycerol stocks are maintained in 96-well plates at −70°.

2. Transformed cells are transferred from glycerol stock onto solid 2× SC-Ura medium with a 96-pronged replicator. Transformants are grown for 2 days at room temperature before being transferred into 96-well microplates containing 500 μl of YPD broth per well. Cells are subsequently grown for 16–24 hr with aeration on a rotary shaker at room temperature.

3. After this growth period, small aliquots of cell suspension (approximately 5 μl) are transferred in an ordered pattern onto various selective test plates as well as a YPD control plate; again, all transfers can be easily accomplished with a 96-pronged replicating tool. We routinely use 22-cm-square plates large enough to accommodate six 96-well microplates; therefore, 576 strains can be simultaneously screened for a given phenotype on one plate (macroarray). Larger plates could be used to screen even more strains in a single array. In addition, any number of phenotypes can be assayed by this approach simply by varying the test growth conditions.

4. mTn-mutagenized strains are scored for disruption phenotypes as follows. Growth of each transformant on a selective plate is compared against growth of the same transformant on a YPD control plate over a period of 2–6 days. Observed differences in growth are classified on a scale of 1–4: 1, corresponding to no difference in growth between the two conditions; 4, indicating a strong difference in growth on selective medium as compared to growth on YPD. As an indication of the practicality of this approach, we have already used this method to analyze nearly 8000 mTn insertion mutants under 21 different test conditions.

Production of HAT-Tagged Strains by Cre–lox-Mediated Recombination

While full-length mTn-3xHA/*lacZ* is useful in generating disruption mutants, the mTn-encoded HAT tag insertion element is an effective means of generating hypomorphic mutants and epitope-tagged strains for immunodetection. Full-length mTn-3xHA/*lacZ* may be reduced to this HAT tag element *in vivo* by Cre–*lox* site-specific recombination. The phage P1 Cre recombinase is expressed exogenously from plasmid p*GAL-cre* (*amp, ori, CEN, LEU2*)[17]; on this construct, the *cre* gene has been placed under transcriptional control of the *GAL* promoter; therefore, induction on galactose can be used to drive *cre* expression. After galactose induction, cells having undergone Cre-mediated recombination (and accompanying loss of the *URA3* marker) may be selected by growth on medium containing 5-fluoroorotic acid (5-FOA).[30] In our hands, galactose induction is capable of inducing Cre-mediated excision of the mTn-encoded *URA3* marker in >90% of cells analyzed.[17] To facilitate the successful generation of HAT-tagged strains, p*GAL-cre* should be transformed into a desired yeast strain prior to transformation with mTn-mutagenized genomic DNA. A diploid *ura3 leu2* strain would be an appropriate genetic background for the subsequent analysis of HAT tag insertions within essential genes; alternatively, a haploid derivative with these mutations could serve effectively as background for immunolocalization studies of HAT-tagged nonessential proteins. In either case, the approach outlined below is a reliable means of deriving strains producing HAT-tagged proteins on a large scale.

1. As noted above, yeast host strains used in this study should carry p*GAL-cre*; if necessary, strains bearing productive mTn insertions may be transformed with this plasmid as described in the previous section. Resulting transformants should be selected on appropriate dropout medium (e.g., synthetic medium lacking leucine and uracil; 2× SC-Leu-Ura).

2. Using a 96-pronged transfer device as before, Leu^+Ura^+ transformants carrying mTn insertions are inoculated into 96-well microplates containing 580 μl of 2× SC-Leu-Ura per well. Cultures are incubated at 30° on a rotary shaker overnight. Subsequently, a small aliquot (approximately 6 μl) of each culture is transferred into 580 μl of 2× SC-Leu-Ura with 2% (w/v) raffinose as a carbon source. Again using 96-well microplates, strains are grown to saturation on a rotary shaker at 30°. To induce galactose-driven expression of the Cre recombinase, cultures are then diluted 100-fold in 600 μl of 2× SC-Leu with 2% (w/v) galactose as a carbon source; as a control, each culture is also diluted 100-fold in 2× SC-Leu with 2% (w/v) glucose as a carbon source. Both sets of samples are grown to saturation (1–2 days) on a rotary shaker at 30°.

3. Cultures are subsequently diluted in sterile water prior to inoculation onto solid medium supplemented with 5-fluoroorotic acid (5-FOA), a compound converted into a toxic metabolic product by the wild-type *URA3* gene.[30] Using a multichannel microdispenser, a 1-μl aliquot from each culture is diluted in 3 ml of sterile water. A small aliquot of this dilute mixture (roughly 5 μl) is transferred onto solid 5-FOA medium (standard medium supplemented with 1 μg of 5-FOA per milliliter), using a 96-

[30] J. D. Boeke, F. LaCroute, and G. R. Fink, *Mol. Gen. Genet.* **197,** 345 (1984).

pronged replicator. FOA plates are incubated at 30° until growth is visible on those plates inoculated with strains exposed to synthetic medium containing galactose as its carbon source. FOA plates carrying strains grown in the presence of glucose should display little or no growth, as expression of the Cre recombinase is repressed by glucose. We have observed a low rate ($<1\%$) of spontaneous mTn excision in the absence of galactose induction.[17]

4. Strains having lost the *URA3* marker (exclusively following galactose induction) may be maintained as stock in 15% (v/v) glycerol at $-70°$ in 96-well microplates. These HAT-tagged strains are now ready for further functional studies.

Analysis of Strains Carrying an HAT Tag within an Essential Gene

HAT tag insertions offer an effective means of determining cellular functions associated with essential genes. Because the HAT tag does not contain any translational stop codons, HAT tag insertion within a gene may cripple, rather than completely disrupt, its function. HAT-tagged essential genes may, therefore, still retain partial function; gene mutations resulting in partially functional proteins are termed hypomorphic.

To generate hypomorphic mutations within essential genes, those open reading frames encoding proteins essential for yeast cell growth must first be identified. Transposon insertions that fail to yield any resulting haploid transformants serve as a rich source of potentially essential genes; in addition, database searches may be used to generate lists of known essential genes. Having established this list, plasmid DNA bearing mTn insertions that target these genes is prepared. Plasmid DNA is subsequently transformed into a diploid *ura3 leu2* strain (specifically, BY4741). BY4741 carries p*Gal-cre,* enabling Cre-mediated reduction of mTn3x-HA/*lacZ* to the HAT tag element. Resulting diploid HAT-tagged strains are sporulated, and tetrads are dissected. On tetrad dissection, a strain in which the HAT tag insertion has not completely disrupted gene function will exhibit a $4^+:0^-$ pattern of viable haploid segregants. Haploid derivatives of strains exhibiting this segregation pattern are subjected to PCR analysis in order to identify those containing HAT tag insertions within essential genes. Once the presence of the HAT tag and absence of the wild-type gene have been verified (Fig. 4), strains bearing these insertions are stored as haploids of both mating types and as homozygous diploids for subsequent phenotypic macroarray analysis. To verify that these genes are indeed essential, diploids containing full-length mTn-3xHA*lacZ* (roughly 6 kb in length) are sporulated and dissected to observe the segregation of viability in resultant haploids: mTn insertion within an essential gene is expected to yield $2^+:2^-$ segregation of viability.

A

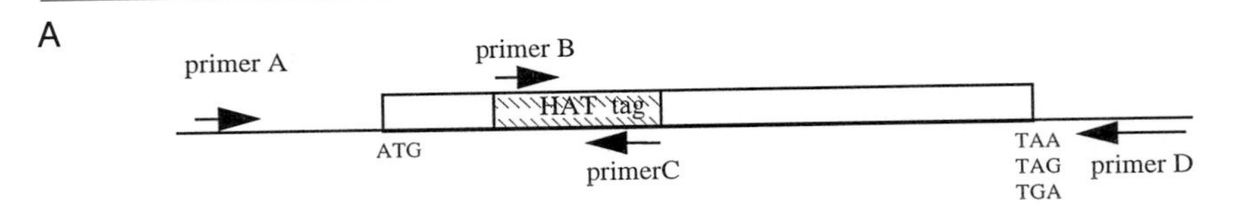

B

Primers	HAT-tagged allele	Wild Type Allele
A+ C	+	-
B+ D	+	-
B+ C	+	-
A+D	+	+

C

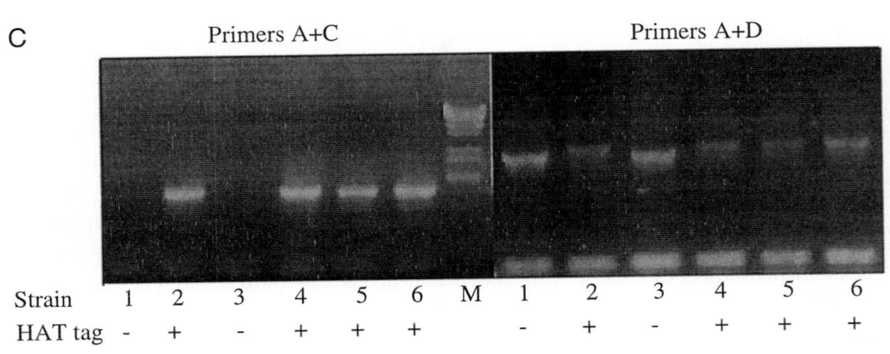

FIG. 4. Identification of HAT-tagged diploids. (A) Colony PCR strategy. Schematic of a HAT-tagged gene with translational start and stop codons indicated. For analysis by colony PCR, primers that anneal to the Watson strand in the upstream untranslated region (primer A) and the Crick strand in the downstream untranslated region (primer D) are used in concert with primers that anneal to the Watson or Crick strand of the HAT tag (primers B and C, respectively). (B) Table of expected results. The A + C or B + D primer sets are used to identify HAT-tagged diploids. The A + D primers verify the absence of the wild-type allele, while confirming the presence of the HAT-tagged allele by a noticeable shift in the size of the band obtained. (C) Examples of typical PCR results. PCR using the primer combinations indicated were performed; half of each reaction mixture was subsequently analyzed by agarose gel electrophoresis. Haploids derived from the same transformant were analyzed with both primer pairs; lanes 1–4 from each gel correspond to members of the same tetrad. The presence or absence of the HAT-tagged allele is indicated by a "+" or "–" respectively. Amplified fragments were sized by comparison against a preparation of *Bst*EII-digested λ DNA (lane M).

Using this approach, we have already identified more than 400 strains carrying in-frame mTn insertions within genes essential for yeast cell growth. Extrapolating from a pilot study of 143 such strains, 28% of all in-frame HAT-tagged essential proteins are expected to retain at least partial function (i.e., generate viable haploids). Therefore, when fully applied on a genome-wide scale, the procedures described below should be sufficient to generate a large collection of hypomorphic or conditional mutants for further phenotypic analysis.

1. Plasmid DNA bearing an mTn insertion allele of interest is prepared and subsequently transformed into yeast strain BY4741 as previously described. After heat-shock treatment and subsequent cell resuspension, a small, 5-μl aliquot of each transformation mixture is streaked onto solid 2× SC-Leu-Ura medium. Transformant cultures are incubated at 30° for 2–3 days. Three independent colonies from each streak are then selected and maintained as stock at −70° in YPAD plus 15% (v/v) glycerol.

2. After Cre-mediated recombination, FOA-resistant isolates are patched onto YPAD and incubated overnight at 30°. For each clone of interest, HAT-tagged derivatives are generated within three independent yeast transformants.

3. Cells from each patch are inoculated into 1.5 ml of sporulation medium [1% (w/v) potassium acetate supplemented with methionine, uracil, leucine, histidine, and lysine]. Cultures are subsequently incubated on a roller drum for 3–5 days at room temperature. Routinely, two independently derived patches are used at this point; the third is preserved and used later only to resolve any conflicting results. Sporulation may be verified by analysis of tetrads on a wet-mount. Sporulated cultures are harvested (3000 rpm, 5 min) and washed prior to resuspension of tetrads in 1 ml of sterile water.

4. An aliquot (100 μl) of resuspended tetrads is digested in 5 μl of β-glucoronidase (134.6 units/μl; available from Sigma, St. Louis, MO) at room temperature for 15–20 min. After incubation, 8 μl of this digestion mixture is gently spread onto a plate; tetrads are dissected, and plates are subsequently incubated at room temperature for 2–3 days. Haploid segregants are scored for cell viability; segregation of the *met, lys,* and *MAT* loci are scored by replica plating onto SC-Met, SC-Lys, and SC-His plates spread with lawns of mating-type testers. For those strains exhibiting $4^+:0^-$ segregation of viability, the necessity of the tagged gene is verified by sporulating and subsequently dissecting the transformants from which the HAT-tagged strains were derived.

5. Four haploids derived from one tetrad, as well as strains that are *MAT***a** *met15* and *MATα lys2,* are selected for further analysis. A complete

tetrad is selected as a control to confirm that the HAT tag is segregating $2^+:2^-$ as expected. For each strain analyzed, two PCRs, are required: the first will identify those strains containing an HAT-tagged essential gene, while the second will serve to verify that the wild-type allele is no longer present. PCR primers are designed as indicated in Fig. 4A. PCRs are performed as follows.

Using a sterile pipette tip, a small cell culture is smeared around the bottom of a thin-walled PCR tube. This sample is heated in a microwave on full power for 2–3 min. Subsequently, 1 μmol of each primer is added to the side of the PCR tube, followed by 20 μl of a PCR master mix consisting of 1× PCR buffer, 200 μM dNTPs, 1.5 mM $MgCl_2$, and *Taq* polymerase (1 U/μl). The contents of this tube are briefly spun prior to thermal cycling under the following conditions.

1 cycle: 94°, 3 min
35 cycles: 94°, 15 sec; 57°, 15 sec; 72°, 1 min 30 sec
1 cycle: 72°, 3 min

Fragments greater than 3 kb in length may be amplified as follows.

1 cycle: 94°, 3 min
35 cycles: 94°, 15 sec; 68°, 3 min
1 cycle: 68°, 7 min

Typical PCR results are illustrated in Fig. 4C.

6. Strains containing HAT-tagged alleles of essential genes are stored as haploids of both mating types as well as homozygous diploids. Independently derived *MAT***a** *met15* and *MATα lys2* strains bearing HAT-tagged essential genes are crossed to generate a homozygous diploid; all three sets of strains are stored as stock in YPAD plus 15% (v/v) glycerol at −70° for subsequent phenotypic analysis.

Large-Scale Immunolocalization of HAT-Tagged Proteins

Strains carrying HAT-tagged proteins may be immunolocalized with monoclonal antibodies directed against the HA epitope. In practice, this approach is an effective and feasible means of determining protein localization on a large scale: we have already immunolocalized HAT-tagged proteins within nearly 1500 mTn-mutagenized strains (A. Kumar and M. Snyder, personal communication, 1999). On occasion, observed staining patterns may be artifacts; however, we estimate the occurrence of such aberrant localizations at under 7% based on a pilot study of HAT tag insertions within proteins exhibiting known subcellular localization.[17] Background staining is typically low with monoclonal anti-HA antibodies. If problematic, however, nonspecific staining can be minimized by preadsorbing primary antibodies against fixed yeast cells (lacking the HA antigen) suspended in phosphate-buffered saline (as described by Burns *et al.*[13]) Using the procedure outlined below, a single researcher can easily prepare 192 samples (two 96-well microtiter plates) for immunofluorescence in a period of 3 days. Examples of immunofluorescence patterns observed in cells carrying HAT-tagged proteins are presented in Fig. 5.[13,31,32]

1. Strains carrying in-frame HAT tag insertions are individually inoculated into 96-well microtiter dishes containing 75 μl of YPAD broth per well. Cultures are grown overnight (roughly 12 hr) at 30° with gentle shaking on an orbital platform shaker (240 rpm). Alternatively, plates may be incubated on a vortex shaker set to its lowest speed. After overnight growth, a 45-μl volume of fresh YPAD is added; cells are grown for an additional 1.5 hr to an OD_{600} of roughly 0.75 to 1.

2. Cells are subsequently fixed in formaldehyde added to a final concentration of 3.75% (v/v). Fixed cells are agitated gently (again using a vortex shaker set to its lowest setting) for 30 min, prior to being collected by centrifugation at 2000 rpm for 4 min in a Sorvall H-1000B rotor. Pelleted cells are washed three times in 100 μl of solution A (1.2 *M* sorbitol, 50 m*M* KPO_4, pH 7) before finally being resuspended in 100 μl of solution A supplemented with 0.1% (v/v) 2-mercaptoethanol, 0.02% (w/v) glusulase (a trademarked preparation of β-glucuronidase/sulfatase, each available from Sigma), and Zymolyase 100T (5 μg/ml; Seikagaku America, Ijamsville, MD).

3. Cells are spheroplasted in this mixture at 37° with gentle shaking for approximately 20 to 30 min. Cell suspension should be checked periodically during this incubation: Once the cells appear translucent, they are recovered and washed once in 100 μl of solution A. Pelleted cells are subsequently resuspended in 100 μl of phosphate-buffered saline (PBS; 150 m*M* NaCl, 50 m*M* $NaPO_4$, pH 7.4) supplemented with 0.1% (v/v) Nonidet P-40 and 0.1% (w/v) bovine serum albumin (BSA). Samples are incubated at room temperature without shaking for 15 min. Cells are then pelleted by centrifugation as before; excess liquid is removed by aspiration, and pellets are resuspended in 100 μl of PBS supplemented with 3% (w/v) BSA. Cells are gently shaken at room temperature for 1 hr.

4. Cells are subsequently collected by centrifugation prior to treatment with 40 μl of mouse anti-HA monoclonal antibody 16B12 (MMS101R; BAbCO, Richmond, CA) used at 1:1000 final dilution in PBS plus 3% (w/v) BSA. Incubation with primary antibodies is continued overnight at 4° with gentle shaking.

[31] E. Hurt, S. Hannus, B. Schmelzl, D. Lau, D. Tollervey, and G. Simos, *J. Cell Biol.* **144,** 389 (1999).

[32] K. G. Hardwick and H. R. Pelham, *J. Cell Biol.* **119,** 513 (1992).

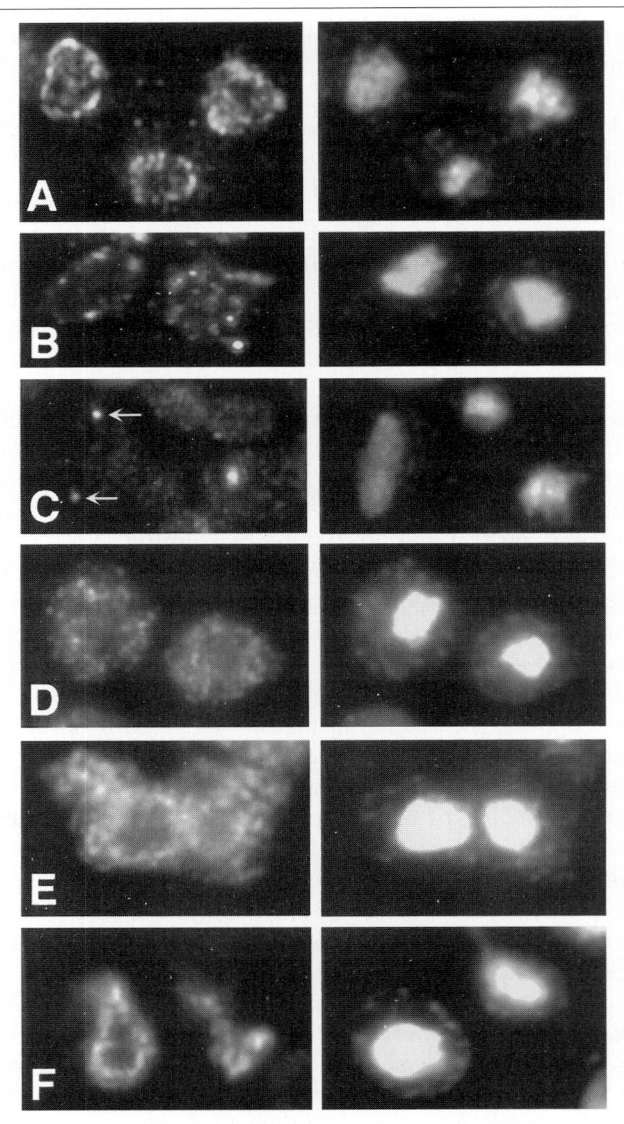

FIG. 5. Large-scale immunolocalization of HAT-tagged proteins. *Left:* Examples of immunofluorescence patterns in vegetative yeast cells stained with monoclonal antibody directed against the HA epitope. *Right:* The same cells stained with the DNA-binding dye DAPI. (A) Staining of the nuclear rim in a transformant carrying an in-frame HAT-tag in Nsp1p, a known nuclear pore protein involved in the binding and translocation of proteins during nucleocytoplasmic transport.[31] (B) Staining of mitochondria in a strain containing a HAT-tagged allele of the NADH-ubiquinone oxidoreductase Nid1p. (C) Localization of HAT-

5. After overnight incubation, cells are washed as follows: once in 150 μl of PBS supplemented with 0.1% (w/v) BSA, once with 100 μl of PBS supplemented with 0.1% (w/v) BSA and 0.1% (v/v) Nonidet P-40, and finally once again in 100 μl of PBS supplemented with 0.1% (w/v) BSA. During this final wash, cells should be gently shaken for 5 min at room temperature prior to centrifugation.

6. Pelleted cells are subsequently treated with Cy3-conjugated affinity-purified goat anti-mouse IgG (Jackson Laboratories, West Grove, PA) used at a 1:200 final dilution in a 40-μl volume of PBS supplemented with 3% (w/v) BSA. Cells are shaken gently at room temperature for 2 hr in the dark (to minimize Cy3 expsoure to light).

7. After treatment with secondary antibodies, cells are again washed thoroughly as follows: once in 150 μl of PBS plus 0.1% (w/v) BSA; once in 100 μl of PBS plus 0.1% (w/v) BSA with gentle shaking for 5 min prior to centrifugation; twice in 100 μl of PBS plus 0.1% (w/v) BSA plus 0.1% (v/v) Nonidet P-40; and, once, in 100 μl of PBS plus 0.1% (w/v) BSA.

8. Cells are finally resuspended in 60 μl of PBS; 10 μl of this cells suspension is subsequently transferred into 12 individual wells on a poly-lysine-treated slide, using a multichannel pipetter. Cells are allowed to sit for a minimum of 15 min before excess solution is aspirated. Cells are immediately mounted in a solution of 70% (v/v) glycerol–30% (v/v) PBS containing 2% (v/v) *n*-propyl gallate and Hoechst (0.25 μg/ml). A coverslip is carefully added and sealed with nail polish. Slides may be stored at −20° in the dark until ready for examination.

Conclusion

The methods outlined above describe a transposon-based system for the large-scale analysis of gene function. *Escherichia coli–S. cerevisiae* shuttle mutagenesis is used to generate a large number of yeast strains containing open reading frames tagged with a specially designed multipurpose Tn*3*-derived minitransposon. Through subsequent manipulation of this multi-functional mTn *in vivo,* we can generate a variety of chimeric protein products for the genomic analysis of gene expression, gene function, and

---

tagged Spc105p to the spindle pole body (indicated with arrows in a mitotic cell). (D) Staining of the Golgi apparatus resulting from HAT tagging of Sed5p, a syntaxin homolog functioning in vesicular transport.[32] (E) Localization of HAT-tagged ERG1 to the endoplasmic reticulum. (F) Vacuolar staining in a strain containing a HAT-tagged allele of the ATPase VPH1. Interestingly, no such vacuolar staining was observed in a previous immunolocalization study of β-Gal fusion proteins.[13] All examples shown here were generated according to the protocol for large-scale immunofluorescence described in text.

protein localization, without construction of any additional gene fusions. As such, our transposon-tagging system offers an economical approach to functional genomics. A single mTn insertion event is sufficient to generate a variety of functional information, drastically reducing the amount of work that would otherwise be required to generate a comparable data set. In addition, our approach does not require any particularly expensive equipment, only common laboratory tools and reagents.

Assuming a reasonable investment in time and effort, a standard-sized laboratory could use these protocols to generate a large collection of defined mutant alleles. If widely available, such collections would constitute an important resource for future studies. Traditionally, genetic studies have utilized the following paradigm: Researchers first generate mutations at random within a test population, then screen for a desired phenotype, and finally identify the affected gene in a mutant of interest by complementation analysis or genetic mapping. These latter procedures, however, are both time consuming and potentially unrewarding. Our mutant collections offer a means of altering this paradigm. Researchers may now immediately identify affected genes within clones of interest by directly screening a collection of defined mutants. These collections, therefore, hold the potential to expedite gene analysis by fundamentally altering the manner in which genetic screens are performed. Similar paradigm shifts will likely take hold in the near future as new genomic methodologies are developed and implemented.

Acknowledgments

A.K. is supported by a fellowship from the American Cancer Society. P.S.R.C. is supported by the Fundacao de Amparo a Pesquisa do Estado de San Paulo, Brazil. This work was supported by NIH Grant CA77808.

[34] GATEWAY Recombinational Cloning: Application to the Cloning of Large Numbers of Open Reading Frames or ORFeomes

By ALBERTHA J. M. WALHOUT, GARY F. TEMPLE, MICHAEL A. BRASCH, JAMES L. HARTLEY, MONIQUE A. LORSON, SANDER VAN DEN HEUVEL, and MARC VIDAL

Introduction

Complete genome sequences are available for three model organisms, *Escherichia coli, Saccharomyces cerevisiae,* and *Caenorhabditis elegans,* and for several pathogenic microorganisms such as *Helicobacter pylori.*[1–4] In addition, complete genome sequences are expected to become available soon for other model organisms and for humans.[5] This information is expected to revolutionize the way biological questions can be addressed. Molecular mechanisms should now be approachable on a more global scale in the context of (nearly) complete sets of genes, rather than by analyzing genes individually. However, most open reading frames (ORFs) predicted from sequencing projects have remained completely uncharacterized at the functional level. For example, of 19,099 ORFs predicted from the *C. elegans* genome sequence,[3] the function of approximately 1,200 has been characterized (see, e.g., Ref. 6). The emerging field of "functional genomics" addresses this limitation by developing methods to characterize the function of large numbers of predicted ORFs simultaneously.

[1] F. R. Blattner, G. Plunkett III, C. A. Bloch, N. T. Perna, V. Burland, M. Riley, J. Collado-Vides, J. D. Glasner, C. K. Rode, G. F. Mayhew, J. Gregor, N. Wayne Davis, H. A. Kirkpatrick, M. A. Goeden, D. J. Rose, B. Mau, and Y. Shao, *Science* **277,** 1453 (1997).
[2] A. Goffeau, *et al., Nature* (*London*) **387**(Suppl.), 1 (1997).
[3] The *C. elegans* Consortium, *Science* **282,** 2012 (1998).
[4] J. R. Tomb, J. F. Tomb, O. White, A. R. Kerlavage, R. A. Clayton, G. G. Sutton, R. D. Fleischmann, K. A. Ketchum, H. P. Klenk, S. Gill, B. A. Dougherty, K. Nelson, J. Quackenbush, L. Zhou, E. F. Kirkness, S. Peterson, B. Loftus, D. Richardson, R. Dodson, H. G. Khalak, A. Glodek, K. McKenney, L. M. Fitzegerald, N. Lee, M. D. Adams, E. K. Hickey, D. E. Berg, J. D. Gocayne, T. R. Utterback, J. D. Peterson, J. M. Kelley, M. D. Cotton, J. M. Weidman, C. Fujii, C. Bowman, L. Watthey, E. Wallin, W. S. Hayes, M. Borodovsky, P. D. Karp, H. O. Smith, C. M. Fraser, and J. C. Venter, *Nature* (*London*) **388,** 539 (1997).
[5] F. S. Collins, A. Patrinos, E. Jordan, A. Chakravarti, R. Gesteland, L. R. Walters, and the members of the DOE and NIH planning groups, *Science* **282,** 682 (1998).
[6] A. J. M. Walhout, H. Endoh, N. Thierry-Mieg, W. Wong, and M. Vidal, *Am. J. Hum. Genet.* **63,** 955 (1998).

0076-6879/00 $30.00

Functional Genomics

The number of uncharacterized ORFs ranges from thousands for *E. coli* and *S. cerevisiae,* to tens of thousands for higher organisms such as *C. elegans, Drosophila melanogaster,* and others. Hence, it is crucial to develop standardized assays and high-throughput procedures that allow the functional characterization of large numbers of genes and/or proteins simultaneously. Such "functional genomics" projects were initially developed for *S. cerevisiae,* the first eukaryotic organism for which a complete genome sequence has been available, and are now in progress for *C. elegans* and other model organisms. The data obtained from genome-wide functional assays such as gene knockouts and microarray or chip analysis are increasingly made publicly available into various databases (see, e.g., Refs. 7–9). Together with new projects under development, such as protein–protein interaction maps, protein localization maps, spatial and temporal expression patterns, and even three-dimensional structures, these data should provide a backbone of annotations from which new hypotheses can be formulated.

Thus far, standardized functional genomics assays have mainly been developed to study the genome (complete chromosomal DNA content) and the transcriptome (complete set of transcripts) (Fig. 1). For example, using microarray or DNA chip analysis, the level of mRNA expression can be measured for large sets of genes.[10–12] DNA chips and/or arrays are now available to measure the level of expression of nearly every yeast gene,[13] 50% of *C. elegans* genes (http://cmgm.stanford.edu/~kimlab/

[7] E. A. Winzeler, D. D. Shoemaker, A. Astromoff, H. Liang, K. Anderson, B. Andre, R. Bangham, R. Benito, J. D. Boeke, H. Bussey, A. M. Chu, C. Connelly, K. Davis, F. Dietrich, S. Whelen Dow, M. El Bakkoury, F. Foury, S. H. Friend, E. Gentalen, G. Giaever, J. H. Hegemann, T. Jones, M. Laub, H. Liao, N. Liebundguth, D. J. Lockhart, A. Lucau-Danila, M. Lussier, N. M'Rabet, P. Menard, M. Mittmann, C. Pai, C. Rebischung, J. L. Revuelta, L. Riles, C. J. Roberts, P. Ross-MacDonald, B. Scherens, M. Snyder, S. Sookhai-Mahadeo, R. K. Storms, S. Véronneau, M. Voet, G. Volckaert, T. R. Ward, R. Wysocki, G. S. Yen. K. Yu, K. Zimmermann, P. Philippsen, M. Johnston, and R. W. Davis, *Science* **285,** 901 (1999).

[8] F. C. P. Holstege, E. G. Jennings, J. J. Wyrick, T. Ihn Lee, C. J. Hengartner, M. R. Green, T. R. Golub, E. S. Lander, and R. A. Young, *Cell* **95,** 717 (1998).

[9] S. Chu, J. DeRisi, M. Eisen, J. Mulholland, D. Botstein, P. O. Brown, and I. Herskowitz, *Science* **282,** 699 (1998).

[10] M. Schena, D. Shalon, R. W. Davis, and P. O. Brown, *Science* **270,** 467 (1995).

[11] D. J. Lockhart, *et al., Nature Biotechnol.* **14,** 1675 (1996).

[12] M. Chee, R. Yang, E. Hubbell, A. Berno, X. C. Huang, D. Stern, J. Winkler, D. J. Lockhart, M. S. Morris, and S. P. A. Fodor, *Science* **274,** 610 (1996).

[13] J. L. DeRisi, V. R. Iyer, and P. O. Brown, *Science* **278,** 680 (1997).

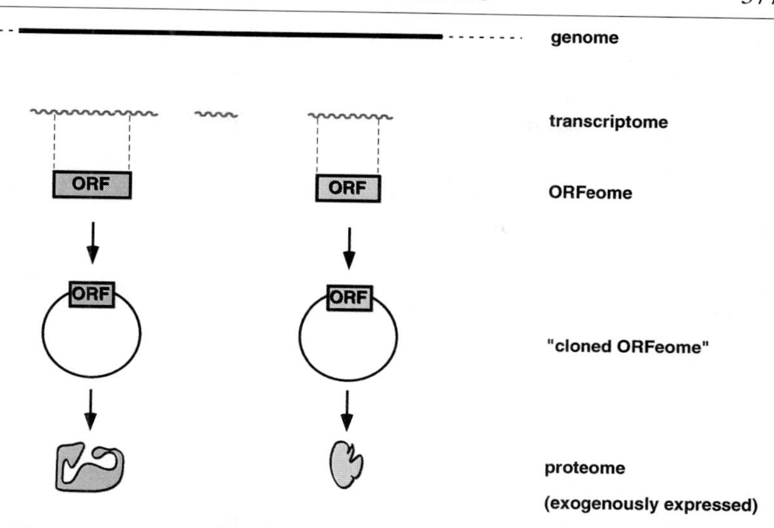

FIG. 1. The ORFeome. As for the genome (complete chromosomal DNA content), transcriptome (complete set of transcripts), and proteome (complete set of proteins), the ORFeome is defined here as the complete set of open reading frames of an organism. Each set of squares and circles represents different ORFs and their corresponding proteins, respectively. Cloning of complete ORFeomes should provide a valuable resource for many functional genomics projects that are currently under development.

wmdirectorybig.html), and approximately 20% of human genes.[14] Similarly, strategies have been developed for genome-wide gene-deletion analyses in yeast[7] and large-scale abrogation of gene function by RNA-mediated interference (RNAi) in *C. elegans*[15] (A. Hyman, personal communication, 1999). The success of such "gene-based" approaches is, in part, attributable to the relative convenience of manipulating DNA and RNA. Most steps in these approaches rely heavily on polymerase chain reactions (PCRs) and/or primer synthesis, which can be readily automated.[16]

Functional genomics projects aimed at characterizing the proteome (the complete set of proteins) have primarily focused on the study of endogenous sets of proteins expressed in an organism of interest. For example, high-throughput approaches to identify the subunits of molecular com-

[14] V. R. Iyer, M. B. Eisen, D. T. Ross, G. Schuler, T. Moore, J. C. F. Lee, J. M. Trent, L. M. Staudt, J. Hudson, Jr., M. S. Boguski, D. Lashkari, D. Shalon, D. Botstein, and P. O. Brown, *Science* **283,** 83 (1999).

[15] A. Fire, S.-Q. Xu, M. K. Montgomery, S. A. Kostas, S. E. Driver, and C. C. Mello, *Nature (London)* **391,** 806 (1998).

[16] D. A. Lashkari, J. H. McCusker, and R. W. Davis, *Proc. Natl. Acad. Sci. U.S.A.* **94,** 8945 (1997).

plexes formed *in vivo* have been developed by a combination of mass spectrometry (MS) and matrix-laser desorption/ionization (MALDI) or "time of flight" (TOF) techniques (see, e.g., Ref. 17). However, it is also important to design approaches in which (nearly) complete sets of ORFs can be expressed exogenously to perform various protein function assays such as large-scale two-hybrid analysis, systematic fusions to the green fluorescent protein (GFP), and protein production and purification for structural analyses. By analogy to the terminology used so far, we refer to the complete set of ORFs of a model organism as its "ORFeome" (Fig. 1).

Projects using predicted ORF sequences as a starting point for the exogenous production and characterization of the corresponding encoded products have been rather limited so far. This can be mainly attributed to two problems. First, it remains technically challenging to precisely clone large numbers of ORFs (i.e., from the start codon to the stop codon) into different expression vectors. Second, it is difficult to produce and manipulate large numbers of different proteins because they usually exhibit different chemical properties, in contrast to DNA and RNA.

The purpose of this chapter is to describe the protocols of a novel cloning technique, referred to as GATEWAY™ recombinational cloning (RC),[18,19] which provides a solution for large-scale cloning of ORFs into many different expression vectors in a standardized manner. GATEWAY is thus compatible with the automation procedures required for the cloning of nearly complete ORFeomes.

ORFeome Cloning

To facilitate various proteome-wide projects such as protein interaction mapping, structure determination, protein localization, and others, nearly complete ORFeomes will need to be cloned multiple times into many different expression plasmids. However, cloning of large sets of ORFs by conventional methods is impractical for numerous technical reasons. Most importantly, these methods involve ORF-specific restriction analyses and rely heavily on the purification of DNA fragments from agarose gels. Altogether conventional methods cannot be automated and thus are not compatible with high-throughput projects. Therefore, alternative systems that allow large numbers of ORFs to be conveniently cloned into different vectors by standardized protocols would be extremely useful.

Several alternative cloning systems based on different recombination

[17] A. Shevchenko, O. N. Jensen, A. V. Podtelejnikov, F. Sagliocco, M. Wilm, O. Vorm, P. Mortensen, A. Shevchendo, H. Boucherie, and M. Mann, *Proc. Natl. Acad. Sci. U.S.A.* **93,** 14440 (1996).

[18] J. L. Hartley, G. F. Temple, and M. A. Brasch, submitted (2000).

[19] A. J. M. Walhout, R. Sordella, and M. Vidal, *Science* **287,** 116 (2000).

reactions have been described to date. In yeast, *in vivo* recombination can be used to clone DNA fragments into linearized vectors provided that the extremities of the fragments and the linearized vectors share at least 40 nucleotides of identical sequence, a method referred to as "gap repair." For example, gap repair has been used to generate an array of nearly all *S. cerevisiae* ORFs fused to the sequence encoding the activation domain (AD) in a two-hybrid plasmid.[20] The main advantage of yeast gap repair is that all steps can be integrated and performed in 96-well plates. However, the main disadvantage is that the method cannot be used for the cloning of ORFs into nonyeast vectors.

Another system, referred to as the "univector-plasmid fusion system" (UPS), has been described. The UPS utilizes a Cre–*lox* site-specific recombination reaction[21] to fuse two different vectors. This in turn moves the 5′ end of ORFs cloned in a starting vector directly downstream of a regulatory sequence, or of an N-terminal fusion encoding sequence, in the new fusion plasmid. In another version of the UPS both 5′ and 3′ ends of ORFs can be transferred into new vectors by a combination of Cre–*lox* and R–*RS* recombination reactions. Compared with gap repair, the main advantage of the UPS is that it can be used with a large set of vectors, not necessarily limited to yeast vectors. However, one disadvantage at this point is that the cloning of ORFs into the starting vector is based on a blunt-end ligation.

A novel cloning system that combines most of the features needed for efficient ORFeome cloning has been developed.[18,19] This recombinational cloning (RC) system, designated GATEWAY™, allows both the initial cloning of ORFs and their subsequent transfer into different expression vectors by site-specific recombination *in vitro*. The features of GATEWAY make it amenable to automation, which is crucial for large-scale ORFeome cloning.

GATEWAY Recombinational Cloning

GATEWAY is based on the recombination reactions that mediate the integration and excision of phage λ into and from the *E. coli* genome, respectively[18] (Fig. 2A; see color insert). The integration involves recombination of the attP site of the phage DNA with the attB site located in the bacterial genome. This generates an integrated phage genome flanked by attL and attR sites. The integration reaction (Fig. 2A, "BP reaction"; see color insert) requires two enzymes: the phage protein integrase (Int) and

[20] P. Vetz, S. Fields, *et al., Nature* **403,** 623 (2000).

[21] Q. Liu, M. Z. Li, D. Leibham, D. Cortez, and S. J. Elledge, *Curr. Biol.* **8,** 1300 (1998).

the bacterial protein integration host factor (IHF) (collectively referred to as "BP Clonase"). The recombination reaction is reversible. The phage DNA can be excised from the bacterial genome by recombination between the attL and attR sites (Fig. 2A, "LR reaction"; see color insert). This excision reaction requires Int, IHF, and an additional phage enzyme, excisionase (Xis) (collectively referred to as "LR Clonase").

The scheme for ORF cloning by GATEWAY is as follows: a PCR product containing the ORF of interest is recombined into a "Donor vector," using the BP reaction. The resulting "Entry clone" is then used to recombine the ORF by the LR reaction into one or more "Destination vectors," generating "Expression clones" (Fig. 2B; see color insert).

For the purpose of GATEWAY, the recombination reactions have been modified in several ways. First, both BP and LR Clonases have been purified, allowing the GATEWAY reactions to take place *in vitro*. Second, the att sites have been mutated to generate pairs of derivatives: attB was modified into attB1 and attB2, attP into attP1 and attP2, attL into attL1 and attL2, and finally attR into attR1 and attR2. These sites were designed such that recombination reactions can take place only between attB1 and attP1, attB2 and attP2, attL1 and attR1, and attL2 and attR2. The duplication of att sites allows two independent recombination reactions to take place in the same molecules, one at the 5′ end of the ORF to be cloned and the other at the 3′ end. Third, the sequence of the attB1 and attB2 sites was selected such that each frame is open and thus both N- and C-terminal fusion proteins can be generated.

GATEWAY vectors have several features required for high-efficiency and accurate RC cloning (Fig. 2C; see color insert). Donor vectors carry a cassette flanked by attP1 and attP2 sites (attP1–attP2) that are compatible for recombination with the attB1 and attB2 sites. Using this configuration a PCR product containing an ORF flanked by attB1 and attB2 (attB1–ORF–attB2) can be readily recombined into a Donor vector. This will generate an Entry clone containing the ORF flanked by attL1 and attL2. The attB1 and attB2 sequences are only 25 nucleotides long and can be added to the primers used for PCR amplication of the ORFs. Destination vectors carry a cassette flanked by attR1 and attR2 sites (attR1–attR2) that can recombine with the attL1 and attL2 sites of the entry clones, respectively. Importantly, the 5′ and the 3′ ends of the ORFs recombine in a direction-dependent manner during both the BP and the LR reactions.

Donor and Destination vectors contain different antibiotic resistance markers. For example, when using a Donor vector that contains the gentamicin resistance marker, Destination vectors that contain the ampicillin resistance marker can be used. Both cassettes of GATEWAY vectors, attP1–attP2 and attR1–attR2, also contain between the att sites the *ccdB*

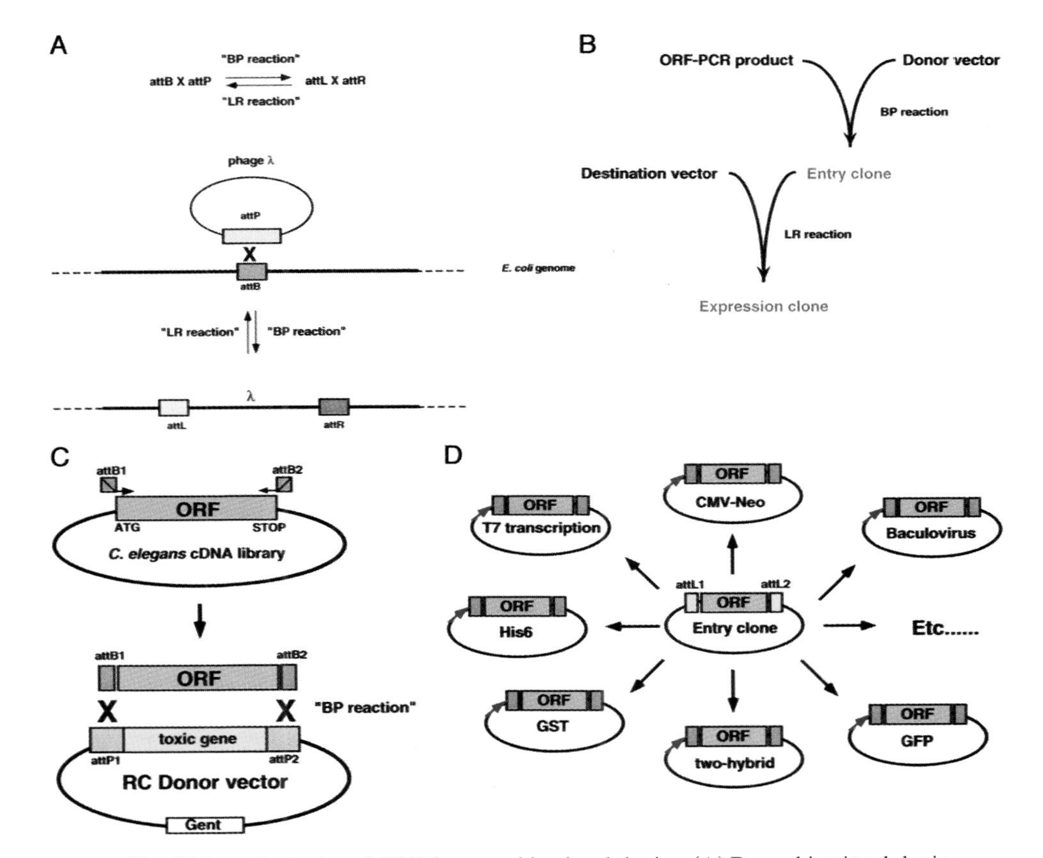

FIG. 2. Versatile cloning of ORFs by recombinational cloning. (A) Recombinational cloning is based on integration of phage λ into the *E. coli* genome. Briefly, the attP site (242 bp, blue box) of the phage recombines with the attB site (25 bp, green box) of the bacterial genome, generating attL (100 bp, yellow box) and attR (168 bp, orange box) sites. The reaction is reversible (BP and LR reaction: recombination between attB and attP or attL and attR, respectively), allowing phage DNA integration into, and excision from the *E. coli* genome. (B) Scheme for RC. PCR products corresponding to the ORFs of interest are cloned into an RC donor vector by the BP reaction, generating entry clones. Subsequently, the ORFs are subcloned, using the LR reaction, into destination vectors, generating expression clones. (C) *Caenorhabditis elegans* ORFs can be obtained by PCR using attB-tailed primers and an appropriate cDNA library as template DNA. Briefly, modified versions of the attB site, attB1 and attB2 (green boxes), are added to the 5′ end of the forward and reverse ORF-specific primers, respectively. This allows specific amplification of the ORFs of interest (orange box) and generates a product in which the ORF is flanked by attB1 and attB2 sites. The presence of these sites allows two RC reactions to occur with a donor vector that contains a toxic gene (yellow box), flanked by attP1 and attP2 sites (indicated in blue). In the resulting entry clone, the ORF is flanked by attL1 and attL2 sites. (D) Once an ORF is cloned into an entry vector it can be transferred by RC into different destination vectors that contain the toxic gene flanked by attR1 and attR2 sites. Examples of destination vectors include two-hybrid, GFP, baculovirus, CMV-neo, His, GST, and T7 transcription vectors. In addition, new destination vectors can easily be generated.[18]

gene, whose protein product interferes with DNA gyrase. Expression of this gene is toxic in many commonly used *E. coli* strains such as DH5α. Thus the absence of *ccdB* provides a selective advantage: After recombination, only the constructs that have lost the *ccdB* gene should confer growth in DH5α cells. It is important to note that propagation of Donor vectors requires a specific *E. coli* strain, DB3.1, which carries a mutation in *gyrA* conferring resistance to *ccdB*.

Subsequent to the LR reaction, ORFs are flanked by the 25-bp attB1 and attB2 sites in the resulting expression clones (Fig. 2D; see color insert). Translation of the attB sites adds an extra eight amino acids at both the N- and C-terminal ends of the proteins. Preliminary observations suggest that the addition of these amino acids does not generally alter the functional properties of the proteins.[19]

Here we describe the cloning of 31 *C. elegans* ORFs by GATEWAY into a Donor vector and the subsequent subcloning of these ORFs into two different two-hybrid destination vectors.

GATEWAY Cloning of 31 Caenorhabditis elegans Open Reading Frames

Many proteins require interactions with other proteins in order to function appropriately. Hence, the identification of interaction partners for a protein of unknown function might provide insight into its molecular role. Combination of such interaction data and results obtained from other functional genomics projects, such as protein localization, developmental expression patterns, and deletion analyses, should allow more valuable functional predictions.[6]

The objective of our laboratory is to initiate a protein interaction mapping project for the *C. elegans* proteome. We have chosen *C. elegans* as a model organism because it is the first multicellular organism for which a complete genome sequence has become available.[3] Furthermore, the biological significance of potential protein–protein interactions in the nematode can be tested by methods such as RNAi. In addition, the interaction map data generated will help to formulate hypotheses about functions of human proteins because many developmental and behavioral molecular mechanisms are conserved (see, e.g., Ref. 6).

The yeast two-hybrid system[22] was selected as a standardized functional assay for protein–protein interaction mapping because it has been demonstrated to be useful for the identification of protein–protein interactions

[22] S. Fields and O. Song, *Nature* (*London*) **340,** 245 (1989).

and is also amenable to automation (see, e.g., Ref. 23). We have developed a novel, semiautomated version of the yeast two-hybrid system, which solves most of the common problems of the method, including the relatively high number of false positives.[19]

We started the *C. elegans* protein interaction mapping project with a subset of ORFs implicated in nematode vulval development (vORFs).[19] Conventional genetic screens have identified approximately 40 genes required for that process.[24,25] We have successfully RC-cloned 31 vORFs into a Donor vector and subsequently into two two-hybrid Destination vectors required for protein interaction mapping.

Protocols

Design of attB-Tailed Primers

The ORF sequences were obtained from either GenBank (http://www.ncbi.nlm.nih.gov/genbank/query form.html) or ACeDB (a *C. elegans* database[26]). Subsequently, the positions of the translational start and stop codons were defined to design the forward and reverse primers. A sequence of at least 18 nucleotides from either the 5′ or the 3′ end of the ORF was used to design the forward and reverse primers, respectively. The length of this sequence was optimized to generate primers with similar annealing temperatures of approximately 60°. The start and stop codons were excluded from the sequence in order to prevent internal translational initiation and allow the generation of N- and C-terminal fusion proteins, respectively. In addition, the primers contain a 5′ tail corresponding to either the attB1 or the attB2 sequence for the forward and reverse primers, respectively. This allows recombination with the attP1 and attP2 sites of the Donor vector (Fig. 2C; see color insert). To automate the design of primers for large-scale ORFeome cloning we have developed a method based on the oligonucleotide synthesis program (OSP) algorithm.[27] The primer sequences used in this study can be found at our website.[19]

1. Define the start codon of the ORFs and select the first 30 nucleotides starting with the TG of the predicted translation initiation codon.

[23] M. Vidal and P. Legrain, *Nucleic Acids Res.* **27,** 919 (1999).
[24] K. Kornfeld, *Trends Genet.* **13,** 55 (1997).
[25] P. W. Sternberg and H. R. Horvitz, *Trends Genet.* **7,** 366 (1991).
[26] R. Durbin and J. Thierry-Mieg, *in* "Computational Methods in Genome Research" (S. Suhai, ed.). Plenum, New York, 1994.
[27] L. Hillier and P. Green, *PCR Methods Appl.* **1,** 124 (1991).

2. Calculate the T_m (see http://alces.med.umn.edu/rawtm.html) and adjust the length of the sequence to obtain a T_m (annealing temperature) of approximately 60°.

3. Add the attB1 tail: 5′-GGGGACAAGTTTGTACAAAAAAGCAGGCT to the 5′ end of the sequence.

4. Define the stop codon and select the last 30 nucleotides 5′ of the stop codon, i.e., excluding the stop codon.

5. Determine the reverse complement of the sequence.

6. Calculate the T_m and adjust the sequence as in step 2.

7. Add the attB2 tail: 5′-GGGGACCACTTTGTACAAGAAAGCTGGGT to the 5′ end of the sequence.

8. Verify primer sequences using the BLASTN program (http://www.ncbi.nlm.nih.gov/cgi-bin/BLAST/nph-newblast?Jform=0).

Polymerase Chain Reaction Amplification of Open Reading Frames

For the purpose of protein interaction mapping we generated a *C. elegans* cDNA library (AD-wrmcDNA) in which poly(dT)-primed reverse transcribed cDNAs are fused to the AD-encoding sequence. This library was made with poly$(A)^+$ RNA isolated from mated populations of wild-type animals of all stages of development including embryonic, larval (L1–L4 stages), adult, and dauer (Fig. 3). Each stage was nonsynchronous to include RNAs expressed within each stage. An approximately equal quantity of RNA from the different populations was acquired. cDNAs were generated and cloned into the two-hybrid vector pPC86.[28] The library contains $\sim 3 \times 10^7$ clones of which 48 of 48 tested contained an insert with insert sizes varying between 0.3 and 3.0 kb (data not shown).

Approximately 75% (28 of 37) of the vORFs were amplified successfully by PCR using the attB-tailed primers and the AD–wrmcDNA library as template DNA (Table I). PCRs were scored as successful when the product migrated as a single band of the expected size in an agarose gel. An example of six attB1–vORF–attB2 PCR products is shown in Fig. 4A. In the control reaction no template DNA was added (Fig. 4A and C). The PCRs were unsuccessful for nine vORFs (Table I: PCR). For two of these, *lin-12*(Notch) and *let-23*(EGF receptor), we chose an alternative approach for amplification. These ORFs are relatively large (4.3 and 4.1 kb, respectively) and both encode transmembrane proteins.[29,30] Such proteins are often difficult

[28] M. Vidal, *in* "The Yeast Two-Hybrid System" (P. Bartels and S. Fields, eds.), pp. 109–147. Oxford University Press, New York, 1997.

[29] R. V. Aroian, M. Koga, J. E. Mendel, Y. Ohshima, and P. W. Sternberg, *Nature* (*London*) **348,** 693 (1990).

[30] I. Greenwald, *Genes Dev.* **12,** 1751 (1998).

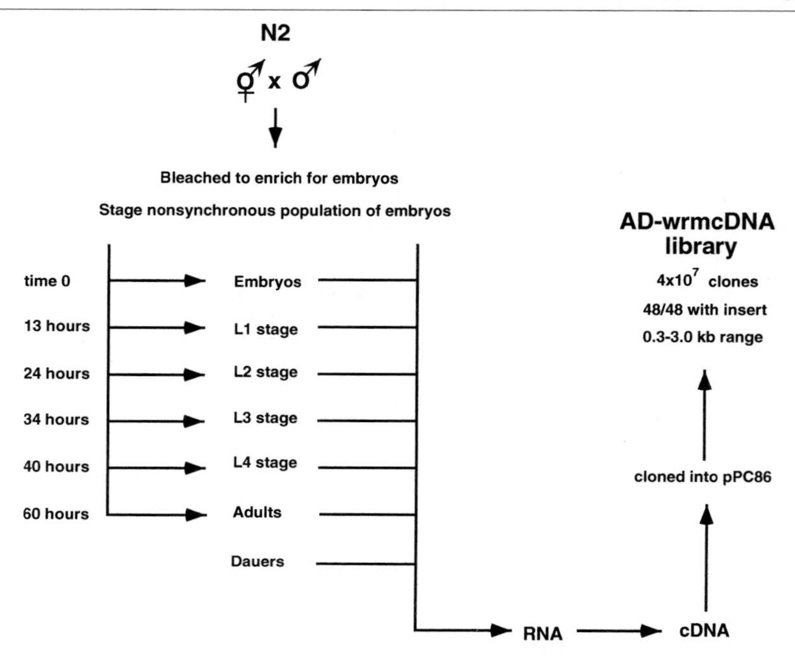

FIG. 3. Generation of a *C. elegans* cDNA library. The AD–wrmcDNA library was generated with poly(A)$^+$ RNA isolated from both hermaphrodite and male N2 worms of all larval stages, embryos, adults, and dauers and the subsequent generation of cDNAs by poly(A) priming. The cDNAs were cloned into pPC86 (28) and 4×10^7 clones were obtained. Forty-eight of 48 clones contained an insert, varying in size between 0.3 and 3.0 kb.

to assay in the context of the two-hybrid system.[23] Therefore, we designed new 5′ attB1-tailed primers covering the 5′ region of the intracellular domain [Table I, *lin-12(I)* and *let-23(I)*] and used these in a similar PCR. Both truncated ORFs have a predicted size of 1.4 kb. While an attB1–*lin-12(I)*–attB2 PCR product of 1.4 kb was readily obtained, we were unable to amplify the correct *let-23(I)* fragment. Therefore, we used as template DNA a plasmid containing the *let-23(I)* sequence. This approach was also used to obtain PCR products for *lin-15B* and *lin-7* (indicated by an asterisk in Table I). In summary, we obtained 32 attB1–vORF–attB2 PCR products.

There are several possible explanations for our failure to amplify a subset of the vORFs. First, the full-length ORF might not be present in the AD–wrmcDNA library. The majority of inserts in this library are between 0.3 and 3.0 kb in length. Thus full-length ORFs are less likely to be present for genes such as *lin-15B* (4.3 kb) and *emb-5* (4.6 kb). Second, some ORFs may be underrepresented due to low gene expression levels.

TABLE I

CLONING OF ORFs INVOLVED IN *Caenorhabditis elegans* VULVA DEVELOPMENT INTO THREE DIFFERENT VECTORS BY RECOMBINATIONAL CLONING[a]

Gene	Size (kb)	PCR	No. of entry colonies	% Correct size	No. of pDB colonies	No. of pAD colonies
bar-1	2.5	##	500	55	150	500
emb-5	4.6	N				
gap-1	1.9	N				
ksr-1	2.3	N				
lag-1	2	###	400	100	250	750
lag-2	1.2	###	300	92	350	550
*let-23 (l)**[b]	1.4	###	200	100	500	500
let-60	0.6	###	1000	100	150	1000
lin-1	1.3	##	500	100	600	500
lin-2	2.9	##	50	80	300	500
lin-3	1.3	##	800	80	250	200
*lin-7**	0.9	###	800	100	500	250
lin-9	1.9	N				
lin-10	2.9	###	300	25	200	500
lin-12 (l)	1.4	##	800	95	80	400
lin-15A	2.1	##	700	83	800	250
*lin-15B**	4.3	###	200	0		
lin-25	3.4	#	20	58	70	60
lin-31	0.7	###	1000	100	150	1000
lin-35	2.9	##	500	75	150	800
lin-36	2.9	##	100	80	400	100
lin-37	0.9	###	500	100	200	800
lin-39	0.8	###	500	100	100	320
lin-45	2.4	N				
lin-53	1.3	###	300	100	70	800
lin-54	1.3	###	600	100	100	100
mek-2	1.1	###	400	100	550	100
ptp-2	2	##	1000	100	500	500
sel-1	2	###	400	100	150	300
sel-10	1.8	###	500	70	240	100
sel-12	1.4	###	500	100	150	400
sem-5	0.7	###	500	100	800	500
sli-1	1.7	##	200	83	300	200
sup-17	2.8	#	100	62	70	30
sur-1	1.3	###	400	100	128	500
sur-8	1.7	##	400	83	600	300
unc-101	1.3	##	800	100	600	200
lin-6 (GR)[c]	2.4	###	0	NA[d]		
No PCR	NA	NA	0	NA	NA	NA
No entry clone	NA	NA	NA	NA	0	0
No clonase	0.6	*let-60*	0	NA	0	0

[a] The expected size of the ORFs is indicated. PCR: the quality of the obtained PCR products (###, a large amount of PCR product of the expected size; ##, less DNA obtained; #, very weak amplification; N, no PCR product detected). No. of entry colonies: The number of colonies obtained after transformation of the BP reaction into DH5α. % Correct size: the percentage of colonies that contain an insert of the expected size, as determined by PCR. No. pDB colonies/No. pAD colonies: the number of colonies obtained after the LR reaction in the DB and AD two-hybrid destination vector.

[b] An asterisk denotes that plasmid DNA was used as template rather than the AD-wrmcDNA library.

[c] GR refers to the fact that yeast gap repair tails were added to the *lin-6* PCR product as a negative control.

[d] NA, Not applicable.

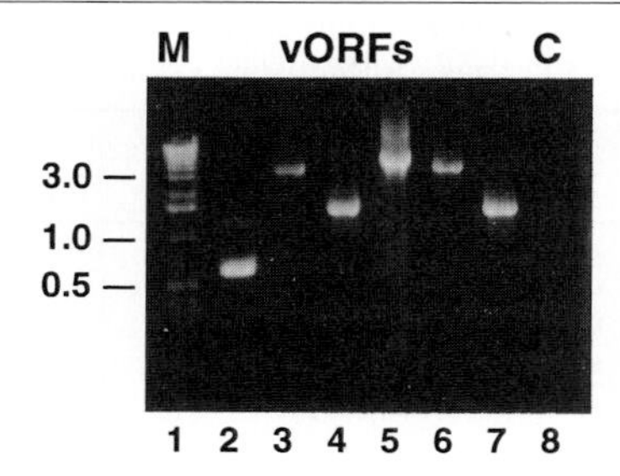

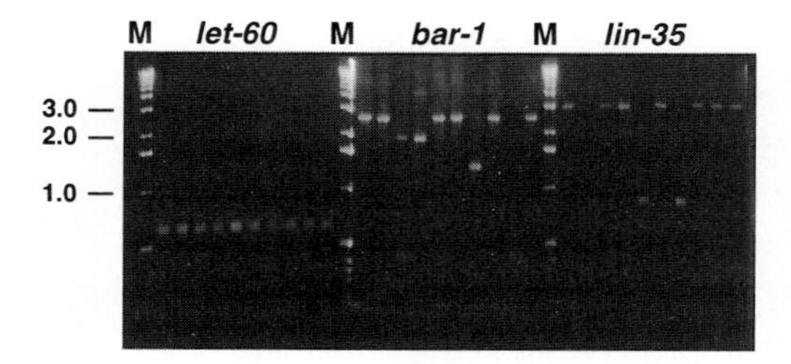

FIG. 4. PCR of ORFs involved in *C. elegans* vulva development. (A) AttB1–vORF–attB2 PCR products. The sequences corresponding to 35 genes involved in vulva development were obtained from GenBank/ACeDB (see Protocols). Primers containing attB tails were designed with a similar T_m (see Protocols) and were used together with the AD–wrmcDNA library as a template for PCR amplifications. The products generated in PCRs were separated by agarose gel electrophoresis. An example of such an experiment is shown. Lane 1 (M): DNA size marker. Lanes 2–7: attB1–vORF–attB2 PCR products containing the following ORFs: *let-60, lin-2, lin-3, lin-15B, lin-36* (sequences were obtained from GenBank), and *lin-54* (R. H. Horvitz, personal communication, 1999). Lane 8 (C): negative control in which no template DNA was added and the *let-60*-specific primers were used. (B) Verification of RC by PCR. The insert size was determined by PCR for at least 10 independent transformants for each entry clone. RC cloning of the attB1–*let-60*–attB2 PCR products resulted in entry clones in which 100% (10 of 10) had the correct size insert. In the case of *bar-1* 55% of the transformants contained the correct size insert (in this experiment 6 of 10) and 70% (7 of 10) of the *lin-35* inserts were of the predicted size.

For example, those ORFs might be expressed only in a small subset of cells and/or at defined stages of development. Here, we have used plasmid DNA as a source of template to amplify a subset of the ORFs described. However, this alternative is not available for the majority of genes that have not been molecularly characterized previously. Alternative sources of template DNA, such as normalized cDNA libraries, will therefore be useful for large-scale ORFeome cloning. As a final consideration, the attB-tailed primers may not anneal properly and might have to be redesigned and resynthesized.

1. Dilute the attB-tailed primers in filter-sterilized H_2O to a concentration of 25 ng/μl.

2. Compose PCRs on ice, using filter tips in either strips containing 8 tubes or 96-well plates. When using 96-well plates, add 50 μl of mineral oil (Sigma, St. Louis, MO) after setting up the reactions. Reactions (50-μl total volume) should contain 60 mM Tris–SO_4 (pH 9.1), 18 mM $(NH_4)_2SO_4$, 2 mM $MgSO_4$, 200 ng of AD–wrmcDNA library, 200 μM dNTPs, 1 μl of Elongase (GIBCO BRL), and 250 ng of each primer.

3. For each reaction, prepare a cocktail containing 10 μl of 5× elongase buffer B [300 mM Tris–SO_4 (pH 9.1), 90 mM $(NH_4)_2SO_4$, 10 mM $MgSO_4$], 5 μl of 2 mM dNTPs (freshly diluted from stocks), 10 μl of each primer (25 ng/μl), 1 μl of Elongase (GIBCO BRL), and 0.2 μl of AD–wrmcDNA library (1 mg/ml). Add filter-sterilized H_2O (so that each reaction is 50 μl) and aliquot the mix in the tubes/plate.

4. Start the program (see below). Place the tubes or 96-well plate in the PCR machine when it reaches a temperature of 90°.

5. After the PCR is completed, analyze 10–20 μl of each reaction on a 1% (w/v) agarose gel in 1× TBE. In addition, run 1 μl of 1-kb DNA marker (GIBCO BRL). Use the PCR products that give a single product of the expected size in the RC BP reaction. (Purification of the PCR products is generally not required.)

6. The PCR program should be as follows (all PCRs were performed in a PTC-100 PCR machine from MJ Research, Watertown, MA):

Step 1: 94°, 2 min
Step 2: 94°, 1 min
Step 3: 56°, 1.5 min
Step 4: 68°, 2 min
Step 5: Repeat from step 2, 35 cycles
Step 6: 68°, 5 min

Propagation of Donor and Destination Vectors

Both donor and destination vectors contain the toxic *ccdB* gene, which interferes with the rejoining step of DNA gyrase. Therefore, these vectors must be propagated in the DNA gyrase mutant strain DB3.1. The BP reaction results in replacement of the *ccdB*-containing cassette by the ORF of interest, allowing a genetic selection of the desired entry clones in DH5α cells.

Preparation of Competent Cells

1. Streak the appropriate *E. coli* strain (DB3.1 or DH5α) on an LB plate without antibiotics and incubate overnight at 37°. Inoculate a single colony in 5 ml of LB and incubate overnight at 37° in a shaking incubator.
2. Transfer the 5-ml culture to 500 ml of LB and incubate for ~2.5 hr or until the OD_{600} is between 0.3 and 0.6.
3. Centrifuge the cells in two sterile 250-ml tubes for 5 min at 4° at 2000 rpm.
4. Gently resuspend the cells in 25 ml of ice-cold, filter-sterilized TSB [LB-HCl (pH 6.1), 10% (w/v) polyethylene glycol (PEG) 3350, 5% (v/v) dimethyl sulfoxide, 10% (v/v) glycerol, 10 m*M* $MgCl_2$, 10 m*M* $MgSO_4$].
5. Incubate for 10 min on ice.
6. Aliquot the cells (500 μl/tube) and quick-freeze in dry ice–ethanol.
7. Store at −80°.
8. Determine the transformation efficiency by transforming (see below) 10 pg of pUC19 DNA (see below). After plating on LB–ampicillin plates and overnight incubation at 37°, this should yield ~50 colonies ($5 \times 10^6/\mu$g).

Transformation of PEG–DMSO Competent Cells

1. Thaw the cells on ice.
2. Add 2 μl of DNA to 20 μl of 5× KCM (0.5 *M* KCl, 150 m*M* $CaCl_2$, 250 m*M* $MgCl_2$) mixed with 80 μl of sterile H_2O.
3. Add 100 μl of cells, mix gently, and incubate for 20 min on ice.
4. Incubate for 10 min at room temperature.
5. Add 1 ml of LB and incubate for 1 hr at 37°
6. Plate the cells on LB plates containing the appropriate antibiotic.
7. Incubate overnight at 37°.

Cloning of attB-Tailed Polymerase Chain Reaction Products in a Recombination Cloning Donor Vector

The attB1–ORF–attB2 PCR products were used in the BP reaction together with the appropriate Donor vector. The number of colonies obtained after transformation of the RC reactions into DH5α cells varied between 20 and ~1000 (Table I, No. of entry colonies). For each entry clone at least 10 independent *E. coli* colonies were directly analyzed in PCRs with att-specific primers (Fig. 4B and see below). Hereby, the length of the insert was determined and thus the accuracy of the first RC step could be judged (Table I, % Correct size). Because cloning of the vORFs was based on PCR, which can be mutagenic, we sequenced 10 independent clones corresponding to 2 ORFs (*lin-53* and *lin-37*). The mutation rate observed was approximately 10^{-4}. Because the average size of vORFs is about 1.5 kb, we estimated that approximately 9 of 10 entry clones generated in the BP cloning step correspond to wild-type vORFs. To make sure that the wild-type ORF was included, we combined 10 colonies containing the correct size insert. On average, this means that 10% of the population of ORFs likely carry a mutation. Our preliminary observations indicate that such mutations are not problematic for functional analyses such as protein interaction mapping.[19] For other downstream analyses, such as protein purification, completely wild-type ORFs might be needed. In such cases, the sequence of a few Entry clones will have to be verified to select a single wild-type clone to be used in subsequent analyses.

We succeeded in cloning 31 of 32 (97%) of attB1–vORF–attB2 PCR products into the Donor vector, indicating that the BP reaction is efficient. The BP reaction of the attB1–*lin-15B*–attB2 PCR product resulted in 200 transformants; however, none of these contained an insert of the correct size (4.3 kb). One possible explanation for this result is that unpurified PCR products were used in the cloning reactions. We are currently testing whether purification of PCR products will allow more efficient recombination of larger ORFs. However, given the cloning efficiency of the majority of ORFs (see above), purification of the attB1–ORF–attB2 PCR product is not generally required.

Fresh attB-tailed PCR products should be used, preferably generated the same day. For GATEWAY reactions, minipreparation quality vector DNA is sufficient. The controls for the RC reaction should be as follows: (1) vector without PCR product, (2) PCR product without vector, (3) no BP Clonase, and (4) no DNA in the transformation. As a positive control we used the attB1–*let-60*–attB2 PCR product, which had worked consistently in the initial test reactions (Table I).

1. Take the BP Clonase from the −80° freezer and place on dry ice.
2. Compose the reactions on ice.
3. Combine in sterile tubes: 125 ng of donor vector DNA (pDNR-gent), 5 μl of 5× BP reaction buffer, and 5 μl of PCR product, and add TE to a final volume of 16 μl.
4. Thaw the BP Clonase on ice; vortex briefly.
5. Add 4 μl of BP Clonase to each reaction and mix by pipetting.
6. Incubate for 1 hr at 25°
7. Add 2 μl of proteinase K (1 mg/ml); mix by pipetting.
8. Incubate for 10 min at 37°.
9. Transform 2μl of each reaction into a 100-μl aliquot of competent DH5α cells, plate the whole transformation reaction on LB–gentamicin

plates (15 μg/ml), and incubate overnight at 37°. The number of colonies obtained varies between 10 and >1000 (Table I).

Polymerase Chain Reaction Verification of Vulval Open Reading Frame Entry Clones

The insert size of the Entry clones generated by the BP reaction can be verified by PCR from *E. coli* colonies, using the attB1- and attB2-specific primers DBB1 and B2Term.

DBB1: 5′-TAGTAACAAAGGTCAAAGACAGTTGACTGTATC-GTCGAGGTTGTACAAAAAAGCAGGCT-3′

B2Term: 5′-GCCGTTACTTACTTAGAGCTCGACGTCTTACT-TACTTAGCTTGTACAAGAAAGCTGGGT-3′

These PCRs can be performed with regular *Taq* polymerase (Fisher, Pittsburgh, PA) because the fidelity of the enzyme is not important for the verification. At least 10 colonies should be tested for every entry clone.

1. Compose the reactions on ice; the final volume should be 50 μl.
2. Prepare a mix containing for each reaction: 5 μl of 10× assay buffer B [100 m*M* Tris–HCl (pH 8.3), 0.5 *M* KCl], 5 μl of $MgCl_2$ solution (FisherBiotech; 25 m*M*), 5 μl of dNTPs (2 m*M*), 5 μl of 10× BSA (1 mg/ml), 100 ng of DBB1 and B2Term primer, 0.25 μl of *Taq* polymerase, and filter-sterilized H_2O to a final volume of 50 μl. Transfer aliquots into tubes or a 96-well plate and store on ice.
3. With a sterile toothpick, pick a single *E. coli* colony from the LB–gentamicin plate and transfer the cells to the appropriate tube or well. *Note:* Do not discard the toothpick. Subsequently, patch the remainder of the cells on a fresh LB–gentamicin plate for future use.
4. Incubate the plate overnight at 37°.
5. Run the PCR according to the following program:
   Step 1: 94°, 3 min
   Step 2: 94°, 1 min
   Step 3: 52°, 1.5 min
   Step 4: 72°, 3 min
   Step 5: Repeat from step 2, 35 cycles
   Step 6: 72°, 5 min
6. Run 10 μl of the reactions on a 1% (w/v) agarose gel in 1× TBE; include a size marker as described above.
7. Identify the colonies containing a PCR product of the expected size. The next day, pick 10 colonies from the LB–gentamicin plate and pool in 5 ml of LB–gentamicin (15 μg/ml). Incubate overnight at 37°. Purify the Entry clone DNA by minipreparations for subsequent destination cloning.

8. Repeat this procedure for ORFs for which fewer than 10 colonies containing an insert of the expected size were obtained.

Recombination Cloning of Vulval Open Reading Frames into Two-Hybrid Destination Vectors

Entry clone plasmids were purified for recombinational cloning of the vORFs into the AD and DNA binding domain (DB) two-hybrid destination vectors (Fig. 2B, "LR reaction"; see color insert). Every vORF (31 of 31) was successfully transferred to both Destination vectors and, in all cases tested, an insert of the correct size was obtained (data not shown). This indicates that the LR reaction is also efficient and accurate. The expression clones were subsequently transformed into the appropriate yeast strains and successfully used in two-hybrid assays.[21a]

In addition to the two-hybrid vectors used here, many other destination vectors are available or currently being developed. These include a baculovirus vector (for expression in insect cells), a GFP vector (for protein localization studies), tagging vectors such as His tag (for expression in bacteria), a T7 transcription vector (for *in vitro* transcription), and a CMV-neo vector (for expression in mammalian cells) (Fig. 2D; see color insert).

1. Linearize the DB– and AD–Destination vectors (minipreparation quality), using *Sma*I. Verify digestion on an agarose gel by comparing with the undigested DNA.
2. Take the LR Clonase from the −80° freezer and place on dry ice.
3. Compose the reactions on ice.
4. Combine in sterile tubes: 100 ng of entry clone, 300 ng of linearized Destination vector, 5 μl of 5× LR buffer, and TE such that the final volume is 14 μl.
5. Thaw the LR Clonase on ice.
6. Add 6 μl of LR Clonase to each reaction and mix by pipetting.
7. Incubate for 4 hr at 25°.
8. Add 2 μl of proteinase K (1 mg/ml); mix by pipetting.
9. Incubate for 10 min at 37°.
10. Transform 2 μl of each reaction in DH5α cells as described above, plate the whole transformation reaction on LB–ampicillin (AD-Dest) or LB–kanamycin (DB-Dest) plates (both 50 μg/ml), and incubate overnight at 37°. The number of colonies obtained varies between 10 and >1000 (Table I).

Conclusions

Using the GATEWAY recombinational cloning system, we have cloned 31 *C. elegans* ORFs into a versatile vector and from that vector into two different expression vectors used for two-hybrid analysis. We have demonstrated that RC can be used to clone dozens of ORFs simultaneously. We are now in the process of automating RC cloning in 96-well plates (our unpublished data, 1999). Our preliminary results indicate that we will be able to clone several hundred ORFs each week by this method. Therefore, the GATEWAY cloning of nearly complete ORFeomes appears feasible in a relatively short period. Because subcloning of ORFs by GATEWAY into different destination vectors is possible and relatively easy, cloned ORFeomes should not only be useful for protein interaction mapping projects but also for many other projects that aim at characterizing the proteome.

In summary, we have outlined and tested a rapid and automatable cloning method that should greatly facilitate the cloning of entire ORFeomes. The generation of cloned ORFeomes will benefit functional genomics projects as well as smaller scale projects in which only a few proteins are studied.

Acknowledgments

We thank L. Matthews and J. Dekker for comments on the manuscript, H. R. Horvitz and S. Kim for reagents and/or unpublished information, and C. Ezuma-Ngwu for generous help with the design of our website. The *C. elegans* ORFeome cloning project is supported by grants from the National Human Genome Research Institute (1 RO1 HG01715-01), the National Cancer Institute (1 R21 CA81658A01), and the Merck Genome Research Institute awarded to M.V.

[35] Gene Trapping Methods for the Identification and Functional Analysis of Cell Surface Proteins in Mice

By WILLIAM C. SKARNES

Introduction

With the sequencing of all the genes in the mammalian genome, the functional analysis of genes will take precedence over gene discovery. Cell surface proteins are a highly desirable class of molecules for study as they delineate signaling pathways used by cells in response to extracellular cues. A number of strategies based on expression cloning have been developed

0076-6879/00 $30.00

specifically to identify secreted products.[1,2] While these methods are useful for cloning new genes, little or no information is gained regarding their function.

The generation of mutations in murine embryonic stem (ES) cells by targeted and random approaches offers a powerful tool for the functional analysis of genes in mice. ES cells can be grown in large numbers, screened in culture for desired genetic changes, and returned to the embryo, where they can contribute to the germ line. ES cells are now widely used for the disruption of cloned genes by homologous recombination en route to creating mutant mice that lack or express an altered form of a specific gene. Gene trapping offers an alternative method to create random insertional mutations that are immediately accessible to molecular characterization.[3–5] By eliminating the time-consuming step of constructing targeting vectors for each gene of interest, the rate at which new mutations may be introduced in the germ line by gene trapping far exceeds that of conventional gene targeting. Thus, gene trapping represents a rapid and cost-efficient method for the identification and functional analysis of new genes in mice.

Gene trap vectors are activated through the production of a reporter gene fusion transcript following insertions of the vector within endogenous transcription units.[6–9] Gene expression is tagged by the reporter and a portion of the target gene may be cloned directly from the fusion transcript.[10,11] Most importantly, gene trap insertions are expected to be mutagenic, allowing for the functional analysis of the target gene after germ line transmission of the gene trap cell line. Thousands of gene trap cell lines can be conveniently isolated and characterized to generate a library of mutant ES cell lines.[12–14] Given the time and resources required for the

[1] K. Tashiro, H. Tada, R. Heilker, M. Shirozu, T. Nakano, and T. Honjo, *Science* **261,** 600 (1993).

[2] R. D. Klein, Q. Gu, A. Goddard, and A. Rosenthal, *Proc. Natl. Acad. Sci. U.S.A.* **93,** 7108 (1996).

[3] W. C. Skarnes, *Curr. Opin. Biotechnol.* **4,** 684 (1993).

[4] D. P. Hill and W. Wurst, *Methods Enzymol.* **225,** 664 (1993).

[5] G. Friedrich and P. Soriano, *Methods Enzymol.* **225,** 681 (1993).

[6] A. Gossler, A. L. Joyner, J. Rossant, and W. C. Skarnes, *Science* **244,** 463 (1989).

[7] D. G. Brenner, S. Lin-Chao, and S. N. Cheon, *Proc. Natl. Acad. Sci. U.S.A.* **86,** 5517 (1989).

[8] G. Friedrich and P. Soriano, *Genes Dev.* **5,** 1513 (1991).

[9] S. Reddy, J. V. DeGregori, H. Von Melchner, and H. E. Ruley, *J. Virol.* **65,** 1507 (1991).

[10] W. C. Skarnes, B. A. Auerbach, and A. L. Joyner, *Genes Dev.* **6,** 903 (1992).

[11] H. Von Melchner, J. V. DeGregori, H. Rayburn, S. Reddy, C. Friedel, and H. E. Ruley, *Genes Dev.* **6,** 919 (1992).

[12] D. J. Townley, B. J. Avery, B. Rosen, and W. C. Skarnes, *Genome Res.* **7,** 293 (1997).

[13] G. G. Hicks, E. Shi, X.-M. Li, C.-H. Li, M. Pawlak, and H. E. Ruley, *Nature Genet.* **16,** 338 (1997).

phenotypic analysis of mutant mice, it is advantageous to tailor screens toward specific genes of interest. A variety of criteria may be used to preselect insertional mutants prior to germ line transmission and phenotype analysis. These include the sequence of the target gene,[12–14] the expression profile of the target gene,[15–23] and the subcellular localization of the fusion product.[24,25] This chapter focuses on a strategy we developed to select insertional mutations specifically in genes encoding cell surface proteins based on the subcellular localization of the fusion protein products in ES cells. An overview of our experience with this technology is presented below, followed by a detailed description of the experimental procedures.

Gene Trapping in Mouse Embryonic Stem Cells

Vector Design

Gene trapping vectors are designed to detect insertions within transcribed regions of the genome. A variety of plasmid- and retrovirus-based vectors have been developed that differ in their individual requirements for reporter gene activation. Promoter trap vectors simply consist of a promoterless reporter gene that is activated after insertions into exons. In contrast, gene trap vectors contain a splice acceptor sequence upstream of a reporter and are activated after insertion into introns. Both promoter

[14] B. P. Zambrowicz, G. A. Friedrich, E. C. Buxton, S. L. Lilleberg, C. Person, and A. T. Sands, *Nature* (*London*) **392,** 608 (1998).

[15] W. Wurst, J. Rossant, V. Prideaux, M. Kownacka, A. Joyner, D. P. Hill, F. Guillemot, S. Gasca, D. Cado, A. Auerbach, and S.-L. Ang, *Genetics* **139,** 889 (1995).

[16] L. Forrester, A. Nagy, M. Sam, A. Watt, L. Stevenson, A. Bernstein, A. Joyner, and W. Wurst, *Proc. Natl. Acad. Sci. U.S.A.* **93,** 3087 (1996).

[17] C. A. Scherer, J. Chen, A. Nachabeh, N. Hopkins, and H. E. Ruley, *Cell Growth Differ.* **7,** 1393 (1996).

[18] R. K. Baker, M. A. Haendel, B. J. Swanson, J. C. Shambaugh, B. K. Micales, and G. E. Lyons, *Dev. Biol.* **185,** 201 (1997).

[19] W. Yang, T. S. Musci, and S. L. Mansour, *Hear. Res.* **114,** 53 (1997).

[20] I. S. Thorey, K. Muth, A. P. Russ, J. Otte, A. Reffelmann, and H. Von Melchner, *Mol. Cell Biol.* **18,** 3081 (1998).

[21] W. L. Stanford, G. Caruana, A. Vallis, M. Inamdar, M. Hidaka, V. L. Bautch, and A. Bernstein, *Blood* **92,** 4622 (1998).

[22] J.-W. Xiong, R. Battaglino, A. Leahy, and H. Stuhlmann, *Dev. Dynamics* **212,** 181 (1998).

[23] A. Stoykova, K. Chowdury, P. Bonaldo, M. Torres, and P. Gruss, *Dev. Dynamics* **212,** 198 (1998).

[24] W. C. Skarnes, J. E. Moss, S. M. Hurtley, and R. S. P. Beddington, *Proc. Natl. Acad. Sci. U.S.A.* **92,** 6592 (1995).

[25] P. Tate, M. Lee, S. Tweedie, W. C. Skarnes, and W. A. Bickmore, *J. Cells Sci.* **111,** 2575 (1998).

and gene tinsertions create a fusion transcript that includes endogenous exons 5′ to the insertion site and, therefore, the endogenous gene may be identified by 5′ RACE (rapid amplification of cDNA ends).[10,12,26] A third type of vector design, the poly(A) trap, incorporates a splice donor sequence downstream of the reporter and is activated by splicing to exon sequences 3′ of the insertion site.[27] In this case, the target gene may be identified by 3′ RACE.[14,27]

Individual vector designs have their own inherent biases that will influence the range of target genes isolated. For example, gene trap vectors are more efficient than promoter trap vectors,[6,8] but will favor the detection of genes composed of large intronic regions and are likely to miss genes possessing few or no introns. Gene trap vectors are also constrained by the reading frame imposed by the splice acceptor sequence. One solution has been to incorporate the splice acceptor derived from the murine leukemia virus (MuLV) *env* gene, which is capable of splicing in all three reading frames simultaneously.[7] However, splice acceptors with this property are predicted to be weak and wild-type transcripts may be produced by splicing around the vector. Alternatively, an internal ribosome entry site (IRES) may be used to initiate translation of the reporter gene independent of the upstream open reading frame.[28,29] Because IRES-containing vectors are inappropriate for screens based on subcellular localization, our solution to this problem has been to construct separate vectors for each of the three reading frames.

Targeting genes that possess few or no introns may be accomplished by using a promoter trap design. However, the ends of plasmid-based vectors are susceptible to nucleases during electroporation and recombination into the genome (Ref. 30; and J. Brennan, B. Avery, and W. C. Skarnes, unpublished results). Deletions from the ends of the vector reduce the efficiency of cloning and make primer selection for 5′ RACE more difficult. In this regard, retrovirus-based promoter trap vectors may be more useful, as the ends of the vector are generally preserved at the site of insertion. Such vectors will require further optimization because a significant proportion of promoter trap insertions occur within introns of genes.[13]

[26] M. A. Frohman, M. K. Dush, and G. Martin, *Proc. Natl. Acad. Sci. U.S.A.* **85,** 8998 (1988).

[27] H. Niwa, K. Araki, S. Kimura, S. Taniguchi, S. Wakasugi, and K. Yamamura, *J. Biochem. (Tokyo)* **113,** 343 (1993).

[28] P. Mountford, B. Zevnik, A. Duwel, J. Nichols, M. Li, C. Dani, M. Robertson, I. Chambers, and A. Smith, *Proc. Natl. Acad. Sci. U.S.A.* **91,** 4303 (1994).

[29] K. Chowdhury, P. Bonaldo, M. Torres, A. Stoykova, and P. Gruss, *Nucleic Acids Res.* **8,** 1531 (1997).

[30] A. K. Voss, T. Thomas, and P. Gruss, *Dev. Dynamics* **212,** 171 (1998).

The choice of sequences included in the vector can be critical to obtain optimal results. For example, splicing around gene trap insertions may occur if either the splice acceptor and/or polyadenylation signal is inefficient. If any part of the vector contains tissue-specific transcriptional enhancers or repressors the pattern of reporter gene activity may not necessarily reflect that of the target gene.[31] For plasmid-based vectors, it is important to include buffer sequences on either side of the reporter to protect the ends from nuclease activity during transfection. The buffer sequence itself must be devoid of splicing signals or regulatory elements that may perturb the proper function of the vector. For these reasons, the use of proven vector designs that have been demonstrated to accurately report endogenous gene expression and effectively mutate the target gene is highly recommended.

ATG-less Vectors

The gene trap vectors used in our studies lack a translation start signal (ATG), such that the activation of the reporter gene depends on the production of an in-frame fusion to the N terminus of the target gene. The *βgeo* gene,[8] a fusion of β-galactosidase (β-Gal) and neomycin phosphotransferase (*neo*), represents a useful reporter for this purpose as it can tolerate most N-terminal fusions without the loss of β-Gal or *neo* activity. Moreover, it provides a convenient histochemical marker and a direct drug selection for desired gene trap events.

The ATG-less vector design has several advantages: (1) the vectors should enrich for insertions in protein-coding genes; (2) homology to other genes is more likely to be found within protein-coding sequences; (3) disrupting the coding sequence, as opposed to untranslated regions, should guarantee a loss of function mutation; and (4) the generation of fusion proteins provides a means to identify particular classes of molecules. For example, nuclear factors can be readily identified among the subset of protein fusions that acquire a nuclear localization signal.[25] Similarly, the secretory trap vector (see below) relies on the differential activity of fusions that capture an N-terminal signal sequence of a membrane or secreted protein.[24]

βgeo-based vectors are designed to detect insertions in expressed genes, and therefore the sensitivity of the drug selection is crucial to accessing genes expressed at low levels in ES cells. Although a bias in favor of highly expressed genes was noted with the original βgeo reporter,[8] this was due to the presence of a mutation in neomycin phosphotransferase.[24] We corrected this mutation in our vectors and, consequently, are able to trap genes

[31] A. Camus, C. Kress, C. Babinet, and J. Barra, *Mol. Reprod. Dev.* **45,** 255 (1996).

expressed at less than detectable levels in undifferentiated ES cells (based on β-Gal staining and Northern blot analysis).[24] These lines generally show the most restricted patterns of expression in the embryo and represent the current limit of recoverable events in genes expressed at low levels. Whether it will be possible to trap genes that are not expressed in ES cells (if such a class of genes exists) remains to be addressed.

Aberrant Events

Gene trap vectors do not necessarily trap genes in a predictable manner: as many as one-half of secretory trap insertions produce unspliced fusion transcripts.[12,24] In many cases, 5′ RACE yields two products, one representing the intron sequence of the vector and a second properly spliced sequence corresponding to the target gene. We hypothesize that these events correspond to insertions in exons of genes, where the splice acceptor of the vector competes with the splice acceptor of the disrupted exon. A second class of inefficiently spliced events represent insertions in rRNA genes, resulting in what can best be described as *trans*-splicing (J. Sleeman, B. Rosen, and W. C. Skarnes, unpublished observations). Cloning of 5′ RACE products from two characterized insertions in rRNA genes detected several different endogenous gene sequences joined to the splice acceptor of the vector. Presumably these uncapped RNA pol I transcripts are poor substrates for *cis*-splicing and splicing factors assembled at the splice acceptor of the vector are free to interact with other splice donors *in trans*.

A second class of aberrant events involved deletions from the ends of the vector. cDNA from these cell lines failed to amplify with the standard primers used for 5′ RACE but could be amplified with nested primers further 3′ in the vector (J. Brennan, B. Avery, and W. C. Skarnes, unpublished observations). Because the splice acceptor site is lost, these events do not represent true gene trap events and are difficult to characterize. The frequency of nonamplifiable events varied with each electroporation, ranging from less than 5% to as high as 30%. In experiments that gave a high proportion of nonamplifiable cell lines, we noted an overall reduction in the absolute number of productive, efficiently spliced events. This may reflect the degree of damage to the vector by exposure to nucleases in the cell suspension prior to electroporation. When we minimized the time the DNA and cells were in contact prior to electroporation, a marked improvement in the proportion of productive events was observed. More rarely, insertions in exons of genes have been found in which cryptic donor sites within the interrupted exon or within the vector are utilized. These insertions do not pose a problem as they are expected to mutate the target gene even if the splice acceptor of the vector is poorly utilized.

Insertional Hot Spots

As the number of characterized gene trap lines increases, it has become clear that these insertions do not occur randomly in the genome as has been previously suggested.[29] Two independent insertions in the same gene were first recorded among a small sample of six secretory trap insertions, raising concerns about the number of genes accessible with this approach.[24] More recently, in a sample of 2000 retroviral gene trap cell lines, approximately 30% of the insertions in known genes were recovered more than once.[14] Similarly, in our current sample of more than 200 secretory trap cell lines, multiple insertions have been found in one-quarter (24 of 100) of the known target genes (Table I). The distribution of repeat events does not follow a simple Poisson distribution, but rather suggests that a small fraction of the target loci correspond to more highly preferred sites of integration.

Secretory Trap Screen

To gain access to genes important for cell–cell interactions in the embryo, we developed a modified gene trap approach to identify and mutate genes encoding cell surface molecules.[24] Most proteins destined for the plasma membrane contain a cleavable N-terminal signal sequence. We found that β-Gal enzyme activity was abolished when fused to the N-terminal signal sequence of the CD4 receptor, presumably due to misfolding or modification the protein in the lumen of the endoplasmic reticulum (ER). To identify insertions in genes that encode N-terminal signal sequences, a type II transmembrane domain was placed upstream of the β-Gal reporter. N-terminal fusions to β-Gal that do not acquire a signal sequence are inserted in a type II orientation by default, exposing β-Gal to the lumen of the ER where its enzymatic activity is lost. In contrast, fusions that acquire an N-terminal signal sequence are inserted in the membrane in a type I orientation and retain β-Gal activity. Thus, the secretory trap vector provides a simple selective assay for insertions in genes that encode a signal sequence. In practice, we find that the secretory trap vector design enriches but does not absolutely select for insertions in genes encoding cell surface proteins (Table I). This enrichment, however, depends on preselecting cell lines that show a characteristic pattern of β-galactosidase staining.

Preselection of Cells that Show a Secretory Pattern of β-Galactosidase Localization

Electroporation of ES cells with the secretory trap vector typically yielded between 200 and 400 G418-resistant colonies, of which 30% express detectable β-Gal activity. Initially (Ex series), all β-Gal-positive colonies

TABLE I

KNOWN GENES ASSOCIATED WITH CELL LINES DISPLAYING A SECRETORY PATTERN[a]

Secreted proteins
- BMP-8a
- Cripto
- IGF-binding protein-2
- Netrin-1 (4)
- Nodal
- Semaphorin C

Extracellular matrix components
- Agrin (2)
- Fibronectin (2)
- Glypican-1
- Glypican-3/OCI-5 (2)
- Glypican-4/K-glypican
- Laminin β-1
- Laminin-γ1 (9)
- Perlecan

Cell adhesion molecules
- E-Cadherin (5)
- Embigin (3)
- Endoglin
- Entactin-2
- FAT (3)
- ICAM
- Integrin β1
- Junctional adhesion molecule
- SDR-1 (2)

Membrane receptors
- APLP-2
- CD10
- CIRL-2
- EphA2/Eck (4)
- EphA4/Sek (3)
- FGF receptor
- gp130
- IGF-1 receptor
- IGF-2 receptor
- Interferon α/β receptor
- LAR (2)
- LDL receptor
- LRP-2/megalin LRP-6 (2)
- Neuropilin-2
- Phosphodiesterase I
- PTPγ
- PTPκ (2)
- Roundabout-1
- SIM
- SorLa/LR11

ER/Golgi/lysosomal proteins
- ATF-6 (3)
- DNAJ-3
- Erp72 (2)
- FuCT
- β-1,4-Galactosyltransferase
- GlucNAc transferase I
- GlucNAc transferase V
- Gfpt-2
- hsp86
- α-Mannosidase II (2)
- β-Mannosidase
- Prolylcarboxypeptidase

Multiple membrane-spanning proteins
- $GABA_B$ receptor 1
- GLUT-1
- TASK-2
- TWIK-1

Nonsecreted proteins
- β-Actin (5′ UTR)
- Atrophin-related protein (5′ UTR)
- B120
- CaRG binding factor
- cdc42 (5′ UTR)
- CtBP-2 (5′ UTR) (6)
- α-Endosulfine (5′ UTR)
- fb19 (5′ UTR)
- FLH-1 (5′ UTR)
- hnRNP C1/C2 (5′ UTR)
- IKAP (5′ UTR)
- LINE-1 (2)
- Lissencephaly protein-1 (5′ UTR)
- MDM-2
- Mym (5′ UTR)
- Nedd-5 (5′ UTR)
- NIPSNAP-2
- Nonmuscle myosin (5′ UTR)
- Oncomodulin (5′ UTR)
- OSF-3 (5′ UTR)
- PKU-β
- Ras GAP (5′ UTR)
- RBP-Jκ (5′ (UTR) (2)
- rRNA (2)
- SVCT-2 (5′ UTR)
- U17 snRNA (5)
- U1 snRNP (70 kDa)
- wbscr-1

[a] Numbers in parentheses indicate the number of independent insertions in the same gene.

were expanded and RNA from 333 cell lines was analyzed by direct sequencing of 5′ RACE products.[12] A total of 122 cell lines (37%) efficiently utilized the splice acceptor of the vector, whereas 169 lines were uninformative (unspliced vector sequence or multiple sequences upstream of the splice acceptor) and 42 failed to amplify altogether (Table II). More disconcerting was the fact that half of the 88 sequences matching known genes encoded nonsecreted gene products. On reexamination of the 5-bromo-4-chloro-3-indolyl-β-D-galactopyranoside (X-Gal) staining patterns in cells, we noted that all of the properly spliced insertions in bona fide secreted and membrane-spanning proteins showed the distinctive "secretory" pattern of β-Gal localization. This pattern is distinguished by staining around the nucleus and in multiple cytoplasmic dots.[24] In contrast, nearly all of the uninformative lines and a majority of insertions in nonsecreted proteins exhibited a localization of β-Gal activity distinct from the secretory pattern. The most common nonsecretory patterns are typified by a single dot of expression per cell or uniform staining throughout the cytoplasm. In this retrospective analysis, we discovered that properly spliced events in cell surface molecules were greatly enriched among those lines that express β-Gal in a secretory pattern. Putting this criterion to the test, a second series of electroporations (GST and NST series) was performed in which cell lines were prescreened for the secretory pattern. From 114 cell lines subjected to 5′ RACE, 97 (85%) represented properly spliced events, of which 51 sequences corresponded to known secreted or membrane proteins and 10 matched nonsecreted gene products. Thus, preselecting cell lines that display a secretory pattern is critical.

Representation of Target Genes in the Database

More than 400 cell lines showing a secretory pattern of β-Gal activity were analyzed by direct sequencing of 5′ RACE products and sequence

TABLE II
EFFICIENCY OF ISOLATING PRODUCTIVE SECRETORY TRAP INSERTIONS[a]

Experiment	Number of lines sequenced	Properly spliced	Inefficiently spliced	Nonamplifiable
1 (Ex001–120)	118	37 (32%)	52	29
2 (Ex121–525)	215	85 (40%)	117	13
3 (GST1–205)	97	82 (84%)	15	0
4 (NST1–35)	17	15 (88%)	2	0

[a] Ex series, unselected lines; GST/NST, lines selected for the secretory pattern.

tags were obtained for 215 cell lines (Table III; sequence information for each line is available at http://socrates.berkeley.edu/~skarnes/resource.html). The majority of these tags (77%) matched a human or mouse gene or EST sequence in the public database and the remainder corresponded to novel sequences. The high degree of correspondence between our sequence tags and sequences in the public database differs from other random gene trap screens, in which the proportion of sequence matches to known genes ranged from 10 to 30%.[13,14] Reasons for this discrepancy are not entirely clear, but may be related in part to the fact that secreted and membrane-spanning molecules have been intensely studied and are perhaps better represented in the current database. Another contributing factor may be that our ATG-less vectors are designed to select for insertions within protein-coding sequences, whereas ATG- or IRES-bearing vectors used in other screens are expected to target noncoding sequences, particularly 5′ untranslated regions (UTRs), which are underrepresented in the current database. The poly(A) trap vector[14] is even less stringent, requiring only a polyadenylation signal, and could lead to the trapping of intronic or intergenic sequences.

The genes associated with cell lines that display a secretory pattern of β-Gal expression are listed in Table I. All classes of secreted and membrane proteins are represented, including growth factors, extracellular matrix components, cell adhesion molecules, membrane receptors, ER/Golgi and lysosomal proteins, and multiple membrane-spanning proteins. The sites of insertion were evenly distributed within the extracellular domains of these target genes: 28% of the insertions occurred within <0.5 kb of the 5′ end of the mRNA, 23% between 0.5 and 1 kb; 33% between 1 and 2 kb;

TABLE III
SEQUENCES ASSOCIATED WITH SECRETORY TRAP CELL LINES[a]

Sequence	Total lines	Total number of genes	Repeated genes	Unique genes	Unique genes/total genes
Known	**137**	**90**	**22**	**68**	**68/90 (76%)**
Secreted genes	97	62	17	45	
Nonsecreted genes	40	28	5	23	
Novel	**78**	**51**[b]	**11**[b]	**40**[b]	
ESTs	28				
No matches	50				
Total	**215**	**141**	**33**	**108**	

[a] Cell lines that display a secretory pattern of β-Gal activity.
[b] Based on the repeat frequency observed among known genes.

and 16% beyond 2 kb. It is noteworthy that the majority of insertions in nonsecreted genes (28 of 35) occurred in the 5′ untranslated region of the target gene. This result implies that reporter gene expression is initiated at one or more of seven internal ATGs within the vector, all located in the vicinity of the transmembrane domain. Initiation at these sites presumably circumvents the selective property of the vector. Insertions in nonsecreted genes often produce short fusion transcripts, which in fact provides a useful criterion to identify and eliminate putative nonsecreted molecules from the novel set of target genes.

Efficient Germ Line Transmission of Feeder-Independent Embryonic Stem Cells

The value of our resource of secretory trap insertions for the functional analysis of secreted and membrane-spanning molecules will be determined by the efficiency of ES cell lines to contribute to the germ line of mice and the ability of these insertions to cause loss of function mutations. We chose to use feeder-independent ES cells,[32] as they effectively halve the amount of tissue culture and thus reduce the cost and effort in a large-scale screen. The rate of germ line transmission of feeder-independent cells was untested and it was important to demonstrate that these cells were equally efficient in colonizing the germ line as feeder-dependent cell lines. One such early-passage cell line established in the absence of feeders (CGR8; generously provided by A. G. Smith, University of Edinburgh) was tested and proved to be satisfactory (1 germ line male per 10 blastocysts injected). However, a significant reduction in the rate of germ line transmission was observed in later passage cells. Sublines of CGR8 cells were then isolated and one of these sublines (CGR8.8) gave an extremely favorable rate of germ line transmission (one germ line male per five injected blastocysts for more than 85% of clonal cell lines tested). In a period of 6 months, we succeeded in transmitting 59 secretory trap cell lines (Ex series) to the germ line of mice by injecting an average of 10–15 blastocysts per cell line. More recently, we have tested a second feeder-independent cell line (E14Tg2a-clone 4; provided by A. G. Smith) with comparable results.

In all, 85 secretory trap cell lines were transmitted to the germ line of mice for expression and phenotype analysis. Many of these cell lines were injected without prior knowledge of the target gene, and regrettably, only 56 of these cell lines represented efficiently spliced insertion events, of which 42 lines were associated with β-Gal expression in a secretory pattern

[32] J. Nichols, E. P. Evans, and A. G. Smith, *Development* **110,** 1341 (1990).

(Table IV). Staining of embryos at midgestation showed restricted patterns of reporter gene expression in approximately half of the transgenic lines, agreeing well with the fraction of developmentally regulated loci reported in other screens using conventional gene trap vectors.[15] We observed β-galactosidase activity in a pattern similar to that of the target gene in cases where the embryonic expression pattern of the target gene has been characterized. These include EphA4, netrin-1, neuropilin-2, glypican-1 and −3, K-glypican, LDL-receptor, fibronectin, agrin, and laminin-γ1 (Ref. 24; and our unpublished results). In contrast, the vast majority of embryos derived from cell lines that display a nonsecretory pattern of β-Gal activity failed to stain or stained weakly in the extraembryonic yolk sac. Because these cell lines represent aberrant events or insertions in 5′ UTRs of nonsecreted genes, we believe that this result reflects poor translation of the reporter gene to produce levels of β-Gal below the threshold of detection. Thus, reporter gene activity associated with aberrant insertions may not accurately report endogenous gene expression patterns and should be interpreted with caution.

Mutagenic Potential of Secretory Trap Insertions

We wished to establish the rate of lethal mutations caused by secretory trap insertions and to compare the mutation rate of insertions within coding sequences of genes (Table IV) to aberrant insertion events (Table V). Among the insertion events that interrupt the coding sequences of genes, we found that 40% (14 of 36 loci) caused recessive embryonic lethal mutations. This rate of mutation compares favorably with the proportion of lethal phenotypes observed with targeted gene knockouts and other conventional gene trap screens. The severity of the phenotypes observed with our insertional mutations in agrin, EphA4, fibronectin, glypican-3, laminin-γ1, LAR, LDL-receptor, and MDM-2 were similar, if not identical, to those generated by conventional gene targeting (our unpublished observations). These data demonstrate that secretory trap insertions within the coding sequence of the target gene cause phenotypic null mutations. In contrast, only 1 in 43 of the aberrant class of insertions resulted in lethality (GST1, an unspliced insertion event). We do not fully understand the nature of the events that lead to inefficient splicing, but clearly, this class of events should be avoided. Moreover, it was surprising to find that none of the nine efficiently spliced insertions in 5′ UTRs caused lethal phenotypes. These results suggest that insertions in 5′ UTRs may be less likely to cause a mutation than insertions that interrupt the coding sequences of the target gene.

In summary, the secretory trap approach offers a powerful tool for the functional analysis of cell surface proteins. The key elements of an effective

TABLE IV
EXPRESSION PATTERNS AND PHENOTYPES ASSOCIATED WITH SECRETORY TRAP INSERTIONS

Cell line	Gene	Expression	Phenotype
ST402	Embigin	Restricted	Viable
ST497	EphA4 (Sek)	Restricted	Viable, locomotor defect
GST2	EphA4 (Sek)	Restricted	Viable, locomotor defect
ST531	PTPκ	Restricted	Viable
Ex125	PTPκ	Restricted	Viable
ST534	LAR	Widespread	Viable, mammary gland defect
ST629	Netrin-1	Restricted	Neonatal lethal; axon guidance defects
JST51	LDL-receptor	Ubiquitous	Viable
JST185	Novel (multi-TM)	Restricted	Neonatal lethal; lung defect
JST213	DNAJ-3	Restricted	Viable
Ex005	α-Mannosidase II	Ubiquitous	Viable
Ex054	BRUCE	Restricted	Neonatal lethal; gut defect
Ex057	APLP-2	Widespread	Viable
Ex061	Novel	Ubiquitous	Implantation lethal
Ex065	MDM-2[a]	Widespread	Embryonic lethal
Ex106	BMP8A	Restricted	Viable
Ex124	Novel	Restricted	Viable
Ex132	RB13-6	Restricted	Viable
Ex136	Glypican-3 (X-linked)	Widespread	Neonatal lethal; overgrowth, kidney defect
Ex151	Novel	None	Viable
Ex153	SDR-1	Widespread	Viable
Ex160	cdc-42[a]	Ubiquitous	Implantation lethal
Ex166	CArG binding factor[a]	Restricted	Viable
Ex177	Novel LIG-1 related	None	Viable
Ex180	Fibronectin	Widespread	Embryonic lethal
Ex183	Laminin γ1	Widespread	Embryonic lethal
Ex186	LDL receptor-related	Widespread	Viable
Ex187	LRP6	Ubiquitous	Neonatal lethal; many developmental defects
Ex192	Agrin	Widespread	Neonatal lethal; neuromuscular junctions[b]
Ex194	K-glypican (X-linked)	Restricted	Viable
Ex255	SorLa/LR11	ND[c]	Viable
Ex503	EST	Restricted	Neonatal lethal
Ex504	human KIAA0573	Restricted	Viable
GST8	AFT-6	Widespread	Viable
GST48	ATF-6	Widespread	Viable
GST11	Entactin-2	ND	Viable
GST36	Novel	ND	Viable
GST39	Neuropilin-2	Restricted	Lethal; strain dependent (not characterized)
GST45	Novel	Widespread	Viable
GST62	Glypican-1	Restricted	Viable
GST68	Novel	Ubiquitous	Viable
GST70	Human FAT-related	Widespread	Neonatal lethal

[a] Nonsecreted gene products.
[b] R. Burgess, W. C. Skarnes, and J. Sanes (2000). Manuscript submitted.
[c] ND, Not determined.

TABLE V
EXPRESSION AND PHENOTYPE ANALYSIS OF CELL LINES SHOWING A NONSECRETED PATTERN[a]

Cell line	Gene	Expression	Phenotype
ST356	ND	None	Viable
ST356	Unspliced	Yolk sac	Viable
ST387	Unspliced	None	Viable
ST397	ND	Restricted	Viable
ST469	Unspliced	Yolk sac	Viable
ST474	Unspliced	Yolk sac	Viable
ST639	Unspliced	Yolk sac	Viable
Ex003	Nonamplifiable	None	ND
Ex009	GFaT (5′ UTR)	Restricted	Viable
Ex011	Adenosine kinase (5′ UTR)	Yolk sac	ND
Ex030	Adenosine kinase (5′ UTR)	Yolk sac	ND
Ex012	Double sequence (intron)	None	ND
Ex014	Double sequence (intron)	None	ND
Ex016	Nck-alpha (5′ UTR)	Widespread (very weak)	Viable
Ex023	B-myb (5′ UTR)	None	ND
Ex024	Enx-1 (5′ UTR)	Widespread (weak)	Viable
Ex025	OSF-3 (5′ UTR)	Yolk sac	Viable
Ex029	Unspliced	None	ND
Ex032	Vector	Yolk sac	Viable
Ex036	GlucNAc transferase V (5′ UTR)	Restricted	Viable
Ex038	Unspliced	None	ND
Ex040	Unspliced	None	ND
Ex044	Nonamplifiable	Yolk sac	ND
Ex045	Novel	None	Viable
Ex047	Nonamplifiable	Restricted	ND
Ex052	GRASP55	Yolk sac	Viable
Ex056	Novel (probable 5′ UTR)	None	ND
Ex059	Vector	Restricted	Viable
Ex060	Unspliced	Yolk sac	ND
Ex063	Double sequence	Widespread (very weak)	ND
Ex071	Nonamplifiable	None	ND
Ex072	Double sequence (vector intron)	ND	ND
Ex073	Double sequence (vector intron)	None	ND
Ex075	Unspliced	Restricted	Viable
Ex076	Double sequence (vector intron)	None	ND
Ex103	Nonamplifiable	None	ND
Ex112	hfb2 cDNA (5′ UTR)	None	ND
Ex140	Novel (probably 5′ UTR)	None	Viable
Ex145	Unspliced	None	ND
Ex151	Novel	None	ND
Ex172	Unspliced	None	ND
Ex173	EST	None	ND
GST1	Unspliced	Widespread	Lethal
GST44	Ring canal protein (5′ UTR)	None	Viable

[a] ND, not determined.

screen are preselection of cell lines that show a "secretory" pattern of β-Gal expression, direct sequencing to identify efficiently spliced insertion events, and the selection of lines that interrupt the coding sequence of the target gene for germ line transmission and phenotype analysis. These lessons may be generally applicable to other gene trap screens. We strongly recommend the use of ATG-less vectors to trap coding sequences of target genes that, as a consequence, are more likely to correspond to sequences in the database and more likely to mutate the gene at the site of insertion. Large-scale screens using the secretory trap technology, if approached with caution, should provide a valuable future source of new mutant phenotypes and a starting point to elucidate novel signaling pathways that impinge on all aspects of embryonic and postnatal development.

Maintenance of Feeder-Independent Embryonic Stem Cells

We use feeder-independent ES cell lines derived from the 129/Ola strain of mice.[32] Feeder-independent ES cells are easy to maintain and significantly reduce the amount of tissue culture required. Parental cell lines (CGR8 and E14Tg2A) were established from delayed blastocysts on gelatinized tissue culture dishes in ES cell media containing leukocyte inhibitory factor (LIF) as described.[32] Sublines were isolated by plating cells at a single cell density, picking and expanding single colonies, and testing several clones for germ line competence. Two sublines (CGR8.8 and E14Tg2A.4) were expanded and large numbers of vials were frozen at early passage as stock.

Tissue Culture Reagents

Dulbecco's phosphate-buffered saline (PBS), without calcium and magnesium (GIBCO-BRL, Gaithersburg, MD)

2-Mercaptoethanol stock solution: Add 70 μl of 2-mercaptoethanol (Sigma, St. Louis, MO) to 20 ml of distilled, deionized water (GIBCO-BRL). Filter sterilize and store at 4° for up to 2 weeks

ES cell medium: 1× Glasgow MEM/BHK12 medium (Sigma) supplemented with 2 mM glutamine (GIBCO-BRL), 1 mM sodium pyruvate (GIBCO-BRL), 1× nonessential amino acids, 10% (v/v) fetal bovine serum (characterized; HyClone, Logan, UT), and a 1:1000 dilution 2-mercaptoethanol stock solution

Freezing medium: Add dimethyl sulfoxide (DMSO, tissue culture grade; Sigma) to ES cell medium to a final concentration of 10% (v/v). Filter sterilize. Make fresh prior to use

Trypsin solution: Add 100 mg of EDTA tetrasodium salt (Sigma) to 500 ml of PBS. Filter sterilize and add 10 ml (for 1×) or 20 ml (for 2×) of 2.5% (w/v) trypsin solution (GIBCO-BRL) and 5 ml of chicken serum (GIBCO-BRL). Store in 20-ml aliquots at −20°. [*Note:* 2.5% (w/v) trypsin solution should be aliquoted and stored at −20° to avoid multiple freeze–thawing]

Gelatin (0.1%, w/v): Add 25 ml of a 2% (w/v) bovine gelatin solution (Sigma) to 500 ml of PBS. Store at 4°

Geneticin (GIBCO-BRL): Dissolve powder in PBS to make a 125-mg/ml stock solution (active concentration). Filter sterilize and store at −20°. Add 0.56 ml to each bottle (560 ml) of ES cell medium. (*Note:* The concentration of Geneticin should be titrated to determine the minimum concentration that will kill nontransfected ES cells in 5 days)

Thawing Embryonic Stem Cells

ES cells are frozen in medium containing 10% (v/v) DMSO. Because DMSO can induce the differentiation of ES cells, we advise thawing the cells late in the day and changing the medium the following morning to minimize the effects of residual DMSO.

1. Coat a 25-cm^2 tissue culture flask with 0.1% (w/v) gelatin and aspirate off immediately prior to use.
2. Thaw ES cells (approximately 5×10^6 cells, equivalent to one confluent six-well plate or half of a confluent 25-cm^2 flask) in a 37° water bath and dilute into 10 ml of prewarmed ES cell medium.
3. Pellet the cells by spinning for 3 min at 1200 rpm in a benchtop clinical centrifuge.
4. Aspirate off the medium and gently resuspend the cells in 10 ml of prewarmed medium.
5. Transfer the cell suspension to a 25-cm^2 flask and grow at 37° in a humidified 6% CO_2 incubator.
6. Change the medium the following day to remove dead cells and residual DMSO.

Passage and Expansion of Embryonic Stem Cell Cultures

ES cells are routinely passaged every 2 days, changing the medium on alternate days. Thus, ES cells require daily attention. In our experience, feeder-independent ES cells grow rapidly and quickly acidify the medium, turning it yellow in color. Allowing the cells to acidify the medium (by not changing the medium every day or by passaging the cells at too low a dilution) will cause the cells to undergo crisis, triggering excess differentiation and cell death, after which their totipotency cannot be guaranteed. Plating the cells at too low a density, insufficient dispersion of cells during passage, or uneven plating can cause similar problems as the cells will form large clumps prior to reaching confluence and the cells within these clumps will differentiate or die. We have observed a significant reduction in germline transmission of cells that have been mistreated even when they appear healthy at the time of injection.

1. For a confluent 25-cm^2 flask of cells, aspirate off the medium and wash with 5–10 ml of prewarmed PBS, pipetting it away from the cells. Rock the flask gently and aspirate off the PBS. Repeat.

2. Cover the cells with 1 ml of 1× trypsin solution and return to the 37° incubator for 1–2 min or until cells are uniformly dispersed into small clumps.

3. Add 9 ml of medium to inactivate the trypsin.

4. Count the cells and add 10^6 cells (usually 1/10 of a 25-cm^2 flask) to a freshly gelatinized flask.

5. To expand ES cells for electroporation (requiring a total of 10^8 cells), seed 3×10^6 cells (one-third to one-fourth of a confluent 25-cm^2 flask) into a gelatinized 75-cm^2 flask and add 30 ml of medium. Add 20 ml of medium on the following day. Once the cells reach confluence, trypsinize the contents of the 75-cm^2 flask and add 5×10^6 cells (one-fifth of a confluent 75-cm^2 flask) to each of three 175-cm^2 gelatinized flasks containing 50 ml of medium. Add an additional 30 ml of medium the following day.

Freezing Embryonic Stem Cells

1. Trypsinize a confluent 25-cm^2 flask of cells (approximately 10^7 cells) as described above.

2. Collect trypsinized cells in 9 ml of medium and pellet for 3 min at 1200 rpm.

3. Aspirate off the medium and resuspend the cell pellet in 1 ml of freshly prepared freezing medium. Aliquot 0.5 ml of cells into two cryotubes.

4. Freeze the vials at −80° overnight and transfer to liquid nitrogen for long-term storage.

Electroporation and Isolation of Secretory Trap Cell Lines

Electroporation is the preferred method for transfecting ES cells with plasmid-based gene trap vectors, as the concentration of DNA and cell density can be adjusted to favor single-copy insertions. The conditions described below have been optimized for feeder-independent ES cells to maximize colony number and minimize multicopy insertions. Occasionally, multiple copies of the vector will integrate at a single site, but these multicopy insertions do not appear to interfere with 5′ RACE, reporter gene expression, or rates of mutation.

DNA Preparation

Vector DNA may be isolated by alkaline lysis and cesium chloride density centrifugation or by Qiagen (Valencia, CA) column purification. Allow 2 days prior to the electroporation for the preparation of the DNA sample.

1. Linearize the vector by digesting 150 μg of pGT1.8TM DNA overnight at 37° with *Hin*dIII restriction enzyme (0.5 U/μg) in a final volume of 0.3 ml.
2. Precipitate the DNA sample with 2 volumes of absolute ethanol on ice for 5 min. Spin down the DNA in a microcentrifuge and wash the pellet several times with 70% (v/v) ethanol.
3. Drain off as much 70% (v/v) ethanol as possible and allow the remainder to evaporate in a sterile laminar flow hood, leaving the lid of the tube open (1 hr is usually sufficient).
4. Resuspend the DNA in 0.1 ml of sterile PBS and vortex occasionally over a period of 4 hr or more to ensure the DNA is in solution.

Electroporation of Embryonic Stem Cells

For optimal results, the ES cells should be approximately 80% confluent on the day of electroporation and the medium changed a few hours prior to harvesting the cells.

1. Trypsinize the contents of three 175-cm^2 flasks, using 5 ml of trypsin per flask. Add 10 ml of medium to each flask and combine the cells in a 50-ml sterile centrifuge tube. Pellet the cells for 5 min at 1200 rpm and resuspend in 20 ml of PBS.
2. Count the cells, repellet, and resuspend the cells at a concentration of 10^8 cells/0.7 ml of PBS.
3. Add 0.7 ml of cells to the tube containing 0.1 ml of linearized vector DNA, quickly mix with a 1-ml plastic pipette, and transfer immediately to a 0.4-cm electroporation cuvette. Electroporate [Bio-Rad (Hercules, CA, Gene Pulser unit] at 3 μF/800 V (time constant, 0.04 msec).
4. Allow cells to recover in the cuvette for 20 min at room temperature, and then transfer to 200 ml of ES cell medium. Plate 10 ml (5×10^6 cells) onto each of 20 gelatinized 10-cm-diameter tissue culture petri dishes.
5. Aspirate the medium the following day and replace with medium containing Geneticin (125 μg/ml). Change the medium daily for the first 4–5 days. Once most of the cells have been cleared from the plate, the

medium may be replaced on alternate days. Eight days after electroporation the colonies should be about 1 mm in diameter and are ready to be picked.

Picking G418-Resistant Colonies and Replica Plating

1. Count and circle the colonies on the bottom of each dish with a marker pen. Gelatinize the appropriate number of 48-well plates.

2. Aspirate culture medium and add 10 ml of room temperature PBS to each dish. If possible, colonies should be picked in a laminar flow hood to minimize contamination. Using a Pipetman (P200) and sterile plugged (aerosol-resistant) tips, break up the colony and collect in 50 μl of PBS. Transfer the cell clumps to a 48-well plate and repeat until all the wells contain cells.

3. Add 50 μl of a 2×-strength trypsin solution to each well and incubate for 10 min at 37°. Tap the plates to disperse the cells into small clumps and add 1 ml of medium to each well. The following day replace with 0.5 ml of medium.

4. When the cells approach confluence, trypsinize with 100 μl of 2×-strength trypsin for 2–4 min, and then add 1 ml of medium. Using a Pipetman (P1000) and sterile-plugged tips, transfer 0.9 ml to one set of gelatinized 48-well plates (replica plate for X-Gal staining) and the remainder to a second set of gelatinized 48-well plates (master plate for expanding selected cell lines). Add 1 ml of medium to each well of the master plate. (*Note:* Individual clones grow at different rates, so cells are harvested in batches over a period of 1 week. As a rule, we replenish the medium at least every 2 days until the wells reach near confluence.)

5. One day after plating, the replica plates are stained with 0.2 ml of X-Gal staining solution and incubated overnight at 37° in a humidified chamber.

6. On the following day, each of the X-Gal stained wells are examined under bright-field illumination for cell lines that display the "secretory" pattern of β-Gal localization: perinuclear staining and multiple cytoplasmic dots as opposed to a single dot of expression commonly seen (see Ref. 24 for examples of the secretory staining pattern). The secretory pattern can be seen best in the subset of flattened, spontaneously differentiated cells in the culture. It is more difficult to distinguish the pattern in lines where β-Gal expression is confined to undifferentiated ES cells. Roughly half of the β-Gal-positive clones should display this pattern.

7. Cell lines that display the secretory pattern of β-Gal staining are selected from the master 48-well plates and expanded for RNA isolation and for storage in liquid nitrogen. Typically, cells in confluent 48-well plates are expanded up to 12-well plates and then split into duplicate 6-well dishes:

one for RNA, one for freezing. It is important to maintain G418 selection during expansion to prevent wild-type (nontransfected) cells from contaminating the cultures.

X-Gal Staining Protocol

1. Prepare 0.1 *M* phosphate buffer by dissolving $NaH_2PO_4 \cdot H_2O$ (3.74g/liter) and anhydrous Na_2HPO_4 (10.35 g/liter) in water (pH 7.3).
2. Remove the medium from each well and fix cells for 5 min at room temperature with 0.1 *M* phosphate buffer containing 5 m*M* EGTA, 2 m*M* magnesium chloride, and 0.2% (v/v) glutaraldehyde (Sigma).
3. Wash the cells twice briefly with 0.1 *M* phosphate buffer containing 2 m*M* magnesium chloride (do not add detergents when staining cultured ES cells).
4. Prepare the required amount of X-Gal staining buffer immediately before use by diluting a 50-mg/ml X-Gal stock solution (dissolved in dimethylformamide) to a final concentration of 1 mg/ml in 0.1 *M* phosphate buffer containing 2 m*M* magnesium chloride, 5 m*M* potassium ferrocyanide (Sigma), and 5 m*M* potassium ferricyanide (Sigma). Filter sterilize the X-Gal staining buffer to prevent crystal formation (*Note:* If X-Gal buffer is to be reused, store frozen, protected from the light, and filter again prior to use.)
5. Stain the cells overnight at 37° in a humidified chamber with 0.2 ml of X-Gal staining buffer per well. Rinse the wells with 0.1 *M* phosphate buffer and add fix buffer prior to scoring the cells for staining patterns. Plates can be stored a 4° for several weeks and it is useful to refer back to the staining patterns once sequence information becomes available.

Generation of Sequence Tags

Given the preponderance of aberrant splicing events and insertional hot spots, an efficient screen will depend on the ability to rapidly sequence a portion of the endogenous gene upstream of the site of insertion. Direct sequencing of 5′ RACE products offers a robust method to generate sequence tags for each insertion, obviating the need to clone individual RACE products.[12] With this method, sequence information will be obtained only for those insertions that undergo efficient splicing. Aberrant events, i.e, vector deletions, inefficient and/or complex splicing patterns, will not produce a readable sequence and can be eliminated from the screen. This protocol is also suitable for automated fluorescent sequencing.

RNA Purification

1. ES cells from a confluent well are washed once with PBS and lysed in RNAzol B (Tel-Test, Friendswood, TX) or Trizol (GIBCO-BRL). Use 1 ml for 6-well plates or 0.25 ml for 24-well plates and transfer the lysate to a microcentrifuge tube (samples may be stored at −80° indefinitely prior to RNA extraction).

2. Add 0.1 volume of chloroform, vortex for 15 sec, and place on ice for 5 min.

3. Spin the sample at top speed in a microcentrifuge for 15 min at 4° and transfer the aqueous phase to a new tube.

4. Add 1 volume of phenol–chloroform (1 : 1, v/v) to the aqueous phase, vortex, and spin. (*Note:* The additional phenol–chloroform extraction is essential to remove contaminating proteins that can interfere with microdialysis.) Extract the aqueous phase once more with chloroform and transfer the aqueous phase to a new tube.

5. Add 1 volume of isopropanol and place on ice for 5 min. Pellet the RNA at top speed in a microcentrifuge and wash the pellet twice with 70% (v/v) ethanol [made with diethyl pyrocarbonate (DEPC)-treated water].

6. Air dry the pellet and resuspend in DEPC-treated water (20 or 50 μl for samples isolated from 24- or 6-well plates, respectively).

5′ RACE Reactions

1. Add 5 μg of total RNA, 10 ng of a primer for first-strand cDNA synthesis, and DEPC-treated water to a volume of 12 μl. Heat the sample at 70° for 5 min, chill on ice, and spin briefly in a microcentrifuge.

2. Add 4 μl of 5× buffer (supplied with Superscript II; GIBCO-BRL), 2 μl of 0.1 *M* dithiothreitol (DTT), 1 μl of dNTPs (10 m*M* each). Incubate at 37° for 2 min, add 1 μl (200 U) of Superscript II, and continue the incubation for 1 hr.

3. Hydrolyze the RNA by adding 2.2 μl of 1 *M* sodium hydroxide and heating at 65° for 20 min. Neutralize with 2.2 μl of 1 *M* HCl.

4. Load the entire volume of the first-strand cDNA reaction onto a 0.025-μm microdialysis filter (VSWP; Millipore, Bedford, MA) floating in a petri dish containing 15 ml of TE buffer [10 m*M* Tris-HCl (pH 7.6) and 1 m*M* EDTA]. Microdialyze for 4 hr. Remove the sample from the filter and adjust the volume to 20 μl with distilled water.

5. To the microdialyzed sample, add 6 μl of 5× terminal deoxytransferase (TdT) buffer (supplied with the enzyme; GIBCO-BRL) and 2 μl of 2 m*M* dATP. Prewarm the tube at 37° for 2 min, add 1 μl (15 U) of TdT, and incubate the sample for 5 min at 37°, and then for 5 min at 70°. Spin briefly.

6. Remove 15 μl of the A-tailed reaction to a new tube and add 2 μl of 10× restriction buffer M (Boehringer Mannheim, Indianapolis, IN), 1 μl of 10 m*M* dNTPs, 1 μl of a T_{17}-tailed anchor primer, and 1 μl of Klenow (2 U; Boehringer Mannheim). Incubate the sample at room temperature for 15 min, shift to 37° for 15 min, and then heat inactivate at 70° for 5 min.

7. Microdialyze the sample on a 0.1-μm filter (VSWP; Millipore) as described above. Pipette 20 μl of distilled water onto the filter, recover the sample, and adjust the volume to 37 μl with distilled water.

8. To the 37-μl sample on ice, add 5 μl of 10× PCR buffer II [500 m*M* KCl–100 m*M* Tris (pH 8.3 at 37°), supplied with AmpliTaq; Perkin-Elmer, Norwalk, CT], 4 μl of 25 m*M* magnesium chloride, 1 μl of 10 m*M* dNTPs, 100 ng (1 μl) of the anchor primer (without the T-tail), and 100 ng of the first nested polymerase chain reaction (PCR) primer. Add 1 μl of AmpliTaq (Perkin-Elmer) and carry out 30 rounds of PCR (90 sec at 94°; 90 sec at 60°; 3 min at 72°).

9. Microdialyze the PCR sample for 4 hr as described in step 7. Remove the sample and adjust the volume to 40 μl with distilled water.

10. Use 5 μl of the first-round PCR sample and repeat the PCR as described in step 8, using a 5′ biotinylated anchor primer and a second nested PCR primer.

11. Microdialyze the PCR for 4 hr as described in step 7. Remove the sample and adjust the volume to 80 μl with distilled water.

Solid-Phase Seqencing Reaction

1. Prepare end-labeled primer by mixing 10 pmol of the seqeuncing primer, 2 μl of 10× polynucleotide kinase buffer (supplied with enzyme; Boehringer Mannheim), 50 μCi of [γ-^{32}P or -^{33}P]ATP and 10 units of polynucleotide kinase (Boehringer Mannheim) in a volume of 20 μl. Incubate at 37° for 10 min, heat inactivate at 95° for 2 min, and store at −20° until ready to use.

2. Prepare 20 μl (200 μg) of Dynabeads M-280 (Dynal, Great Neck, NY), following the protocol described by the supplier.

3. Add 40 μl of the second PCR to the beads and incubate for 15 min (flick the tube periodically to keep beads from settling).

4. Using a magnet, remove the supernatant from the beads. Wash the beads once with 40 μl of 1× B&W buffer [5 m*M* Tris-HCl (pH 7.5), 1 m*M* EDTA, 1 *M* NaCl], and then resuspend the beads in 40 μl of freshly prepared 0.1 *M* sodium hydroxide. Incubate the sample at room temperature for 10 min.

5. Wash the beads twice with 40 μl of 1× B&W buffer, once with 40 μl of TE, and resuspend the beads in 12.5 μl of distilled water.

6. To 12.5 μl of the single-strand cDNA immobilized on beads, add 2 μl of 10× Thermo Sequenase buffer (US 78500 kit; United States Biochemical, Cleveland, OH), 1 μl of end-labeled primer (0.5 pmol), and 2 μl of Thermo Sequenase. Add 4 μl of this mix to four 0.2-ml PCR tubes, each containing 4 μl of a termination mix (ddATP, ddCTP, ddGTP, or ddTTP). If not using a heated lid, overlay each sample with 20 μl of mineral oil and place in a PCR machine.

7. Carry out 40 rounds of cycle sequencing (30 sec at 90°; 30 sec at 60°; 60 sec at 72°). Add 6 μl of stop solution (supplied with kit), heat at 90° for 3 min, and load half of each reaction on a sequencing gel. After electrophoresis, rinse the gel in water, transfer to 3MM Whatman (Clifton, NJ) paper, dry the gel in a gel dryer, and expose the gel to autoradiographic film. [*Note:* For fluorescent automated sequencing, substitute 5–10 pmol of 5′Cy5 sequencing primer in the reaction mix (step 6) and run the sequencing reactions on an ALF Express sequencing machine.]

Blastocyst Injections and Breeding Protocols

ES cell lines are thawed and passaged for 6 days prior to injection in ES cell medium in the absence of Geneticin. On the day of injection, the medium should be changed several hours before harvesting the cells. The cells in a confluent 25-cm^2 flask are trypsinized for 3–4 min and diluted into 9 ml of cold ES cell medium without LIF, pelleted, and resuspended in 0.8 ml of ES cell medium minus LIF in a sterile 1.5-ml screw-top microcentrifuge tube. To prevent the cells from clumping, they are kept on ice (for up to several hours) prior to adding them to the injection chamber. Blastocysts are flushed from pregnant C57B1/6 females (Jackson Laboratories, Bar Harbor, ME) and collected into CO_2-independent medium (GIBCO-BRL) containing 10% (v/v) fetal bovine serum. Blastocysts are expanded for 1–2 hours in ES cell medium in a 37°/6% (v/v) CO_2 incubator, transferred to a hanging drop chamber, and cooled to 4°. ES cells are added to the hanging drops and the blastocysts are injected with enough cells (20 or more) to fill the blastocoele. Injected blastocysts are then transferred to pseudo-pregnant recipient females (10–15 blastocysts/uterine horn). Typically, injection of 10 blastocysts will yield an average of 2 male chimeras with ES cell contribution to the germ line. [*Note:* The 129/Ola cells carry the recessive pinkeye (p) and chinchilla c^{ch}) mutations. Strong chimeras will exhibit patches of cream-colored fur and, occasionally, pink eyes.]

Genotyping by Dot-Blot Analysis of Tail DNA

Male chimeras are test bred to C57B1/6 females and agouti offspring are genotyped by dot-blot hybridization of tail DNA. We use a rapid one-tube method to prepare genomic DNA. To obtain uniform hybridization signals, it is essential to saturate the membrane with DNA. This protocol was worked out empirically with a 96-well dot-blot manifold (Bio-Rad) and Hybond N^+ hybridization membrane (Amersham, Arlington Heights, IL). With this method we can reliably distinguish animals that are homozygous or heterozygous for gene trap vector insertions based on signal intensity. Either radioactive or nonradioactive probes may be used.

1. Prepare fresh tail buffer by dissolving proteinase K powder (Sigma) in 10 *M* Tris-HCl (pH 8), 100 m*M* NaCl, 50 m*M* EDTA (pH 8), 0.5% (w/v) sodium dodecyl sulfate (SDS) to a final concentration of 1 mg/ml.
2. Tail biopsies (approximately 1.5 cm in length) are taken at weaning age and digested overnight at 65° in 0.4 ml of tail buffer.
3. Add 0.1 ml of 5 *M* NaCl to digested tails while still warm and vortex at high speed for 10 sec. Add 0.5 ml of chloroform and vortex as described previously. Spin in a microcentrifuge at full speed for 15 min.
4. Transfer 50 μl of the aqueous (top) phase to a 96-well plate. Denature the DNA by adding 150 μl of 0.53 *M* NaOH and incubating for 30 min at 37°.
5. Prepare Hybond N^+ hybridization membrane by prewetting it briefly in water and soaking it in 0.4 *M* NaOH for 10 min. Wet one piece of Whatman paper in 0.4 *M* NaOH and place on the manifold. Lay the membrane on top, assemble the apparatus, and briefly apply vacuum to empty wells of excess liquid.
6. Apply samples and wait for 30 min before applying vacuum.
7. Draw samples through the manifold, using a gentle vacuum. Disassemble the apparatus and wash the membrane in 150 m*M* sodium phosphate buffer containing 0.1% (w/v) SDS at 65° for 1 hr.
8. Hybridize the membrane overnight in Church and Gilbert prehybridization/hybridization buffer [0.5 *M* sodium phosphate buffer, 7% (w/v) SDS, and fraction V bovine serum albumin (2.5 mg/ml; Sigma)] with a random-prime labeled βgeo probe.
9. Wash the filter three times with 30 m*M* sodium phosphate buffer containing 0.1% (w/v) SDS at 60° and expose the filter to autoradiographic film.

Acknowledgment

I thank Paul Wakenight and Barry Rosen for helpful discussions and Peri Tate for help in compiling the protocols. I gratefully acknowledge the following contributors to this screen, past and present: Brian Avery, Jane Brennan, Sarah Elson, Andrew Jeske, Sue Monkley, Kathy Pinson, David Townley, and Joel Zupicich. This work was funded in part by the Biotechnology and Biological Sciences Research Council (United Kingdom), Exelixis Pharmaceuticals, Inc., and the Searle Scholars Program of the Chicago Community Trust W.C.S. is a 1998 Searle Scholar.

Author Index

Numbers in parentheses are footnote reference numbers and indicate that an author's work is referred to although the name is not cited in the text.

A

B

C

D

E

F

G

I

J

K

L

M

Q

R

S

T

U

V

W

X

Y

Z

Subject Index

A

B

C

D

E

H

I

L

M

N

O

R

S

U

V

W

Y

ISBN 0-12-182229-X